AF530712

Insect Physiology

About the Author

Dr. K.P. Sanjayan is an Associate Professor and Head, in the Department of Advanced Zoology and Biotechnology, Guru Nanak College, Chennai, India. Since 1981 he has been researching in the field of insect plant interactions with primary focus on digestive and reproductive physiology of bugs and grasshoppers. His research has been funded by grants from the Government of India funding agencies like the DST, DBT, ICAR, CSIR, MoEF and UGC. He has more than 150 research publications in national and international peer reviewed journals. During the span of more than three decades of teaching and research, twenty students have obtained their doctorate degrees and twelve student received their M.Phil degree from the University of Madras. He is a member of two prestigious scientific organization such as the Madras Science Foundation and the Indian Academy of Entomology besides being a member for various National Scientific Bodies. He is a life member of the Association for Advancement of Entomology, Trivandrum, the National Institute of Ecology, New Delhi and the Indian Academy of Environmental Sciences, Haridwar.

Insect Physiology

Dr. K.P. SANJAYAN

Department of Advanced Zoology and Biotechnology,
Guru Nanak College, Chennai–600042
Tamil Nadu, India.

Published by
SCIENTIFIC PUBLISHERS (INDIA)

Jodhpur –
5 A, New Pali Road
P.O. Box 91
Jodhpur - 342 001 INDIA
INDIA

Delhi –
4806/24, Ansari Road
Daryaganj
New Delhi - 110 002

ISBN: 978-93-86652-27-0

Visit the Scientific Publishers (India) website at
http://www.scientificpub.com
Printed in India

In Memory of

Our Son

Preetham

Of all the precious gifts in our lives
However great or small
To have you as our son
Was the greatest gift of all.
We miss you with every beat
Of our broken Heart.
You have just gone ahead of us.
Hope to see you when we get there!

Acha and Amma

Preface

Studies on insects have interested mankind from time in memorial. Their occurrence in various shapes, sizes and colours have attracted several Naturalists who began studying their behavior, especially those of social insects such as species of wasps, ants, bees and termites. Social insects also served to provide early evidence of the close and evolving association between plants and insects. Naturalists recognized that bees were attracted to flowers because of their fragrances. However, the concept of chemical attraction was rather slowly extended to solitary insects and host plants, especially crops. Studies on host preference of phytophagous insects paved the way for it to become the very centre of agricultural entomology. The secondary plant metabolites served not only as a means of protecting plants from insects, but also facilitated to guide insects for food. The cultivation of food plants on a large scale to meet the burgeoning human population and storage of the subsequent harvest has provided favorable conditions for the emergence of phytophagous and stored product pests respectively. Man's crops are attacked in the field by locusts, an enormous array of Lepidoptera, and by Diptera and Coleoptera among others. Major sources of food and fiber, thus damaged include cereals, rice, vegetables and fruit as well as cotton and timber. This paved the way for a focussed study of insects under the field of economic entomology.

While several generations of pesticides have been used to protect the plants from insect attack, we are still not able to control the insect pests. A better understanding of the physiology of the insects would facilitate fine tuning of our strategies in Insect control. The physiology of all systems functioning in an insect could be used as targets for insect control. Applied entomologists, being confronted with the ravages of insects in agriculture and public health, have begun to recognize the need for thoroughly understanding the physiology of insect nutrition, the laws governing their responses to sensory stimuli, about their reactions to parasites, on the precise way in which insects adapt to diverse climatic conditions, on the reproductive physiology of insects, to name a few, in order to refine their control strategies. Although several books on insect physiology have been written by several authors, rapid advances in entomological studies and physiology in particular demand the need to provide an up to date account on the subject. With this in view, this textbook contains important, comprehensive and in-depth account of all aspects of insect physiology, providing wherever necessary also the fundamental knowledge of the various systems. Although it is aimed as a resource material for postgraduate students of entomology, it would serve as an essential reference source for invertebrate physiologists and neurologists, entomologists, zoologists and insect biochemists. To achieve this goal, extensive references have been made to several textbooks and reviews, to a few research papers dealing with applied aspects of insect physiology and the resources available over the net.

The first chapter deals with the anatomical and physiological attributes of the integument conferring insect success with a discussion on the use of the chemical properties of the cuticle to design novel molecules to control insect pests. The chapter also indicates that the structural design of the cuticle could itself be applied in the field of material science to develop hard structures which can withstand the harshness of the environment. Chapter two discusses the diversity in growth and life cycle patterns in insects. Chapters three and six deals with the digestive and excretory systems as potential targets for pest management. Aspects of the circulatory system of insects are presented along with an account on the new frontiers in insect immunity in chapter four. This would appraise the reader on the possible improved use of entomopathogens in biological control, in the discovery of antimicrobial molecules that can be exploited by humans, and of new strategies for management of insect vectors of human and animal disease. While the dynamism of the respiratory system (Chapter five) is presented as a key to their success, the use of the knowledge thus gained in fluid dynamics and biomechanical research is mentioned. An up to date account on the insect nervous system is presented in Chapter seven, together with a note on learning, memory and intelligence in insects. Chapter eight deals with the reproductive system of insects while chapter nine deals with hormones and regulation of metabolism, moulting and diapause. General protein, carbohydrate and lipid metabolism and their energetic are presented in chapter ten along with the physiology of regulation in cold hardiness and flight. Chapter eleven deals with muscular coordination while an in depth account on the sensory physiology and behaviour is presented in chapter twelve.

This book has been written with the encouragement and support given to me by Dr. Anantanarayanan Raman, Charles Sturt University, Australia.

He believed that I had the knowledge and potential to undertake this fruitful exercise. His critical review of the manuscript has undoubtedly facilitated in improving the content and presentation. I am deeply indebted to him. Two of my research students- Dr. N. Senthilkumar, Scientist D, Tropical Forest and Tree Breeding Research Institute, Coimbatore and Dr. V.A. Vijayashanthi, Asst. Professor, Agricultural College and Research Institute, Madurai, helped me immensely in providing resource materials which enabled me in writing a few chapters. I thank them for their assistance.

I deem it a privilege to thank Dr. K.G. Sivaramakrishnan, Retd. Professor, Madura College, Dr. R. Varatharajan, Professor, Manipur University, Imphal, Dr. Sabu K Thomas, Associate Professor, St. Xavier's College, Calicut, and Dr. S. Subramanian, Principal Scientist, IARI, New Delhi, for peer reviewing several of the chapters. Their suggestions and critical reading have added to the quality of the book.

I would also like to place on record my sincere thanks to Sardar Manjit Singh Nayar, Secretary and Correspondent and Dr. M. Selvaraj, Principal, Guru Nanak College for their support in all my endeavours at the Guru Nanak College.

Lastly, I would like to thank my family members- my wife Er. Manonmani, K.R, my sons Utham and Preetham for tolerating me and giving me the much needed environment to accomplish this work. Their unstinted support gave me the impetus to sit for several hours into the night working to beat the deadline.

Dr. K.P. Sanjayan
email: kpsanjayan@yahoo.co.in

Series Preface

With great pleasure, the Scientific Publishers (Jodhpur) and I launch the fourth volume of the postgraduate textbook series under the general title *21st Century Biology and Agriculture* for use in the Indian subcontinent.

This series aims at fulfilling the knowledge needs of postgraduate learners in agriculture and biology, focussing on self-directed learning. Keeping this point in full view, these volumes would dilate on contemporary information in chosen themes in a pertinent, but brief backdrop of historical knowledge, with appropriate textual information laced with relevant illustrations. Most importantly, these books aim to cater to the self-learning needs in passionate and committed learners. By self-learning, the teacher's role turns into mentoring rather than tutoring. Vital details have been 'box'ed so that learners can internalize them easily, swiftly, and forever. Care has been exercised to integrate examples from the Indian subcontinent, so that the learners can relate to concepts and principles quickly. The most critical aspect is the inclusion of specific case studies and interactive-mode of learning, so that learners can learn about the day-to-day issues and application of theory that surround the nominated theme effectively. Every effort has been meticulously made to see that the books launched under this series are easily readable and user friendly.

Tanay Sharma of Scientific Publishers (Jodhpur) readily agreed to my demand to identify and recruit an efficient copy editor, who has done a neat job.

This book entitled *Insect Physiology* executed by K.P. Sanjayan is the fourth book of this series. Sanjayan is a well-known entomologist and insect physiologist—ecologist of India. Presently he is working as a professor of Zoology, Guru Nanak College, Chennai. He has contributed to diverse aspects of insect biology, physiology, and ecology. Scientific Publishers (Jodhpur) and I consider his acceptance to write this book a great honour.

Many more similar titles have been enlisted and will be appearing in the near and far future. Scientific Publishers (Jodhpur) and I thank Sanjayan for readily and willingly accepting to our request by executing this task, and importantly doing it speedily.

India is a land of varied landscapes. One of the remarkable natural icons of India is the Great Indian Bustard, the populations of which are restricted to Western India. Because this series is being published from Jodhpur, I thought it will be appropriate to remind ourselves of the uniqueness of this splendid bird and make every effort of conserve its diminishing population. All books slated to appear in the series *21st Century Biology and Agriculture* will celebrate this Indian icon.

I am confident that learners will benefit from the information and knowledge shared by K.P. Sanjayan; I am also confident that post-graduate teachers of biology and agriculture in the Indian subcontinent would find this material appropriate to prescribe this book as a learning resource.

Anantanarayanan Raman

Editor-in-Chief, Textbook Series
21st Century Biology and Agriculture
School of Agricultural & Wine Sciences
Charles Sturt University
Orange, NSW 2800, Australia

Contents

12. Insect Sensory Physiology and Behaviour 631

The Integument: Anatomical and Physiological Attributes Contributing to Insect Success

1

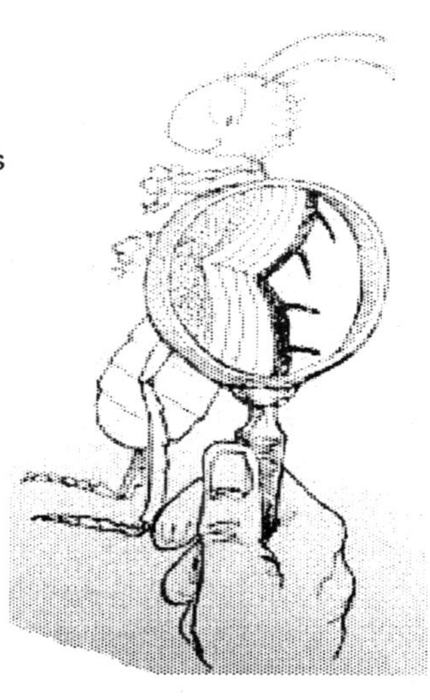

Synopsis

1. Insect integument structure
   - 1.1 The insect cuticle
     - 1.1.1 The cuticular components
     - 1.1.2 Schmidt's layer
     - 1.1.3 Epidermis
     - 1.1.4 Basement membrane
     - 1.1.5 Pore canals
   - 1.2 Chemical components of cuticle
   - 1.3 Endoskeleton
   - 1.4 Cuticular appendages
   - 1.5 Functions of cuticular surfaces
   - 1.6 Cuticular glands
2. Physiology of moulting
   - 2.1 Steps in moulting
   - 2.2 Control of moulting:
3. Cuticle sclerotisation
   - 3.1 Mechanism of sclerotisation
   - 3.2 The control of the sclerotisation process: hormones
   - 3.3 Sclerotisation in larvae and pupae
4. Colouration and melanisation
5. Learning from cuticle

INTRODUCTION

Insects exhibit a great diversity of form and within each order and family, they exist in a myriad of different ways that make it seem that their anatomy is very complex. However, they have the same basic plan, which rather make them very successful animals as a group. An insect body has a hard exoskeleton formed of the integument which protects the soft interior of the insect. In addition to acting as a physical barrier providing protection against entry of pathogens, parasites, predators and pesticides, the integument also serves are a base for muscle attachment and reduces the loss of water to a very low level in most insects, thereby being of great evolutionary significance to this predominantly terrestrial class of invertebrates. The integument acts as sensory windows to the outside world for the insect. It also lines the tracheae, tracheoles, salivary glands and portions of the reproductive tract, as well as the fore and hind gut. The cuticle is, preternaturally, multifunctional; it not only supports the insect, and gives it its shape, and facilitates locomotion, but also serves as a waterproofing device. The integument has a range of localized mechanical specializations such as high compliance, adhesion, wear resistance and diffusion control. It can also serve as a temporary food store (Box 1.1).

The structural complexities of the integument in terms of its biochemical nature and the physiological changes accompanying growth, development, and moulting have contributed to the insect success.The structure of the integument is not rigid as it shows dynamic changes correlated to the physiology of the insect. Both the cellular and acellular components of the integuments may be differentiated in a variety of ways to suit the insect's needs. Some epidermal cells of the integument may form specialised gland cells or may develop into particular parts of sense organs. Further, at different regions of the body, the cuticle is variously differentiated according to the function it is required to perform. In some regions they are thick and rigid, while at other regions they are flexible and elastic and over sensory structures it may be extremely thin.

From an evolutionary perspective, the integument provides several positive advantages to the insect; however, as the size of the insect increases, problems with the cuticle being too heavy crop up and also the cuticle impede gaseous exchanges and are therefore factors of significance in evolution.

BOX 1.1. FUNCTIONS OF THE INTEGUMENTARY SYSTEM

1. Protection of internal organs and tissues.
2. Protective barrier against entry of pathogens, parasites, predators and pesticides.
3. Preventive barrier against water loss.
4. Provides for the insect the sensory windows to the outside world.
5. Lines the tracheae, tracheoles, salivary glands and portions of reproductive tract.
6. Protective barrier for foregut and hindgut.
7. Has a range of localised mechanical specialisations.

I. INSECT INTEGUMENT STRUCTURE

The amazing success of insects on planet earth can be attributed in no small measure to the incredible mixture of flexibility and strength of the insect integument. The combination of flexibility and hardness of the insect integument allows insects to move freely without loss of defense and protection. A typical insect integument is mainly composed of three parts (Fig1.1). The outer layer, the **cuticle**, is most visible together with the attendant bristles and hairs. It gives the insect integument most of its colour and hardness. Below this is the **epidermis**, which is a cell layer responsible for synthesis and release of various materials for the cuticle, and the **basement membrane**, which is a non-cellular layer that serves as a barrier between the integument and the inside tissues. The basement membrane is a bilayer of amorphous mucopolysaccharides and collagen fibers.

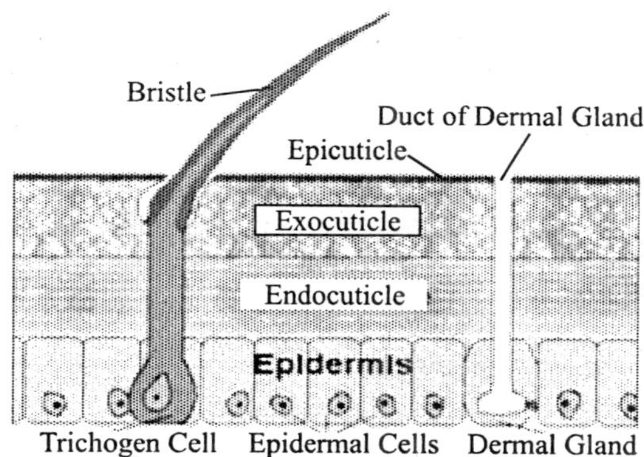

Fig. 1.1. The general structure of the insect integument (*http://www.nazarian.ir/biodb/en/insects_anatomy.asp*).

1.1 The Insect Cuticle

The insect cuticle is a relatively thin, non-cellular and hard layer of exoskeleton. The hardened, or sclerotised, cuticle has several important functions: to provide protection against desiccation (or from imbibing excessive water in case of aquatic insects), external physical injuries and microorganisms and to serve as muscle attachment for insect movement and flight. Yet, these functions are inherently dependent upon the occurrence of sclerotization of the cuticle during moulting. Insects are especially vulnerable between moults, when the cuticle is soft and flexible. A better understanding of the physiological events involved in moulting, specifically sclerotisation of the cuticle, could possibly provide ways to exploit these vulnerabilities in controlling pest populations.

Structurally, the insect cuticle at some regions are hard and stiff and are called the **sclerites**, while in some regions they are more flexible and pliable and are called the **arthrodial membranes**. The arthrodial membranes connect among the sclerites and facilitate locomotion. The properties of the individual regions of the cuticle correspond to their different functions. The regions corresponding to flight are characterised by near-optimal balance between stiffness and flexibility, while those regions corresponding with protection are mostly covered by hard and rigid sclerites. Marked differences in mechanical properties of the cuticle can be present at the microscopic level; two neighbouring epidermal cells can produce cuticle with contrasting properties, indicating that cuticular composition is precisely controlled at the cellular level. The mechanical properties of the various cuticular regions are related to the function they perform in the life of the insect. For example, proper flight can only be sustained when the various wing regions have a near-optimal balance between stiffness and flexibility. If the wing material is locally too soft or too stiff, the varying air pressure during the wing strokes will not cause the wings to bend into the shapes needed for generating optimal lift. It would therefore suffice to mention here that cuticle throughout the body would be present as a mosaic of structures that vary in their mechanical properties based on the physiological requirement. However, the description that follows below is a generalised structure.

1.1.1 The Cuticular Components

A typical insect cuticle consists of three layers: the **endocuticle**, the **exocuticle**, and the **epicuticle** (Fig 1.2) and are bestowed with specific functions (Box 1.2 and 1.3)

The **endocuticle** and the **exocuticle**, together known as **procuticle**, are mostly composed of a large number of layers of protein and chitin fibers laying down in a laminated pattern such that the individual strands in each layer cross each other, creating an extremely tough and flexible substance. The relative amounts of chitin and proteins, chitin architecture, protein composition, water content, intra-cuticular pH and the degree of sclerotisation are the main factors affecting the hardness and rigidity of the insect cuticle. In some cuticles, the border between the endo and exo-cuticle is not clear and an intermediate area, the mesocuticle is visible.

The **epicuticle** contains no chitin and is highly resistant to water and other solvents. It blocks the invasion of foreign matter. This layer is differentiated into the following layers:

1. **Inner epicuticle** that is often called the cuticulin layer. It is a stratum composed of lipoproteins and chains of fatty acids embedded in a protein-polyphenol complex.
2. **Outer epicuticle:** It is an oriented monolayer of wax molecules that lies just above the cuticulin layer and serves as a chief barrier to movement of water into or out of the insect's body.
3. **Cement layer:** It is the outer most layer formed by lipid and tanned protein. It protects the wax layer from abrasion.

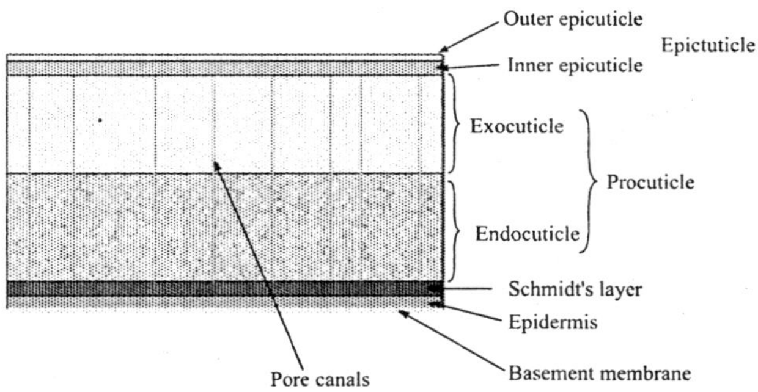

Fig. 1.2. Schematic representation of insect integument

1.1.2 *Schmidt's Layer*

Below the endocuticle and adjacent to the epidermal cells, a narrow amorphous layer, Schmidt's layer may be seen. It was suggested originally that this is an adhesive layer holding the cuticle to the cells. It is now believed to represent newly secreted, less well stabilised endocuticle

1.1.3 *Epidermis*

The epidermis is an inner unicellular layer resting on basement membrane. It is one cell thick. During periods of inactivity, its cells are flattened and intercellular boundaries are indistinct. When active, cells are more or less cuboidal, and their plasma membranes are readily apparent. The density of the cells in a particular area varies and is correlated with the moulting cycle. It has the following functions: 1. Cuticle secretion, 2. Digestion and absorption of old cuticle, 3. Wound repairs and 4. Provide a surface look.

Epidermal cells may be differentiated into sense organs or specialised glandular cells. Oenocytes are large, ductless, often polyploidy cells, upto 100 μm in diameter. They occur in pairs or small groups and show signs of secretary activity that can be correlated to the moulting cycle. It has been suggested that they produce the lipoprotein component of the epicuticle. Epidermal cells are also differentiated into dermal glands of various types. Typically, they are unicellular and have long duct that penetrate the cuticle to the exterior. The gland cells exhibit cyclic activity associated with new cuticle formation and it has been proposed that they secrete the cement layer of epicuticle.

The sex pheromone in most female Lepidoptera is secreted by a small patch of tall, columnar epidermal cells located beneath the cuticle of the ventral inter-segmental membrane in the 8-9th segment of the abdomen. Golgi complex is prominent in epidermal cells, and probably serves several functions including 1) processing of secretory substances necessary to synthesise cuticle, 2) production of material for the plasma membrane of the cell and 3) processing and packaging of lysozymes needed for autophagy.

BOX 1.2. FUNCTION(S) OF EACH MAJOR LAYER OF CUTICLE

1. **Cement layer:** Thin outside layer. Closely associated with wax layer and may serve to protect it. Not found in all insects.
2. **Wax layer:** Hydrocarbons constitute 90% of this layer. Important to insects for water loss, thus water proofing of cuticle.
3. **Outer epicuticle:** First layer formed following the moult and is the layer that protects the new procuticle from digestion by moulting enzymes (i.e., chitinases and proteinases). Also called cuticulin layer.
4. **Inner epicuticle:** Function not that clear but it is a much thicker layer than the outer epicuticle.
5. **Wax canal filament:** A filament of wax that is produced by the hypodermal cells and extends to the inner part of the epicuticle. Probably is a filament that continually is moved towards the surface of the cuticle.
6. **Exocuticle:** Region in which cross-linking of proteins occurs to give cuticle hardness. Region not broken down by proteases at moult. Melanin and other pigments found in this area.
7. **Endocuticle:** Region directly above the hypodermal cells. Is continually synthesised (in a dark/light way-24 hrs) and can often be used to age-grade some insects. Contains chitin, which is broken-down at the moult by chitinase. Little cross-linking of proteins, thus most is broken-down at moulting and reabsorbed.

1.1.4 Basement Membrane

It is non-cellular and about 0.05μm thick. It is composed of amorphous granular material, most likely neutral mucopolysaccharides. It provides support for epidermal cells. It is penetrated by nerves and tracheae going to hypodermis, and is a selective barrier between hemolymph and epidermal cells. Hormones and other nutrients can pass through this selectively permeable layer to reach the hypodermal cells. Is important in the recognition of 'self'; thus, the insect's blood cells do not recognise it as foreign. Molecules in this layer are charged and probably act like a molecular sieve.

1.1.5 Pore Canals

These are tiny pores that run from the hypodermal cells to the inner part of the inner epicuticular layer. Inside the canals are wax filaments that extend upto the epicuticular layer. They probably serve as a transport passage for wax from the hypodermal cells upto the wax layer (Box 1.3)

BOX 1.3. FUNCTIONS OF CUTICLE AND EPIDERMAL LAYER

1. **Pore canals:** Narrow pores that extend from the hypodermal cells to the inner most part of the epicuticle. Involved in transport of lipids from hypodermal cells to the epicuticle or outer layer.
2. **Schmidt's layer:** The zone of deposition of new cuticle or procuticle.
3. **Hypodermis or epidermis:** A single layer of cells that underlies the cuticle and is responsible for its production. These cells can be modified to form dermal glands, sensory receptors and oenocytes.
4. **Basement membrane or basement envelope (Basal lamina):** Mucopolysaccharide layer that is secreted by the hemocytes, is penetrated by nerves and tracheae going to hypodermis, and is a selective barrier between hemolymph and epidermal cells. Hormones and other nutrients can pass through this selectively permeable layer to reach the hypodermal cells. Is important in the recognition of 'self'; thus, the insect's blood cells do not recognize it as 'foreign.' Molecules in this layer are charged and probably act like a molecular sieve.

The functions of the integument can be summarised as follows:

1. Acts as external armour and strengthen external organs like jaws and ovipositor.
2. Protects the organs against physical aberration, injurious chemicals, parasites, predators and pathogen.
3. Internally protects the vital organs, foregut, hindgut and trachea.
4. Provides space for muscle attachment and gives shape to the body.
5. Prevents water loss from the body.
6. Cuticular sensory organs help in sensing the environment.
7. Cuticular pigments give color.

1.2 Chemical Components of Cuticle

Chemically, insect cuticle contains carbohydrates, proteins and lipids (Box 1.4).

Chitin

Chitin is an important constituent of the insect cuticle. Chitin is a polymer of N-acetyl-β-D-glucosamine residues linked together by β-(1,4)-glycosidic linkages (Fig 1.3). Chitin occurs in long chains in nature and is packed into a crystallite structure that is embedded in a protein matrix. Within the body of the cuticle the chitin is assembled into nanofibres about 3 nm in diameter and about 0.3 mm long, each containing 19 molecular chains. Although it has never been specifically measured, the stiffness of these nanofibres isat least 150 GPa, based on the observation that cellulose isabout 130 GPa and the extra bonding in the chiti crystallite is going to stiffen it further. In suspension in water, chitin nanofibres are highly thixotropic and liquidcrystalline. Stiffness and thixotropy are crucial properties of the self-assembly of the components of the cuticle.

The content of chitin can constitute up to 40% of the dry mass of insect exuviate (Kramer *et al.*, 1995). Chitin is found in the exo- and endocuticle, but not in the epicuticle. It functions as a light, but mechanically strong scaffold material and is always associated with several cuticle proteins that mainly determine the mechanical properties of the cuticle.

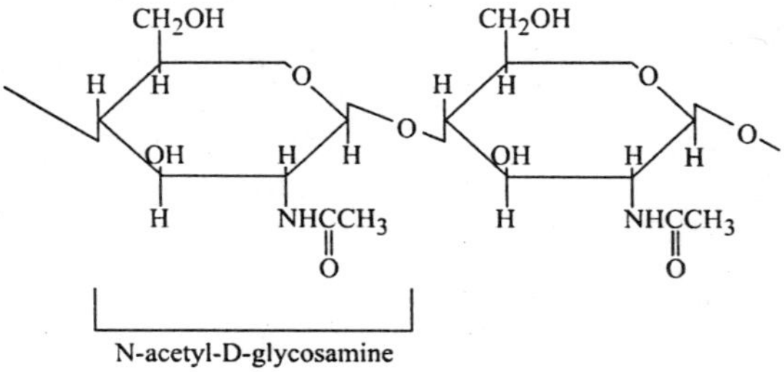

Fig 1.3. Chitin

BOX 1.4. CHEMICAL COMPOSITION OF THE CUTICLE

1 **Carbohydrates**

a. **Chitin:** Polysaccaride with repeating units of *N*-acetyglucosamine residues. Chitin gives the cuticle its strength, not its hardness.

2 **Proteins**

a. **Arthropodin:** Untanned protein that during sclerotisation is crosslinked to produce sclerotin and give the cuticle its hardness.

b. **Sclerotin:** Tanned protein that gives cuticle hardness.

c. **Chitinase:** Enzyme involved in digestion of chitin at the moult. Is released and produced by the epidermal cells.

d. **Proteinases:** Enzymes other than chitinase that aid in the digestion of the endocuticle at the moults.

e. **Resilin:** Colourless rubber-like protein. Can be stretched and stores energy due to tension. Places found: Wing hinges; food pump of reduviid bugs; hind legs of jumpers; aids in inspiration in beetles, which lack inspiration muscles.

3 **Lipids**

a. **Waxes:** Wax blooms produced by several groups of insects.

b. **Cuticular hydrocarbons:** It is also used by systematists.

X-ray diffraction studies indicate chitin exhibits a crystalline structure. Three types of chitin, namely, α, β and γ-chitin that differ in the crystal cell size, number of chitin chains per unit cell, and degree of hydration have been described. Differences in orientation of the three chitin chains in chitinized structure lead to differences in physical packing of chains and in overall physiological properties of chitin (Fig 1.4). In α-chitin, adjacent chains run antiparallel to each other, which allow them to pack closer together and maximizes the number of within- and between-chain hydrogen bonds. Typically 18-10 α-chitin chains are packed together in a roughly circular rod or crystallite that is embedded in a protein matrix. The extensive hydrogen bonding contributes to the rigidity, strengthand waterproofing of the cuticle. Adjacent chains in β-chitin run parallel to each other, and the relative large N-acetyl groups projecting from the chain act like spacers to hold chains further apart, thus reducing tightness of the packing and the number of hydrogen bonds that can form between chains. Chains in γ-chitin may be oriented in various ways, but one common orientation is a repeating pattern of two parallel chains adjacent to an antiparallel chain, again reducing packing tightness and interchain hydrogen bonds.

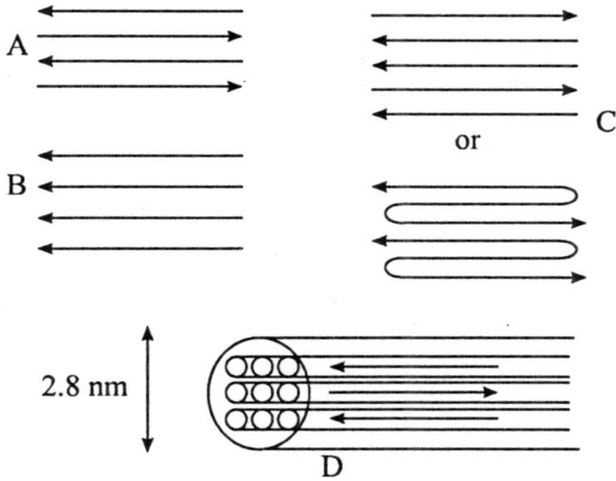

Fig 1.4. Orientation of chitin chains in the cuticle

A. The orientation of α chitin, the most common form in insects;
B. The orientation of β chitin;
C. Two possible orientation of γ chitin;
D. The location of the chains of chitin in α single chitin microfibril

(After Marc J Klowden (2013) Physiological systems in Insects, Academic Press)

The starting point in the biosynthesis of chitin is β-D-glucose which may come from a storage form as trehalose or glycogen. Glucose is phosphorylated at the expense of ATP and then isomerised to fructose-6-phosphate. An amino group is transferred from glutamine to fructose-6-phosphate to form glucoseamine-6-phosphate. The latter molecule is acetylated, probably with acetyl CoA contributing the acetyl group, to form N-acetyl-glucosamine-6 phosphate. A transfer of a phosphate group from carbon-6 to carbon-1 is necessary to form N-aceyl-glucosamine-1-phosphate, which reacts with uridine triphosphate and forms uridine diphospho-N-acetyl-glucosamine (Fig 1.5). This latter compound, then undergoes polymerisation to form chitin.

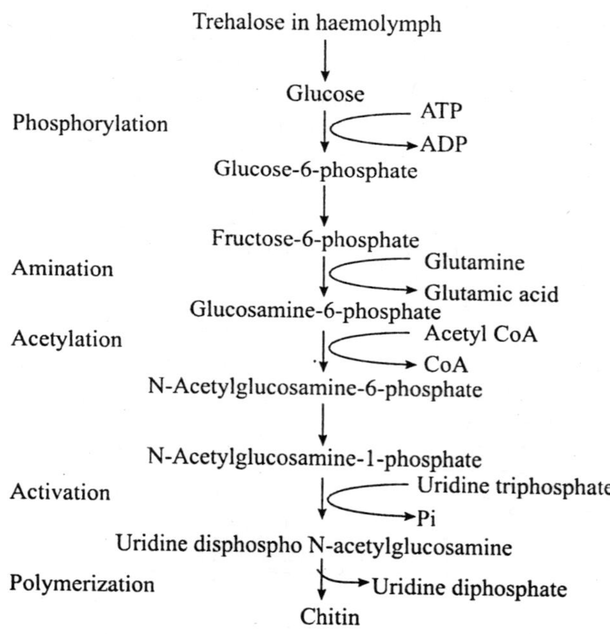

Fig 1.5. Steps in the biosynthesis of chitin

Proteins

Proteins are another major and important cuticular component. Cuticular proteins are synthesised in the epidermal cells and are then secreted into the cuticle from the apical surface of these cells. The protein has to produce a matrix of varying mechanical properties, which will also interact with and stabilize the chitin. It interacts with the chitin in a narrow zone (the Schmidt layer) where there is a relatively large amount of water (about 90%), presumably to drive the hydrophobic interactions between chitin and protein. The proteins can interact with each other giving well defined complexes in the absence of H-bond breakers, suggesting that hydrophobicity is an important factor in their stabilisation; in general proteins from more highly evolved insects tend to be less hydrophobic (Hillerton andVincent, 1983).

Cuticular proteins are generally classified into two classes: non-structural proteins and structural proteins (Willis *et al.*, 2005).

Non-structural Proteins

Pigment

So far proteins from three classes of pigments have been identified and sequenced in insects: insecticyanins, which are blue pigments, and two anonymous yellow proteins (Willis *et al.*, 2005). These proteins, in cooperation with carotene, confer the green coloration. One of the yellow proteins, encoded by the gene *yellow*, has been reported to be responsible for the melanisation of the cuticle (Kornezos and Chia, 1992), and cis- and trans- regulation of the yellow gene is responsible for the color differences in different species and different patterns within a species (Wittkopp *et al.*, 2002). The other yellow pigment protein has putative carotene and/or JH binding activity and was proposed to be involved in transporting carotene into the cuticle (Wybrandt and Andersen, 2001).

Enzymes

Different types of enzymes are located in the insect cuticle having different functions during cuticular events, such as moulting and sclerotisation. Several of the enzymes involved in sclerotisation, such as prophenoloxidase, diphenoloxidase, laccases, peroxidases, isomerases, and tautomerases, have been identified in insect cuticle (Andersen, 1979a, b).

Defense proteins

Defense proteins, such as cecropin (Lee and Brey, 1994) and scolexin (Molnar *et al.*, 2001), are involved in insect immune response to pathogens like bacteria and viruses.

Arylphorins

Arylphorins are proteins with a high content of aromatic amino acids and some lipid; the function of arylphorin proteins is not clear.

Arthropodin

Arthropodins are an untanned cuticular protein, which is water soluble.

Sclerotin

These are tanned cuticular protein, which is water insoluble.

Resilin

This is an elastic cuticular protein responsible for the flexibility of sclerites, e.g., wing articulatory sclerites. It provides elastic storage energy and also prevents or delay material damage and failure by fatigue due to the repeated folding and unfolding as observed in the wings of beetles. Resilin contains various amino acids, including the ones that provide links between protein chains. The amino acid sequence in the protein is such that it prevents other cross-links occurring, since this would impair the rubber-like properties.

Structural Proteins

There are a number of structural proteins present in insect cuticle. Over one hundred structural proteins have been sequenced, and in addition, almost 200 more proteins were predicted based on the available genomic sequences in the fruit fly *Drosophila melanogaster* and the mosquito *Anopheles gambiae* (Nation, 2002). Many of these structural proteins have common consensus sequences, such as Rebers and Riddiford Consensus (R&R Consensus) and Pfam00379 consensus (Andersen *et al.*, 1995; Finn *et al.*, 2008), so their presence in the cuticle can sometimes be predicted simply based on sequence analysis.

Lipids

Lipids in the cuticle not only promote water conservation, but they also have additional behavioral, pheromonal, ecological and taxonomic significance (Box 1.5). Lipids in the cuticle include true chemical waxes, hydrocarbons, alcohols, fattyacids, glycerides, sterols, ketones, aldehydes and esters. The lipid layer may lie below the cement layer, but in many insects there is a mosaic pattern consisting of patches of lipids and cement on the cuticle like the one seen in tenebrionid beetle as a **wax bloom**. The type and quantity of lipid on the cuticle seem to be an adaptation to the environment in which the insect lives. For examples insect larvae which live in wet environments have only small quantities of cuticular hydrocarbons on the cuticle. Diapausing pupae of *Manduca sexta* have a thicker lipid layer on the cuticle than the non-diapausing ones. Season and apparently temperature, are correlated with the type and quantity of hydrocarbons on the cuticle of the desert tenebrionid *Eleodes armata* (Hadley, 1977). In social insects there are a number of hydrocarbons which act as a sex pheromone and cues for caste recognition. Worker ants recognise the eggs laid by their queens by the surface hydrocarbons. Some insects secrete wax and deposit them as bloomus and use it as protection from predators and from desiccation. The wax produced from the abdominal epidermal cells of honey bees are used to build the honeycomb of the hives.

BOX 1.5. FUNCTIONS OF CUTICULAR LIPIDS

1. Hydrophobicity of cuticle.
2. Inter- and intraspecific signal molecules for behavior.

Other Components of the Cuticle

The cuticle also has several other components like the phenols derived from tyrosine metabolism, and trace amounts of inorganic components such as calcium, manganese and zinc. The pupae of the dipteran *Musca autumnalis* have large amounts of calcium deposits. High levels of Mn and Zn are seen in the mandibles of herbivorous insects such as locusts and caterpillars to increase the hardness and reduce the wear from abrasions.

1.3 Endoskeleton

In certain regions of the body the integument becomes invaginated and greatly hardened, forming rigid processes which serve for the attachment of muscles and the support of certain other organs. This internal framework is termed the **endoskeleton** and its individuals parts are known as **apodemes** or **phragmata**. They arise as invagination of the body wall between adjacent sclerites, or at the edge of a sclerite or segment. In some insects, the invagination remains open throughout life, but they usually become completely solid through deposition of cuticular material. There are two types:

1. **Apodeme:** These are hollow invagination of the body wall that meet internally and form a brace for the head and for muscle attachment and found in all insects. In pterygote insects, however, these ingrowths start at the tentorial pits on the face and have evolved to form the head endoskeleton or tentorium. Apodemes can take the form of being just a point-like invagination or a line (or sulcus or suture). **Phragma** are transverse partitions of the endoskeleton used mainlyfor muscle attachment
2. **Apophysis:** These are solid invagination of the body wall. They are plates of the endoskeleton that provide both extra strength and attachment sites for muscles in the thorax.

1.4 Cuticular Appendages

Insects are well endowed with cuticular extensions, varying from fine and hair-like to robust and spine-like. Four basic types of protuberance (Fig 1.6), all with sclerotised cuticle, can be recognised on morphological, functional, and developmental grounds:

1. Spines are multicellular with undifferentiated epidermal cells;

2. Setae, also called hairs, macrotrichia, or trichoid sensilla, are multicellular with specialised cells;
3. Acanthae are unicellular in origin;
4. Microtrichia are subcellular, with several to many extensions per cell.

Setae sense much of the insect's tactile environment. Large setae may be called bristles or chaetae, with the most modified being scales, the flattened setae found on butterflies and moths (Lepidoptera) and sporadically elsewhere. Three separate cells form each seta, one for hair formation (trichogen cell), one for socket formation (tormogen cell), and one sensory cell. There is no such cellular differentiation in multicellular spines, unicellular acanthae, and subcellular microtrichia. The functions of these types of protuberances are diverse and sometimes debatable, but their sensory function appears limited. The pattern of production, including colour, may be significant for some of the microscopic projections. Spines are immovable, but if they are articulated, then they are called **spurs**. Both spines and spurs may bear unicellular or subcellular processes.

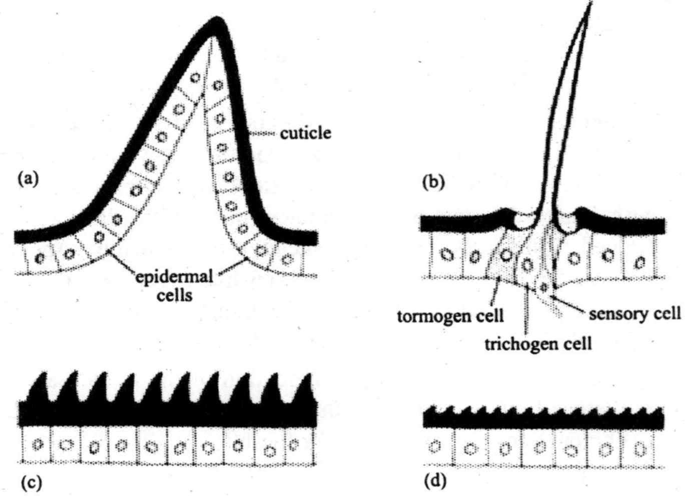

Fig 1.6. The four basic types of cuticular protuberances: (a) a multicellular spine; (b) a seta, or trichoid sensillum; (c) acanthae; and (d) microtrichia. (After Richards & Richards 1979)

The cuticular appendages are also classified into:

1. **Non-cellular:** Non-cellular appendages have no epidermal association, but rigidly attached. e.g. minute hairs and thorns .

2. **Cellular:** Cellular appendages have eipdermal association. They may be unicellular or multicellular.

A. Unicellular

- Scales - flattened outgrowth of body wall e.g. Moths and butterflies
- Glandular seta. eg. caterpillar
- Sensory setae - associated with sensory neuron or neurons
- Seta - hair like outgrowth (Epidermal cell generating site is known as **trichogen**, while the socket forming cell housing trichogen is known as **tormogen**. Study of arrangement of seta is known as **Chaetotaxy**).

B. Multicellular: e.g. Spur-movable structure, Spine- Immovable structure.

The different kinds of setae are as follows

1. **Clothing hairs:** These are hair like structures that cover the entire body, legs and wings. Eg: honey bee.
2. **Scales:** These are varyingly pigmented plate like structures covering the body as well as wings of adult moths and butterflies.
3. **Glandular setae:** Setae that functions as outlet for the secretion of hypodermal glands. If they are stout they are called glandular bristles as in case of certainlarvae of Lepidoptera.
4. **Sensory setae:** These are the setae or trichomes provided with sensory receptors or sensory nerve cells and respond to the external stimuli. These are present on antennae, legs and mouth parts.

1.5 Functions of Cuticular Surfaces

Most parts of the exoskeleton are covered with protuberances ranging from relatively flat scales in the micrometer range to long hairs and brush-like structures of several mm in size which themselves have a complex geometry : long, stiff spines for defence, scales and bristles for generation of sound, structural colour (camouflage, mimicry, species and sex recognition), prevention of water loss and thermoregulation (reflection of sunlight, insulation), aero- and hydrodynamic efficiency through the creation of turbulence which act as air and water bearings, respectively, long bristles as tools for grooming, food sampling and food filtering, and spines forfood grinding, both in the mouth-parts and also in the digestivesystem. The essential water and dirt repellence of the exoskeleton is achieved through the combination of a hydrophobic surface chemistry with cuticular protuberances in the range from 0.1 to 1.0 mm. There is also a large range of attachment devices, both hairy and smooth. In the form of hooks, brushes, snap and lock-and-key mechanisms they allow the organism to attach to the substratum, to interlock body parts, to attach during copulation and to associate one organism with another for phoresy, predation and parasitism.

1.6 Cuticular Glands

Insects possess two fundamental types of cuticular exocrine glands: **type 1 glands** are unicellular, that is they are small glands made up of single (epidermal) cells, and **type 2 glands** which are larger multicellular organs. Both types of gland secrete chemicals to the outside world via cuticular ducts that carry the secretion across the cuticle to the outside world, opening via pores on the surface of the cuticle. The rove beetles *Aleochara bilineata* and *Aleochara bipustulata* have type I glands in the antenna where the epidermal cells make up the antennae by forming a hollow cylinder and they secrete the cuticle. The hollow interior is occupied by blood, nerves and tracheae. Each of these type 1 glands opens via a rosette cluster of pores, each pore is about 100 nm in diameter, whilst the whole pore cluster is about one micrometre.

Antennal glands in beetles have several possible functions. It has been suggested that glands concentrated on the antennae and at points of articulation of the joints secrete a lubricant to reduce friction in moving parts of the cuticle. Some produce enzymes that seem to have an antimicrobial function. Another possibility is that these glands secrete pheromones. Pheromones are chemicals that carry signals, encoded as odours, from one animal to another. Many insects produce sex pheromones which identify the sex of the individual and their ripeness for reproduction. Other pheromones are used by social insects to communicate with one another.

Following are some of the examples of cuticular glands.

1. Wax gland - e.g. Honey bee and mealy bug.
2. Lac gland - e.g. Lac insects.
3. Moulting gland secreting moulting fluid.
4. Androconia or scent scale - e.g.moth .
5. Poison gland - e.g. slug caterpillar.

2. PHYSIOLOGY OF MOULTING

Insects' success on the earth is partly due to the hard exoskeleton or cuticle. The hard cuticle provides insects muscle attachment for movement and protects insects from water loss, external physical injury and pathogens. However, the cuticle also prevents insects from growth due to lack of the ability to expand. To grow bigger, insects must periodically shed the old cuticle and replace the old one with a new bigger one. The shedding of the old cuticle and replacement of a new one is called **moulting**.

2.1 Steps in Moulting

Moulting in insects begins with the separation of the cuticle from the underlying epidermal cells (apolysis) and ends with the shedding of the old cuticle (ecdysis). During apolysis, the old cuticle (including the endocuticle, exocuticle and epicuticle) is detached from the underneath newly synthesized endocuticle and the epidermis. Then after apolysis, moulting fluid is secreted into the space between the old cuticle and the epidermis (the exuvial space). The fluid contains inactive enzymes that are activated when the new epicuticle is secreted. This prevents the enzymes from digesting the new procuticle as it is laid down. After the synthesis of the new epicuticle, the lower region of the old cuticle - the endocuticle - is then digested by the enzymes and subsequently absorbed. The exocuticle and epicuticle resist digestion, and are hence shed at ecdysis (Fig 1.7).

Several changes accompany the moulting process and these can be summarized as follows:

1. **Behavioural changes:** The larva stops feeding and become inactive. These changes are probably a consequence of elevated ecdysteroid titre as observed in locusts. However, in the case of the final stage of lepidopteran larvae preparing to moult into the pupal stage, there is increased locomotor activity. In these insects, at the conclusion of the larval growth, the larva stop feeding and moves away from its food plant, seeking a pupation site. The larvae bury themselves in the soil prior to pupation, digging down under ground before constructing a pupation cell in which moulting will take place. During this time, the insect empties its gut, and covers its body with saliva. This complex of behavioral events is called **wandering behaviour**.
2. **Changes in epidermis:** The epidermis cell size, its activity, protein content and enzyme level increases. Cells divide mitotically and increase the tension, which results in loosening of cells of cuticle.
3. **Apolysis:** This is detachment of cuticlefrom epidermis. As a result of the changes in the cell shape, a tension is generated at the surface of the epidermal cells which results in their separating from the cuticle.
4. **Formation of sub cuticular space.**
5. Secretion of **moulting gel** in the sub cuticular space which is rich with chitinase and protease.

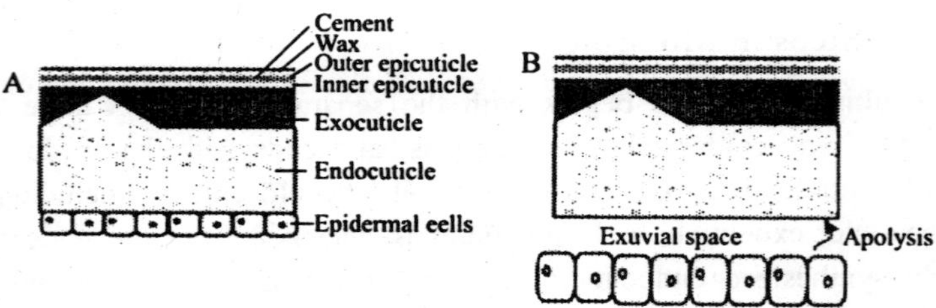

A: Cuticle layers B: Exuvial space- the area between the cuticle and epidermis; fills with a molting gel that contains inactiveenzymes including a chitinase and proteases for digesting the old cuticle

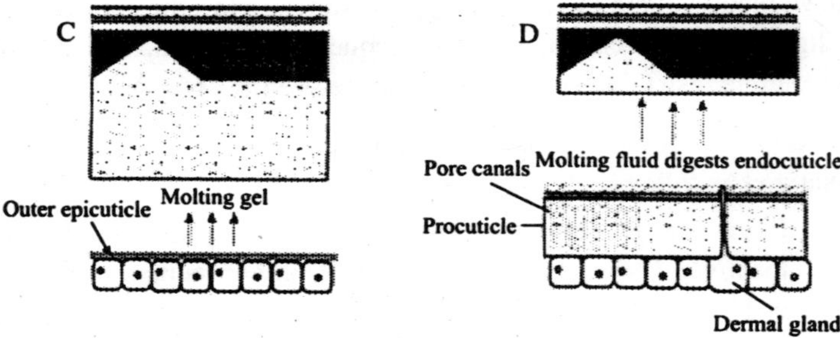

C: The epidermal cells secrete a new outer epicuticle (lipoprotein: cuticulin);The activation of the enzyme in the molting gel, now called the moulting fluid; D: The molting fluid begin the digestion of the old unsclerotized endocuticle;The epidermal cells begin to secrete the new procuticle;

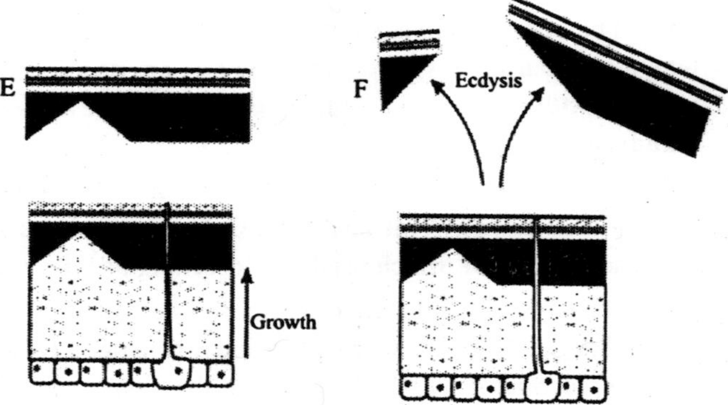

E: Formation of the new epicuticle; Absorption of the moulting fluid; F: Ecdysis: induced by eclosion hormone

Fig 1.7. Sketch of the steps in moulting
(*Source: entomol.nchu.edu.tw*)

6. **New epicuticle formation:** Lipoprotein layer (cuticulin) is laid over the epidermis. At first the cuticulin layer is smooth, but then it increases in area and this produces complex surface patterns.
7. **Procuticle formation:** Procuticle is formed below the epicuticle.
8. **Activation of moulting gel:** Moulting gel is converted into moulting fluid rich in enzymes. This activates endocuticle digestion and absorption. The enzymes when first secreted are inactive. The secretion of cuticulin follows the secretion of moulting fluid, but precedes its activation. This important as otherwise the moulting fluid would digest the newly formed layers of the procuticle and the old endocuticle.
9. **Wax layer formation:** Wax threads of pore canals secrete wax layer. The wax is secreted shortly before ecdysis on the surface of the new cuticle. It is probably transported to the surface as lipid-water liquid crystals.
10. **Cement layer formation:** Dermal glands secretes cement layer (Tectocuticle) soon after ecdysis.
11. **Ecdysis:** This involves two steps:
    A. **Rupturing of old cuticle:** Insect increases its body volume through intake of air or. water, which enhances the blood flow to the head and thorax. There by the old cuticle ruptures along predetermined lines of weakness known as **ecdysial line.**
    B. **Removal of old cuticle:** Peristaltic movement of the body and the lubricant action of molten fluid helps in the removal of the old cuticle. During each moulting the cuticular coverings discarded are the cuticular of legs, internal linings of foregut and hindgut and trachea.
12. **Formation of exocuticle:** The upper layer of procuticle develops as exocuticle through the addition of protein and tanning by phenolic substance.
13. **Formation of endocuticle:** The lower layer of procuticle develops as endocuticle through the addition of chitin and protein. This layer increases in thickness.

2.2 Control of Moulting

Three types of hormones are involved in the process of moulting. They are:

Juvenile hormone (JH): Produced from corpora allata of brain that helps the insects to be in the immature stage.

Moulting hormone (MH): Produced from prothoracic glands of brain that induces the process of moulting.

- **Eclosion hormone:** Released from neurosecretory cells in the brain that help in the process of ecdysis or eclosion.

Endocrine glands are activated by prothoracicotropic hormones produced by neurosecretory cells of the brain. Ecdysone and 20-hydroxyecdysone stimulate moulting and are thought to be the principal moulting hormones. It is common to assume that the prothoracic glands control moulting by synthesising ecdysone, even though there is no direct evidence for this. Experimental evidences show that the head and thorax of *Calpodes* larvae isolated with intact activated prothoracic glands do not moult; they can only do so if the moulting hormone is injected or if at least part of the abdomen is present. This suggests that more than an activated prothoracic gland is needed for moulting.

3. CUTICLE SCLEROTISATION

Right after moulting, since the newly synthesised cuticle is very soft, insects are extremely vulnerable. They have limited mobility which makes them easy prey for predators. The cuticle is unable to provide enough protection from desiccation (or from imbibing excessive water in the case of aquatic insects) and pathogens. Thus, for the survival of insects, a cuticle hardening and darkening process, the sclerotisation (or tanning), must occur.

Sclerotisation is a chemical process by which certain regions of the insect cuticle undergo an irreversible change from a soft layer into a stiffer and harder structure. Sclerotisation mainly occurs in the exocuticle and epicuticle, while in the endocuticle, although a high amount of chitin and proteins are present, very little sclerotisation occurs and this layer of cuticle always remains soft and flexible.

During sclerotisation the colour of the cuticle may change; some cuticles remain nearly colorless, and some become lighter or darker shades of brown or black. The term **tanning** is often used synonymously with sclerotisation, but sometimes it is specifically used for the processes whereby brownish (tan) cuticles are formed. Sclerotisation often takes place in connection with moulting, starting just after the new, as yet unsclerotized, cuticle has been expanded to its final size and shape, but some specialisedcuticular regions are sclerotised while the insect is still in its pharate state inside the old cuticle. Such pre-ecdysiallysclerotised regions cannot be expanded post-ecdysially, but help the insect to escape from the exuvium. The dipteran puparium is an example of a soft larval cuticle which is sclerotised at the end of the last larval instar to form a hard protective case, in which metamorphosis to pupa and adult can take place.

3.1 Mechanism of Sclerotisation

The exact metabolic pathway involved in the insect cuticle sclerotisation process is not yet clear. Sclerotisation mostly involves post-translational modifications that lead to covalent cross-linking between cuticular proteins and catecholamines.

There are two currently proposed models as to how catecholamines are transformed during the final sclerotisation process. The first and most widely accepted model is the quinone tanning hypothesis. This model was first proposed by Pryor (1940 a, b), modified by many researchers and well-reviewed by Andersen (2005). The quinone-tanning hypothesis proposes that N-acylcatecholamines are oxidised by a phenoloxidase to quinones and quinone methides, which serve as electrophilic cross-linking agents to form covalent cross-links between cuticular proteins. As shown in Fig. 1.8, the amino acid tyrosine is hydrolated to 3,4-dihydroxyphenylalanine (DOPA) through the action of tyrosine dehydrogenase. Then DOPA is transformed into dopamine by decarboxylation under dopa decarboxylase activity. Dopamine is the key intermediate compound for the formation of sclerotisation and another very likely related agent called melanin. The synthesised dopamine is then N-acylated to either N-acetyldopamine (NADA) or N-β-alanyldopamine (NBAD), both of which can serve as precursors in the sclerotisation process. As described in Fig 1.9, NADA (or NBAD) can be oxidized to *o*-quinone, which in turn reacts with nucleophilic groups, whereby catecholic structure is regained and the nucleophile is linked to the aromatic ring. The *o*-quinones of NADA and NBAD can also be enzymatically isomerised to *p*-quinonemethide. The *p*-quinone methides are then enzymatically isomerized to unsaturated catechol derivatives, dehydro-NADA and dehydro-NBAD, which after oxidation to un-saturated quinones can react with catechols to form dihydroxyphenyl-dihydrobenzodioxine derivatives and with other nucleophilic groups to form other unidentified compounds. These derivatives, are further linked with other proteins and perhaps chitin, to provide the final colour and stiffness of the sclerotised cuticle. The appearance and properties of cuticle from different body regions of the same animal can vary widely, and a considerable part of this variation, such as the different coloration and mechanical properties, is probably due to quantitative or qualitative differences in the sclerotisation process. The proposed sclerotisation reactions reflect studies in only a few insect species, such as the cuticle of *C. erythrocephala* larvae, the pupae of the tobacco hornworm *Manduca sexta,* and the *Locusta migratoria* femurs (Anderson, 2005).

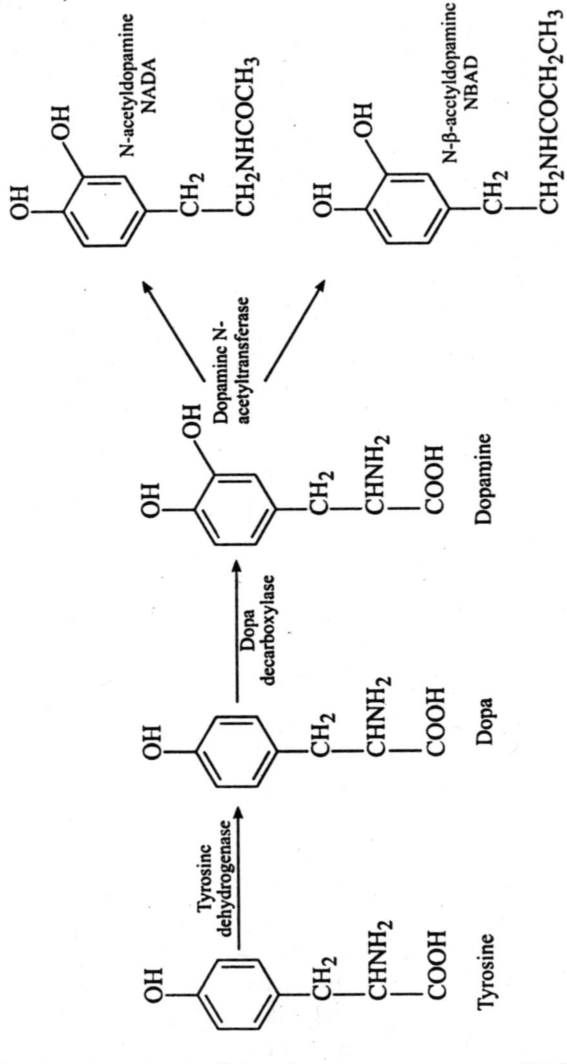

Fig 1.8. Biosynthesis of the sclerotisation precursors NADA and NBAD from tyrosine.

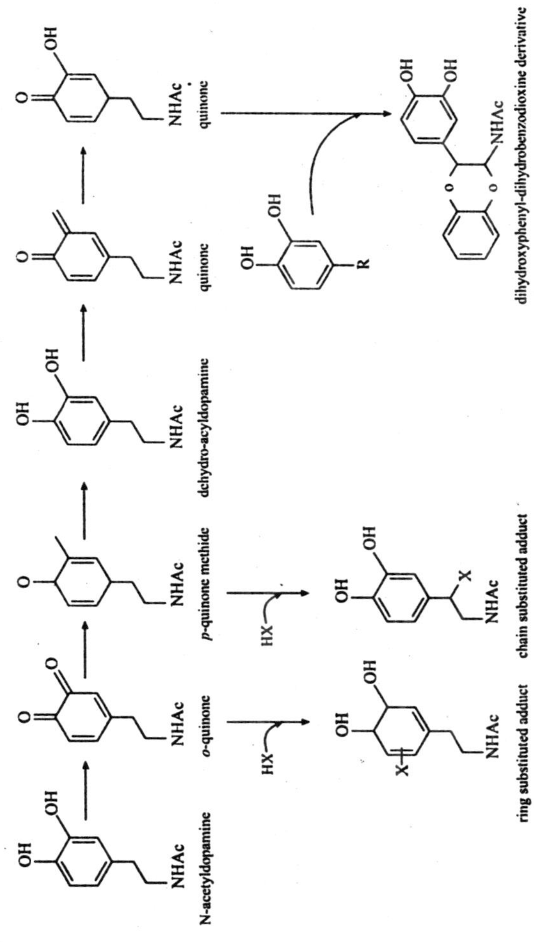

Fig 1.9. Cuticular oxidation of NADA and NBAD to various quinone derivatives

The second hypothesis for the mechanism of sclerotisation is that incorporation of large quantities of oxidised N-acylcatechols displaces water and dehydrates the cuticle to such an extent that cuticular proteins become trapped in a hydrophobic matrix of polymerized N-acylcatechols (Vincent and Hillerton, 1979). However, due to lack of evidence and experimental data, fewer scientists now support this hypothesis.

The proposed hypotheses for insect cuticle sclerotisation indicate the possible chemical reactions taking place to make the final rigid cuticle. However, these are very likely the processes that take place at the final level within the whole sclerotisation pathway. So far, it has been proposed and believed that the cuticle sclerotisation process is under control of a series of hormones.

The enzyme tyrosine dehydrogenase hydrolases tyrosine to 3,4-dihydroxyphenylanaline (DOPA), which is then decarboxylated to dopamine by the enzyme dopa-decarboxylase. Dopamine can then be acylated to either N-acetyldopamine (NADA) or N-β-alanyldopamine (NBAD).

NADA and NBAD are oxidated to various quinone isomers (*o*-quinone, *p*-quinone), which then react with catechol's to form various known and/or unidentified derivatives. These derivatives, after further linking with proteins and possibly chitin, form the final colour and stiffness of the cuticle. (The model is taken from Anderson, 2005).

3.2 The control of the sclerotisation process: hormones

At the end of the insect moulting process, insects shed their old cuticle by performing the ecdysis sequence, an innate behaviour consisting of three steps: **pre-ecdysis**, **ecdysis**, and **post-ecdysis**. The coordination of these processes is regulated by at least six hormones, prothoracicotropic hormone (PTTH), ecdysteroid 20-hydroxyecdysone (20-E), eclosion hormone (EH), ecdysis triggering hormone (ETH), crustacean cardioactive peptide (CCAP), and bursicon (Ewer and Reynolds, 2002).

The possible sequential control was proposed and summarised by Davis *et al.* (2007), as cited in Fig 1.10. The very first hormone, PTTH, initiates the whole process and is responsible for controlling the release of 20-E. The change in the titre of 20-E stimulates the release of ETH and EH, and gives the signal for the initiation of ecdysis and the synthesis of a new cuticle (Truman, 1981). ETH and EH act in a positive feedback loop to cause further releases of each other (Ewer *et al.*, 1997). ETH and EH together are responsible for the first step of moulting-**pre-ecdysis** (Baker *et al.*, 1999). It was recently reported that a subset of bursicon containing neurons in the thoracic ganglia in the larvae of the hawkmoth *M. sexta* express ETH subtype A receptors (ETHR-A), indicating that these cells are the targets of ETH, and ETH may directly regulate the expression and/or release of bursicon (Dai *et al.*, 2008).

The first hormone, PTTH, controls the release of ecdysone (20-E). A decrease in the level of 20-E triggers the release of ecdysis-triggering hormone (ETH) and eclosion hormone (EH), which further stimulate the release of each other in a manner of positive feedback. ETH and EH together regulate pre-ecdysis behavior. EH causes the release of crustacean cardioactive peptide (CCAP), which shuts off pre-ecdysis and causes the release of the hormone bursicon for the final cuticle sclerotisation process (Modified from Davis *et al.*, 2007).

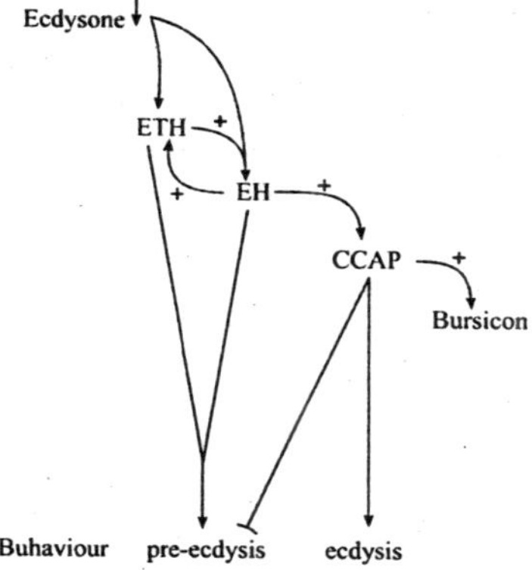

Fig 1.10. The neuropeptide signaling pathway at eclosion.

At the last step after ecdysis, the last hormone in this signal cascade, bursicon, is released and causes the sclerotisation of the newly synthesised cuticle and the expansion of insect wings (Dewey *et al.*, 2004; Huang *et al.*, 2007; Luo *et al.*, 2005; Mendive *et al.*, 2005). Another hormone, crustacean cardioactive peptide (CCAP), which behaves as a cardioaccelerator, is co-localised in all neurosecretory cells that produce bursicon (Kostron *et al.*, 1996; Luan *et al.*, 2006). The major function of CCAP hormone in moulting and sclerotisation process is not well understood, but it was proposed that CCAP might be responsible for activating the PKA signaling pathway (Luo *et al.*, 2005) and initiating tyrosine hydroxylase (TH) translation, which is thought to be a critical enzyme involved in the synthesis of tanning materials following eclosion (Davis *et al.*, 2007).

Although it is widely accepted that the bursicon hormone alone controls insect sclerotisation, the exact signal transduction pathway that it acts through is yet not well characterised.

3.4 Sclerotisation in larvae and pupae

Sclerotisation occurs not only in insects after the pupa to adult moulting, but also takes place in early developmental stages. In insects that undergo larval to larval and/or larval to pupal moulting, the sclerotisation process also occurs in various types of tissues, such as mandibles and puparium. The control of these types of sclerotisation has not been very well characterised, but it is very likely that they are different from the sclerotisation that occurs after pupal to adult moulting.

In higher holometablous orders, hardening of insect cuticle during puparium is controlled by a set of hormones called the **pupariation factors**. In these orders, insects discard their whole thoracic and abdominal epidermis during metamorphosis and build a new adult body wall from imaginal discs and abdominal histoblasts. At the end of the last larval stage, the cuticle contracts to a smooth barrel shape puparium, which sclerotises and provides a safe place where adult development takes place. Sclerotisation of the puparium is controlled by a hormone called **puparium tanning factor** (PTF). Two additional hormones, anterior segment retraction factor (ARF) and puparium-immobilising factor (PIF), are also likely involved in puparium formation (Zdarek, 1985).

Anterior segment retraction factor causes anterior segments of the larva to retract and form the larval integuments into the barrel-shaped puparium. Puparium-immobilising factor causes the larvae to become immobile for a period of time. The three hormones seem to be secreted simultaneously to cause the hardened puparium (Denlinger and Zdarek, 1994). Recently a group of peptides, which all belong to the pyrokinin hormone family, were identified to be potential pupariation factors in several insect species, including the grey flesh fly *Neobellieria bullata* (formerly called *Sarcophaga bullata*), the cockroach *Leucophaea maderae*, and the locust, *L. migratoria* (Zdarek *et al.*, 1997, 1998). One of these factors, a SVQFKPRL amide designated as Neb-pyrokinin-2 (Neb-PK-2), is believed to be the primary pupariation factor (Verleyen *et al.*, 2004). These factors are similar in sequences and are homologous to the products of two *Drosophila* genes, *capa* and *hugin*, which are too closely related FXPRL amide encoding genes (Verleyen *et al.*, 2004).

4. COLOURATION AND MELANISATION

The colour of the exoskeleton is produced in several ways. The brown colour of most insects is produced by the quinone tanning process. Melanisation is the process of producing true black pigments in the cuticle. Most insect colours are due to the pigments found in the epidermis or cuticle. For example, insecticyanin is found in the epidermal cells of *Manduca sexta.* The colour of the tomato hornworm larvae is a combination of blue pigment, biliverdin and yellow pigment, lutein obtained in the diet. A lot of the bright colours aredue to the reflective properties of the cuticle (Box 1.6). Bright iridescent colours are produced as the result of interference by multiple thin films separated by material of slightly different refractive index results in the bright metallic colours. Scattering of light produces white colours and some greens and blues. Diffraction is the splitting of light into its component spectral colours. This is done by a series of fine grooves or ridges separated by spaces corresponding to a certain wavelength of light. The latter is structural colors produced by the microarchitecture of the cuticle.

BOX 1.6. COLOURATION AND MELANISATION

Insect colours are due to either:

1. Pigments in the exocuticle such as melanin and they are usually lacking in endocuticle.
2. Physical structure of the cuticle to form defraction gradients that defract light in various ways. This results in the iridescence of the blue morpho butterflies and gold of beetles, gold on monarch chrysalis, and colour of tiger beetles, etc.

Melanogenesis involves the formation of melanin pigments and toxic by-products from the action of phenoloxidase (PO) on quinone precursors, and occurs primarily in the cuticular structures, midgut epithelium and haemolymph, where it performs dual roles in hardening and darkening the cuticle and in immune defence. Such defences include non-self-recognition and humoral encapsulation of invading organisms. The presence of melanin and melanogenesis intermediates in the insect cuticle not only limits the growth of certain fungi, possibly by acting as a physical barrier, but also suppresses synthesis of fungal cuticle-degrading enzymes, thus further impeding cuticle penetration.

Melanins are derived from the catecholamine precursors dopa and dopamine, which are synthesised from tyrosine. Tyrosine hydroxylase, encoded by the gene *pale* and Dopa decarboxylase convert tyrosine to dopa and dopamine, respectively, which are then processed by phenoloxidases and co-factors to form melanin during cuticle development. The differential production of pigments among cells generates pigment patterns. In *Drosophila,*yellow (*y*) and Ebony (*e*)genes play critical, reciprocal roles in the formation of melanin patterns. Yellow protein is required for the production of black melanin, Ebony protein is required to produce tan pigment, and in the absence of both genes, a brown pigment remains. The distribution of the yellow protein correlates with the pattern and intensity of black melanisation in the adult, and changes in this expression pattern alter melanin patterns in the absence of Ebony. The Ebony protein is not spatially regulated, despite its function in pigment patterning. Its low level, uniform expression is necessary for proper pigmentation. Differences in the level of Ebony expression between cells affect pigment patterns.

5. LEARNING FROM CUTICLE

An understanding of the biochemistry of the cuticle and its molecular structural design can have immense applied value. The chemical composition of the cuticle can be used to design novel molecules to control insect pests, while the structural design of the cuticle could itself be applied in the field of material science to develop hard structures which can withstand the harshness of the environment. Success in the development of biomimeticmaterials suitable for industrial production may be found in the application of the principles of design optimisation identified in biological materials and in the awareness that material and shape always go hand in hand, that the one needs to be analysed in parallel with the other. Bearing this in mind, cuticle with its great variety of structures and functions presents an immense source of inspiration. Adhesive-free and reversible attachment systems, wear resistant articulations with variable frictional properties, functional surfaces, mechanosensors, hardening and stiffening of polymers through metal incorporation, aspecific fiber alignment for wear resistant surfaces, fatigue and fracture resistance, and the structure and properties of the fiber composite which allow the cuticle to fulfil all the complex requirements imposed on it by the environment in which the organism lives, are examples. The use of nanofibers, resembling chitin, in composite design for improved mechanical properties is another. However, in order to derive the maximum benefit it is probably necessary to unwind our analysis so that we can generate the structures of interest by defining the chemistryand chemical environment of the components.

In recent decades, a greater knowledge of chitin chemistry, and the increased availability of chitin-containing waste materials from the seafood industry, has led to the testing and development of chitin-containing products for a wide variety of applications in the agriculture industry. A number of modes of action have been proposed for how chitin and its derivatives can improve crop yield. In addition to direct effects on plant nutrition and plant growth stimulation, chitin-derived products have also been shown to be toxic to plant pests and pathogens, induce plant defenses and stimulate the growth and activity of beneficial microbes. The effectiveness of chitin-based treatments has been found to be comparable to those achieved with current synthetic pesticides and fertilizers.The cationic properties of the chitosan oligosaccharide bestow it with unique properties that can be exploited by biotechnologists; including applications in the fields of medicine, material science, and crop science. Chitosan nanoparticles have found use in oral drug and gene delivery systems and glucosamine therapy has been used for treating osteoarthritis. It has found use as an immunoadjuvant in antitumour, antifungal and blood coagulant studies besides being used in wound healing films and absorbable surgical sutures. This effectiveness combined with the low cost, low concentration required, ample supply (recycled waste) and health/environmental safety lead to a forecast that a range of chitin-based/ augmented products will become a more common feature in agriculture and medicine in the near future.

REFERENCES

Andersen, S.O. (1979a). Biochemistry of insect cuticle. *Annu. Rev. Entomol.* 24, 29-61.

Andersen, S.O. (1979b). Characterization of the sclerotization enzyme(s) in locust cuticle. *Insect Biochem.* 9, 233-239.

Andersen, S.O. (2005). Cuticular sclerotization and tanning. In: Gilbert,L.I., Iatrou, K., Gill, S. (Eds.), *Comprehensive Molecular Insect Science.* Elsevier Press, Oxford, UK, pp. 145-170.

Andersen, S.O., Hojrup, P., Roepstorff, P. (1995). Insect cuticular proteins. *Insect Biochem. Mol. Biol.* 25, 153-176.

Baker, J. D., McNabb, S. L., Truman, J. W. (1999). The hormonal coordination of behaviour and physiology at adult ecdysis in *Drosophila melanogaster*. *J. Exp. Biol.* 202, 3037-3048.

Dai, L., Dewey, E.M., Zitnan, D., Luo, C.W., Honegger, H.W., Adams, M.E. (2008) Identification, developmental expression, and functions of bursicon in the tobacco hawkmoth, Manduca sexta. *J. comp. neurol.* 506, 759-774.

Davis, M.M., O'Keefe, S.L., Primrose, D.A., and Hodgetts, R.B. (2007). A neuropeptide hormone cascade controls the precise onset of post-eclosion cuticular tanning in *Drosophila melanogaster*. *Development* 134, 4395-4404.

Denlinger, D.L. and Zdarek, J. (1994). Metamorphosis behavior of flies. *Annu. Rev. Entomol.* 39, 243-266.

Dewey, E.M., McNabb, S.L., Ewer, J., Kuo, G.R., Takanishi, C.L., Truman, J.W., Honegger, H.W. (2004). Identification of the gene encoding Bursicon, an insect neuropeptide responsible for cuticle sclerotization and wing spreading. *Curr. Biol.* 14, 1208-1213.

Ewer, J., Gammie, S. C., Truman, J. W. (1997). Control of insect ecdysis by a positive feedback endocrine system: roles of eclosion hormone and ecdysistriggering hormone. *J. Exp. Biol.* 200, 869-881.

Ewer, J. and Reynolds, S. (2002). Neuropeptide control of molting in insects. In: Pfaff, D.W., Arnold, A.P., Fahrbach, S.E., Etgen, A.M., Rubin, R.T. (Eds.), Hormones, Brain and Behavior. Academic Press, San Diego, CA, pp. 1-92.

Finn, R.D., Mistry, J., Schuster-Böckler, B., Griffiths-Jones, S., Hollich, V., Lassmann, T., Moxon, S., Marshall, M., Khanna, A., Durbin, R., Eddy, S.R., Sonnhammer, E.L.L., Bateman, A. (2008). The Pfam protein families database. *Nucleic Acids Res.* 36, D281-D288.

Hadley, N.F. (1977) Epicuticular lipids of the desert tenebrionid beetle, *Eleodes armata*: seasonal and acclamatory effect on composition. *Insect Biochem.* 7, 277-283

Hillerton, J.E., and Vincent,J.F.V. (1983) Consideration of the importance of hydrophobic interactions in stabilizing insect cuticle. *International Journal of Biological Macromolecules*, Vol 5, 163-166.

Huang, J., Zhang, Y., Li, M., Wang, S., Liu, W., Couble, P., Zhao, G., Huang, Y. (2007). RNA interference-mediated silencing of the bursicon gene induces defects in wing expansion of silkworm. *FEBS Lett.* 581, 697-701.

Kornezos, A. and Chia, W. (1992). Apical secretion and association of the Drosophila yellow gene product with developing larval cuticle structures during embryogenesis. *Mol. Gen. Genet.* 125, 397-405.

Kostron, B., Kaltenhauser, U., Seibel, B., Braunig, P., Honegger, H.W. (1996). Localization of bursicon in CCAP-immunoreactive cells in the thoracic ganglia of cricket *Gryllus bimaculatus*. *J. Exp. Biol.* 199, 367-377

Lee, W.J. and Brey, P. T. (1994). Isolation and identification of cecropin antibacterial peptides from the extracellular matrix of the insect. *Anal. Biochem.* 217, 231-235.

Luan, H., Lemon, W.C., Peabody, N.C., Pohl, J.B., Zelensky, P.K., Wang, D., Nitabach, M.N., Holmes, T.C., White, B.H. (2006). Functional dissection of a neuronal network required for cuticle tanning and wing expansion in *Drosophila*. *J. Neurosci.* 26, 573-584.

Luo, C.W., Dewey, E.M., Sudo, S., Ewer, J., Hsu, S.Y., Honegger, H.W., Hsueh, A.J. (2005). Bursicon, the insect cuticle-hardening hormone, is a heterodimeric cystine knot protein that activates G protein-coupled receptor LGR2. *Proc. Natl. Acad. Sci.* 102, 2820-2825.

Marc J Klowden (2013) Physiological systems in Insects, Academic Press

Mendive, F.M., Van Loy, T., Claeysen, S., Poels, J., Williamson, M., Hauser, F., Grimmelikhuijzen, C.J.P., Vassart, G., Vanden-Broeck, J. (2005). *Drosophila* molting neurohormone bursicon is a heterodimer and the natural agonist of the orphan receptor DLGR2. *FEBS Lett.* 579, 2171-2176.

Molnár, K., Borhegyi, H.N., Csikós, G., Sass, M. (2001). The immunprotein scolexin and its synthesizing sites, midgut and epidermis. *Acta Biol. Acad. Sci. Hung.* 52, 473-488.

Nation, J. (2002). Insect physiology and biochemistry. CRC Press, Boca Raton, Fla.

Pryor, M.G.M. (1940a). On the hardening of the ootheca of *Blatta orientalis. Proc. R. Soc. B* 128, 378-392.

Pryor, M.G.M. (1940b). On the hardening of cuticle of insects. *Proc. R. Soc. B* 128, 393-407.

Richards, A. G., and P. A. Richards. (1979). The cuticular protuberances of insects. *Int. J. Insect Morphol. Embryol,* 8143-157.

Truman, J.W. (1981). Interaction between ecdysteroid, eclosion hormone, and bursicon titers in *Manduca sexta. Am. Zool.* 21, 655-661.

Verleyen, P., Clynen, E., Huybrechts, J., Van Lommel, A., Vanden Bosch, L., De Loof, A., Zdarek, J., Schoofs, L. (2004). Fraenkel's pupariation factor identified at last. *Dev. Biol.* 273, 38-47.

Willis, J.H., Iconomidou, V.A., Smith, R.F., Hamodrakas, S.J. (2005). Cuticular proteins. In: Gilbert, L.I., Iatrou, K., Gill, S. (Eds.), *Comprehensive Molecular Insect Science.* Elsevier Press, Oxford, UK, pp. 79-110.

Wittkopp, P.J., Vaccaro, K., Carroll, S.B. (2002). Evolution of *yellow* gene regulation and pigmentation in *Drosophila. Curr. Biol.* 12, 1547-1556.

Wybrandt, G.B. and Andersen, S.O. (2001). Purification and sequence determination of a yellow protein from sexually maure males of the desert locust, *Schistocera gregaria. Mol. Biol.* 31, 1183-1189.

Zdarek, J. (1985). Regulation of pupariation in flies. In: Kerkut, G.A. and Gilbert, L.I. (eds.), *Comprehensive Insect Physiology, Biochemistry and Pharmacology*, Vol 8. Pergamon, New York, pp. 301-333.

Zdarek, J., Nachman, R., Hayes, T. (1997). Insect neuropeptides of the pyrokinin/PBAN family accelerate pupariation in the fleshfly (*Sarcophaga bullata*) larvae. *Ann. N. Y. Acad. Sci.* 814, 67-72.

Zdarek, J., Nachman, R., Hayes, T. (1998). Structure-activity relationships of insect neuropeptides of the pyrokinin/PBAN family and their selective action on pupariation in fleshfly (*Neobellieria bullata*) larvae (Diptera: Sarcophagidae). *Eur. J. Entomol.* 95, 9-16

Diversity in Growth and Life Cycle Patterns in Insects 2

Synopsis

1. Introduction
2. Growth
3. Life history patterns
   - 3.1 Embryonic development and hatching
   - 3.2 Larval development
     - 3.2.1 Types of larvae
     - 3.2.2 Number of instars
     - 3.2.3 Larval growth
   - 3.3 Pupal development
   - 3.4 Metamorphosis
   - 3.5 The Imaginal or Adult Stage
     - 3.5.1 Emergence
     - 3.5.2 Hormonal control of adult emergence
     - 3.5.3 Maturation
     - 3.5.4 Reproduction
     - 3.5.5 Senescence

4. Hormonal control of metamorphosis
   - 4.1 Hormones in development of ametabolous insects
   - 4.2 Hormones in nymph to adult transformation in hemimetabolous insects
   - 4.3 Hormones in larva to pupa to adult transformation in holometabolous insects
5. Voltinism
6. Polymorphism and polyphenism
   - 6.1 Genetic polymorphism
   - 6.2 Environmental polymorphism or polyphenism
   - 6.3 Phase dimorphism or kentromorphism
   - 6.4 Caste or social polymorphism
   - 6.5 Polyethism
7. Environmental effects on development
   - 7.1 Temperature
   - 7.2 Photoperiod
   - 7.3 Humidity
   - 7.4 Mutagens and toxins
   - 7.5 Biotic effects

1. INTRODUCTION

Insects are the most abundant and diverse group of organisms on Earth. They have maintained a position of ecological pre-eminence for over 400 million years. They have witnessed the rise and fall of the dinosaurs and have survived at least four major cataclysms that resulted in planet-wide extinctions. They have continued to thrive despite mans' best efforts at eradication. Their unparalleled success is due to the unique combination of characteristics which, as a whole, have given them an unusual survival advantage over others. Attributes such as diversity in growth and life cycle, metamorphosis and adaptability in an ever changing environment, have contributed in a great detail to their success. Most insects undergo significant developmental changes from their immature stage to adults stage. These changes may involve physical, biochemical and/or behavioural alterations that promote survival, dispersal and reproduction of the species. This chapter dwells into the physiological aspects of growth and life history of insects.

2. GROWTH

The soft body of the insect is covered over by an exoskeleton made up of rigid cuticle which limits growth. Insects overcome this limitation by periodically shedding the old cuticle through a phenomenon of **ecdysis** and subsequently forming a new cuticle of greater surface area that provides for size increase. Thus, for sclerite-bearing body segments and appendages, increases in body dimensions are confined to the postmoult period immediately after moulting, before the cuticle stiffens and hardens. Hence, the sclerotised head capsule of a beetle or moth larva for example, increases in dimensions in a saltatory manner (in major increments) during development, whereas the membranous nature of body cuticle allows the larval body to grow more or less continuously.

Insect development comprises two components of growth. The first is **moult increment**, which is the increment in size, measured in terms of length or width of some sclerotised body part occurring between one instar stage and the next. The second component is the **intermoult period** or **stadium** or **instar duration**, which is the time between two successive moults. These two components may be affected by food, temperature, larval density and sex.

In some insects, such as collembolans, diplurans and apterygotes, growth is **indeterminate** and is continuous till death. In a vast majority of insects, growth is **determinate** in which case there is a distinctive instar that marks the cessation of growth and moulting. All insects with determinate growth become reproductively mature in this final instar, called the **adult** or **imaginal instar**. The imago or adult is the final stage of insect with well developed organs of reproduction, which emerges out from the pupal body. In Ephemeroptera, there is a pre-adult stage, the sub-imago, which has fully developed wings but without reproductive organs.

3. LIFE HISTORY PATTERNS

As the insect moults and grows, subtle changes take place in the body form. The external morphological changes are more apparent than the internal changes during ontogeny. Three broad patterns of morphological changes during ontogeny can be distinguished- **ametabolous, hemimetabolous, holometabolous**. In addition, two other life history patterns are also noticed in insects- **intermediate metabolous** and **hypermetabolous.**

1. Ametabolous: Insects do not undergo any metamorphosis. When the insect hatches from the egg, it resembles the adult in all the characters except the small body size, which later increases, until they reach sexual maturity with well developed reproductive organs. E.g. Apterygotes e.g. silver fish, springtails.

2. Incomplete metabolous or hemimetabolous or direct development or simple metamorphosis: The life cycle includes egg, nymph and adult stages. The nymph resembles the adult in all the characters except wings. Nymphs possess wing buds which transform into fully developed wings in the adult stage (Fig. 2.1). In these insects, wings develop externally and hence are also called **exopterygota**. Pupal stage is absent hence, development is said to be **direct** and **simple**.

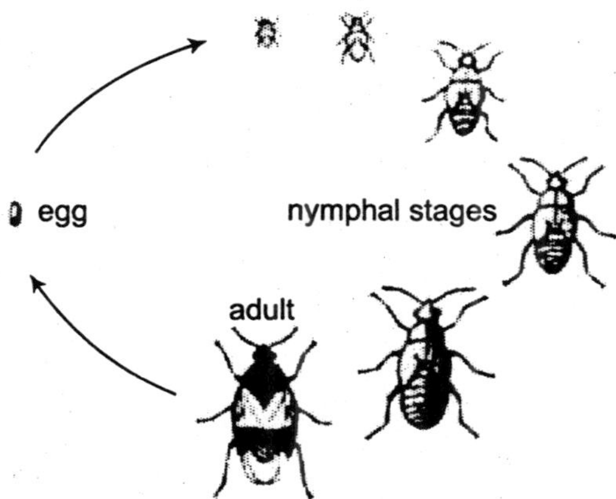

Fig 2.1. Incomplete metamorphosis as seen in bugs

3. Complete (or) holometabolous or indirect development: The life cycle includes four stages; **egg, larva, pupa and adult** (Fig 2.2). Larva differs from the adult both in body structure and habits. The larva has both thoracic and abdominal legs, sometimes legs may be absent in larva, whereas adult has only thoracic legs. Compound eyes are absent in larva. The larva undergoes moulting to enter in tothe pupal stage from which the adult insect emerges. Wings develop **internally** during the pupal stage and hence, they are called **endopterygotes.**

4. Intermediate metabolous: In this case, insects may be either **hemi** or **holometabolous.** If they are holometabolous, there is a short pupal stage. E.g.: coccids, thrips etc.

5. Hypermetabolous: This is a peculiar type of development which consists of two or more types or forms of larvae in the life cycle of insects. In majority of the cases the first larval instar is **campodeiform** and the subsequent larval forms depends on the type and mode of life of the larva. E.g.: In blister beetle (Meloidae; Coleoptera), the first larval instar is **campodeiform** followed by **scarabeiform** larval type.

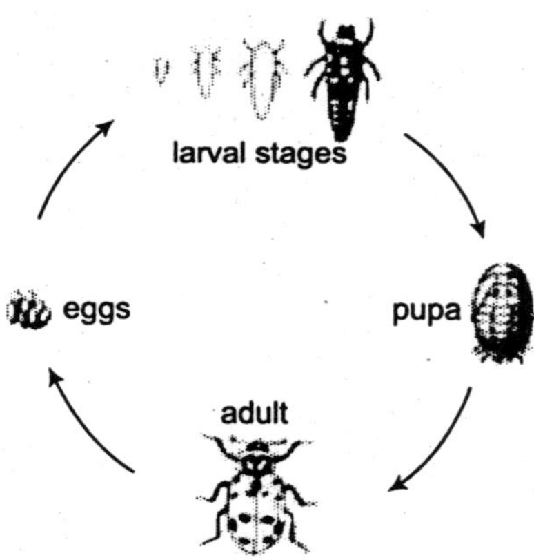

Fig 2.2. Complete metamorphosis as seen in beetles

Insect life cycle has several phases which correspond to distinct stages in insect ontogeny- embryonic period, the larval period, metamorphosis and imaginal life.

3.1 Embryonic development and hatching

All insect eggs are enclosed by a chorion which is a product of the ovarian follicle cells. The chorion usually consists of two layers, which are penetrated by one or more micropyles near the anterior egg pole. In many terrestrial and some aquatic insects, the chorion contain a respiratory system of channels which are continuous with the external environment by means of tiny holes called **aeropyles**.

Most insect eggs are ovoid in shape, while in some they are compressed to a disc, elongated to an ellipsoid, or modified to another shape. Some eggs are found on stalks or have long flanges and horns that serve for the attachment or respiration. During deposition, the eggs are covered by products of the accessory collateral glands. Production of a cement by which the eggs are attached to a substrate is very common and when, eggs are deposited in batches, they are also cemented to one another.

An individual insect's ontogeny actually begins when embryonic development within the egg is triggered by activation. This trigger usually results from fertilization in sexually reproducing insects, but in parthenogenetic species appears to be induced by various events at oviposition, including the entry of oxygen to the egg or mechanical distortion. Following activation of the insect egg cell, the zygote nucleus subdivides by mitotic division to produce many daughter nuclei, giving rise to a syncytium. These nuclei and their surrounding cytoplasm, called **cleavage energids**, migrate to the egg periphery where the membrane infolds, leading to cellularisation of the superficial layer to form the one-cell thick blastoderm. This distinctive superficial cleavage during early embryogenesis in insects is the result of the large amount of yolk in the egg. The blastoderm usually gives rise to all the cells of the larval body, whereas the central yolky part of the egg provides the nutrition for the developing embryo and will be used up by the time of eclosion, or emergence from the egg.

Regional differentiation of the blastoderm leads to the formation of the germ Anlageor the germ disc (Fig 2.3a), which is the first sign of the developing embryo, whereas the remainder of the blastoderm becomes a thin membrane, the serosa, or embryonic cover. Next, the germ Anlage develops an infolding in a process called **gastrulation** (Fig 2.3b) and sinks into the yolk, forming a two-layered embryo containing the amniotic cavity (Fig 2.3c). After gastrulation, the germ Anlage becomes the germ band, which externally is characterized by segmental organization(Fig 2.3d). The germ band essentially forms the ventral regions of the future body, which progressively differentiates with the head, body segments and appendages becoming increasingly well defined. At this time, the embryo undergoes movement called **katatrepsis**, which brings it into its final position in the egg. Later, near the end of embryogenesis the edges of the germ band grow over the remaining yolk and fuse mid-dorsally to form the lateral and dorsal parts of the insect—a process called **dorsal closure**.

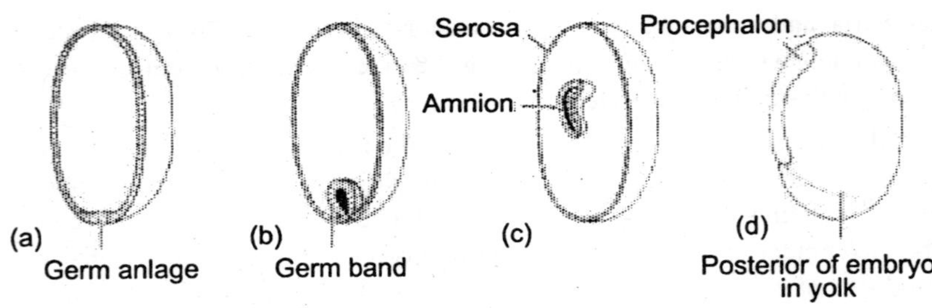

Fig 2.3. Schematic drawing of early embryonic development of insect (Scorpionfly) (a) 32 h; (b) 2 days; (c) 7 days; (d) 12 days (Modified from Suzuki 1985)

In the well-studied *Drosophila*, the complete embryo is large, extending across the egg, andbecomes segmented at the cellularisation stage, termed **long-germ** (as in most studied Diptera, Coleoptera and Hymenoptera). This long-germ band contains the primordial (rudiments) of all segments that will form the embryo. In contrast, in **short-germ band** insects (such as locusts and the beetle *Tribolium*), the embryo derives from only a small region of the blastoderm, and the germ band mostly forms only the anterior parts of the headwith the posterior segments added post-cellularisation, during subsequent growth. In **intermediate-germ band** insects (such as damselflies and crickets), the germ band forms from two ventrolateral cell aggregations, which fuse ventrally to become primordia of the head and thorax, with abdominal segments budded off later, as in short-germ band insects. In the developing long-germ embryo, the syncytial phase is followed by cell membrane intrusion to form the blastoderm phase.

Functional specialisation of cells and tissues occurs during the latter period of embryonic development, so that by the time of hatching, the embryo is a tiny proto-insect packed into an eggshell. In ametabolous and hemimetabolous insects, this stage may be recognised as a **pronymph**—a special hatching stage.

Many eggs have a special line of weakness when the pressure from inside increases and the shell ruptures at the time of hatching. An opening is formed in this way in many Diptera and Heteroptera. In Mantodea, the opening is present from the time of chorion secretion but is plugged by a stopper which is expelled at hatching. The chorion ruptures when the larva inside expands by swallowing amniotic fluid and air and exerting muscular activity. During hatching, the larvae of most exopterygotes, also ecdyse from an embryonic cuticle which is left behind in the egg shell.

The length of the incubation period between egg deposition and hatching varies from a couple of days to many months. The extreme prolongations are largely due to intercalation for a period of quiescence in the embryonic development. Without the developmental arrest, the embryogenesis of most insects lasts 1-3 weeks, but in many dipterans it takes only 2 days. Embryogenesis is also short in some Hymenoptera and Coleoptera. In some Mantodea and Phasmida embryonic development continues for several months.

3.2 Larval development

Eclosion conventionally marks the beginning of the first stadium, when the young insect is said to be in its **first instar**. This stage ends at the first ecdysis, when the old cuticle is cast off and the insect enters into its **second instar**. Third and often subsequent instars generally follow. Thus, the development of the immature insect is characterised by repeated moults separated by periods of feeding, with hemimetabolous insects generally undergoing more moults to reach adulthood than do holometabolous insects. Most immature holometabolous insects are called **larvae.** Larval forms of coleopterans are called **grubs**. Immature terrestrial insects with hemimetabolous development, such as cockroaches and termites (Blattodea), grasshoppers and crickets (Orthoptera), mantids (Mantodea) and bugs (Hemiptera), always are called **nymphs** (Fig 2.4). However, immature individuals of aquatic hemimetabolous insects (Odonata, Ephemeroptera and Plecoptera), although possessing external wing pads at least in later instars, also are frequently, but incorrectly, referred to as **larvae** (or sometimes **naiads**). True larvae look very different from the final adult form in every instar, whereas nymphs more closely approach the adult appearance at each successive moult. Larval diets and lifestyles are very different from those of their adults. In contrast, nymphs often eat the same food and coexist with the adults of their species (Box 2.1). Competition thus is rare between larvae and their adults, but is likely to be prevalent between nymphs and their adults. There are exceptions with nymphal forms having pupal stage or puparia as in some Thysanoptera and Hemiptera.

Fig. 2.4. Grasshopper life cycle

Source: Hill, M.A. (2016) Embryology Grasshopper lifecycle.jpg.Retrieved from https://embryology.med.unsw.edu.au/embryology/index.php/File:Grasshopper_lifecycle.jpg

Box 2.1 General differences between larva and nymph

Larva	Nymph
1. It is an immature stage of endopterygotes	Immature stage of exopterygotes
2. It undergoes holometamorphosis	It undergoes hemimetamorphosis
3. Body is vermiform which differs from the adult both in structure and feeding habits	Body resembles the adult in all the characters except wings
4. Consists of ocelli and reduced antennae	Generally have compound eyes and antennae
5. Possess both thoracic and abdominal legs	Possess only thoracic legs
6. The larva is different from adult in feeding habits and behaviour	Nymph resembles the adult in feeding habits and behaviour
7. The larva enters pupal stage Eg.Lepidoptera, Coleoptera	Generally no pupal stage Eg.Hemiptera, Orthoptera.

3.2.1 *Types of larvae*

1. Protopod larva: The larva emerges (hatch) from an egg, which is still in an early embryonic phase as the egg contain less yolk material. The larvae partially develop. They possess a well developed head and thoracic segments, but lacks segmentation in the abdomen. They possess rudimentary cephalic and thoracic appendages, but no abdominal appendages. They have partially developed digestive system and under developed respiratory and nervous systems. Eg: endoparasitic Hymenoptera.

2. Oligopod larva: These are characterised by the presence of well developed thoracic legs, head capsule and without any abdominal legs. These are classified into two types viz.,campodeiform and scarabaeiform. The differences between these two types are given in Box 2.2.

Box 2.2 Differences between Campodeiform and Scarabaeiform

Campodeiform	Scarabaeiform
1. The body is long and fusiform in shape	1. Body is 'C' shaped
2. Body is dorso-ventrally compressed with sclerotised cuticle	2. Body is cylindrical or sub cylindrical, stout and fleshy in nature
3. Prognathous type of head	3. Hypognathous type of head
4. Long thoracic legs	4. Short thoracic legs
5. A pair of terminal abdominal processes (anal cerci) are present	5. Absent
6. These are active	6. Inactive
7. Predatory in nature E.g: Neuroptera, Trichoptera	7. Phytophagous E.g.: Scarabidae of Coleoptera

3. Polypod larva (Eruciform larva): The larva possesses well defined segmentation of the body with three pairs of thoracic legs, and 2-5 pairs of abdominal legs (3rd, 4th, 5th,6th and 10th abdominal segment. They are phytophagous and destructive. Some forms of polypod larvae are:

A. **Hairy caterpillar:** Larval body is fully covered with hairseg: red hairy caterpillar (*Amsacta albistriga*), castor hairy caterpillar (*Pericaliaricini*).

B. **Sphingid caterpillar / larva:** The larva consists of a horn (or) hook on the dorsal surface of 8th abdominal segment. Eg: *Acherontia styx* (Gingelly death's head moth).

C. **Looper:** Only two pairs of abdominal legs present on 6thand last abdominalsegment. During walking, the insect body forms a complete loop like structure, hence, the name **looper**. Eg: Mango looper, *Thalassodes quadraria.*

D. **Semilooper:** e.g. Castor semilooper (*Achoea janata*). First two pairs of abdominal legs (on 3rd and 4th segments) are reduced, hence a part of the insect body forms a small loop during its movement.. Eg: Castor semilooper *Achoea janata.*

4. Apodous larva: These are characterised by the absence of trunk appendages (or) legs.They possess 3 pairs of sensory papillae in the place of thoracic legs. They are usually derived from oligopod type. Based on the degree of development of the head capsule and its appendages, these larvae are divided into three types (Fig 2.5):

A. **Eucephalous:** E.g. Sub order Nematocera of Diptera, Mosquito (Culcidae). The larva consists of a well sclerotised head capsule.
B. **Hemicephalous.** E.g. Brachycera of Diptera, robberflies (Asilidae). Larva possess partially developed head capsule.
C. **Acephalous.** E.g. Cyclorrhapa of Diptera, Muscidae (house flies). The larva is characterised by the absence of head capsule and the mouth parts are represented by mouth hooks.

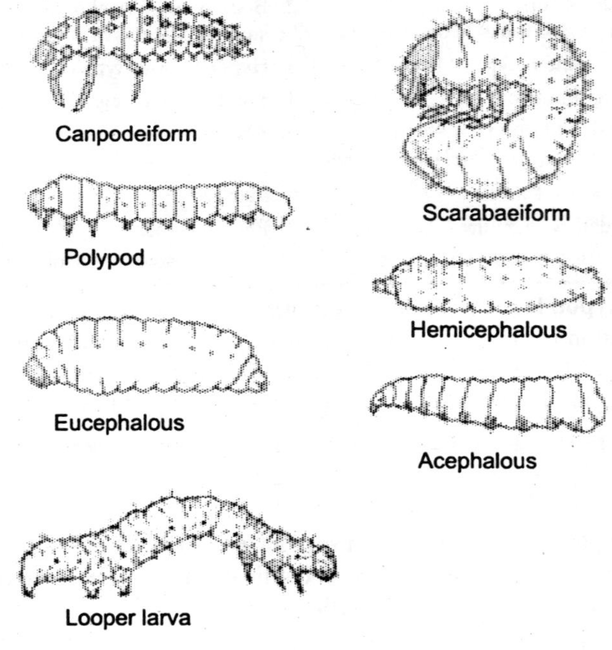

Fig. 2.5. Types of insect larvae

(Source: Compiled from drawings available on the web)

3.2.2 *Number of instars*

The post embryonic development of insects is characterised by a series of moults during which the old cuticle is replaced by a new one. The detachment of the epidermis from the cuticle is called **apolysis**. Apolysis usually occurs first on the extremities from where it spreads to all over the body surface. The development phase of complete apolysis when a new instar is externally defined, but the animal remain within the remnants of its old cuticle is called **pharate phase**. This term is applied only to insects with an externally obvious exuvial space filled with copious amounts of moulting fluid or air. Ecdysis marks the end of one and the beginning of the next interecdysial period, which is conventionally called **instar**. Exopterygote larvae exhibit at each moult some differentiation of the final imaginal features and this requires time. Rapid progress of the moulting process can be viewed as an apomorphic developmental characteristic of endopterygote larvae which is associated with lack of considerable modification in shape. Short moulting time permits the prolongation of the feeding period and allows for larger size increase in the instar.

The number of larval instars varies over broad limits. Just prior to moulting, the larvae stops feeding and physiologically has reduced endurance to environmental hazards. This has led to an evolutionary pressure for reducing the number of moults. The larvae of phylogenetically primitive orders usually moult more frequently than those of Paraneoptera and Oligoneoptera.

The number of larval instars is genetically fixed. For example, different strains of *Bombyx mori* have four, five or six larval instars, and in some species the larvae of one sex, usually the females, have an additional instar than the males as observed in some species of grasshoppers. In most insects, the number of larval instars moults varies under the influence of environmental conditions. For example, the cloth moth *Tineola*, which may moult upto 40 times, whereas normally it, has just five larval instars. Primitive insects are more prone to environmentally influenced fluctuations than the advanced groups. The number of larval instars is fairly constant in most Paranoptera and Oligoneoptera. In cyclorrhaphous Diptera, some Hymenoptera and Coleoptera there are always just three larval instars.

3.2.3 *Larval Growth*

Since insects are endowed with a heavy sclerotised cuticle. Growth can take place only during the brief post-ecdysal period when the cuticle is stretchable. At this time insects are expanding its body volume by swallowing air or water and by inhaling air to the full expansion of the tracheal sacks. Once the cuticle stiffens, growth is limited to space available when cuticular folds and wrinkles are pressed out. The unfolding of the pliable intersegmental membrane can also facilitate growth in exopterygotes between ecdysis. In most endopterygote larvae the number and size of rigid sclerites is reduced and most of the body surface is covered with soft extensible cuticle.

Integumental folds and cuticular extensibility permit a nearly continuous increase in the overall body size and the larval body weight. It has been proposed that insects growing under optimal conditions, the growth is regular and the weight increments are predictable from the equation $W=a + bt + c\log(t + x)$ (Gier 1947) where W =body weight, a,b, c are species specific constants, t is time in days, and x is length of embryonic incubation period in days.

The weight increase is briefly suspended around each ecdysis when the insect stops feeding. At this time, the body weight has dropped due to the purging of the gut, sloughing off of the exuviae and loss of water. Weight begins to rise within a few hours after ecdysis when insects start to feed. Daily coefficients of food consumption and the output of excrement fluctuate with maximums in the first third or middle of each instar and minimum at the time of ecdysis. A similar trend is also noticed for the oxygen consumption. Respiration is minimal around ecdysis, increases concomitantly with the resumption of feeding and culminates when the gut is fully extended with ingested food.

The relative growth of various body parts can be compared with Huxley's equation

$$y = bx^{\alpha}$$

where y is the dimension of the organ studied, b and α are constants, and x is the body size or dimension of a **reference organ**. The growth of an organ is isometric with the body or reference organ when $\alpha = 1$, but this is rather an exception. The growth of different organs is usually allometric and α is positive and different from 1. Negative values of α are obtained in case of regressive growth. Transformation of Huxley's equation into a logarithmic form results in a linear relationship, $\log y = \log b + \alpha \log x$. The growth constants of two organs can also be compared in a derived equation $K_1 = K_2^{\alpha}$ (K_1 and K_2 are growth constants of two organs) which is often also used in the logarithmic form, $\log K_1 = \alpha K_1 = \alpha \log K_2$.

3.3 Pupal development

Pupa is a resting, inactive stage of the holometabolous insects and a transitional phase during which the wings are developed and the insect attains matured sexual organs (Fig 2.6). The pupa is incapable of feeding, locomotion except in some cases where they crawl (Neuroptera) (Aphid lion), or can swim e.g.: mosquitoes. Pupa represents apharate adult stage and later it emerges as an adult. Pupation may take place either in the soil, or on the plant surface or within the webs. Pupae are divided on the following bases:

1. Based on the presence or absence of powerful mandibles

A. Decticous pupae: It possess relatively powerful mandibles which are used for escaping of the adult from the cocoon i.e. to break the cocoon. E.g: Neuroptera.

B. Adecticous pupae: It do not possess the mandibles, but with the help of other appendages the adults escape from the cocoon eg: Lepidoptera, Diptera.

2. Based on the attachment on the appendages (or) shape of the pupae

A. Exarate pupa: e.g. most of the Coleoptera. The pupae have appendages which are free without any secondary attachment to the body.

B. Obtect pupa: eg: Lepidoptera (moths). The pupae have appendages which are firmly pressed against the body and the pupa is highly chitinised.

C. Coarctate: e.g. Cyclorrhapha of Diptera (housefly). The pupa remains enclosed in a puparium formed by the last larval skin and the pupa looks like a capsule or barrel.

D. Chrysalis: It is an obtect type of pupa which has golden colouration and a stalk eg: butterflies.

In the life cycle of an insect, the pupal stage follows the larval stage and precedes adulthood (imago). It is during the time of pupation that the adult structures of the insect are formed while the larval structures are broken down. Pupae are inactive, and usually sessile. Some have a protective covering. Most moth caterpillars spin a cocoon made of silk within which they metamorphose into the pupal stage. Most butterflies, on the other hand, form an exposed pupa, which is also termed as a **chrysalis**.

Pupation may be brief, for example, twoweeks, as in monarch butterflies, or the pupa may enter dormancy or diapause until the appropriate season for the adult insect. Intemperate climates, pupae usually stay dormant during winter, while in the tropics pupae usually do so during the dry season.

Insects emerge from pupas by splitting the pupal case, and the whole process of pupation is controlled by hormones. Most butterflies emerge in the morning. In mosquitoes, the emergence typically is in the evening or night. In fleas, the process is triggered by vibrations that indicate the possible presence of a potential host.

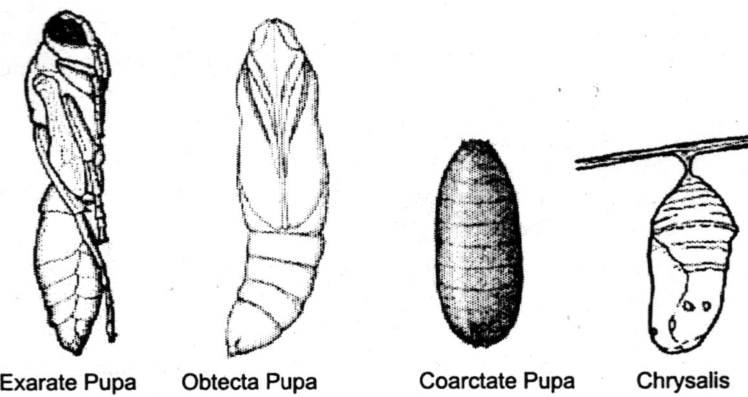

Fig 2.6. Types of pupa

(Source: Compiled from drawings available on the web)

During the pupal stage, the insect's larval tissues break down and groups of cells known as **imaginal disks** develop into the adult tissues, using the nutrients from the broken down larva. This process of cell death is called **histolysis**, and cell re-growth **histogenesis**.

In some insect orders, the appendages such as legs and proboscis are free and visible in the pupal stage. Such pupae are termed as **exarate** and examples are seen in the Hymenoptera. In many other orders, the pupa is a tight and compact shell with all the appendages tightly packed within and these pupae are termed as **obtect**. The familiar lepidopteran chrysalis is obtect.

Another form has the appendages visible, but covered within a shell. In some cases the covering is formed by the integument of the last larval instar. Such pupae are termed as **coarctate** and are found in many of the Diptera. Some exarate pupae, such as those of the Neuroptera, also have movable mandibles attached to the head. Such pupae are termed **decticous**. In most other insects the mandibles are immovable and such pupae are termed **adecticous**.

The wings or wing pads, which are not visible on the outside of the larva, become apparent in the pupal stage. However, tiny developing wing disks actually can be found in the larval stage. In butterfly caterpillars near pupation, the wings are forced outside the epidermis under pressure from the hemolymph, and although they are initially quite flexible and fragile, by the time the pupa breaks free of the larval cuticle they have adhered tightly to the outer cuticle of the pupa (in obtect pupae). Within hours, the wings form a cuticle so hard and well-joined to the body that pupae can be picked up and handled without damage to the wings. To transform from the miniature wings visible on the outside of the pupa into large structures usable for flight, the pupal wings undergo rapid mitosis and absorb a great deal of nutrients.

Pupae are usually immovable and are largely defenseless. To overcome this, a common feature is concealed placement. Some species of Lycaenid butterflies are protected in their pupal stage by ants. Some species of pupae are capable of making sounds or vibrations to scare potential predators. A few species use chemical defences, including toxic secretions. The pupae of social hymenopterans are protected by adult members of the hive.

A **chrysalis** or **nympha** is the pupal stage of butterflies. Because chrysalids are often showy and are formed in the open, they are the most familiar examples of pupae. Most chrysalids are attached to a surface by a velcro-like arrangement of a silken pad spun by the caterpillar and a set of hooks (cremaster) at the tip of the pupal abdomen. Like other types of pupae, the chrysalis stage in most butterflies is one in which there is little movement. However, some butterfly pupae are capable of moving the abdominal segments to produce sounds or to scare away potential predators. Within the chrysalis, growth and differentiation occur. The adult butterfly emerges from this and expands its wings by pumping haemolymph into the wing veins.When the butterfly emerges from the chrysalis, usually it will sit on the empty shell in order to expand and dry its wings.

Moth pupae are usually dark in colour and either formed in underground cells, loose in the soil, or their pupa is contained in a protective silk case called a **cocoon**. Very few butterflies larvae spin a cocoon. A cocoon is a casing spun of silk by many moth caterpillars and numerous other holometabolous insect larvae as a protective covering for the pupa. Cocoons may be tough or soft, opaque or translucent, solid or meshlike, of various colours, or composed of multiple layers, depending on the type of insect larva producing it. Many moth caterpillars shed the larval setae and incorporate them into the cocoon; if these are urticating hairs then the cocoon is also irritating to the touch. Some larvae attach small twigs, fecal pellets or pieces of vegetation to the outside of their cocoon in an attempt to disguise it from predators. Others spin their cocoons in a concealed location—on the underside of a leaf, in a crevice, down near the base of a tree trunk, suspended from a twig or concealed in the leaf litter (Scoble 1992).

Insects that pupate in a cocoon must escape from it, and they do this either by the pupa cutting its way out, or by secreting fluids that soften the cocoon. Some cocoons are constructed with built-in lines of weakness along which they will tear easily from inside, or with exit holes that only allow a one-way passage out. Such features facilitate the escape of the adult insect after it emerges from the pupal skin.

3.4 Metamorphosis

Metamorphosis is a period of intensive morphogenesis by which the structural difference between the larva and imago are bridged. Insects attain maximum size by undergoing a succession of moults or ecdyses. The number of moults that an insect passes through is quite constant for the species, and the form assumed by the animal between any two ecdyses is called **instar**. The animal's existence is thus made up of a succession of instars (growth) during which the insect is immature, followed by the attainment of the final adult instar (metamorphosis). In the simplest and most generalised insects the instars resemble one another and only differ from the adult in the absence of wings and the incomplete development of the reproductive system. Where the adult is primitively wingless, as in silverfish and springtails the change from young to adult is so slight as to be ignored, and metamorphosis, involving only a development of the reproductive system, is usually considered to be absent. The insect orders in this category may be grouped under the Ametabola.

However, in winged insects, the adult is in sharp contrast to the young stage not only because of its wings, but also on account of the sexual appendages that in the adult have become prominent. Such forms are said to undergo a **metamorphosis**. Here the degree of metamorphosis varies considerably, irrespective of wings, depending on whether the young stages resemble their adults or not. A growth stage of a cockroach, eg., has the general appearance of the adult. But the young stage of a housefly is a grub and has no resemblance to the final stage with its wings, complex body form and mouthparts.

Metabolous insects pass through a distinctive metamorphosis, and classifications have been based on this.

The heterometabola are insects whose young stages, called **nymphs**, closely resemble the adult in body form and type of mouthparts. These young stages have compound eyes and, where wings develop, their growth is external and clearly visible throughout most or all of the nymphal instars. Included are the following orders: Isoptera, Zoraptera, Orthoptera, Embioptera, Dermaptera, Homoptera, Heteroptera, Mallophaga, Anoplura, Thysanoptera, Ephemeroptera, Odonata and Plecoptera. The last three orders, having nymphs adapted for aquatic life, have sometimes been grouped accordingly as hemimetabola.

The Holometabola, including the orders Coleoptera, Neuroptra, Mecoptera, Trichoptera, Lepidoptera, Diptera, Hymenoptera, Strepsiptera and Siphonaptera, have young stages known as **larvae** that differ noticeably from the adult in body form and mouthparts. These young stages do not have compound eyes, but in their place are lateral ocelli. Also, their wings develop from within pockets of the hypodermis and are not visible from the outside during growth. So great is the difference between the larva and the adult that an instar known as the **pupa** has been developed to connect the gulf between them.

The pupa is described as the **resting stage**. However, it is really one of many physiological and developmental activity, and it is here that many larval tissues, e.g. the muscles and the alimentary canal, are broken down by phagocytic or other processes and the new adult tissue is constructed from many growth centers, usually called **imaginal discs**. The change from larva to pupa is often accompanied by a period of inactivity at the end of the last larval instar.

In the Heterometabola, development of adult form is a gradual process and the appendages, including mouthparts, antennae and legs, grow directly into those of the adult. Wings develop gradually as external dorsolateral extensions of the meso- and metathoracic body wall. All the Heterometabola have this kind of wing development and for that reason the alternative name **Exopterygota** is sometimes given to the group.

Larvae of the Holometabola have, for the most part, mouthparts with a form and mode of working different from that of their adults, their legs are reduced in size and complication or even absent, and wing growth is hidden from view; thus, the alternative name **Endopterygota**. It is in the pupal stage of Holometabola that adult appendages show up on the surface.

The development of adult appendages in the larva is only one of the many aspects of metamorphosis. The wings that suddenly appear in the pupa of the butterfly grow gradually through each of the five larval instars, but instead of growing externally as in the Heterometabola (Exopterygota) they appear as outgrowths from the bottom of invaginations of the body wall. This obliging folds of the body wall forming a sac, opening at the surface by a minute pore, hides the growing wing bud within it, making it the main difference between **endopterygote** and **exopterygote** development.

At pupation, the sac carrying the wing disc or bud at its base straightens out by contraction of its walls and the wing bud is in this way brought to view. Similar limb buds occur for the adult legs and mouthparts that always grow in association with the corresponding larval organs. These buds are also called **imaginal discs** and their existence are typical of all endopterygote insects.

Thus, there are two post-embryonic processes that occur in the lives of all insects. Both require the process of moulting, the shedding of the old and the production of the new cuticle. The first of these, growth, is the attribute of the young insect; the second, metamorphosis, is the attainment at the final moult of the condition of sexual maturity. These processes are under hormonal control from **neurosecretory cells** of the brain, the **corpora allata** and the **prothoracic glands**. The interchange of these is complex, depending on a precisely timed activation. The result of this interplay is the suppression of adult characters from the appearance during youth and the final attainment of sexual maturity when the juvenile features are lost. The prothoracic gland produces the hormones necessary for moulting but only when activated to do so by the neurosecretory cells of the brain.

Moulting is the route by which either further juvenile stages or the final adult stage can be attained. In the presence of the secretion of the corpora allata the juvenile condition results. When these glands stop functioning, usually at the end of the larval period, the moulting process, still activated by the prothoracic gland, gives rise to the adult.

3.5 The Imaginal or Adult Stage

The adult is the perfect or sexual stage of an insect. In this stage the sexual organs mature and locomotory appendages are sufficient for propagation. The storage of energy varies considerably among species. Some adults can only take water or liquid; some adults use fat bodies that were stored in a previous immature stage, and some, like Chironomid midges, cannot even imbibe liquids. Much of metamorphosis is directed toward producing an adult that can propagate the species.

Except for the mayflies, pterygote insects do not moult again once the adult phase is reached. The adult, or imaginal, the stage has a reproductive role, and is often the dispersive stage in insects with relatively sedentary larvae. The imago that emerges (ecloses) from the cuticle of the previous instar may be capable of reproduction almost immediately or a period of maturation may precede sperm transfer or oviposition. Depending on species and food availability, there may be from one to several reproductive cycles in the adult stadium. The adults of certain species, such as some mayflies, midges and male scale insects, are very short-lived. These insects have reduced or no mouthparts, and fly for only a few hours or at the most a day or two—they simply mate and die. Most adult insects live at least a few weeks, often a few months and sometimes for several years; termite reproductives and queen antsand bees are particularly long-lived. The evolution of eusociality is associatedwith a 100-fold increase in adult lifespan, based on a comparison of the mean average longevity of ant, termite and honeybee queens with that of adult solitary insects.

3.5.1 *Emergence*

Adult life begins at eclosion from the pupal or last-nymphal cuticle. Metamorphosis, however, may have been complete for some hours, days or weeks previously, and the pharate adult may have rested within the pupal cuticle until the appropriate environmental trigger foremergence. Changes in temperature or light, and perhaps chemical signals, may synchronize adult emergence in most species. Perceived signals are interposed over the entrained endogenous circadian rhythm. Environmental and probably also intrapopulation signals (pheromones) synchronise emergence of the species to a short period of time. This is particularly important in species with short-lived adults.

A complex of behavioural patterns is observed in preparation for emergence. Aquatic pharate adults of Odonata, Ephemeroptera, Plecoptera, Trichoptera and other insects swims to the surface or crawl out of water before emergence. Pharate adult of caddisflies first uses the pupal mandibles to bite their way out of the larval case. Some insects burrow through the soil and come to the ground surface.

In *M. sexta*, a combined action of external and internal stimuli triggers the release of a peptide eclosion hormone that turns on a behavioural programme encoded in the abdominal ganglia. The programme ends with the sloughing off the old exuviae. The hormone further conditions the inflation of the wings, release of bursicon and post-ecdysial degeneration of some muscles.

Imaginal ecdysis is regularly followed by the involution of muscles that facilitated emergence behaviour, but has no function in the adult. Degeneration of intersegmental abdominal muscles in *Antherae apernyi* is affected by the eclosion hormone and takes place shortly after the emergence.

3.5.2 *Hormonal control of adult emergence*

Our understanding of hormonal control of emergence derives substantially from studies of the tobacco hornworm, *Manduca sexta* (Sphingidae). Despite the huge range of insect body plans, highly conserved systems of eclosion control in other taxes, resemble that found in *M. sexta*. At least six hormones are involved in eclosion. A few days pre-eclosion, a rise in the neuropeptide **corazonin** and decline in ecdysteroid level initiates physiological and behavioural preparation for ecdysis, including the release of neuropeptide hormones and transmitters. **Ecdysis triggering hormones** (PETH and ETH), secreted from **Inka cells** clustered in epitracheal glands and/or dispersed throughout the tracheal system, and **eclosion hormones** (EH), from neurosecretory cells in the brain, stimulate pre-eclosion behaviour,such as seeking a site suitable for ecdysis and inducing contractions that aid extrication from the old cuticle. PETH (pre-edysis triggering hormone) is released first and ETH and EH then stimulate each other's release, forming a positive feedback loop. The build-up of EH releases **crustacean cardioactive peptide** (CCAP) from cells in the ventral nerve cord. CCAP induces a switch from pre eclosion to eclosion behaviours, including body contractions and wing-base movements, and accelerates heartbeat. EH appears also to permit the release of further neurohormones **cardiopeptides** and **bursicon**—that are involved in wing expansion and cuticle hardening after ecdysis. **Cardiopeptides** stimulate the heart, pumping haemolymph into the thorax and thus into the wings. **Bursicon** induces a brief increase in cuticle plasticity, allowing wing expansion, followed by sclerotisation of the cuticle.The newly emerged, or **teneral**, adult has soft cuticle, allowing expansion of the body surface by swallowing air, by taking air into the tracheal sacs, and by locally increasing haemolymph pressure by muscular activity. The wings normally hang down (Fig 2.7), which aids their inflation. Pigment deposition in the cuticle and epidermal cells occurs just before or after emergence, and is either linked to, or followed by, sclerotisation of the body cuticle under the influence of the neurohormone bursicon. A newly moulted insect is often white. Following emergence from the pupal cuticle, many holometabolous insects void a faecal fluid called the **meconium**. This represents the metabolic wastes that have accumulated duringthe pupal stadium. Sometimes, the teneral adult retains the meconium in the rectum until sclerotisation is complete, thus aiding increase in body size.

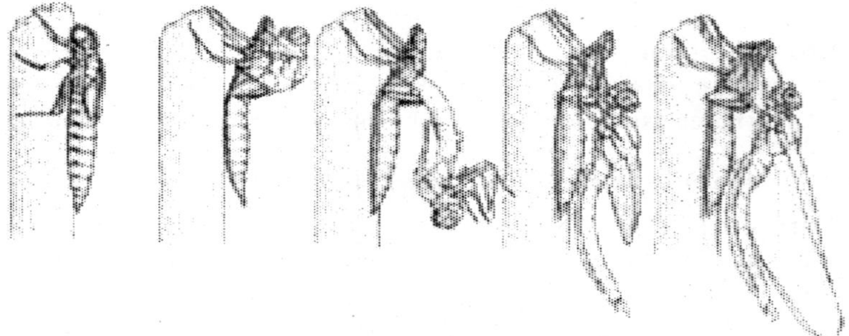

Fig. 2.7. The nymphal-imaginal moult of a male dragonfly, *Aeshna cyanea* (After Blaney 1976)

3.5.3 *Maturation*

Adult emergence is followed by cuticular sclerotisation, only after which the imago can feed and fly around. However, to attain sexual maturity considerable time has to pass. In a few individuals of Ephemeroptera, Phasmida and Plecoptera, females emerge with eggs in the oviduct and are able to mate and deposit eggs within a few hours after emergence. Most emerging females contain previtellogenic egg chambers and differentiating cells in the germarium. During maturation the chambers grow and only afterwards become competent to sequester yolk proteins. Males in contrast already contain ripe sperms and their maturation primarily is concerned with the activation of accessory glands.

Maturation of gonads is accompanied by morphogenesis of other tissues. Degeneration of the unnecessary remnants of the larval organs such as abdominal intersegmental muscles and larval fat body takes place. Continuous formation of imaginal organs like the flight muscles and growth of the brain take place. Differentiation of the reproductive system sometimes also continues in the adult. JH stimulates differentiation of the colleterial glands in females and of accessory glands, tubular glands and the ejaculatory duct in the males of the butterfly *Danaus plexipus* during the first day after eclosion.

Various processes of imaginal maturation are under hormonal control. Elevation of JH titre stimulates the previtellogenic development of ovarian follicles. Vitellogenin sequestration occurs in the mature insect when the JH concentration increases further. In *Rhodnius prolixus* the activation of early follicles by JH is clearly separated from the subsequent stimulation of yolk deposition by JH. Similarly, in the previtellogenic females of *Schistocerca gregaria* and *Periplaneta americana*, an increased activity of corpora allata which was clearly separated from the following burst of JH secretion affecting the vitellogenic cycles was reported.

3.5.4 *Reproduction*

Reproduction is the main function of adult life, and the length of the imaginal stadium, atleast in the female, is related to the duration of egg production. Senescence correlates with termination of reproduction, and death may be predetermined in insect's ontogeny. Females may die after egg deposition and males may die after mating. An extended post-reproductive life is important in distasteful, aposematic insects as it allows predators to learn distastefulness from expendable prey individuals.

Egg production and related reproductive events are triggered by numerous hormones, which get released in relation to feeding, photoperiod, temperature, etc. Adult females having oocytes in the terminal previtellogenic stage wait for a long time for the stimulus effecting vitellogenesis. However, if hormone stimulation does not occur within a certain period, the follicles are resorbed. The resorption of previtellogenic follicles is ensured by special mechanisms in the workers of some social insects. For example, the mandibular glands of honey bee queen produce a secretion that inhibits the corpora allata and consequently prevents development of eggs in the worker bees.

Although many follicles are exposed to the same hormonal conditions, the development of eggs is initiated only in some of them, usually synchronously in different ovarioles. Oocytes arranged within each ovariole become engaged in yolk deposition successively. This pattern of egg formation is found in larva (Ephemeroptera), at metamorphosis (Plecoptera, certain Lepidoptera) and in adults (most insects). The mechanism by which the development of proximal follicles is restrained until the more distant follicles reach a certain stage of egg formation is not known. It is proposed that the primary vitellogenic follicles of flies secrete an oostatic hormone which prevents the neuro-hormonal stimulation of the more proximal follicles. The arrest of vitellogenesis in these secondary follicles is lifted when the primary follicles reach a certain size.

In most insects, the desire to mate is evoked by environmental and social signals and is mostly independent of the stage of gonadal differentiation. In some species it may occur during the maturation period. Signals used to locate or lure the mate are visual (dancing swarms of mayflies, caddisflies), auditory (chirping of cicadas, crickets) and chemical (pheromones). Mating in some insects is preceded by courtship. Impregnation is brought about by the insertion of free sperm or spermatophore into the vagina or bursa copulatrix of the female, but the males of apterygotes drop the spermatophore during courting and the female picks it up into her vagina. In Strepsiptera and a few bugs, the penis penetrates the body wall and in the bed bug it is inserted into an integumental invagination on the ventral body side. Depending on the species, copulation either occurs once in a lifetime or repeatedly.

Adult insects show various types of reproduction. A detailed account of the various modes of reproduction is given in Chapter 8. In **oviparous** reproduction, an egg is formed and the female lays the egg covered by a chorion. In **polyembryony,** there is a lot of division in the egg to give many individuals from one egg. In **vivipary,** the insect is born alive, although its origin was still from an egg. In **parthenogenesis,** the young are produced from infertile eggs, as in aphids. Some species do not have males, as in the white-fringed beetle. Some may have a combination of fertile and infertile periods. **Paedogenesis** involves reproduction in immature stages. The maggot young fasten themselves on the parent and consume them. Several generations may pass in this manner.

The various kinds of sculpturing found on insect eggs, are formed by epithelial cells. Many insects have the means of fixing their eggs so that they will have a proper environment for development. Examples are nits, the eggs of lice, that are glued to the hair of their host and the ovipositor can cut holes in wood where the eggs are laid. Lacewings lay their eggs on a stalk, which is a piece of silk with the egg attached to the end. However, some insects depend wholly on large numbers of eggs for survival of the species.

3.5.5 *Senescence*

Senescence of insects is correlated with the termination of reproduction, although the contingent of primordial cells in the germarium is not exhausted. At death, many female insects still contain egg chambers at different pre-vitellogenic and vitellogenic stages. The onset of ageing and timing of death is pre-determined in the ontogenetic cycle. Some males die after mating and some females after egg deposition. Imaginal life last for a single day in some mayflies, several years in certain beetles and a quarter of a century in termite reproductives. Loss of cells (neurons are lost in the brain of the honey bee), accumulation of pigments and urates, decreased respiration and other metabolic changes are features that determine the nearing of death.

4. HORMONAL CONTROL OF METAMORPHOSIS

Wingless insects continue to moult after attaining sexual maturity and there is no predetermined final instar. This is considered as **primitive character**. In advanced insects, i.e. pterygotes (excepting mayflies), there is always only one flying instar that never moults again. The greatest morphological change occurs between the last larval and adult stages in hemimetabolous insects. This occurs in two steps – larva to pupa and pupa to adult in holometabolous insects, i.e. always towards the end of development. Three major events occur during this period:

1. Structural metamorphosis (reorganisation of morphology)
2. Obligatory termination of moulting cycles (attainment of predetermined final instar)
3. Achieving reproductive competence (reaching adult stage or sexual maturity).

Originally there may have been some temporal difference in attaining these three events, but in modern insects it is difficult to distinguish between them, especially with respect to endocrinological regulation of metamorphosis. It is possible that mechanisms associated with structural metamorphosis may have been primarily connected with reproductive maturity or termination of moulting, and has secondarily been involved in morphological changes.

4.1 Hormones in development of ametabolous insects

Amongst wingless insects, the firebrat *Thermobia domestica* (Zygentoma) has been found to produce juvenile hormone (JH) III and ecdysteroids similar to pterygotes. It was found that JH secretion by the corpora allata was highest in the **first** and **second instars**, declined in the **third instar** and then rose again, and corresponding changes in volume of corpora allata cells were also observed. It was inferred that the decline in JH in the third instar allowed the appearance of integumental scales that are absent in earlier instars and appear for the first time in the **fourth instar** – possibly an effect of absence of JH during moulting. This suggests that the occurrence of scales is controlled by JH in a process similar to larva–adult metamorphosis in pterygotes. It has also been found in the same species that the haemolymph shows a peak in ecdysteroids concentration during each moulting cycle that canbe correlated with apolysis (separation of epidermis from the old cuticle) and the onset of new cuticle deposition.This is analogous to the cyclic increase in ecdysteroids during moulting in hemimetabolous and holometabolous insects. Unfortunately, the physiological bases of moulting remain largely unexplored in ametabolous insects.

4.2 Hormones in nymph to adult transformation in hemimetabolous insects

Juvenile hormone (JH) and ecdysteroids are the two key hormones regulating metamorphosis in insects. JH titre remains high from hatching till the last nymphal moult, i.e. during the juvenile moults, and its level falls abruptly in the pre-adult stage finally becoming almost undetectable. This may be a primitive situation in metamorphic pterygotes. JH is secreted from the corpora allata, and its production declines rapidly during the last nymphal instar, a low JH level being acrucial requirement for metamorphosis. By artificially increasing the JH titre by implanting corpora allata or by application of exogenous JH, metamorphosis can be prevented, and conversely, on reducing the JH level by removing corpora allata from the penultimate larval instar, precocious metamorphosis can be induced to give rise to miniature adults. Production of JH is regulated by stimulatory allatotrophins and inhibitory allatostatins, that are neuropeptides that act on receptors in corpora allata and corpora cardiaca. Their mode of action can depend on the age and intrinsic level of JH production of the individual, but the exact mechanisms of how these peptides are regulated and how in turn they regulate JH production remains to be explored.

Although the major focus on endocrinology of metamorphosis has been through the study of holometabolous insects, recent work has unravelled some of the molecular mechanisms of mode of action of metamorphosis hormones in hemimetabolous insects. JH appears to act through its putative receptor methoprene-tolerant (*Met*) to regulate Kruppel homolog 1 (Kr-h1) and **broad-Complex (BR-C) genes**. *Met* conveys the signal induced by JH to maintain a high expression level of *Kr–h1*, and knocking down *Met* or *Kr–h1* by RNAi induces the precocious appearance of adult colouration, wings and genitalia. *Kr–h1* expression falls naturally during the last nymphal instar, permitting adult development. Thus, Met and *Kr–h1* serve as JH dependent repressors of metamorphic changes.

Juvenile hormone may further act by modifying the expression of genes activated by ecdysteriods, or may also directly influence gene expression through positive and negative genetic control. However, such evidences from hemimetabolous insects are lacking. *BR-C* transcription factors also appear to govern progressive morphogenesis through successive nymphal moults, as wing bud growth through different stages can depend on *BR-C* factors. *BR-C* factors are expressed throughout all nymphal instars and the expression falls on the last nymphal stage. Thus, hemimetabolous insects undergo growth (especially wing bud, development) in the nymph stage with high JH level, which may induce the high expression of *BR-C*, and moult into the adult stage after both JH and *BR-C* levels decline in the last nymphal instar.

Another key regulatory factor for successful metamorphosis is the **ecdysteroid titre.** Ecdysteroids are typically produced by the prothoracic gland, and the ecdysteroid level increases rapidly towards the end of each instar stage, reaching a moult-inducing peak, and then again decreases and remains low till the next molting cycle. The increase is first associated with epidermal cell division, then with apolysis, i.e. separation of the old cuticle from the body, and finally with ecdysis, i.e. exfoliation of the exuviae or remnants of the old cuticle. Ecdysteroids are secreted mostly by the prothoracic gland, or sometimes by the integument. 20 Hydroxy ecdysone is the more active form of ecdyosteroids and it binds to a heterodimeric nuclear receptor (formed by two receptors, EcR and RXR/USP), in turn regulating a cascade of nuclear hormone receptor genes (found to be *BgE75*, *BgHR3*, *BgHR4* and *BgFTZ-F1* in cockroach *Blatella germanica*) that display regulatory crosstalk to regulate molting. Different phases of the ecdysteroid cycle– increase, peak or decrease, activate or inhibit different players of this network of nuclear receptors.

Other peptidic hormones, especially neurohormones, may also be studied properly in hemimetabolous insects. Recent studies have revealed the role of a new level of regulation in hemimetabolous insects – microRNAs (miRNA), which are small RNA molecules around 22 nucleotides long, generally playing a role in repressing action of messenger RNA (mRNA). Experimental down regulation of miRNA production prevents molting into adults and instead causes insects to retain nymphoid features, suggesting that miRNAs may perhaps regulate genes that are JH or ecdysteroid targets.

4.3 Hormones in larva to pupa to adult transformation in holometabolous insects

The role of JH and ecdysteroids in inhibiting and stimulating metamorphosis respectively, remains conserved in holometabolous insects. Peaks of ecdysteroids trigger moulting, whereas the presence of JH during moulting maintains the juvenile characters. Thereby broad JH titre peaks may be found overlapping with the moult inducing peaks of ecdysteroids. When the JH level remain slow during moulting, a metamorphosis to adult state ensues. In some insects like honey bees (*Apis*: Hymenoptera), a derived condition is found where JH levels remain at a steady level from embryos till early last larval instar in worker larvae, and in queen larvae it goes up to several folds higher than worker larvae. The dependence of sexual maturation on the decrease in JH and increase in ecdysteroids may be an ancient characteristic, as similar control of previtellogenesis (egg yolk formation) has been found in the ametabolous insect *T. domestica*.

As in hemimetabolous insects, JH (or sometimes carboxylic acids of JH) are secreted by the corpora allata and JH secretion can be governed by neurosecretory cells releasing allato regulatory peptides and amines. JH titres in haemolymph may be a result of balance between JH secretion and its catabolism by JH specific epoxide hydrolases and esterases, and it may be protected from degradation by binding with JH-binding proteins (Fig 2.8). An important function of JH in holometabolous insects is to modulate the action of ecdysteroids, and the lowering of the JH level prior to pupation induces inhibition of feeding and the onset of wandering, where the larva looks for a suitable location to undergo pupation. Like in hemimetabolous insects, JH acts through the *Met* and *Kr-h1* to regulate JH-inducible genes,and in Drosophila activates *FTZ-F1* after binding with the *Met* and *Tai* to initiate regulation of nuclear genes. Theantagonistic effect of JH on metamorphosis is through *Kr-h1*, which suppresses ecdyosteroid-induced switching on of *BR-C*, in turn permitting larval growth and development (Fig 2.8). Thus the interplay between JH and *BRC* is quite different in hemimetabolous and holometabolous insects, and the acquisition of an inhibitory effect of JH on *BR-C* expression in juvenile instars may have been a key step in the evolutionary transition from hemimetaboly to holometaboly. Since *BR-C* permits changes in gene expression required for change in morphology between instars, it plays a vital role in regulating metamorphosis.

Ecdysteroids have numerous functions in holometabolous insects – regulation of neuronal growth, reproductive roles, stimulation of growth and development of imginal discs, initiation of breakdown of larval structures during metamorphosis, elicitation of deposition of cuticle by epidermis, etc. Ecdysteroids are produced in the prothoracic gland of holometabolans, which typically secretes ecdysone, which in turn is converted to the principal moulting hormone 20-hydroxyecdysone (20E) by *P*450 monooxygenase in mitochondria of peripheral tissues. The production of ecdysone is controlled by the prothoracicotrophic hormone (PTTH) secreted by corpora allata (Fig 2.8). 20E binds to a heterodimeric nuclear receptor composed of the ecdysone receptor (EcR) and ultraspiracle (USP) or RXR, in turn activating a hierarchy of transcription factors that regulate the expression of target genes governing cellular changes associated with moulting and metamorphosis.

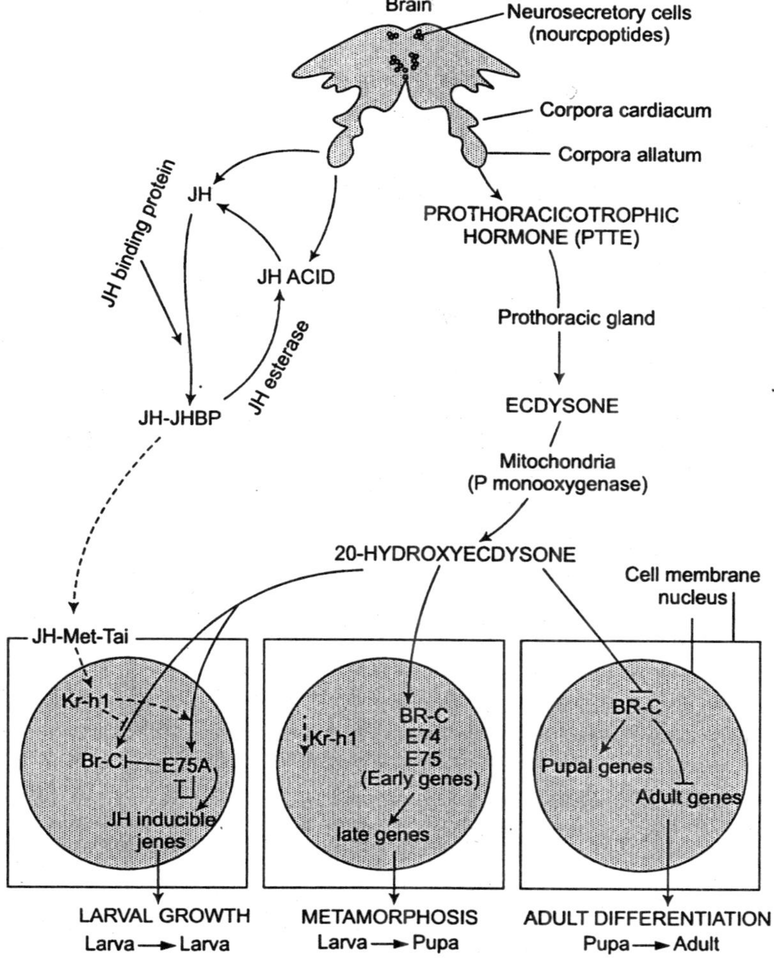

Fig 2.8. Physiological pathway controlling metamorphosis in holometabolous insects (After A.Mitra, 2013)

Some of the nuclear receptors described for *B. germanica*, that ultimately modulate β*FT-Z1* and *HR4* remain conserved in both hemimetabolous and holometabolous insects, but their regulatory interplay differs. *HR4* appears to be involved in all post-embryonic transitions in hemimetabolous insects, which may be the ancestral function, but it appears to have acquired a new holometabolous-specific role in coordinating growth and maturation in the last larval instar.

There appears to be some crosstalk between JH and 20E in modulating various transcription factors and genes that are crucial for execution of metamorphosis with JH sometimes facilitating and sometimes inhibiting 20E-induced gene expression (Fig 2.8). Along with various other genes, the genes associated with apoptosis (*diap2*, *rpr*, *hid*) are expressed in the pupal stage and this is important for breakdown and cell death of larva specific structures. Changes in ecdysone titres translate into waves of gene expression during holometabolan metamorphosis. Indeed in *Drosophila*, probably the most well studied holometabolan insect, about 872 target genes have been found to change their expression during metamorphosis, out of which around 200 are induced in the larva, more than 400 in the prepupa and around 350 in the pupa. Interestingly, the different sets of genes can be modulated by the same set of master regulatory genes, like the *Hox* genes. Recently, it has been found that some transcription factors like E93 can act as a temporal identity determinant, bestowing responsiveness to other signalling pathways on the target genes in a time dependent manner. Thereby, the same genes and signalling pathways that were employed in a different way during embryonic development can be redeployed in another way during metamorphosis to bring about large scale morphological changes. Some peptidic hormones are involved in regulating the process of molting or ecdysis, and emergence of the molted insect from the exuvia. Ecdysis itself is a complicated process comprising pre- and post-ecdysis behaviours, controlled by a cascade of peptide hormones from Inka cells and neuropeptides of the central nervous system. Inka cells in tracheas produce pre-ecdysis and ecdysis triggering hormones that activate ecdysis initiated by the molting hormone, acting through receptors on specific neurons.

5. VOLTINISM

Voltinism refers to the numbers of generations insects produce per year. Most insects take a year or less to develop, with either one generation per year (**univoltine** insects), or two (**bivoltine** insects), or more than two (**multivoltine**, or polyvoltine, insects). Generation times in excess of one year (**semivoltine** insects) are found, for example amongst some inhabitants of the polar extremes, where suitable conditions for development may exist for only a few weeks in each year. Large insects that rely upon nutritionally poor diets also develop slowly over many years. For example, periodical cicadas feeding on sap from tree roots may take either 13 or 17 years to mature, and beetles that develop within dead wood have been known to emerge after more than 20 years of development.

Most insects do not develop continuously throughout the year, but arrest their development during unfavourable times by quiescence or diapause. Many univoltine and some bivoltine species enter diapause at some stage, awaiting suitable conditions before completing their life cycle. For some **univoltine insects**, many social insects, and others that take longer than a year to develop, adult longevity may extend to several years. In contrast, the adult life of multivoltine insects may be as little as a single evening for many Ephemeroptera, or even a few hours at low tide for marine midges such as *Clunio* (Diptera:Chironomidae). **Multivoltine insects** tend to be small and fast-developing, using resources that are more evenly available throughout the year. Univoltinism is common amongst temperate insects, particularly those that use resources that are seasonally restricted. These might include insects whose aquatic immature stages rely on spring algal bloom, or phytophagous insects using short-lived annual plants. **Bivoltine insects** include those that develop slowly on evenly spread resources, and those that track a bimodally distributed factor, such as spring and autumn/fall temperature. Some species have fixed voltinism patterns, whereas others may vary with geography, particularly in insects with broad latitudinal or elevational ranges.

Major factors influencing insect development and therefore also voltinism are ambient temperature and photoperiod. Latitude is correlated with both these factors and may therefore be used as a **surrogate variable**. The number of generations per year decreases with latitude. In low latitudes (i.e. in the Tropics) most species are multi- or bi-voltine, whereas in high latitudes most species are semi- or parti-voltine. In mid-latitudes and the subtropics the frequencies are intermediate. In odonates voltinism differs significantly among perennial lotic, perennial lentic and temporary lentic habitats. This is to be expected, particularly due to the lifecycle constraints caused by drying out of temporary waters. Temporary waters usually require rapid growth to permit successful development (Wellborn et al. 1996). This pattern is especially evident in arid countries, such as Namibia (Johansson & Suhling 2004; Suhling *et al.*, 2004, 2005). Species occupying temporary waters there develop more rapidly than those occupying permanent waters, and those occupying lentic waters there develop more rapidly than those occupying lotic waters. **Voltinism** is a resultant of the interaction of exogenous and endogenous factors that, acting in concert, maintain a viable, seasonally appropriate life cycle under varying environmental conditions.

6. POLYMORPHISM AND POLYPHENISM

The existence of several generations per year is associated with morphological change between generations. Similar variation may occur contemporaneously within a population, such as the existence simultaneously of both winged and flightless forms (**morphs**). Sexual differences between males and females, and the existence of strong differentiation in social insects such as ants and bees, are further obvious examples of the phenomenon. The term **polymorphism** encompasses all such discontinuities, which occur in the same life history phase at a frequency greater than might be expected from repeated mutation alone. It is defined as the simultaneous or recurrent occurrence of distinct morphological differences, reflecting and often including physiological, behavioural and/or ecological differences among conspecific individuals.

6.1 Genetic polymorphism

The distinction between the sexes is an example of a particular polymorphism, namely **sexual dimorphism**, which in insects is almost totally under genetic determination. Environmental factors may affect sexual expression, as in castes of some social insects or in feminisation of genetically male insects by mermithid nematode infections. Aside from the dimorphism of the sexes, different genotypes may co-occur within a single species, maintained by natural selection at specific frequencies that vary from place to place and time to time throughout the range. For example, adults of some gerrid bugs are fully winged and capable of flight, whereas other coexisting individuals of the same species are brachypterous and cannot fly. Intermediates are at a selective disadvantage and the two genetically determined morphs coexist in a balanced polymorphism. Some of the most complex, genetically based, polymorphisms have been discovered in butterflies that mimic chemically protected butterflies of another species for purposes of defence from predators. Some butterfly species may mimic more than one model and, in these species, the accuracy of the several distinct mimicry patterns is maintained because inappropriate intermediates are not recognised by predators as being distasteful and are eaten. Mimetic polymorphism predominantly is restricted to the females, with the males generally monomorphic and non-mimetic. The basis for the switching between the different mimeticmorphs is relatively simple Mendelian genetics, which may involve relatively few genes or supergenes. It is a common observation that some individual species with a wide range of latitudinal distributions show different life-history strategies according to location. For example, populations living at high latitudes (nearer the pole) or high elevation may be **univoltine**, with a long dormant period, whereas populations nearer the equator or lower in elevationmay be **multivoltine**, and develop continuously, without dormancy. Dormancy is environmentally induced, but the ability of the insect to recognise and respond to these cues is programmed genetically. In addition, at least some geographical variation in life histories is the result of genetic polymorphism.

6.2 Environmental polymorphism or polyphenism

A phenotypic difference between generations that lacks a genetic basis and is determined entirely by the environment often is termed **polyphenism**. The expression of a particular phenotype depends upon one or more genes that are triggered by an environmental cue. An example is the temperate to tropical Old World butterfly *Eurema hecabe* (Lepidoptera:Pieridae), which shows a seasonal change in wing colour between summer and autumn/ fallmorphs. Photoperiod induces morph change, with a dark-winged summer morph induced by a long day of greater than 13 h. A short day of less than 12 h induces the paler-winged autumn/fall morph, particularly at temperatures of under 20°C, with temperature affecting males more than females. A second example is the **colour polyphenism** of the caterpillars of the American peppered moth *Biston betularia cognataria* (Lepidoptera: Geometridae), which are generalist herbivores with a body colour that varies to match the green or brown twigs of their host plants. Experiments have shown that the caterpillars change the pigments in their epidermal cells in response primarily to their visual experience, rather than to their diet. This larval colour polyphenism is not related to the genetic polymorphism for melanic forms seen in adult moths. In another genus of geometrid moths, caterpillars of *Nemoria arizonaria* of the spring generation, feed on and develop to resemble oak catkins (flowers), whereas the summer-generation caterpillars feed on leaves and resemble oak twigs. These catkin and twig morphs are induced solely by the larval diet.

Phenotypic plasticity refers to the emergence of phenotypic variation in a developing organism in response to changes in environmental conditions. An extreme example of phenotypic plasticity is **polyphenism**, in which discrete alternative phenotypes are produced from the same genotype in response to extrinsic factors (Nijhout, 1999, 2003). How a single genotype can produce such markedly different phenotypes is an important question in developmental biology. In many cases, the developmental switches are regulated by physiological factors, like hormones, that mediate environmental (or genetic) factors and differential developmental mechanisms. In all known cases of polyphenisms in insects, the switching mechanisms of developmental pathways leading to alternative phenotypes are mediated by either the timing of hormone secretion, the timing of a hormone sensitive period, or the threshold of hormone sensitivity (Nijhout, 1999).

Amongst the most complex polyphenisms are those seen in the aphids (Hemiptera:Aphidoidea). In addition to wing (winged/macropterous, brachypterous or wingless) and reproductive (sexual reproduction or parthenogenesis) polymorphisms, aphids also exhibit body-colour and caste polymorphisms (Fukatsu, 2010; Tsuchida *et al.*, 2010), most of which are also recognised as being polyphenisms. Many aphid species show wing polyphenism in which winged morphs appear in response to changes in environmental factors in order to facilitate migration to new host plants or habitats (Brisson, 2010). Reproductive polyphenism, in which sexual reproduction and parthenogenesis are switched depending on seasonal conditions, is also exhibited by many aphid lineages (Le Trionnaire *et al.*, 2008; Davis, 2012), and in social aphids, caste polyphenism results in the production of soldier aphids which appear to defend their gall (nest).

Within parthenogenetic lineages (i.e. those in which there is absolute geneticidentity), the females may show up to eight distinct phenotypes, in addition to polymorphisms in sexual forms. These female aphids may vary in morphology, physiology, fecundity, offspring timing and size, development time, longevity, and host-plant choice and utilisation. Environmental cues responsible for alternative morphs are similar to those thatgovern diapause and migration in many insects, including photoperiod, temperature and maternal effects, such as elapsed time (rather than number of generations) since the winged founding mother. Overcrowding triggers many aphid species to produce a winged dispersive phase (Box 2.3).

Box 2.3 Polyphenism in Aphids

Determined by season, food quality, crowding and predator pressure. Mediated by hormones. In many spp., involves asexual and sexual reproductive phase, apterous and winged phases

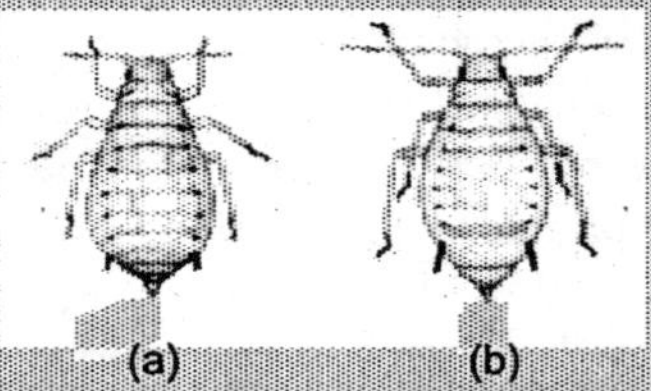

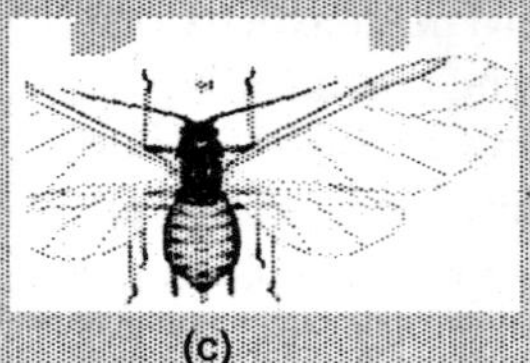

(a) (b) (c)

(a) & (b) ovoviviparous apterous form (during summer with plentiful rich food); (c) sexual alate (during fall with decreasing food quality, crowding

Crowding also is responsible for one of the most dramatic examples of polyphenism, the phase transformation from the solitary young locusts (hoppers) to the gregarious phase. Studies on the physiological mechanisms that link environmental cues to these phenotype changes have implicated JH in many aphid morph shifts. In reproductive polyphenism, morph determination (male, oviparous or viviparous female) within mother aphids is regulated by juvenile hormone (JH) titers in the mothers. In wing polyphenism, although JH is considered to play an important role in phenotype determination (winged or wingless), the role is still controversial. In both cases, the acquisition of viviparity in Aphididae is considered to be the basis for maternal regulation of these polyphenisms, and through which environmental cues can be transferred to developing embryos through the physiological state of the mother.

If aphids show the greatest number of polyphenisms, the social insects are a close second, and undoubtedly have a greater degree of morphological differentiation between morphs, termed **castes**. Maintenance of the phenotypic differences between castes as different as queens, workers and soldiers includes physiological mechanisms such as pheromones transferred with food, olfactory and tactile stimuli, and endocrine control including JH and ecdysone. Superimposed on these polyphenisms are the dimorphic differences between the sexes, which impose some limits on variation.

Several parasitoids show developmental interactions with their hosts. Sex ratio, imaginal morphology, length of development and other characteristics of parasitic hymenoptera are modified by their host. For example, alate males are produced when *Trichogramma semblidis* develops in the eggs of Lepidioptera but apterous males emerge when it is reared in the eggs of *Sialis* (Megaloptera). The males also differ in size and structure appendages. It is assumed that extreme variation is based on genetic heterogeneity. Natural selection and epigenetic factors (nutrition from host) determine which form will be realized.

6.3 Phase dimorphism or kentromorphism

Growth and development of insects proceeds optimally under certain population densities. Some species grow best when reared singly or in small groups, others prefer crowding. Some locusts, phasmids, and crickets adjust to wide range of population densities by being able to appear in two distinct phenotypes or phases. Under low densities insects form solitary phase which possess a cryptic pigmentation. The gregarious phase usually has a warning colouration. The phases differ from one another in colouration, morphological features, rate of development, mortality, behaviour etc. In general the solitary phase is geared to survival within the original habitat, while the gregarious phase serves for dispersal. The occurrence of phases is affected not only by population density, but other factors such as temperature, humidity and possibly photoperiod and food are important.

The phenomenon of phase dimorphism was discovered in locusts. The solitary phase of *Locusta migratoria* is robust. The body colour varies from white to yellow and green, and it often blends with the background. When the population build up, the locusts become smaller in size, darker in colour and exhibit restlessness. The nymphs often migrate over the ground and the adults fly away in swarms.

Phase dimorphism in orthopterans and lepidopterans is controlled by hormones. Melanization of integument is stimulated by neurohormones. In *Leucania* separate neurohormone was identified as a peptide called **melanisation and reddish pigmentation hormone**. Solitary pigmentation is believed to be caused by elevated JH titre in locusts. Ecdysteroids is also believed to be involved in control of pigmentation.

6.4 Caste or social polymorphism

Social insects have several castes which differ by their pattern of growth. Each colony includes female(s), the queen, and at least in certain seasons also the male(s). Majority of the members of the colony are workers, and in some termites and ants, the soldiers. The non-reproductive castes are of both sexes in termites, but exclusively of females in Hymenoptera. Morphological differences between castes are largely due to different growth patterns but the loss of wings in the termite and ant reproductive occurs abruptly in the adults.

Termite colonies contain a long-lived royal pair and its progeny, i.e. larvae, nymphs with wing pads, workers and soldiers. Older larvae and nymphs help with various duties in the colony. In the lower termites the fully grown larvae represent the worker caste (pseudergates) which maintains full capacity to differentiate into reproductives or soldiers. In higher termites the developmental fate of the larvae is determined in the early instars. Workers, soldiers and secondary termites are always wingless.

Ants resemble termites by their winglessness and other adaptations to life in underground galleries. The workers of some ants lay few male eggs while the queen produces only female eggs. In other species, the workers deposit trophic eggs which are devoured by the queen and other colony members. The effect of larval food on the development of castes has been demonstrated in a number of species.

Social polymorphism in bees and wasps is determined exclusively postnatally. In the most primitive social species, a colony includes only mated females with a distinct social hierarchy. One of the females dominates over the others and become the egg laying queen and the status of others is reduced to a non-reproducing worker. In other species the reproductive capacity of the future workers is restrained by the food they receive during larval stage.

Determination of caste has been well documented in honey bees. Any female larvae becomes a queen if it is continuously nourished by royal jelly. Larvae which are fed with a slightly modified worker jelly for the first three days after hatching and then on pollen and honey produce the worker caste.

6.5 Polyethism

Polyethism is the division of labour observed in eusocial insects like ants, termites, or bees. Biologists differentiate between at least two means of dividing roles amongst workers in natural insect colonies namely caste polyethism and age polyethism. Other types of polytheism are also observed (e.g. elitism).

Caste polyethism occurs when distinct types of individuals are bred by the colony. An individual is effectively born into its role, often displaying morphological differences from individuals from other castes. The clearest example of castes is the division between the reproductive caste and the worker caste in eusocial insects. A single or small group of reproductive females (called queens) are responsible for all reproductive tasks in the colony while non-reproductive workers carry out all other tasks required by the colony (brood care, nest constructions and maintenance, waste removal, foraging, and defense).

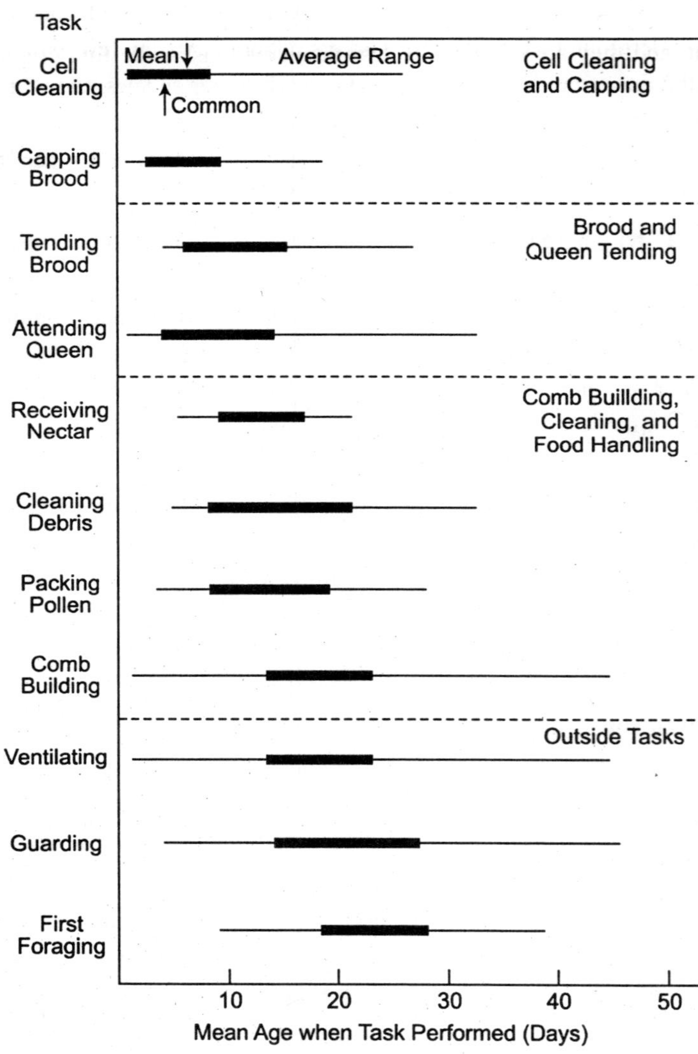

Fig. 2.9. Age-related polyethism in the honey bee *Apis mellifera* (Adapted from Winston 1987)

In some species workers are further divided into sub-castes. Differences among workers from different castes are particular to the worker's role. For instance in some species of ants the workers can be divided into majors and minors (occasionally with an intermediate caste as well) where the majors are larger than the minors, this size being helpful in the task they carry out (primarily colony defense). Minors are smaller, making them more energy efficient, and they are relegated to less dangerous tasks like foraging and nest maintenance. Only in rare occasions will a worker do a task that is typically assigned to a different caste.

Age or **temporal polyethism** is a type of division of labour where the worker's role is correlated with its age or changes over time (Fig 2.9). Age polyethism is more common than caste polyethism in natural insect colonies. In colonies displaying age polyethism younger workers commonly carry out less risky tasks (nursing or nest maintenance allowing them to stay in the nest) whereas older workers carry out more dangerous tasks (foraging, defense, or raiding where the agent must leave the nest). It is hypothesised that this division of labour allows the colony to maximise the work carried out by each individual worker (i.e. young workers will be less likely to die and thus can live longer to carry out more work). This will be beneficial to the colony since it will have to breed fewer workers if each worker's longevity (and thus productivity) is maximized (Tofilski (2002, 2009)). In certain cases this progressive role assignment may also allow younger and less experienced workers to gain the experience necessary to carry out more difficulttasks (say at the very least allowing them to become familiar with the layout of the nest and surrounding environment before having to venture far from the nest) (Tofts and Franks (1992); Tofts (1993); Franks and Tofts (1994)). Many mechanisms have been suggested as the underlying reason for observed age polyethism.

7. ENVIRONMENTALEFFECTSONDEVELOPMENT

The rate or manner of insect development or growth may depend upon a number of factors.These include the type and amount of food, the amount of moisture (for terrestrial species) and heat (measured as temperature), or the presence of environmental signals (e.g.photoperiod), mutagens and toxins, or other organisms, either predators or competitors. Two or more of these factors may interact to complicate interpretation of growth characteristics and patterns.

7.1 Temperature

Insects belong to an ectothermic group of animals. They are highly dependent on thermal conditions of the surrounding environment. Most insects are poikilothermic, that is with body temperature more or less directly varying with environmental temperature, and thus heat drives the rate of growth and development when food is unlimited. A rise in temperature, within a favourable range, will speed up the metabolism of an insect,and consequently increase its rate of development. In higher temperature conditions, the development of egg, larva and pupa shortens. Faster development of preimaginal stages implies shorter time of exposure to adverse environment conditions such as low temperature, too high or insufficient humidity, attacks of predators and parasitoids, and entomopathogen's activity. It can result in reproductive success of many insect species. Temperature influence on a length of larval development has been observed under laboratorial conditions for two significant species of native foliophages: the nun moth, *Lymantriamonacha* (L.) and the gypsy moth *Lymantriadispar* (L.) (Karolewski *et al*., 2007). In both cases, temperature increase has had an influence on reducing growth period, from egg phase to pupa. Different results have been obtained regarding larva survivability of both species. When the average environment temperature has increased, higher mortality has been observed for caterpillars of *L. monacha.* Whereas the survivability of *L. dispar* larvae has increased. These differences probably result from two different thermal optima for both species reflected in varied environmental preferences. This indicates that a variety of climatic parameters influence insect development, even when closely related species are compared.

Each species and each stage in the life history may develop at its own rate in relation to temperature. Thus, **physiological time**, a measure of the amount of heat required over time for an insect to complete development or a stage of development, is more meaningful as a measure of development time than age in calendar time. Knowledge of temperature–development relationships and the use of physiological time allow comparison of the life cycles and/or fecundity of pest species in the same system, and prediction of the larval feeding periods, generation length and time of adult emergence under variabletemperature conditions that exist in the field. Such predictions are especially important for pest insects, as control measures must be timed carefully in order to be effective.

Physiological time is the cumulative product of total development time (in hours or days) multiplied by the temperature (in degrees) above the **developmental (or growth) threshold**, or the temperature below which no development occurs. Thus, physiological timeis commonly expressed as **day-degrees** (also degree-days) (D°) or hour-degrees (h°). Normally, physiological time is estimated for a species by rearing a number of individuals ofthe life-history stage(s) of interest under different constant temperatures in several identicalgrowth cabinets. The developmental threshold is estimated by the linear regression –axis method, although more accurate threshold estimates can be obtained by more time-consuming methods.

In practice, the application of laboratory-estimated physiological time to natural populations may be complicated by several factors. Under fluctuating temperatures, especially if the insects experience extremes, growth may be retarded or accelerated compared with the same number of day-degrees under constant temperatures. Furthermore, the temperatures actually experienced by the insects, in their often sheltered microhabitats on plants or in soil or litter, may be several degrees different from the temperatures recorded at a meteorological station even just a few metres away. Insects may select microhabitats that ameliorate cold night conditions or reduce or increase daytime heat. Thus, predictions of insect life-cycle events based on extrapolation from laboratory to field temperature records may be inaccurate. For these reasons, the laboratory estimates of physiological time should be corroborated by calculating the hour-degrees or day-degrees required for development under more naturalconditions, but using the laboratory-estimated developmental threshold.

Another problem with laboratory estimation of physiological time is that insect populations maintained for lengthy periods under laboratory conditions frequently undergo acclimation to constant conditions, or even genetic change in response to the altered environment or as aresult of population reductions that produce **genetic bottle-necks**. Therefore, insects maintained in rearing cages may exhibit different temperature–development relationshipsfrom individuals of the same species in wild populations. For all of the above reasons, any formula or model that purports to predict insect response to environmental conditions must be tested carefully for its fit with natural population responses.

7.2 Photoperiod

Many insects do not develop continuously all year round, but avoid some seasonally adverse conditions by a resting period or migration. Summer dormancy (aestivation) and winter dormancy (hibernation) provide two examples of avoidance of seasonal extremes. The most predictable environmental indicator of changing seasons is **photoperiod**—the length of the daily light phase or, more simply, day length. Near the equator, although sunrise to sunset of the longest day may be only a few minutes longer than on the shortest day, if the period of twilight is included then total day length shows more marked seasonal change. The photoperiod response is to duration rather than intensity and there is a critical threshold intensity of light below which the insect does not respond; this threshold is often as dim as twilight, but rarely as low as bright moonlight. Many insects appear to measure the duration of the light phase in the 24 h period, and some have been shown experimentally to measure the duration of dark.

Others recognise long days by light falling within the dark half of the day. Most insects can be described as long-day species, with growth and reproduction in summer, and with dormancy commencing with decreasing day length. Others show the reverse pattern, with short-day (often autumn/fall and spring) activity and summer aestivation. In some species, the life-history stage in which photoperiod is assessed is inadvance of the stage that reacts, as is the case when the photoperiodic response of the maternal generation of silkworms affects the eggs of the next generation. The ability of insects to recognise seasonal photoperiod and other environmental cues requires some means of measuring time between the cue and the subsequent onset or cessation of diapause. This is achieved through a **biological clock,** which may be driven by internal (endogenous) or external (exogenous) daily cycles, called **circadian rhythms**. Interactions between the short time periodicity of circadian rhythms and longer term seasonal rhythms, such as photoperiod recognition, are complex and diverse, and haveprobably evolved many times within the insects.

7.3 Humidity

The high surface area : volume ratio of insects means that loss of body water is a serious hazard in a terrestrial environment, especially a dry one. Low moisture content of the air can affect the physiology and thus the development, longevity and oviposition of many insects. Air holds more water vapour at high than at low temperatures. The relative humidity (RH) at a particular temperature is the ratio of actual water vapour present to that necessary for saturation of the air at that temperature. At low relative humidity, development may be retarded, for example in many pests of stored products; but at high relative humidity or insaturated air (100% RH), insects or their eggs may drown or be infected more readily by pathogens. The fact that stadia may be greatly lengthened by unfavourable humidity has serious implications for estimates of development times, whether calendar or physiological time is used. The complicating effects of low, and sometimes even high, air moisture levelsshould be taken into account when gathering such data.

7.4 Mutagens and toxins

Stressful conditions induced by toxic or mutagenic chemicals may affect insect growth and form to varying degrees, ranging from death at one extreme to slight phenotypic modifications at the other end of the spectrum. Some life-history stages may be more sensitive to mutagens or toxins than others, and phenotypic effects may not be measured easily by crude estimates of stress, such as percentage survival. One measure of the amount of genetic or environmental stress experienced by insects during development is the incidence of **fluctuating asymmetry**, or the quantitative differences between the left and right sides of each individual in a sample of the population. Insects should be bilaterally symmetrical, with left and right halves of their bodies as mirror images—except for obvious differences in structures such as the genitalia of some male insects. Under stressful (unstable) developmental conditions, the degree of asymmetry tends to increase. However, the underlying developmental genetics is poorly understood, and the technique is unreliable, with doubts having been raised about interpretation, such as variation in asymmetry response between different organ systems measured.

7.5 Biotic effects

In most insect orders, adult size has a strong genetic component, and growth is strongly determinate. In many Lepidoptera, for example, final adult size is relatively constant within aspecies; reduction in food quality or availability delays caterpillar growth rather than causing reduced final adult size, although there are exceptions. In contrast, in flies that have limited or ephemeral larval resources, such as a dung pat/pad or temporary pool, cessation of larval growth would result in death as the habitat shrinks. Thus, larval crowding and/or limitation of food supply tend to shorten development time and reduce final adult size. In some mosquitoes and midges, success in short-lived pool habitats is attained by a small proportion of the larval population developing with extreme rapidity relative to their slower-growing siblings. In paedogenetic gall midges, crowding with reduced food supply terminates larva-only reproductive cycles and induces the production of adults, allowing dispersal to more favourable habitats. Food quality appears important in all these cases, but there may be related effects, forexample as a result of crowding. Clearly, it can be difficult to segregate out food effects from other potentially limiting factors. In the California red scale, *Aonidiella aurantii* (Hemiptera), development and reproduction on orange trees is fastest on fruit, intermediateon twigs, and slowest on leaves. Although these differences may reflect differing nutritional status, a microclimatic explanation cannot be excluded, as fruit may retain heat longer than do stems and leaves, and such slight temperature differences might affect the development of the insects.

The effects of crowding on development are well understood in some insects, as in locusts inwhich two extreme phases, termed **solitary** and **gregarious**, differ in morphometrics, colour and behaviour. At low densities, locusts develop into the solitaryphase, with a characteristic uniform-coloured hopper, and large-sized adult withlarge hind femora. As densities increase, induced in nature by high survivorship of eggs andyoung nymphs under favourable climatic conditions, graded changes occur and a darker-striped nymph develops into a smaller locust with shorter hind femora. The most conspicuous difference is behavioural, with more solitary individuals shunning each other's company but making concerted nocturnal migratory movements that result eventually inaggregations in one or a few places of gregarious individuals, which tend to form enormous and mobile swarms. The behavioural shift is induced by crowding, as can be shown bysplitting a single locust egg pod into two: rearing the offspring at low densities induces solitary locusts, whereas their siblings reared under crowded conditions develop into gregarious locusts. The response to high population density results from the integration of several cues, including the sight, odour and touch of conspecifics (perhaps via cuticular hydrocarbons acting as contact primer pheromones and detected by the antennae), which lead to endocrine and neuroendocrine (ecdysteroid) changes associated with developmental transformation. Under certain circumstances, biotic effects can override growth factors. Across much of the eastern United States, 13- and 17-year periodical cicadas (*Magicicada* spp.) emerge highly synchronously. At any given time, nymphal cicadas are of various sizes and in differentin stars, according to the nutrition they have obtained from feeding on the xylem sap fromroots of a variety of trees. Whatever their growth condition, after the elapse of 13 or 17 years since the previous emergence and egg-laying, the final moult of all nymphs prepares them for synchronous emergence as adults. In a very clever experiment, host plants were induced to flush new foliage and flowers twice in one year, inducing adult cicada emergence one yearearly compared to controls on the roots of single flushing trees. This implies that synchronised timing for cicadas depends on an ability to "count of" annual events—the predictable flush of sap with the passing of each spring once a year.

REFERENCES

Brisson, J.A.(2010) Aphid wing dimorphisms: linking environmental and genetic control of trait variation. *Philos Trans R Soc Lond B Biol Sci.*; 365(1540):605-16.

Davi,s G.K. (2012) Cyclical parthenogenesis and viviparity in aphids as evolutionary novelties. *J. Exp. Zool. B Mol. Dev. Evol.* 318, 448–459

Franks, N. and Tofts, C. (1994). Foraging for work: how tasks allocate workers. *Animal Behaviour*, 48:470–472.

Fukatsu, T (2010) Evolution. A fungal past to insect colour. *Science.;* 328(5978):574-5.

Johansson, F. and F. Suhling, (2004). Behaviour and growth of dragonfly larvae along a permanent to temporary water habitat gradient. *Ecological Entomology* 29: 196-202.

Karolewski P., Grzebyta J., Oleksyn J., and Giertych M. J. (2007). Effects of temperature on larval survival rate and duration of development of *Lymantria monacha* (L.) on needles of *Pinus silvestris* (L.) and of *L. dispar* (L.) on leaves of *Quercus robur* (L.). *Polish Journal of Ecology*, 55 (3): 595–600.

Le Trionnaire G, Hardie J, Jaubert-Possamai S, Simon, J.C., and Tagu, D. (2008) Shifting from clonal to sexual reproduction in aphids: physiological and developmental aspects. *Biol Cell.* 100(8):441-51.

Mitra A (2013) Cinderella's new shoes- how and why insects remodel their bodies between life stages. *Current Science*, Vol 104(8) 1028-1036

Nijhout H. F. (1999). Control mechanisms of phenotypic development in insects. *Bioscience* 49, 181–192

Nijhout, H F 2003 Development and evolution of adaptive polyphenisms. *Evol Dev.* 5(1):9-18

Scoble, M. J. (1992). *The Lepidoptera: Form, Function and Diversity.* Oxford: Oxford University Press. ISBN 0198540310.

Suhling, F., G. Sahlén, J. Kasperski and D. Gaedecke, (2005). Behavioural and life history traits in temporary and perennial waters: comparisons among three pairs of sibling dragonfly species. *Oikos* 108: 609-617.

Suhling, F., K. Schenk, T. Padeffke and A. Martens (2004). Larval development of some dragonflies from experimental temporary ponds in the Pro-Namib desert. *Hydrobiologia* 526: 75-85.

Tofilski, A. (2002). Influence of age polyethism on longevity of workers in social insects. *Behaviour Ecology and Sociobiology*, 51:234–237.

Tofilski, A. (2009). Shorter-lived workers start foraging earlier. *Insectes Sociau,* 56(4):359–366.

Tofts, C. (1993). Algorithms for task allocation in ants (a study of temporal polyethism theory). *Bulletin of Mathematical Biology,* 55:891–918.

Tofts, C. and Franks, N. (1992). Doing the right thing: ants, honeybees and naked mole rats. *Trends in Ecology and Evolution*, 7:346–349.

Tsuchida T, Koga R, Horikawa M, Tsunoda T, Maoka T, Matsumoto S, Simon, J.C., and Fukatsu, T. (2010) Symbiotic bacterium modifies aphid body colour. *Science.* 330(6007):1102-4.

Wellborn, G.A., D.K. Skelly and E.E. Werner, (1996). Mechanisms creating community structure across a freshwater habitat gradient. Annual Review of Ecology and Systematics 27: 337-363.

Winston, M. L. (1987) *The Biology of Honey Bee*. Harvard University Press, Cambridge, MA.

Insect Digestive System as a Potential Target for Pest Management 3

Synopsis

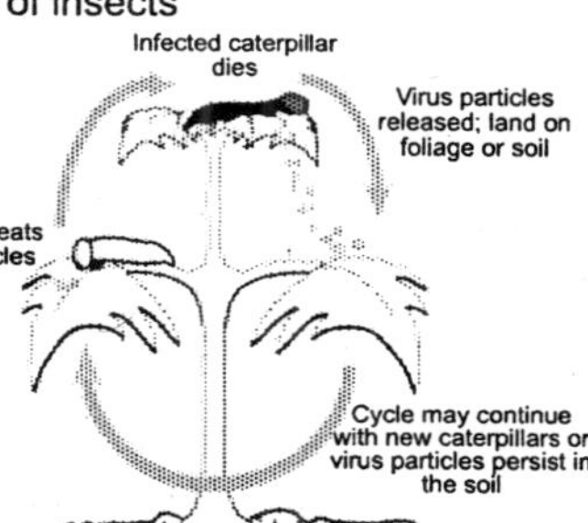

1. INTRODUCTION

Losses of agricultural production are mainly due to pest and diseases. The methods available today to protect plants from insects rely heavily on the use of aggressive chemicals. The adverse effect of chemical insecticides on the environment has resulted in the search for other methods for control of insect pests. Studies on the midgut of insects as a target for pest control got importance after the realisation that the gut is a very large and relatively unprotected interface between the insect and the environment. Plant natural defense against herbivores has been mediated through chemicals that impair the digestive process in the insect gut. To take advantage of this mechanism, a thorough understanding of the physiology of digestion in insects is essential, which this chapter attempts to provide.

The digestive system is the most prominent system in insects. It facilitates the extraction of nutrients and other substances from the food it consumes. All insects have a **complete digestive system**. This means that food processing occurs within a tube-like enclosure, the **alimentary canal**, running lengthwise through the body from mouth to anus. The length of the alimentary canal and the complexity of its structure depend on the food habits of the insects. It is the part of insect anatomy which discharges the functions of digestion, absorption and egestion of food materials. The complete digestive system permits **functional specialisation** - different parts of the system may be specially adapted for various functions of food digestion, nutrient absorption, and waste excretion. This chapter describes the structural organisation of the alimentary canal, digestion and absorption of nutrients with emphasis on the molecular processes underlying the digestive phenomena and highlighting the potential in the development of new insect control techniques.

1. ALIMENTARY CANAL MORPHOLOGY AND FUNCTION

In insects, food processing occurs within a tube-like enclosure, the **alimentary canal**, running lengthwise through the body from mouth to anus. Alimentary canal (gut) is formed by a one-cell thick layer of epithelial cells, a non-cellular layer of basement membrane, circular muscles, longitudinal muscles and peritoneal membrane. The length of the alimentary canal and the complexity of its structure depend on the food habits of the insects. It is longer in insects feeding on a high - carbohydrate diet and shorter in insects feeding on a high - protein diet. It is the part of insect anatomy which discharges the functions of digestion, absorption and egestion of food materials. The alimentary canal in insects is divided into three functional regions *viz.*, foregut, midgut and hindgut (Fig 3.1). Insects have paired **salivary glands** and **salivary reservoirs** which are associated with the foregut. These structures are usually located in the thorax (adjacent to the foregut). **Salivary ducts** start from the glands to the reservoirs and then forward, through the head, to an opening, the salivarium, behind the **hypopharynx**. Movement of the mouthparts helps in mixing of saliva with food in the buccal cavity.

1.1 Foregut (Stomodaeum)

Foregut is ectodermal in origin. It is lined with a cuticle (impermeable layer) that is continuous with body surface linings. Cuticular lining (endocuticle and epicuticle) of the foregut, known as the **intima**, protects foregut from abrasion by food particles. Cuticular lining inside the foregut is shed as exuviae during moulting and then a new cuticular lining is formed. In primitive insects like Collembolans and Silver fish, foregut is an undifferentiated simple tube. However, in highly evolved forms, it is further divided into the i) mouth, ii) pharynx, iii) esophagus (gullet),iv) crop, and v) proventriculus (Box 3.1).

1. **Mouth,** also called **buccal cavity,**is a muscular valve (sphincter) that marks the front of the foregut. It is located at the center of the mouthparts. The mouth parts are modified to suit different types of feeding habits. It is generally made up of the labrum, labium, maxilla, mandibles and hypopharynx. The pre-oral cavity between labrum and hypopharynx is called **cibarium**. In insects with biting and chewing type of mouth part, themandibles break the food into small particles so that more surface area is available for the action of digestive enzymes

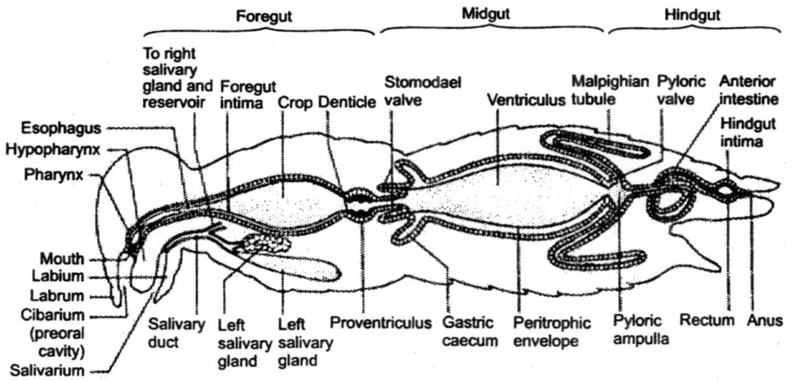

Fig 3.1. Generalised structure of insect alimentary canal

The pearl cavity between hypopharynx and salivary duct is called **salivarium**. Salivary gland secretes enzymes into the mouth, which also aid in the breakdown of food and add moisture that aid in movement of food through esophagus more easily. In bugs with piercing and sucking type of mouth parts, the cibarium acts as the sucking pump. In Lepidoptera with the siphoning type of mouth parts,the pharynx is developed into a sucking pump. Salaivarium acts as a saliva syringe or salivary pump. In Lepidoptera larvae the salaivarium works as a **silk regulator**.

Box. 3.1 Foregut

1. Starts at mouth and ends just after proventriculus but before cardiac sphincter.
2. Little or not digestion.
3. Involved in ingestion of food and its mechanical breakdown by teeth or muscle sets.
4. In some insects has a dilated crop that serve as food reservoir.
5. Next to crop is proventriculus where food is ground.

2. **Pharynx** is the region provided with dorsal dilator-muscles originating from the frontal and dorsal area of the head wall. Food in the buccal cavity is sucked through the mouth opening into the pharynx by contractile action of cibarial muscles. Pharynx is not well developed in mandibulate insects.
3. **Oesophagus** is the undifferentiated part which passes food from the pharynx to the crop. It is variable in length and their inner walls are longitudinally folded. It is commonly enlarged to form the crop and proventriculus.
4. **Crop** serves as a **food reservoir** until it can be pushed into the midgut. This allows insects to survive for a longer period of time without eating. The crop is wide and thin walled in grasshoppers and is well developed in honey bees. In case of ants, they use their crops to carry food, which they can then feed to other ants. While in the crop, some digestion of food may occur as a result of salivary enzymes that were added in the buccal cavity and/or other enzymes regurgitated from the midgut.
5. **Proventriculus** or **Gizzard**: Next to crop is the proventriculus where the grinding of food particles takes place (Fig 3.2). Proventriculus is the narrow posterior end of stomodaem. In most insects, there is a valve which controls the entry of food into the midgut.The gizzard is usually absent or reduced in sucking insects of the orders Diptera and Hemiptera and well developed in insects of orders Orthoptera, Coleoptera, Dictyoptera that feed on hard substances.
6. **Stomodaeal** or **oesophageal valve** is a circular fold of stomodaeal wall projecting into the ventriculus. The valve is composed of two cellular lamellae and is covered on each side by stomodaealintima.The valve regulates the transport of food from proventriculus to the mesenteron and prevents the movement of food from mesenteron to stomodaeum.

Types of Proventriculus (Gizzard)

Based on the type of food habits the internal cuticle of the gizzard is variously modified.

1. In insects with biting and chewing type of mouth parts inner walls of gizzard develop into strong cuticular plates or teeth to grind and strain food e.g. Blattidae, Gryllidae, Tettigonidae..
2. In honey bees, it is plate like to separate pollen grains from nectar.
3. In flea, it is spine like to break the blood corpuscles.

Functions of foregut

1. It is the conducting tube, carrying food from the cibarial cavity to the midgut.
2. It is the site of temporary food storage.
3. It is not involved in digestion and absorption of nutrients.
4. Sometimes, partial digestion may occur by salivary enzymes.

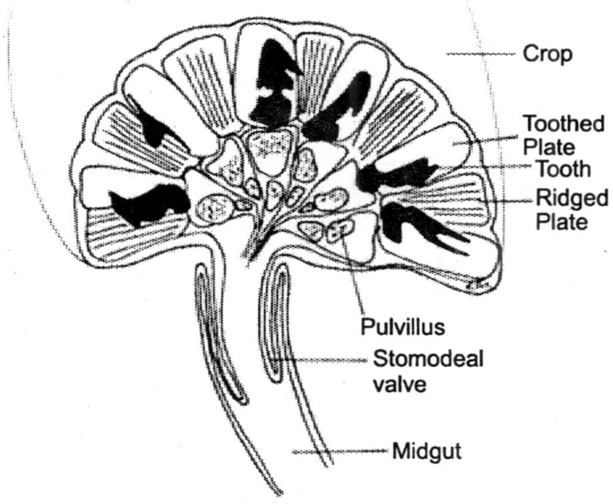

Fig 3.2. Schematic diagram of the gizzard of blattids

1.2 Midgut (Mesenteron)

Midgut is endodermal in origin and is also called **ventriculus.** This part contains no cuticular lining. In general form, the ventriculus is tubular and uniform in diameter. The epithelial walls are thickened and the cells are spongy than those of other parts of the alimentary canal. It is the middle part of the alimentary canal and functions for all the practical purposes as a **stomach**. Hence it is also called **functional part of alimentary canal** where the maximum amount of digestion, absorption of food and secretion of digestive enzymes occur in this region. Midgut starts at the cardiac sphincter and ends at the pyloric sphincter (Box 3.2).

The midgut is composed of a single layer of epithelium resting on a continuous basal lamina or basement membrane. On the haemolymph side of the epithelium is a discontinuous latticework of inner circular and outer longitudinal muscles. Typically in fluid-feeding insects the muscle layers are poorly formed, whereas in insects with solid diets, extensive muscle layers and connective tissues are present. These muscles are relatively poor in mitochondria, possess sparse tracheation, have a reduced sarcoplasmic reticulum and T system and have 12 actin filaments around each myosin filament. Thus, they are designed for slow contraction and perhaps supercontraction. The nervous supply to the midgut is also poor. Often there are small number of nerves stretching back from the stomatogastric system over the proventriculus and extreme anterior end of the midgut.

Insects do not produce mucin, instead the meal in the midgut lumen is separated from the epithelium by a peritrophic matrix or membrane. This may be produced either by all cells along the midgut (type I peritrophic matrix) or by specialised cells just posterior to the proventriculus (type II peritrophic matrix).

Midgut epithelial cells are distinguished into **secretory cells** and **regenerative cells**, which replaces secretory cells. Secretory cells include columnar cells and goblet cells (aged secretory cells). Columnar cells possess microvilli (*micro*-small; *villi*- bushy hair), forms the striated border adjacent to the lumen, which provides an extensive surface area for absorption. It contains absorptive and secretory cells, hence actively involved in the secretion of enzymes and the absorption of the products of digestion. Regenerative cells replace the degenerated cells. Laterally the cells are linked by simple or septate desmosomes.

Box 3.2 MIDGUT

1. Starts at cardiac sphincter and ends at pyloric sphincter.
2. Contains different cell types:
   a. Digestive cells
   b. Regenerative cells
   c. Endocrine cells
   d. Goblet cells
   e. Cuprophilic cells
3. Peritrophic membrane covers microvilli and separates food.

Important structures present in midgut are as follows:

1. **Peritrophic membrane:** This is the internal lining of midgut, secreted by a cluster of cells, called the **cardial epithelium.** It lies just behind the stomodael valve. Peritrophic membrane (*Peri-* around; *trophic-* food) is present in the majority of the insects, both larval instars and adults. It is well developed in solid feeders and absent in sap feeders (Hemiptera and adult Lepidoptera) with few exceptions. The main functions of peritrophic membrane are:
   A. Lubricate and facilitate food movement.
   B. Envelop the food and protects the midgut epithelial cells against harder food particles. It serves as a sort of a covering for waste products and is excreted with faeces.
   C. Protect the delicate midgut digestive cells from abrasion by food particles, without inhibiting absorption of nutrient molecules and also act as a barrier to micro-organisms.
   D. It is also permeable to enzymes and dissolved nutritive material resulting from digestion.

   The peritrophic membrane is made up of proteins (peritrophins) and chitin to which other components (enzymes, food molecules) may associate. Two types of membranes have been reported. The formation of the peritrophic membrane may occur in part of the midgut or in the entire organ (type I), or only at the entrance of the midgut (cardia) (type II). Peritrophic membrane type 1 is seen in *Aedes aegypti* adult and *Tribolium castaneum* and peritrophic membrane 2, seen in larvae of *Aedes aegypti* larvae and *Drosophila melanogaster*.The type 1 and type 2 peritrophic membranes have varied molecular architectures determined by different peritrophins, which may be partly modified by diet.

   Peritrophic membranes are not seen in Hemiptera and Thysanoptera. These insect orders have perimicrovillar membranes in their cells. The larvae and adults of Bruchidae have a peritophic gel instead of a peritrophic membrane separating the midgut epithelium from the food. Most of the pores of the peritrophic membrane measure 7 to 9 nm, although some may be as large as 36nm. This therefore hinders free movement of molecules. The function of the peritrophic membrane includes protection against food abrasion and invasion by microorganisms and several other roles associated with compartmentalisation of the midgut.

2. **Gastric caecae (Enteric caecae or Hepatic caecae):** Finger like projections (usually from 2 to 10 projections) found in the anterior

or posterior ends of midgut known as **gastric caecae**. This structure increases the functional area of midgut for secretion of enzymes or absorption of water (and other substances) from the alimentary canal. Also, theymay shelter symbiotic bacteria in some insects that help digestion.

3. **Ventriculus:** The posterior part of the midgut is ventriculus. It is the primary site for enzymatic digestion of food and absorption of nutrients. There are microscopic projections (microvilli) on the walls of the ventriculus which increase the surface area for nutrient absorption.

Functions of midgut

1. Secretes enzymes for the chemical breakdown of food.
2. Helps in digestion of food materials.
3. Absorption of nutrients produced takes place in the midgut.

1.3 Hindgut (Proctodaeum)

It is ectodermal in origin and lined by a thin cuticle layer similar to foregut which is permeable than that of foregut to salts, ions, amino acids and water. Cuticular lining inside the hindgut is shed during moulting and new cuticular lining is formed. In hindgut, the circular muscles are developed both inside and outside the layer of the longitudinal muscles. It is marked by the pyloric valve and the insertion of the Malpighian tubules. The hindgut consists of the pylorus, the anterior intestine (ileum and colon), rectum or posterior intestine and anus (Box 3.3).

Box 3.3 Hindgut

1. Major refuse dumping area for wastes from midgut and Malpighian tubules
2. In fluid feeders, involved in getting rid of excessive water
3. Houses symbonts
4. Aid in resorption of certain salts and amino acids
5. Helps in maintaining osmotic pressure of the haemolymph
6. In male Scolytid beetles, produces aggregation pheromone
7. Respiration in larvae of Anisoptera

1. **Pyloric valve:** Midgut opens into the hindgut through a pyloric valve, which regulate food flow from the mesenteron to the proctodaeum. The Malpighian tubules often arise from it. In certain immature stages of insects like Honey bee grubs, the midgut is not connected to hindgut till pupation.

2. **Malpighian tubules:** In the cockroach, *Periplaneta americana*, the Malpighian tubulues has about 150 tubules in each insect. Each tubule consists of at least three parts. The short distal region is thinner than the other parts and is highly contractile. The middle region comprises most of the tubule length and is composed of primary and stellate cells. Primary cells contain numerous refractile mineral concretions, while stellate cells have smaller nuclei, fewer organelles, simpler brush border, and numerous multivesicular bodies. Symbiont protozoa are sometimes present within the lumen of the middle region near where it opens into the proximal region of the tubule. The latter is a short region that drains the tubular fluid into one of the six ampullae. These are contractile diverticula of the intestine located at the midgut-hindgut junction. The ampulla is highly contractile, and consists of a layer of epithelial cells surrounding a cavity that opens into the gut via a narrow slit lined by cells of unusual morphology. The proximal region of the tubule and the ampulla resemble the midgut in that they have similar micromal origin and reabsorptive function for the proximal region of the tubule and for the ampulla.

   In some insects the distal end of these tubules is found in contact with the wall of the rectum through the perinephric membrane, an arrangement known as **cryptonephric complex** (Box 3.4). This system facilitates the reabsorption of water and, therefore, is more highly developed in insects that feed on dry substances or live in a relatively dry environment and therefore need to retain water.

   The main function of the Malpighian tubules of insects is the excretion of nitrogenous products and other remains of metabolites. The Malpighian tubules also secrete fluids within the posterior intestine where they are reabsorbed and certain substances are transported into the haemolymph.

   The Malpighian tubules work as follows: potassium is actively secreted into the lumen of the tubules and water follows passively due to osmotic forces, resulting in the formation of an abundant amount of potassium-rich liquid. This fluid is isotonic relative to haemolymph, but has a quite different composition that differs in surprising ways with the mammalian kidney in which the initial urinary fluid is an ultrafiltrate of blood plasma.

   Within the Malpighian tubule fluid is modified through processes of secretion and resorption, then there is passage of fluid into the posterior intestine where most of the solutes and water are reabsorbed, occurring the precipitation of uric acid (part the fluid in the form of water-soluble potassium urate). This facilitates subsequent removal of water, because the precipitated uric acid does not contribute to the osmotic activity of the rectal contents. Finally, the remaining content is expelled into the rectum feces mixed with urine

3. **Ileum:** It is a narrow tube running to the rectum. In larval Scarabaeidae, it is called as the fermentation chamber. It is expanded to house the symbionts.
4. **Colon:** Posterior part of the ileum is sometimes called as colon.
5. **Rectum**: It is an enlarged sac with a thin epithelium except in the rectal pad region. Rectal pads, bears epithelial thickenings, called rectal papillae (usually six), and project into the lumen. They are important in reabsorbtion of water, salts, amino acids from the urine and help in dehydration of faeces. It opens out through the anus. Wastes containing nitrogen are excreted as uric acid, which is relatively nontoxic and is excreted dry. It conserves water for terrestrial organisms. Rectal papillae are absent in Hemiptera and the larvae of endopterygotes.
6. **Anus** is the posterior opening of the alimentary canal. It is situated at the caudal end of the abdomen. Waste is packaged as pellets and excreted.

Box 3.4 Cryptonephridia

1. Some desert insects have intimate association between Malpighian tubules and hindgut.
2. Distal end of these tubules are in contact with the wall of the rectum.
3. Bounded by perinephricmembrane.
4. Allows extreme conservation of water, including absorption of water from humid air in the rectum.

Functions of hindgut

The main functions of the hindgut are the absorption of water, salt and other useful substances from the faeces. Other functions associated with the hindgut are:

1. Filters haemolymph and helps in maintaining the osmotic pressure.
2. Recovers or reabsorbs lost metabolites, water and salts before expelled as it passes towards the anus.
3. Waste products are concentrated and packaged for excretion.
4. Allows indigestible material to pass out of the digestive system as waste.

5. Houses symbionts that use the waste products.
6. In male Scolytid beetles, produces aggregation pheromone.
7. Respiration in larvae of Anisoptera..

1.4 Modifications in alimentary canal of insects

1. In sucking insects, **pharynx** is well developed into pharyngeal pump and used to draw up fluids especially in Lepidopterans and Hymenopterans.
2. **Proventriculus** of the honey bee (Honey stomach) is very specialised. An anterior invagination into the crop ends in four muscle-controlled mobile lips each armed with a number of spines. The lips make snapping movements in such a way that the spine strain of the pollen grains from a suspension in nectar inside the crop while the nectar is retained, which is regurgitated and processed to form honey.
3. **Ileum** of hindgut in wood feeding termites forms a pouch called **fermentation chamber** to harbour cellulase secreting flagellate protozoans connected with the digestion of cellulose. In Scarebaeiodea (dung roller) the intima of the hindgut house cellulose fermenting bacteria where decaying wood and microorganisms are retained to accelerate fermentation of cellulose. In some insects, these symbionts are free in the lumen of the gut, but occur in cells called **mycetocytes**. They may be transmitted between individuals through an oral food exchange called **trophallaxis** and through egg called **transovarial transmission.**In wax moth, bacteria help in wax digestion. Bed bug and cockroach acquire vitamin and amino acids from microbes. Leafhoppers harbour yeast to digest starch and sucrose.
4. **Filter chamber in midgut:** Homopterans and some hemipterans ingest a large volume of liquid food, in order to extract enough nutrients for growth, development and reproduction. If the entire liquid food is passed through the gut, there would be a dilution of digestive enzymes and ultimately dilution of haemolymph. The rapid removal of water to the rectum is achieved by the anterior midgut forming a large thin walled bladder, which is closely bound to the anterior hindgut and malphighian tubules by its own basement membrane. This chamber formed within the folds of the anterior midgut is called **filter chamber** (Fig 3.3).Water passes directly from the midgut to the hindgut, which prevents dilution of digestive enzymes and concentrates food for efficient digestion. This also helps in osmoregulation by preventing dilution of haemolymph.

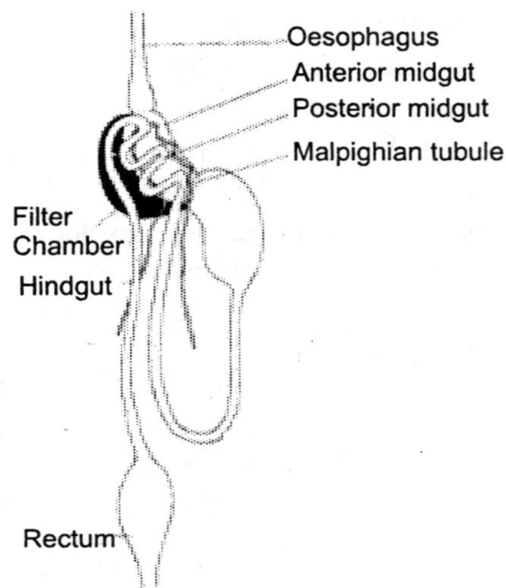

Fig 3.3. Diagram showing the filter chamber in a cercopid

5. **Rectal chamber** is modified as **branchial chamber** in larval Anisoptera (Dragonflies). Here the water is pumped in and out of the rectum so that the water round the gills in the anterior part of the rectum is constantly renewed and by forcible ejection of water, the insect is able to propel forward. Exchange of gases takes place inside this rectal chamber. Thus propulsion assists in locomotion.

1.5 Histologyof midgut epithelium

The gut epithelium is simple and rests on a basal lamina that is surrounded by conspicuous circular and a few longitudinal muscles. Wave like contractions of the circular muscles causes peristalsis, propelling the food bolus along the gut. The entire gut is oxygenated by the tracheal system. The foregut and hindgut is well innervated but not the midgut. The gut is connected to the body wall through extrinsic visceral muscles. These act as dilators of the midgut mainly at the foregut, where they form a pump (cibarium pump) that is highly developed in fluid feeders. However, it is also present in chewing insects (pharyngeal pump), which enables them to drink water and pump air into the gut during moult.

The epithelium contains columnar cells. It also has regenerative cells that are often collected together as nests at the base of the epithelium, besides cells believed to have endocrine function, and also specialised cells like the goblet cells, oxyntic cells.

1.5.1 The digestive epithelial cells

The apical plasma membrane of the midgut digestive cells is highly folded, usually regularly, into microvilli. The microvilli have an internal cytoskeleton composed of actin-like microfibrils and may be formed early in the maturation of the midgut cells. On the luminal surface of the microvilli is a **glycocalyx**, composed of a regular carbohydrate-rich electron-dense layer. Microvilli may possess blebs or protasome-like structures associated with secretory activity.

The basolateral plasma membrane is usually highly and irregularly convoluted, the infoldings sometimes stretching into the apical third of the cells. The status of these folded membranes is important to cell function. When tightly folded so that they remain juxtaposed they appear to play no role in the transport mechanism. However, when the membranes become separated after a meal, they appear to function in ion and water transport out of the lumen.

1.5.2 *Absorptive and storage epithelial cells*

In general, the absorptive and storage epithelial cells display greater elaborations of the basolateral membrane, contain such storage products as lipids and glycogen and has little or no apparent secretory activity. Absorptive cells of the midgut sometimes have multilayered mineral concentrations of cations in the form of phosphates, carbonates and chlorides. They may be produced by the rough endoplasmic reticulum or Gogi body. Glycogen deposits are often found associated with absorptive cells. These tend to form large, electron-dense structures of no particular shape near the central part of the midgut cells.

1.5.3 *Endocrine cells*

Distributed throughout the epithelium are cells which have an endocrine function. These cells are of variable shapes being pyramidal, bowl-shaped, oval or fusiform and are often of different electron density from neighbouring cells. Some endocrine cells, type the so-called **open-type** are in direct contact with the midgut lumen via a narrow extension from the broader basal part of the cell, suggestive of a receptor-secretory role. Other endocrine cells, the **closed-type**, do not extend through the epithelial layer as far as the luminal surface. Midgut endocrine cells are most common in the posterior midgut.

1.5.4 *Regenerative cells*

Scattered throughout the epithelium is undifferentiated cells found singly, paired or in groups. These cells have a regenerative role, and are characterised by a dense cytoplasm with few differentiated organelles, and none of the elaborations of the apical and basal plasma membranes. In cockroaches, the fully formed midgut is renewed continuously from regenerative cells and in *Tenebrio molitor*, the whole midgut epithelium is replaced on a four day cycle.The cells in the regenerative Nidi may also be used in replacement following injury, or can be triggered into mitosis and epithelial replacement by specific physiological events, such as blood feeding in fleas.

1.5.5 Goblet cells

These cells are typically seen in ephemeropteran and lepidopteran midgets and are so called because of their invaginated shape. Each goblet cell encloses a large extracellular lumen that is continuous with the midgut lumen, via a labyrinthine apical valve formed of interdigitating microvilli. The enclosed lumen of the goblet cell is filled with flocculent material formed of about 20% solids, probably sulphated glycosaminoglycans. In the anterior and middle portions of the midgut the irregular microvilli of the goblet cells contain mitochondria, a feature unique to arthropod epithelia, and characteristic of active ion-transporting tissues. In keeping with their absorptive function, goblet cells may also have a basolateral plasma membrane that folds right into the body of the cell. As a consequence, goblet cells have very little cytoplasm and forms the thinnest of cellular layers between the midgut lumen and the haemocoel.

1.5.6 Cuprophilic cells

These cells are seen in dipteran larvae and are associated with midgut zones that are acidic. The apical invaginations of these cells are lined with microvilli that are poorly structured unlike goblet cells. The invaginations may be almost completely occluded by irregular microvilli, just below which is a zone of apical cytoplasm which is less dense and has no mitochondria. The cells possess carbonic anhydrase activity and the microvilli are studded on the cytoplasmic side with portasome-like particles.

2. MICROVILLAR MEMBRANE- STRUCTURE AND ENZYMOLOGY

Midgut cells are associated with one another by junctions that separate the plasma cell membranes into an apical and a basolateral domain. The apical domain is usually modified into finger-like projections, the microvilli. The insect midgut cell microvillus is homologous to that described in vertebrates. Thus, a bundle of parallel actin filaments cross-linked by actin-bundling proteins like fimbrin and villin form the core of a microvillus.

Lateral side arms (composed of myosin I and calmodulin) connect the sides of the actin bundle to the overlying plasma membrane (Fig 3.4).

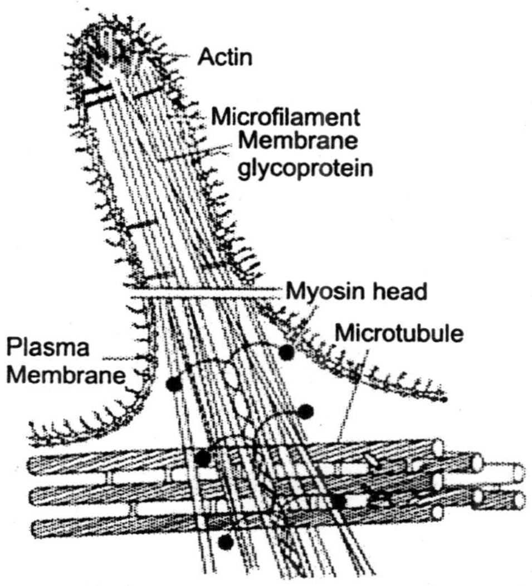

Fig 3.4 Structure of microvilli

Insect midgut microvillar membrane densities vary widely, with insects appearing later in evolution (more derived insects) having denser membranes (and hence a higher protein content) than insects appearing earlier in evolution (less derived insects) (Terra et al., 2006). Although higher protein content does not necessarily mean a richer variety of proteins, there is evidence supporting this.

Insect midgut cell apices are involved in the transport of water, and organic compounds. They are also known to play a role in digestive events since disaccharidases were observed firmly bound to cell membrane covering the enterocyte microvilli. Digestive enzymes performing final digestion were found associated with microvilli.

The physiological role of midgut microvillar membranes may change along the midgut and among insect taxa, and should include surface (terminal) digestion, absorption, ion homeostasis, signaling, and unique digestive enzyme secretion mechanisms. The microvillar integral digestive enzymes vary among different taxa. Most frequently, they are amino-peptidase, alkaline phosphatase, carboxypeptidase, dipeptidase, and α-glucosidase (Terra and Ferreira, 1994).

Three groups of predicted microvillar proteins were recognised in *T. molitor* (Coleoptera) (Ferreira *et al.*, 2007): digestive enzymes (aminopeptidase and α-mannosidase), putative peritrophic membrane ancillary protein (PMAP)-like proteins; and peritrophic membrane proteins (peritrophins). The aminopeptidase corresponds to that described previously (Cristofoletti and Terra, 1999, 2000). PMAP may be involved in PM formation, and is homologous to proteins having insect-allergen repeat-related domains.

Lepidopteran midgut microvillar proteins are grouped into six (Pauchet *et al.*, 2009): (1) digestive enzymes (aminopeptidases with GPI-anchors, carboxypeptidase, alkaline phosphatase, astacin-like protein, dipeptidyl peptidase A, maltase-like protein; (2) midgut protection (thioredoxin peroxidase; protein disulfide isomerase; aldehyde dehydrogenase, serpin); (3) peritrophic membrane formation (peritrophins); (4) membrane-tight-bound cytoskeleton proteins (fimbrin, actin, afadin, desmocollin, cadherin-like proteins); (5) proteins associated with microapocrine secretion (annexin, calmod-ulin, gelsolin); and (6) others (V-ATPase, chlorophylide A binding protein, ABC transporters).

Larval mosquito midgut microvillar proteins fall into four categories (Bayyareddy *et al.*, 2009; Popova-Butler and Dean, 2009): (1) digestive enzymes (alkaline phosphatase, trypsin, serine proteinase, zinc-metalloprotease, a-amylase); (2) midgut protection (protein disulfide isomerase, aldehyde dehydrogenase, peroxiredoxin); (3) membrane-tightly-bound cytoskeleton proteins (flotillins, prohibitin, actin); and (4) others (V-ATPase, calmodulin).

Lepidopteran and dipteran midgut microvillar membranes are denser and show a greater variety of proteins than those of coleopterans, because there are a host of microvillar proteins that assist the larvae in dealing quickly with huge amounts of food derived from a variety of plants. There are proteins involved in counteracting plant chemical defenses, in protecting the midgut surface against the larval serine proteinases, and in promoting peculiar secretory mechanisms. In addition to those proteins, there are still others that lepidopterans and dipterans share with coleopterans; namely, those forming the peritrophic membrane, a few digestive enzymes, and those not found among predicted proteins, like receptors and ion and organic compound transporters. Coleopterans attack a variety of plants, but they usually deal with their food more slowly than lepidopterans and dipterans. Perhaps because of this, their digestive physiology strategy seems to rely less on midgut microvillar proteins than is the case in lepidopterans and dipterans.

3. PERITROPHIC MEMBRANE- STRUCTURE, FORMATION AND FUNCTION

There is a film surrounding the food bolus in most insects that is occasionally fluid (peritrophic gel), but is more frequently membranous (peritrophic membrane, PM). The PM is made up of a matrix of proteins (mainly peritrophins) and chitin to which other components (e.g., enzymes, food molecules) may associate. This anatomical structure is sometimes called the **peritrophic matrix**, but this term should be avoided because it does not convey the idea of a film and rather suggests that it is the fundamental substance of some structure, usually filling a space, as in the mitochondrial matrix. The argument that "membrane" means a lipid bilayer is not valid, because the PM is not a cell part but an anatomical structure, like the nictitating membrane of birds and reptiles.

There are two types of insect PMs. **Type 1 PM** is synthesised from the digestive epithelial cells that line the gut of the insect, whereas **type 2 PM** is synthesised as a continuous tube from a specialised organ, the cardia (or proventriculus) situated in the anterior midgut region of the insect. The PM is the first barrier in insects encountered by a number of ingested viral and protozoal organisms, some of which use blood- or tissue-feeding insects as vectors for transmission to vertebrate hosts, where these organisms cause considerable morbidity and mortality. There is evidence that the PM of insect vectors can affect the survival and virulence of these pathogenic organisms (Box 3.5).

Box 3.5 Peritrophic membrane

1. Secreted by microvillate columnar epithelial cells.
2. Made up of an amorphous sheet of polysaccharide, chitin, glycoprotein, and protein.
3. Tubular film that surrounds the bolus and within which considerable digestion occurs.
4. Numerous insect pathogens center activity on peritrophic membrane including Bt.

Peritrophic membraneproteins have been classified by Tellam *et al.* (1999) according to the ease with which they can be extracted: **class 1 proteins** can be removed with physiological buffers, **class 2** with mild detergents, and **class 3** with strong denaturants (e.g., urea), while **class 4 proteins** are the remaining residue. Class 1 proteins are thought to be digestive enzymes and food proteins loosely adsorbed at the PM surface. Class 2 proteins, at least in lepidopteran PMs, may correspond to proteins enclosed in membrane vesicles entrapped between PM sheets. These membrane vesicles bud off from the cell microvilli. The numerous hydrolases found in the PM proteomes may correspond to those entrapped enzymes. The same might be true for the so-called β-glucan binding **like proteins** (GNBP) like proteins (Campbell *et al.*, 2008), which might correspond to active laminarinases. Class 3 proteins are the integral proteins of PM. Class 4 proteins are likely the same as those in class 3, since Campbell *et al.*, (2008) were able to completely solubilise PM with anhydrous trifluoromethanesulfonic acid, and found the same proteins previously described as class 3. The major class 3 proteins are named peritrophins, and are characterized by the presence of chitin-binding domains (CBDs) (called peritrophin, pfam 01607, CBM_14) (CDART database) (Geer *et al.*, 2002), and may also have mucin-like domains (Tellam *et al.*, 1999).

Peritrophins are made of several domains. The major domain (peritrophin A-domain) is a cysteine-rich domain with chitin-binding properties having the consensus sequence CX13_2qCX5_6CX9_19CX10 _14CX4_14C (where X is any amino acid except cysteine), which includes several conserved aromatic amino acids. Variations of this chitin-binding domain are the peritrophin-B and peritrophin-C domains, with consensus sequences CX12J3 CX20-21CX10CX12CX2CX8CX7-12C and CX8-9CX17-21 CX10_11CX12_13CXnC, respectively. These variations may be absent from many insects, exemplified by the predicted sequences of *T. castaneum* proteins (Jasrapuria *et al.*, 2010). The mucin-like domains occurring in peritrophins are proline/threonine-rich domains that are heavily glycosylated and similar to mucins (Lang *et al.*, 2007). Peritrophins may have one (e.g., Cb-peritrophin-15 from *Lucilia cuprina*) to several (e.g., peritrophin-44 from *L. cuprina*) chitin-binding domains or chitin-binding domains with small (e.g., Ag-Aperl from *Anopheles gambiae*) or very large mucin-like domains (e.g., IIM from *Trichoplusia ni*).

Other PM integral proteins are chitin deacetylases (CDA) (Dixit *et al.*, 2008). This kind of protein was first found in the PM of *Trichoplusia ni* by Guo *et al.* (2005). The authors did not detect any chitin deacetylase activity, and suggested that binding might be the only interaction of those proteins with chitin. The interaction was suggested by its extraction from PM with calcofluor. Taking into account that some CDA-like proteins found associated with PM are active in producing chitosan from chitin, Campbell *et al.*, (2008) suggested several functions for chitosan: (1) it might be present to modify the flexibility of chitin fibers; (2) it may regulate redox condition, as it has antioxidant properties; (3) it may have antimicrobial properties; and, finally, (4) modification of chitin may be required for the binding of other proteins.

PMs are classified into two types (Peters, 1992). **Type I PM** is found in cockroaches (Dictyoptera), grasshoppers (Orthoptera), beetles (Coleoptera), bees, wasps, and ants (Hymenoptera), moths and butterflies (Lepidoptera), and in hematophagous adult mosquitoes (Diptera). **Type II PM** occurs in larval and adult (except hematophagous ones) mosquitoes and flies (Diptera), and in a few adult Lepidoptera.

Type I PM is formed either by the whole midgut epithelium, or by part of it (anterior or posterior regions). The formation of these PMs is frequently induced by the distension of the gut caused by food ingestion. PM produced by the whole or anterior midgut epithelium envelops the food along the whole midgut. When PM is produced only by the posterior part, the anterior midgut epithelium is usually covered with a viscous material, the peritrophic gel, as observed in carabid beetles and bees. This gel is also observed in the anteriorly placed midgut ceca of some insects, and along the whole midgut of others (Terra, 2001).During formation of type I PM, chitin is synthesised outside the cells by a chitin synthase bound to microvillar membranes using precursors formed inside the cells. Once chitin is self-organised in chitin fibers, it interlocks with peritrophin molecules that are released by exocytosis (Bolognesi *et al*., 2001).

Peritrophins are soluble proteins that are extensively immobilized at the surface of *T. molitor* cells due to cell glycocalyx association (Ferreira *et al*., 2007). This arguably facilitates chitin-peritrophin association, and it is conceivable that ancillary proteins help this process. *T. molitor* PMAP may be one of those ancillary proteins. PMAP is homologous to proteins having insect allergen-related repeat domains like AEG12 (Shao *et al*., 2005). The role of PMAP is based on circumstancial evidence: (1) PMAP is a soluble protein that, like peritrophins, occurs partially immobilized at the cell surface; and (2) PMAP is found associated with PM, but, in contrast with peritrophins, it seems to be removed from the PM as it moves along the midgut.

Type II PM is secreted by a few rows of cells at the entrance of the midgut (cardia), and is usually found in insects irrespective of food ingestion. Peritrophins are secreted by exocytosis (Eisemann *et al*., 2001).

Although a PM is found in most insects, it does not occur in Hemiptera and Thysanoptera, which have peri-microvillar membranes in their cells. The other insects that apparently do not have a PM are adult Lepidoptera, Phthiraptera, Psocoptera, Zoraptera, Strep-siptera, Raphidioptera, Megaloptera, adult Siphonaptera, bruchid beetles, and some adult ants (Hymenoptera) (Peters, 1992). These insects may have a peritrophic gel instead of PM, one example being bruchid beetles (Terra, 2001), or may have had their PMs overlooked because the insects were unfed. Another possibility is that minute hematophagous insects (like Siphonaptera and Phthiraptera) have lost their PM because the blood clot ensures countercurrent flows and their small size makes for easy and efficient diffusion of digestion products up to the midgut surface.

The PM may have a wide range of pore sizes, with some being small or very large, but most of them in the middle range. As a consequence of its small pores, the PM hinders the free movement of molecules, dividing the midgut lumen into two compartments with different molecules. The functions of this structure include those of the mucus (protection against food abrasion and invasion by microorganisms), and several roles associated with the compartmentalization of the midgut. These roles result in improvements in digestive physiology efficiency, thereby leading to decreased digestive enzyme excretion, and restrict the production of the final products of digestion close to their transporters, thus facilitating absorption.

3.1 The Physiological Role of the Peritrophic Membrane

A multitude of functions is associated with the peritrophic membrane. They are:

1. Protection against food abrasion
2. Act as a barrier against invasion by microorganisms
3. Facilitate increased growth rate
4. Enhancing digestive efficiency
5. Prevention of non-specific binding of undigested material onto midgut cell surface,
6. Prevention of enzyme excretion
7. Increase in the efficiency of digestion of polymeric food favoured by oligomer removal
8. Increase in the efficiency of oligomeric food hydrolysis
9. Restriction of food monomer production at the cell surface
10. Enzyme immobilisation
11. Toxin binding

4. PHYSIOLOGY OF DIGESTION

Digestion is the chemical transformation of large and complex molecules in food to smaller molecules that can be absorbed across the gut wall for insect nutrition. All insects have intrinsic digestive enzymes like carbohydrases, proteinases and lipases. Enzymes concerned with digestion are present in the saliva and in the midgut secretions. In the midgut, enzymes are secreted by hepatic caecae and midgut epithelial cells. Midgut enzyme secretion may be **holocene** (disintegrate epithelial cells in the process of enzyme secretion) and **merocrine** (secrete enzymes without breakdown of the cells). In addition, digestion may be facilitated by micro-organisms in the gut.

Most of the food is ingested in the form of macromolecules and other complex polymer substances (such as proteins, polysaccharides, fats, nucleic acids, etc.) which must be broken down by catabolic reactions into smaller molecules (i.e. Amino acids, simple sugars, etc.) before being used by cells of the body for energy, growth, or reproduction. This process of break-down is known as **digestion** and it involves three phases. **Primary digestion** is the dispersion and reduction in molecular size of the polymers and results in oligomers. During **intermediate digestion**, these undergo a further reduction in molecular size to dimers, which in **final digestion** form monomers. Digestion usually occurs under the action of digestive enzymes from the midgut, with minor or no participation of salivary enzymes. In most insects, midgut pH is either mildly acidic or neutral. Lepidopteran and trichopteran larvae, scarabaeid beetles, and nematoceran flies have alkaline midguts, whereas cyclorrhaphous flies have a very acidic section in the middle of the midgut. The mid-gut is, as a rule, an oxidizing site, although in some wool-digesting insects, it is a reducing site, a condition necessary to break disulfide bonds in keratin, thus facilitating enzymatic hydrolysis. Microbes may provide insects with nutrients or help detoxify noxious substances with a negligible role in digestion.

4.1 Digestive Glands

The primary function of the gut is to digest the ingested food and to absorb the metabolites. The digestion process is enhanced with the help of enzymes, which are produced by digestive glands and special cells as well as by symbiotic microbes.

Salivary glands

In insects, the labial glands are commonly known as **salivary glands**. They are usually paired, tubular branched structures with a number of lobes situated on either side of the foregut in the head, secreting saliva. Two types of salivary glands are present in insects- (1) Acinous (alveolar) type which are multicellular sac like and directly innervated by the central nervous system and (2) the tubular type which is not innervated by the CNS (Box 3.6). In Orthoptera and Dictyoptera, they are very large and composed of a number of lobes, each lobe consisting of salivary reservoir. In Hemiptera, they are differentiated into 1-4 lobes. In Lepidoptera, it produces silk for cocoon construction, open in salivarium. Salivarium is modified to form a pump. Hymenoptera, consists of two pairs of racemose organs and their four ducts form a common canal. It contains clear, neturàl liquid and salivary enzymes. The major salivary enzymes are amylase, invertase, protease and lipase. Other salivary enzymes are hyaluronidase, phospholipase and pectinase. Starch is digested by amylase into sucrose. Proteases are helpful in digesting proteins into peptones and aminoacids and lipases break down fat into free fatty acids.

Functions of saliva

1. Dissolve the ingested food.
2. Saliva contains salivary enzymes (aids in partial digestion outside the mouth or midgut, called extra-oral digestion), various anticoagulants and agglutinins.
3. Secrete silk in larval Lepidoptera and Hymenoptera.
4. Used to glue puparial cases to the substrate in certain flies.
5. Mosquito saliva contains anticoagulant (to prevent the clotting of blood during bloodsucking) and anesthetic (to make the victim unaware of biting).
6. Ant saliva contains formic acid.
7. Saliva secreted by gall midges has a plant growth regulator, indole acetic acid (IAA), which aids in gall formation at the point of injection of saliva.
8. In plant and animal vectors, saliva assists in the transmission of pathogens.
9. Tea mosquito bug saliva contains toxins, causing necrosis and phytotoxemia symptoms.

Box 3.6 Types of salivary glands

1. Acinous (Alveolar) type
   - Multicellular with sac like units
   - Directly innervated by CNS
   - Stomatogastric NS produces serotonin for enzyme production
   - Suboegophageal ganglion produces dopamine for fluid secreton
2. Tubular type
   - Not directly innervated
   - serotonin from haemolymph regulates enzyme production

Mandibular glands

These are small sac like glands that opens near the base of mandibles in Apterygota, Isoptera, Coleoptera and Hymenoptera. In *Apis* queen, a big mandibular gland produce **queen substance** and other pheromones concerned with colony control. In larval Lepidoptera, it is large and functional as salivary gland, having become specialised in silk production.

Maxillary glands

It opens near maxillae and lubricatethe mouth parts. They are present in Protura, Colembola, some larval Neuroptera and Hymenoptera. In Carnivorous Heteroptera, it produces a toxin to kill the prey.

Pharyngeal gland

It is a long coiled tube with large number of solid lobules on each side of the head and it opens at the hypopharynx. In Hymenoptera, it is well developed in worker bees. In early days they produce brood food the **royal jelly** with which young larvae are fed and which plays some role in caste determination. Later, it also secretes invertase and amylase enzymes.

4.2 Types of digestion

4.2.1 Extra-oral/ Extra-intestinal digestion

Complete or partial digestion outside the intestine is termed as **extra-intestinal** or **extra-oral digestion**.This occurs before ingestion with the help of salivary enzymes or sometimes digestive enzymes. It may constitute a major part of the total digestion, as found in all heteropteran insects, in adult robber fly, larval neuropterans (antlions, *Chrysoperla*), water beetles and firefly. In honeybees, sucrose in the nectar is digested by the action of the invertase enzyme into glucose (dextrose) and fructose (levulose). In honey bee, invertase is present in the nectar and the bees add more of their hypopharyngeal glands. In silkworm, the moth on emergence secretes protease, which attacks sourced of silk, thus facilitating the emergence of moth from the cocoon.

In Aphids and plant bugs, digestion of the plant cells by pectinase enzyme, aid in penetrating their Stylets into the plant. The saliva of Assassin bugs, *Platymeris* contains the hyaluronidase enzyme, which attacks the mucopolysaccharids of connective tissues. It also acts as a spreading agent, facilitating the entry of the other salivary enzymes, pectinase and phospholipase, into the host tissues. Pectinase helps to break down the prey's tissue. Phospholipase helps to dissolve the lipid layers in the cells of nervous tissue, before feeding.

The predatory larval water beetles regurgitate digestive enzymes from the gut and inject them into the prey through their mandibles. The tissues of the prey get digested and become a soupy material, which is drawn into the predatory larva through the mandibles.

4.2.2 *Internal/ Intestinal digestion*

Based on the diet on which they feed, digestive enzymes are secreted in the midgut and its caeca. The major digestive enzymes are protease, lipase, carbohydrases (amylase, maltase, invertase). Omnivorous, phytophagous insects and honeybees have protease, lipase and carbohydrases. Carnivorous and necrophagous insects have protease and lipase enzymes. In insects with symbiotic microorganisms, digestion of cellulose takes place in the hindgut.

4.3 Digestion of Carbohydrates

Polysaccharides are major constituents of cell walls and energy reserves, such as starch granules within plant cells and glycogen within animal cells. For phytophagous insects, disruption of plant cell walls is necessary in order to expose storage polymers in cell contents to polymer hydrolases. Cell wall breakdown may be achieved by mastication, but more frequently it is the result of the action of digestive enzymes. Thus, even insects unable to obtain nourishment from the cellulosic and non-cellulosic cell wall biochemicals would profit from having enzymes active against these structural components. Cell walls are disrupted by β-glucanases, xylanases, and pectinases (plant cells), lysozyme (bacterial cells), or chitinase and β-1,3-glucanase (fungal cells). The carbohydrates associated with cellulose, are frequently named **hemicelluloses** and the enzymes that attack them hemicellulases. Thus, the hemicellulases include β-glucanases other than cellulases, xylanases, and pectinases. Following the loss of cell wall integrity, starch digestion is initiated by amylases. A complex of carbohydrases converts the oligomers resulting from the action of the polymer hydrolysing enzymes into dimers (such as sucrose, cellobiose, and maltose), and finally into monosaccharides like glucose and fructose (Box 3.7).

Box 3.7 Major insect carbohydrases in carbohydrate digestion

1. Amylases
2. β-Glucanases
3. Xylanases and Pectinases
4. Chitinases and Lysozymes
5. α-Glucosidases
6. β-Glucosidases, β-Galactosidases, and Myrosinases
7. Trehalases
8. Acetylhexosaminidases, β-Fructosidases, and α-Galactosidases

4.3.1 Amylases

α-Amylases (EC 3.2.1.1) catalyse the endohydrolysis of starch and glycogen. Most insect amylases have molecular weights in the range 48-60 kDa, pi values of 3.5-4.0, and Km values with soluble starch around 0.1%. pH optima generally correspond to the pH prevailing in midguts from which the amylases were isolated. Insect amylases are calcium-dependent enzymes, and are activated by chloride with displacement of the pH optimum. Activation also occurs with anions other than chloride, such as bromide and nitrate, and it seems to depend upon the ionic size.

The best-known insect α-amylase, is the midgut α-amylase of *Tenebrio molitor* larvae. The enzyme has three domains (Fig 3.5). The central domain (domain A) is a (P/a)8-barrel that comprises the core of the molecule and includes the catalytic amino acid residues (Asp 185, Glu 222, and Asp 287). Domains B and C are almost opposite each other, on each side of domain A. The substrate-binding site is located in a long V-shaped cleft between domains A and B. There, six saccharide units can be accommodated, with the sugar chain being cleaved between the third and fourth bound glucose residues. A calcium ion is placed in domain B, and is coordinated by Asn 98, Arg 146, and Asp 155. This ion is important for the structural integrity of the enzyme, and also seems to be relevant due its contact with His 189. This residue interacts with the fourth sugar of the substrate bound in the active site, forming a hinge between the catalytic site and the Ca^{2+}-binding site. A chloride ion is coordinated by Arg 183, Asn 285, and Arg 321 in domain A (Strobl *et al.*, 1998). Domain C is placed in the C-terminal part of the enzyme, contains the so-called **Greek key** motif, and has no clear function. These structural features are shared by all known a-amylases (Nielsen and Borchert, 2000), although not all α-amylases have a chloride-binding site (Strobl *et al.*, 1998).

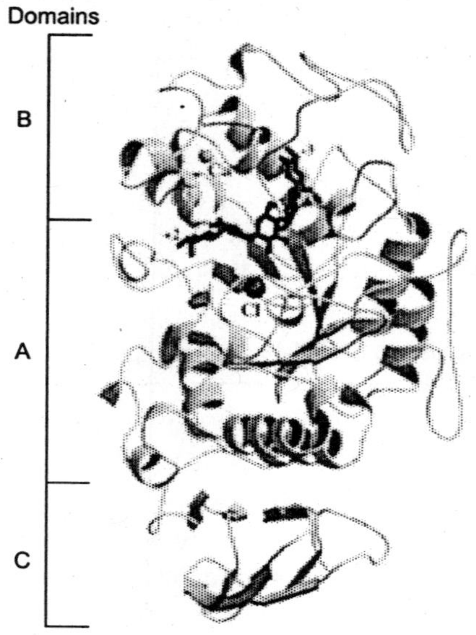

Fig. 3.5. Ribbon diagram of α-Amylase

(Robert Maurus et al 2005 in *Protein Science* 14 (3): 743-55)

4.3.2 β-Glucanases

β-Glucanases are enzymes acting on β-glucans. These are major polysaccharide components of plant cell walls, and include β-1,4-glucans (cellulose), β-1,3-glucans (callose), and β-1,3;1,4-glucans (cereal β-glucans). The cell walls of certain groups of fungi have β-1,3;1,6-glucans (Bacic *et al*., 1988).

Cellulose is by far the most important β-glucan. It occurs in the form of β-1,4-glucan chains packed in an ordered manner to form compact aggregates which are stabilized by hydrogen bonds. Cellulose is digested by a combined action of two enzymes. An endo-β-1,4-glucanase (EC 3.2.1.4), with an open substrate-binding cleft, cleaves bonds located within chains in the amorphous regions of cellulose, creating new chain ends. An exo-β-1,4-glucanase (EC 3.2.1.91) processively releases cellobiose from the ends of cellulose chains in a tunnel-like active site.

Cellulose digestion in insects is rare, presumably because the dietary factor that usually limits growth in plant feeders is protein quality. Nevertheless, cellulose digestion occurs in several insects that have, as a rule, nutritionally poor diets such as wood or humus. The role of symbiotic organisms in insect cellulose digestion is less important than initially believed, although symbiotic nitrogen-fixing organisms are certainly involved in increasing the nutritive value of diets of many insects (Terra, 1990).

Insect cellulases are known mainly from the lower termite *Reticulitermes speratus*, the higher termite *Nasutitermes takasagoensis*, and the woodroach *Panesthia cribrata* (Lo *et al.*, 2000; Watanabe and Tokuda, 2010). Alignments of the sequences of termite and woodroach endo-glucanases from data banks showed that they belong to family 9 of glycoside hydrolases (Coutinho and Henrissat, 1999).

The phytophagous beetle *Phaedon cochleariae* has cellulase activity in its midgut that was shown to belong to the glycoside hydrolase family 45 and to consist only of a catalytic domain. Similar enzymes occur in the midgut of the beetles *Psacothea hilaris* (Sug-imura *et al.*, 2003) and *Apriona germari* (Wei *et al.*, 2006). The cellulase from the cricket *Teleogryllus emma*, however, is from family 9 (Kim *et al.*, 2008).

Licheninases (EC 3.2.1.73) digest only β-1,3;1,4-glucans, whereas laminarinases may hydrolyse β-1,3;1,4-glucans and also β-1,3-glucans (EC 3.2.1.6). Three laminarinases produced by the salivary glands were isolated from *Periplaneta americana* (LAM, LIC1 and LIC2; Genta *et al.*, 2003). LIC1 and LIC2 hydrolyze laminarin and lichenin, whereas LAM is active only against laminarin. *Abracris flavolineata* (Genta *et al.*, 2007), *T. molitor* (Genta *et al.*, 2009), and *Spodoptera frugiperda* (Bragatto *et al.*, 2010) have only one major laminarinase (ALAM, TLAM, and SLAM, respectively) produced by the midgut. ALAM and SLAM have only laminarin as substrate, and TLAM hydrolyzes, besides laminarin, yeast β-1,3-β-1,6-glucan. The enzymes have pH optima from 5.0 to 6.5, except for SLAM (9.0), Km values for laminarin hydrolysis from 0.074 to 0.36%, and molecular masses from 23.4 to 46.2 kDa.

Some laminarinases are processive - that is, they perform multiple rounds of catalysis when the enzyme remains attached to the substrate. The exo-β-1,3-glucanase of *A. flavolineata* has a high-affinity accessory site that, on substrate binding, causes active site exposure, followed by the transference of the substrate to the active site. Processivity results in this case from consecutive transferences of substrate.

Insect laminarinases belong to glycoside hydrolase family 16, and are evolutionarily related and strikingly similar in sequence to Gram-negative bacteria-binding proteins (GNBPs) and other glucan-binding proteins that are active in the insect innate immune system. Because of this, many true laminarinases had been wrongly annotated as glucan-binding proteins in sequence databases.

The role of laminarinases in the insect gut is not yet clear. Insects eating contaminated food, such as *T. molitor* and *P. americana*, may use β-1,3 glucanases to digest fungi from their diet. In Lepidoptera that eat a less contaminated food, β-1,3- glucanase may be important to digest callose. Since the larvae eats on and in the same plant, callose deposition can impair nutrient digestibility in the absence of a β-1,3- glucanase. Pauchet *et al.* (2009) correlate the increase in β-1,3-glucanase activity in *Helicoverpa armigera* after fungi or bacteria ingestion with a role of the enzyme in immunity.

4.3.3 Xylanases and Pectinases

Xylans constitute the major non-cellulosic polysaccharides (hemicelluloses) of primary walls of grasses and secondary walls of all angiosperms, accounting for one-third of all renewable organic carbon available on earth. Chemically, xylans are a family of linear β-1,4-xylans with a few branches. Endo-β-1,4-xylanase activities (EC 3.2.1.8) were found in several insects (Terra and Ferreira, 1994). One of these enzymes was cloned from a beetle and shown to correspond to a protein of 22 kDa, with high sequence identity to fungal xylanases, and conserving the usual two catalytic regions (Girard and Jouanin, 1999). An exo-β-1,4-xylanase (EC 3.2.1.37) was partially purified from termites (Matoub and Rou-land, 1995), and was thought to act synergistically with an endo-β-1,4-xylanase originating from fungus ingested by the termites. Recent work showed that xylanase genes were expressed in the hindguts of termites by symbiotic flagellates (Arakawa *et al.*, 2009).

Pectin is a linear chain of a D-galacturonic acid units with α-1,4-linkages in which varying proportions of the acid groups are present as methyl esters. It is the main component of the rhamnogalacturonan backbone of the structure formed by the pectin polysaccharides. Pectin is hydrolyzed by pectinases (polygalacturonases, EC 3.2.1.15) described in many insects (Terra and Ferreira, 1994).

Pectinases are thought to be important for hemipterans, as they would facilitate penetration of their stylets through plant tissues into sap-conducting structures, and for insects boring plant parts. Accordingly, pectinases have been found in hemipteran saliva, and have been isolated and characterised from two weevils (Shen *et al.*, 1996) and cloned from a phytophagous beetle (Girard and Jouanin, 1999a).

4.3.4 Chitinases and Lysozymes

Chitin, the simplest of glycosaminoglycans, is a β-1,4-homopolymer of N-acetylglucosamine.Chitinolytic enzymes include chitinase (EC 3.2.1.14), which catalyses the random hydrolysis of internal bonds in chitin forming smaller oligosaccharides, and β-N-acetyl-D-glucosaminidase (EC 3.2.1.52), which liberates N-acetylglucosamine from the non-reducing end of oligosaccharides (Arakane and Muthukrishnan, 2010). Lysozyme, also has some chitinase activity, whereas chitinase rarely has lysozyme activity. These enzymes, besides being active in the ecdysial cycle, may also have a digestive role. Chitinase assays with mid-guts of several insects showed that there is a correlation between the presence of chitinase and a diet rich in chitin (Terra and Ferreira, 1994).

The best-known insect chitinase is the **moulting fluid** chitinase from the lepidopteran *Manduca sexta*, which pertains to family 18 of glycosyl hydrolases. The enzyme has a multidomain architecture that includes a signal peptide, an N-terminal catalytic domain with the consensus sequence (F/L)DG(L/I)DLDWEYP, and a C-terminal cysteine-rich chitin-binding domain, which are connected by the interdomain serine/threonine-rich O-glycosylated linker (Tellam *et al.*, 1999). The three disulfide bonds in the domain may constrain the polypeptide to present the aromatic amino acids on the protein surface for interactions with the ring structures of sugars. Thus, the chitin-binding domains enhance activity toward the insoluble polymer, whereas the linker region facilitates secretion from the cell and helps to stabilize the enzyme in the presence of proteolytic enzymes (Lu *et al.*, 2002; Arakane *et al.*, 2003). In some cases, proteins with chitin-binding domains, but lacking catalytic activity aid in the degradation of chitin (Vaaje-Kolstad *et al.*, 2005).

The *Anopheles gambiae* gut chitinase is secreted upon blood-feeding as an inactive proenzyme that is later activated by trypsin. Sequencing a cDNA coding for the gut chitinase showed that the enzyme comprises a putative catalytic domain at the N-terminus, a putative chitin-binding domain at the C-terminus, and a serine/threonine/proline-rich amino acid stretch between them (Shen and Jacobs-Lorena, 1997). The mosquito chitinase seems to modulate the thickness and permeability of the chitin-containing peritrophic membrane. Supporting this conjecture, the authors found that the peritrophic membrane is stronger and persisted longer when the mosquitoes were fed diets containing chitinase inhibitor.

The beetle *P. cochleariae* has one group of chitinases of 40-70 kDa active at pH 5.0 and detected in guts. This enzyme has an active site centered on the catalytic residues Asp 146 and Glu 150 (*M. sexta* chitinase numbering), but lacks the C-terminal chitin-binding domain and the serine/threonine-rich interdomain (Girard and Jouanin, 1999). *T. molitor* midgut chitinase (Genta *et al.*, 2006b) is similar to that of *P.cochleariae*, and both pertain to group IV chitinases (Arakane and Muthukrishnan, 2010) that as a rule lack a chitin-binding domain. The function of the T. molitor enzyme may be digestion of the cell walls of fungi usually present in its food, without damaging the peritrophic membrane.

Lysozyme (EC 3.2.1.17) catalyses the hydrolysis of the 1,4-P-glycosidic linkage between N-acetylmuramic acid and N-acetylglucosamine of the peptidoglycan present in the cell wall of many bacteria, causing cell lysis. Lysozyme is part of an immune defence mechanism against bacteria, and has been described in most animals, including insects. Lysozyme has also been implicated in the midgut digestion of bacteria by organisms which ingest large amounts of them, such as marine bivalves, or that harbor a bacterial culture in their guts (exemplified by Callewaert and Michiels, 2010).

Among insects, the capacity for digesting bacteria in the midgut seems to be an ancestral trait of Diptera Cyclorrhapha, which agrees that most Diptera Cyclorrhapha larvae are saprophagous, feeding largely on bacteria. These insects have midgut lysozymes (Lemos *et al.*, 1993; Regel *et al.*, 1998) similar to those of vertebrate fermenters. Thus, these enzymes have low pi values, are more active at pH values 3-4, and are resistant to the cathepsin D-like aspartic proteinase present in midguts (Fujita, 2004).

Lysozyme is also found in the salivary glands of *Reticulitermes speratus*. This insect is a termite that feeds mainly on dead wood, which tends to lack nitrogen. Fujita *et al.* (2001) suggested, on the basis of the distribution and activity of lysozyme in this termite, that wood-feeding termites use lysozyme secreted from the salivary gland to digest their hindgut bacteria, which are transferred by proctodeal trophallaxis..

4.3.5 α-Glucosidases

α-Glucosidases (EC 3.2.1.20) catalyse the hydrolysis of terminal, non-reducing α-1,4-linked glucose residues from aryl (or alkyl) -glucosides, disaccharides, or oligosaccharides. α-Glucosidases are frequently named **maltases**, although some of them may have weak activity on maltose. Insect α-glucosidases occur as soluble forms in the midgut lumen, or are trapped in the midgut cell glycocalyx. They also bind to microvillar membranes (Terra and Ferreira, 1994), to perimicrovillar membranes (lipoprotein membranes ensheathing the midgut cell microvillar membranes in most hemipterans) (Silva and Terra, 1995), or to the modified perimicrovillar membranes of aphid midgut cells (Cristofoletti *et al.*, 2003). α-Glucosidase is a major protein in dipteran mid-gut microvilli (Terra and Ferreira, 1994), and probably because of that it is the receptor of endotoxins, similar to that observed with aminopeptidase N in lepidopteran midgut cells.

Insect α-glucosidases hydrolyse oligosaccharides up to at least maltopentaose (Terra and Ferreira, 1994), although there are exceptions. The perimicrovillar α-glucosidase from *Dysdercus peruvianus* (Silva and Terra, 1995) and *Quesada gigas* (Fonseca *et al.*, 2010) prefers oligosaccharides up to maltotetraose, and the midgut bee α-glucosidase, up to maltotriose

(Nishimoto *et al.*, 2001). The purified midgut α-glucosidase of the pea aphid *Acyrthosiphon pisum* catalyzes transglycosylation reactions in the presence of excess sucrose, thus freeing glucose from sucrose without increasing the osmolarity of the medium (Cristofoletti *et al.*, 2003). This phenomenon, associated with a quick fructose absorption (Ashford *et al.*, 2000), explains why the midgut luminal osmolarity decreases as the ingested sucrose-containing phloem sap passes along the aphid midgut. *A. pisum*α-glucosidase, pertains to family 13 glycosidase.

4.3.6 β-Glucosidases, β-Galactosidases, and Myrosinases

β-Glycosidases (EC 3.2.1) catalyse the hydrolysis of terminal, non-reducing β-linked monosaccharide residues from the corresponding glycoside. Depending on the monosaccharide that is removed, the β -glycosidase is named β -glucosidase (glucose), β -galactosidase (galactose), β -xylosidase (xylose), and so on. Frequently, the same β -glycosidase is able to hydrolyse several different monosaccharide residues from glycosides. In this case, β -glucosidase (EC 3.2.1.21) is used to name all enzymes that remove glucose efficiently. The active site of these enzymes is formed by subsites numbered from the glycosidic linkage to be broken, with negative (towards the non-reducing end of the substrate) or positive (towards the reducing end of the substrate) integers (Davies *et al.*, 1997). The non-reducing monosaccharide residue binds at the glycone (-1) subsite, whereas the rest of the molecule accommodates at the aglycone subsite, which actually may correspond to several monosaccharide residue-binding subsites.

Some insects have three or four digestive β -glycosidases with different substrate specificity. In others, only two of these enzymes are found, which are able to hydrolyse as many different P-glycosides as the other three or four enzymes together (Ferreira *et al.*, 1998; Azevedo *et al.*, 2003).

Insect β-glycosidases best characterised have molecular masses of 30-150 kDa, pH optima of 4.5-6.5, and pi values of 3.7-6.8, whereas K_m values, with cellobiose as substrate, are in the range of 0.2-2 mM. Based on relative catalytic efficiency on several substrates, insect β -glycosidases can be divided into two classes. **Class A** includes the enzymes that efficiently hydrolyse substrates with hydrophilic aglycones, such as disaccharides and oligosaccharides. **Class B** comprises enzymes that have high activity only on substrates with hydrophobic aglycones, such as alkyl-, p-nitrophenyl-, and methylumbelliferyl-glycosides. Enzymes from class A are more abundant than β -glycosidases from class B. Class A β -glycosidases hydrolyse di- and oligosaccharides, and have four subsites for glucose binding in the active site: one in the glycone (-1) and three in the aglycone (+1, +2, +3) positions (Ferreira *et al.*, 2003; Azevedo *et al.*, 2003). Some enzymes seem to be adapted to use disaccharides as well as oligosaccharides as substrates.

Class A β-glycosidases are able to hydrolyse β -1,3, β -1,4, and β -1,6 glycoside bonds from di- and oligosac-charides. These enzymes are likely to be involved in the intermediate and terminal digestion of cellulose, hemicellulose, and glycoproteins present in food.

Class B β-glycosidases with high activity against hydrophobic substrates may have the physiological role of hydrolysing glycolipids, mainly galactolipids that are found in high amounts in vegetal tissues. The main galactolipids in plants are 2,3-diacyl β -galactoside D-glycerol (mono galactosyl diglyceride) and 2,3-diacyl 1-(a-galactosyl 1,6 P-galactosyl)-D-glycerol (digalacto-syl diglyceride). These enzymes may act directly against the monogalactosyl diglyceride, or on digalactosyl diglyceride after the removal of one of the galactose residues by α-galactosidase. The activation by amphiphatic substances may be a mechanism to maintain high enzyme activity only in the neighborhood of plant cell membranes undergoing digestion in the insect midgut. These membranes are the source of glycolipid substrates and activating detergent-like molecules. Distant from membranes, the β-glycosidase would be less active, thus hydrolyzing plant glucosides ingested by the insect with decreased efficiency.

Besides having a role in digestion, β glycosidases are important in insect-plant relationships. To avoid herbivory, plants synthesise a large number of toxic glucosides which may be present in concentrations from 0.5% to 1% (Spencer, 1988). The presence of these glucosides in some insect diets may explain the variable number of β-glycosidases with different specificities present in their guts. Most plant glucosides have a hydrophobic aglycone and are β-linked O-glycosyl compounds. Since aglycones are usually more toxic than the glycosides themselves, intoxication may be avoided by decreasing the activity of the enzyme most active on toxic glucosides, without affecting the final digestion of hemi-cellulose and cellulose carried out by the other enzymes. This is exemplified by *D. saccharalis* larvae, which have three β-glycosidases in their midgut, feeding on diets containing the cyanogenic glucoside amygdalin. In this condition, the activity of the enzyme responsible for the hydrolysis of prunasin is decreased. Prunasin is the saccharide resulting after the removal from amygdalin, and that forms the cyanogenic mande-lonitrille upon hydrolysis. Resistance to toxic glucosides may also be achieved by the lack of enzymes able to hydrolyse toxic β-glucosides, as observed by *S. frugiperda* larvae, which have two β-glycosidases unable to efficiently hydrolyse prunasin (Marana *et al.*, 2001).

Myrosinase (EC 3.2.147) is a member of the glycosyl hydrolase family 1 that hydrolyses glucosinolates, which are β-δ-thioglucosides. The aphid *Brevicoryne brassicae* has a midgut myrosinase for which coding cDNA was cloned and sequenced (Jones *et al.*, 2002), and its 3D structure resolved (Husebye *et al.*, 2005). The enzyme has two catalytic Glu residues (plant myrosinase has only one), and its sequence is more similar to animal β-O-glucosidases than to plant myrosinases. The data suggest that myrosinase has twice arisen from β-glucosidases in plants and animals.

4.3.7 *Trehalases*

Trehalase (EC 3.2.1.28) hydrolyses α,α'-trehalose into two glucose molecules, and is one of the most widespread carbohydrases in insects, occurring in most tissues. Trehalase is very important for insect metabolism, since trehalose is the main circulating sugar in these organisms.

Apical and basal trehalases can be distinguished in insect midguts. The apical trehalase may be soluble (glycocalyx-associated or secreted into the midgut lumen) or micro-villar, whereas the midgut basal trehalase is an integral protein of the basal plasma membrane. The apical midgut trehalase is a true digestive enzyme. The midgut basal trehalase probably plays a role in the midgut utilization of haemolymph trehalose (Terra and Ferreira, 1994). Trehalases, partially or completely purified from insect guts, have pH optima from 4.8 to 6.0, Km from 0.33 to 1.1 mM, pI around 4.6, and molecular masses from 60 to 138 kDa (Terra and Ferreira, 1994).

Plant toxic β-glucosides and their aglycones can inhibit, with varying efficiency, some or all trehalases from Malpighian tubules, fat body, midgut, and body wall of *P. americana, M. domestica, S. frugiperda,* and *D. saccharalis* (Silva *et al.*, 2006). Toxic β-glucosides are produced by many plant species, and are present in high concentrations. It is not known whether the glucosides or aglycones are absorbed by the insect gut and interact with trehalases in tissues other than the midgut. It would be interesting to know how insects resistant to toxic β-glucosides avoid the damage they can cause.

4.3.8 *Acetylhexosaminidases, β-Fructosidases, and α-Galactosidases*

An enzyme related to chitinolytic enzymes is β-N-acetyl-D-hexosaminidase (EC 3.2.1.52), which differs from β-N-acetyl-D-glucosaminidase in having a rather wide substrate specificity. The enzyme is found in many insects, and its presumed physiological role is the hydrolysis of N-acetylglucosamine β-linked compounds such as glycoproteins (Terra and Ferreira, 1994). Detailed studies of this digestive enzyme are lacking.

Sucrose hydrolysis is catalysed by enzymes that are specific for the α-glucosyl (α-glucosidase, EC 3.2.1.20) or for the β-fructosyl residue (β-fructosidase, EC 3.2.1.26) of the substrate. β-Fructosidase is characterised by its activity toward sucrose and raffinose, and lack of activity upon maltose and melibiose. In insect midguts, sucrose hydrolysis generally occurs by action of the conspicuous α-glucosidase rather than by β-fructosidase. However, larvae and adults of the moth *Erinnyis ello* have a midgut β-fructosidase with pH optimum 6.0, Km 30 mM (sucrose), pi 52, and molecular mass of 78 kDa. The physiological role of this enzyme is to hydrolyse sucrose, the major leaf (larvae) or nectar (adults) carbohydrate, which is not efficiently digested by *E. ello* midgut α-glucosidase (Santos and Terra, 1986). The presence of this enzyme is thought to be an adaptation to the presence of alkaloid sugars in latex that inhibits the α-glucosidase, but not the β-fructosidase (Daimon *et al.*, 2008). As there seem to be exceptions for this inhibition, more data should be gathered to settle the hypothesis.

-Galactosidase (**α**-D-galactoside galactohydrolase, EC 3.2.1.22) catalyses the hydrolysis of α-D-galactosidic linkages in the non-reducing end of oligosaccharides, galactomannans, and galactolipids, and is widely distributed in nature (Dey and Pridham, 1972). Galactooligosaccharides, such as melibiose, raffinose, and stachiose, are common in plants, mainly in lipid-rich seeds (Shi-Roya, 1963), whereas galactolipids are widespread among the leaves. The major lipids in chloroplast membranes are monogalactosyldiglyceride and digalactosyldiglyceride.

There have been few attempts to resolve insect mid-gut α-galactosidases. Gel filtration and heat inactivation suggested that there is a single α-galactosidase activity (30 kDa, pH optimum 5.0) in *D. peruvianus* midgut that is more efficient on raffinose than on melibiose (Silva and Terra, 1997). There are two α-galactosidases in *A. flavolineata* midguts: the major (112 kDa, pH optimum 5.4) is active on melibiose and raffinose in addition to NPaGal, whereas the minor (70 kDa, pH optimum 5.7) hydrolyses only NPαGal. In the case of *Psacothea hilaris*, gel filtration gave evidence of the presence of multiple overlapping α-galactosidases more active on NPαGal (p-nitrophenyl α-D-galactopyranoside) than on melibiose (Scrivener *et al.*, 1997). There are three midgut luminal α-galactosidases (TG1, TG2, and TG3) from *T. molitor* larvae that are partially resolved by ion-exchange chromatography (Grossmann and Terra, 2001). The enzymes have approximately the same pH optimum (5.0), pi value (4.6), and molecular mass (46-49 kDa). TG2 hydrolyses α-1,6-galactosaccharides, exemplified by raffinose, whereas TG3 acts on melibiose and apparently also on digalactosyldiglyceride, the most important compound in thylacoid membranes of chloroplast, converting it into monogalactosyldiglyceride.

4.4 Digestion of Proteins

The initial digestion of proteins is carried out by proteinases (endopeptidases) that break internal bonds in proteins. Different proteinases are necessary to do this because the amino acid residues vary along the peptide chain. There are three subclasses of proteinases involved in digestion, classified according to their active site group (and hence by their mechanism): serine, cysteine, and aspartic proteinases (Box 3.8). In each of the mentioned subclasses, there are several proteinases differing in substrate specificities. The oligopeptides resulting from proteinase action are attacked from the N-terminal end by aminopeptidases and from the C-terminal end by carboxypeptidases, both enzymes liberating one amino acid residue at each catalytic step. Finally, the dipeptides formed are hydrolysed by dipeptidases. For an efficient management of pest control through proteinase inhibitor transgenes, it is imperative to know the type of enzymes present in the gut of insects and pests. The two major proteinase classes in the digestive systems of phytophagous insects are the serine and cysteine proteinases (Haq *et al.*, 2004).

Box 3.8 Enzymes in protein digestion

1. Serine proteinases
   A. Trypsins
   B. Chymotrypsins
   C. Elastases
2. Cysteine proteinase
3. Aspartic proteinases
4. Aminopeptidases
5. Carboxypeptidases and dipeptidases

4.4.1 *Serine Proteinases*

Serine proteinases (EC 3.4.21) have serine, histidine, and aspartic acid residues (called the **catalytic triad**) in the active site. The family of enzymes homologous to chymotrypsin (Barrett *et al.*, 1998) includes the major digestive enzymes trypsin, chymotrypsin, and elastase. These enzymes differ in structural features that are associated with their different substrate specificities. The numbering of residues in enzyme polypeptide chains is referred to that of bovine chymotrypsin.

Trypsins

Trypsins (EC 3.4.21.4) preferentially cleave protein chains on the carboxyl side of basic L-amino acids such as arginine and lysine. Most insect trypsins have molecular masses in the range 20-35 kDa, pi values 4-5, and pH optima 8-10. These enzymes occur in the majority of insects, with the remarkable exception of hemipteran species and some taxa belonging to the series Cucujiformia of Coleoptera like Curculionidae (Terra and

Ferreira, 1994). Nevertheless, some heteropteran Hemiptera have trypsin in the salivary glands (Zeng *et al*., 2002).

Trypsin is usually identified in insect midgut homogenates using benzoyl-arginine p-nitroanilide (B-R-pNA, often referred to as BApNA) or benzoylarginine 7-amino-4-methyl coumarin (B-R-MCA) as substrates, and with N-a-tosyl-l-lysine chloromethyl ketone (TLCK), phenylmethylsulfonyl fluoride (PMSF), or diisopropyl-fluorophosphate (DFP) as **inactivating compounds**. In spite of having structural features resembling vertebrate trypsins, insect trypsins differ from these because they are not activated or stabilised by calcium ions, and frequently are unstable in acidic pH (Terra and Ferreira, 1994). Finally, other differences between vertebrate and insect trypsins include their substrate specificities and their interaction with protein inhibitors.

Trypsins from different insects also differ in the strength with which their sub-sites bind the substrate or the transition state (high-energy intermediate of the reaction). In other words, trypsin sub-sites differ in how they favour substrate binding or catalysis (Marana *et al*., 2002).

Plants have protein inhibitors (PIs) of insect midgut serine proteinases that affect insect development (Ryan, 1990). Insects may adapt to the presence of PIs in the diet by over-expressing proteinases, by proteolytical inactivation of PIs mediated by the insect's own proteinases, or by expressing new proteinases that are resistant to the inhibitor (Mazumdar-Leighton and Broadway, 2001a, 2001b). PIs produced by plants have a region, named the **reactive site,** which interacts with the active site of their target enzymes. The reactive sites of many PIs are hydrophilic loops with a lysine residue at P1 (Lopes *et al*., 2004). As lepidopteran trypsins have hydrophobic subsites and prefer lysine rather than arginine at P1, they are usually more resistant to PIs than the other insect trypsins. In this respect, it is interesting to note that PI-insensitive trypsins from *Heliothis virescens* bind more tightly to a hydrophobic chromatographic column than do sensitive trypsins (Brito *et al*., 2001). These observations led to the hypothesis that the molecular differences between sensitive and insensitive trypsins must rely on the interactions of PIs with residues in and around the enzyme active site.

Chymotrypsins

Chymotrypsin (EC 3.4.21.1) preferentially cleaves protein chains at the carboxyl side of aromatic amino acids. Insect chymotrypsins usually have molecular masses of20-30 kDa and pH optima of 8-11, and they differ from their vertebrate counterparts in their instability at acidic pH, inhibition pattern with SBTI (Terra and Ferreira, 1994), and, finally, in reacting with N-a-tosyl-l-phenylalanine chloromethyl ketone (TPCK). The distribution of chymotrypsin among insect taxa is similar to that of trypsin (Terra and Ferreira, 1994), including the occurrence in the salivary glands of some heteropteran bugs (Colebatch *et al.*, 2002). Insect chymotrypsins prefer Phe at P1, and are almost inactive if Tyr is at that position (Lopes *et al.*, 2009). Furthermore, insect chymotrypsins have an extended active site and larger substrates, like succinyl-AAPF-p-nitroanilide (Suc-AAPF-pNA), are usually necessary for their detection (Lee and Anstee, 1995; Lopes *et al.*, 2009).

The insect digestive chymotrypsin that has been most thoroughly studied is that of *M. sexta* (Lepidoptera: Sphingidae) (Peterson *et al.*, 1995). In this enzyme, the activation peptide is longer and has a net charge different from that of bovine chymotrypsinogen, leading the authors to suggest that the insect enzyme is activated by a peculiar mechanism. The mammalian chymotrypsin has a pH optimum around 8, with two catalytic important pK^s of 6.8 and 9.5, corresponding to the active-site histidine and N-terminal leucine, respectively. In contrast, the *M. sexta* chymotrypsin has pH optimum 10.5-11, and a single kinetically significant pKa at pH 9.2. This pK may represent the active-site histidine in an appropriate environment, although several other hypotheses were discussed (Peterson *et al.*, 1995).

The molecular mechanism of chymotrypsin PI inhibition was investigated. Two chymotrypsins were purified from the midgut of *Helicoverpa punctigera*, one PI-sensitive and the other PI-insensitive. After their corresponding cDNAs were cloned and sequenced, molecular modeling revealed that a Phe — Leu substitution at position 37 in the chymotrypsin results in the loss of important contacts with the PI. This was confirmed by site-directed mutagenesis of chymotrypsin molecules, followed by inhibition tests (Dunse *et al.*, 2010). Chymotrypsins from insects that routinely ingest ketone-releasing compounds (like several plant glycosides) are not affected much by these compounds and others that react with His 57. Thus, in comparison with bovine chymotrypsin, the chymotrypsin from polyphagous lepidopteran insects reacts slowly with chloromethyl ketones, whereas those of oligophagous pyralid insects react rapidly (Lopes *et al.*, 2009). Modeling *Spodoptera frugiperda* (Noctuidae) chymotrypsin, based on its sequence and on crystallographic data of bovine chymotrypsin, showed that the neighborhood of His 57 differs from bovine chymotrypsin, thus affecting His reactivity (Lopes *et al.*, 2009). These adaptations are new examples of the interplay between insects and plants during their evolutionary arms race.

Elastases

Since Christeller *et al.* (1990) described an elastase (EC 3.4.21.36)-like enzyme in the cricket *Teleogryllus commodus*, this enzyme has been described in many other insects, including in homogeneous form (Terra and Ferreira, 1994; Whitworth *et al.*, 1998). Usually elastase is identified with the substrate Suc-AAPL-pNA, combined with the observation of lack of activity on B-Y-pNA or B-Y-ee and resistance to TPCK. Since the mentioned substrate may also be hydrolysed by chymotrypsin, and lack of activity on B-Y-pNA and/or resistance to TPCK are usual among chymotrypsins, most described elastases may actually be chymotrypsins. True elastases were isolated from gypsy moth midguts and from whole larvae of *Solenopsis invicta* (Whitworth *et al.*, 1998). The last-mentioned enzymes hydrolyse Suc-APA-pNA, but not substrates with phenylalanine at P1. Although the specific substrate for elastase (Suc-AAA-pNA) (Bieth *et al.*, 1974) was not tested, the hydrolysis of Suc-AAAPV-pNA and the lack of hydrolysis of substrates with phenylalanine in P1 discount a chymotrypsin. Further work is necessary to evaluate the extent of this enzyme in insect midguts, and its importance in digestion.

4.4.2 *Cysteine Proteinases*

Cysteine proteinase is usually assayed in insect midgut contents or midgut homogenates at pH 5-6 with B-R-pNA, B-R-NA, casein, or hemoglobin as substrate. Activation by sulfhydryl agents (dithiothreitol (DTT) or cysteine) and inhibition by trans-epoxysuccinyl-l-leucyl-amido (4-guanidinobutane) (E-64) are usually indicative of the presence of the enzyme. The identification of cysteine proteinase was made easier with the substrate e-amino-caproyl-leucyl-(S-benzyl)-cysteinyl-MCA, which is hydrolysed by cysteine proteinase but not by serine proteinases (Alves *et al.*, 1996). Using such criteria, cysteine proteinases were described in Hemiptera, Heteroptera and in species belonging to the series Cucujiformia of Coleoptera (Terra and Ferreira, 1994). Exceptions to this rule are the identification of cysteine proteinase in Hemiptera Sternorrhyncha (aphids) (Cristofoletti *et al.*, 2003; Deraison *et al.*, 2004), and the lack of this enzyme in cucujiform cerambycid beetles (Johnson and Rabosky, 2000).

Insect midgut cysteine proteinases were at first denoted as cathepsin B (EC 3.4.22.1) -like enzymes, because cathepsin B was the first animal cysteine proteinase described. Later on it became known that cathepsin B is more important as a peptidyl dipeptidase, rather than as an endopeptidase, because of the existence of an extended loop that forms a cap to the active-site cleft, and carries a pair of histidine residues that are thought to bind to the C-terminal carboxylate of the substrate (Barrett *et al.*, 1998). Cathepsin L (EC 3.4.22.15) is a true endopeptidase that preferentially cleaves peptide bonds with hydrophobic amino acid residues in P2 (cathepsin B prefers arginine at the same position) (Barrett *et al.*, 1998). Thus, by using substrates like carbobenzoxy (Z) -FR-MCA and Z-RR-MCA it is possible to distinguish between the two enzymes. Current research has revealed that cathepsin L-like enzymes are the only insect mid-gut cysteine proteinase that is quantitatively important. Much more difficult to ascertain is that a cathepsin L-like enzyme assayed in insect midguts has been secreted into midgut contents, and hence may be considered as a truly digestive enzyme. As in other animals (Barrett *et al.*, 1998), cathepsin L-like enzymes in insects are expected to occur in lysosomes and never leave the cells. The same difficulties arise in trying to relate digestion to cathepsin L-like enzymes encoded by cDNAs cloned from midgut cells.

The problems that may arise during cathepsin L-like enzyme characterisation are well illustrated in a study with *T. molitor* larvae. Three cathepsin L-like sequences were recognised in a cDNA library prepared from T. molitor midguts. One sequence, after being expressed and used to raise antibodies, was found to correspond to a lysosomal cathepsin L immunolocalised mainly at haemocytes and fat body cells. The second sequence was not related yet with any enzyme active in the midgut. Finally, the third sequence corresponds to a cathepsin L-like enzyme purified from midgut contents (Cristofoletti *et al.*, 2005).

There are many cathepsin L-like sequences corresponding to coleopterans and hemipterans (those known to have digestive cathepsins) registered in GenBank. The sequences demonstrate the features characteristic of family 1 of cysteine proteinases (Barrett *et al.*, 1998): an N-ter-minal propeptide that must be removed to activate the enzyme and the catalytic triad, Cys 25, His 169, and Asn 175 (papain numbering), and the ERFNIN motif. The sequences form a monophyletic grouping, and a polyphyletic array seems to correspond to lysosomal and digestive enzymes, respectively (Cristofoletti *et al.*, 2005). The data suggest that the Hemiptera digestive cathepsin is close to the lysosomal one from *T. molitor* (CAL 1), whereas those from the coleopterans (except from *Sitophilis oryzae*) is similar to the *T. molitor* digestive enzyme (CAL 3). It is probable that the *S. oryzae* enzymes are not true digestive but lysosomal, like *T. molitor* CAL (Cristofoletti *et al.*, 2005). More work is needed to clarify the role of cathepsin L-like enzymes in insect digestion.

4.4.3 Aspartic Proteinases

Aspartic proteinases are active at acid pH, hydrolyse internal peptide bonds in proteins, and attack some synthetic substrates – either chromophoric or internally quenched fluorescent substrates (Pimenta *et al.*, 2001). The first report of aspartic proteinases in insects was made by Greenberg and Paretsky (1955), who found a strong proteolytic activity at pH 2.5-3.0 in homogenates of whole bodies of *Musca domestica*. Lemos and Terra (1991) showed that the enzyme occurs in midguts, and is cathepsin D-like. An aspartic proteinase similar to cathepsin D was found in families of Hemiptera and Heteroptera, and in several families belonging to the cucujiform series of Coleoptera (Terra and Ferreira, 1994). Thus, it is possible that aspartic proteinases occur together with cysteine proteinase in Hemiptera and in most Coleoptera. The aspartic proteinase isolated from *Callosobruchus maculatus* (pH optimum 3.3, 62 kDa) (Silva and Xavier-Filho, 1991) and *Tribolium castaneum* (pH optimum 3.0, 22 kDa) (Blanco-Labra *et al.*, 1996) were partially purified and shown to be similar to cathepsin D. *M. domestica* midguts express 3 cathepsin D-like enzymes (CAD1, CAD2, and CAD3). All CADs have catalytic Asp 33 (together with T34 and G35) and Asp 229 (together with T230 and G231) (bovine cathepsin D numbering), and the conserved substrate binding pockets (Padilha *et al.*, 2009). CAD3 is a luminal digestive enzyme, and lacks the proline loop. CAD2 also lacks the proline loop and may be another digestive enzyme. CAD1 is expressed in all *M. domestica* tissues (not only in midgut as CAD2 and CAD3), and shows the proline loop as the vertebrate lysosomal cathepsin D (Padilha *et al.*, 2009). The data suggest that, on adapting to deal with a bacteria-rich food in an acid midgut region, *M. domestica* CAD resulted from the same archetypical gene as the intracellular cathepsin D.

4.4.4 *Aminopeptidases*

Aminopeptidases sequentially remove amino acids from the N-terminus of peptides, and are classified on the basis of their dependence on metal ions (usually Zn^{2+} or Mn^{2+}) and substrate specificity. Aminopeptidase N (EC 3.4.11.2) has a broad specificity, although it preferentially removes alanine and leucine residues from peptides, whereas aminopeptidase A (EC 3.4.11.7) prefers aspartyl (or glutamyl) peptides as substrates. Both are metallo-enzymes (Noren *et al.*, 1986).

In insect midguts, major amounts of soluble amino-peptidases are found in less evolved insects (e.g., Orthoptera, Hemiptera, Coleoptera Adephaga), whereas in more evolved insects (e.g., Coleoptera Polyphaga, Diptera, and Lepidoptera) aminopeptidase is found mainly bound to the microvillar membranes of midgut cells (Terra and Ferreira, 1994). Insect midgut aminopeptidases are metalloenzymes (ethylenediaminetetraacetic acid [EDTA] inhibition) and have pH optima of 7.2-9.0, irrespective of the pH of the midgut lumen from the different species, Km values (L-pNA) of 0.13-0.78 mM, and molecular masses of 90130 kDa. With a single exception, all known insect aminopeptidases have a broad specificity, hydrolysing a variety of amino acyl P-naphthylamides (except acidic amino acyl P-naphthylamides), indicating they are aminopeptidases N (Terra and Ferreira, 1994). The exception is a soluble glycocalyx-associated midgut aminopeptidase from *R. americana*. This enzyme is an aminopeptidase removing N-terminal aspartic acid or glutamic acid residues from peptides that are not efficiently attacked by the other aminopeptidases (Klinkowstrom *et al*., 1994).

In addition to a midgut aminopeptidase A, the dipteran *R. americana* has three midgut aminopeptidase Ns (one soluble and two membrane-bound). The soluble aminopeptidase N (115.7 kDa) prefers tetrapeptides over tripeptides (Ferreira and Terra, 1984), like the minor 107-kDa membrane-bound enzyme, whereas the contrary is true for the major 169-kDa membrane-bound amino-peptidase (Ferreira and Terra, 1986a, 1986b). The single midgut aminopeptidase N of the coleopterans *Attagenus megatoma* (Baker and Woo, 1981) and *T. molitor* (Cristofoletti and Terra, 1999) resembles the 115.7-kDa and 107-kDa aminopeptidases of *R. americana*. Approximately the same substrate specificity was observed with the two midgut aminopeptidases of the lepidopteran *Tineola bisselliella* that from the aphid *A pisum* (Cristofoletti *et al*., 2006), and that of the cerambycid beetle *Morimus funereus* (Bozic *et al*., 2008). The data suggest that panorpoid insects (Diptera and Lepidoptera) present multiple aminopeptidases with different substrate specificities, in contrast with the single aminopeptidase of coleopterans.

There have been few attempts to characterise the active site of insect midgut aminopeptidases. Using multiple inhibition analysis and observing the protection against EDTA inactivation that different competitive inhibitors conferred to the enzyme, two subsites were proposed to occur in the active center of *P. americana* microvillar aminopeptidase: a hydrophobic subsite, to which isoamyl alcohol binds, exposing the metal ion, and a polar subsite, to which hydroxylamine binds. The effect of pH at different temperatures on kinetic parameters of *T. molitor* midgut aminopeptidase and its inactivation by different compounds were studied (Cristofoletti and Terra, 2000). The data showed that *T. molitor* aminopeptidase catalysis depends on a metal ion, a carboxylate, and a protonated imidazole group, and is somehow influenced by an arginine residue in the neighborhood of the active site. The catalytic metal binding depends on at least a deprotonated imidazole. In addition to the above-mentioned groups involved in catalysis, at least one phenol group and one carboxylate are associated with substrate binding. Thus, *T. molitor* amino-peptidase shares common features with those of other zinc metallopeptidases, especially with mammalian aminopeptidase N, but it differs in some details. An imidazole group seems to be involved in catalysis in T. molitor aminopeptidase; this is not observed in mammalian aminopeptidase N, which has an imidazole group participating in substrate binding.

Probably associated with the fact that aminopeptidases are major proteins in some microvillar membranes (55% of *T. molitor* midgut microvillar proteins) (Cristofoletti and Terra, 1999), they are targets of insecticidal *Bacillus thuringiensis* crystal S-endotoxins. These toxins, after binding to aminopeptidases and receptor molecules called **cadherins**, form channels through which cell contents leak, leading to the death of the insect (Knight *et al.*, 1995). Although data on substrate specificity for lepidopteran aminopeptidase isoforms are lacking, there is evidence that the isoforms may have differences in toxin binding (Nakanishi *et al.*, 2002; Rajagopal *et al.*, 2003).

4.4.5 *Carboxypeptidases and Dipeptidases*

Carboxypeptidases hydrolyse single amino acids from the C-terminus of the peptide chain, and are divided into classes on the basis of their catalytic mechanism. There are two digestive metallocarboxypeptidases in mammals: **carboxypeptidase A** (EC 3.4.17.1), which hydrolyses (in alkaline medium) C-terminal amino acids, except arginine, lysine, and proline; and **carboxypeptidase B** (EC 3.4.1.7.2), which releases (in alkaline conditions) C-terminal lysine and arginine preferentially. Insect digestive carboxypeptidases have been classified as carboxypeptidase A or B depending on activity in alkaline medium against Z-GF (or hippuryl P-phenyllactic acid) or Z-GR (or hippuryl-l-arginine), respectively. Digestive insect carboxypeptidase A-like enzymes are widespread among insects, and most of them have pH optima of 7.5-9.0 and molecular masses of 20-50 kDa (Terra and Ferreira, 1994).

The features typical of carboxypeptidase As, including the residues His 69, Glu 72, and His 196, which bind the catalytic zinc ion; Arg 71, Asn 144, Arg 145, and Tyr 248, responsible for substrate binding; and Arg 127 and Glu 270, responsible for catalysis. In spite of the overall similarity of *H. armigera* procarboxypeptidase with human procarboxy-peptidase A2, there are differences in the loops between the conserved secondary structures, including the loop where the activation processing occurs. Another important difference is the residue 255 (bottom of the S'1 pocket) that defines the enzyme specificity. In mammalian sequences, Asp 255 is found in carboxypeptidase B and Ile 255 in carboxypeptidase A. In insect carboxypeptidase As, this residue varies (but never is an acid residue)

Carboxypeptidase B-like enzymes have been detected in insect midguts (Terra and Ferreira, 1994), and the enzyme from H. zea (CPBz) has been described in 3D detail (Bayes *et al.*, 2005). CPBhz has Glu 255 in the specificity pocket and differs from other carboxypeptidase Bs in some loops, which renders it insensitive to the potato carboxypeptidade inhibitor. Furthermore, the enzyme is unable to hydrolase substrates ending in Arg because of its bulky size, whereas it cleaves quite well substrates ending in Lys (Bayes *et al.*, 2005).

Dipeptidases hydrolyse dipeptides, and are classified according to their substrate specificities. Dipeptidases comprise the least known of the insect peptide hydrolases. There have been few studies in which dipeptidase assays were performed, and even fewer attempts to characterise the enzymes (Terra and Ferreira, 1994). The larval mid-gut of *R. americana* has three dipeptidases (two soluble, of 63 kDa and 73 kDa, respectively, and one membrane-bound) that hydrolyse Gly-Leu, resembling dipeptide hydrolase (dipeptidase, E.C. 3.4.13.18), although, in contrast to the mammalian enzyme, they are very active upon Pro-Gly. *R. americana* also seems to have an amino acyl-histidine dipeptidase (carnosinase, EC 3.4.13.3) (Klinkowstrom *et al*, 1995).

4.5 Digestion of Lipids and Phosphates

Lipids that contain fatty acids comprise storage lipids and membrane lipids. Storage lipids, such as oils present in seeds, and fats in adipose tissue of animals, are triacylglycerols (triglycerides), and are hydrolysed by lipases. Membrane lipids include phospholipids and glycolipids, such as mono- and digalactosyldiglycerides. Phospholipids are digested by phospholipases. A combination of α- and β-galactosidases may remove galactose residues from mono- and digalactosyl-diglyceride to leave a diacylglycerol which may be hydrolysed by a triacylglycerol lipase.Phosphate moieties need to be removed from phosphorylated compounds prior to absorption. This is accomplished by non-specific phosphatases. The phosphatases may be active in an alkaline (alkaline phosphatase, EC 3.1.3.1) or acid (acid phosphates, EC 3.1.3.2) medium.

4.5.1 *Lipases*

Triacylglycerol lipases (EC 3.1.1.3) are enzymes that preferentially hydrolyse the outer links of triacylglycerols and act only on the water-lipid interface. Activity of the lipase is increased as the interface becomes larger due to lipid emulsification caused by emulsifiers (surfactants). Insects lack emulsifiers comparable to the bile salts of vertebrates, but surfactant phospholipids, including lysolecithin, occur in their midguts in sufficient concentration to alter the surface tension of midgut contents (De Veau and Schultz, 1992). Lysolecithin, and other surfactants, may be formed by the action of phospholipase A on ingested phospholipids.

Insect midgut triacylglycerol lipases have been studied in a few insects, and only in crude preparations. The data suggest that the enzyme preferentially releases fatty acids from the a-positions, prefers unsaturated fatty acids, and is activated by calcium ions, thus resembling the action of mammalian pancreatic lipase. The resulting 2-monoacyl-glycerol may be absorbed or hydrolysed (Terra *et al.*, 1996). Hydrolysis of 2-monoacylglycerol may be accomplished by triacylglycerol lipase, following migration of the fatty acid to the 1-position, which seems to be favoured by the alkaline midgut pH, at least in *M. sexta* (Terra *et al.*, 1996).

Current research on insect midgut lipases is focused on classifying the different enzymes in the families already described for mammals. Thus, a large group of proteins is identified as neutral (or pancreatic) lipase. They have the catalytic triad (Ser 152, Asp 176, His 263, human pancreatic lipase numbering) and two important features that are associated with triacylglycerol hydrolysis; the P9 loop and lid, which cover the active site and are implicated in substrate recognition (Horne *et al.*, 2009). The C-terminal domain is absent in insect neutral lipases, which is consistent with the lack of a colipase, the protein associated with the vertebrate pancreatic lipases. Also differing from the pancreatic lipases, insect neutral lipases may have the lid structure severely reduced and the P9 loop partially deleted, resembling phospholipases (Christeller *et al.*, 2010). In accordance with this, the major triacyl-glycerol lipase from *the Manduca sexta fat body* is also an active phospholipase A1 (Aires *et al.*, 2006).

Another large group of lipases corresponds to the acid (gastric) lipases. They have conserved catalytic triad and lid similar to those of the neutral lipases (Horne *et al.*, 2009). Alignment of the sequences of insect midgut lipases with those of mammalian lipases, for which substrate specificity information is available, suggests that the lepidopteran neutral lipases are galactolipases and phospholipases in vivo, whereas the lepidopteran acid lipases are triacylglycerol lipases (Christeller *et al.*, 2010).

Insect midgut lipases may have a primary role in the acquisition of dietary lipid, but may also have strong antiviral activity by disrupting viral envelopes, as observed in *B. mori* larvae (Ponnuvel *et al.*, 2003).

Esterases, which are usually named the carboxylesterases (aliesterases, EC 3.1.1.1) catalyse the hydrolysis of carboxyl ester into alcohol and carboxylate. This enzyme, in contrast to lipases, attacks molecules that completely dissolve in water. It also hydrolyses water-insoluble long-chain fatty acid esters in the presence of surfactants, but at a rate much slower than that of triacylglycerol lipase. A role for esterases in digestion is unclear.

4.5.2 *Phospholipases*

Phospholipase A2 (EC 3.1.1.4) and **phospholipase A1** (EC 3.1.1.32) remove the fatty acids from phosphatides attached to the 2- and 1-positions, respectively, resulting in a lysophosphatide. Lysophosphatide is more stable in micellar aggregates than on membranes. Thus, the action of phospholipase A on the membrane phosphatides causes the solubilisation of cell membranes, rendering the cell contents free to be acted upon by the appropriate digestive enzymes. **Phospholipase** is widespread among insects (Terra *et al*., 1996). Phospholipase A2 partially purified from the midgut of adult beetle *Cincindella circumpicta* has a molecular mass of 22 kDa and pH optimum 9.0, is calcium dependent, and is inhibited by the site-specific inhibitor oleyoxyethyl phosphoryl-choline. Unfed beetles did not express the phospholipase in the midgut contents (Uscian *et al*., 1995).

Although lysophosphatide may be further hydrolysed by a lysophospholipase (phospholipase B, EC 3.1.1.5), evidence suggests it is absorbed intact by insects (Terra *et al*., 1996). Phosphatides may also be hydrolysed by phospholipase C (EC 3.1.4.3), yielding the phosphoryl base moiety and diacylglycerol, or by phospholipase D (EC 3.1.4.4), resulting in phosphatidate and the base. Both enzymes have been found in the insect midgut (Terra *et al*., 1996).

4.5.3 *Phosphatases*

Alkaline phosphatase is usually a midgut microvillar membrane marker in dipteran and lepidopteran species, although it may also occur in midgut basolateral membranes, and even as a secretory enzyme. **Acid phosphatase** is usually soluble in the cytosol of midgut cells in many insects, and may also appear in midgut contents or be found membrane-bound in midgut cells (Terra and Ferreira, 1994).The best-known alkaline phosphatases are those from *B. mori* (Lepidoptera: Bombycidae) larval midgut. The major membrane-bound and the minor soluble alkaline phosphatases were purified and shown to be monomeric enzymes with the following properties: (1) soluble enzyme, molecular mass of 61 kDa, pH optimum 9.8; (2) membrane-bound enzyme, molecular mass of 58 kDa, pH optimum 10.9. Both enzymes have wide substrate specificity and are inhibited by cysteine. The membrane-bound alkaline phosphatase occurs in the microvillar membranes of columnar cells, whereas the soluble enzyme is loosely attached to the goblet cell apical membrane facing the cell cavity.

Acid phosphatases have been characterised in some detail only in *Rhodnius prolixus* (Hemiptera: Reduvidae). The major enzyme activity is soluble, and has the following properties: wide specificity, a molecular mass of 82 kDa, Km for p-nitrophenyl phosphate 0.7 mM, and is inhibited by fluoride, tartrate, and molybdate. The minor enzyme activity is membrane-bound, and is resolved into two enzymes (123 and 164 kDa) that are resistant to fluoride and tartrate (Terra *.et al.*, 1988).

5. EVOLUTIONARY TRENDS OF INSECT DIGESTIVE SYSTEM

Insects may be grouped relative to their digestive physiology, assuming they have common ancestors. Those putative ancestors correspond to basic gut plans from which groups of insects may have evolved by adapting to different diets.The basic plan of digestive physiology for most winged insects (Neoptera ancestors) ancestors consist of the major part of digestion being carried out in the crop by digestive enzymes propelled by anti-peristalsis forward from the midgut. Saliva plays a variable role in carbohydrate digestion. Shortly after ingestion, the crop contracts, transferring digestive enzymes and partly digested food into the ventriculus. The anterior ventriculus is acidic and has high carbohydrase activity, whereas the posterior ventriculus is alkaline and has high proteinase activity. This differentiation along the midgut may be an adaptation to the instability of ancestral carbohydrases in the presence of proteinases. The food bolus moves backward in the midgut of the insect by peristalsis. As soon as the polymeric food molecules are digested and become sufficiently small to pass through the peritrophic membrane, they diffuse with the digestive enzymes into the ectoperitrophic space. The enzymes and nutrients are then displaced toward the caeca with a countercurrent flux caused by secretion of fluid at the Malpighian tubules and its absorption back by cells at the caeca, where final digestion is completed and nutrient absorption occurs. When the insect starts a new meal, the caeca contents are moved into the crop. As a consequence of the countercurrent flux, digestive enzymes occur at a decreasing gradient in the midgut, and their excretion is lowered.

The Neoptera basic plan gave origin to that of the Polyneoptera orders, which include Dictyoptera (cockroaches, termites and mantids), Orthoptera and Phasmatodea, and evolved to the basic plans of Paraneoptera and Holometabola. The characteristics of the Paraneoptera ancestors cannot be inferred, because midgut function data are available only for Hemiptera.

The basic gut plan of the Holometabola (which include Coleoptera, Megaloptera, Hymenoptera, Diptera, and Lepidoptera) is similar to that of Neoptera, except that fluid secretion occurs by the posterior ventriculus, instead of by the Malpighian tubules. Because the posterior midgut fluid does not contain wastes, as is the case for Malpighian tubular fluid, the accumulation of wastes in the caeca is decreased. Caeca loss probably further decreases the accumulation of noxious substances in the midgut, which would be more serious in insects that have high relative food consumption rates, as is common among Holometabola. ·

The basic plan of Coleoptera did not evolve dramatically from the holometabolan ancestor, whereas the basic plan of the Hymenoptera, Diptera, and Lepidoptera ancestor (hymenopteran-panorpoid ancestor) presents important differences. Thus, hymenopteran-panorpoid ancestors have countercurrent fluxes like holometabolan ancestors, but differ from them in the lack of crop digestion, midgut differentiation in luminal pH, and in which compartment is responsible for each phase of digestion. In holometabolan ancestors all phases of digestion occur in the endoperitrophic space, whereas in hymenopteran-panorpoid ancestors only initial digestion occurs in that region. In the latter ancestors, intermediate digestion is carried out by free enzymes in the ectoperitrophic space, and final digestion occurs at the midgut cell surface by immobilised enzymes. The free digestive enzymes do not pass through the PM, because they are larger than the PM's pores. As a consequence of the compartmentalisation of digestive events in hymenopteran and panorpoid insects, there is an increase in the efficiency of digestion of polymeric food

6. MECHANISM OF DIGESTIVE ENZYME SECRETION

Insects are continuous (e.g., Lepidoptera and Diptera larvae) or discontinuous (e.g., predators and many hematophagous insects) feeders. The synthesis and secretion of digestive enzymes in continuous feeders seem to be constitutive – that is, they occur continuously (at least between moults) – whereas in discontinuous feeders they are regulated (Lehane *et al.*, 1996). Digestive enzymes, as with all animal proteins, are synthesised in the rough endoplasmic reticulum and processed in the Golgi complex, and are packed into secretory vesicles (Fig 3.6). There are several mechanisms by which the contents of the secretory vesicles are freed in the midgut lumen. In **holocrine secretion**, secretory vesicles are stored in the cytoplasm until they are released, at which time the whole secretory cell is lost in the extracellular space. During exocytic secretion, secretory vesicles fuse with the midgut cell apical membrane, emptying their contents without any loss of cytoplasm (Fig 3.6A). In contrast, **apocrine secretion** involves the loss of at least 10% of the apical cytoplasm following the release of secretory vesicles (Fig 3.6B). These have previously undergone fusions, leading to larger vesicles that, after release, eventually free their contents by solubilization (Fig3.6B). When the loss of cytoplasm is very small, the secretory mechanism is called microapocrine. Microapocrine secretion consists in releasing budding double-membrane vesicles (Fig. 3.6C) or, at least in insect midguts, pinched-off vesicles that may contain a single or several secretory vesicles (Fig 3.6D). In both cases, the secretory vesicle contents are released by membrane fusion and/or by membrane solubilisation caused by high pH contents or by luminal detergents.

Holocrine secretion is usually described on histological grounds mainly in midgut of insects other than higher Holometabola. These insects have large numbers of regenerative cells in their midguts. Thus, it is probable that cell renewal in these insects is being misinterpreted as holocrine secretion (Terra and Ferreira, 1994). In spite of this, immunocytochemical data showed that trypsin-containing vesicles along with cell organelles are discharged by opaque zone cells of adult stable flies, suggesting holocrine secretion (Jordao *et al.*, 1996a).

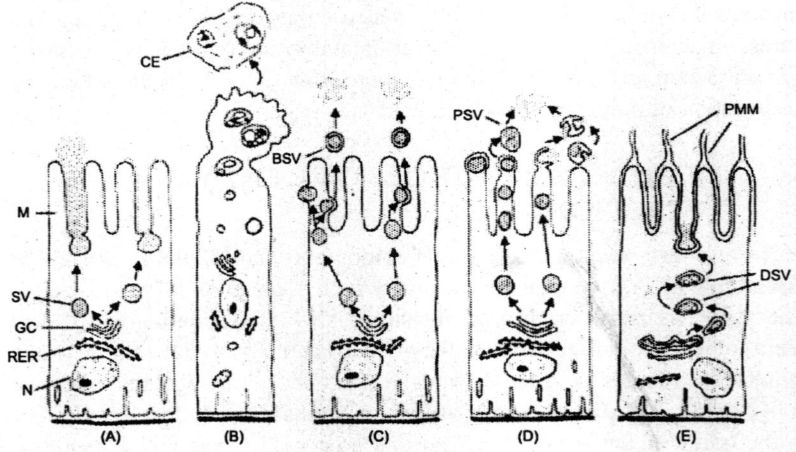

Fig 3.6. Models for secretory processes of insect digestive enzymes.

(A) Exocytic secretion; (B) apocrine secretion; (C) microapocrine secretion with budding vesicles; (D) microapocrine secretion with pinched-off vesicles; (E) modified exocytic secretion in hemipteran midgut cell.

BSV, budding secretory vesicle; CE, cellular extrusion; DSV, double-membrane secretory vesicle; GC, Golgi complex; M, microvilli; N, nucleus; PMM, perimicrovillar membrane; PSV, pinched-off secretory vesicle; RER, rough endoplasmatic reticulum; SV, secretory vesicle.

(Taken from http://what-when-how.com/insect-molecular-biology-and-biochemistry/biochemistry-and-molecular-biology-of-digestion-insect-molecular-biology-part-1/)

Exocytic, apocrine, and microaprocrine secretory mechanisms depend largely on midgut regions. Digestive enzymes are usually secreted by exocytosis in the posterior midgut, whereas alternate mechanisms may be observed in anterior midgut. Thus, trypsin is secreted by the posterior midgut of adult mosquitoes, larval flies (Jordao *et al.*, 1996b), and caterpillars (Jordao *et al.*, 1999) by exocytosis, as well as P-glycosidase by *T. molitor* middle midguts. Trypsin is secreted by the anterior midgut of caterpillars using a microapocrine route (Santos *et al.*, 1986; Jordao *et al.*, 1999), whereas in the anterior midgut of *T. molitor* amylase secretion occurs by an apocrine mechanism (Cristofoletti *et al.*, 2001). Based only on morphological evidence, it appears that, in addition to *E. ello* and *Spodoptera frugiperda*, microapocrine secretion occurs in other lepidopteran species, such as *Manduca sexta*, whereas apocrine secretion is observed in some Orthoptera and in many coleopteran species other than *T. molitor* (Terra and Ferreira, 1994).

Immunocytolocalisation data (Silva *et al.*, 1995) showed that secretion by hemipteran midgut celldisplays, special features, as the cells have perimicrovillar membranes in addition to microvillar ones (Fig3.6E). In this case, double-membrane vesicles bud from modified (double membrane) Golgi structures (Fig 3.6E). The double-membrane vesicles move to the cell apex, their outer membranes fuse with the microvillar membrane, and their inner membranes fuse with the perimicrovillar membranes, emptying their contents (Fig 3.6E).

7. ROLE OF SYMBIONTS IN DIGESTION

Although insects produce a wide variety of digestive enzymes, many species harbour an intestinal microbiota that converts a substantial portion of the dietary components to fermentation products before they are resorbed by the intestinal epithelia. The occurrence of such associations is always correlated with a dietary specialisation, which indicates that the symbiosis provides metabolic capacities that are normally not available to the host.

Digestive symbioses are most common among insects feeding on wood or other lignified plant materials. The most prominent example is that of wood-feeding termites (Isoptera), which represent the only group of insects whose interactions with intestinal microorganisms has been studied in detail. Here, they serve to illustrate principles that most likely govern also other cases of symbiotic associations in which detailed information on the gut microbiota and its function in digestion is lacking.

The major components of plant cell walls are cellulose and hemicelluloses. Although the glycosidic linkages between the sugar subunits of these structural polysaccharides are relatively easy to hydrolyse, the sheer size of the molecules, their primary and secondary structures, and the intimate physical contact of the different components, especially in lignified cell walls, impart a considerable recalcitrance to enzymatic digestion.

Cellulose depolymerisation is catalysed by endoglucanases (endo-(3-1,4-glucanases), which cleave randomly within the extremely long polyglucose chains, and by exoglucanases (exo-f3-1,4-cellobiohydrolases and (3-glucosidases), which require free nonreducing ends for their catalytic activity. Because native cellulose contains only one nonreducing end per many thousand glucose units, and because the action of endoglucanases is restricted to the amorphous regions of the microfibrils, an efficient cellulolytic system requires the synergistic action of both types of enzymes.

Enzymatic activity of cellulases is further impeded by the insolubility of cellulose, and depolymerization is usually the kinetically limiting step even in the digestion of pure cellulose. In the cell wall, the cellulose fibrils are intimately associated with hemicelluloses, which comprise a wide variety of homopolymers and heteropolymers of different sugars and sugar acids. Hemicellulose chains are often branched and lack the highly ordered structure of cellulose. The variety of primary structures require an equal variety of digestive enzymes, and it is not astonishing that the ability to degrade hemicelluloses is a prerequisite for an efficient cellulose degradation.

For the reasons outlined above, the degradation of plant cell walls requires the synergistic action of many different enzymes and, in the case of lignified substrates, also a mechanism to break up the ligno-cellulose complex. The most efficient cellulose and hemicellulose degraders in nature are microorganisms, i.e., bacteria, protozoa, and fungi. Fungi and certain filamentous bacteria (actinomycetes) are also the only organisms that have developed a strategy for the chemical breakdown of lignin. Not surprisingly, insects and other animals have made use of these capacities by using microorganisms as symbionts in the digestion of lignocellulosic food.

In addition to being difficult to degrade, lignocellulose is an extremely nutrient-poor substrate. The C-to-N ratio of sound wood is up to 100-fold higher than that of the insect body. Moreover, a lignocellulosic diet typically lacks most of the essential nutrients required by the animal, such as amino acids, vitamins, and sterols.

In contrast to higher animals, many microorganisms are capable of fixing dinitrogen, assimilating nitrate and ammonia, and synthesising those amino acids and vitamins essential for the host. Many animals, including insects, have developed means of exploiting these biosynthetic capacities of microorganisms.

One of the key contributions of the termite to wood digestion is the action of the mandibles and proventriculus, which grinds the wood particles down to a microscopic size. This mechanically destroys many lignin-carbohydrate complexes and creates an enormous surface area for the digestive enzymes provided by host and symbionts, thereby relieving much of the kinetic limitations of cellulose digestion. In lower termites, communication is also a prerequisite for the ingestion (phagocytosis) of the wood particles by symbiotic protozoa.

Insects lacking a pronounced gut microbiota usually have a short hindgut that serves mainly in recovering water and useful electrolytes from the residual material before the feces are voided. The digestive tract of termites, however, is characterised by one or more proctodeal enlargements and may reach enormous dimensions in length and volume. The dilatations increase the residence time of the digesta, thereby prolonging the exposure to the activities of the intestinal microbiota. Moreover, the increased diameter also affects the oxygen status of the fermentation chambers because it reduces the proportion of the gut volume rendered oxic by the diffusion of oxygen into the periphery, thus providing a favorable environment for the oxygen-sensitive microbiota and reducing the inevitable loss of fermentation products to aerobic processes.

In **lower termites**, the solid food particles entering the hindgut are immediately phagocytized by the intestinal protozoa. These oxygen-sensitive flagellates, which make up a large fraction of the hindgut volume, are essential for wood digestion. They represent a major source of cellulolytic and xylanolytic activities in the hind-gut. It appears that the different flagellate species are nutritionally specialised and each species might fill a specific niche in symbiotic digestion. The consecutive action of the host endoglucanases secreted in the salivary glands and the full complement of endoglucanases and exo-cellobiohydrolases of their eukaryotic symbionts forms the so-called **dual cellulolytic system**.

In phylogenetically **higher termites** (family Termitidae), the site of secretion of host endoglucanases has shifted from the salivary glands to the midgut epithelium. Although all families of lower termites sport a hindgut packed with flagellates, the higher termites contain a largely prokaryotic microbiota. Apparently, other symbionts took over the cellulolytic function of the protozoa in the course of termite evolution. In the case of the fungus-cultivating termites (subfamily Macrotermitinae), which cultivate basidiomycete fungi (Termitomyces spp.) on predigested food in "fungus gardens" located within their nests, the key activities of the symbiotic partner are extensive delignification and conversion of lignocellulose to fungal biomass in the fungus combs. It has been proposed that fungal cellulases ingested together with the comb material are essential for cellulose digestion in the gut of fungus-cultivating termites, but this "acquired enzyme hypothesis" is controversial. Moreover, evidence is accumulating that fungal biomass became an important nutritional basis for certain species.

In all other higher termites cellulolytic activities in the hindgut are probably produced by symbiotic bacteria. Two major bacterial lineages (Fibrobacteres and Spirochetes) seem to be involved in this process. While the former are related to an important group of fiber-degrading bacteria in the cow rumen, the exact role of the latter is still enigmatic.

8. PHYSIOLOGY OF ABSORPTION OF DIGESTED FOOD

Absorption is the uptake of nutrients from the gut lumen across the gut wall to the hemocoel. In many insects absorption of nutrients occurs through the microvilli of midgut epithelial cells by diffusion. Water and ions are absorbed in the hindgut. The crop is impermeable to most nutrients except some lipids.

Absorption involves first the passage of molecules and ions from the gut lumen into the gut cells, thus traversing the cuticle (if present) and the cell plasma membrane. Absorption depends on the permeability of those barriers and on the concentration ratio of a compound in gut lumen and inside gut cells. The permeability of cuticles is variable, whereas that of the plasma membrane is greater for water and for hydrophobic compounds. Thus, absorption of hydrophilic compounds requires special devices (transporters) to help the molecules find their way through the cell membrane. These **transporters** are transmembrane proteins that bind the molecule to be transported in a membrane face and, after suitable conformational changes, deliver the molecule from the other face. A **uniporter** is a transporter that carries a single solute, whereas symporters and antiporters are transporters that carry two solutes into the same and opposite directions, respectively. To transport molecules against a concentration gradient, the process must be energised by coupling with ATP hydrolysis or with cotransport of another molecule down its concentration gradient. Transporter-mediated absorption may be inhibited by molecules resembling those of the transported solute, and its velocity attains a maximum (transporter becomes saturated) at a high solute concentration. This behaviour is not observed in the case of simple diffusion, exemplified by lipid absorption.

In insects, as is usual for all animals, most nutrient absorption occurs in the midgut through symporters, with the ions being cotransported down the concentration gradient. Favorable ion gradients are maintained by ion pumps. The most ubiquitous of these pumps is the ATP-driven Na+, K+-antiporter (Na+-K+-ATPase) localised in the midgut cell basal membrane. Another important ion pump is the H+, K+-ATPase found in the goblet cell microvillar membranes of lepidopteran larvae. Those pumps maintain cell Na+ and cell K+ low in insects with Na+-rich and K+-rich diets, respectively. The midgut of lepidopteran larvae contains K+ -dependent symporters for amino acids, consistent with the K+ -rich plant diets of these larvae. There are usually symporters for neutral, acidic, and basic amino acids. Insects with high-Na+ diets seem to have midgut amino acid symporters dependent on Na+, as shown in cockroaches.

8.1 Process of Absorption

Absorption of nutrients may be a passive or an active process in insects. Passive absorption depends on the relative concentrations of a substance inside and outside the gut, diffusion taking place from the higher (midgut) to the lower concentration (hemocoel) (Fig 3.7).

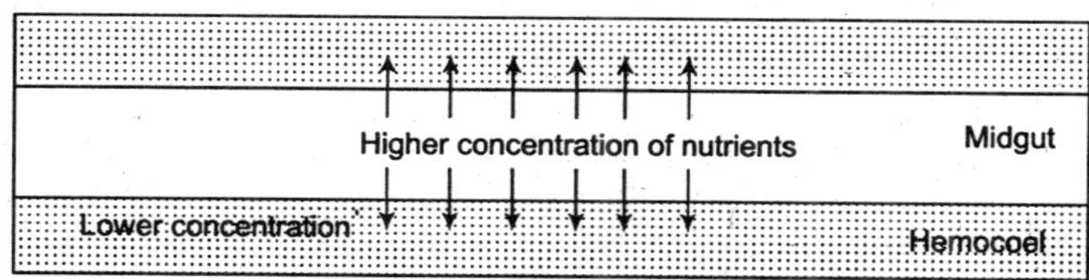

Fig 3.7 Passive absorption

Active absorption depends on some metabolic process for movement of a substance against a concentration gradient, that is, from lower concentration (midgut) to higher concentration (hemocoel) (Fig 3.8).

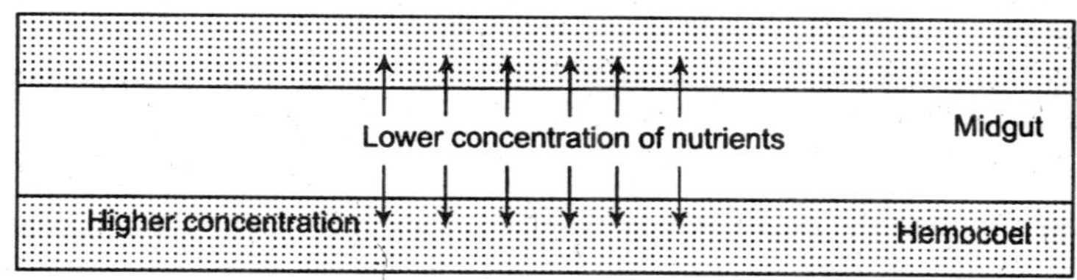

Fig 3.8 Active absorption

8.2 Absorption of water

In solid feeders, resorption of water from the faeces occurs in the rectum and the faecesisexpelled as pellets. In sap feeders (liquid feeders) the faeces is liquid like. The liquid faeces of homopteran bugs (aphids, mealy bugs, scales and psyllids) with soluble sugars and aminoacids is known as **honey dew**, which attracts ants for feeding.

Water absorption in the midgut occurs associated with midgut fluid fluxes, but in large amounts it is characteristic of insects feeding on dilute diets, of blood feeders, and also of insects in which salivation is important (e.g., grasshoppers and seed-sucker bugs). Water absorption in the hindgut is part of the water conservation mechanism that is important in all terrestrial insects. Frequently, special cell aggregates (rectal pads) are involved in this process. Water uptake is thought to depend on salt being pumped into spaces enclosed by the basolateral infoldings of the absorbing cell. This creates an osmotic pressure that moves water into these restricted spaces. The resulting hydrostatic pressure drives water into the haemolymph, with salts being absorbed from the water on its way out. Water is also cotransported with sugars.

8.3 Absorption of digested Carbohydrate

It occurs in the midgut as monosaccharides. Sugar absorption may occur in the hindgut also as in the case of larvae of *Aedes sp*. For glucose, there is no evidence of active transport. Glucose diffuses rapidly into the hemolymph and it is quickly taken up by the fat body and converted into trehalose. The rate of absorption appears to be limited by the rate of conversion of glucose to trehalose and by the rate at which fluid containing glucose is released from the crop into the midgut.

8.4 Absorption of Protein

Proteins are digested into amino acids, which are absorbed. Absorption of amino acids in certain insects like *Aedes* larvae is in the posterior portion of the midgut, while in *Rhodnius* it is absorbed in the middle part of the midgut. In Tenebrionids larvae amino acids are absorbed in the rectum. In the larvae of few lepidopterans that have been studied, amino acids are actively absorbed by transport proteins in the apical membranes of midgut columnar cells. Energy for absorption comes from the high K+ concentration in the gut lumen and the high transepithelial potential created by the proton –ATPase pump active in the midgut goblet cells

8.5 Absorption of lipids

In insects, there are few studies on lipid absorption. Absorption occurs in midgut region. However, in *Blatta orientalis* crop is the site of absorption of fats. In termites and white grubs absorption occurs through ileum. Absorption of fat by hindgut has alsobeen reported in some adult Hymenopterans.

9. INSECT DIGESTIVE ENZYME INHIBITORS IN THE CONTROL OF INSECT PEST

The possible role of protease inhibitors (PIs) in plant protection was investigated as early as 1947 when, it was observed that the larvae of certain insectswere unable to develop normally on soybean products.Subsequently the trypsin inhibitors present in soybean wereshown to be toxic to the larvae of flour beetle,*Tribolium confusum* (Lipke *et al.*, 1954). Following these early studies,there have been many examples of protease inhibitors active against certain insect species, both in *in vitro* assays against insect gut proteases and in *in vivo* artificial diet bioassays. The term **protease** includes both endopeptidases and exopeptidases whereas, the term **proteinase** is used to describe only endopeptidases (Ryan, 1990). Several non-homologous families ofproteinase inhibitors are recognised among the animal,microorganisms and plant kingdom. The majority of proteinase inhibitors studied in the plant kingdom originates from threemain families, namelyLeguminosae, Solanaceae and Gramineae. These inhibitor families that have been found are specific for each of the four mechanistic classes of proteolytic enzymes, and based on the active amino acid in their **reaction center** (Koiwa *et al.,* 1997), are classified as serine, cysteine, aspartic and metallo-proteases.

9.1 Serine proteinase inhibitors

The role of serine PIs as defensive compounds against predators is particularly well established, since the major proteinases present in plants, used for processes such as protein mobilization in storage tissues, contain a cysteine residue as the catalytically active nucleophile in the enzyme active site. The serine class of proteases such as trypsin, chymotrypsin and elastase, which belong to a common protein superfamily, is responsible for the initial digestion of proteins in the gut of higher animals (Garcia Olmedo *et al.*, 1987). *In vivo* they are used to cleave long, essentially intact polypeptide chains into short peptides which are then acted upon by exopeptidases to generate amino acids, the end products of protein digestion. These three types of digestive serine proteinases are distinguished based on their specificity, trypsin specifically cleaving the C-terminal to residues carrying a basic side chain (Lys, Arg), chymotrypsin showing a preference for cleaving C-terminal to residues carrying a large hydrophobic side chain (Phe, Tyr, Leu), and elastase showing a preference for cleaving C-terminal to residues carrying a small neutral side chain (Ala, Gly) (Ryan, 1990).

Serine proteinases have been identified in extracts from the digestive tracts of insects from many families, particularly those of Lepidoptera (Houseman *et al.,* 1989) and many of these enzymes are inhibited by proteinase inhibitors. The order Lepidoptera, which includes a number of crop pests, the pH optima of the guts are in the alkaline range of 9-11 (Applebaum, 1985) where, serine proteinases and metalloexopeptidases are most active. Additionally, serine proteinase inhibitors have anti-nutritional effects against several lepidopteran insect species (Applebaum, 1985).

9.2 Cysteine proteinase inhibitors

Isolation of the midgut proteinases from the larvae of cowpea weevil, *C. maculatus* (Kitch and Murdock, 1986; and bruchid *Zabrotes subfaceatus* (Lemos *et al.,* 1987) confirmed the presence of a cysteine mechanistic class of proteinase inhibitors. Similar proteinases have been isolated from midguts of the flour beetle *Tribolium castaneum*, Mexican beetle *Epilachnavarivestis* (Murdock *et al.,* 1987) and the bean weevil *Ascanthoscelides obtectus*. Cysteine proteinases isolated from insect larvae are inhibited by both synthetic and naturally occurring cysteine proteinase inhibitors (Wolfson and Murdock, 1987). In a study of the proteinases, from the midguts of several members of the order coleopteran, 10 of 11 beetle species representing 11 different families, had gut proteinases that were inhibited by p-chloromercuribenzene sulfonic acid (PCMBS), a potent sulphydryl reagent (Murdock *et al.,* 1988) indicating that the proteinases were of the cysteine mechanistic class. The optimum activity of cysteine proteinases is usually in the pH range of 5-7; which is the pH range of the insect gut that use cysteine proteinases (Murdock *et al.* 1987). Advances in enzymology has revealed the existence of a variety of cysteine proteinases resulting in their classification into several families namely papain, calpin and asparagines specific processing enzyme (Turk and Bode, 1991).

The rice cysteine proteinase inhibitors are the most studied of all the cysteine PIs, which is proteinaceous in nature (Abe and Arai, 1985) and highly heat stable (Abe *et al.* 1987). Recent three dimensional structure analysis of oryzacystatin OC-I by Tanokura's group (Nagata *et al.*, 2000), using NMR has shown a well defined main body consisting of amino acids from Glu 13 - Asp 97 and an alpha helix with five stranded anti parallel beta-sheet, while the N terminus (Ser 2-Val 12) and C terminus (Ala 98-Ala 102) are less defined. Further, analysis has demonstrated OC-I to be similar to chicken cystatin which belongs to type-2 animal cystatin. Another rice cystatin named as OC II, with a putative target binding motif Gln-x-Val-x-Gly shares similar motif with OCI but has a different inhibition constant (Ki) value (Arai *et al.*, 1991; Kondo *et al.*, 1991).

9.3 Aspartic and metalloproteinase inhibitors

Knowledge on the role of aspartic proteinases in insect digestion is more limited than that of cysteine proteinases. In species of six families of the order Hemiptera, aspartic proteinases (cathepsin D-like proteinases) were found along with cysteine proteinases (Houseman and Downe, 1983). The low pH of midguts of many members of coleopteran and Hemiptera provides more favourable environments for aspartic proteinases (pH optima ~ 3 - 5) than the high pH ofmost insect guts (pH optima ~ 8 - 11) (Houseman *et al.,* 1987) where the aspartic and cysteine proteinases would not be active. No aspartic proteinases have been isolated from coleoptera but Wolfson and Murdock 1987 demonstrated that pepstatin, a powerful and specific inhibitor of aspartyl proteinases, strongly inhibited proteolysis of the midgut enzymes of Colorado potato beetle, *Leptinotarsa decemlineata* indicating that an aspartic proteinase was present in the midgut extracts. Potato tubers possess an aspartic proteinase inhibitor, cathepsin D (Mares *et al.,*1989) that shares considerable amino acid sequence identity with the trypsin inhibitor SBTI from soybeans. Plants have also evolved at least two families of metallo-proteinase inhibitors, the metallocarboxypeptidase inhibitor family in potato (Rancour and Ryan, 1968), and tomato plants (Graham and Ryan, 1981) and a cathepsin D inhibitor family in potatoes (Keilova and Tomasek, 1976).

The cathepsin D inhibitor (27 kDa) is unusual as it inhibits trypsin and chymotrypsin as well as cathepsin D, but does not inhibit aspartyl proteases such as pepsin, rennin or cathepsin E. The inhibitors of the metallo carboxypeptidase from tissue of tomato and potato are polypeptides (4 kDa) that strongly and competitively inhibit a broad spectrum of carboxypeptidases from both animals and microorganisms,but not the serine carboxypeptidases from yeast and plants(Havkioja and Neuvonen, 1985). The inhibitor is found in tissues of potato tubers where it accumulates during tuber development along with potato inhibitor I and II families of serine proteinase inhibitor. The inhibitor also accumulates in potato leaf tissues along with inhibitor I and II proteins in response to wounding (Hollander-Czytko *et al.* 1985). Thus, the inhibitors accumulated in the wounded leaf tissues of potato have the capacity to inhibit all the five major digestive enzymes, i.e.trypsin, chymotrypsin, elastase, carboxypeptidase A andcarboxypeptidase B of higher animals and many insects(Hollander-Czytko *et al.*, 1985).

9.4 Mechanism of toxicity

The secretion of proteases has been attributed to two mechanisms, involving either a direct effect of food components (proteins) on the midgut epithelial cells, or a hormonal effect triggered by food consumption (Applebaum, 1985). Models for the synthesis and release of proteolytic enzymes in the midguts of insects proposed by Birk and Applebaum, 1960, Brovosky, 1986 reveal that ingested food proteins trigger the synthesis and release of enzymes from the posterior midgut epithelial cells. The enzymes are released from membrane associated forms and sequestered in vesicles that are in turn associated with the cytoskeleton. The peptidases are secreted into the ectoperitrophic space between the epithelium, as a particulate complex (Eguchi *et al.,* 1982), from where the proteases move transversely into the lumen of the gut, where the food proteins are degraded. PIs inhibit the protease activity of these enzymes and reduce the quantity of proteins that can be digested, and also cause hyperproduction of the digestive enzymes, which enhances the loss of sulfur amino acids (Shulke and Murdock, 1983) as a result of which, the insects become weak with stunted growth and ultimately die.

Coleopteran and hemipteran species tend to utilise cysteine proteinases (Murdock *et al.*, 1987) while lepidopteran, hymenopteran, orthopteran and dipteran species mainly use serineproteinases (Ryan, 1990; Wolfson and Murdock, 1990). Examples from both of these classes of proteinases havebeen shown to be inhibited by their cognate proregions (Taylor *et al.*, 1995). The effect of class specific inhibitors on the pest digestive enzymes is not always a simple inhibition of proteolytic activity, but recent studies haveindicated the reverse may happen. It would appear that there are often two populations of digestive enzymes intarget pests, those that are susceptible to inhibition and those that are resistant (Michaud *et al.*, 1996; Bown *et al.*, 1997), and some insects respond to ingestion of plant PIs such as soybean trypsin inhibitor (Broadway and Duffey, 1986) and oryzacystatin (Michaud *et al.*, 1996) by hyperproducing inhibitor-resistant enzymes. Coleopteran and hemipteran species tend to utilize cysteine proteinases (Murdock *et al.*, 1987) while lepidopteran, hymenopteran, orthopteran and dipteran species mainly use serine proteinases (Ryan, 1990; Wolfson and Murdock, 1990). Examples from both of these classes of proteinases have been shown to be inhibited by their cognate proregions (Taylor *et al.*, 1995). The effect of class specific inhibitorsofthe pest digestive enzymes is not always a simple inhibition of proteolytic activity, but recent studies have indicated the reverse may happen. It would appear that there are often two populations of digestive enzymes in target pests, those that are susceptible to inhibition and those that are resistant (Michaud *et al.*, 1996; Bown *et al.*, 1997), and some insects respond to ingestion of plant PIs such as soybean trypsin inhibitor (Broadway and Duffey, 1986) and oryzacystatin (Michaud *et al.*, 1996) by hyperproducing inhibitor-resistant enzymes.

The mechanism of binding of the plant protease inhibitors of the insect proteases appears to be similar with all the four classes of inhibitors. The inhibitor binds to the active site on the enzyme to form a complex with a very low dissociation constant (107 to 1014 M at neutral pH values), thus effectively blocking the active site. A binding loop on the inhibitor, usually "locked" into conformation by a disulphide bond, projects from the surface of the molecule and contains a peptide bond (reactive site) cleavable by the enzyme (Terra *et al.*, 1996; Walker *et al.*, 1998). This peptide bond may be cleaved in the enzyme inhibitor complex, but cleavage does not affect the interaction, so that a hydrolysed inhibitor molecule is bound similar to an unhydrolysed one. The inhibitor, thus directly mimics a normal substrate for the enzyme, but does not allow the normal enzyme mechanism of peptide bond cleavage to proceed to completion ie., dissociation of the product (Walker *et al.*, 1998). It would also appear that insect digestive trypsins do not fall into the classification of peptidase hydrolases, as defined by inhibition spectra. It has been shown, notably, that the trypsin like digestive proteases of several lepidopteran species are inhibited by (l- 3-trans carboxiran-2 carbonyl)-l-leu-agmatin (E-64) (Lee and Anstee, 1995; Novillo *et al.*, 1997) an inhibitor, generally considered to be specific to cysteine proteinases (Dunn, 1989). Thus, true interactions will become clear only when we have protein crystals and X-ray diffraction data for the structure of insect enzyme/inhibitor complexes. Further, the specificity of the inhibitor enzyme interaction is primarily determined by the specificity of proteolysis determined by the enzyme (Blancolabra *et al.*, 1996).

Do you kniow

Box 3.9 provides an interested aspect of insects

Hundreds of thousands of years before mankind's first efforts, the world's first paper makers were collecting weathered fibers from plants that were chewed and mixed with saliva to make paper for nests. These paper makers were the yellow jackets, umbrella or paper wasps, hornets, and the social wasps of the Central and South American tropics, the polybiines (Akre 1982)(all Hymenoptera: Vespidae). The fiber they gather is made into a hard paper, called carton, used for constructing cells for their brood and the envelope which encloses the nest. The nest envelopes often are of interesting colour patterns depending on the source of the original plant fibers. Some wasps even incorporate bits of sand into their paper for strength and hardness. The envelope covering most nests is an excellent insulator. It is laid down in layers with air spaces between. Wasps make maximum use of dead air spaces in their nest construction to help regulate the internal temperature of the nest. In addition, some species of yellow jackets build these paper nests in a cavity below the surface of the soil which also tends to protect the nest and brood from fluctuations in temperature.

REFERENCES

Abe, K. and Arai, S. (1985) Purification of a cysteine proteinase inhibitor from rice, *Oryza sativa* L. *japonica. Agriculture Biology and Chemistry.* vol. 49, p. 3349-3350.

Abe, K., Emori, Y., Kondo, H., Suzuki, K. and Arai, S. (1987). Molecular cloning of a cysteine proteinase inhibitor of rice (*Oryzacystatin*). Homology with animal cystatins and transient expression in the ripening process of rice seeds. *Journal of Biological Chemistry*, vol. 262, no. 35, p. 16793-16797.

Akre R D. (1982) Social wasps. In: H.Rermann (Ed) *Social Insects* Vol. IV New York AP Inc. XIII pp 105

Alves,L C, P.C. Almeida, L. Franzoni, L. Juliano, and M.A. Juliano (1996) Synthesis of N-α-protected aminoacyl 7-amino-4-methyl-coumarin amide by phosphorus oxychloride and preparation of specific fluorogenic substrates for papain. *Peptide Research*, 9, pp. 92–96

Applebaum, S.W. (1985). Biochemistry of digestion. In: Kerkut, G.A., Gilbert, L. (Eds.), *Comprehensive Insect Physiology, Biochemistry and Pharmacology*, vol 4. Pergamon Press, New York, pp. 279–311.

Arai,S., Watanabe, H., Kondo, H., Emori, Y. and Abe, K. (1991). Papain inhibitory activity of oryzacystatin, a rice seed cysteine proteinase inhibitor, depends on the central Gln-Val-Val-Ala-Gly region conserved among cystatin superfamily members. *J. Biochem,* vol. 109, p. 294-298

Arakane,Y., Q. Zhu, M. Matsumiya, S. Muthukrishnan, K.J. Kramer (2003). Properties of catalytic, linker and chitin-binding domains of insect chitinase. Insect Biochem. Mol. Biol., 33 (2003), pp. 631–648

Arakane, Y., and Muthukrishnan, S. (2010) Insect chitinases and chitinase-like proteins. *Cellular Molecular Life Sciences*, 67, 201-216.

Arakawa G, Watanabe H, Yamasaki H, Mekawa H, and Tokuda, G. (2009). Purification and molecular cloning of xylanases from the wood-feeding termite, *Coptotermes formosanus* Shiraki. *Biosci. Biotechnol. Biochem.*, 73: 710-718.

Arrese, E.L., Patel, R.T., and Soulages, J. L. (2006). The main triglyceride-lipase from the insect fat body is an active phospholipase A(1): identification and characterization. *J Lipid Res.* ;47:2656–67

Ashford D.A., Smith, W.A and Douglas A.E. (2000) Living on a high sugar diet: The fate of sucrose ingested by a phloem-feeing insect, the pea aphid *Acyrthosiphon pisum. J.Insect Physiol.* 46: 335-341

Azevedo T.M., Terra, W.R and Ferreira, C (2003). Purification and characterization of three β-glycosidases from midgut of the sugar cane borer, *Diathraea saccharalis. Insect Biochem. Mol.Biol. 33*: 81-92

Bacic, A., Harris P.J and Stone B.A (1988) Structure and functions of plant cell walls. In P.K.Stumpf and E.E Conn (Eds) *Biochemistry of Plants* Vol 14, pp 297-372 N.York, Academic Press

Barrett,A.J, N.D. Rawlings, and J.F. Wasner (1998) Handbook of Proteolytic Enzymes. Academic Press, London

Bayes, A., Comellas-Bigler, M., Rodriguez de la Vega M., Maskors, K., Bode, W., Aviles, F.X., Jongsma, M.A., Beekwilder, J and Vendrell, J. (2005). Structural basis of resistance of an insect carboxypeptidase to plant protease inhibitors. *Proc. Natl.Acad.Sci.* 102: 16602-16607

Bayyareddy, K., Andacht, T.M., Abdullah, M.A., and Adang, M.J. (2009). Proteomic identification of Bacillus thuringiensis subsp. israelensis toxin Cry4Ba binding proteins in midgut membranes from Aedes (Stegomyia) aegypti Linnaeus (Diptera, Culicidae) larvae. *Insect Biochem. and Mol. Biol.* 39, 279–286.

Bieth, J., Spiess, B and Weermuth C.G. (1974). The synthesis and analytical use of a highly sensitive and convenient substrate of elastase. *Biochem. Med. 11*: 350-357

Blancolabra, A., Martinez-Gallardo, N.A., Sandoval-Cardoso, L. and Delano-Frier, J.(1996) Purification and characterization of a digestive cathepsin D proteinase isolated from *Tribolium castaneum* larvae (Herbst). *Insect Biochem. and Mol. Biol.* vol. 26, p. 95-100.

Bolognesi, R., Ribeiro, A.F., Terra, W.R., and Ferreira, C. (2001). The peritrophic membrane of *Spodoptera frugiperda*: secretion of peritrophins and role in immobilization and recycling digestive enzymes. *Archives of Insect Biochemistry and Physiology* 47, 62–75.

Bown, D.P., Wilkinson, H.S., and Gatehouse, J.A. (1997) Differentially regulated inhibitor sensitive and insensitive protease genes from the phytophagous pest, *Helicoverpa armigera*, are members of complex multigene families. *Insect Biochem. Mol. Biol.* 27: 625-638.

Bozicc, N., Ivanovic, J., Nenadovic, V., Berdstrom, J., Larsson, t., and Vujcic, Z. (2008). Purification and properties of a major leucyl aminopeptidase of *Morimus funereus* (Coleoptra: Cerambycidae) larvae. *Comp. Biochm.Physiol. B. 149: 454-462*

Bragatto I , Fernando A. Genta A. B, Alberto F. Ribeiro , C., Walter R. Terra and Clélia Ferreira (2010). Characterization of a b-1,3-glucanase active in the alkaline midgut of Spodoptera frugiperda larvae and its relation to b-glucan-binding proteins . *Insect Biochem. and Mol. Biol.* 40 861-872

Brito L., Lopes, A.R., Parra, J.R.P., Terra, W.R., and Silva-Filho, M.C.(2001) Adaptation of tobacco budworm *Heliothis virescens* to proteinase inhibitors may be mediated by synthesis of new proteinases. *Comp. Biochem. Physiol.* 128B: 365-375.

Broadway, R. M., and S. S. Duffey (1986) Plant proteinase inhibitor: mechanism of action and effect on the growth and digestive physiology of larval *Heliothis zea* and *Spodoptera exigua J. Insect Physiol*, 32 pp. 827–833

Callewaert L and Michiels, C.W. (2010). Lysozymes in the animal kingdom. *J. Biosciences* 35(1):127-60

Campbell P.M., Cao, A.T., Hines, E.R., East., P.D and Gordon K.H.J. (2008) Proteomic analysis of the peritrophic matrix from the gut of the caterpillar, *Helicoverpa armigera. Insect Biochem. Mol. Biol.* 38: 950-958.

Christeller, J. T., Poulton, J., Markwick, N. M., and Simpson, R. M. (2010). The effect of diet on the expression of lipase genes in the midgut of the lightbrown apple moth (*Epiphyas postvittana* Walker; Tortricidae). *Insect Mol. Biol.* 19, 9–25

Colebatch, G., Cooper, P., and East P. 2002 cDNA cloning of a salivary chymotrypsin-like protease and the identification of six additional cDNAs encoding putative digestive proteases from the green mired, *Creontiades dilutus* (Hemiptera: Miridae). *Insect Biochem.Mol. Biol.32:* 1065-1075

Coutinho, P.M and Henrissat, B. (1999) Carbohydrate-active enzyme server. http://www.afmb.cnrs-mrs.fr/~cazy/CAZY/index.html.

Cristofoletti, P.T., Terra, W.R.(1999). Specificity, anchoring, and subsites in the active center of a microvillar aminopeptidase purified from Tenebrio molitor (Coleoptera) midgut cells. *Insect Biochem. and Mol. Biol* 29, 807–819.

Cristofoletti, P.T., and Terra, W.R. (2000). The role of amino acid residues in the active site of a midgut microvillar aminopeptidase from the beetle *Tenebrio molitor. Biochimica et Biophysica Acta* 1479, 185–195.

Cristofoletti PT, Ribeiro AF, Deraison C, Rahbé Y, and Terra WR (2003) Midgut adaptation and digestive enzyme distribution in a phloem feeding insect, the pea aphid ***Acyrthosiphon pisum***. *J Insect Physiol* 49:11–24

Cristofoletti P T, A.F. Ribeiro, and W.R. Terra (2005) The cathepsin L-like proteinases from the midgut of *Tenebrio molitor* larvae: sequence, properties, immmunocytochemical localization and function. *Insect Biochem. and Mol. Biol.* 35, pp. 883–901

Cristofoletti,P.T., F.A.M. Mendonça, Y. Rahbé, and W.R. Terra (2006) Characterization of a membrane-bound aminopeptidase purified from *Acyrthosiphon pisum* midgut cells. A major binding site for toxic mannose lectins. *FEBS J.*, 273 pp. 5574–5588

Daimon, T., Taguchi, T., Meng, Y., Katsuma, S., Mita, K., and Shimada, T. (2008). β-fructofuranosidase genes of the silkworm, *Bombyx mori*: insights into enzymatic adaptation of *B. mori* to toxic alkaloids in mulberry latex. *J. Biol. Chem.* 283, 15271e15279

Davies, G. J ., Wilson K.S., and Henrisat, B. (1997) Nomenclature for sugar-binding subsites in glycosyl hydrolases. *Biochem. J. 321:* 557-559

De Veau, E.J. I. and Schultz, J.C (1992) Reassessment of interaction between gut detergents and tannins in Lepidoptera and significance for gypsy moth larvae. *J. Chem. Ecol.* 18:1437-1453

Deraison, C., Darboux. I., Duportets, L., Gorojankina, T., Rahbe, Y., and Jouanin (2004). Cloning and characterization of gut-specific cathepsin L from the aphid *Aphis gossypii. Insect Mol. Biol.* 13: 165-

Dey, P. M. and Pridham, J. B. (1972), Bi0Chemistry of a-galactosidase: *Adv. Enzymol.*, 36, 911-930.

Dunse, K.M., Stevens, J.A., Lay, F.T, Gaspar, Y.M., Heath RL, *et al.* (2010) Coexpression of potato type I and II proteinase inhibitors gives cotton plants protection against insect damage in the field. *Proc Natl Acad Sci* USA 107: 15011–15015.

Eguchi, M.; Iwamotoa, A. and Yamaguhi, K. (1982) Interaction of proteases from the midgut lumen, epithelial and peritrophic membrane of the silkworm *Bombyx mori* L. *Comparative Biochemistry and Physiology-A*, vol. 72, p. 359-363.

Eisemann, C., Wijffels, G., Tellam, R.L.,(2001). Secretion of type 2 peritrophic matrix protein, peritrophin-15, from the cardia. *Arch. Insect Biochem. Physiol.* 47, 76-85

Ferreira,C. and W.R. Terra (1986) Substrate specificity and binding loci for inhibitors in an aminopeptidase purified from the plasma membrane of midgut cells of an insect (*Rhynchosciara americana*). *Arch. Biochem. Biophys.*, 244, pp. 478–485

Ferreira C, Ribeiro AF, Garcia ES and Terra WR. (1988). Digestive enzymes trapped between and associated with the double plasma membranes of *Rhodnius prolixus* posterior midgut cells. Insect Biochem 18: 521– 530.

Ferreira L., J.C. Molina, C. Brasil, and G. Andrade (2003). Evaluation of *Bacillus thuringiensis* bio- insecticidal protein effects on soil microorganisms. *Plant and Soil* 256: 161–168.

Ferreira A H P., P.T. Cristofoletti, D.M. Lorenzini, L.O. Guerraa, P.B. Paiva, M.R.S. Briones, W.R. Terra, and C. Ferreira (2007) Identification of midgut microvillar proteins from Tenebrio molitor and Spodoptera frugiperda by cDNA library screenings with antibodies *J.Insect Physiol.* 53 1112–112

Fonseca, M. G., Vidal, D. M., and Zarbin, P. H. G. (2010). Male-produced sex pheromone of the cerambycid beetle *Hedypathes betulinus*: Chemical Identification and biological activity. *J. Chem. Ecol.* 36:1132–1139.

Fujita, A., Shimizu, I. and Abe, T. (2001) Distribution of lysozyme and protease, and amino acid concentration in the guts of a wood-feeding termite, *Reticulitermes speratus* (Kolbe): possible digestion of symbiont bacteria transferred by trophallaxis. *Physiological Entomology*, 26, 116–123.

Fujita, A. (2004), Lysozymes in insects: what role do they play in nitrogen metabolism?. *Physiol Ent*, 29: 305–310.

Geer, L. Y., Domrachev, M., Lipman, D. J. and Bryant, S. H. (2002). CDART: protein homology by domain architecture. *Genome Res.* 12,1619 -1623

Genta, F.A., Dumont, A.F., Marana, S.R., Terra, W.R., Ferreira, C. (2007). The interplay of processivity, substrate inhibition and a secondary substrate binding site of an insect exo-beta-1,3-glucanase. *Biochem. Biophys. Acta* 1774, 1079-1091

Genta, F.A., Bragatto, I., Terra, W.R., and Ferreira, C. (2009). Purification, characterization and sequencing of the major beta-1,3-glucanase from the midgut of Tenebrio molitor larvae. Insect Biochem. Mol. Biol. 39, 861-874

Girard, C and Jouanin, L. (1999a) Molecular cloning of cDNAs encoding a range of digestive enzymes from a phytophagous beetle *Phaedon cochleariae. Insect Biochem. Mol.Biol. 29: 1129-1142*

Girard, C and Jouanin, L. (1999b) Molecular cloning of gut specific chitinase cDNA from the beetle *Phaedon cochleariae. Insect Biochem. Mol. Biol. 29:549-556*

Grossmann, G. A. and Terra, W. R. (2001) α-galactosidases from the larval midgut of *Tenebrio molitor* (Coleoptera) and *Spodoptera frugiperda* (Lepidoptera) *Comp. Biochem. Physiol.*128B:109–122

Guo, W., G. X. Li, Y. Pang, and P. Wang. (2005). A novel chitin binding protein identified from the peritrophic membrane of the cabbage looper, *Trichoplusia ni. Insect Biochemistry and Molecular Biology* 35: 1224–1234.

Havkioja, E., and Neuvonen, L. (1985). Induced long-term resistance to birch foliage against defoliators: defense or incidental. Ecology 66, 1303-1308.

Hollander-Czytko, H., Andersen, J.L., and Ryan, C.A. (1985). Vacuolar localization of wound-induced carboxy peptidase inhibitor in potato leaves. Plant Physiology 78, 76-79.

Horne, I., Haritos V.S., and Oakeshott, J. G. (2009) Comparative and functional genomics of lipases in holometabolous insects. *In Insect biochemistry and molecular biology* 39(8):547-67

Houseman J.G, and Downe A.E.R. (1983) Activity cycles and the control of four digestive proteinases in the posterior midgut of *Rhodnius prolixus* Stal (Hemiptera: Reduviidae). *J Insect Physiol* 29: 141–148.

Houseman, J.G., Campbell, F.C. and Morrison, P.E. (1987) A preliminary characterization of digestive proteases in the posterior midgut of the stable fly *Stomoxys calcitrans* (L.) (Diptera: Muscidae). *Insect Biochem.* vol. 17, p. 213-218.

Husebye H, Arzt S., Burmeister W.P., Hartel, F,V, Brandt, A., Rossiter, J.T and Bones A M. (2005). Crystal

structure at 1.1 A resolution of an insect myrosinase from *Brevicoryne brassicae* shows its close relationship to β-glucosidases. *Insect Biochem. Mol. Biol.* 35: 1311-1320.

Jasrapuria, S., Arakane, Y., Osman, G., Kramer, KJ., Beeman, RW., and Muthukrishnan, S. (2010) Genes encoding proteins with peritrophin A-type chitin binding domains in *Tribolium castaneum* are grouped into three distinct families based on phylogeny, expression and function. *Insect Biochem Mol. Biol.,* 40: 214-227.

Johnson, K. S., and D. Rabosky. (2000). Phylogenetic distribution of digestive proteinases in beetles: evidence for an evolutionary shift to an alkaline digestive strategy in Cerambycidae. Comparative Biochemistry and Physiology B 126:609-619.

Jones , A.M..E., Winge, P., Bones, A.M., Cole, R., and Rossiter, J.T . (2002) Characterization and evolution of a myrosinase from the cabagge aphid *Brevicorine brassicae. Insect Biochem. Mol.Biol.* 32: 275-284

Jordaoeiro A., Terra, W.R., Ribeiro, A.F, Lehane M.J., and Ferreira, C. (1996) Trypsin secretion in *Musca domestica* larval midgets: A biochemical and immunocytochemical study. *Insect Biochem.Mol. Biol.*26: 445-453.

Jordao lla, ,B.P., Capella, A.N., Terra, W.R., Ribeir A.F. and Ferreira, C. (1999) Nature of the anchiors of membrane-bound aminopeptidase, amylase and trypsin and secretory mechanisms in *Spodoptera frugiperda* (Lepidoptera) midgut cells. *J.Insect Physiol. 45:* 29-37.

Keilova, H. and Tomasek, V. (1976). Isolation and properties of cathespin D inhibitor from potatoes. *Collection of Czechoslovak Chemical Communication* 41, 489-497.

Kim, N., Choo, Y.M., Lee, K.S., H, Kong, S.J., Seol, K.Y. et al. (2008) Molecular cloning and characterization of a glycosyl family 9 cellulase distributed throughout the digestive tract of the cricket *Teleogrllus emma. Comp.Biochem.Physiol. 150B:* 368-376.

Kitch, L.W. and Murdock, L.L. Partial characterization of a major gut thiol proteinase from larvae of *Callosobruchus maculatus* (F.). *Archives of Insect Biochem. and Physiol*, 1986, vol. 3, p. 561-575.

Klinkowstrom, A.M., Terra., W.R ., and Ferreira, C. (1994) Aminopeptidase A from *Rh*ynchosciara *americana* (Diptera) larval midgets: Properties and midgut distribution. *Arch.Iinsect B*iochem. Physiol. 27: 302-305.

Klinkowstrom, A.M., Terra, W.R., and Ferreira, C. (1995) Midgut dipeptidases from *Rhynchosciara americana* (Diptera) larvae: Properties of soluble and membrane bound forms. *Insect Biochem. Mol. Biol.*25: 303-310

Knight P.K.J., Knowles, B.H., and Eller D.J. (1995) Molecular cloning of an insect aminopeptidase N that serves as a receptor of *Bacillus thuringiensis* Cry 1a (c) toxin. *J.Biol.Chem. 270:* 17765-17770.

Kondo, H., Abe, K.., Emori, Y. and Arai, S. (1991) Gene organization of oryzacystatinII, a new cystatin superfamily member of plant origin, is closely related to that of oryzacystatinI but different from those of animal cystatins. *FEBS Letters*, vol. 278, p. 87-90.

Lange, T., Hansson, G.C and Samuelsson, T.(2007) Gel-forming mucins appeared early in metazoan evolution.*Proc. Natl.Acad. Sci. USA, 104:* 16209-16214.

Lee, M.J. and Anstee, J.H. (1995) Endoproteases from the midgut of larval *Spodoptera littoralis* include a chymotrypsinlike enzyme with an extended binding site. *Insect Biochemistry and Molecular Biology*, vol. 25, p. 49-61.

Lemos, F.J.A. and Terra, W.R (1991) Properties and intracellular distribution of a cathepsin D-like proteinase active at the acid region of *Musca domestica* midgut. *Insect. Biochem. 21:* 457-465.

Lemos, F.J.A , Xavier-Filno, J. and Campos, F.A.P.(1987) Proteinases of the midgut of *Savrotes subfasciatos* larvae. *Agricultural Biology Tecnoi*, vol. 80, p. 46.

Lemos, F.J.A, Ribeiro, A.F., and Terra, W.R. (1993) A bacteria digesting midgut lysozyme from *Musca domestica* (Diptera) larvae: Purification, properties and secretory mechanism. *Insect Biochem. Mol.Biol. 23:* 533-541.

Lo, N., Toduda, G., Watanabe, H., Rose, H., Slaytor, M et al., (2000) Evidence from multiple gene sequence indicates that termites evolved from wood-feeding cockroaches. *Current Biol. 10:* 801-804.

Lopes, A.R, Juliano, M.A., Juliano, L., and Terra, W.R.(2004) Coevolution of insect trypsins and inhibitors. *Arch. Insect Biochem. Physiol.* 55: 140-152.

Lopes A R, P.M. Sato, and W.R. Terra (2009) Insect chymotrypsins: chloromethyl ketone inactivation and substrate specificity relative to possible coevolutional adaptation of insects and plants. Archives of Insect Biochemistry and Physiology, 70 , pp. 188–203

Lu, Y., Zen, K.C., Muthukrishnan, S., and Kramer, K.J. (2002) Site-directed mutagenesis and functional analysis of active site acidic amino acid residues D142,D144 and E146 in *Manduca sexta* (tobacco hornworm) chitinase. *Insect Biochem.Mol.Biol. 32:* 1369-1382

Marana, S.R., Jacobs-Lorena, M., Terra, W.R., and Ferreira, C. (2001) Amino acid residues involved in substrate binding and catalysis in an insect digestive β-glycosidase. Biochem. Biophys. Acta 1545: 41-52

Marana, S.R., Lopes, A.R., Juliano, L., Juliano, M.A., Ferreira, C. (2002) Subsites of trypsin active site favour catalysis or substrate binding. *Biochem.Biophys. Res.Commun. 290:* 494-497.

Mares, M., Meloun, B., Pavlik, M., Kostka, V., and Baudys, M. (1989). Primary structure of Cathepsin-D inhibitor from potatoes and its structural relationship to trypsin inhibitor family. *FEBS Letters* 251, 94-98.

Matoub, M and Rou-land, C. (1995) Purification and properties of the xylanses from the termite *Macrotermes bellicosus* and its symbiotic fungus *Termitomyces* sp. *Comp. Biochem. Phsysiol. B 112:* 629-635.

Mazumdar-Leighton, S. and Broadway, R.M. (2001a) Transcriptional induction of diverse midgut trypsins in larval *Agrotis ipsilon* and *Helicoverpa zea* feeding on the soybean trypsin inhibitor. *Insect Biochem. Mol. Biol. 31:* 1881-1889.

Mazumdar-Leighton, S. and Broadway, R.M. (2001b) Identification of six chymotrypsin cDNAs from larval midgut of *Helicoverpa zea* and *Agrotis ipsilon* feeding on the soybean trypsin inhibitor.*Insect Biochem. Mol. Bi*ol. 31: 633-644

Michaud, D., Nguyen-Quoc, B., Vrain, T.C., Fong, D. and Yelle, S. (1996) Response of digestive cysteine proteinases from the Colorado potato beetle (*Leptinotarsa decemlineata*) and the black vine weevil (*Otiorynchus sulcatus*) to a recombinant form of human stefin A. *Archieves of Insect Biochemistry and Physiology*, vol. 31, p. 451-64.

Murdock, L.L., Brookhart, G., Dunn, P.E., Foard, D.E. and Kelley, S. (1987) Cysteine digestive proteinases in Coleoptera. *Comparative Biochemistry and Physiology-B*, vol. 87, p. 783-787.

Nagata, K., Kudo, N., Abe, K.., Arai, S. and Tanokura, M. (2000) Three dimensional solution structure of oryzacystatin-I, a cysteine proteinase inhibitor of the rice, *Oryza sativa* L. japonica. *Biochemistry*, vol. 39, p. 14753-14760.

Nakanishi, K., Yaoi, K., Naginao, Y., Hara, H., Kitami, M., et al., (2002) Aminopeptidase N isoforms from the midgut of *Bombyx mori* and *Plutella xylostella*: Their classification and the factors that determine their binding specificity to *Bacillus thuringiensis*Cry 1A toxin. *FEBS Lett.* 519: 215-220.

Nielsen, J.E and Borchert, T.U. (2000) Protein engineering of bacterial α-amylase. *Biochem. Biophys.Acta.* 1543: 253-274.

Nishimoto, M., Kubota, M., Tsuji, M., Mori, H., Kimura, A. et al., (2001) Purification and substrate specificity of honey bee, *Apis mellifera* L., alpha glucosidae III. *Biosci. Biotechnol. Biochem.* 65: 1610-1616.

Novillo, C., Castanera, P. and Ortego, F. (1997) Inhibition of digestive trypsin like proteases from larvae of several lepidopteran species by the diagnostic cysteine protease inhibitor E64. *Insect Biochemistry and Molecular Biology*, vol. 27, p. 247-254.

Noren, O., Sjostrom, H., Danielsen, E.M., Cowell G.M. and Skovbjerg, H. (1986) The enzymes of the enterocyte plasma membrane. In P.Desnuelle, H. Sjostrom and O. Noren (Eds) *Molecular and cellular basis of digestion* (pp 355-365) Amsterdam: Elsevier.

Padilha, M.H.P., Pimentel, A.C, Ribeiro, A.F., and Terra, W.R . (2009) Sequence and function of lysosomal and digestive cathepsin D-like proteinase of *Musca domestica* midgut. *Insect Biochem. Mol. Biol.* 39: 782-791.

Pauchet, Y., Freitak, D, Heidel-Fischer, H.M., Heckel, D, G., and Vogel, H. (2009a) Immunity or digestion: Glucanase activity in a glucan-binding protein family from Lepidoptera. *J. B*iol. Chem. 284: 2214-2224.

Pauchet, Y., Muck, A., Svatos A and Heckel D.G (2009b) Chromatographic and electrophoretic resolution of proteins and protein complexes from the larval midgut microvilli of *Manduca sexta. Insect Biochem. Mol. Biol. 39: 467-474*

Peters, W. (1992) *Peritrophic membranes.* Berlin: Springer.

Peterson, A.M., Fernando, G.J.P., and Wells, M. (1995) Purification, characterization and cDNA sequence of an alkaline chymotrypsin from the midgut of *manduca sexta. Insect Biochem. Mol. Biol. 25:* 765-774

Pimenta, D.C., Oliveira, A., Juliano, M.A and Juliano L. (2001) Substrate specificity of human cathepsin D using internally quenched fluorescent peptides derived from reactive site loop of kallistatin. *Biochem. Biophys. Acta, 1544:* 113-122.

Popova-Butler, A., and Dean, D.H. (2009) Proteomic analysis of the mosquito *Aedes aegypti* midgut brush border membrane vesicles. *J. Insect Physiol*. 55: 264-272

Rajagopal , Agrawal, N., Selvapandiayn, A., Sivakumar S., Ahmad, S. And Bhatnagar, R.K.(2003). Recombinantly expressed isoenzymic aminopeptidases from *Helicoverpa armigera* (American cotton bollworm) midgut display differential interaction with closely realated *Bacillus thuringiensis* insecticidal proteins. *Biochem. J.*370: 971-978.

Rancour, J.M., & Ryan, C.A. (1968). Isolation of a carboxypeptidase B inhibitor from potatoes. Archives of Biochemistry and Biophysics 125, 380-382

Robert, M.,, Anjuman, B., Hsin-Hen Kuo, Andrew R., Shin, N., Carsten, A., Jeppe W. Tams, Jesper, V., Christopher, M. O., Stephen, G. Withers, and Gary D. Brayer (2005) Structural and mechanistic studies of chloride induced activation of human pancreatic α-amylase *Protein Science* (2005), 14:743–755.

Regel, R., Matioli, S.R and Terra, W.R (1998) Molecular adaptation of *Drosophila melanogaster* lysozymes to a digestive function. *Insect Biochem. Mol. Biol.* 28: 309-319.

Ryan CA. (1990) Protease inhibitors in plants: genes for improving defenses against insects and pathogens. *Annu. Rev. Phytopathol.* 28: 425-449.

Santos, C.D., Ribeiro, A.F., Terra, W.R.,(1986). Differential centrifugation, calcium precipitation and ultrasonic disruption of midgut cells of *Erinnyis ello* caterpillars. Purification of cell microvilli and inferences concerning secretory mechanisms. *Can J Zool* 64: 490– 500.

Santos, C.D., and Terra, W.R. (1986) Midgut alpha glucosidase and beta-fructosidase from *Erinnyis ello* larvae and imagoes: Physical and kinetic properties. *Insect Biochem. 16:* 819-824.

Scrivener, A.M., Watanabe, H., and Noda, H. (1997) Diet and carbohydrate digestion in the yellow spotted longicorn beetle *Psacothea hilaris. J.Insect Physiol. 43:* 1039-1052.

Shao, L., Devenport, M., Fujioka, H., Ghost, A., Jacobs-Lorena, M., (2005). Identification and characterization of a novel peritrophic matrix protein, Ae-Aper50, and the microvillar membrane protein, AEG12, from the mosquito, *Aedes aegypti. Insect Biochem. and Mol. Biol.* 35, 947–959

Shen, Z., and Jacobs-Lorena, M. (1997) Characterization of a novel gut-specific chitinase gene from the human malaria vector *Anopheles gambiae. J.Biol. Chem*. 272: 28895-28900

Shen, Z., Reese, J.C and Reeck, G.R. (1996) Purification and characterization of polygalacturonase from the rice weevil, *Sitophilus oryzae.* (Coleoptera: Curculionidae). *Insect Biochem. Mol. Biol. 26: 427-433.*

Shiroya, T. (1963) Metabolism of raffinose in cotton seeds. *Phytochemistry 2:* 23-46

Silva,C.P. and W.R. Terra (1995) An α-glucosidase from perimicrovillar membranes of *Dysdercus peruvianus* (Hemiptera: Pyrrhocoridae) midgut cells. Purification and properties Insect Biochem. Mol. Biol., 25, pp. 487–494

Silva, C.P and Terra, W.R. (1997) Alpha-galactosidase activity in ingested seeds and in the midgut of *Dysdercus peruvianus* (Hemiptera: Pyrrhocoridae). *Arch. Insect Biochem. Physiol.* 34: 443-460.

Silva, C.P and Xavier-Filho, J. (1991) Comparison between the levels of aspartic and cysteine proteinase of the larval midguts of *Callasobruchus maculates* (F) and *Zabrotes subfasciatus* (Boh.) (Coleoptera: Bruchidae). *Comp. Biochem. Physiol. B 99: 529-533*

Silva CP, Ribeiro AF, Gulbenkian S, Terra WR (1995) Organization, origin and function of the outer microvillar (perimicrovillar) membranes of *Dysdercus peruvianus* (Hemiptera) midgut cells. J Insect Physiol 41:1093–1103

Silva, M.C.P., Terra. W.R., and Ferreira, C. (2006) Absorption of toxic beta-glucosides produced by insects and their effect on tissue trehalases from insects. *Comp.Biochem. Physiol.* 143 B: 367-373

Spencer, K.C (1988) Chemical mediation of coevolution in the *Passiflora-Heliconius* interaction. In K.C Spenser (Ed.) *Chemical Mediation of Coevolution* (pp 167-240). London, UK, Academic Press.

Strobl, S., Maskos, K., Betz, M., Wiegand, G., Huber, R. et al., (1998). Crystal structure of yellow meal worm α amylase at 1.64 A resolution. *J.Mol.Biol. 178: 617-628.*

Sugimura, M., Watanabe, H., Lo, N., and Saito, H. (2003) Purification, characterization, cDNA cloning and nucleotide sequencing of a cellulose from the yellow spotted longicorn beetle, *Psacothea vilaris. Eur.J.Biochem. 270: 3455-3460.*

Taylor, M.A.J., Baker, K.C., Briggs, G.S., Connerton, I.F., Cummings, N.J., Pratt, K.A., Revell, D.F., Freedman, R.B. and Goodenough, P.W. (1995) Recombinant proregions from papain and papaya proteinase IV are selective high affinity inhibitors of the mature papaya enzymes. *Protein Engineering,* January vol. 8, no. 1, p. 59-62.

Tellam, R.L., Wijffels, G., Willadsen, P. (1999). Peritrophic matrix proteins. Insect Biochemistry and Molecular Biology 29, 87–101.

Terra, W.R., and Ferreira, C. 1994. Insect digestive enzymes: properties, compartmentalization and function. Com Biochem Physiol 109B: 1– 62.

Terra, W.R. Ferreira, C. and Jordao, B.P. (1996) Digestive enzymes. In: Lehane M. J., ed. Billin London, Chapman and Hall, p.153-194.

Terra, W.R., Espinoza-Fuentes, F.P., Ribeiro, A.F., and Ferreira, C. (1988). The larval midgut of the housefly (*Musca domestica*): ultrastructure, fluid fluxes and ion secretion in relation to the organization of digestion. *J Insect Physiol* 34: 463– 472.

Terra, W.R., Costa, R.H., Ferreira, C., (2006). Plasma membranes from insect midgut cells. *Anais da Academia Brasileira de Ciencias* 86, 1–15.

Terra, W.R. (1990). Evolution of digestive systems of insects. *Annu Rev Entomol* 35: 181– 200.

Terra, W.R. (2001). The origin and functions of the peritrophic membrane and peritrophic gel. *Arch Insect*

Biochem Physiol 47: 47– 61.

Uscian, J M, Miller, J., Sarath, G and Stanley D W. (1995) A digestive phospholipase A2 in the Tiger beetle *Cicindela circumpicta. J. Insect Physiol.* 41(2):135-141

Vaaje-Kolstad G, Houston DR, Riemen AH, Eijsink VG, van Aalten DM. (2005). Crystal Structure and Binding Properties of the Serratia marcescens Chitin-binding Protein CBP21. *J Biol Chem* 280:11313-11319.

Walker, A.J., Ford, L., Majerus, M.E.N., Geoghegan, I.E., Birch, A.N.E., Gatehouse, J.A. and Gatehouse, A.M.R. (1998). Characterisation of the midgut digestive proteinase activity of the two-spot ladybird (*Adalia bipunctata* L.) and its sensitivity to proteinase inhibitors. *Insect Biochemistry and Molecular Biology*, vol. 28, p. 173-180.

Wolfson,J L, and L.L. Murdock (1990) Diversit y in digestive proteinase activity among insects. Journal of Chemical Ecology, 16 (1990), pp. 1089–1102

Watanabe H and Tokuda, G (2010) Cellulolytic systems in insects. Annu Rev Entomol. 2010;55:609-32

Wei Z, Xue Y, Xu H, and Gong W (2006) Crystal structure of the C-terminal domain of S.cerevisiae eIF5. *J Mol Biol* 359(1):1-9

Wolfson, J.L. and Murdock, L.L.(1990) Diversity in digestive proteinase activity among insects. *Journal of Chemical Ecology*,vol. 16, p. 1089-1102.

Wolfson, J.L. and Murdock, L.L. (1987) Suppression of larval Colorado beetle growth and development by digestive proteinase inhibitors. *Entomologia Experimentalis et Applicata*, 1987, vol. 44, p. 235-240.

Zeng F, Y. Zhu, and A. Cohen (2002) Partial characterization of trypsin-like protease and molecular cloning of a trypsin-like precursor cDNA in salivary glands of Lygus lineolaris. *Comparative Biochem. and Physiol.* Part B 131, 453–463

The Circulatory System and Immune Responses in Insects 4

Global warming could lead to lighter coloured insects with waning immune defenses

Synopsis

1. Introduction
2. Structural pattern of the open circulatory system
    - 2.1 Dorsal diaphragm
    - 2.2 Ventral diaphragm
    - 2.3 Alary muscles
    - 2.4 Ostia
3. Components of the circulatory system
    - 3.1 Dorsal vessel: Heart and aorta
    - 3.2 Sinuses
    - 3.3 Accessory pulsatile organs
4. Process of blood circulation
    - 4.1 Diastole
    - 4.2 Systole
    - 4.3 Diastasis
    - 4.4 Heartbeat
5. Path of circulation
6. Extracardiac pulsations (ExP)
7. Autonomic nervous system for haemolymph circulation
8. Haemolymph
    - 8.1 Functions of haemolymph
    - 8.2 Composition of haemolymph
    - 8.3 Role of haemolymph in thermoregulation
9. Haemocytes
    - 9.1 Origin and types of haemocytes
    - 9.2 Functions of haemocytes
10. Insect immunity
    - 10.1 The cellular immune response
    - 10.2 The systemic immune response (Humoral response)
    - 10.3 The local immune response
    - 10.4 Physical barriers to invasion
11. Recognition of non-self by insect immune system
    - 11.1 Pattern recognition receptors
    - 11.2 Signaling pathways
    - 11.3 Antiviral defence
12. Synthesis of antifungal and antibacterial peptides
    - 12.1 Toll pathway for synthesis of antimicrobial peptides

1. INTRODUCTION

Insects, like all other arthropods, have an open circulatory system which differs in both structure and function from the closed circulatory system found in mammals and other vertebrates. In a **closed system**, blood is always contained within vessels (arteries, veins, capillaries, or the heart itself). In an **open system**, blood (usually called haemolymph) spends much of its time flowing freely within body cavities where it makes direct contact with all internal tissues and organs. In insects, blood is confined to vessels during only a portion of its circuit through the body. The remainder of its journey takes place within the body cavity called the hemocoel. For this reason, insect blood is called **haemolymph**. The internal organs and tissues are bathed in the haemolymph, which is propelled actively to all internal surfaces by specialized pumps, pressure pulses, and body movements.

The circulatory system in insects has multifaceted functions (Box 4.1). It is responsible for the movement of nutrients, salts, hormones, and metabolic wastes throughout the insect's body. In addition, it plays several critical roles in defence: it seals off wounds through a clotting reaction, it encapsulates and destroys internal parasites or other invaders, and in some species, it produces (or sequesters) distasteful compounds that provide a degree of protection against predators. Hydrostatic pressure generated internally by muscle contraction is used to facilitate hatching, moulting, expansion of the body and wings after moulting, physical movements (especially in soft-bodied larvae), reproduction (e.g. insemination and oviposition), and evagination of certain types of exocrine glands. In some insects, the blood aids in thermoregulation: it can help cool the body by conducting excess heat away from active flight muscles or it can warm the body by collecting and circulating heat absorbed while basking in the sun.

Box 4.1 Functions of the circulatory system

1. Transport of minerals or food materials
2. Storage of water for tissues
3. Lubrication of internal organs
4. Facilitate hatching, moulting, expansion, reproduction
5. Defence- encapsulation, phagocytosis
6. Immunity and wound healing
7. Provides lipoproteins for connective tissue formation.
8. Detoxification
9. Protection from natural enemies.

Insects are exposed to all sorts of infectious agents, and to survive in the world insects developed a potent defence mechanism that recognises and removes microbial threats. Insects have a rapid and effective immune system against microbial infections. Insects have a well-developed innate immune system that allows general and rapid responses to infectious agents while they lack an acquired immune system. This innate immunity includes the cellular response and the systemic response. An essential component of immunity is the recognition of nonself which is the first step in humoral defence that stimulates defensive responses. These responses are often accompanied by proteolytic activation of the phenoloxidase zymogen that is present in the haemolymph. A component of insect immune responses to bacteria is the synthesis by fat body and haemocytes of a variety of antibacterial proteins and peptides, which are secreted into the haemolymph to fight off pathogens. These molecules attack bacteria by several mechanisms. Inducible antifungal proteins have also been recently discovered in insect haemolymph. The promoters of several antibacterial protein genes in insects are regulated by transcription factors similar to those involved in mammalian acute phase responses. Some of the immune reactions have potential for autotoxicity to the insect itself.

2. STRUCTURAL PATTERN OF THE OPEN CIRCULATORY SYSTEM

Insects have an open circulatory system in comparison to the closed system in mammals (Box 4.2). The internal organs and tissues are bathed in the haemolymph, which is propelled actively to all internal surfaces by specialized pumps, pressure pulses, and body movements and is directed by vessels, tubes, and diaphragms. To facilitate circulation of haemolymph, the body cavity is divided into three compartments (called **blood sinuses**) by two thin sheets of muscle and/or membrane known as the dorsal and ventral diaphragms (Fig 4.1). The **dorsal diaphragm** is formed by Alary muscles of the heart and related structures; it separates the pericardial sinus from the perivisceral sinus. The **ventral diaphragm** usually covers the nerve cord; it separates the perivisceral sinus from the perineural sinus. In some insects, pulsatile organs are located near the base of the wings or legs. These muscular pumps do not usually contract on a regular basis, but they act in conjunction with certain body movements to force haemolymph out into the extremities.

Box 4.2 Comparison of the mammalian and insect circulatory system

Mammals	Insects
1. Closed system with blood Vessels and heart	1. Open system
2. Heart- ventral	2. Heart/aorta – dorsal
3. RBC carry O_2 and CO_2	3. No RBC
4. Haemoglobin is carrier molecule	4. Some insects have Hb. O_2 is delivered by trachea
5. Immune system with long-term Recognition system	5. No long-term immune recognition system
6. Acquired immunity present	6. No acquired immunity

2.1 Dorsal diaphragm

The dorsal diaphragm present in most insects consists of multiple layers of thin sheets of connective tissue with small dorsal transverse muscle fibers which are called **Alary muscles**, enclosed within the sheets, which separates the upper pericardial sinus from the lower perivisceral sinus. It is a fenestrated membrane and incomplete laterally, so that haemolymph readily passes between the upper pericardial sinus from the lower perivisceral sinus. In some insects, for example, cockroaches, they resemble wings projecting laterally from the dorsal vessel of each abdominal segment. Although sometimes mistakenly thought to play a key role in a heartbeat, the Alary muscles are more properly called **muscles of the dorsal diaphragm**. Whereas the myocardium is specialised to contract rapidly and constantly, the ultrastructure of the Alary muscles indicates infrequent and slow contractions. The muscles associated with the dorsal diaphragm may likewise be arranged in a loose network, as in Lepidoptera, or arranged like a weave surrounding the dorsal vessel, as in some Diptera; the functional role of these muscles is difficult to determine (Miller and Pass 2009). In Hemiptera, it is greatly reduced and a series of muscles near the posterior end of the heart, along with tracheal connections, support the heart. The dorsal diaphragm usually does not extend into the thorax, although it extends into the thorax to a limited extent in some insects (for example, some orthopterans). Loose clusters and strings of cells, called **pericardial cells**, often occur on the external surface of the heart, and they also are attached to the dorsal diaphragm at various places in the pericardial sinus. They phagocytise injected particles, such as India ink, some dyes, and other small particles, and are presumed to serve a protective phagocytic function. The alimentary canal, reproductive organs, and some of the fat body lay in the central or perivisceral sinus below the dorsal diaphragm (Nation, 2008).

2.2 Ventral diaphragm

Generally, it is present in larvae and adults of Odonata, Orthoptera, Hymenoptera, Ephemeroptera, Lepidoptera, and Neuroptera. It is present only in adults of Mecoptera and lower Diptera, and not present in the higher Diptera (the Cyclorrhapha). The ventral diaphragm plays a prominent role in perfusing the ventral nerve cord of insects. Nearly 40 years ago Glenn Richards surveyed the ventral diaphragms in insects and found that insects with a well-defined ventral nerve cord in the abdomen also have a well-developed ventral diaphragm. In contrast, insects with the ventral nerve cord condensed into a complex ganglion structure in the thorax invariably lack a ventral diaphragm. This correlation suggests that the role of the ventral diaphragm is inexorably tied to perfusion of the ventral nerve cord in the abdomen. When present, the ventral diaphragm loosely defines a perineural sinus below and the peri-visceral sinus above containing the gut. The ventral diaphragm usually does not extend into the thorax and not even present in the abdomen of all insects. In some insects, the ventral diaphragm is a strong muscular structure with a great deal of contractile activity. The activity of the ventral diaphragm is dictated by innervation from the central nervous system. The ventral diaphragm is fenestrated, allowing haemolymph to circulate. Undulations of both the dorsal and ventral diaphragms due to intrinsic muscle fibers, Alary muscle action, gut movements, ventilatory movements, and general body muscle action aid circulation of haemolymph and keep it mixed and moving within the body, especially in active insects. In some large flying insects, the ventral diaphragm assists in the haemolymph flow during thermoregulation by facilitating the removal of warm haemolymph from the hot thoracic muscles to the abdomen for cooling. The intimate association between the ventral diaphragm in insects and perfusion of the ventral nerve cord is strengthened by considering the structure in cockroaches that takes the place of a proper diaphragm. In these insects, four strips of muscle, together called **hyperneural muscle** are near the back of each of the abdominal ganglia, and contract slowly but not in a rhythmic order. The muscles are electrically inexcitable, which means that they do not contract myogenically, as the myocardium does, but instead are neurally driven by motor neurons located in the ventral ganglia. Thus, each of the ventral nerve cords in cockroaches has its own muscle supply that pulls it back and forth along the midline of the abdomen upon demand, thereby increasing the contact and mixing between the ganglia and the haemolymph (Miller and Pass, 2009).

2.3 Alary muscles

Alary muscles are the series of small muscles found in pericardial wall of insects. The Alary muscles, so named because of their general wing or delta shape in many insects, lay immediately on top of the dorsal diaphragm. The muscles probably aid the dorsal diaphragm in providing support to the heart, the part of the dorsal vessel in the abdomen. Some Alary muscle fibers pass beneath the heart and extend from the lateral side to side. Their contraction causes blood to flow into the pericardium from the perivisceral cavity and then into the heart. Each chamber has a pair of Alary muscles which may assist in the expansion (diastole) of the heart after a contraction wave and, thus, aid in pulling haemolymph into the incurrent Ostia.The pairs of alary muscles tend to agree with the number of pairs of ostia. The Alary muscles, then contract to close the Ostia and their valve-like structure prevents the haemolymph from returning to the body cavities. In places, the fibers may also run parallel to the long axis of the heart for a short distance (Nation, 2008).

2.4 Ostia

Ostia is a small valve- like paired openings which separates the chambers of the heart and allow haemolymph to enter or leave the dorsal vessel. Incurrent ostia allows haemolymph to enter during diastole and excurrent ones permit haemolymph to exit. During diastole, haemolymph is drawn through lateral incurrent ostial valves, facilitated by contraction of alary muscles, and during systole, contractions of the cardiac muscle close the ostia and force haemolymph anteriorly towards the head capsule. In a limited number of species, excurrent Ostia is also present, allowing haemolymph to pass out of the dorsal vessel into the main body cavity from the anterior abdominal/thoracic region. Ostial openings tend to occur at the base of shallow pockets or at deeper. Incurrent and excurrent ostia may be difficult to distinguish (Jones, 1977). Excurrent ostia most often occur in the thoracic portion of the vessel, but occur in the abdomen of some insects. Some ostia do not have developed valves that control haemolymph flow. Some insects have valve flaps in incurrent ostia that open inward; haemolymph readily passes the valves into the lumen of the heart. During a contraction the valve flaps are forced together, preventing backflow, which may be their main function (Nation, 2008).

3. COMPONENTS OF THE CIRCULATORY SYSTEM

3.1 Dorsal vessel: Heart and Aorta

The principal organ of haemolymph propulsion is the **dorsal vessel** which is the most visible organ specialised in haemolymph movement in insects. It is a hollow tube which runs along the midline for the whole length of the body. The dorsal vessel of most insects is not uniform in its course through the body and shows a differentiation into two regions which is reflected by their traditional denomination. The dorsal vessel consists of two parts; the posterior part in the abdomen is referred to as the **heart**, whereas the anterior part in the thorax and the head is the **aorta**. The heart is the abdomen portion of the vessel, but it may extend into the thorax in some insects. The presence of alary muscles and incurrent ostia in the heart portion are the major criteria for deciding where the heart ends and the aorta begins. The aorta does not have alary muscles and incurrent ostia, but it may have excurrent ostia. Both alary muscles and incurrent ostia occur in the thoracic portion of the vessel in many Orthoptera and consequently, this thoracic portion is still the heart and only a short aorta leads into the head. Anteriorly the aorta sometimes gets divided into two or more **cephalic arteries** in the head.

The heart remains in position with the help of Alary muscles that are attached to the tergum of the abdomen on one side and to the dorsal diaphragm on the other side. These Alary muscles appear to be distributed fan like over the heart (Fig 4.1). The heart consists of a number of chambers marked by constrictions and the presence of the opening called the **incurrent ostia** which allow the entry of blood from the pericardial sinus into the heart. The number of Ostia depends upon the number of heart chambers which is usually 9. The wall of the heart also consists of muscles. Heart mainly functions as the **pumping organ**, pumping haemolymph into the aorta, Insects also have sac-like structures at the base of appendages such as wings, legs and antenna called **accessory pulsatile organs**. They pulsate independently and supply adequate blood to the appendages.

3.2 Sinuses

In insects theblood flows through the body cavity, that is, the haemocoel irrigating various tissues and organs. The haemocoel is divided into 3 sinuses or regions due to the presence of two fibrous muscular septa or diaphragms composed of connective tissue. These sinuses are:

1. **Dorsal or pericardial sinus:** The area lying in between the tergum and dorsal diaphragm. It contains the heart.
2. **Ventral or perineural sinus:** The area lying in between the sternum and ventral diaphragm. It contains nerve cord.
3. **Visceral sinus:** The area in between dorsal and ventral diaphragms. It harbours the visceral organs like alimentary canal and gonads.

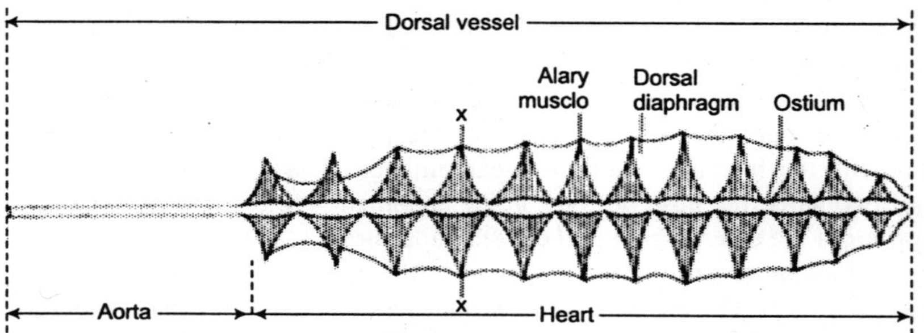

Fig. 4.1. Schematic sketch of the dorsal vessel in insects

3.3 Accessory pulsatile organs

The dorsal vessel and the two large diaphragms are responsible for pumping haemolymph through the main body cavity. However, these organs are incapable of achieving a circulation in body appendages such as the antennae, legs, wings, and various long abdominal processes (e.g., cerci and ovipositors). To supply these appendages, insects rely on special, small circulatory pumps, known collectively as **accessory pulsatile organs** or **accessory hearts**. As a rule, accessory hearts are separated from the dorsal vessel and function autonomously. The pumping organ is generally located at the base of the appendage and is connected to vessels or special diaphragms which guide the flow of haemolymph through the appendage. Accessory pulsatile organs are present in an astounding array of functional constructions. They are evolutionary novelties of higher insects and are absent in the phylogenetically ancestral insects. Antennal circulatory organs are nearly universal in insects, lacking only in groups with extremely short antennae, such as fleas and lice. The ancestral state of antennal circulation in insects is antennal vessels connected to the dorsal vessel as in certain apterygotes. The connection between the two vessels was probably lost during the further evolution of insects and replaced by autonomous pulsatile organs at the base of the antennal vessels. The anatomy of these organs differs widely in various insects. The best investigated antennal heart in terms of morphology and physiology is that of the American cockroach, *Periplaneta americana*. Remarkably, the antennal heart of the cockroach functions not only as a circulatory pump but also as a neurohemal organ: hormones released into the ampulla lumen are pumped into the antennae where they most likely modulate the sensitivity of the numerous sensilla. In leg circulatory organs, the flow of haemolymph is guided by longitudinal diaphragms instead of vessels. A diaphragm of connective tissue divides the inner cavity of each leg into two channels permitting a counter-current flow. Some insects have pulsatile leg organs with muscular attachment to the diaphragm. When the muscle contract, one channel is compressed forcing haemolymph toward the thorax; at the same time, the other channel expands drawing haemolymph into the leg. Other insects utilise changes in the volume of an elastic tracheal sac in the legs as the driving force for haemolymph exchange. A common misconception is that insect wings are dead cuticular structures. However, the veins of the wings are filled with haemolymph to maintain living tissues, such as nerves and tracheae. Circulation is achieved by pumping organs in the thorax, which in ancestral winged insects are ampullary enlargements of the dorsal vessel. The wing hearts of most holometabolan insects, however, are muscular diaphragms which are separated from the dorsal vessel and which are either paired or unpaired.

Recently, a further vital function of the wing hearts was discovered in *Drosophila*, namely that they are essential for the proper maturation of wings. Toward the completion of the wing formation process, they suck epidermal cells out of the wings, which are necessary before the two cuticular surfaces of each wing bond together. Flies lacking functional wing hearts never develop flight ability! Finally, it should be noted that accessory hearts may also partake in the hydraulic movements of body appendages. For example, the long sucking mouthparts of butterflies are uncoiled by action of special pumping organs in the head, and the lamellae of the antenna in scarabaeid beetles are spread out by action of the antennal heart. The antennal heart of the cockroach, *Periplaneta americana*, consists of a pulsatile ampulla at the base of each antennal vessel. The two ampullae expand by action of the interconnecting dilator muscle. Haemolymph rushes into each ampulla through a small ostium. When the muscle relaxes, the ampullae collapse by their own elasticity forcing the haemolymph into the antennal vessels. The haemolymph is conveyed to the end of each vessel and empties into the lumen of the antenna before returning to the head capsule; the two small muscles which extend to the anterior opening of the aorta are responsible for the suspension of the apparatus (Miller and Pass, 2009).

4. PROCESS OF BLOOD CIRCULATION

Heart mainly function as a pulsatile organ whose expansion and contraction leads to blood circulation. It takes place generally in an **anti clock** manner starting from posterior end to the anterior end in a forward direction. Circulation of blood takes place in two phases due to the action of the Alary muscles as well as the muscles of the walls of the heart. The two phases are:

1. **Diastole:** Expansion of heart takes place.
2. **Systole:** Contraction of the heart takes place.

4.1 Diastole

Expansion of heart (diastole) occurs, when the Alary muscles that are spread fan like over the heart and connected to the tergum get contracted. It results in **increase of volume** of heart and **decrease in the area** of pericardial sinus. This creates a pressure on the blood in pericardial sinus forcing the blood to enter into the heart through the incurrent ostia. The incurrent ostia allows only the entry of blood from the sinus into the heart and prevent its backflow from the heart to the sinus.

4.2 Systole

Contraction of the heart (systole), is brought about by the **expansion** of the Alary muscles as well as contraction of the muscles of the heart wall. This creates pressureon the blood within the heart, leading to its **forward movement** into the aorta. From the aorta blood enters into the head and flows back bathing the visceral organs in the visceral sinus and neural cord in the perineural sinus.

4.3 Diastasis

In between diastole and systole there will be a short period of rest which is known as **diastasis.**

During the process of backward flow of blood, after entering in to the visceral and perineural sinuses, blood flows to the posterior part of the body and in to the pericardial sinus, due to the undulating movementsof the dorsal and ventral diaphragms. During the process of circulation throughout the body some part of the blood flows into the accessory pulsatile organ that are present at the base of appendages like legs, wings and antennae.

4.4 Heartbeat

The heartbeat is a wave of contractions (systole) that generally originates at the posterior end of the heart and travels anteriorly. The heart beat is highly variable and depends upon physiological conditions, temperature, species, stage of development, nervous activity, and neurosecretions. In the larva of *Lucanus cervus* (Coleoptera), heart beat rate is 15 beats/min and near or higher than 100 beats/min in several insects (Jones, 1977). The beat is considered to be myogenic (originate in the muscle itself) even though the heartbeat of many other invertebrates is neurogenic. Various parts of the dorsal vessel are capable of acting as a pacemaker unlike the human heart (Box 4.3). However, the one that usually dominates, is located at the posterior of the heart. One of the unusual and little understood features of insect hearts is that the beat can reverse, and originate at or near the anterior and travel posteriorly. Reversals are unpredictable, but usually the back-directed beats are slower, and duration of the reversal is usually brief.

Certain experimental treatments, such as amputations of parts of the heart, lateral pressure on the dorsal vessel, and ligation near the middle of the heart may alter conduction rates and cause reversal of contraction waves in some insects. Dorsal vessel contraction rates and direction of contraction wave are variable in adult moths, with fast forward heart rates of 47.6 beats/min, slow forward rates of 32.8 beats, and reversal of contraction wave (anterior origin) of 32.2 beats/min. Heart activity may sometimes be initiated by abdominal movements *via* cardiac reflex or mechanical excitation. Sometimes, when heart function is weakened by histolysis, the abdomen may temporarily take over the main circulatory function or occasionally contribute to acceleration of low-speed haemolymph flow (Tartes *et al.*, 2002).

Box 4.3 Human vs insect pacemaker

Human	Insect
1. More than one pacemaker	1. May be few pacemakers but not in each Individual segment
2. Hierarchy of pacemaker activity:	2. Each segment can maintain or assume
a. Sinoatrial node	pacemaker role with last posterior one
b. Atrioventricular node	setting the pace
c. His bundle	
3. Fastest pacemaker drives heart	3. Not driven by one single pacemaker

The heart receives innervation through lateral neurons from ventral ganglia in the abdomen and from a chain of cardiac neurons that lie alongside the heart in some insects, notably Odonata, Orthoptera, Dictyoptera, and Hemiptera. Insects in the orders Diptera and Lepidoptera, and probably other orders, do not have cardiac neurons and a nerve chain paralleling the heart. The heart may still receive a nerve supply from the segmental ganglion.Various secretions and neurosecretory peptides act upon the heart to change the rate and amplitude of the heartbeat (Nassel, 1993). Proctolin, a neuropeptide, produced in motoneurons, interneurons, and neurosecretory cells located typically in the brain, but sometimes in other ganglia, stimulates heart rate. Whether this is one of its functions in the normal course of insect life, however, is not clear. 5-Hydroxytryptamine (serotonin) a neurotransmitter in the nervous system causes a vigorous increase in the heartbeat rate at very low concentrations in isolated or semi-isolated hearts of some insects, but not in others (Nation, 2008).

The basal heartbeat rate of most insects is around 60 beats min-1 at room temperature and at rest (e.g., in the American cockroach, *Periplaneta americana*, and the locust, *Locusta migratoria*). In adult house flies (*Musca domestica*), however, it ranges between 300 or more beats per minute during flight to zero beats for a while when at rest. The central nervous system of the adult house fly is composed of the brain and a thoracic ganglion mass, but lacks abdominal ganglia. Because of this unusual anatomy, the dorsal vessel of the abdomen can be separated from the central nervous system input by experimentally severing the thorax from the abdomen. After this operation, the heartbeat of the fly becomes quite regular at around 60 beats min- 1. This indicates that the heart of the house fly is innervated by both inhibitory and excitatory motor neurons from the central nervous system (Miller and Pass, 2009).

5. PATH OF CIRCULATION

The pumping force that propels the haemolymph is provided by the pulsation of the heart. The respiratory movements of the Alary muscles, increase this force. From the pericardial sinus, the haemolymph enters into the heart (the dorsal vessel) through the ostia (Fig 4.2). The valve like ostia closes, thus preventing back flow of haemolymph into the pericardial sinus. The valves which project into the vessel lumen close the ostia at the systolic contraction. Haemolymph emerges from the dorsal vessel at the open anterior end, and in insects with heartbeat reversal it usually also flows out the open posterior end. In Lepidoptera, there is two-way ostia, which permits influx during forward phases and outflow during backward pass. The anterior end of the dorsal vessel opens just beneath or in front of the brain. Most of the haemolymph is poured into the head sinus through the terminal opening at the anterior aorta. This arrangement ensures a constant supply of nutrients and removal of waste products to and from the brain mass. In addition, the dorsal vessel is often intimately associated with the retrocerebral nervous system (hypocerebral ganglion, corpora cardiaca, and corpora allata complex) just behind the brain, which delivers neurohormones and possibly other hormones into the aorta at specialised release sites.

Some insects have in addition outflow openings which are paired or unpaired and located more ventrally. In cockroaches, mantids, and some orthopterans, the outflow openings are outfitted with sphincter-like valves and associated with segmental vessels laterally diverging from the heart. These vessels are formations of connective tissue with no inherent musculature, thus providing a simple channel to ensure lateral perfusion of the pericardial sinus.

From the head sinus, the haemolymph flows backwards into the thorax and abdomen. While flowing backwards from head sinus, the haemolymph remains in the ventral part due to presence of oesophagus in dorsal part and so it fills into the perineural sinus. From the perineural sinus, the haemolymph now flows into the perivisceral sinus through pores of ventral diaphragm in the abdominal region. Then from the perivisceral sinus, it flows into the pericardial sinus through the pores of dorsal diaphragm. During the heart's diastole, it fills the heart through the Ostia.

The dorsal vessel is comprised of two rows of opposing pairs of muscle cells (collectively called **myocardium**). The cells of each row are offset, so that they cause a spiral-like peristaltic wave of contraction. The myocardium in all insects is spontaneously active, usually beginning in the embryonic stages. This type of heart control is termed **myogenic**, because the electrical activity underlying contractions arise in the myocardium itself. This is in contrast to a neurogenic heart control present in crustaceans, such as crabs and lobsters, in which a barrage of nerve impulses drives the heartbeat from a discrete cardiac ganglion center. In insects, the myogenic pacemaker may be neurally or hormonally modulated.

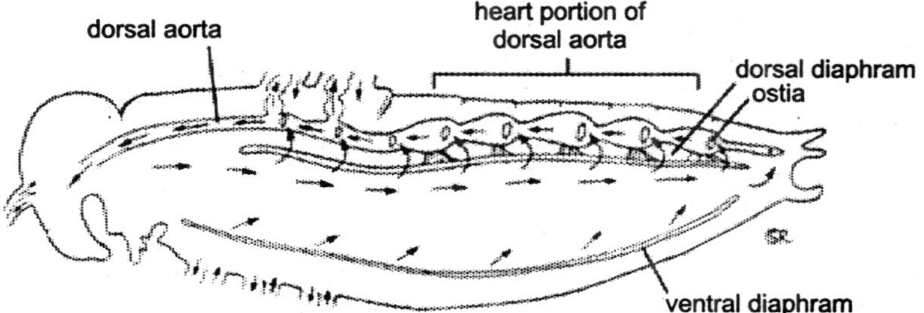

Fig 4.2. Schematic sketch of the circulation of the haemolymph

6. EXTRACARDIAC PULSATIONS (EXP)

Extracardiac pulsations of insects are the simultaneous contractions of inter-segmental muscles, usually of the abdomen, that cause a sharp increase in the pressure in the insect body. The amount of movement accompanying each pulse is too small to be seen, but it can be readily measured as a slight shortening or telescoping of the abdomen as measured from its tip. Terrestrial insects exhibit extracardiac pulsations in haemocoelic pressure, similar in some respect to the human blood pressure pulse. The pulsations are produced by large intersegmental abdominal musculature (abdominal pressure pump). The dorsal vessel of insects is a relatively weak organ which is unable to pump haemolymph against an increased gradient of pressure. The weak cardiac pulsations (myogenic nature) and strong ExP (neurogenic nature) occasionally occur hand in hand during similar periods with similar, but not identical frequencies (Slama, 2008).

The larger over movements of the abdomen, especially in bees and bumble bees, that accompany ventilation during times of high activity or exertion such as flight larger over movements of the abdomen, especially in bees and bumble bees, that accompany ventilation during times or high activity or exertion such as flight. The larger over movements of the abdomen, especially in bees and bumble bees, that accompany ventilation during times of high activity or exertion such as flight are not extracardiac pulses. Either the extracardiac pulsations occur in coordination with openings of certain of the spiracles, and therefore can play a role in ventilation, or they occur when all the spiracles are tightly closed, hence affecting haemolymph movement. The extracardiac pulsations become suspended only in quiescent stages of insect development, such as during diapause, but they can be evoked immediately upon disturbance or stimulation. The extracardiac pulsations are driven by a part of the nervous system for which Karel Slama coined the name **coelopulse nervous system**. The pressures induced by extracardiac pulsations are 100-500 times greater than pressures caused by contractions of the dorsal vessel and are transmitted by the haemolymph throughout the entire body of the insect (Miller and Pass, 2009).

The described coelopulse system controls a number of important physiological functions (Slama, 2008):

1. Expo in haemocoelic pressure causes rapid circulatory inflow and outflow of haemolymph between thoracic and abdominal parts of the body.
2. The relatively strong pressure changes caused by ExP can vigorously move tissue and organs against each other, thus preventing occlusion of haemolymph among densely packed organs.

3. Large extracardiac peaks in haemocoelic pressure open or close passively, one-way valves or tissue folds and promote circulation of haemolymph to destinations that cannot be reached by the heartbeat, i.e. ventral perineural sinus, appendages.
4. Strong ExP in haemocoelic pressure producea rhythmic, up and down compressions of tracheal tubes and air sacs, resulting in actively regulated inspirations or expirations of air through individual spiracles, i.e. actual insect breathing.
5. ExP controlled by the coelopulse neuroendocrine system causes unidirectional ventilation of the determined spiracles during emergency hypoxia, or during enzymatically produced outbursts of CO_2.
6. The coelopulse system effectively controls various homeostatic physiological functions, like respiratory water loss, water retention, isoosmosis, optimum body volume, or economic gaseous exchange.
7. ExP in haemocoelic pressure plays an important role in execution of special developmental events, like ecdysis, oviposition or pupariation.

7. AUTONOMIC NERVOUS SYSTEM FOR HAEMOLYMPH CIRCULATION

The tidal flow of haemolymph, the extracardiac pulsations, heartbeat reversal, and thermoregulation all imply a very sophisticated control of circulation by the central nervous system. The central nervous system also plays a role in regulation of the respiratory system. The activities of circulatory and respiratory systems are coordinated by the central nervous system, perhaps to an extent not fully appreciated. It would be convenient and satisfying to be able to point out a particular part of the central nervous system and related peripheral nerves in insects that might comprise this regulatory system; however, beyond evidence that the meso- and/or metathoracic ganglia play a major role in some of these functions, entomologists know of no such discrete structure or structures, possibly because these inter-regulatory functions have been undertaken by different parts of the nervous system in different insects. It is known that insects have a number of regulatory mechanisms that can be recruited to achieve such control, from motor and sensory neurons to neurosecretory neurons and neurohormonal organs located throughout the insect hemocoel (Miller and Pass, 2009).

8. HAEMOLYMPH

Haemolymph is the fluid plasma that circulates in the interior of the insect body and is in direct contact with the tissues. Several different types of nucleated cells, the hemocytes, are suspended in the haemolymph. These cells are involved in several functions that include phagocytosis, wound healing and even storage and intermediate metabolism. The haemocytes, plasma, and dissolved inorganic and organic molecules constitute the haemolymph. Nutrients, nitrogenous waste products, hormones, and a variety of other substances are also transported in the haemolymph for distribution to various tissues. The haemolymph make up 15-30% of the total body weight and 15-70% of the total body volume. The volume of the haemolymph may vary at different stages in the life cycle of the insect and according to its physiological condition. The haemolymph has very many functions, but it serves primarily as a means of transport of nutrients, waste and hormones and plays little part in respiration. It may also provide as a store of substances such as sugars and proteins, while its water acts as a reservoir for maintenance of tissue fluid. Its hydrostatic pressure aids in movement and expansion.

8.1 Functions of Haemolymph

1. Haemolymph is important as a sink for carbon dioxide (GO_2) where it gets soluble as bicarbonate ion (HCO_3). In *Hyalophora cecropia* pupae, for example, CO_2 in the gas phase built up slowly because most of the CO_2 goes into solution in the haemolymph. This allows the spiracles to remain closed for many minutes until CO_2 in the gas phase reaches a critical level and the spiracles open. Keeping the spiracles closed as much as possible prevents excessive water loss, a critical factor for a closed system pupa.
2. The circulatory system transports nutrients to cell and tissues. Haemolymph is a major storehouse of trehalose, the major insect sugar used particularly for flight energy.
3. The circulatory system delivers waste products, excess water absorbed from food, such as a blood meal or plant sap, and ingested allelochemicals or metabolites to excretory organs, or in some cases for the storage structure.
4. The circulatory system is a reservoir of fluid, nutrients, and enzymes. Trehalose, some storage proteins, and amino acids are typical nutrients present in large quantity in the haemolymph. Some large proteins (cecropins) with antibacterial and antifungal activity are induced and transported in the haemolymph after an insect is exposed to bacteria and fungi, or some of their chemical byproducts.

5. Haemolymph transports hormones from neurohemal organs to target tissues. Many neuropeptides, which have good water solubility, are transported in the haemolymph from their release points in the nervous system to target systems such as the Malpighian tubules (diuretic hormone), prothoracic glands (PTTH), pheromone glands (PBAN), and others.
6. Haemolymph is a lubricant and hydraulic support that assists in maintaining body shape and movement, especially of soft-bodied forms such as caterpillars, and expansion of the wings in newly emerged adults.
7. Pumping of haemolymph into the wings and, in some cases, secretion of plasticization factors is critical to proper expansion of the wings in newly emerged adults. The mesotergal and metatergal accessory hearts are important in directing haemolymph into the wings.
8. Haemolymph and the hemocytes provide protection from invading bacteria, eggs of parasitoids, and other foreign substances by biochemical immune-type reactions, phagocytosis and encapsulation by haemocytes, and wound healing by coagulation in some insects.
9. The circulatory system is important in some insects as a means of heat transfer to prevent excessively high temperatures in the thorax from flight muscle activity. Control of thoracic temperature is probably important to many insects in flight. Many insects do not initiate flight until the temperature in the thorax reaches some critical temperature above the ambient. They warm the thorax by wing whirring before attempting to take flight.

8.2 Composition of haemolymph

Haemolymph contains many dissolved inorganic and organic substances, colloidally suspended proteins, and lipoproteins. It is about 90% water and 10% solids. Probably most components contribute to the osmotic pressure and specific gravity of haemolymph. Specific gravity is usually slightly greater than 1. Haemolymph generally has an osmotic pressure, expressed in freezing point depression, of about 0.7 to slightly over 1°C. This is less than some Crustacea (Cancer [crab] and Homarus [lobster]), and slightly higher on average than that in humans.

Inorganic Ions

The ionic composition of haemolymph plasma is highly variable in different insects. Sodium, potassium, calcium, and magnesium are typical cations, and chloride, phosphate, amino acids, and sometimes bicarbonate are present as anions. Chloride and phosphate are the major anions in haemolymph of honeybee larvae, but these are not usually the major anions balancing the cations in haemolymph of most insects. Amino acids and organic acids account for a substantial part of the anions in some groups, particularly Neuroptera and Lepidoptera. Most insects have rather high levels of amino acids in the haemolymph, and some amino acids can contribute to the cation load.

Free Amino Acids

Insect haemolymph contains very large amounts of free amino acids, much more than the body fluids of other animals. These amino acids contribute to the osmotic value of haemolymph, and account for a substantial portion of the cations and anions of haemolymph. A good physiological explanation for such large quantities of amino acids in haemolymph is not available, but the haemolymph probably is a reservoir of amino acids for protein synthesis, much of which occurs in fat body cells.

Proteins

Haemolymph contains a wide variety of different proteins (Kanost *et al.*, 1990), which may change with physiological state, age, sex, and other factors. Some insects (larvae of some Diptera and Lepidoptera) synthesise large quantities of one or more storage proteins during late larval life. Calliphorin accounts for up to 60% of the protein in mature blowfly larvae. It is synthesised by the fat body, released into the haemolymph, and finally reabsorbed back into the fat body and stored for use during pupation and adult development. Similar proteins have been described from other insects. Haemolymph proteins include many different enzymes, such as lysozyme, antibacterial and antifungal proteins, and transport proteins that carry hydrophobic substances, such as juvenile hormone, cholesterol, diglycerides, hydrocarbons, and other lipoidal substances through the aqueous medium of haemolymph. One haemolymph enzyme, phenoloxidase (PO), plays an important role in sclerotization of the cuticle and in protecting the insect from foreign invaders (Ashida and Yamazaki, 1990: Nappi *et al.*, 1991; Soderhall and Aspan, 1993). It exists in the haemolymph and in haemocytes as a proenzyme that can be converted to active phenoloxidase. PO converts a variety of phenols to quinones. The enzyme and some of its products are important in tanning the cuticle. PO activity often causes the haemolymph to melanize and eventually become dark brown or black at wound sites.

Other Organic Constituents

Many organic compounds have been found in the haemolymph of insects. For example, triacylglycerols stored in fat body cells are transported by lipoproteins (lipophorin) to thoracic flight muscles in Lepidoptera and some other insects and uric acid is transported by haemolymph from the main site of synthesis, the fat body, to the Malpighian tubules. Possibly potassium and sodium urate account for much of the uric acid transported in haemolymph. The uric acid is transported for excretion in many insects or for storage in various sites by some insects. Trehalose is the principal haemolymph sugar in most insects and is rapidly converted to glucose for immediate metabolism in glycolysis and the Krebs cycle. Ecdysteroids, juvenile hormone, PTTH, PBAN, AKH, HTH, diuretic hormone, and a host of other neuropeptides and hormones are transported from the site where they are released into the haemolymph to their targets. The titer of these in the haemolymph varies continuously with the physiological state of the insect.

8.3 Role of haemolymph in Thermoregulation

Insects, through evolution, have developed their own means of thermoregulation. Thermoregulation refers to an insect's ability to regulate body temperature through behaviour, physiology, and morphology resulting in a temperature suitable for survival and reproduction. On the basis of body weight, most flying insects have higher rates of metabolism, and hence heat production, than other animals. However, the rapid rate of cooling because of small body size in most cases precludes appreciable endothermy. The use of the haemolymph in thermoregulation of flying insects was firstly described by Bernad Heinrich (1971). The body temperature of small flies in flight is probably close to ambient temperature, and that of flying butterflies and locusts is 5 to 10°C above ambient temperature. Many moths and bumble bees are insulated with scales and hair, and their metabolism during flight can cause the temperature of the flight muscles to increase 20°C to 30°C above ambient temperature. Curiously, those insects which (because of size, insulation) retain the most heat in the thorax during flight, also require the highest muscle temperature in order to maintain sufficient power output to continue flight. The minimum muscle temperature for flight varies widely between different species, while the maximum temperature varies over the relatively narrow range of 40°C to 45 °C.

As a consequence, those insects that necessarily generate higher muscle temperatures during flight must maintain their thoracic temperature within a relatively narrow range during flight. Active heat loss from the thorax to the abdomen prevents overheating of the flight motor and allows some large moths to be active over a wide range of ambient temperatures (Heinrich 1974). The optimum temperature for flight muscle contraction in many insects, such as the tobacco hornworm, *Manduca sexta*, is surprisingly high, up to 45°C. Before this moth can fly, it must warm the thorax to near this temperature, which it accomplishes by means of a series of simultaneous isometric contractions of the antagonistic pairs of flight muscles that appear to the casual observer as **shivering**, or vibrations of the wings. A **thermometer** in the thoracic ganglia detects the proper temperature. When the thoracic temperature is below optimum, the central nervous system signals the dorsal vessel to circulate haemolymph slowly. When the thoracic temperature rises above optimum, the central nervous system brings about maximal amplitude and rate of heartbeat to drive haemolymph through the thoracic muscles. The increased haemolymph flow pulls heat away from the flight muscles in the thorax and eventually delivers hot haemolymph to the abdomen, where the heat is dissipated. Then relatively cool haemolymph is redelivered to the thoracic muscles by the dorsal vessel, completing the thermoregulation cycle. The warm haemolymph is then delivered to the head and percolates back past the ventral ganglia in the thorax to the abdomen, where the heat is dissipated. The cooler haemolymph is then delivered again to the thorax. The dorsal vessel and the ventral diaphragm in the tobacco hornworm act together for haemolymph circulation. When the thorax is too warm, both the amplitude and the frequency of heartbeat contractions are increased, and the rate of delivery of haemolymph increases. When the thorax is too cool, amplitude and frequency of contraction of the dorsal vessel are decreased. The activity of the ventral diaphragm acts in concert with that of the dorsal vessel. Thermoregulation of the flight muscles of the tobacco hornworm implies a sophisticated nervous control. The overall nervous control can be easily demonstrated by severing the ventral nerve cord between the thorax and abdomen. Then the moth can no longer thermoregulate because the feedback loop of temperature detection by the thoracic ganglia has been destroyed, and control over ventral diaphragm and dorsal vessel contractions have been lost (Miller and Pass, 2009). Thermoregulation is a key factor in the energetics of foraging of some of the flower-visiting insects. The higher their muscle temperature the more flowers they can visit per unit time. When food supplies are ample, bees may invest relatively large amounts of energy for thermoregulation. While shivering to maintain high body temperatures during the short intervals they are perched on flowers (as well as while in the nest), bumblebees often expend energy at rates similar to the rates of energy expenditure in flight.

9. HAEMOCYTES

Haemocytes are free moving blood cells that circulate in the plasma, within the hemoceleof insects. Hemocytes originate from mesodermally derived stem cells that differentiate into specific lineages. The most common types of haemocytes are granulocytes, plasmatocytes, spherulocytes, and oenocytoids (Lavine and Strand, 2002). The number and types of haemocytes vary with insect species, developmental stage, and physiological state. In general, total haemocyte count (the number of haemocytes per unit volume) increases throughout larval development, reaching a maximum at each larval ecydsis but probably not during pupal/adult ecdysis. Cellular defences require haemocyte-mediated responses like phagocytosis and encapsulation. Gillespie *et al.*, (1997) reviewed haemocyte functions, with particular emphasis on their role in immunity. Haemocytes participate in wound healing by aggregating at the wound site, where some cells phagocytize cellular debris or foreign organisms, such as bacteria. Entrapment of haemocytes in the coagulum helps plug the wound. Some of the trapped cells may also play an active role in defence at the wound site. Some haemocytes contain enzymes that aid in detoxification, including detoxification of some insecticides. Haemocytes participate in nodule formation and encapsulation of foreign objects.

Invasions of bacteria may be attacked by nodule formation in which the bacteria are aggregated into nodules containing haemocytes and coagulum. Larger objects or invading parasites may be encapsulated. Encapsulated objects often become attached to various tissues or organs in the body. The ability to encapsulate parasitoids seems to be dependent on an evolutionary race between the parasitoid and its host. Sometimes encapsulated objects also become melanized by the action of the phenoloxidase cascade of enzymatic action on tyrosine or other phenolic compounds. Often, but not always, the whole mass becomes melanized, a process that also may be toxic to the offending organism and assist in killing it (Nation, 2008). The number of haemocytes in the circulation of insects is quite variable from species to species, and even within the same individual at different times depending on its physiological state. Sex, age, stage of development, and activity are known to influence the observed number of cells in some insects. Also, some haemocytes are sessile, attached to various tissues, at least for much of the time, but may be released into the circulation under certain circumstances. The absolute number and kind of haemocytes and, possibly, the temporal sequence of their appearance and increase in concentration may be important to the ability of an insect to defend itself from foreign invaders. Haemocytes also have many important roles during embryonic development and are crucial for successful embryogenesis to occur. During embryogenesis (eg: *Drosophila*), a proportion of the mesodermal cells originating from the head region migrate along specific pathways and subsequently disperses throughout the body (Tepass *et al.*, 1994). These embryonic haemocytes (EH) either differentiate into small spherical cells with phagocytic capacities, so-called **plasmatocytes**, or into crystal cells that are involved in the melanization of pathogens (Alfonso and Jones, 2002).

9.1 Origin and types of haemocytes

Origin of haemocytes

In most insect larvae, the haemocytes are produced within haematopoietic organs where they replicate and differentiate. *Drosophila* blood consists of several types of haemocytes, which originate from the mesoderm. By lineage analysis of transplanted cells, it was shown that two separate anlagen gives rise to different populations of haemocytes: embryonic haemocytes and lymph gland haemocytes. The Anlage of the embryonic haemocytes is restricted to a region within the head mesoderm between 70 and 80% egg length and a second analog, which gives rise to additional haemocytes at the onset of metamorphosis, is located within the thoracic mesoderm at 50 to 53% egg length. The cells of the anlage of the embryonic haemocytes are already determined at the blastoderm stage whereas the anlage of the lymph gland-derived haemocytes is not determined at the blastoderm stage. The haemocytes of pupae and adult flies consist of a mixture of embryonic haemocytes and lymph gland-derived haemocytes, originating from two distinct anlagen that are determined at different stages of development (Holz *et al.*, 2003). In *Gryllus* the haemopoietic organs are represented by organized tissues connected to each side of the dorsal vessel in the second and third abdominal segments. The organs are delimited by connective tissue and fibroblast-like cells and the cortical region consist of a meshwork of reticular cells. Some of the dividing reticular cells become thehemocyitoblast which Plasmatocytes, Granulocytes or Coagulocytes. In *Periplaneta, Schistocerca* and *Locusta*, the haemopoetic organs are found as irregular accumulations of cells along the dorsal vessel.

Types of haemocytes

Themostcommontypesofhaemocytesfoundininsectsareprohaemocytes, plasmatocytes, granulocytes, spherulocytes, adipohaemocytes, oenocytoids, thrombocytoids and coagulocytes (Gupta, 1979a, 1979b). However, majority of insects does not possess all types of haemocytes. Prohaemocytes are small circular cells with a large nucleus and a basophilic cytoplasm that can differentiate into a number of cell types. Plasmatocytes are the most common haemocyte, they are leaf like in shape and their cytoplasm contains lysosomal enzymes. They are usually larger than granular cells and are involved in capsule formation. Granular cells have a small nucleus and many granules in the cytoplasm. Spherulocytes display a variety of differing shapes, with numerous small spherical inclusions. Oenocytoids are large, bi-nucleate, non-phagocytic cells. Coagulocytes participate in the clotting process(Browne *et al.*, 2013). Sessile haemocytes are found amongst lobes of fat body or as accumulations within the hemocoel or transiently adherent to hemocoelic surfaces. Most sessile haemocytes are pericardial cells alongside the dorsal vessel and hematopoietic organs.

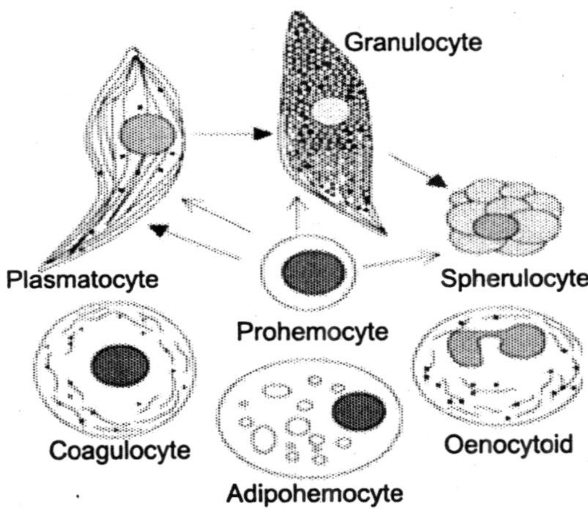

Fig 4.3 A possible scheme for the types and relationships among insect haemocytes
[http://igtrcn.org/hemocyte-specific-promoters/]

The main classes of circulating haemocytes are:

1. **Prohaemocytes (PR):** Small, round cells, more common in larvae than adults. They are considered to be stem cells from which certain other types develop.
2. **Plasmatocytes (PL):** Agranular (Lepidoptera) or few too many granules (Orthoptera, Dictyoptera), round or spindle shaped in suspension, but adhesive, flattened and often motile over substrata. Within this class can also be included Lamellocytes, which are large flattened cells into which the plasmocytes transforms.
3. **Granular cells (GR):** Cells that are adherent by fine filopodia but are rounded and phase-bright in vitro
4. **Coagulocytes (CO):** Granular cells that lyse to produce a gel or coagulum and become phase-bright and non-adherent, with the nucleus surrounded by thin rim of vacuolated cytoplasm.

Other types of circulating haemocytes include:

1. **Crystal cells:** Found in larval Drosophila and contain a prominent crystalline array that is rich in prophenoloxidase known to play a role in melanisation
2. **Spherule cells (SP):** Containing large, spherical phase-bright inclusions that stain for acid mucopolysaccharides
3. **Oenocytoids (OE):** Very large cells, fairly uncommon and apparently contain prophenoloxidase.
4. Thrombocytoids of Calliphora are important in wound healing and encapsulation.

9.2 Functions of Haemocytes

1. Haemocytes play a crucial role in both development and immunity.
2. Haemocytes also have many important roles during embryonic development and are crucial for successful embryogenesis to occur. In the embryo, the haemocytes phagocytose cells secrete extracellular matrix and other proteins that promote organogenesis.
3. Haemocytes are involved in the clotting reaction, phagocytosis, nodule formation andencapsulation of foreign bodies. Some haemocytes are phagocytic and.function to eliminate microbial pathogens and tissue debris generated during embryonic development, metamorphosis and by wounding, while others protect against large foreign bodies by encapsulation.
4. Some haemocytes contain enzymes that aid in detoxification, including detoxification of some insecticides.
5. Haemocytes consist of adhesive reactions against microbes or parasites microorganism, which directly attack the foreign object and are also important in the elimination of larval tissues that have undergone programmed cell death at metamorphosis.
6. Among these various haemocytes, the granulocytes and plasmatocytes are thought to be the most important haemocytes involved in insect cellular defence.
7. They play a central role in wound healing and protection against infection in the embryo and throughout the life of the insect.

10. INSECT IMMUNITY

Millions of insect species live in practically every known habitat and ecological niche, though marine environments are an important exception. This diversity exposes insects to all sorts of infectious agents, such as viruses, bacteria, fungi and protozoa. To survive in a world full of microorganisms and parasites, insects developed a potent defence mechanism that recognises and removes microbial threats. Insects for survival have a well-developed innate immune system that allows general and rapid responses to infectious agents while they lack an acquired immune system.The innate immune system of insects consists of physical barriers, humoral responses and cellular responses (Lavine and Strand, 2002; Kanost *et al.*, 2004).The protection against pathogens begins primarily with certain barriers such as cuticle, gut and trachea, tissues that are difficult to be penetrated, while immune response is originated by the fat body and the haemocytes. Insects combat invading microbial organisms by several innate mechanisms, including, (1) phagocytosis of small objects and encapsulation of larger objects with layers of haemocytes, (2) localised coagulation of haemolymph at wound sites, (3) melanisation reactions at wound sites and usually at encapsulated objects, and (4) synthesis of antimicrobial peptides. Insects lack the complement system of acquired immunity with memory that occurs in vertebrates.This innate immunity can be divided into two classes, the cellular response and the systemic (humoral) response (Box 4.4). This classification is artificial since many of the humoral factors are produced by the cellular components.

Box 4.4 Defence mechanism in insects

1. Hemocytic (cellular) defence	2. Humoral (systemic) defence
a. Phagocytosis	a. Antibacterial proteins
b. Encapsulation	b. Serum lectins
c. Melanization	
d. Nodulation	

10.1 The cellular (hemocytic) immune response

Cellular reactions are initiated immediately upon invasion of a foreign microorganism and involve direct attack of the foreign object by haemocytes in the circulating haemolymph. Cellular immune responses are immediate after an invasion of the hemocoel, while humoral responses appear several hours after an infection. Haemocytes are responsible for a variety of defence responses in insects. Cellular events are initiated by the cells that encounter the invading organisms, usually epithelial cells beneath the cuticle, haemocytes in the haemolymph, fat body cells, and epithelial cells lining the gut. The cellular immune reactions involve mainly thehaemocytes and consist of adhesive reactions of haemocytes against microbes or parasites. Among the various haemocytes, the granulocytes and plasmatocytes are thought to be the most important haemocytes involved in insect cellular defence.In general three cell types of the haemolymph are involved, namely: (1) Plasmatocytes that accounts for 90-95% of the homicides, (2) Crystal cells (storage of antimicrobial particles) accounting for 5% of cells and (3) Lamellocytes (can handle larger microorganisms than plasmatocytes) but are few in number.

The crystal cells contain particles that are lethal for microorganisms and are released upon infection. The plasmatocytes and lamellocytes are cells that eat microorganisms in a process called **phagocytosis**. The plasmatocyte surrounds for example a bacterium and consumes it. The bacterium will be on the inside of the plasmatocyte and will be broken down there. These cells rapidly respond to the invasion by secreting pattern recognition proteins that have a variety of functions, including eliciting synthesis of antimicrobial peptides. Haemocyte proliferation occurs making increased numbers available to attack the invaders by phagocytosis, encapsulation, melanisation and nodule formation.

Phagocytosis

Phagocytosis is the process by which cells recognise, bind and ingest relatively large particles (usually larger than 0.5 μm in diameter). Phagocytosis is probably the oldest defence mechanism against microorganisms and a fundamental cellular process performed by unicellular organisms and by particular cell types in multicellular organisms. In insects, phagocytosis is performed by a subset of haemocytes in the haemolymph (Strand, 2008). The main steps for phagocytosis are recognition, endocytosis, ingestion and killing (Gupta, 1999). These processes have been studied in different insect species. Generally, granulocytes and plasmatocytes are the major haemocytes involved in phagocytosis. The surface of the haemocytes contains binding sites that recognise cell wall components of foreign objects. Thus, when the foreign object encounters the haemocytes, the surface components of its cell wall will be recognized by the haemocytes. The bound objects are then surrounded by pseudopods and ingested by coated vesicles. Finally the engulfed particles are lysed by lysozyme and antibacterial peptides (Götz, 1985). Many of the initial interactions leading to coagulation and phagocytosis reactions in insects are sugar-lectin binding reactions, involving homeostasis- and defence-related receptors, such as scavenger receptors (Franc *et al.*, 1999) and mucin-like immune receptors (Theopold *et al.*, 1996). A range of sugar-lectin interactions have been documented to be involved in phagocytosis events (Wilson *et al.*, 1999).

Encapsulation

Encapsulation is the response of haemocytes to large targets such as parasites, protozoa, and nematodes. Haemocytes bind to the target in multiple cell layers until they form a capsule around the invader. The capsule is normally melanised at the end. Inside the capsule the invading organism is killed by reactive cytotoxic products or by asphyxia (Rosales, 2011).

Melanisation

Melanisation involves the action of phenoloxidase (PO) on phenolic compounds (tyrosine and dihydroxyphenylalanine) to produce quinines that autopolymerize and produce melanin (Chase *et al.*, 2000). Activation of PO is one of the principal defences of insects against bacterial and fungal invasion (Cerenius and Soderhall, 2004). The thick layer of melanin and haemocytes formed around encapsulated organisms may help suffocate as well as be toxic to them. A zymogen, prophenoloxidase, is stored in cells, but mainly circulates in the open haemolymph system of insects. It is converted to active phenoloxidase by a series of serine proteases. Inhibitors help regulate the active enzyme and serve to restrict its activity to the invasion site. Thus, excessive systemic damage is avoided by highly toxic and reactive compounds that result from the PO action on a variety of substrate molecules (phenolic compounds) (Nation, 2008).

Nodulation

Nodulation, a predominant cellular response in insects to large bacterial infections is a formation of multicellular hemocyte aggregates that entrap large numbers of bacteria. It begins when haemocytes, together with antimicrobial molecules in the haemolymph, surround bacteria. Haemocytes with their entrapped bacteria form small aggregates that grow by joining with additional haemocytes to form large nodules. The nodule is completed by covering it with layers of flattened haemocytes. In some cases, but not always, nodules are also melanised. These melanin-covered nodules are an effective way to isolate the invading bacteria from the haemolymph, since they have an impermeable wall between the nodule and the rest of the insect organism (Rosales, 2011).

10.2 The systemic immune response (Humoral response)

Next to the cellular immune response, insects synthesise massive amounts of antimicrobial peptides (AMPs) in response to infection which is called the **systemic immune response**. These AMPs are produced in the fat body in insects andsecreted into the haemolymph where the AMPs fight the infection. But to make these AMPs, first the infection needs to be recognised; the recognition of infection induces a chain reaction which will eventually lead to the production of AMPs. First the proteins in the haemolymph recognise microorganisms. When they recognise a microbe, these will activate proteins (receptors) that are in the cell membrane of fat/ liver cells. This protein in turn will initiate a cascade of several different proteins which will eventually send a transcription factor (a protein that induces the production of other proteins through DNA and RNA) into the cell nucleus and initiate the production of RNA. This RNA will be secreted into the cytoplasm and here it will be translated into protein. These proteins (AMPs) are secreted into the haemolymph (or other tissue depending where the infection is) and attack and kill the microorganisms. A detailed account on the synthesis of antifungal and antibacterial peptides, serum lectins and serpins is provided elsewhere in this chapter.

10.3 The local immune response

Insects can also synthesise AMPs in epithelia. Epithelia line the outside of the insect (skin/integument) and the intestine, so these are exposed first to microorganisms. Although there are differences in the systemic immune response the general idea is the same. After recognition of microorganisms by extracellular (outside the cells) proteins the chain reaction will be initiated and induce the production of AMPs. These will fight the infection (Lemaitre and Hoffmann, 2007).

10.4 Physical Barriers to Invasion

The primary physical defence is the chitinous cuticular lining, which cover most of the external surface of an insect and extends through the foregut, the hindgut, the tracheal tubes and the most of reproductive tissues. The external cuticle is virtually impenetrable to microorganism. The thickness of the external cuticle of insects is the main natural physical barrier to many microorganisms and fungi (Siva-Jothy *et al.*, 2005), but in many insects, there are numerous thin areas of the cuticle, particularly in inter-segmental boundaries between segments in the abdomen and in the lining of the tracheal system. Insect ingests many microorganisms with their food and, for some pathogens, the oral route is the main entry point into the insect. The foregut and hindgut have an attached cuticular lining on the lumen surface of the gut epithelial cells that protects the epithelial cells, but the midgut is more vulnerable. The midgut may have an unattached protective layer, the peritrophic matrix (Lehane, 1997) composed of chitin along with glucosaminoglycans, glycoproteins and a variety of other proteins that protects the delicate brush border on midgut cells from harsh food particles as well as ingested microorganisms. Lining the midgut, protecting the epithelial cells from damage and serving as a barrier against infection is the peritrophic matrix or membrane (PM). Some insects, however, do not have a peritrophic matrix and yet their line has survived for hundreds of millions of years. The peritrophic matrix has holes or pores through which digestive enzymes secreted by the midgut cells pass into the food bolus and through which small digested molecules pass for absorption by midgut cells. Invading microorganism finds these pores a potential site of entry to the midgut cells.

In the *tse-tse* flies (*Glosina* sp., transmitters of the parasite for sleeping sickness to humans), the pores are about 9 nm in diameter and probably are too small to allow invasion of theparasite through the pores. Some insects (notably some mosquitoes that are disease transmitters) secrete the peritrophic matrix only after food is ingested, and orally ingested microorganisms might contact the surface of midgut epithelial cells before the matrix is completely formed (Lehane, 2005). The mature ookinete stage of the malaria parasite produces the enzyme chitinase, which may aid in penetrating the midgut peritrophic matrix (Shahabuddin, 1998; Sieber *et al.*, 1991). Penetration, however, may depend on how complete the secretion of the type I peritrophic matrix characteristic of the mosquito adults is, and whether the blood meal of the mosquito female contains immature malarial gametocytes or mature ookinetes, which are the infective stage (Janse *et al.*, 1985; Lehane, 2005).

II. RECOGNITION OF NON-SELF BY INSECT IMMUNE SYSTEM

The basis of any immune response is the ability to discriminate as non-self, foreign invaders of the body. Such discrimination may rely solely on the physicochemical properties of the foreign surface or else recognition and receptor molecules may be present in the plasma and/or the surfaces of the haemocytes. An essential component of immunity is a mechanism for surveillance, by which an organism can detect foreignness or the presence of non-self molecules (or perhaps the absence of "self" molecules). The recognition of non-self stimulates defensive responses. The first step in mounting a humoral defence is the recognition of non-self by fat body cells, haemocytes, midgut epithelium, and cuticular epithelium. Insects may use the basement membrane that lies on the basal surface of all insect cells (except haemocytes) as an indicator of self against which they direct non-self reactivity (Siva-Jothy *et al.*, 2005). When stimulated by invading microorganisms, epithelial cells in various parts of the body secrete pattern recognition proteins (PRPs) which are also called **pattern recognition receptors** (PRRs) (Werner *et al.*, 2000). Some of the PRPs are released into the circulating haemolymph while others are attached to the surface of the cells that produce them.

The PRPs recognise and bind to particular carbohydrate or carbohydrate-peptide linkages in the structure of invading fungi orbacteria by acting as receptors for characteristic bacterial or fungal wall components. The bacterial and fungal structures recognized by PRPs are called **pathogen-associated molecular patterns** or **PAMPs**. Identified PAMPs include β-1,3-glucans as a part of the fungal cell wall, and lipopolysaccharides (LPS) and peptidoglycans (PGNs) as part of the cell surface of bacteria. Binding of the PRPs to invading microbial cells marks them for destruction. Lipopolysaccharide (LPS) is a bacterial surface component that can stimulate synthesis of haemolymph proteins associated with antibacterial responses and lectins are possible recognition molecules because of their ability to bind to oligosaccharide structures present on cell surfaces. Also, many lectins have multiple binding sites, resulting in lectin-linked aggregates of target cells. Such cell agglutination activity, inhibited by various carbohydrates, has been detected in haemolymph of many insect species. β-1,3-Glucans are components of fungal cell walls that can serve as a pattern recognition signal. The open circulatory system of insects makes PRRs especially well suited to communicate rapidly the presence of non-self because the PPR-marked microorganisms are conveyed directly to haemocytes which may encapsulate or phagocytise them, and to fat body cells which are stimulated to synthesize antimicrobial peptides (Ferrandan*et al*., 2004). *Drosophila* has 13 genes coding for peptidoglycan recognition proteins (PGRPs) (Royet, 2004).

Some of the PGRPs are membrane-bound molecules, while others circulate. The presence of a constitutively low level of some PRRs may serve to facilitate an immediate response to foreign invasion, and allow time for the induction of additional PRPs to assist in combating the infection. β-GRP isolated from *Plodia interpunctella* had the broad binding capability and multiple recognition capability, binding to β-1,3-glucan, lipopolysaccharide, and the lipoeichoic acid of fungi and bacteria, and it agglutinates yeast and bacteria (Fabrick*et al*., 2003). PRPs or PGRPs additionally have a critical role in initiating the cascade of reactions leading to synthesis of antifungal and antibacterial peptides (Nation, 2008). Haemolin, a haemolymph protein (composed of repeated immunoglobulin domains), have a role in immune recognition and in modulation of defensive responses in *Halophora cecropia and Manduca sexta* (Kanost and Zhao 1996). Hemolin inhibits the aggregation of haemocytes there by regulates the adhesive properties of haemocytes during defensive responses.

11.1 Pattern recognition receptors

Insects have an array of pattern-recognition receptors that bind to molecules (lipopolysaccharides, peptidoglycans, and glucans) associated with microbes and trigger signaling cascades to activate immune cells or the transcription of AMPs to isolate or kill invaders. There are two main families of pattern-recognition receptors, the peptidoglycan-recognition proteins (PGRPs) and the Gram-negative binding proteins (GNPBs), which have homology to enzymes (amidase or glucanase).

PGRPs were initially identified as extracellular sensors of bacterial infection, but recent results show that at least one PGRP function intracellularly. There are two isoforms of PGRP-LE (Kaneko *et al.*, 2006). One is a short version that functions extracellularly as a coreceptor of PGRP-LC in the recognition of diaminopimelic acid-type peptidoglycan, whereas the longer version functions intracellularly and recognises trachealcytotoxin (TCT), a small peptidoglycan fragment released by bacteria. It is thought that the long version of PGRP-LE might defend against intracellular bacteria

Another subgroup of PGRPs, called **catalytic PGRPs**, has amidase activity that removes peptides from the glycan chains and thereby reduces peptidoglycan biological activity. Research shows that catalytic PGRPs negatively regulate the immune response. Two catalytic molecules, such as PGRP-SC1 and PGRP-SC2, have been described that degrade peptidoglycan (Bischoff *et al.*, 2006). Flies that lack these proteins have an overactive Imd pathway after infection with *Escherichia coli*, causing developmental defects and larval death. A similar immune-regulatory role has been attributed to another catalytic family member, PGRP-LB (Zaidman-Remy *et al.*, 2006). Importantly, both PGRP-SC and PGRP-LB are primarily expressed in the gut, and their main function might be to prevent innocuous peptidoglycan in the diet from initiating an immune response.

Peptidoglycans are long molecules that sometimes need to be processed to be recognised. The detection of Gram-positive bacteria through their Lys-type peptidoglycan leads to the activation of the Toll pathway and requires the pattern-recognition receptors PGRP-SA and GNBP1. GNBP1 is responsible for hydrolysing the Gram-positive peptidoglycan, and PGRP-SA binds to the peptidoglycan fragments, leading to activation of Toll.GNBPs contain a glucanase-like domain and are important in detecting fungal infections. GNBP3 detects fungal β-1,3-glucan and leads to the activation of the Toll pathway and the production of antifungal peptides.

11.2 Signalling Pathways

Several pathways have been postulated in the immune response of insects, of which the Toll and Imd pathways are important. Both the Toll and the Imd pathways result in the transcription of AMPs (Fig 4.4). Some AMPs are specific to one pathway, and others are activated by both, but little is known about how this specificity is translated into a gene-expression profile. Both the Toll and the Imd pathways culminate in the activation of NF-κB family transcription factors—the Toll pathway activates Dorsal and Dif, and the Imd pathway activates Relish.

Although the Toll and Imd pathways are separate, knocking out both pathways can have a greater phenotypic effect than knocking out either Toll or Imd alone. It is believed that a crosstalk occurs at the level of Relish, Dif, and Dorsal and their interaction with the promoters of immune genes. Cooperation between these NF-κB factors (including the formation of heterodimers) results in the synergy between the Toll and Imd pathways.

Several important immune mechanisms, such as Toll and Imd-mediated defence and phenol oxidase (PO) activation, involve proteolytic cascades. These are mediated by serine proteases and controlled by the serine protease inhibitors (serpins). Studies of serine proteases and serpins are made difficult by the large number of serine proteases encoded in the genome (more than 200 in *D. melanogaster*), but genome-sequence information and RNAi have recently boosted this field. In the *Anopheles gambiae,* knocking down the genes SRPN2 and SRPN6, compromises the ability of mosquitoes to clear plasmodium parasites through melanisation or lysis (Abraham *et al.*, 2005, Michel *et al.*, 2005). In *D. melanogaster*, Serpin27A regulates the processing of pro-PO in the melanization response. Two proteases, MP1 and MP2, have been identified to be involved in infection specific roles-MP1 activates melanisation in response to bacteria and fungi, but MP2 is involved in an antifungal response.

11.3 Antiviral Defence

Little is known about antiviral responses; indeed, it is not yet clear whether there is a dedicated antiviral pathway in insects. Microarray analysis of DCV infection revealed genes controlled by a Jak kinase and STAT transcription factor (Dostert *et al.*, 2005). The Jak-STAT pathway has a role in interferon signaling in mammals and could represent an ancient conserved mechanism for dealing with viral infections. However, the stimuli that trigger the Jak-STAT pathway and the nature of induced antiviral molecules remain to be determined.

A role for RNAi in antiviral defense in animals was first described in 2002, when *D. melanogaster* S2 cells were infected with flock house virus (FHV). For FHV to infect and kill adult flies, it must express a protein called B2, which suppresses RNAi. Conversely, flies with a loss-of-function mutation in the gene encoding Dicer-2 (*Dcr-2*), which is an essential component of the RNAi system, are more susceptible to infection by members of three families of RNA viruses: FHV, DCV, and Sindbis virus (Galiana-Arnoux *et al.*, 2006).

12. SYNTHESIS OF ANTIFUNGAL AND ANTIBACTERIAL PEPTIDES

Antimicrobial peptides (AMPs) are an essential component of innate immunity, which can rapidly respond to diverse microbial pathogens. Antimicrobial peptides, also called **host defence peptides** (HDPs) are found among all classes of life. Humoral immune responses mainlyinvolve the rapid synthesis and release of different types of antimicrobial peptides (AMPs) in the insect haemolymph upon microbial infection (Bulet *et al.*, 1999). Collectively these AMPs are known to mount effective immune defence against invading pathogens. Bacteria injected into the hemocoel of an insect elicit the synthesis of a numberof antibacterial peptides and proteins, which are secreted into the haemolymph.The fat body is the tissue, primarily responsible for synthesis of these proteins. AMPs are part of the armament that insects have developed to fight off pathogens, which are low molecular weight heat stable, typically cationic and often made of less than 100 amino acid residues (Wang and Lai, 2010). The fat body of insects is mainly responsible for the production of AMPs (Breyand Hultmark, 1998; Hoffmann *et al.*, 1999; Ganz, 2003; Schmid-Hempel, 2005).

Although their structures are diverse, most of the AMPs can be assigned to a limited number of families. The most common structures are represented by peptides assuming an alpha-helical conformation inorganic solutions or disulfide-stabilised beta-sheets with or without the presence of alpha-helical domains (Bulet and Stocklin, 2005). Synthesis of the antimicrobial peptides involves activation of genes in the Toll, IMD (Immune deficiency pathway), and JAK/STATpathways in fat body cells.

12.1 Toll Pathway for Synthesis of Antimicrobial Peptides

Gram (+) bacteria and fungi each activate serine-based proteolytic cascades that converge on and activate the Toll pathway. Gram (+) bacteria induce fat body cells to synthesize the antimicrobial peptide drosomycin (possibly drosomysins) and fungi elicit the secretion of metchnikowin. Both antimicrobial compounds are synthesised after activation of the Toll pathway. The protein Toll and the Toll pathway are ancient parts of a conserved innate immunesystem that extends from lower invertebrates to humans, understandably with numerous variations and differences in detail along the evolutionary path. For example, the Toll protein is a trans-membrane protein in fat body cells of insects, but in vertebrates it is a cytoplasmic component, not the trans-membrane receptor, in the pathway. *Drosophila* produces a number of peptidoglycan recognizing proteins (PRPs), but only three, namely, PGRP-SA, PGRP-LC, and PGRP-LE have been characterised functionally. PGRP-SA recognises the peptidoglycan structure of Gram (+) bacteria and PGRP-LC and PGRP-LE recognise the peptidoglycan structure of Gram (–) bacteria. Gram (+) bacteria invasion of *Drosophila* results in the release of PGRP-SA and Osiris, another recognition molecule. The peptidoglycan recognition protein that binds an invading fungus in a *Drosophila* larva may be GNBP3. Identified genes that function in the Toll pathway, include Späetzle, Toll, tube, pelle, and Cactus operating in the antifungal response of *Drosophila*. The genes Späetzle, Toll, and Cactus are expressed in the adult fat body and are upregulated when *Drosophila* is immune-challenged (Lemaitre *et al*., 1996).

The Toll pathway is activated mainly by fungi and bacteria. In this pathway, the Toll receptor is activated upon binding a cleaved form of Spätzle. Proteolytic cascades is initiated by secreted recognition molecules (PGRP-SA and Gram-negative bacteria-binding protein 1 (GNBP1) for Gram-positive bacteria, GNBP3 for β-glucans) or by direct cleavage of serine proteases (by fungal virulence factors) converge on Spätzle processing enzyme (SPE), which cleaves Spätzle. Spätzle binding induces Toll dimerisation and subsequent recruitment of the adaptor proteins MyD88, and Tube and the kinase Pelle, leading to phosphorylation and proteasomal degradation of Cactus. Cactus degradation allows the Rel transcription factors Dif and Dorsal to translocate to the nucleus where they bind NF-κB-response elements and activate transcription of genes including Drosomycin (Figs 4.4, 4.5).

The Toll pathway is conserved in mosquitoes and also plays a key role in antiviral defense in these insects. Dengue virus (DENV) infection of the *Ae. aegypti* midgut, carcass, and salivary gland activates the transcription of Toll pathway components and putative effectors such as Späetzle, Toll, Rel1A, and multiple AMPs (Xi *et al.,* 2008; Sim *et al.,* 2012). The DENV-infected mosquito transcriptome and that of Cactus-silenced (or Rel1-activated) mosquitoes also overlap considerably in terms of the magnitude and direction of gene regulation. Functional assays provide further evidence for the role of the Toll pathway in DENV control: Transient activation of Rel1 through RNAi-mediated gene silencing of Cactus significantly reduces midgut DENV titers, whereas silencing of the adaptor protein MyD88 leads to significantly increased DENV titers. The Toll pathway controls DENV replication in the midgut as early as 3 days post-infection, and the Toll pathway-regulated defense is active against multiple DENV serotypes as well as in field-derived mosquitoes (Sim *et al.,* 2013).

12.2 IMD Pathway for Synthesis of Antimicrobial Peptides

Gram (–) bacteria and some Gram (+) bacilli induce the synthesis of the antibacterial peptides, diptericin, drosocin, cecropins, and attacins. These peptides are synthesised by fat body cells after genes in the nucleus have been turned on by signalling through the IMD pathway. IMD stands for immune deficiency, so named because earlier discovery (Lemaitre *et al.*, 1995) of a mutation in the gene IMD reduced the resistance of *Drosophila* to Gram (–) bacterial infection. Gram (–) bacteria likely induce pattern recognition receptors, but they have not been identified. A serine proteinase cascade activates the trans-membrane IMD receptor. PGRP-LE seems to be involved near or at the fat body cell membrane, possibly acting in conjunction with the IMD receptor.PGRP-LC and PGRP-LE recognise peptidoglycan (PGN) with diaminopimelic acid (DAP) in the short cross-linking peptide bridges in the peptidoglycan molecule. Binding of DAP–PGN activates the IMD pathway in the cell cytoplasm. The intracellular portion of the IMD pathway involves an additional cascade of intracellular proteolytic kinase reactions that lead to the movement of Relish (Rel) into the nucleus. Rel is a transcription factor, and Diptericin is transcribed and translated in the cytoplasm with the synthesis of diptericin and related antibacterial peptides.These peptides are exported into the circulating haemolymph to attack the invading bacteria. The IMD pathway controls the transcription of several hundred genes in *Drosophila*, but only a few are known to function in immune defence.The IMD pathway has a branch in the cytoplasm downstream of the IMD membrane receptor named the JNK path (Delaney *et al.*, 2006). JNK is a c-Jun N-terminal kinase cascade that leads to activation of cytoskeletal genes important in tissue repair. Thus, the upstream activation of IMD receptor may allow linking of antimicrobial defence with tissue repair due to the invasion or damage from invasion (Nation, 2008).

Imd is recruited by PGRP-LC upon direct binding to monomeric (LCx/LCa heterodimer) or polymeric (LCx heterodimer) diaminopimelic acid (DAP)-type peptidoglycan. Imd then recruits dFADD and the caspase Dredd, which might be responsible for the cleavage of phosphorylated Relish. This phosphorylation is thought to be mediated by an inhibitory κB (IκB) kinase complex (IRD5 and Kenny), itself activated by TAK1. TAK1 activation of IκB kinase is dependent on its adaptor TAB2 and Imd and possibly dFADD and DIAP2. The precise nature of these relationships is unclear. Phosphorylation and cleavage of its ANKyrin repeats allows the Rel domain of Relish to translocate to the nucleus where it binds different NF-κB-response elements and activates transcription of genes including diptericin.

In *Drosophila,* activation of the IMD pathway, like that of the Toll pathway, is initiated by PRR-mediated recognition of microbial pathogen-associated molecular patterns (PAMPs). Intracellular signaling through the adaptor

IMD protein and various caspase-like proteins and kinases then leads to a functional split in the pathway into two downstream branches. One branch, similar to the mammalian c-Jun/JNK pathway, activates the transcription factor AP-1 via JNK signaling (Fig 4.5), while the other branch culminates in the processing and activation of the NF-kB transcription factor Relish (Rel2 in mosquitoes) via caspase-mediated cleavage of its carboxy-terminal end. Activated Relish is then translocated to the nucleus to promote the transcription of anti-microbial effectors (Fig 4.5). The human Fas-associated factor 1 ortholog Caspar negatively regulates Relish activation, possibly by interfering with the enzymes involved in its cleavage. In mosquitoes, the IMD pathway also plays important roles in the antibacterial defense, and it also directs immune responses against *Plasmodium* parasites.

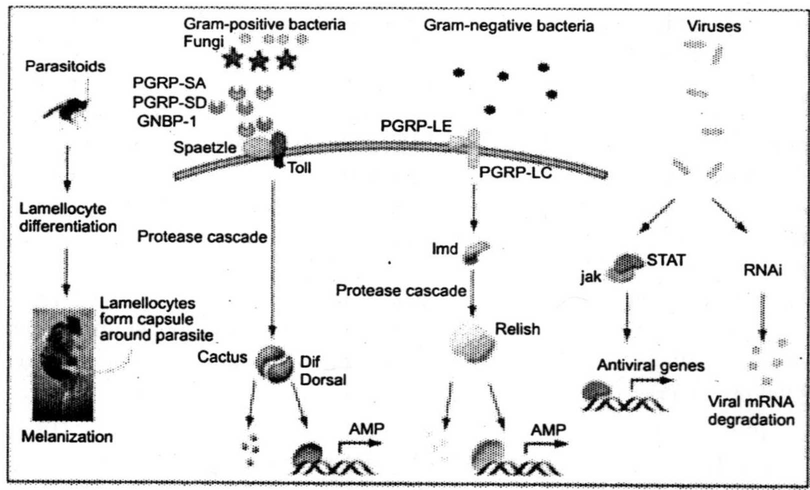

Fig 4.4 Simplified description of four of the immune responses of *Drosophila melanogaster* (Bangham *et al.*, 2006)

The antiviral role of the IMD pathway has more recently been investigated, and in flies it has been found to be active against SINV and cricket paralysis virus (CrPV) (Costa *et al.*, 2009; Avadhanula *et al.*, 2009). In mosquitoes, up-regulation of IMD components and effectors in response to DENV and SINV infection has been observed, but transient activation of the pathway by RNAi-mediated gene silencing of Caspar has no effect on mosquito midgut DENV titers. A more recent study, however, has found thatcompromising the pathway in DENV-refractory strains of *Ae. aegypti via* IMD silencing results in increased midgut DENV titers (Sim *et al.*, 2013), suggesting that the IMD pathway may be required for anti-

DENV defense, but that in susceptible strains it may already be operating at maximum capacity. The situation may be different during Semliki Forest Virus (SFV, family *Togaviridae)* infection of *Ae. albopictus* cells, however, where the addition of heat-inactivated Gram-negative bacteria (which stimulates the IMD and JAK-STAT pathways simultaneously) prior to virus challenge results in reduced SFV replication.

12.3 JAK/STAT pathway

The Janus kinase/signal transducers and activators of transcription (JAK-STAT) pathway was first identified as an interferon (IFN) -induced signaling pathway in vertebrates and it plays a key role in antiviral immunity in mammals. It is conserved in invertebrates and was first characterized in *Drosophila* in the context of its role in several aspects of development.

The JAK/STAT pathway is activated by stress/injury sensed by haemocytes. JAK (Hopscotch), pre-associated with dimmers of receptor Domeless, is activated upon receptor binding of the cytokines Upd-3. Upd -3 itself is secreted into the haemocytes at sites of septic injury. JAK activation leads to recruitment of STATs (Stat92E), which are phosphorylated and dimerize, translocating to the nucleus to activate transcription of targets including Turandot genes. The stress/injury response may also be regulated by the c-Jun N-terminal kinase (JAK) pathway, which is activated by TAK1 downstream of Imd.

The Janus kinase protein and the signal transducer and activator of transcription represent two core components of the JAK/STAT signal transduction cascade. The core components of the canonical JAK/STAT signalling pathway present in *Drosophila* include three related ligands called **unpaired (upd), unpaired 2 (upd2)** and **unpaired 3 (Upd3)**. Binding of ligand to the predimerised Domeless (Dome) receptor stimulates signalling and causes the receptor associated JAK tyrosine kinase Hopscotch (Hop) to phosphorylate both itself and the cytoplasmic tail of the Dome receptor to create docking sites for the latent STAT92E proteins. STAT92E is in turn phosphorylated, dimerises and translocates into the nucleus where it is capable of binding to a palindromic response element to induce target gene expression.

The canonical *Drosophila* JAK-STAT pathway is triggered by the attachment of the Unpaired (Upd) peptide ligand to the extracellular region of the transmembrane receptor Dome. Ligand recognition leads to conformational modification and dimerization of Dome, resulting in self-phosphorylation of the receptor-associated Janus kinases (JAKs). Activated JAKs then phosphorylate the cytoplasmic tail of the receptor, generating docking sites for the recruitment of STAT proteins. The recruited STATs are then phosphorylated by the Dome/JAK activated complex, resulting in STAT activation and dimerization. The activated STAT dimers are translocated to the nucleus and induce the expression of effector genes (Fig 4.5).

The first evidence for JAK-STAT pathway involvement in insect immunity came from studies in the malaria vector mosquito *An. gambiae*, in which bacterial challenge resulted in nuclear translocation of STAT (Barillas-Mury 1999). It was later found that the JAK-STAT pathway modulates viral infection and survival of flies infected with the *Drosophila* C virus (DCV), suggesting an evolutionarily conserved antiviral mechanism in insects and humans.

The antiviral role of the JAK-STAT pathway is conserved in the *Ae. aegypti* defense against DENV. DENV replication in the mosquito midgut is significantly increased when the pathway is transiently suppressed by RNAi-mediated depletion of the receptor Dome or the JAK ortholog Hop, and the opposite effect on virus replication is observed when the pathway is activated by silencing of protein inhibitor of activated STAT (PIAS), a negative regulator. As mentioned above, prior stimulation of the IMD / JAK-STAT pathways with Gram-negative bacteria reduces SFV replication in *Ae. albopictus* cells (Fragkoudis *et al.,* 2008), but it is difficult to distinguish between the roles of these two pathways in this case.

JAK-STAT pathway-activated antiviral mechanisms are poorly understood. Two DENV-induced, JAK-STAT-regulated putative effector genes that restrict DENV replication in midgut tissues have been identified but remain uncharacterised. These were termed dengue virus restriction factors (DVRFs) 1 and 2. **DVRF1** is a putative transmembrane protein that presumably acts as a pathway receptor; **DVRF2** contains antifreeze and allergen domains and may be involved in virus recognition. Immunity-related genes comprise only a small proportion of the mosquito's JAK-STAT-regulated transcriptome, suggesting that this pathway may not fight DENV infection through a classical innate immune response.

In contrast to *Aedes* and *Culex* mosquitoes, which transmit many arboviruses, the *Anopheles* mosquitoes are known to be the primary vector for only one—O'nyong-nyong virus (ONNV, family *Togaviridae*). Differences between the immune responses of culicine and anopheline mosquitoes lead to one hypothesis for this distinction. Intriguingly, ONNV infection of *Anopheles gambiae* does not appear to transcriptionally activate the Toll, IMD, or JAK-STAT pathways, and knockdown of Toll or JAK-STAT pathway components also had no effect on ONNV replication. In this study, however, mosquitoes were intrathoracically inoculated with ONNV instead of orally infected, and so it remains possible that these immune pathways are active against ONNV in the midgut, while other responses target it upon dissemination. Indeed, several ONNV antagonists have been identified, including an MD2 co-receptor-like gene, a galectin, two lysozymes, and a heat-shock protein (Waldock *et al.*, 2012). The mechanisms by which these gene products mediate virus killing remain unknown, but may involve processes such as immune recognition, immune signaling, membrane perturbation, and modulation of cellular stress responses. ONNV infection also activates the transcription of numerous LRIMs and thioester-containing proteins (TEPs), suggesting that complement-mediated lysis or opsonisation may also act against virus (Waldock *et al.*, 2012).

12.4 The RNA Interference (RNAi) Pathway

The RNAi antiviral mechanism is not a classical pathogen-stimulated immune response, but instead appears to use enzymes that are constitutively expressed in the cell cytoplasm. The insect exogenous small interfering RNA (siRNA) response is the best studied of the RNAi pathways, and is initiated when long, virus-derived double-stranded RNA (dsRNA) in the cytoplasm of infected cells is recognised and cleaved by Dicer-2 (Dcr2) into siRNAs, usually 21 base pairs (bp) in length. siRNAs are then loaded onto the multi-protein RNA-induced silencing complex (RISC), which unwinds the duplex RNA and degrades one of the siRNA strands, using the other for targeted degradation of single-stranded viral RNA with sequence complementary to the siRNA (Fig 4.5).

RNAi had earlier been shown to influence RNA virus replication and pathogenic outcomes in *D. melanogaster* but recently been recognised as an antiviral mechanism in mosquitoes. Studies in transgenic mosquito lines with inducible midgut and salivary gland expression of DENV-specific dsRNA have shown that RNAi inhibits DENV infection in these organs (Mathur *et al.*, 2010). The use of dsRNA to transiently deplete key RNAi components has provided direct evidence for a natural antiviral role of RNAi in *Ae. aegypti*: Knockdown of the effector enzymes Dcr2, R2D2, and Argonaute-2 allows DENV to replicate more efficiently after oral challenge. Knockdown of these components also shortens the extrinsic incubation period of DENV in mosquitoes and increases the efficiency of virus transmission, suggesting that RNAi is a key mediator of vector competence. RNAi has also been found to control SINV and CHIKV replication in *Ae. aegypti*. It is also a key component of anti-ONNV defense in *An. gambiae*, in which the Toll, IMD, and JAK-STAT pathways do not appear to play antiviral roles during systemic infection(Waldock *et al.*, 2012).

Arbovirus infection of mosquitoes is generally asymptomatic and persists throughout the life of the vector, but the virus appears to be continually targeted by the RNAi response. DENV2 and WNV siRNAs are detected at both 7 and 14 days after oral infection, and more siRNAs accumulate at later times after infection. Interestingly, DENV2-derived small RNAs make up less than 0.05% of the total small RNAs sequenced, a proportion similar to that observed for West Nile virus (WNV) in *Culex* mosquitoes but much lower than that observed for alphaviruses (10% for SINV). It has been suggested that sequestration of flavivirus replication complexes in membrane-enclosed vesicles limits the access of Dcr2 to dsRNA replicative intermediates, and that given the low abundance of DENV2-specific small RNAs, Dcr2 cleavage of dsRNA alone may be sufficient to control DENV infection.

Although the origin of the initiating dsRNA for RNAi activation is not yet fully defined in mosquitoes, replicative intermediates formed between the positive and negative viral RNA strands or secondary structure within an individual viral RNA molecule are likely sources; these two possibilities are not mutually exclusive. Deep sequencing of small RNAs from DENV-infected mosquitoes has revealed an almost equal ratio of positive- to negative-sense DENV-derived small RNAs, suggesting that most small RNAs are derived from dsRNA replicative intermediates instead of intra-strand secondary structures.

While 21-bp virus-derived siRNAs are the dominant species during the middle and late stages of DENV infection, another study found a predominance of longer (24–30-bp) DENV2-derived small RNAs during early infection. These RNAs are most likely derived from the PIWI RNA (piRNA) pathway, suggesting a role for this Dcr2-independent pathway in antiviral defense. Virus-derived piRNAs have also been detected during CHIKV infection of *Ae. aegypti* and *Ae. albopictus*. SFV infection of mosquito cell lines also triggered virus-derived piRNA production, and knockdown of piRNA pathway components enhanced SFV infection in these cells, further implicating this small RNA pathway in antiviral defenCe.

Studies in flies may provide further insight into the antiviral mechanisms of RNAi. Systemic spread of RNAi via cellular dsRNA uptake has been reported to be necessary for *D. melanogaster* survival after RNA virus challenge. Since the flies in this study were challenged via septic injury with a pathogenic virus (DCV) and a virus that does not typically infect drosophilids (SINV), lysis of infected cells may have released dsRNA for uptake by other cells. Although the spread of RNAi has been described in cultured mosquito cells infected with SFV, it is unclear whether systemic RNAi is necessary for the mosquito defense against arboviral infections, which have classically been regarded as non-pathogenic in the vector.

Another study from the *Drosophila* model suggests a role for insect-encoded reverse transcriptase (RT) enzymes and the RNAi machinery in maintaining the persistence of RNA viruses. Here, viral genome fragments are reverse-transcribed and inserted into the insect genome by retrotransposon elements; these insertions later serve as templates for RNAi responses against the virus (Goic *et al.*, 2013).

12.5 Pathway Crosstalk

The Toll and IMD pathways have been reported to dually regulate subsets of immune effector genes, and there is growing evidence for a functional overlap between insect antiviral, anti-bacterial, and anti-parasite immune responses (Rmirez *et al.*, 2012). Although it is still poorly understood, crosstalk between pathways is likely to occur so that an effective immune response can be coordinated.

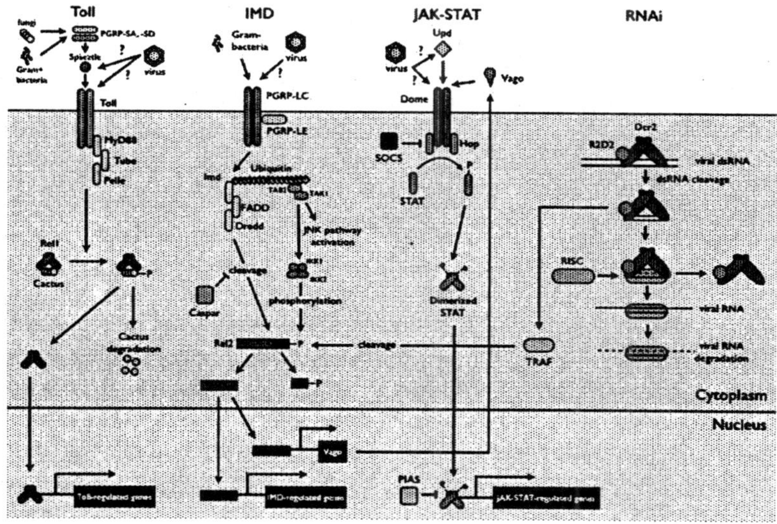

Fig. 4.5 Mosquito immune signalling and RNAi pathways (Sim *et al.*, 2014)

The best-studied example of pathway crosstalk in insect antiviral immunity involves Dcr2, the RNAi pathway DExD/H-box helicase that recognises viral dsRNA and cleaves it into siRNAs. In addition to its role in RNAi, Dcr2 has been reported to mediate the induction of antiviral genes, similar to the action of mammalian cytoplasmic RIG-I-like dsRNA sensors: DCV infection of *Drosophila* induces the expression of the antiviral polypeptide *Vago* in a Dcr2-dependent manner . Other components of the RNAi machinery do not appear to be involved in the modulation of *Vago* gene expression, suggesting that its induction through Dcr2 is independent of the RNAi pathway. Recently, this mechanism has also been characterised in the *Cx. Quinque fasciatus* response to WNV. Here, sensing of viral dsRNA by Dcr2 activates TNF receptor-associated factor (TRAF), which in turn leads to Rel2 cleavage and activation. Activated Rel2 binds to a conserved NF-kB binding site on the *Cx*Vago promoter, inducing its transcription. *Cx*Vago, a secreted peptide, restricts WNV by activating the JAK-STAT pathwa (Fig 4.5). These findings are in contrast to those of the *Drosophila* study, in which the induction of fly *Vago* was not mediated by NF-kB transcription factors, and *Vago* did not activate the fly JAK-STAT pathway. Interestingly, the *Aedes* ortholog of *Vago* was also found to restrict WNV in *Culex* cells, suggesting that the antiviral action of this molecule may be conserved in mosquito vectors of DENV. These findings characterise a distinct pathway leading to Rel2 activation and suggest potential crosstalk mechanisms between the mosquito RNAi, JAK-STAT, and IMD pathways.

Otherpeptideswithcytokine-likeantiviralactivityhavebeenisolatedfrom arbovirus-infected mosquito cells: The anionic septapeptideviprolaxikine, for example, has been isolated from the cell-free medium supernatants of DENV-infected *Ae albopictus* C6/36 cells; this septapeptide protects against DENV infection in naive mosquito and vertebrate cells. The molecular mechanisms of this protection, however, have yet to be studied.

13. C-Type lectins

C-type lectins are important factors in insect defence against invading microorganisms.In insect humoral responses, many haemolymph proteins are involved. These proteins include a group of antimicrobial proteins/ peptides, lectins, and cell adhesion proteins. C-type lectins (CTLs) are a family of proteins that share a common structural motif, the carbohydrate recognition domain, and may act as receptors in pathogen recognition. Lectins are sugar binding proteins that participate in the innate immune system. C-type lectins are calcium-dependent proteins that bind to particular carbohydrate sequences and,thus, function as pattern recognition proteins (Kanost*et al.*, 2004). Four C-type lectins found inthe haemolymph of *M. sexta* and named immunolectins (IMLs) bind to bacterial LPS, and IML-1and IML-2 stimulate activation of proPO in plasma (Yu *et al.*, 2002; Yu and Kanost, 2000). IML-2 appears to have activity against Gram (–) bacteria (Yu and Kanost, 2003). IMLs may help in localising PO response to the surface of invading bacteria (Yu *et al.*, 2003) and, thus, avoid widespread action of phenoloxidase that might be detrimental to tissues of the insect. Insect lectins have different functions in the innate immune systems, including haemocyte nodule formation (Koizumin *et al.*, 1999),prophenol oxidase activation (Yu *et al.,* 1999) and opsonisation (Jomori and Natori, 1992).

14. SERPINS

Serpins are ubiquitously distributed serine protease inhibitors that covalently bind to target proteases to exert their activities. Serpins regulate a wide range of activities, particularly those in which protease-mediated cascades are active. Serine proteases play critical roles in the regulation of the invertebrate innate immune responses. Serpin is a coined name that stands for (ser)ine (p)roteinase (in)hibitor. Serpins act like a brake to regulate the cascade of reactions occurring in the immune response and innumerous other cascade reactions involving serine proteinases (Pelte *et al.*, 2006). Serpins are proteins about 45 kDa in size forming a family of serine proteinase inhibitors that function in both invertebrates and mammals (Kanost, 1999). Necrotic (Nec) is a serpin that regulates Toll activation by inhibiting a serine proteinase involved in cleavage of Späetzle. They perform their inhibition of serine proteinases by trapping and distorting the proteinase enzyme in a covalently linked serpin–proteinase complex, which is then targeted for destruction. *Manduca sexta* synthesizes a mixture of 12 serpins, differing mainly in the carboxyl-terminalreactive site loop that is critical to the mechanism of inhibition of serine proteinases. Serpin-1 is not upregulated in response to challenge and ecdysteroid negatively regulates serpin-1 in fat bodycells (Kanost *et al.*, 1995). Serpin-2 from *M. sexta* is upregulated by bacterial challenge (Gan *et al.*, 2001), but its specific target proteinase is unknown. Serpins 3 to 5 are present in fat body cells andare upregulated by challenge. Serpin-3 might help prevent excessive melanisation in the haemocoel of insects by localising proPO activation to the site of infection i.e.,serpin-3 might inhibit the prophenoloxidase activating protein, a serine proteinase that is upregulatedin an infection (Kanost *et al.*, (2004). Systemic hyper activation of the prophenoloxidase system, excessive formation of quinines and inappropriate excessive melanin synthesis are delicious to the host, hence all should be tightly regulated by serpins or melanisation regulatory molecules or inhibitors.

15. Evolutionary ecology of insect immune defences

Evolutionary ecology seeks to understand the selective reasons for the immune defence. Failure to recognise an antigen, response specificity, the cost of defence, and the risk of autoimmunity are the limitations in the molecular processes and sex, resource availability, and interference by parasites also affect a response. Evolution of an effective immune defence can result in someother fitness-related parameter becoming reduced because of **pleiotropic effects** or **genetic covariance** (Schmid-Hempel, 2003). Evolutionists are interested in how defences are selected and how they adapt insects for success in their ecological setting. Synthesis of the new compounds involved in animmune response is likely to be costly and may divert energy away from other activities, with possible trade-offs operating during evolution of defensive mechanisms that negatively impact fitnessof insects. Benefits of defence are generally obvious, such as survival of an attack, continued growthand development, and success in reproduction. The costs of defence are less clear-cut, but may involve reduced function in some other biological process, especially when nutritional resources are limited.

Nutritional resources may be limited for many insects at certain times, and especially when an insect feeds on one or only a few food sources. For example, *Rhodniusprolixus*, a blood feeder, takes only one blood meal in each instar as an immature, and adult females need blood to produce eggs. Feder *et al.* (1997) found that Rhodnius was less able to resist challenge with bacteria if fed only plasma instead of whole blood, which suggested that the plasma was not nutritionallyas good as whole blood. On plasma, the bugs had reduced production of antimicrobial peptides, reduced nodule formation, and synthesised less lysozyme than those getting whole blood. Addition of α-ecdysone aided these plasma-fed bugs during infection, and although the mechanism by which α-ecdysone might work is not clear, it might signal ecdysteroid-related control of some defence mechanisms. Mounting an immune response is reduced in the mealworm, *Tenebrio molitor*, if it is deprived of food for short periods of time (Siva-Jothy and Thompson, 2002). Welburn *et al.*, (1989) found that lectins in the midgut of various species of Glossina (tsetse flies that also feed on vertebrate blood) were less able to kill trypanosomes when the insects were fed with modified blood. There are consistent and predictable differences between individuals, sexes and species in terms of parasite defence mechanism which can be understood through ecology and evolutionary background of the species concerned. The prospects for ecological immunology making further significant basis and applied contribution to the study of insect host –parasite interactions look promising.

16. AUTOIMMUNE CONSEQUENCES OF SOME DEFENCE REACTIONS

Some of the immune reactions have potential for autotoxicity to the insect itself, a subject arearecently reviewed by Zuk and Stoehr (2002), Schmid-Hempel (2003), and Rolff and Siva-Jothy (2003). Some of the reactive oxygen and nitrogen radicals generated by phenoloxidase activity maydamage the host's own tissues in some cases. Nevertheless, melanin, a product of the action ofphenoloxidase, may be beneficial as a scavenger of reactive free radicals (Schmidt-Hempel, 2005). Green *et al.*, (2003) found mutants of *D. melanogaster* that experienced necrotic tissues after microbial challenge, apparently because the mutants were not able to mobilise serpin inhibitors of the prophenoloxidase activation cascade. Thus, it appeared that phenoloxidase activity was not localised and excessive tissue damage occurred. The potential for autoimmune reactions is another potential cost of activating the immune system. Rahman *et al.*, (2006) present evidence that lipopharin particles may assemble into cage-like coagulation products that protect tissues and cells from somepathogens and PO products (Nation, 2008).

17. GENDERDIFFERENCES IN IMMUNE RESPONSES

Since males and females differ in their life histories, males tend to have weak immune responses than female and as a result tend to have higher prevalence of parasitism. These sex-specific differences in immunity have been related to Bateman's principle, i.e. females gain fitness by maximising their lifespan through immunity investment, assuming that higher immunity increases longevity, which in turn increases the time for egg production and oviposition (Rolff, 2002). Males gain fitness by increasing their mating frequency, and should thus invest in increasing their mating rates instead of immunity. Male insects typically invest less in the next generation and their role is usually simply related tomating, although some males offer a nuptial gift to the female and some provide nutrients in theseminal fluid. Many male insects have less immune capacity than females (Schmid-Hempel, 2005). Meylaers *et al.*, (2007) found that male adults of the wax moth, *Galleria mellonella*, were smaller than females when the larvae were immune-challenged. There were also stage-specific effects, with pupae having the most effective immune response, the larvae next, andadults with the lowest immune response in the wax moth. Kageyama *et al.*, (2007) found that infectionof *D. melanogaster* with spiroplasms had an adverse effect on longevity of male flies, and could kill them early or late, depending on maternal host age when infected. Male *D. Melanogaster* that were allowed to repeatedly mate were found to have reduced immune capability (McKean and Nunney, 2001), but Ryder and Siva-Jothy (2000) and Siva-Jothy (2000) found positive correlation between sexually selected traits and immune function.

Vigorous courtship behaviour may be indicative of general vigour and the fact that the individual has escaped from parasites, predators, and pathogens (Contreras-Garduno *et al.*, 2007). Territorial males of the American ruby spot damselfly, *Hetaerina americana*, have larger red spot areas on the wings, greater phenoloxidase activity, more hydrolytic enzyme activity, and higher survival after challenge with *Serratia marcescens* than males with smaller spots (Contreras-Garduno *et al.*, 2007).

18. COMPARISON OF THE IMMUNE SYSTEM OF INSECTS *VIS-A-VIS* THE VERTEBRATES

Contrary to the vertebrates, insects depend solely on innate defence mechanisms. The vertebrate adaptive immunity mediated by lymphoid lineage derived B-cells and T-cells are absent in insects (Lavine and Strand 2002). Insects have a powerful innate immune system consisting of cellular and humoral mechanisms. Humoral factors that are secreted into the haemolymph include anti-microbial peptides (AMPs), reactive oxygen species (ROS), reactive nitrogen species (RNS), proteolytic cascades that lead to haemolymph clotting and melanisation. Cellular defence mechanisms include phagocytosis, nodulation and encapsulation. Generally insect immunity is mediated by the fat body which is analogous to the vertebrate liver and via patrolling leuckocyte-like haemocytes. Sometimes the expression of immune effectors can also be observed in the gut, trachea, testis and salivary glands (Ferrandon *et al.,* 1998, Ha *et al.,* 2005ab, Senger *et al.,* 2006). The plasmatocytes share similarities with mammalian monocyte/macrophage lineage (Williams 2007)

After crossing the physical barriers such as larval cuticle, adult exoskeleton, the peritrophic membrane of the mid gut, or the chitinous lining of the trachea, the pathogen has to face an effective clotting system (Wang *et al.,* 2010) and microbes that escape the clotting may be phagocytosed or encapsulated. The immune insults usually elicit transcription and secretion of humoral effector molecules such as AMPs and opsonins. Most AMPs are specific against certain types of microbes. For example, in *Drosophila* Drosomycin is active against fungi, Diptericin against Gram-negative bacteria, and Defensin against Gram-positive bacteria (Meister *et al.,* 1997; Imler and Bulet 2005; Lemaitre and Hoffmann 2007).

Drosophila and vertebrates share many key factors of immune signalling pathways including Toll/Toll-like Receptor (TLR), Immune Deficiency (IMD)/ Tumour Necrosis Factor Rector (TNFR), Janus Kinase /Signal Transducer and Activator of Transcription (HAK?STAT) and c-Jun N-terminal Kinase (JNK) as well as effector molecules such as antimicrobial peptides (AMPs) that are involved in innate immune response (Lemaitre and Hoffmann 2007).

Plasmatocytes share similarities with mammalian monocyte/ macrophage lineage (Williams 2007). In general, higher vertebrates use a broader repertoire of specialised innate immune cells compared to insects. The crystal cells and lamellocytes of insect blood do not have clear homologues in vertebrates, but signalling pathways that control their differentiation have in parts mammalian counterparts (Meister 2004).

19. INSECT IMMUNITY AND GLOBAL WARMING

Insects are among the groups of organisms most likely to be affected by climatic changes because climate has direct influences on their development, reproduction and survival (Bale *et al.*, 2002; Savage *et al.*, 2004; Frazier *et al.*, 2006; Menéndez, 2007). Nevertheless, insects have short generation times and high reproductive rates, so that they can respond quicker to climate change than long-lived organisms, such as plants and vertebrates (Bale *et al.*, 2002; Menéndez, 2007). Warming can therefore potentially affect several aspects of insect life-cycle and ecology, and potential responses could include changes in phenological patterns and habitat selection and the expansion/contraction of geographic and altitudinal ranges (Bale *et al.*, 2002; Menéndez, 2007; Berg *et al.*, 2010).

Physiologically with the global warming, insects will have to maintain an efficient immune response to survive which require a shift of resources to the immune system from other functions (such as reproduction). In view of its cost, the immune response is influenced therefore by both biotic and abiotic factors (such as food availability and temperature) (Karl *et al.*, 2011). The predicted increase in thermal stress due to global warming is likely to induce cascading effects on other functions such as the individual fitness and favouring the distribution and prevalence of infectious diseases (Lafferty, 2009; Travers *et al.*, 2009).

In order to verify this hypothesis, Karl *et al.* (2011) investigated in the tropical butterfly *Bicyclusanynana* the effects of temperature changes on fitness-related adult traits (such as body mass and fat content) and on phenoloxidase (PO) activity and haemocyte numbers that are two key parameters for evaluating the immune function at both cellular and humoral levels. Interestingly, results on body mass and fat content suggest that global warming could be beneficial, whereas haemocyte numbers and PO activity decreased at increasing temperatures (Karl *et al.*, 2011) supporting the hypothesis that immune parameters were negatively affected by global warming. At higher temperatures, insects may therefore increase their rate of growth and reproduction, but may become more susceptible to diseases, leading to reduced lifespan and possibly reduced fitness in the field.

At the same time, Karl *et al.* (2011) observed that the decrease in PO activity with increasing temperature was more evident in food-deprived individuals, in respect to butterflies having access to food showing a trade-off relating energy shortage and immune response. Even if the molecular mechanism at the basis of this change in the PO activity has not been studied, a direct link between the expression of heat shock proteins and a decrease in immune parameters could be suggested.

The study of the trade-off between immune response and thermal tolerance is absolutely relevant in medicine and agriculture since if both immune function and reproduction are simultaneously enhanced, then climate change will result in the reproduction of more insects that will be more resistant to diseases. As a consequence, we will have more pest insects in our fields with more damages for agriculture, together with larger numbers of insects relevant for the transmission of human diseases. On the contrary, if higher temperatures induce or exacerbate trade-offs between reproduction and immune function, in order to face the global warming insects have a choice to put their resources mainly on reproduction or on the immune systems. A possible reply to this question can be obtained from studies performed in the cricket *Gryllus texensis*, where empirical evidence suggested that warmer temperatures lead to a decline in some immune functions (e.g. melanisation) (Suwanchaichanda and Paskewitz, 1998; Adamo and Lovett, 2004). In particular, it emerged that *G. Texensis* may become more susceptible to some pathogens at higher temperatures suggesting that reproductive rate and immune function are not simultaneously enhanced at higher temperatures.

Adamo and Lovett (2004) reported that elevated temperatures resulted in increased egg laying, faster egg development and greater mass gain in *Gryllus texensis*. In the same experiments a reduction in the resistance to the Gram-positive bacterium *Bacillus cereus* was observed both above or below the average field temperature (26°C) suggesting that increased temperatures induce trade-offs between reproduction and disease resistance for some species–pathogen interactions (Adamo and Lovett, 2004).

20. NEW FRONTIERS IN INSECT IMMUNITY

Nutrition is critical to immune defence and parasite resistance, which not only affects individual organisms, but also has profound ecological and evolutionary consequences. Theimportance of adopting an integrative and multi-dimensional approach to nutritional immunology is a new frontier in insect immunity. Continued research in insect immunity can be expected to result in improved use of entomopathogens in biological control, in discovery of antimicrobialmolecules that can be exploited by humans, and in new strategiesfor management of insect vectors of human and animal disease. In the near future,discovery of a wealth of antibacterial molecules from other insect orders should be expected, as investigators extend efforts to diverse insect groups. Likewise,research on antifungal peptides and proteins from insects has begun to yield interesting new results.The cell biology and biochemistry of haemocyte responses, haemocyte research, molecular basis of recognition, adhesion, and killing responses·by insect haemocytes,communication between haemocytes, and interactions between haemocytes and other immune-responsive tissues are the topics should receive significant efforts in the future. Signalling molecules that function as cytokines are likely to be discovered and their function in cell recruitment and regulation of gene expression elucidated (Gillespie and Kanost, 1997).

REFERENCES

Abraham, E.G., Pinto, S.B., Ghosh, A., Vanlandingham, D.L., Budd, A., Higgs, S., Kafatos, F.C., Jacobs-Lorena, M., and Michel, K. (2005). An immune-responsive serpin, SRPN6, mediates mosquito defense against malaria parasites. *Proc. Natl. Acad. Sci.* USA 102, 16327–16332.

Adamo SA, and Lovett MME.(2004). Some like it hot: the effects of climate change on reproduction, immune function and disease resistance in the cricket *Gryllus texensis*. *J. Exp. Biol.* 214: 1997-2003.

Agaisse, H., and N. Perrimon. (2004). The roles of JAK/STAT signaling in Drosophila immune responses. *Immunol. Rev.* 198: 72–82.

Alfonso, T.B., and B.W. Jones. (2002). gcm2 promotes glial cell differentiation and is required with glial cells missing for macrophage development in ***Drosophila***. *Dev. Biol.* 248:369–383

Ashida, M and H, I.Yamazaki (1990) Biochemistry of the phenoloxidase system in insects: With special respect to its activation, in E.Ohnishi and H.Eshizaki (eds) *Moting and metamorphosis.* Japan Scientific Society Press, Tokyo Japan: Springer Verlag, Berlin Germany, pp 239-256

Avadhanula, V.; Weasner, B.P.; Hardy, G.G.; Kumar, J.P.; Hardy, R.W. (2009). A novel system for the launch of alphavirus RNA synthesis reveals a role for the IMD pathway in arthropod antiviral response. *PLoS Pathog.*, *5*, e1000582

Bale JS, Masters GJ, Hodkinson ID, Awmack C, Bezemer TM, Brown VK, *et al.* (2002) Herbivory in global climate change research: direct effect of rising temperature on insect herbivores. Global Change Biol. 8: 1-16

Barillas-Mury, C.; Han, Y.S.; Seeley, D.; Kafatos, F.C.(1999). *Anopheles gambiae* Ag-STAT, a new insect member of the STAT family, is activated in response to bacterial infection. *EMBO J. 18*, 959–967

Berg MP, Kiers ET, Driessen G, van der Heijden M, Kooi BW *et al.* (2010).. Adapt or disperse: understanding species persistence in a changing world. Global Change Biol. 16: 587-598Bischoff, V., Vignal, C.,

Duvic, B., Boneca, I.G., Hoffmann, J.A., and Royet, J. (2006). Downregulation of the Drosophila immune response by peptidoglycan-recognition proteins SC1 and SC2. *PLoS Pathogens* 2, e14.

Brey, P.T. and Hultmark, D. (1998). Molecular Mechanisms of Immune Responses in Insects. (Eds.), Chapman & Hall, London/New York, 1998.

Browne, N, Heelan M, and Kavanagh K (2013). An analysis of the structural and functional similarities of insect hemocytes and mammalian phagocytes. Virulence, 4(7): 597–603.

Buchon, N., Silverman, N., and Cherry, S. (2014). FromImmunity in Drosophila melanogaster — from microbial recognition to whole-organism physiology. Nature Reviews Immunology, 14, 796–810.

Bulet, P. and R. Stocklin (2005) Insect antimicrobial peptides: structures, properties and gene regulation. Protein Pept Lett 12, 311.

Bulet, P., Hetru, C., Dimarcq, J.L., Hoffmann, D.(1999). Antimicrobial peptides in insects: structure and function. *Dev. Comp. Immunol.* **23:** 329–344.

Cerenius, L., and Soderhall, K.(2004) The prophenoloxidase-activating system in invertebrates. *Immunological reviews.* 198: 116-126

Chase, M., K. Raina, J. Bruno, and M. Sugumaran (2000). Purification, characterization and molecular cloning of prophenoloxidase from Sarcophaga bullata. *Insect Biochem. Mol. Biol.* **30:** 953–967.

Contreras-Garduño, J., H. Lanz-Mendoza, and A. Córdoba-Aguilar. (2007). The expression of a sexually selected trait correlate with different immune defense components and survival in males of the American ruby spot. *J. Insect Physiol.* 53: 612–621.

Costa, A.; Jan, E.; Sarnow, P.; Schneider, D.(2009) The Imd pathway is involved in antiviral immune responses in *Drosophila. PLoS One*, *4*, e7436.

Delaney, J.R., S. StÖven, H. Uvell, K.V. Anderson, Y. EngstrÖm, and M. Mlodzik. (2006). Cooperative control of Drosophila immune responses by the JNK and NF-kappa B signaling pathways. *Embo J.* **12:** 3068–3077.

Dostert, C., Jouanguy, E., Irving, P., Troxler, L., Galiana-Arnoux, D., Hetru, C., Hoffmann, J., and Imler, J. (2005). The Jak-STAT signaling pathway is required but not sufficient for the antiviral response of Drosophila. *Nat. Immunol.* 6, 946–953.

Fabrick, J.A., J.E. Baker,· and M.R. Kanost. (2003). cDNA cloning, purification, properties, and function of a β-1,3-glucan recognition protein from a pyralid moth, *Plodia interpunctella. Insect Biochem. Mol. Biol.* 33: 579–594.

Feder, D., C.B. Mello, E.S. Garcia, and P. Azambuja. (1997). Immune response in *Rhodnius prolixus*: Influence of nutrition and ecdysone. *J. Insect Physiol.* 43: 513–519.

Ferrandon D1, Jung AC, Criqui M, Lemaitre B, Uttenweiler-Joseph S, Michaut L, Reichhart J, Hoffmann JA (1998). A drosomycin-GFP reporter transgene reveals a local immune response in Drosophila that is not dependent on the Toll pathway. *EMBO J.* 17(5):1217-27.

Ferrandon, D., J.-L. Imler, and J.A. Hoffman. (2004). Sensing infection in Drosophila: Toll and beyond. *Sem. Immunol.* 16: 45–53.

Fragkoudis, R.; Chi, Y.; Siu, R.W.; Barry, G.; Attarzadeh-Yazdi, G.; Merits, A.; Nash, A.A.; Fazakerley, J.K.; Köhl, A. Semliki (2008) Forest virus strongly reduces mosquito host defence signaling. *Insect Mol. Biol.*, *17*, 647–656.

Franc NC, Hèitzler P, Ezekowitz RAB, White K. (1999). Requirement for croquemort in phagocytosis of apoptotic cells in Drosophila. *Science.* 284:1991-1994.

Frazier M, Huey R, Berrigan D. (2006) Thermodynamics constrains the evolution of insect population growth rates: "Warmer is better". *Am. Nat.* 168: 512–520

Galiana-Arnoux, D., Dostert, C., Schneemann, A., Hoffmann, J.A., and Imler, J.-L. (2006). Essential function in vivo for Dicer-2 in host defense against RNA viruses in Drosophila. *Nat. Immunol.* 7, 590–597

Gan, H., Y. Wang, H. Jiang, K. Mita, and M.R. Kanost. (2001). A bacteria-induced, intracellular serpin in granular hemocytes of *Manduca sexta. Insect Biochem. Mol. Biol.* 31: 887–898.

Ganz, T. (2003). Defensins: antimicrobial peptides of innate immunity. *Nat. Rev. Immunol.* 3: 710-720.

Gillespie J. P. and Kanost M. R (1997). Biological mediators of insect immunity. *Annu. Rev. Entomol.* 1997. 42:611–43

Gillespie, J.P., Kanost, M.R., Trenczek, T. (1997) Biological mediators of insect immunity. *Annu. Rev. Entomol.* 42: 611-643.

Goic, B.; Vodovar, N.; Mondotte, J.A.; Monot, C.; Frangeul, L.; Blanc, H.; Gausson, V.; Vera-Otarola, J.; Cristofari, G.; Saleh, M.-C.; *et al.*(2013) RNA-mediated interference and reverse transcription control the persistence of RNA viruses in the insect model *Drosophila. Nat. Immunol. 14*, 396–403

Götz P and Boman H.G. (1985). Comprehensive Insect Physiology, Biochemistry and Pharmacology. In *Comprehensive Insect Physiology, Biochemistry and Pharmacology*; Volume 3: Integument, Respiration and Circulation (Eds, Kerkut, G. A. and Gilbert, L. I.) Pergamon Press, Oxford, pp. 453-485.

Green, C., G. Brown, T.R. Dafforn, J.-M. Reichart, T. Morley, D. Lomas, and D. Grubb. (2003). *Drosophila necrotic* mutations mirror disease-associated variants of human serpins. *Development* 130: 1473–1478.

Gupta A.P. (1991). Insect immunocytes and other hemocytes: roles in humoral and cellular immunity. In A. P. Gupta (Ed.), *Immunology of Insects and other Arthropods*, (pp. 19-118). Boca Raton, Florida: CRC Press.

Gupta, A.P. (1979a). Insect hemocytes development, forms, functions and techniques. Cambridge University Press, New York.

Gupta, A.P.(1979b). Arthropod hemocytes and phylogeny. In: Arthropod phylogeny (ed. A.P. Gupta), Van Nostrand Reinhold, pp. 669-735.

Ha E .M., Oh C.T., Bae Y.S and Lee W.J (2005a). A Direct Role for Dual Oxidase in Drosophila Gut Immunity. *Science*, 310: 847-850.

Ha E.M., Oh C.T., Ryu J.H., Bae Y.S., Kang S.W., Jang I.H., Brey P.T and Lee W.J (2005b). An antioxidant system required for host protection against gut infection in Drosophila. *Dev Cell.* 8(1):125-32.

Heinrich , B. (1971) . Temperature regulation in the sphinx moth, Manducasexta. II: Regulation of heat loss by control of blood circulation . *J. Exp. Biol.* 54 , 153 – 166 .

Heinrich, B (1974). Thermoregulation in endothermic insects. Science. 185 (4153):747-56.

Hoffman, J. A., F. C. Kafatos, C. A. Janeway and R. A. Ezekowitz (1999) Phylogenic perspectives in innate immunity. *Science* 284, 1313-1318.

Holz A, BossingerB, StrasserT, Janning W and Klapper R (2003). The two origins of hemocytes in Drosophila. *Development* 130, 4955-4962.

Imler J.L and Bulet P (2005). Antimicrobial peptides in Drosophila: structures, activities and gene regulation. *Chemical Immunology and Allergy,* **86:**1–21.

Janse, C.J., R.J. Rouwenhorst, P.F.J. van der Klooster, H.J. van der Kay, and J.P. Overdulve.(1985). Development of Plasmodium berghei ookinetes in the midgut of Anopheles atroparvus mosquitoes and in vitro. *J. Parasitol.* 91: 219–225.

Jomori T, and Natori S. (1992). Function of the lipopolysaccharide-binding protein of *Periplaneta americana* as an opsonin. *FEBS Lett.* 296:283–86

Jones, J. C. (1977). "The Circulatory System of Insects". Thomas, Springfield, IL

Kageyama, D., Anbutsu, H., Shimada, M. and Fukatsu. T. (2007). Spiroplasms infection causes either early or late male killing in Drosophila, depending on maternal host age. *Naturwissenschaften* **94:** 333–337.

Kaneko, T., Yano, T., Aggarwal, K., Lim, J.H., Ueda, K., Oshima, Y., Peach, C., Erturk-Hasdemir, D., Goldman, W.E., Oh, B.H., et al. (2006). PGRP-LC and PGRP-LE have essential yet distinct functions in the Drosophila immune response to monomeric DAP-type peptidoglycan. *Nat. Immunol.* 7, 715–723.

Kanost, M.R, and Zhao L. 1996. Insect hemolymph proteins from the Ig superfamily. *Adv. Comp. Environ. Physiol.* 23:185–97

Kanost MR, Jiang H, Yu XQ. (2004) Innate immune responses of a lepidopteran insect, *Manduca sexta. Immunol. Rev.* 198: 97 - 105.

Kanost, M.R., Kawooya, J.K, Law, J.H et al (1990) Insect hemolymph proteins In P.D. Evans and V.B.Wigglesworth, eds, *Advances in Insect Physiology,* pp 299-396 Academic Press, San Diego, CA

Kanost, M.R., Prasad, S.V., Huang, Y., and Willot, E (1995) Regulaltion of serpin gene-1 in *Manduca sexta, Insect Biochemistry and Molecular Biology* 25: 285-291

Kanost, M.R (1999). Serine protein inhibitors in arthropod immunity. *Development and comparative Immunology* 23, 291-301

Karl L, Stokcs I, De Block M, Janowitz SA, and Fischer K. (2011). Temperature extremes and butterfly fitness: conflicting evidence from life history and immune function. Global Change Biol. 17: 676-687

Koizumin, A N., Imai, Y., Morozumi A., et al. (1999). Lipopolysaccharide-binding protein of *Bombyx mori* participates in a hemocyte-mediated defence reaction against Gram-negative bactrria. *Journal of Insect Physiology* 45: 853-859

Lafferty, K.D. (2009) The ecology of climate change and infectious diseases. Ecology 90: 888-900

Lavine, M.D and Strand M R. (2002). Insect hemocytes and their role in immunity. *Insect Biochem. Mol. Biol.* 32: 1295-1309.

Lehane, M.J (1997). Peritrophic matrix structure and function. *Annu. Rev. Entomol.* 42: 525–550.

Lehane, M (2005). The Biology of Blood-Sucking in Insects, 2nd ed. Cambridge University Press, Cambridge, U.K.

Lemaitre B., and Hoffmann J. (2007) The Host Defense of Drosophila melanogaster. *Annual Review of Immunology* 25(1), 697-743.

Lemaitre, B., E. Kromer-Metzger, L. Michaut, E. Nicolas, M. Meister, P. Georgel, J.-M. Reichhart, and J.A. Hoffmannn. (1995). A novel mutation, immune deficiency, defines two distinct control pathways in the Drosophila host defense. *Proc. Nat. Acad. Sci.* USA 92: 9465–9469.

Lemaitre, B., Nicolas, E., Michaut L., Reichhart J.-M and Hoffmann J.A (1996). The dorsoventral regulatory gene cassette spätzle/Toll/cactus controls the potent antifungal response in Drosophila adults. *Cell* 86:973–983.

Mathur, G.; Sanchez-Vargas, I.; Alvarez, D.; Olson, K.E.; Marinotti, O.; James, A.A. (2010) Transgene-mediated suppression of dengue viruses in the salivary glands of the yellow fever mosquito, *Aedes aegypti*. *Insect Mol. Biol.*, *19*, 753–763

McKean, K.A., and Nunney L. (2001). Increased sexual activity reduces male immune function in *Drosophila melanogaster*. *Proc. Nat. Acad. Sci. USA* 98: 7904–7909.

Meister, M., Lemaitre, B. and Hoffmann,J.A. (1997).Antimicrobial peptide defense in Drosophila.BioEssays 19,1019 -1026.

Meister, M. (2004). Blood cells of Drosophila: cell lineages and role in host defense. Curr Opin Immunol 16:10–15

Menéndez, R. (2007) . How are insects responding to global warming? Tijdsch. Entomol. 150: 355-365

Meylaers, K., Freitak D and Schoofs L. (2007). Immunocompetence of *Galleria mellonella*: Sex- and stage specific differences and the physiological cost of mounting an immune response during metamorphosis. *J. Insect Physiol.* 53: 146–156.

Michel, K., Budd, A., Pinto, S., Gibson, T.J., and Kafatos, F.C. (2005). Anopheles gambiae SRPN2 facilitates midgut invasion by the malaria parasite Plasmodium berghei. EMBO Rep. 6, 891–897.

Miller, T. A and Pass , G. (2009). Circulatory System. In Encyclopedia of Insects, (Eds.). Vincent H. Resh, Ring T. Cardé, London: Academic Press, 169-173.

Nappi, A.J., Y. Carton, and F. Frey. (1991). Parasite induced enhancement of haemolymph tyrosinase in a selected immune reactive strain of *Drosophila melanogaster. Arch. Insect Biochem. Physiol.* **18:** 159–168.

Nässel, D.R. (1993). Neuropeptides in the insect brain: A review. *Cell Tissue Res.* 273: 1–29

Nation, J. L. (2008). Circulatory system. In *Insect Physiology and Biochemistry*, pp. 339-365. Boca Raton, FL: CRC Press.

Pelte, N., A.S. Robertson, Z. Zou, D. Belorgey, T.R. Dafforn, H. Jiang, D. Lomas, J.-M. Reichart, and D. Grubb. (2006). Immune challenge induces N-terminal cleavage of the *Drosophila serpin* Necrotic. *Insect Biochem. Mol. Biol.* **36:** 37–46.

Rahman, M. M., G. Ma, H.L.S. Roberts, and O. Schmidt. (2006). Cell-free immune reactions in insects. *J. Insect Physiol.* 52: 754–762.

Ramirez, J.L.; Souza-Neto, J.; Torres Cosme, R.; Rovira, J.; Ortiz, A.; Pascale, J.M.; Dimopoulos, G. (2012) Reciprocal tripartite interactions between the *Aedes aegypti* midgut microbiota, innate immune system and dengue virus influences vector competence. *PLoS Negl. Trop. Dis.*, *6*, e1561.

Rolff, J., and M.T. Siva-Jothy. (2003). Invertebrate ecological immunity. *Science* 301**:** 472–475.

Rolff J. (2002) Bateman's principle and immunity. Proc. R. Soc. B: Biol Sci. 269(1493):867–872.

Rosales, C . (2011). Phagocytosis, a cellular immune response in insects. ISJ 8: 109-131,

Ryder, J., and M.T. Siva-Jothy. (2000). Male calling song provides a reliable signal of immune function in a cricket. *Proc. Roy. Soc. London Sci. Ser. B* 263**:** 1171–1175.

Savage VM, Gillooly JF, Brown JH, West GB, Charnov EL.(2004). Effects of body size and temperature on population growth. Am. Nat. 163: 429-441

Schmid-Hempel, P. (2003). Variation in immune defense as a question of evolutionary ecology. *Proc. Roy. Soc. London Sci. Ser. B* 270: 357–366.

Schmid-Hempel, P. (2005).Evolutionary ecology of insect innate defenses. *Annu. Rev. Entomol.* **50:** 529–551.

Senger K, Harris K, Levine M (2006) GATA factors participate in tissue-specific immune responses in Drosophila larvae. *Proc Natl Acad Sci* USA 103:15957–15962.

Shahabuddin, M. (1998). Plasmodium ookinete development in the mosquito midgut: A case of reciprocal manipulation. *J. Parasitol.* 116: S83–S93.

Sieber, K.-P., M. Huber, D. Kaslow, S.M. Banks, M. Torii, M. Aikawa, and L.H. Miller.(1991). The peritrophic membrane as a barrier: Its penetration by *Plasmodium gallinaceum* and the effect of a monoclonal antibody to ookinetes. *Exp. Parasitol.* **72:** 145–156.

Sim, S.; Ramirez, J.L.; Dimopoulos, G. (2012) Dengue virus infection of the *Aedes aegypti* salivary gland and chemosensory apparatus induces genes that modulate infection and blood-feeding behavior. *PLoS Pathog.* *8*, e1002631

Sim, S.; Jupatanakul, N.; Ramirez, J.L.; Kang, S.; Romero-Vivas, C.M.; Mohammed, H.; Dimopoulos, G.(2013). Transcriptomic Profiling of Diverse *Aedes aegypti* Strains Reveals Increased Basal-level Immune Activation in Dengue Virus-refractory Populations and Identifies Novel Virus-vector Molecular Interactions. *PLoS Negl. Trop. Dis.*, *7*, e2295.

Siva-Jothy, M.T. (2000). A mechanistic link between parasite resistance and expression of sexually selected trait in a damselfly. *Proc. Roy. Soc. London Sci. Ser. B* 267**:** 223–2527.

Siva-Jothy, M.T., and J.J.W. Thompson (2002). Short-term nutrient deprivation affects immune function. *Physiol. Entomol.* 27**:** 206–212.

Siva-Jothy, M.T., Y. Moret, and J. Rolff (2005). Insect immunity: An evolutionary ecology perspective. *Adv. Insect Physiol.* 32: 1–48.

Sláma, K (2008). Extracardiachaemocoelic pulsations and the autonomic neuroendocrine system (coelopulse) of terrestrial insects. *Terrestrial Arthropod Reviews* 1: 39–80.

Söderhäll, K., and A. Aspán (1993). Prophenoloxidase activating system and its role in cellular communication, pp. 113–129, in J.P.N. Pathak (Ed.), Insect Immunity. Oxford & IBH Publishing Co., New Delhi.

Strand M.R. (2008) The insect cellular immune response. *Insect Science* 15: 1-14

Suwanchaichinda C, and Paskewitz SM. 1998 Effects of larval nutrition, adult body size, and adult temperature on the ability *Anopheles gambiae* (Diptera: Culicidae) to melanize sephadex beads. J. Med. Entomol. 35: 157-161

Tartes U1, Vanatoa A, and Kuusik A (2002). The insect abdomen--a heartbeat manager in insects? Comp BiochemPhysiol A MolIntegr Physiol. 133(3):611-23.

Tepass, U., Fessler L.I., Aziz, A and Hartenstein V (2004). Embryonic origin of hemocytes and their relationship to cell death in Drosophila. *Development,* 120: 1829–1837.

Theopold U, Samakovlis C, Tempst P, Dillon N, Axelsson B, Schmidt O, Hultmark D. 1996. *Helix pomatia* Lectin, an Inducer of Drosophila Immune Response, Binds to Hemomucin, a Novel Surface Mucin. *J. Biol. Chem* **271**:12708-12715

Travers MA, Basuyaux O, and Le Goic N. 2009 Influence of temperature and spawning effort on *Haliotis tuberculata* mortalities caused by *Vibrio harveyi*: an example of emerging vibriosis linked to global warming. Global Change Biol. 15: 1365-1376

Waldock, J., Olson, K.E., and Christophides, G.K.(2012) *Anopheles gambiae* antiviral immune response to systemic O'nyong-nyong infection. *PLoS Negl. Trop. Dis.*, *6*, e1565.

Wang, YP, and Lai R. (2010). Insect antimicrobial peptides: structures, properties and gene regulation. *Dongwuxue Yanjiu*. 31(1): 27-34.

Welburn, S.C., I. Maudlin, and D.S. Ellis. (1989). Rate of trypanosome killing by lectins in midgets of different species and strains of Glossina. *Med. Vet. Entomol.* 3: 77–82.

Werner, T., G. Liu, D. Kang, S. Ekengren, H. Steiner, and D. Hultmark. (2000). A family of peptidoglycan recognition proteins in the fruit fly *Drosophila melanogaster*. *Proc. Nat. Acad. Sci. USA* 97: 13772–13777.

Williams, M. (2007). Drosophila hemopoiesis and cellular immunity. *The journal of immunology*, 178(8):4711.

Wilson R, Chen CW, Ratcliffe NA. (1999). Innate immunity in insects: The role of multiple, endogenous serum lectins in the recognition of foreign invaders in the cockroach, *Blaberus discoidalis*. *J. Immunol.* 162:1590-1596.

Xi, Z.; Ramirez, J.L.; Dimopoulos, G. (2008) The *Aedes aegypti* Toll pathway controls dengue virus infection. *PLoS Pathog.*, *4*, e1000098.

Yu, X.-Q., H. Gan, and M.R. Kanost. (1999). Immulectin, an inducible C-type lectin from an insect, *Manduca sexta*, stimulates activation of plasma prophenol oxidase. *Insect Biochem. Mol. Biol.* **29:** 585–597.

Yu, X.-Q., Y. Zhu, C. Ma, J.A. Fabrick, and M.R. Kanost. (2002). Pattern recognition proteins in *Manduca sexta* plasma. *Insect Biochem. Mol. Biol.* 32: 1287–1293.

Yu, X.-Q., Kanost, M.R., (2000). Immulectin-2, a lipopolysaccharide-specific lectin from an insect *Manduca sexta* is induced in responseto Gram-negative bacteria. *J. Biol. Chem.* 275, 37373–37381.

Yu, Q-Y., and M.R. Kanost. (2003). *Manduca sexta* lipopolysaccharide-specific immulectin-2 protects larvae from bacterial infection. *Dev. Comp. Immunol.* **27:** 189–196.

Yu, X.-Q., H. Jiang, Y. Wang, and M.R. Kanost. (2003). Nonproteolytic serine proteinase homologs are involved in prophenoloxidase activation in the tobacco hornworm *Manduca sexta. Insect Biochem. Mol. Biol.* 33: 197–208.

Zaidman-Remy, A., Herve, M., Poidevin, M., Pili-Floury, S., Kim, M.S., Blanot, D., Oh, B.H., Ueda, R., Mengin-Lecreulx, D., and Lemaitre, B. (2006). The Drosophila amidase PGRP-LB modulates the immune response to bacterial infection. *Immunity* 24, 363–366.

Zuk, M., and A.M. Stoehr. (2002). Immune defense and host life history. *Am. Natural.* 160: 9–22.

Dynamics of the Respiratory System as a Key to Insects' Success 5

Synopsis

1. Structure and diversity of tracheal system
   - 1.1 Spiracles
   - 1.2 Tracheae
   - 1.3 Tracheoles
   - 1.4 Air sacs, aeriferous trachea and tracheal lungs
   - 1.5 The tracheal system in Locusts- A case study
2. Classification of the tracheal system
3. Other types of respiration
4. Development of the tracheal system
5. Plasticity in tracheal organisation as a key trait for insect's success
6. Gaseous exchange in insects
   - 6.1 Gas exchange pattern
   - 6.2 Discontinuous gas exchange
7. Gas exchange in aquatic insects
   - 7.1 Oxygen intake from air
   - 7.2 Oxygen intake from water
   - 7.3 Physics of bubble gas exchange in collapsible gas gills
8. Respiration in endoparasites
9. Respiratory pigments in insects
10. Respiration in eggs and water conservation.
11. Control of respiratory patterns
    - 11.1 Spiracle control and innervations
    - 11.2 Metabolic rate determines patterns of respiration
12. Using the knowledge gained from the insect's tracheal system

INTRODUCTION

Most insects respire *via* an air-filled tracheal respiratory system that is represented by a series of branched tubules which ramify throughout the body of the insect carrying oxygen directly to the cells. The tracheal system generally consists of cells joined to form walls of hollow tubes that run throughout the body, terminating in blind-ended tracheoles in close proximity to the cells.In contrast to using liquid (blood) as a respiratory medium, the use of air for delivering oxygen has several advantages such as (1) It is light-weight and requires less energy for production, maintenance, and transport compared to blood-based systems; (2) It supports aerobic metabolism during active flight and is a very important key to the great biodiversity and the ecological success of insects; (3) It provides an internal reservoir for oxygen, which allows prolonged periods of asphyxia; (4) If insects eventually become paralysed by exposure to anoxia, they can recover quickly when re-exposed to oxygen, because oxygen can be delivered by passive diffusion, which is rarely true in vertebrates.

The insects' unique physiology and possession of a tracheal system have been fundamental to their evolutionary history and consequent success in a changing environment down the ages. As insects grow from hatchlings to adult, the body mass increases several folds and increase in body mass must be supported by more oxygen supply. Insects achieve this by several mechanisms. For example, larger insects may have more or larger tracheae or they may breathe faster. In addition, larger insects may change cellular or gene expression patterns that alter oxygen supply or demand. Studies on the dynamics of the respiratory system in insects may help to develop non-pesticide methods of insect control, as manipulation of atmospheric gases is increasingly important for controlling insects in grain storage, greenhouses, and museum collections.

I. STRUCTURE AND DIVERSITY OF THE TRACHEAL SYSTEM

Insects have a tracheal respiratory system in which oxygen and carbon dioxide travel primarily through air-filled tubes called **tracheae**. Usually the tracheal system penetrates the cuticle via closable valves called **spiracles** and ends near or within the tissues in tiny tubes called **tracheoles** (Fig 5.1). The spiracles occur on the pleural surface of the body, one on either side of each segment. The tracheal system with functional spiracles is called **open tracheal system** and with non-functional spiracles are called **closed tracheal system.** The tracheae primarily serve as pipes that transport gases between the spiracles and the tracheoles, whereas the thin-walled tracheoles are thought to be the main sites of gas exchange with the tissues. However, in many insects, the tracheae are compressible, and dilations of the tracheae form thin-walled **air sacs** that together serve as bellows for enhancing the flow of gases through the tracheal system. The organisation of the tracheal system varies dramatically among insects, with spiracle number ranging from 0 to 20 and with tracheal branching patterns varying widely across species, between body regions, and during the developmental stages of holometabolous insects. This variation in the tracheal system organisation is partially at least correlated with species ecology. Some aquatic insects lack spiracles so as to prevent water entry into the trachea. Many dipteran larvae have only one pair of spiracles on their abdominal tip, which they extend into the air as they feed head-down in the water. In contrast, many terrestrial insects have the full complement of 10 pairs of spiracles. Insects also vary in the number of air sacs, the location and size of tracheal branches, and the number of longitudinal trachea.

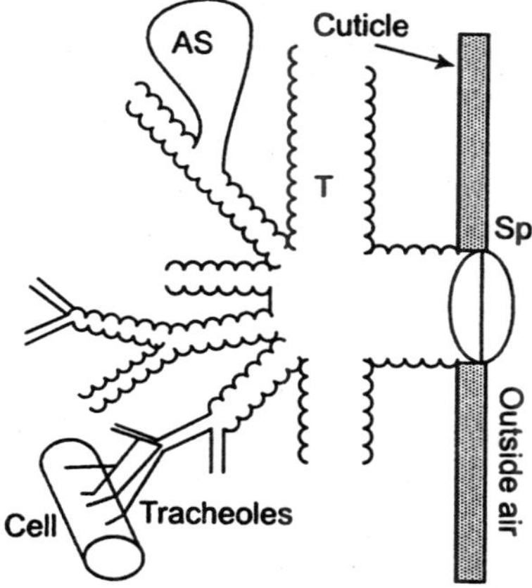

Fig. 5.1 An overview of the tracheal system.
Sp –Spiracles; T- tracheae;AS- air sacs

1.1 Spiracles

Spiracles are the openings of the tracheal system on the integument of the insect. In all the insect orders except Diplura, spiracles are absent in prothorax and are distributed in meso, methathorax and abdomen. A total of ten pairs of spiracles is present generally, with two pairs in the thorax and eight pairs in the abdomen. Each spiracle consists of a small ring like sclerite at opening called **peritreme** leading to a small cavity known as **atrium** (Fig 5.2).The closing and opening of spiracles are accomplished by **atrial valve, lined** with fibrous processes and form the so called **felt chamber** which reduces water loss in the absence of closing mechanism. Some apterygote and larval insects lack valves in their spiracles and therefore have trachea that are always open to the environment, although these are often covered with sieve-like plates containing large number of small apertures through which gas exchange takes place. This modification is to prevent entry of water, especially in aquatic forms. Often the outside of the spiracle is fringed with hairs or covered with a filter to help resist entry of dust, water, or parasites.

The valve that closes the spiracle takes various forms, but can appear as two lips of cuticle or as a bar of cuticle that pinches the trachea shut. The movement of the valve usually depends on the action of muscles that insert on or near it, and elastic elements in the cuticle. Some thoracic spiracles have only a closer muscle, and elastic elements tend to open the spiracular valve. Often the valve is recessed behind the atrium. The most common situation (true for all abdominal spiracles studied to date) is for the spiracle to be recessed behind an atrium and to have both opener and closer muscles.

In most of the terrestrial insects, water loss through spiracles is controlled by the closing mechanism which consists of one or two valves or a constriction ofthe trachea or by muscular activity. The hydrophobic nature of spiracles is also dueto the presence of modified epidermal glands known as **peristigmatic glands,** which secrete a hydrophobe material preventing the wetting of these organs.

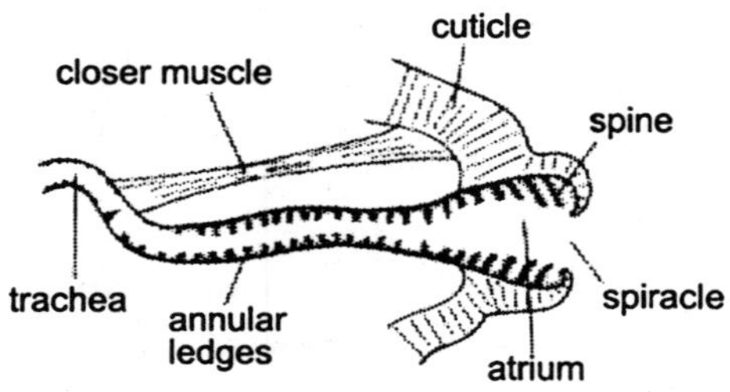

Fig. 5. 2 Structure of spiracle
(Source: Taken from A General Textbook of Entomology (1960) – A.D. Imms

1.2 Tracheae

Tracheae are the largest tubes of the tracheal system. They range in diameter from a few millimetres to about 1 μm. The tracheal wall contains an epithelial cell layer that secretes a basement membrane and forms the outer most layer of the tracheal wall. Internally, the tracheal cells secrete the intima which contains protein and chitin fibres. The intima forms spiral folds known as **taenidia**. It is thought that the tenidia prevent the trachea from collapsing inward, yet allow the trachea to flex. However, in the tracheae of some insect, the taenidia are widely spaced within the tracheal wall, and the intima is quite flexible so that the trachea is compressible. Because tracheal walls are relatively thick, wherein oxygen is not believed to pass out the large trachea in appreciable amounts. Thus the primary function of the trachea is to transport gases between the spiracles and the tracheoles. However, because carbon dioxide can diffuse more easily through tissues than oxygen, it is possible that a significant amount of carbon dioxide emitted may pass through the tracheal walls.

1.3 Tracheoles

The trachea ramify into very fine branches known as **tracheoles** which are about 0.1 – 1 μm in diameter. These tracheoles are formed in to cells called **tracheoblast** or **tracheolar end cell**, which are derived from epidermal cells, lining the trachea. Tracheoles form a network over the visceral organs including the alimentary canal as well as the gonads and penetrate in to the tissues of the organs and become intracellular and supply oxygen directly to the tissues.

Tracheole walls are capable of transporting oxygen at high rates of diffusion because they are thin (usually <0.1μm) and have a large surface area to volume ratio. Thus, they are likely the major site of gas exchange between the tissues and the tracheal system. Tracheoles are generally dense in tissues such as the flight muscles where there is high metabolic activity. Within the flight muscles, the tracheoles occur within the cells and it is believed that they enter the flight muscles via infoldings of the muscle plasma membrane and so they are actually extracellular. Box 5.1 provides the basic difference between the trachea and tracheoles.

Box 5.1 Differences between Trachea and Tracheole

Trachea	Tracheoles
1. These are large tubes running from Spiracles	1. Fine tubes arising distallyfrom trachea
2. Taenidiapresent	2. Absent
3. Intima layer is shedduring moulting	3. Intima layer is retained, unchanged
4. Never become intracellular	4. Intracellular
5. The intima layer consist of protein – chitin matrix with resilin,	5. Chitin – protein matrix present, resilin absent

The trachea coming from spiracles throughout the body joins with those of neighbouring spiracles forming **longitudinal trunks**. Likewise, transverse commisures and longitudinal connectives are formed by the fusion of trachea coming from dorsal, lateral and ventral sides of the body.

All these in total form into dorsal trunk, lateral trunks which are two in number and one ventral trunk (Fig 5.3). The dorsal trunk supply oxygen to proximal part of the body as well as to heart where as the ventral supplies to thecentral nervous system. The two lateral longitudinal trunks spread tracheoles toalimentary canal, legs, gonads and wings. As the head does not contain spiracles, airis supplied through the first pairspiracles by means of two main branches of the dorsal longitudinal trunk, where onebranch supply oxygen to the eyes, antenna, brain; another branch to mouthparts and the muscles of the head.

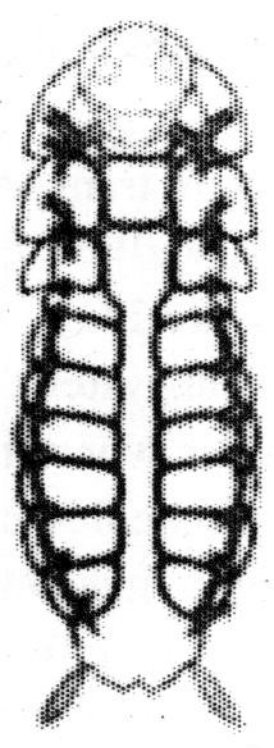

Fig. 5.3 The dorsal tracheal trunks

(Source: Taken from A General Textbook of Entomology (1960) – A.D. Imms)

1.4 Air sacs, aeriferous trachea and tracheal lungs

In insects that are very active and have a rigid cuticle, portions of the tracheal system get enlarged to form thin-walled air sacs. These air sacs lack taenidia and therefore can collapse and expand based on the haemolymph pressure. They serve as bellows to enhance the movement of air through the tracheal system. The air sacs account for the majority of the volume of the tracheal system in Diptera, Hymenoptera, Lepidoptera, Odonata and Orthoptera. Insects such as cockroaches, tenebrionid or carabid beetles lack air sacs and have compressible trachea that may also function as bellows. Box 5.2 lists the functions of the air sacs

Box 5.2 Functions of the air sacs

1. Increase the volume of tidal air
2. Reduce the diffusion path to the tissues
3. Aid in ventilation of the tracheal system
4. Can be used to provide sectioning of the thorax from the abdomen
5. Form tympanic cavities for hearing organs and sound producing organs
6. Lower the specific gravity of insect, thus aiding in flight
7. In some chironomids, it aids to determine the depth they based on how much air is in the sac.

Tracheal lungs or aeriferous tracheae, are similar to air sacs in that they are thin-walled expansions of the tracheal system with few taenidia that float freely in the haemolymph. Their primary function is to deliver oxygen to tissues that lack tracheoles, such as haemocytes, some ovaries and Malphighian tubules. In such cases the oxygen must diffuse through the thin wall of the tracheal lung, through the hemolymph and to the tissues. These tracheal lungs are seen in lepidopteran larvae which apparently do not have haemocyanin in the haemolymph.

1.5 The tracheal system in Locusts- A case study

The tracheal system of the American locust, *Schistocerca americana*, has a complex morphology. There are 10 pairs of spiracles of which 2 are on the thorax, and 8 on the abdominal segments I–VIII. The tracheae are supported by rings of thickened cuticle (taenidia) that resist compression and expansion. The air sacs lack taenidia and are very soft and flexible. The system can be conceptualised as a collection of subsystems that serve the major regions and tissues of the locust: the head, the thorax/flight muscle, the legs, the ventral nerve cord, the digestive tract/renal system, and the dorsal heart.

Within the abdomen, longitudinal tracheal trunks connect the abdominal segments. Pairs of longitudinal trunks run along the dorsal heart and the ventral nerve cord, whereas others run dorsally, laterally, and ventrally along the gut or connect the abdominal spiracles. Regular transverse tracheae connect the spiracles of each segment to these longitudinal tracheae. The dorsal and lateral gut longitudinal tracheae connect anteriorly to spiracle 3, whereas the ventral gut longitudinal tracheae connect to spiracle 4. Finer tracheae branch from both the main transverse tracheae and the longitudinal tracheae into the tissues where gas exchange occurs. The abdominal segments have large dorsal air sacs on each side that connect near the spiracles within that segment. In anterior regions of the abdomen, these dorsal air sacs are interconnected, in the posterior regions, these connect to the dorsal gut longitudinal trunk.

The dorsal air sacs of **abdominal segment I** and **II** are particularly large and paired. **Abdominal segments III–VIII** also have smaller air sacs that connect near the spiracles and lay blind-ended between the gut and lateral abdominal wall (Fig 5.4). Multiple air sacs extend beyond the longitudinal trunks into the most posterior region of the abdomen, **segments IX** and **X**, and the epiproct and paraproct. Anastomoses connect the left and right sides between the paired neural longitudinal tracheae, between the gut longitudinal tracheae, and between the large air sacs in **abdominal segments I** and **II**, but not between the paired cardiac longitudinal tracheae.

The tracheal system of the thorax is separated from that of the abdomen. The cardiac longitudinal tracheae and spiracular tracheae do not extend into the thorax. The spiracular longitudinal trachea bypasses spiracle 3 and connects to the ventral longitudinal tracheae anterior to spiracle 4. The ventral neural longitudinal trachea runs through the thorax between the head and the abdomen, but has minimal connections with the thoracic tracheal system. Thus the primary connection between the abdominal and thoracic tracheal systems occurs where longitudinal tracheae of the dorsal and lateral gut meet tracheae from the thorax at spiracle 3.

In the thorax, tracheae branch from thoracic spiracles 1 and 2 and abdominal spiracle *3* to the flight muscles, as well as to the various air sacs, of which the largest are the mesial tergo-pleural air sacs that occur ventral to the dorso-longitudinal flight muscles. There are also multiple pleuro-coxal air sacs connected to tracheae that feed the dorso-ventral flight muscles and the legs. There are two major tracheae that feed each leg, and these arise from different branches and often different spiracles. For example, the metathoracic (jumping leg) receives one trachea directly from *spiracle 4* and is also serviced from the neural longitudinal tracheae.

The head is supplied dorsally by large tracheae branching from dorsal atrium of the first spiracle and ventrally by the neural longitudinal. Tracheae branch to the brain, mandibular muscles, and alimentary canal, and anastomose. There are many air sacs, with the largest in the mandibles. These accommodate the movements of the mandibles during chewing, allowing the mandibles to move in and out of the rigid head capsule and likely providing autoventilation to support the extra oxygen need associated with chewing.

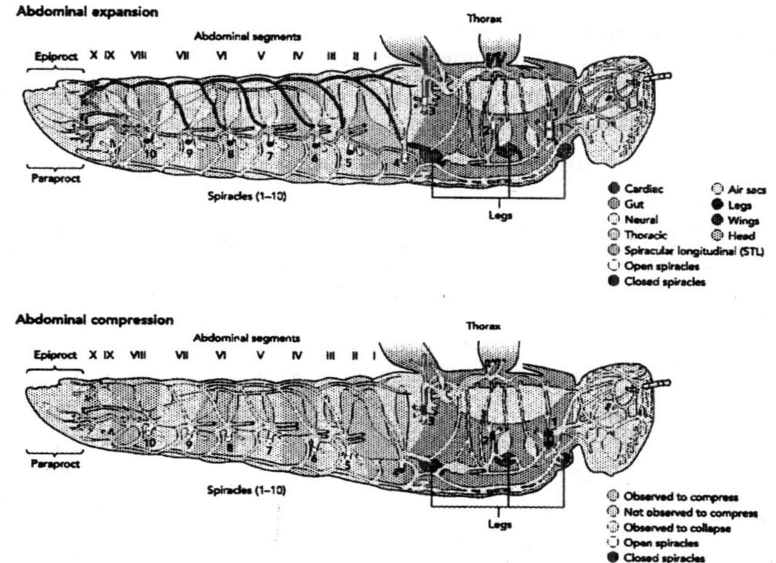

Fig. 5.4 Schematic sketch of the tracheal system of Locusts

(Taken from Jon F. Harrison, James S. Waters, Arianne J. Cease, John M. Vanden Brooks, Viviane Callier, C. JacoKlok, Kimberly Shaffer, John J. Socha Physiology Published 1 January 2013 Vol. 28 no. 1, 18-27)

2. CLASSIFICATION OF THE TRACHEAL SYSTEM

The tracheal system in insects can be classified based on the number and arrangement of functional spiracles. Most insects have ten pairs of spiracles and their modifications are as follows:

1. **Holopneustic:** These are primitive types with 2 pairs of spiracles in the thorax and 8 pairs in the abdomen (1+1+8). All the spiracles are functional. Eg. Dragonflies, grasshoppers and cockroach.
2. **Hemipneustic:** One or more pairs of spiracles become non-functional. They are:
   - A. **Peripneustic:**Metathoracic spiracles are closed (1+0+8) Eg. Larvae of Lepidoptera, Hymenoptera, Coleoptera.
   - B. **Amphineustic:** Only mesothoracic and last pair of abdominal spiracles are open (1+0+1). Eg. Larva of cyclorrhaphan Diptera.
   - C. **Propneustic:** Only the mesothoracic spiracles are open (1+0+0). Eg Mosquito pupa.

D. **Metapneustic:** Only last pair of abdominal spiracles is open. (0 + 0 + 1). E.g.mosquito larvae.

E. **Apneustic:**No functional spiracles. e. g: mayfly larva, nymph of Odonata.

3. **Hypopneustic:** 1 or 2 pairs of spiracles may completely disappear or absent E.g.: Siphunculata, Mallophaga.
4. **Hyperpneustic:** More than 10 pairs of spiracles are present e.g.*Japyx* sps.

3. OTHER TYPES OF RESPIRATION

1. **Cutaneous respiration**: When the spiracles are absent, respiration occurs through the body wall which forms the main sourceof gaseous exchange. E.g. Protura, Collembola and endoparasitic insects.
2. **Tracheal gills**: Also called **abdominal gills**, which occur as the outgrowths of the trachea in theform of gills distributed on the lateral sides of the body. They are useful for absorption of dissolved oxygen. They may vary in shape as lamellate orfilamentous. E.g.larva of Trichoptera, nymphs of Ephemeroptera.
3. **Spiracular gills:** In some aquatic pupae,peritreme or atrium of spiracles is drawn out into a long filament like structure known as **spiracular gills**. These gills are adapted for both aquatic and aerial respiration, enabling the insect to live on air and moist places or completely in water or at the edges of water structures.
4. **Blood gills:** These are tubular or digitiform or eversible structures present at the anal end of body ranging from 4-6 in larva of Trichoptera. In chironomid larva of Diptera, 2 pairs of blood gills are present on penultimate segment and a group of 4 shorter anal gills are present. These are called **blood gills** as they contain blood, but sometimes have trachea. The function of these structures is the absorption of water and inorganic ions rather than respiration.
5. **Rectal gills:** In dragonfly nymphs (naids), the rectum modifies into a barrel like chamber where the rectal wall forms in to basal thick pads and distal gill filaments which are richly supplied with tracheoles. They help in respiration.
6. **Air sacs:** In many winged insects, the trachea gets dilated at some points to form thin walled air sacs which do not contain the taenidia. These can be seen asglistening sac like structures mainly function as storage structures of air which change their volume with respiratory movement.

7. **Plastron respiration:** E.g.aquatic beetles.The plastron is a special type of air store in the form of a thin film held by a system of hydrofuge hairs, scales or other cuticular processes whose volume remains constant. If there is adequate oxygen dissolved in water, the plastron can act as apermanent physical gill.The trachea opens into plastron.

4. DEVELOPMENT OF THE TRACHEAL SYSTEM

Unlike vertebrates which have developed two distinct yet interconnected systems, one for oxygen uptake (lungs) and the other for distribution (circulatory system), insects rely solely on the tracheal system for both these physiological functions. Our knowledge on the development of the tracheal system in insects is mainly from the work carried out on *Drosophila melanogaster*. The development begins around 4h after oocyte, fertilisation, to form an interconnected network of tubes during embryogenesis. This network becomes functional at the beginning of the first larval instar and ramifies dramatically during the 2nd and 3rd larval stages to accompany the increase in body size of the larva. Finally, it is rebuilt at the pupal stage to acquire the ultimate pattern of the adult (Ghabrial *et al.*,, 2003; Weaver and Krasnow, 2008).

The tracheal system arises at mid-embryogenesis as 10 independent ectodermal placodes of around 80 cells each at either lateral side of the embryo. Each placode then invaginates to form a sac, and immediately afterwards, tracheal cells migrate in highly stereotyped directions to give rise to six tracheal branch primordial. After tracheal branches have extended, most of them meet and fuse to branches arising from the contra-lateral hemisegment, or from the adjacent segments, to generate an interconnected and continuous network of epithelial-like tubes. Primary branching of the tracheae is induced by secretion of a local chemoattractant termed **fibroblast** growth factor secreted by local tissues. The primary branches are formed of 3-20 cells organised into tubes. Some cells become terminal cells and ramify in response to fibroblast growth factor to become tracheae, with densities of tracheoles modulated by local oxygen levels.

In general, the size of the tracheal system increases with age in order to support the increased gas exchange need of the larger insect. During the moults, the cuticular lining of the trachea is drawn out of the spiracle with the old integument. Major changes in tracheal structure, including changes in spiracle number and tracheal system organisation, can occur at each moult and during the pupal period for endopterygote insects. In *Drosophila*, the adult tracheal system is formed from both imaginal disc cells and larval tracheal cells that activate and proliferate.

Changes in the tracheal system structure are not limited to moulting periods, because the tracheoles can change structure within an instar. In the event of injury or oxygen deprivation, local tracheoles grow and increase in branching. If no undamaged tracheoles are nearby, damaged tissues produce cytoplasmic threads that extend toward and attach to healthy tracheoles. These threads then contract, dragging the tracheole and its respective trachea to the region of oxygen deficient tissues

Both tracheoles and trachea are fluid filled in newly hatched insects, and fluid fills the space between the old and new trachea at each moult. Usually this fluid is replaced with gas shortly after hatching or moulting. In some cases, the spiracles must be open to the air for gas-filling of the trachea to occur, suggesting that gas filling occurs as fluid is actively absorbed by the tracheal and tracheole epithelia, with air entering through the spiracle to replace the absorbed fluid. However, in aquatic insects that lack spiracles, the trachea also become gas-filled, indicating that this gas can be generated by the tissues or haemolymph.

5. PLASTICITY IN TRACHEAL ORGANISATION AS A KEY TRAIT FOR INSECT'S SUCCESS

Insect are diverse in terms of the ecological niches they occupy and this is reflected in the modifications in the organisation of the tracheal system. Aquatic dipterans have only one pair of spiracles while the true terrestrial forms have 10 pairs of spiracles. Grasshoppers have spiracles on every segment while fly larvae have functional spiracles only on the head and abdominal tip. Variations in the number of air sacs, location and size of tracheal branches and number of longitudinal trachea are also seen in relation to species ecology.

In grasshoppers, the transverse trachea leads from the spiracles to longitudinal tracheae that run along the heart, ventral nerve cord, and several positions of the gut. In these insects, the air sacs are prominent. Flies on the other hand have two longitudinal trunks that connect the two functional spiracles on each side, with branches extending from the dorsal trunk to the tissues and one other longitudinal trachea. In these insects, air sacs are totally absent.

In most insects, there are no spiracles on the head. Therefore the trachea extends forward from the prothoracic spiracle to supply the tissues of the head. Similar extensions are seen in the posterior direction from the 10th abdominal segment. The legs, antennae, and wings that extend from the main body, also do not have spiracles and the trachea extend from the spiracle corresponding to the local segment. These regions have accessory pulsatile structures that pump haemolymph and help circulate air through their long trachea. The distribution of tracheae in the thorax of flying insects is different from the abdomen. The thorax contains extensive air sacs associated with the flight muscles.

The tracheal system is strongly modified during the developmental stages, especially in holometabolous insects. To suit the aquatic life of dipteran larvae, there are only one or two pairs of spiracles with no air sacs. But during the adult stage, to meet the high oxygen demand, especially during flight, adult dipterans have 10 pairs of spiracles and extensive air sacs. Undoubtedly, flexibility in the tracheal structure and function is a key trait in the success, adaptability and diversity of insects. As insects increase in size, the demand for oxygen supply to the tissues is more and therefore the insect must proportionally invest more of their body in tracheal structures. Therefore insects' maximum body size may be limited by the need of larger insects for proportionally more trachea.

The respiratory patterns are also different based on whether the insect is sluggish or active. Even within a species, there is a continuum of respiratory responses driven by metabolic demand experienced by the insect. This may vary from being a **discontinuous respiratory pattern** when temperatures are low and metabolic activity low, to being **cyclic** at moderate activity and at the increased metabolic activity, the pattern is **continuous**.

6. GASEOUS EXCHANGE IN INSECTS

The tracheoles are usually filled with liquid and are intimately associated with the tissues. In the flight muscles where there is great oxygen demand, the tracheoles indent the plasma membrane and penetrate deeply into the muscle fibre. Gaseous exchange follows the principle of diffusion, convection and ventilation. Diffusion is the movement of gases from a region of higher concentration/partial pressure to lower concentration/partial pressure. Convection means mass movement of the medium and ventilation refers to convection that is assisted by some form of pumping activity usually associated with muscular contraction and relaxation (Box 5.3).

Box 5.3 General principles of gas exchange

1. **Diffusion:** Movement of substance down a gradient in concentration or partial pressure without bulk movement of medium.
2. **Convection:** Movement of a substance by being carried in a moving fluid (water or air).
3. **Ventilation:** Convection assisted by pumping activity.

Typically it is thought that oxygen moves along the pathway starting from the spiracles through the main tracheal tube by convection or diffusion and then into the tracheoles from where it diffuses into the mitochondria. Diffusion from the large trachea to the surrounding tissues or haemolymph is thought to be negligible because of the low partial pressure of oxygen between the tracheal lumen and the surrounding tissues. Carbon dioxide is thought to enter the tracheal system at all points from the tissues and haemolymph where it is buffered as bicarbonate.

6.1 Gas exchange pattern

Different patterns of gas exchange have been recognised when the insect is at rest. These ranges from the continuous gas exchange by diffusion-convection or convection, with spiracles held open to **discontinuous gas exchange cycles** (DGCs) where gas exchange is either diffusive-convection kind or active ventilation where in ventilation and spiracle opening are highly coordinated. Between these extremes insects show different patterns, ranging from being barely cyclic through to strongly cyclic gas exchange.

During rest some species show active ventilation accompanied by gas exchange. In these insects ventilatory movements can include active abdominal pumping, head protraction and retraction, and prothoracic pumping. Haemolymph circulation also plays a considerable role in tracheal ventilation, especially in adult insects. This is as a consequence of the presence of many large air sacs and the compartmentalisation of the haemocoel into an anterior section comprising of the thorax and often the first abdominal segment, and a posterior section, consisting of the abdominal segments. In flies (*Calliphora*), haemolymph flow reversal, associated with heartbeat reversal, interacts with the air sacs to produce inspiration via the abdominal spiracles during forward heart beat. During the retrograde heart beat, collapse of the abdominal air sac via the abdominal spiracles produce expiration. Accessory pulsatile organs also play a significant role in circulation and the interaction between circulation and gas exchange.

The central nervous system is thought to control the ventilatory movements via the action of abdominal ganglia and a central pacemaker located in the fused abdominal or metathoracic ganglia, or in abdominal ganglia, as well as by afferent feedback from abdominal ganglia and proprioceptors. There appears to be a central chemoreception of oxygen and carbondioxide in the CNS.

Gas exchange pattern and efficacy are markedly affected by the degree of synchronisation between the spiracles and abdominal pumping. Spiracular opening and closing are affected by the partial pressures of both oxygen and carbon dioxide. Spiracular movements are also affected by the hydration state of the insect and the humidity of the surrounding environment. There is much variation in the degree of spiracular coordination, ventilatory movements, and airflow, among species, among individuals, and even within individuals. Thus, gas exchange may take place through either one or all of the spiracles, spiracular coordination can be either tightly regulated or completely unsynchronised, and airflow may be unidirectional, tidal, or in the form of cross-currents.

A large portion of the variation in gas exchange patterns within and among individuals is clearly a function of external factors (water availability) or the physiological status of the individual (such as levels of dehydration, starvation, or activity).

Locusts and some other insects (bees, cockroaches, beetles) have been shown to have unidirectional air flow. In locusts, inspiration often occurs via the first four pairs of spiracles, which open during the expansion of the abdomen, whereas expiration usually occurs via the posterior six pairs of spiracles that open during abdominal compression. During flight, spiracles 1 and 4–10 remain synchronised with abdominal pumping, but spiracles 2 and 3 are kept wide open, allowing tidal ventilation of the flight muscles due to autoventilation associated with thoracic volume changes driven by the flight muscles. However, this basic pattern can be altered. Some variation is graded. The number of open spiracles (both inspiratory and expiratory) increases with activity and respiratory drive, as does the percent of time during each respiratory cycle that each spiracle is open. Spiracular patterns also vary substantially. Many locusts exhibit discontinuous gas exchange when very quiescent, closing spiracles and ceasing external gas exchange for periods of time. Spiracles 3, 4, and 10 have been reported to exhibit both inspiratory and expiratory functions (McArthur 1929). If the abdominal spiracles are sealed, the locusts can both inspire and expire from their thoracic spiracles. Coordination of spiracular timing depends on central interneurons, but local effects of O_2 and CO_2 on spiracles also influence opening time.

6.2 Discontinuous gas exchange cycles

Discontinuous gas exchange cycles (DGC) are characterised by the repetition of three major periods, during which the exchange of carbon dioxide and oxygen is partially separated. In terms of spiracular behaviour these are (1) an open period (O)when the spiracles are fully open, (2) a closed period (C) during which the spiracles are tightly shut and (3) a flutter period (F) during which the spiracles open and shut by means of reduced fluttering movements. During the closed period, no gas exchange takes place through the tightly shut spiracles. Oxygen in the endotracheal space is depleted by respiration, but carbon dioxide is buffered in the haemolymph, leading to a slow decline in endotracheal pressure and to a steady decline in partial pressure of oxygen. A CNS-mediated to oxygen set point, then causes the spiracles to partially open and close in rapid succession, and the flutter period is initiated. Convective gas transport restricts outward movement of both carbon dioxide and water and is thought to be one of the major ways in which DGC is responsible for water conservation. During the F-period, endotracheal pressure increases rapidly towards ambient in a series of steps caused by spiracular opening and closing. During the open period, gas exchange takes place predominantly by diffusion, whereas when the spiracles are more constricted convective gas exchange predominate.

6.2.1 Internal and external factors that influence the DGC

Not all insects show DGC, in fact, most of the investigated species do not (Marais *et al.*,, 2005). Several factors have been identified that are necessary for DGC or at least favour it. First of all, DGC is a phenomenon typical of resting insects. When they become active, the pattern vanishes and is replaced by continuous gas exchange to meet the increased metabolic demand. The aerobic scope of insects is unsurpassed and reaches to more than 100 fold increased metabolic rates during flight in honey bees (Stabentheiner *et al.*,, 2003). Temperature, body mass and metabolic rate are factors that influence DGC. An increase in temperature increases metabolic rate, which is accompanied by an increase in DGC frequency. An increase in temperature results both in a decline in carbon dioxide solubility and a decline in haemolymph pH, thereby affecting haemolymph buffering capacity. The change in buffering capacity and increased metabolic demand presumably lead to an increase in DGC frequency.Again the crucial factor seems to be metabolic rate. As ectothermic animals, insects show lower resting metabolic rates at lower temperatures. During dormancy, there is likewise a typical depression of metabolic rate (U-shaped curve; Schneiderman and Williams, 1953).

The observations of metabolic rate as a key factor in the DGC lead Bradley (2008)to the conclusion that the respiratory pattern might solely be a result of a control that faces achange in metabolic rate. This idea is supported by a number of observations that asystematic change in the pattern occurs if the temperature is artificially modified (Contreras and Bradley, 2009, 2010).

However, other factors also determine the DGC pattern. Through the occlusion of spiracles the conductance of the spiracles and consequently the whole tracheal system can be artificially modified (Hetz, 2007). The results show that typically the spiracular conductance is high in moth pupae and the DGC pattern is retained for a large number of blocked spiracles. However, the proportions of the three sub-phases change with the number of sealed spiracles. Bursts tend to increase in duration at the expense of the flutter phase (Hetz, 2007; Mörbitz and Hetz, 2009). At one point subsequent bursts fuse and the pattern essentially become continuous. One of the main explanations for the occurrence of DGC is the reduction of respiratory water loss. The influence of environmental relative humidity is well researched with inconclusive results. Whereas earlier studies concluded DGC to occur primarily in dry conditions (Cloudsley-Thompson, 1975; Edney, 1977), newerwork casts some doubt on this simple water DGC interaction. Hadley and Quinlan (1993) found an inverse relationship between hydration state and the occurrence of DGC in grasshoppers. Dehydrated individuals were more likely to give up DGC even though this would have incurred an additional water cost. In a comparative study, Terblanche *et al.,* (2008) likewise found only little evidence of a humidity effect on DGC. However, other authors found the opposite (White *et al.,*, 2007). In conclusion, the question whether environmental humidity directly influences the regulation of gas exchange in insects still is unclear.

For many years, research on DGC had concentrated on few key species often from closely related groups (moth pupae (e.g. Buck, 1958; Levy and Schneiderman, 1966), ants (Lighton and Berrigan, 1995), beetles (Duncan and Byrne, 2000) or roaches (Kestler, 1985)). A broad scale phylogenetic study (Marais *et al.,*, 2005) showed that only a few orders of insects show the typical CFO-type DGC as described earlier. Most others either show periodic fluctuations in the CO_2 release without ever fully closing their spiracles (cyclic gas exchange), or, alternatively, they simply use the continuous gas exchange all the time.

6.2.2 Evolution of the DGC

Given the large diversity of gas exchange modes in the different orders and life stages of insects, it is not surprising that a large number of explanations have been developed regarding the origin of DGC. Chown *et al.,* (2006) sum up the current hypotheses that explain the evolution of discontinuous gas exchange. Most of them are adaptive in that they argue that DGC were the consequence of the selective pressure on a single trait. One hypothesis (the emergent property hypothesis) is non-adaptive with respect to the DGC pattern.

Hygric hypothesis

Like all terrestrial animals, insect also faces the problem of increased water loss in comparison to aquatic animals. Depending on local climate and humidity, a large range of adaptations is seen to minimise water loss across all possible paths. The respiratory system is no exemption. However, as the primary function of a respiratory system is the exchange of respiratory gases, that is oxygen and carbon dioxide; all animals have to trade-off delivery capacity*vs.* possible water loss along the same pathway. Given the generality of the above points it does not surprise that the first explanation which was developed for the DGC is that the pattern serves to minimise respiratory water loss by maximizing the time the spiracles are closed (Buck, 1958).

Strolling arthropods hypothesis

Miller (1974) attributed the prevention of tracheal mites as one of the functions of the spiracles. The dense sieve plate and the closing valve should play a role in this process. However, Miller (1974) did not postulate a causal role of mite prevention for the evolution of the DGC. Observations of Harrison *et al.,* (2001), showed that tracheal mite infestationreduces the flight metabolism, safety margins for oxygen uptake in bees.Though usually treated as an exotic alternative, this hypothesis on the origin of DGChas not been disproved so far (Chown *et al.,* 2006).

Chthonic hypothesis

The first direct competitor to the long standing hygric hypothesis had been developed when Lighton and Berrigan (1995) discovered that all insect species that show DGC spend at least some life stage subterranean. Often these habitats share hypoxic and hypercapnic conditions. The Lighton's idea was that the DGC pattern maximizes the external gas exchange by steepening the partial pressure gradients across the spiracle. The diffusive gas exchange will thus be enhanced with spiracle opening. Since the mechanism in itself provides no adaptive explanation, two different additionalpropositions were put forward. First that gas exchange rate had to be enhanced to minimize prolonged spiracle opening which would incur high respiratory water loss. This is basically the same assumption as the hygric hypothesis, so this particular interpretation of the chthonic hypothesis was later refined as the chthonic-hygric hypothesis (Chown *et al.,* 2006). The second proposition is that the increase of the CO_2 gradient per se is adaptive (Lighton,1998; Chown *et al.,* 2006). It's balanced, especially in hypoxic conditions, the disparaterates of diffusion between O_2 and CO_2. A puzzling consequence of this idea is that apattern would adaptively evolve, which results in the abandonment of the very same pattern, as balanced exchange rates would be equivalent to continuous gas exchange.

Oxidative damage hypothesis

When Hetz and Bradley (2005) found that the oxygen partial pressure is tightly controlled to lower levels during the flutter phase, irrespective of the ambient pO_2. They proposed that the adaptive value of the DGC was to keep internal oxygen levels right to the level that was needed for sufficient oxygenation of the tissues. It has long been known that oxidative damage occurs in tissues if they are well oxygenated (Fridovich, 1977; Landis and Tower, 2005). Animals therefore have developed oxygen detoxification mechanisms. That these are crucial is shown by experiments where a knock out or an over-expression of the particular proteins resulted indirect effects on adult lifespan (Sun *et al.*, 2002).In this interpretation the fluttering phase was the visible sign of the control processfor O_2. The accumulation of CO_2 is deemed a side effect of keeping the oxygen levels low.Because eventually the CO_2 has to be released the open phase occurs with a massiveburst of CO_2. Its purpose is to rid the animal of carbon dioxide as fast as possible because the endotracheal partial pressure of O_2 rise to nearly ambient values after a to less harmful levels.

Emergent property hypothesis

A completely different approach was taken by Chown and Holter (2000). Instead of trying to explain some adaptive value of the pattern they attributed it to the interaction of multiple different control systems. The complex pattern of DGC therefore could be an emergent phenomenon with no immediate advantagein itself. Because the hypothesis is not easily testable, it has often been neglected in current research that comparatively tested multiple hypotheses (White *et al.*, 2007; Terblanche *et al.*, 2008). However, on second look the emergent property hypothesis allows to disintegratespiracularcontrol into simpler **modules**. The pattern is obviously dependent on O_2 and CO_2 (Schneiderman, 1960). Also a number of authors locate the sensing of CO_2 and O_2 in different regions of the body (e.g. Hoyle, 1960; van der Kloot, 1963; Burkett and Schneiderman, 1974). From an evolutionary viewpoint, it might be much easier to find the driving forces that shaped each of these low complexity modules with much simpler behaviour thanto try to find a single cause for the complex pattern of the DGC.

6. GAS EXCHANGE IN AQUATIC INSECTS

Aquatic insects have developed two ways by which they can obtain oxygen, namely, (1) from air or (2) from water (Box 5.4). In order to take oxygen from the air, they must have some semi-permanent connection with the air, or they must frequently come to the surface, the frequency of visit to the surface can be reduced if they are capable of increasing the size of the store of air. Insects which take in oxygen from water depend on gaseous exchange through thin walled gills, which are well supplied with tracheae.

Box 5.4 Mechanism of gas exchange in aquatic insects

1. Oxygen intake from air
   A. Breathing tubes (Siphon)
   B. Air bubble (Physical gills)
      - Compressible gas gill
      - Incompressible gas gill
2. Oxygen intake from water
   C. Cuticular respiration
   D. Biological gill
   E. Plastron
   F. Spiracular gill

7.1 Oxygen intake from air

Aquatic insects that live under water get air straight from the surface through hollow **breathing tubes**, sometimes called **siphons**. In mosquito larvae for example, the siphon tube is an extension of the posterior spiracles. An opening at the end of the siphon is guarded by a ring of closely spaced hairs with a waterproof coat. At the air-water interface, these hairs break the surface tension of the water and maintain an open airway. When the insect dives, the water pressure pushes the hairs close together so that they seal off the opening and keep water out. Water scorpions (Nepidae) and rat-tailed maggots (larvae of Syrphid fly) are examples of insects that have siphons. Some insects make use of oxygen stored in special vacuoles which aquatic plants have to maintain buoyancy. For example the larvae of *Mansonia* spp of mosquito insert their breathing tubes into these air stores of plant and obtain a rich supply of oxygen without ever swimming to the surface of water.

Some aquatic insects (diving beetles, for example) carry a **bubble of air** with them whenever they dive beneath the water surface. This bubble may be held under the elytra (wing covers) or it may be trapped against the body by specialised hairs. The bubble usually covers one or more spiracles so the insect can breathe air from the bubble while submerged.

An air bubble provides an insect with only a short-term supply of oxygen, but its unique physical properties, permits a bubble to also collect some of the oxygen molecules dissolved in the surrounding water. In effect, the bubble acts as a **physical gill**, replenishing its supply of oxygen through the physics of passive diffusion. The larger the surface area of the bubble, the more efficiently this system works. An insect can remain under water as long as the volume of oxygen diffusing into the bubble is greater than or equal to the volume of oxygen consumed by the insect. Unfortunately, the size of the bubble shrinks over time as nitrogen slowly diffuses out into the water. When the bubble's surface area decreases, its rate of gas exchange also decreases. Eventually, the bubble becomes too small to keep up with metabolic demands and the insect must renew the entire bubble by returning to the water's surface.

There are two kinds of physical gill. A **compressible gas gill** is a simple bubble that clings to the animal and its surface is not supported, so the pressure inside is essentially equal to the atmospheric pressure plus the hydrostatic pressure that depends on depth. An **incompressible gas gill** (plastron) is a bubble that adheres to the surface of the insect, but its external surface is supported by hydrophobic structures that suspend it, so the internal pressure can be independent of depth and in fact lower than that of the atmosphere. Physical gills occur chiefly among diving bugs and beetles, many of which spend most of their adult lives under water. They also occur among the Arachnida, including some true spiders (Araneae) and whip spiders (Amblypygi) that forage underwater or at least survive submersion.

7.2 Oxygen intake from water

Many aquatic insects depend on **cuticular respiration** to meet their oxygen requirement. These insects have relatively thin integument that is permeable to oxygen and carbon dioxide. Diffusion of gases through this body wall may be sufficient to meet the metabolic demands of small, inactive insects, especially those living in cold, fast-moving streams where there is plenty of dissolved oxygen. Larger insects, more active ones, or those living in less oxygenated water may need to rely on other adaptations to supplement cuticular respiration.

Some insects have developed **biological gills**, which are usually outgrowths of the tracheal system. They are covered with a thin layer of cuticle that is permeable to both oxygen andcarbon dioxide. In mayflies and damselflies, the gills are leaf-like in shape and located on the sides or rear of the abdomen. Fanning movements of the gills keep them in contact with a constant supply of freshwater. Stoneflies and caddis flies have filamentous gills on the thorax or abdomen. Dragonflies have internal gills associated with the rectum. Water is circulated in and out of the anus by the muscular contractions of the abdomen.

Some insects have a **plastron,** which is a special array of rigid, closely-spaced hydrophobic hairs (setae) that create an airspace next to the body. Air trapped within the plastron operates as a physical gill, but this airspace cannot shrink in volume because the fortress of setae prevents encroachment of surrounding water. When the insect consumes oxygen, it creates a partial pressure deficit inside the plastron. This deficit is made good by dissolved oxygen that diffuses from the water. As nitrogen gradually diffuses out of the bubble, it creates a similar partial pressure deficit. But there is very little dissolved nitrogen present in water. Therefore some of the nitrogen partial pressure deficit is corrected by oxygen. In effect, the plastron trades some of the nitrogen for oxygen, keeping a constant volume of gas that may slowly become enriched with oxygen. The constant volume of a plastron's air supply eliminates the periodic need to come to the surface and replenish the bubble.

Spiracular gills are seen in pupae of many Diptera and of some beetles of the family Psephenidae which live in aquatic habitats that dry up.These are extensions of the spiracle or the cuticle immediately surrounding it, to form long, partly hollow process.

7.3 Physics of bubble gas exchange in collapsible gas gill

The air bubble which aquatic insects carry with them during dive have a dual respiratory significance by providing oxygen at the beginning of the dive and by allowing oxygen uptake from the water. In the bubble, oxygen could decrease from 21% (21 kPa at sea level) to nearly zero, while carbon dioxide does not increase above approximately 2-3%. However, the fraction of N_2 could rise from 78% (78 kPa at sea level) originally in the bubble to above 95%, causing it to diffuse out into the water and eventually collapse the bubble. Also, the bubble is under additional hydrostatic pressure determined by its depth. This pressurisation increases the pressure of all gases in the bubble, further enhancing N_2 loss and restricting O_2 gain.

The movement of gases between a bubble and the surrounding water is determined primarily by diffusion across a theoretically stationary layer of water around the bubble, known as the **boundary layer**. This layer is generally defined by hydrodynamicists as a distance equal to 90 or 99% of the distance from the surface to freely moving water (Pinder and Feder, 1990). Water flows at all levels within the boundary layer, but the velocity at each level decreases as the stationary surface is approached, where velocity is zero (Vogel, 1981). However, if the surface is the air–water interface of a bubble, water velocity does not necessarily become zero, because it can skim along on the air to some extent. In this case, both the outer and inner limits of the hydrodynamic boundary layer are unknown.

The rate of gas flux can be approximated in terms of Fick's first law of diffusion :

$$M_G = K_G\,(A/L)\,(\Delta P_G),$$

where M_G is the rate of diffusion, K_G is Krogh's coefficient of diffusion (=the product of solubility and diffusivity), A is surface area, L is thickness of the effective boundary layer and ΔP_G is the difference in partial pressure of the gas in question.

This equation is useful in understanding the relative exchanges of O_2 and N_2, because both take the same pathway. The partial pressure difference is the **driving force** for diffusion, and the movement of each gas is nearly independent of movements of the other. According to Dalton's law of partial pressures, the total pressure of a gas is the sum of the partial pressures of each gas in the mixture. The total pressure inside a submerged bubble is determined by the sum of three pressures: the atmospheric barometric pressure, the hydrostatic pressure due to depth and the pressure derived from the surface tension of the bubble. At sea level, barometric pressure is 101.3 kPa. Hydrostatic pressure increases 9.8 kPa per meter of depth in freshwater. The surface tension of a 1 mm diameter bubble is 0.3 kPa and decreases in larger bubbles.When insects dive in shallow fresh waters, the bubbles are larger than 1mm. Therefore the atmospheric pressure has the overwhelming effect on total pressure in the bubble. Pure dry air is 78% N_2, 21% O_2 and 1% trace gases (mainly argon; CO_2 is only 0.039%). A bubble taken to 1 m depth at 20°C therefore contains 108.8 kPa total pressure, made up of P_{N2} = 84.9 kPa, P_{O2} = 22.8 kPa, P_{trace} = 1.9 kPa and P_w = 2.3 kPa (P_w is the saturated vapour pressure of water and is constant at a given temperature).

The partial pressures of dissolved gases in equilibrium with the atmosphere are independent of depth and are: P_{N2}=77.2 kPa and P_{O2}=20.8 kPa. Thus simply taking a naked bubble of air from the surface to 1 m depth causes an outward pressure difference for N_2 (84.9–77.2=7.7 kPa) and O_2 (22.8–20.8=2.0 kPa). Both gases continue to diffuse into the water until the bubble disappears completely. If the water body is large, the addition of gas to it does not significantly affect its dissolved gas tensions.

When an insect takes a bubble from the surface, the same conditions apply initially, but because the bubble is connected to the animal's tracheal system, O_2 is consumed from the bubble and P_{O2} decreases. This quickly reverses the outward O_2 diffusion to an inward one. As P_{O2} in the bubble decreases, PN_2 increases practically to the same extent, according to Dalton's law. The reason for this is that the CO_2 produced by the insect's respiration is rapidly taken up by the water, because CO_2 is approximately 28 times more soluble in water than O_2. The P_{CO2} measured in the bubbles of several species of diving insects is greater than that of the water (ca. 1–3 kPa), causing outward diffusion, but the P_{CO2} in the bubble quickly becomes independent of dive duration. Therefore, P_{CO2} can be considered to be a constant, like P_w.

There are two models of the dynamics of volume and pressure changes in a collapsible gas gill that depend on whether the surface area *A* is constant or changing. If the edges of the bubble can move, then area can decrease as volume decreases, but if they are fixed, then area can remain almost constant as the bubble gets thinner. A decreasing area predicts a non-linear decline in P_{O2} until it reaches zero. A constant area predicts that P_{O2} stabilises at a level dependent on the surface area and the O_2 uptake rate of the insect.

Observations of the real collapsible gas gill on the ventral surface of water boatmen, *Agraptocorixa eurynome* (Hemiptera, Corixidae), indicate that a constant-area model is appropriate for this species, because the edges of the bubble are mainly fixed (Matthews and Seymour 2010). Records of bubble P_{O2} during water boatmen dives apparently plateau, but this may in part be related to active regulation of the effective thickness of the boundary layer by leg movements that ventilate the surface of the bubble. Quiescent water boatmen produce P_{O2} values that are well below the critical P_{O2} at which M_{O2} is forced to decline, and the dives are clearly unsustainable.

8. RESPIRATION IN ENDOPARASITIC INSECTS

Most of the endoparasitic insects are parasites only in the immature stages, eg. parasitic wasps. These insects have respiratory problems similar to aquatic insects. They have a tracheal system that is non-functional and therefore respiration is cutaneous. Gasesous exchanges occur directly between the tissues of the parasite and the body fluid of the host. Some endoparasites have tracheal gills. The larvae of *Cotesia* (Hymenoptera) possess caudal vesicle which is an everted structure of the hind gut. The wall of the vesicle is very thin and is associated with the heart and so oxygen passing in is quickly carried around the body. Others depend on aerial oxygen, obtaining it either by tubes or other structures that communicate with the tracheal system and that extend out of the host to the atmosphere. The larvae of Tachinid flies have an open metapneustic respiratory system and the parasites perforates the integument of the host or one of the tracheal trunks. By inserting the spiracular region of the body into the opening thus formed, it can draw on a supply of atmospheric oxygen.

9. RESPIRATORY PIGMENTS IN INSECTS

Very few insects have respiratory pigments in their haemolymph or cells. The larvae of *Chironomous* have haemoglobin composed of two chains with a molecular weight of 31,400. The haemoglobin is present in the haemolymph and not in the haemocytes. It has high affinity for oxygen and is 50% saturated at pO_2 of 0.6mmHg at 17°C. Its loading curve is not influenced by temperature and CO_2 tension unlike vertebrate haemoglobin. Due to its high affinity for oxygen, it unloads oxygen only at places of extremely low oxygen deficit. Its primary function may be to aid recovery from anaerobic conditions and provide limited oxygen to some critical tissues such as the nervous system. In contrast to *Chironomous*, the haemoglobin of the beetle *Anisops pellucens* is only 50% saturated at a pO_2 of 28mmHg at 24°C, giving it much more potential function.

10. RESPIRATION IN EGGS AND WATER CONSERVATION

Insects have a more or less solid chorion in its egg. However, since oxygen enters and carbon dioxide leaves through the shell, it follows that the shell must have holes large enough to exchange respiratory gases. The shell should also have provisions to prevent desiccation if they are laid in the terrestrial environment, and if in the aquatic medium prevention of water entry. The shell of *Leptohy lemyia* has chorion made up of fibrils that form a meshwork with interstices of about 20-25 Å wide that permits gas exchange. Aquatic and semi-aquatic eggs are imbedded in a jelly-like substance which is hygroscopic. When the eggs are subjected to some desiccation, the jelly loses water slowly and protects the egg from drying. Majority of the terrestrial eggs have meshwork in the chorion that holds a layer of gas. These eggs also have aeropyles or holes that extend through much of the shell and bring about the continuity of chorionic layer of gas and the ambient atmosphere.

The respiratory systems in insect eggs are nevertheless very different from systems in other insect life stages.The principle idea is still the same; there must be openings on the eggshell to allow theexchange of gases with the atmosphere and mechanisms to move gases throughout the eggshell. Typically, insect embryos respire through tiny pores called **aeropyles** located on the outer eggshell, which allows for efficient gas exchange and reduces water loss during respiratory activities (Fig. 5.5).

To reduce water loss, many terrestrial insect eggs have a reduced numbers of aeropyles. The aeropyles connect to an inner air-filled space that is referred toas the pillar system, which moves oxygen from the outer eggshell layers to the inner embryo. Insect eggs may have their aeropyles located on the outer rim of the operculum, as seen onmany hemipteran eggs, or, if there is no defined rim, they will be located on the border of theoperculum. Alternatively, aeropyles can also be scattered across the egg surface.

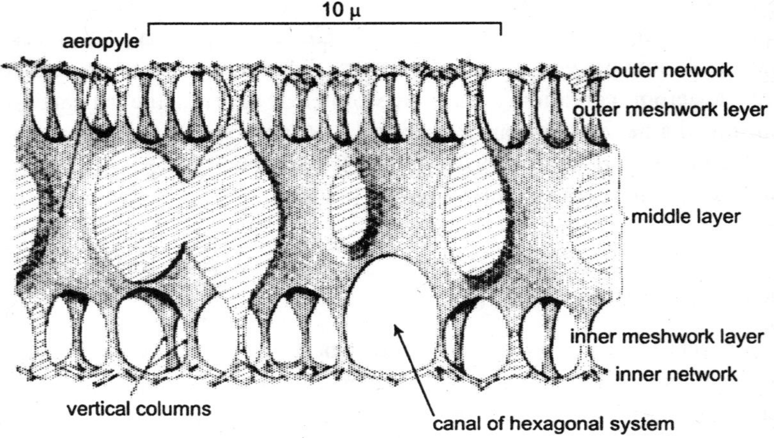

Fig. 5.5 Respiratory system of the common house fly egg. (After Hinton 1967).

There are numerous modifications to the respiratory systems of insect eggs that is dependent upon the environment and respiratory requirements of individual species. While terrestrial insects have an innerpillar system, aquatic insect eggs primarily respire with a plastron. The plastron is a gas filled air layer below the outer chorion of the egg shell. The plastron acts as a physical gill that allows eggs to respire under water. Some terrestrial insect eggs that are oviposited in environments flooded with water may alsohave a plastron. In eggs with a large plastron membrane, only small parts of the plastron are permeable, limiting water loss. Terrestrial eggs typically do not have a plastron but do have a gas layer directly under the outer surface of the chorion and connected to the aeropyles. Aeropyles of terrestrial insects occasionally flooded by rainwater are located on the ends ofrespiratory horns. These horns extend above the water surface to allow for more efficient uptake of oxygen when the egg is surrounded by a layer of water. These respiratory horns are essential to allow the insect to breathe during periods of heavy rain when an insect egg lacks the plastron respiratory system.

The eggshell layers and elaborate respiratory systems of insect eggs are modified to conservewater for the insect egg. Terrestrial insect eggs are provided from their mother all of the water necessary for survival and development at the time of oviposition. Therefore, the embryos must conserve this limited amount of water during their development while also respiring. Water loss occurs across the chorion and is correlated to oxygen consumption requirements of the embryo. The more gas exchange that occurs, the more vulnerable the insect embryo isto water loss.The waxy layer in the insect eggshell most probably is the main barrier in the eggshell thatprevents water loss. There was a significant increase in water loss in *Manduca sexta* eggs whenthe waxy layer was dissolved by organic solvents (Woods, 210). In addition to the waxy layer, the serosal membrane that envelopes the embryo has been shown to also protect insect eggs against desiccation (Jacobs *et al.*, 2013). RNAi technology has been used to prevent the development of the serosa inthe beetle *Tribolium castaneum*. Following RNAi treatment, the serosa-less eggs were subjected to a range of different humidities. Eggs exposed to the lowest humidities could notretain water and fewer eggs hatched compared to eggs that still had their serosal membraneintact.

Environmental factors and embryonic development can exacerbate water loss in insect eggs. Elevated environmental temperatures and low relative humidity can increase water loss greatly. Insect eggs have adapted to these conditions by having an altered density or numberof eggshell chorionic layers. The developing embryo also has different metabolic requirementsas it progresses in development. As insect embryos develop into larvae, the metabolic ratesand water loss rates increase. These metabolic requirements peak shortly before the embryo hatches from the eggshell.

II. CONTROL OF RESPIRATORY PATTERNS

In insects, the exchange of gases between the atmosphere and tracheal system takes place *via* valve-like spiracles. In many quiescent species, this exchange is discontinuous, and can be divided into three phases based on spiracle behaviour. During the **closed (C) phase**, the spiracles are closed, oxygen is depleted from the endotracheal space and pressure within the tracheal system declines. The next, **flutter (F) phase** is triggered by declining pO_2 levels and, via partial spiracular opening, oxygen levels are typically regulated at 2–4 kPa. Eventually, accumulation of CO_2 results in complete spiracular opening, known as the **open (O) phase**, which may be accompanied by ventilation of the tracheal system. The pattern is then repeated, leading to the DGC. Not all insects show evidence of DGC at rest and several major hypotheses have been proposed to explain the evolutionary origins and maintenance of these cycles and why they are restricted in their taxonomic distribution. While some authors prefer a single evolutionary explanation, others have suggested that DGC may have arisen and are maintained for several reasons, among which is the limitation of oxidative damage.

The oxidative damage hypothesis for DGC (Marais *et al.*, 2005) proposes that this pattern has arisen to reduce oxidative damage. During the C phase, oxygen partial pressure is rapidly reduced, and, in the moth *Attacus atlas*, during the F phase, it is closely regulated at *ca* 4 kPa irrespective of variation in external oxygen partial pressures ranging from 5 to 50 kPa (normoxic conditions at sea level is *ca* 21 kPa). It is this tight regulation of intratracheal

pO_2, along with the known cellular-level damage caused by normoxic concentrations of O_2, which underlie the oxidative damage hypothesis.

However, whether insects do actually limit oxidative damage using higher-level mechanisms is unknown. While some studies have provided evidence for the oxidative damage hypothesis based on gas exchange dynamics, no studies have shown that spiracular control can limit oxidative damage via regulation of reactive oxygen species (ROS) production, independently of the cellular-level mechanisms that already exist to do so. Thus, although the pattern of maintenance of internal partial pressure is consistent with the oxidative damage hypothesis, the chain of reasoning from DGC to oxidative damage is missing a critical, causal link. Indeed, it has been suggested that current oxygen toxicity data cannot be used to support the oxidative damage hypothesis for DGC.

11.1 Spiracle control and innervations

Given the particular morphological arrangement of the lepidopteran closing apparatus with only a single closing muscle it is amazing how such a complex and coordinated pattern of opening as is represented by the DGC can be generated at the level of single spiracles. As the muscle is multi-terminally innervated (Chapman, 1998), its action is most probably all-or-nothing. It becomes clear that neural or other forms of coordination between multiple spiracles are necessary to generate the DGC pattern by modulating their opening time.

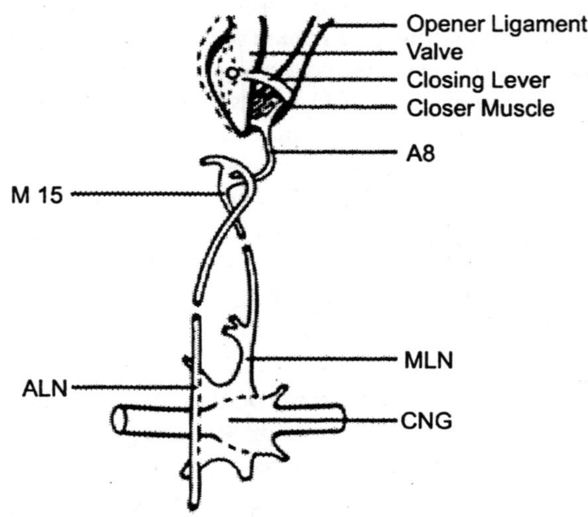

Fig 5.6 Innervation of an abdominal spiracular closer muscle (Redrawn from Backel 1958).

CNG- ganglion, ALC- anterior lateral nerve, MLN-mid lateral nerve, A8- branch of ALN innervating the muscle; M15 – branch of MLN innervating the muscle.

During the three phases of DGC it has been briefly mentioned that mainly endotracheal *p*CO2 and *p*O2 trigger the transition between the phases. Therefore it has to be assumed that these gases somehow exert an influence directly on the spiracle muscle or via neural mechanisms. A morphological study of the innervation of the spiracle muscle in *Hyalophora* (Beckel, 1958; Beckel and Schneiderman, 1957) revealed that the muscle is innervated by two branches of unpaired median nerves, the antero-lateral (ALN) and the medio-lateral (MLN) nerve respectively (Fig. 5.6). Both nerves originate in segmental ganglia, the ALN in the segment preceding the spiracle, MLN in the same segment. The unpaired axons bifurcate outside the ganglia and innervate both spiracles of a single segment.

From this arrangement of unpaired neurons and bifurcating axons it has been concluded that the opening of both spiracles in a single segment is synchronised though slight differences in latency due to differences in the length of the propagation pathway might occur (van der Kloot, 1963). However, on a behavioural basis, the spiracles in asegment might still be independent (Sláma, 1988). So far no simultaneous records of electrical and mechanical activity of both spiracles in a segment exist.Trans-section of the median nerves resulted in prolonged constriction of the denervated spiracle (Beckel and Schneiderman, 1957). It has been concluded that the muscle is spontaneously active (myogenic contraction) and no central nervous input is necessary to close the valve. This was later confirmed by reports that the muscle fibres of *Hyalophora* pupae possess a pacemaker potential, which is possibly unique to this stage (van derKloot, 1963). Apart from the pacemaker potentials in the muscle membrane tonic activity has been described in the axons of the median nerves (ALN, MLN; van der Kloot, 1963; Hoyle,1959, 1960). Perfusion experiments with separate tracheal and ganglionic flows revealedthat the tonic firing most probably is related to the oxygenation state of the ganglion,that is, it provides an O_2 dependent signal to the spiracle (Burkett and Schneiderman,1974). Direct electrical recording demonstrated a stopping of the electrical activity in ALN due to relatively high concentrations of CO_2 (10%; van der Kloot, 1963), although nocentral behavioural effect of CO_2at physiological concentrations has been found later inthe perfusion experiments (Burkett and Schneiderman, 1974).However, a number of experiments revealed a direct effect of CO_2on the neuromuscular transmission. First, no morphological evidence for a peripheral reflex network has been found (Beckel and Schneiderman, 1957). Further the sensitivity of the muscle to CO_2 persists after denervation and even after dissection of most of the fibres of the muscle (Beckel and Schneiderman, 1957; van der Kloot, 1963).

Measurements of the post-synaptic activity and the mechanical force generation under the influence of CO_2 (Hoyle, 1960) led to the conclusion that the effect of CO_2 is direct and on the transmission at the neuromuscular junction. A possible mechanistic explanation for this direct effect probably has been found in the properties of the post-synaptic receptors of the neurotransmitter. Like all insect skeletal muscles, the transmitter at the neuromuscular junction of the spiracle muscle is L-glutamate (Chapman, 1998). Investigating the narcotic effects of CO_2, Badre *et al.*, (2005) found that the post-synaptic glutamate receptors are blocked by CO_2 in the heart and the intersegmental muscles in larvae of the fruit fly *Drosophila*. It is likely that this CO_2 inhibition also occurs in the neuromuscular junction of the spiracle muscles. In conclusion, it can be said that, within a single segment, the action of CO_2 is directly on the spiracle muscle and therefore local, whereas the reaction to O_2 requires central nervous control of the spiracle.

11.2 Metabolic rate determines patterns of respiration

The DGC is characterised by a period in which the spiracles open (the open phase) and gases are free to diffuse and/or exchange between the tracheal space and the external atmosphere. It has been demonstrated (Hetz and Bradley, 2005) that the partial pressure of oxygen in the tracheae during the open phase approaches a value very similar to the external P_{O2}. The open phase is followed by the closed phase. During this period of spiracular closure, the P_{O2} drops substantially due to the consumption of oxygen through aerobic metabolism. Eventually, the P_{O2} reaches a critical low level and the spiracles begin to flutter. During the flutter phase the level of oxygen is closely regulated.

Following the above logic, the length of the closed phase should be negatively correlated with the metabolic rate of the insect. The increased metabolic rate of the insect leads to a shortening of the closed phase because the P_{O2} inside the insect drops more rapidly. In an insect employing the DGC, as the insect's metabolic rate begins to increase, the closed phase will get shorter and shorter until it can no longer be discerned (Levy and

Schneiderman, 1966; Lighton, 1996). At this point the insect will have transitioned to the cyclic pattern, in which bursts of CO_2 release continue to be observed but the closed phase is missing. Further increases in metabolic rate will require the spiracles to open more fully in the flutter phase, allowing more CO_2 to escape in this phase. Finally, when the metabolic rate is sufficiently high, oxygen entry will match CO_2 release, resulting in a continuous pattern of respiration in which no large, rhythmic bursts of CO_2 are observed.The transitions described above are consistent with observations in *Rhodnius* (Contreras and Bradley 2009).

In summary, the respiratory pattern employed by an insect is affected by three factors. The length of the closed phase is (1) positively correlated with the partial pressure of oxygen in the atmosphere surrounding the insect, (2) positively correlated with the amount of oxygen stored in the insect at the end of the open phase (presumably the sum of the gaseous oxygen in the tracheae and the dissolved oxygen in the tissues and hemolymph), and (3) negatively correlated with the metabolic rate of the insect. The external oxygen concentration in the atmosphere varies very little in most habitats, but it can be substantially lower for insects underground. The amount of oxygen stored in the insect at the end of the open phase is probably most significantly affected by the volume of air contained in the trachea upon spiracular closure. This will vary from one insect species to another and may be a major factor in explaining species-specific respiratory patterns. It has been shown, however, that growth within an instar and feeding can reduce tracheal volume (Greenlee and Harrison, 2003). Despite the importance of the above factors influencing oxygen delivery, the major variable affecting respiratory patterns in insects is the metabolic rate of the insect. This can vary over more than an order of magnitude in many insects and therefore is the major variable affecting respiratory pattern.

The hygric hypothesis, which has been the leading explanation for the DGC in textbooks of insect physiology for the past 50 years, has now been questioned by several researchers. Many insects do not show DGC under conditions where the hygric hypothesis would suggest they should (Hadley and Quinlan, 1993; Chappel and Rogowitz, 2000; Chown and Holter, 2000), while other studies indicate that the DGC does not reduce respiratory water loss compared with other respiratory patterns (William and Bradley, 1998; Gibbs and Johnson, 2004; Lighton and Turner, 2008).

The oxidative damage hypothesis not only explains the presence of the DGC in insects at low rates of metabolism but, now also provides a rationale for the otrespiratory patterns observed in insects. The hypothesis is testable through the manipulation of external P_{O2}, humidity and the metabolic rate of the insect.

12. USING THE KNOWLEDGE GAINED FROM THE INSECT'S TRACHEAL SYSTEM

The properties of insect respiratory and cardiac systems differ greatly from typical mammalian physiological systems. These differences include the capacity to deliver oxygen from air to cell by either diffusive or convective mechanisms, convection driven by relatively low pressures, and dramatic morphological and structural differences in the components. The high performance of insect fluidic systems, their unique properties, and the ecological success of insects all suggest that insects offer potential bioinspiration for a variety of biomedical engineering processes including microfluidics, nanofabrication, and tissue engineering

Even very large insects weighing multiple grams can recover from exposure to anoxia when returned to room air, demonstrating that they can attain sufficient oxygen delivery to initiate movement by diffusion. This is a major benefit of a gas-phase oxygen transport system. Someday we may culture these cells or build synthetic models that allow engineering of artificial tracheal systems that could offer a passive system for supporting gas exchange of cultured tissues of at least a few centimeters in thickness. Study of the morphology of insect tracheal systems (e.g., branching patterns, relative diameters at different branching levels) may aid development of such artificial systems.

The topic of microscale flow transport is of interest to engineers because there are many applications that need precise flow control of just microliters of gas or fluid. Microscale flow devices or microfluidic devices are already around us and are assumed to change research tools in the life sciences, aid drug discovery and chemical analyses, and revolutionise health diagnostics. Modern DNA sequencing systems rely on microfluidics. **Droplet** analyses also rely on microfluidic technologies. Cheap paper-based microfluidic immunoassay tests will become important for healthcare in the developing world. And those are just a few of the applications.

Insect tracheal systems have been of particular interest to engineers now that we are able to visualise the physiological process of moving fluid and air through tubes in action. Insects are able to move fluid and air with precision with minimal energy requirements and little input from outside of the body.

Insects use contractions at very specific places along the tracheal system, which probably affords them more precise control than if just relying on the more crude pressure difference approach. Through these small, but targeted, compressions the airflow to specific areas of the body can be achieved. This is apparently not an easy task to achieve in small-scale devices, and one can appreciate that by imagining larger tubes.

Researchers at the fluid dynamics research group and biomechanics research lab at Virginia Tech's Department of Engineering Science and Mechanics have shown by studying the insect tracheal system that fluids can be transported and flow velocity, pressure and direction can be controlled. They say that they have learned from insects that localized rhythmic tracheal contraction is an effective way to move gas through micro-tubes. They claim to have been inspired by the insect's tracheal system to model **selective pumping in a network**. They show, by using theoretical (low Reynolds number) flow analyses and computation fluid dynamics, that the network might also enable fluids to be transported precisely into particular tubes without the use of mechanical valves – at least through an 8-armed tracheal system. It will be interesting to see if this research group will continue to follow the insects' lead by adding more branches/channels, by adding air sacs (which seem to be very important in insects), by making the model 3D, by adding actively opening and closing ports/spiracles to the outside, and by incorporating the diversity of insect respiratory systems into different models. If done correctly, and by using micro fabrication techniques that can make flexible micro- and nano-scale tubes, it is likely insects can make a contribution to the rapidly growing technological field of microfluidics.

REFERENCES

Beckel, W. E. (1958). The morphology, histology and physiology of the spiracular regulatory apparatus of *Hyalophora cecropia* (L.) (Lepidoptera). *Proc 10th Intern Congr Entomol 2*: 87-115.

Beckel, W E. and H. A. Schneiderman. (1957). Insect spiracle as an independent effector. *Science*, 126 (3269): 352–353.

Badre, N.H., M. E. Martin, and R. L. Cooper. (2005). The physiological and behavioural effects of carbon dioxide on *Drosophila melanogaster* larvae. *Comparative Biochemistry and Physiology - Part A: Molecular & Integrative Physiology*, 140(3):363–376.

Bradley, T.J. (2008). Control of the respiratory pattern in insects. In R. C. Roach, P. D. Wagner, and P. H. Hackett, editors, *Hypoxia And The Circulation*, number 618 in *Advances in Experimental Medicine and Biology*, pages 211–220. Springer New York.

Buck, J. (1958). Cyclic CO2 release in insects IV. a theory of mechanism. *Biological Bulletin*, 114(2):118–140.

Burkett, B.N and H. A. Schneiderman. (1974). Roles of oxygen and carbon dioxide in the control of spiracular function in cecropia pupae. *Biological Bulletin*, 147(2):274–293.

Chapman, R.F. (1998). *The insects - structure and function*. Cambridge University Press, 4th Edition.

Chappell, M. A. and Rogowitz, G. L. (2000). Mass, temperature and metabolic effects on discontinuous gas exchange cycles in eucalyptus-boring beetles (Coleoptera: cerambycidae). *J. Exp. Biol.* 203,3809 -3820.

Chown, S.L., and P. Holter. (2000). Discontinuous gas exchange cycles in *Aphodius fossor* (Scarabaeidae): a test of hypotheses concerning origins and mechanisms. *Journal of Experimental Biology*, 203 (2):397.

Chown, S.L., A. G. Gibbs, S. K. Hetz, C. J. Klok, J. R. Lighton, and E. Marais. (2006). Discontinuous gas exchange in insects: A clarification of hypotheses and approaches. *Physiological and Biochemical Zoology*, 79(2):333–343.

Cloudsley-Thompson, J.L.(1975) Adaptations of Arthropoda to arid environments. *Annual Review of Entomology*, 20 (1):261–283.

Contreras, H.L., and T. J. Bradley. (2009). Metabolic rate controls respiratory pattern in insects.*Journal of Experimental Biology*, 212(3):424–428.

Contreras, H. L and T. J. Bradley. (2010). Transitions in insect respiratory patterns are controlled by changes in metabolic rate. *Journal of Insect Physiology*, 56(5): 522-528.

Duncan, F.D. and M. J. Byrne. (2000). Discontinuous gas exchange in dung beetles: patterns and ecological implications. *Oecologia*, 122 (4):452–458.

Edney.E. (1977) *Water Balance in Land Arthropods*. Springer Verlag, Berlin.

Fridovich, I. (1977). Oxygen is toxic! *BioScience*, 27(7):462–466.

Ghabrial, A., Luschnig, S., Metzstein, M.M., Krasnow, M.A., (2003). Branching morphogenesis of the Drosophila tracheal system. *Annual Reviews of Cellular Developmental Biology* 19: 623–647.

Gibbs, A. G. and Johnson, R. A. (2004). The role of discontinuous gas exchange in insects: the chthonic hypothesis does not hold water. *J. Exp. Biol.* 207: 3477 -3482.

Greenlee, K. J. and Harrison, J. F. (2003). Development of respiratory function in the American locust *Schistocerca americana*: II Within-instar effects. *J. Exp. Biol.* 207: 509 -517.

Hadley, N.F and M. C. Quinlan. (1993). Discontinuous carbon dioxide release in the eastern lubber grasshopper *Romalea guttata* and its effect on respiratory transpiration. *Journal of Experimental Biology*, 177(1):169

Harrison, J.F., S. Camazine, J. H. Marden, S. D. Kirkton, A. Rozo, and X. Yang. (2001) Mite not make it home: tracheal mites reduce the safety margin for oxygen delivery of flying honeybees. *Journal of Experimental Biology*, 204(4):805

Harrison, J F, James S. Waters, Arianne J. Cease, John M. Vanden Brooks, Viviane Callier, C. Jaco Klok, Kimberly Shaffer, and John J. Socha. (20130. How Locusts breathe. *Physiology.* 28 (1), 18-27

Hetz S.K. 2007. The role of the spiracles in gas exchange during development of *Samia cynthia* (Lepidoptera, Saturniidae). *Comparative Biochemistry and Physiology – Part A: Molecular & Integrative Physiology*, 148 (4):743–754.

Hetz, S.K., and T. J. Bradley. (2005). Insects breathe discontinuously to avoid oxygen toxicity. *Nature*, 433(7025):516–519.

Hinton, H. E. (1967). Structure and protective devices of the egg of the mosquito, *Culex pipens.* J.Insect Physiol., 14, 145-161

Hoyle, G. (1959). The neuromuscular mechanism of an insect spiracular muscle. *Journal of Insect Physiology*, 3(4):378–394

Hoyle, G. (1960) The action of carbon dioxide gas on an insect spiracular muscle. *Journal of Insect Physiology*, 4(1):63–79.

Imms, A.D. (1960) A general text book of Entomology. Butler & Tanner Ltd, Frome and London.

Jacobs C, Rezende G, Lamers G, van der Zee M. (2013) The extraembryonic serosa protects the insect egg against desiccation. *Proc R Soc Lond B Biol Sci.*280;1–8.

Kestler, P (1985) Respiration and respiratory water loss. In K. Hoffmann, editor, *Environmental Physiology and Biochemistry of Insects*, pages 137–183. Springer Verlag, Berlin, Heidelberg

Landis G.N and J. Tower. (2005). Superoxide dismutase evolution and life span regulation. *Mechanisms of Ageing and Development*, 126(3):365–379 .

Levy, R I and H. A. Schneiderman. (1966). Discontinuous respiration in insects – II. The direct measurement and significance of changes in tracheal gas composition during the respiratory cycle of silkworm pupae. *Journal of Insect Physiology*, 12(1):83–104

Lighton, J. R. B. (1996). Discontinuous gas exchange in insects.*Annu. Rev. Entomol.* 41,309 -324.

Lighton.J. R B (1998) Notes from underground: Towards ultimate hypotheses of cyclic, discontinuous gas-exchange in tracheate arthropods. *Amer. Zool.*, 38(3):483–491, 1998.

Lighton, J R. B., and D. Berrigan. 1995. Questioning paradigms: caste-specific ventilation in harvester ants, *Messor pergandei* and *M. julianus* (Hymenoptera: Formicidae). *Journal of Experimental Biology*, 198(2):521.

Lighton, J. R. B. and Turner, R. J. (2008). The hygric hypothesis does not hold water: abolition of discontinuous gas exchange cycles does not affect water loss in the ant ***Camponotus vicinus***.*J. Exp. Biol.* 211,563 -567.

Marais,E., C. J. Klok, J. S. Terblanche, and S. L. Chown. (2005). Insect gas exchange patterns:a phylogenetic perspective. *Journal of Experimental Biology*, 208(23):4495.

McArthur J. (1929). An experimental study of the functions of the different spiracles in certain orthoptera. *J Exp Zool* 53: 117–128.

Miller PL. (1974). Respiration: aerial gas transport. In: *Physiology of Insecta, Volume IV* , edited by Rockstein M. New York: Academic, p. 345–402

Mörbitz C., and S. K. Hetz. (2009). Tradeoffs between metabolic rate and spiracular conductance in discontinuous gas exchange of *Samia cynthia* (Lepidoptera, Saturniidae). *Journal of Insect Physiology*, . 56(5):536-42

Pinder, A. W. and Feder, M. E. (1990). Effect of boundary layers on cutaneous gas exchange. *J. Exp. Biol.* 143: 67 -80.

Schneiderman. H. A. (1960). Discontinuous respiration in insects: role of the spiracles. *Biological Bulletin*, 119(3):494.

Schneiderman H. A., and C. M. Williams.(1953). The physiology of insect diapause. VII. The respiratory metabolism of the cecropia silkworm during diapause and development. *The Biological Bulletin*, 105(2):320.

Sláma.A. (1988) A new look at insect respiration. *Biological Bulletin*, 175(2):289–300, 1988.

Stabentheiner,A., J. Vollmann, H. Kovac, and K. Crailsheim. (2003) Oxygen consumption and body temperature of active and resting honeybees. *Journal of Insect Physiology*, 49 (9):881–889

Terblanche,J.S., E. Marais, S. K. Hetz, and S. L. Chown. (2008) Control of discontinuous gas exchange in *Samia cynthia*: effects of atmospheric oxygen, carbon dioxide and moisture. *J Exp Biol*, 211(20):3272–328.

van der Kloot, V.G (1963) . The electrophysiology and the nervous control of the spiracular muscle of pupae of the giant silkmoths. *Comparative Biochemistry and Physiology*, 9 (4):317–333

Vogel, *S. (*1981). ***Life in Moving Fluids***. Boston: Willard Grant Press.

Weaver, M., and Krasnow, M.A. (2008). Dual origin of tissue-specific progenitor cells in Drosophila tracheal remodeling. Science 321, 1496–1499.

White,C.R., T. M. Blackburn, J. S. Terblanche, E. Marais, M. Gibernau, and S. L. Chown.(2007) Evolutionary responses of discontinuous gas exchange in insects. *Proceedings of the National Academy of Sciences*, 104(20):8357–8361.

Williams, A. E. and Bradley, T. J. (1998). The effect of respiratory pattern on water loss in desiccation-resistant ***Drosophila melanogaster.*** *J. Exp. Biol.* 201,2953 -2959.

Insect Excretory System and Mechanism of Homeostasis 6

Synopsis

1. Introduction
2. Excretory organs of insects
   - 2.1 Malpighian tubule
   - 2.2 Integument
   - 2.3 Tracheal system
   - 2.4 Alimentary canal
   - 2.5 Nephrocytes
   - 2.6 Oenocyte
   - 2.7 Urate cells
   - 2.8 Labial glands
   - 2.9 Chloride cells

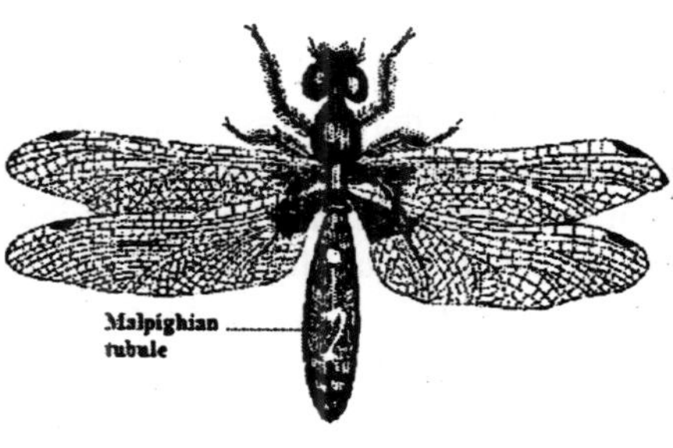

3. Morphology and ultrastructure of Malpighian tubules (MT)
   - 3.1 Cell types
   - 3.2 The basal lamella
   - 3.3 Basal infolds
   - 3.4 Junctions
   - 3.5 Intracellular storage granules
   - 3.6 Microvilli
   - 3.7 Secondary specialisations in malpighian tubules
4. Anatomical specialisation of hindgut epithelial cells
5. Formation of primary urine in Malpighian tubules
6. The basic transepithelial transport system
7. Proton pump is the driving mechanism for urine formation and homeostasis
8. Selective reabsorption in the hind gut
   - 8.1 Reabsorption in the Ileum
   - 8.2 Reabsorption in the Rectum
9. Role of the excretory system in maintaining homeostasis
   - 9.1 Electrolyte homeostasis
   - 9.2 Water homeostasis
   - 9.3 Acid-base homeostasis
   - 9.4 Nitrogen homeostasis
10. Hormonal control of urine formation
    - 10.1 Diuretic hormones
    - 10.2 Antidiuretic hormones
    - 10.3 Hormonal regulation of malpighian tubule transport
    - 10.4 Hormonal regulation of hindgut transport
    - 10.5 Hormonal regulation of midgut transport

1. INTRODUCTION

Excretion is the process through which an organism eliminates metabolic wastes and excess or unwanted chemicals that are either toxic or not useful from the system. Some products of the metabolic reactions are toxic and must be processed or eliminated from the body. Others are simply materials that are present in excess and so must be eliminated as waste. The process of excretion is quite different to defecation, which is the removal of undigested food wastes fromthe gut. However, the gut of many animals also has a role in excretion as some materials may be excreted into the gut and eliminated with the faeces. In insects most excretory products are excreted into the gut lumen and eliminated along with faecal matter. Excretion is also important in eliminating excess water and other unwanted chemicals that may be ingested and enter the body fluids, such as plant poisons and excess salts. The waste materials may be in solid, semi-solid, liquid or gaseous form. The principal excretory product in gaseous form is CO_2, liquid form is honey dew, solid form is urea/uric acid and semi-solid form is allantoin.

Excretion includes, among others, elimination of nitrogenous metabolites, removal of ingested toxic chemicals (allelochemicals) and maintenance of ions and water balance (Box 6.1). The toxic materials are nitrogenous products of metabolism (mainly ammonia), pigments, salts etc. For the efficient maintenance of water and the ionic balance in the haemolymph, the waste products of the metabolism are to be removed or eliminated.

The process of excretion is to remove substances from the metabolic pool so that they do not interfere in metabolism. This process is related to three separate aspects of excretion. The first is the excretion of molecules which are undesirable, perhaps even poisonous, at all except exceedingly low concentrations. The second is the excretion of molecules which are ordinarily useful or, indeed, essential to metabolism but which are present to excess. Included under this category is excretion of water and ions, and this process is of course the outwardly directed part of osmoregulation. The third process is the excretion of substances that are not toxic but merely useless. Their excretion is worthwhile because if allowed to accumulate, they would become obstructive. The storage of water in the rectum by larvae of *Dysdercus* towards the end of the feeding cycle in an instar is surprisingly the only case known of **storage excretion** in insects. **Deposit excretion**, in contrast, seems to be widespread. It appears that some insects have difficulty in eliminating uric acid from the body through the Malpighian tubules and rectum complex. Maddrell (1971) defined storage excretion to include materials with future potential use, such as glycogen as a storage form of glucose or amino acids stored as proteins. He viewed deposit excretion as waste material of no further use that needed to be removed from harmful interaction with the tissues. The distinction between stored and deposit excretion, however, is very subtle and not agreed on by everyone.

Box 6.1 Role of the excretory system

1. Eliminates metabolic wastes and unwanted chemicals from its system.
2. Elimination of nitrogenous metabolites.
3. Removal of ingested toxic chemicals (allelochemicals).
4. Maintenance of ions and water balance.
5. Removal of molecules which are undesirable or toxic.
6. Excretion of molecules which are not toxic but merely useless, and which would become obstructive if allowed to accumulate.
7. Excretion of molecules which are useful or essential such as glucose, amino acids, and certain ions, but are present to excess.

Several organs play a subsidiary role in the excretion of ions and water. The anal papillae of a salt-water mosquito may act to extrude ions when the animal is in a hyper-osmotic medium. Malpighian tubules secrete fluid containing many organic molecules, some of which enter the lumen passively, while others appear actively to be secreted there.

2. EXCRETORY ORGANS OF INSECTS

The processes of excretion in insects are carried out largely by the organs of the insect gut. These include the Malpighian tubules, and the structures in the hindgut (ileum and rectum). Several other organs are also involved in the process of excretion in insects. The organs involved in excretion in insects are:

1. Malpighian tubules
2. Integument or body wall
3. Tracheal system
4. Alimentary canal
5. Nephrocytes
6. Urate cells

7. Oenocytes
8. Labial glands and
9. Chloride cells

2.1 Malpighian tubule

Malpighian tubule is the main excretory organ of the insect. The Malpighian tubule and rectum, function as a unit in most insects. Malpighian tubules are tubular outgrowths of the gut. They typically develop as pouches emerging from the junction between the midgut and the hindgut, though there actual final position varies - they may be attached to the midgut, hindgut or the midgut-hindgut junction.They usually enter the gut individually but may fuse to form a common sac or ureter that lead into the gut. The tubules do not just hang around in the air, they are also typically loosely or firmly anchored in place by the tracheae which attach to them.

Each Malpighian tubule consists of a single-layer of squamous epithelial cells enclosed by an elastic membrane (basement membrane - a fibrous and porous protein mesh). In most insects there is a thin layer of striated muscle around this membrane. Typically, muscle cells spiral around the distalend (the end furthest from the gut) of the tubule, causing it to twist and turn in gentle writhing movements as the muscles contract.The proximal end (near the gut) may be coated in circular and longitudinal muscle fibres, giving rise to peristalsis or squeezing movements which empty the contents of the tubule into the gut. In some cases, such as in caterpillars, the Malpighian tubules on each side empty into a small bladder, this then empties into the gut. In this case only the bladder may be muscular and its lumen is lined by cuticle. Outside the muscle layer is a **peritoneal covering** of cells with embedded tracheoles, which carry oxygen to the Malpighian tubules which their mitochondria use to generate the needed ATP by aerobic respiration.

The Malpighian tubule forms a blind-ended tube whose lumen is continuous with the lumen of the gut. Tubules range from 2 to 70 mm in length, 2–250 in number and up to 100 mm in diameter (Phillips, 1981). The tubules vary in number from 2 to 150 or even more depending on the genus and do not appear to be closely related to either the phylogenetic position or the excretory problem of an insect.In species having few Malpighian tubules (blood sucking Hemiptera, *Rhodnius prolixus* and higher Diptera, *Drosophila melanogaster*), they are long and coiled; in species with numerous (up to 150) tubules (Desert Locust, *Schistocerca gregaria*), they are short. In *Aedes aegypti* each tubule begins with a blind end at its distal extreme (relative to the gut) and connects with the digestive tract at the junction of the midgut and hindgut. The **distal portions** of the tubule secrete fluid into the lumen to form a urinary compartment into which can be dumped nitrogenous wastes, toxins, and other solutes. In **proximal portions** of the tubule and in the hindgut and rectum, life-essential solutes and water may be reabsorbed before the final urine is ejected from the rectum.

The Malpighian tubules produce an isosmotic fluid termed the **primary urine**. The filtration of haemolymph and the production of primary urine take place by an active secretory process involving the active transport of ions into the tubule lumen. This filtered fluid or primary urine carries excretory waste products and toxic materials to the hindgut where the fluid is modified by selective absorption of useful substances. This selective modification of primary urine occurs in the anterior hindgut and rectum. Unwanted materials are secreted into the rectum constituting the secondary urine which is voided out through the anus (Fig 6.1).

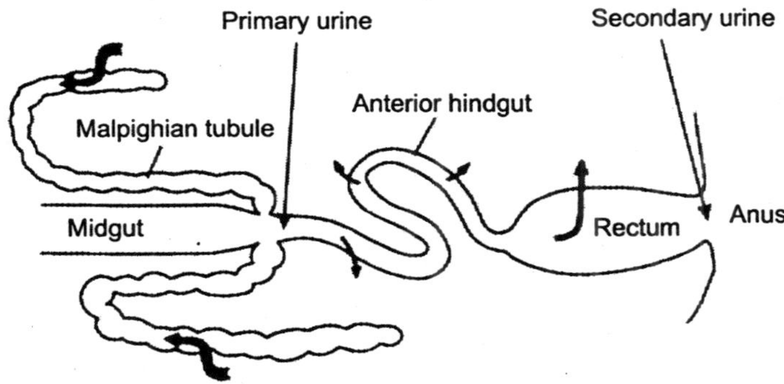

Fig. 6.1 The excretory system of an insect

(www.entomol.nchu.edu.tw/dept/teachers/khlu/insect.../13_excretion)

The Malpighian tubules function is removing toxins, nitrogenous wastes, and ions to maintain ionic concentrations within the haemolymph.They absorb wastes from the haemocoel either by diffusion across a concentration gradient or by active transport, remove wastes from the haemolymph and then excrete them into the gut lumen.The Malpighian tubules are bathed in the insect's blood, but since they are not rigid it is impossible for any hydrostatic pressure to be developed across their walls, such as could bring about filtration. This primary urine is then modified as it passes through more proximal regions of the Malpighian tubules and through the hindgut. The Malpighian tubules and the hindgut each frequently consist of several functionally distinct regions and/or distinct cell types, and each region contributes by specific mechanisms to the overall formation and processing of the urine.

2.2 Integument

Through the process of moulting, insects remove the waste nitrogenous products, i.e. they are deposited in the form of exuviae. In some insects, where respiration occurs through body wall, CO_2 is removed through integument as waste product (cutaneous respiration).

2.3 Tracheal system

The respiratory tubes, the trachea which are distributed throughout the body, function in elimination of CO_2 through spiracles.

2.4 Alimentary canal

The gut of the insects also play a major role in excretion by removing the unwanted material, dead cells formed during enzyme secretion (holocrine) and intima layer during moulting. Rectum plays an important role inexcretion by reabsorbing the water from faeces.

2.5 Nephrocytes

These are the special cells that are distributed in the body cavity and scattered. Nephrocytes are cells that take up foreign chemicals of relatively high molecular weight which Malpighian tubules may be incapable of dealing with.They are of two types:

1. **Dorsal** or **pericardial** nephrocytes on either side of the heart in the pericardial sinus. They are present in immature and adult stages of most of the insects.
2. **Ventral nephrocytes**, arranged as a chain below the foregut and attached byits two ends to the salivary glands. e.g.: dipterous larvae.

Nephrocytes helps in the removal of ammonia, chlorides, dyes, colloidal particles etc.

2.6 Oenocytes

These are large cells and are usually present near the abdominal spiracles. They arise from the ectoderm or hypodermis. These cells are thought to secrete cuticulin layer of the epicuticle and in cockroach, surface grease which covers the integument is believed to be involved in excretion.

2.7 Urate cells

Some of the fat body cells which store urea or uric acid in the form of granules are known as **urate cells**. Preserved uric acid can be utilised subsequently. These are present when Malpighian tubules are absent or may become non-functional. In some of the insects such as cockroach, the waste material in the form of urea or uric acid is stored throughout its life in the fat body cells without anyharmful effect. This phenomenon of storage of urea / uric acid in the fat body cells is called **storage excretion** which is useful for supply of nitrogen, when insect feedson nitrogen deficient food.

2.8 Labial glands

In Collembola, Diplura, Thysanura the labial glands serves in excretion. They consists of a sac like structures called **ampulla** that leads to a long coiled labyrinth that open at the base of labium in the head. These glands help to remove ammonia.

2.9 Chloride cells

These are the cells distributed on the body of aquatic insects such as larva of mayfly or stone fly. They absorb ions from salt water (body) and then excrete into surrounding medium to compensate the changes in the ionic concentration of haemolymph.

3. MORPHOLOGY AND ULTRASTRUCTURE OF MALPIGHIAN TUBULES (MT)

The Malpighian tubule consists of a single layer of one or more types of cells, surrounding a blind-ending lumen. Histologically, the MTs possess pyramidal or cubical cells, with nuclei in the center or the base, containing 1–3 nucleoli. In these species, the transitional portion between the midgut and the ileum (pylorus) is closed on one side by the pyloric valve, and on the other by a glandular epithelium, the MTs being empty in this region. According to insect groups, tubules are divided into distinct regions, based on function and morphology. In MTs of *Solenopsissae vissima* (Hymenoptera, Formicidae), three segments were distinguished- **proximal**, **middle**, and **distal**, and this distinction is based on the shape and structure of the epithelial cells lining the different segments of the tubules. However, this regionalisation is not observed in *Bombus mori*, where the epithelium has two cell types that are interspersed throughout the tubule.

3.1 Cell types

The ultrastructure of the cells of MT is variable. In *Musca domestica* (Diptera), *Frankliniella occidentalis* (Thysanoptera), and *Calpodes ethlius* (Lepidoptera), the MT layer is uniform, while, in other insects, the cells are of different types both functionally and morphologically (colour due to intracellular crystals or tube diameter). In adults, MTs are composed of two cell types: the principal cells and the stellate cells, which play distinct roles in the process of excretion.

. The **principal cells** are characterised by the outer surface being covered by a thin basement membrane. The tracheoles supplying the tubules are located on the outer surface of the basement membrane. There are extensive basal membrane infoldings having mitochondria in cytoplasmic compartments. Multiple spherical cell inclusions are found in the central cytoplasm, including clear vacuoles, lipid droplets, and large vacuoles having double membranes, spherocrystals and lysosomes. In the central cytoplasm there is a large euchromatic nucleus with a nucleolus. The deep basal membrane infoldings are associated with endoplasmic reticulum. The apical surfaces of the principal cells havelumina microvilli containing mitochondria running up to the tips of the microvilli. The lateral plasma membranes are differentiated into septate junctions, frequently with one or more gap junctions. In adult female *Aedes aegypti*, the principal cells are fusiform in shape and curl upon themselves to form a tubule lumen.

Besides their large size, principal cells exhibit distinct ultrastructural features (Box 6.2). First, their luminal side consists of a dense array of elongated microvilli that form a brush border. Within each microvillus is a mitochondrion, in convenient proximity to the consumer of ATP, the H+ V-ATPase. Second, the cytoplasm of principal cells is filled with membrane-bound mineralised concretions that serve as storagesites for ions, metals and uric acid.

The **stellate cells** are thin, having few cell inclusions and with many short luminal microvilli devoid of mitochondria. This fits with its proposed function of allowing the passive flow of water through the cell driven by osmotic gradient set by the principal cells. The basal membrane foldings are extremely reduced and peripheral tracheoles are visible. In mosquito, stellate cells are found sporadically along the distal segment where they intercalate between principal cells. Stellate cells have a small, thin cell body from which 3 or 4 arm-like projections branch out, giving the cell a star-like or stellate appearance. Although the length of a stellate cell can extend over 100 μm (from arm to arm), the cells are very thin compared to principal cells and are usually less than 5 μm thick.

In addition to their smaller size, stellate cells contrast from principal cells in their ultrastructure. The apical microvilli of stellate cells are not nearly as long or as densely packed as those of principal cells, and they do not contain mitochondria. Furthermore, cytoplasmic concretions are absent in stellate cells. Lastly, the basal membranes of stellate cells are characterised by extensive infoldings that appear to form a labyrinth.

Malpighian tubules of the yellow fever mosquito differ morphologically from other epithelia in two ways: the abundance of intracellular concretions (metallo-organic aggregates of Ca^{2+}, Mg^{2+}, and K^{+}) in principal cells and the presence of a long slender mitochondrion in the apical brush border of every microvillus.

Box 6.2 Distinct features of cell types in the Malpighian tubules

	Principal cells		Stellate cells
1.	Large size	1.	Small and thin size
2.	Dense elongate luminal Microvilli with brush border	2.	Short luminal microvilli
3.	Mitochondria present within	3.	Devoid of mitochondria Microvilli
4.	Cytoplasm with membrane	4.	Cytoplasmic concretions absent
5.	Basal membrane infolding With endoplasmic reticulum	5.	Extensive infolding forming a labyrinth

The intracellular concretions in the MTs of ants can be visualised as granules containing electron-dense concentric rings. According to some authors, these granules consist of metabolic remnants of filtrates from the haemolymph and indicates that cells with high metabolic activity store large quantities of granules. These granules are characteristic not only of ant and other insect MTs but also some cells of the midgut epithelium.

The ultrastructure of the MT epithelium can vary among insects, but the basic features are similar. Thus, the apical surfaces of the MT epithelium possess microvilli of varying descriptions, depending upon the cell type. The basal portion of the epithelium has invaginations in its plasma membrane that are perpendicular to the basal lamina and that externally cover the epithelium. Excretory cells have mitochondria, golgi complexes, and granular and agranular endoplasmic reticulum. Most of the mitochondria are associated with basal membrane invaginations that indicate the energetic need of the transport process of material from the haemolymph to the tubule epithelium.

In *Bombus morio*, two types of cells are present. **Type I cells** have mitochondria associated with both, the apical microvilli and the invaginations of the basal plasma membrane, indicating a specialised epithelium in active transport of fluids and ions. They assume the role of the rectal pads, which are absent in this insect. **Type II cells** are more related to the selective endocytosis, because they show a little amount of mitochondria associated with an apical membrane with shorter and irregular projections.

In many insects, the Malpighian cells are polytene, and the nuclei of the cells are large.In *Dinoponera australis*, organs related to digestion and excretion present the much higher levels of endopolyploidy when compared to other organs such as brain and muscle.This was also observed in *Apis mellifera*, although no statistically significant differences were observed. In addition to differences with regard to other organs, in *Apis mellifera,* MT showed higher ploidy values in an age-dependent manner. The higher ploidy in this organ may be related in their function and metabolic capacity.

In a large number of insect species, various types of intracellular crystals can be found. The organic and inorganic composition of intracellular crystals determines the colour of the tubules. The transport properties of the cells are determined by many factors that control their activity besides their position in the membranes. These crystals are always associated with a specific cell type, such as the upper tubule cells of *Rhodnius prolixus* or the primary cell of *Aedes taeniorhynchus* and *Periplaneta americana*. However, in *Bombus morio*, the two cell types I and II have crystals in their cytoplasm crystals. These crystals are generally supposed to serve as a form of storage excretion, and the most common are the **laminate spheres** or **spherocrystals**, which have unclear origin.

3.2 The Basal Lamella

All Malpighian tubules are covered by a non-cellular, fibrous matrix termed **basal lamella**. Collagen fibres have been observed in the basal lamella of some insects but not all. The basal lamella probably serves as a filter on the outside of the tubule to keep particles and cells from clogging the infolds of the basal cellular membrane. It also appears to be impermeable to large proteins and has a role in ion selectivity.

3.3 Basal Infolds

The basal plasma-membrane of Malpighian tubules is highly folded into deep slits running perpendicular to the basal lamella. There is extensive cytoskeletal network inside the folds consisting of both microtubules and microfilaments. The folds allow a substantial increase in surface area, thereby increasing the quantity of transport proteins, which can be inserted and providing a proper geometry to maximise solute-water coupling during fluid transport.

3.4 Junctions

Individual cells of the Malpighian tubules are large and often extend halfway around the tubule. Therefore, very little of the external surface of the tubule is occupied by the intercellular cleft. However, the intercellular space is thought to play a very significant role as pathways of passive movement for non-transported low molecular weight solutes.

The intercellular plasma membranes of Malpighian tubules show differentiation into four types of junctions: (1) Belt desmosomes, (2) Septate junctions, (3) Gap junctions and (4) Scalariform junctions

1. **Belt desmosomes** run as a belt around the cell apex. In insects, these are fibrous material associated with the cytoplasmic face consisting of short fuzz rather than a long,densely intertwined fibres as in the vertebrates.
2. **Septate junctions** are located apically in the epithelium, immediately below the belt desmosomes if present or at the luminal surface if belt desmosomes are absent. Septate junctions consists of two plasma membranes running parallel in a precise distance apart and spanned by septa of uniform thickness (8-9μm) which runs as shelves in the space between the cells. The smooth septa junction run in a gently curving pattern, roughly parallel to the apical surface of the cell, while the pleated septate junction runs closely parallel to each other looking like wavy ribbons. The sepatate junction functions to hold adjacent cells together. It is also thought to reduce the transepithelial movement of water and solute.
3. **Gap junctions** are located basally, either in the basal portion of the smooth septate junction or below it. They are presumed to mediate metabolic and electrical coupling.Occasionally, the basal plasma membranes are the site of membrane specialisation known as **scalariform junctions**. Here the two plasma membranes run closely parallel to one another and are held in place by thin, fibrous columnar processes and are closely associated with mitochondria. They seem to provide a fixed and precise geometry to the basal folding and may be a localized site of active ion transport.

3.5 Intracellular storage granules

Malpighian tubule cells contain storage crystals which serve as a form of storage excretion. In *Musca domestica*, there are three types of crystals. **Type A crystals** are spherical vacuoles. **Type B** generally contains fragments of cell organelles and regions of paracrystalline inclusions. **Type C** contains electron-opaque clumps of material, which aggregate into ever-growing clusters. They are membrane bound and contain β-glycerophosphatase activity.

3.6 Microvilli

The luminal surface of Malpighian tubule cells is covered with microvilli. They increase the surface area available for salt and water transport between the cell cytoplasm and the lumen. They are bounded by a unit membrane containing a coat of glycoprotein. They are organised into a core bundle of parallel filaments. Fine cross-bridges oriented perpendicular to the long axis of the microvilli link the core microfilaments to each other and to the plasma membrane.

3.7 Secondary Specialisations in Malpighian Tubules

In most insects, the Malpighian tubules are all the same as described above, although in some Gryllotalpa (Orthoptera), there are three types of tubules that are completely different based on the total composition. There are white, yellow and embryonic tubules, each seems to have distinct function. However, other roles for each type of tubule have not been conclusively demonstrated.

The Malpighian tubule epithelium of firefly, *Arachnocamp aluminosa*, has four distinct parts, each made up of different cell types:**Part I** – a short segment near the opening of the midgut, **Part II**- a long highly convoluted region containing large and granulosa cells, **Part III**- a crytonephridial region and **Part IV**- a region differentiated into a light organ.

In Malpighian tubules of *Solenopsiss aevissima* (Hymenoptera), three segments are distinguished: **proximal**, **middle** and **distal**, and this distinction is based upon the shape and structure of the epithelial cells lining the different segments of the tubules.

Aedes aegypti have five Malpighian tubules formed by the primary and stellate cells. Morphological differences involving Malpighian tubules of males and females were also reported, the diameter of the tubules in females is twice that of males. In *Aedes taeniorhynchus*, the Malpighian tubules do not exhibit distinct regions, although an arrangement of two types of cells is observed. The **primary cells** comprise the majority of tubule cells and are the site of production of primary urine. The **stellate cells** may be involved in the reabsorption of Na^+ from the primary urine

Larvae of Diptera, *Ephydrahians*, are found in hypersaline lakes of northwestern North America and physiological specialisations must occur for these larvae to live in an unusual environment. *Ephydrahians* larvae are capable of maintaining osmoregulation in the haemolymph with an osmotic concentration ten times higher than the blood. They possess a pair of Malpighian tubules containing large amounts of a white substance, compared with the other pair that does not contain the substance. It was observed that these white granules dissolved in uric acid and release gas bubbles.

Secondary specialisations are also observed in the Malpighian tubules of two superfamilies of the order Hemiptera (Cercopoidea and Membracoidea), where the **distal portions** of the Malpighian tubules are involved in the production of proteins, while the **proximal portions** which are more inflated, are responsible for producing proteins and glycosaminoglycans. These various substances have a role in the composition of the cocoon.

Special regions of the Malpighian tubules of certain coleopteran larvae, and Neuroptera also produce silk. Thus, the silk production in insects is not unique to labial gland and other tissues may also produce this material.

The Malpighian tubules are almost exclusively engaged in secretory mechanisms. Products most commonly found within the lumen of these organs are uric acid, ammonia, salts, riboflavins, pterines and ions. However, a few exceptions are reported for the preparation of special secretions such as glycoproteins and glycosaminoglycans.

The Malpighian tubules of larvae of certain chrysomelids exudate a sticky substance that seems to help procreation in some species and are used to cover the eggs' chambers in other species. Nymphs Cercopidae produces a glycoprotein secretion, possibly related to silk production.

4. ANATOMICAL SPECIALISATION OF HINDGUT EPITHELIAL CELLS

The alimentary tract posterior to the midgut is the cuticle lined hind gut. The anterior portion is narrower than the midgut and is termed the ileum while the posterior portion is expanded into a rectum. The ileum is said to be active in ion transport and is a site of water resorption. The rectum is the major site of osmoregulatory control.

In many insects, the ultra-structure of cells composing the ileum varies in accordance with the transport role of tissues. In the ileum of saline water mosquito, *Aedes compestris*, the cells are small, contain relatively few mitochondria and do not show extensive apical and basal membrane folding. Therefore, this region has lesser role in ion transport. In those insect, where ileum has a role in ion transport, the ileum contains primary cells, which possess numerous mitochondria and highly folded apical membranes, which are locations of high ATPase activity.

The rectum or the posterior hindgut is primarily responsible for final modification of the urine which is a product of the filtration of the haemolymph. An important function of the rectum in terrestrial insects is the removal of water from the faeces, which in some insects are in equilibrium with an atmosphere of 75% relative humidity or less when they are voided. Structural complexity in the insect rectum appears to have evolved along two main lines: (1) The **cryptonephric line**; the ends of the Malpighian tubules are enclosed together with the rectum in a perirectal space. (2) The **rectal gland line**; areas of thickened epithelium have a system of intercellular channels and spaces. These two lines have in common that the perirectal fluid of (1) and the intercellular fluid of (2) higher osmolarity than the haemolymph, and it has been suggested that the reabsorption of water from the faeces is thereby facilitated.

The hindgut, lined by the cuticle, completes the excretion process by selectively reabsorbing some substances into the haemolymph, leaving others in the lumen, and actively secreting somesubstances into the hindgut lumen. The rectal cuticular lining has greater permeability than the cuticular lining on foregut cells, and the epithelial cells of the hindgut are specialised for both active secretion and active reabsorption.

The rectum often contains specialised structures (small groups of cells) called **papillae** or **rectal pads** that are enlarged epithelial cells (Box 6.3). Rectal pad cells are common in many insects, and typically are enlarged, columnar to cuboidalcells arranged in six clusters separated by smaller, squamous cells between the pads of absorptive cells (Fig 6.2).The rectal pad cells in the cockroach, *P. americana* (Phillips, 1981), have highly folded cell membranes on the surface of the rectal lumen. The extensive infoldings of membranes create many intracellular channels that collect fluid (with dissolved solutes) from the rectumand direct it toward channels between adjacent cells (intercellular channels). The intercellular channels lead to the basal membrane of pad cells where water and useful solutes re-enter the haemolymph. The small group of cells have special modifications for reabsorption, and the rectum transports water and ions from the material within the gut lumen into the haemolymph. Fluid that crosses the rectal papillae cells and enters the central channel is returned to the haemolymph (Nation, 2008).

Box 6.3 Rectal cells and the ion transport

1. Papillae or rectal pads are enlarged epithelial cells of the rectum.
2. The rectum transports water and ions from the material within the gut lumen into the haemolymph.

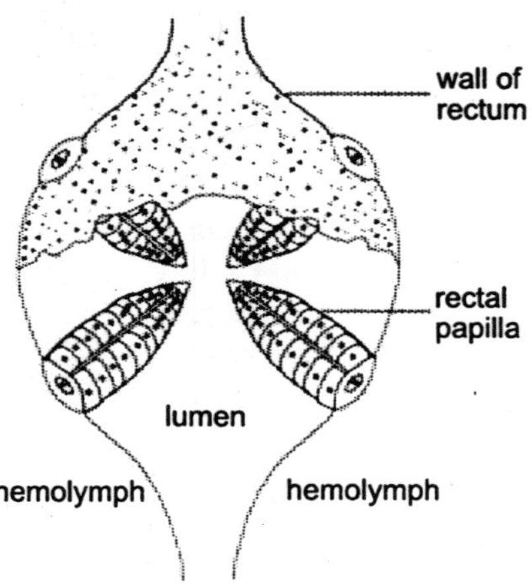

Fig 6.2. Rectal papillae in adult dipterans

5. FORMATION OF PRIMARY URINE IN MALPIGHIAN TUBULES

Primary urine is a fluid produced in the malpighian tubules to carry excretory substances to the hindgut. Because there is little or no blood pressure in the insect circulatory system, primary urine formation is bysecretion rather than by filtration.The primary urine formed in the lumen of the Malpighian tubules is a filtration of the haemolymph and it contains most of the small ions and molecules that occur in the haemolymph. The urine formation mainly relies on a proton pump in the apical membrane of Malpighian tubule. Malpighian tubes are responsible for forming a filtrate from the haemolymph and the formation of the filtrate is not selective. The cells of the malpighian tubules actively transport potassium from the haemolymph and concentrate it in the lumen. The volume of urine formation in insects is proportional with the concentration of K^+ in the haemolymph. The active transport of K^+ is primarily responsible for generating urine flow in the excretory organs of insects.

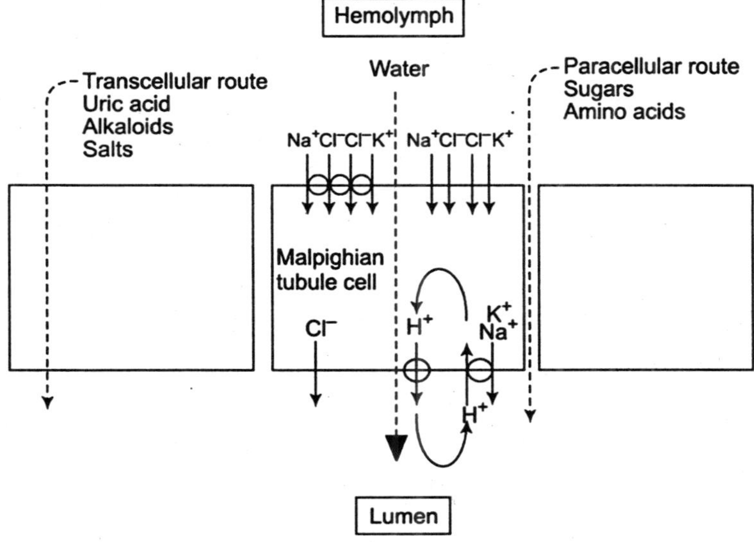

Fig. 6.3. Transcellular and paracellular pathway of the formation of primary urine

(www.entomol.nchu.edu.tw/dept/teachers/khlu/insect.../13_excretion)

The primary urine is formed by a process of secretion in the following way (Fig 6.3): **Potassium ions** are actively transported from the blood into the cavity of the tubule and are necessarily followed by negatively charged ions so as to maintain electroneutrality. In turn, water follows the ions, probably by osmosis, and various other substances sugars, amino acids, and urate ions also enter the primary urine by diffusion from the blood. The primary urine, together with soluble products of digestion and insoluble indigestible matter from the midgut, then passes into the rectum. There (or in some insects at an earlier stage) the urine is acidified, and the soluble urate is thereby converted to insoluble uric acid, which comes out of solution. Water is then reabsorbed together with the soluble products of digestion and other useful substances, including the bulk of the ions that entered the primary urine. In insects that live in dry surroundings the rectum has remarkable powers of reabsorption, its contents finally being voided as hard, dry pellets containing solid uric acid. The activity of the excretory system in insects is controlled by diuretic hormones and certain non peptide compounds such as 5-hydroxy tryptamine (5-HT or serotonin). This has been most clearly demonstrated in the case of *Rhodnius*, a bloodsucking bug. Immediately after the ingestion of a blood meal, there is a rapid flow of urine whereby most of the water taken in with the blood meal is eliminated. The distension of the body after ingestion is the stimulus that causes certain cells in the central nervous system to release a hormone that acts upon the malpighian tubules to promote a brisk flow of primary urine. The primary urine formed by the malpighian tubules is isosmotic or sometimes slightly hyposmotic to the haemolymph . Malpighian tubules are not capable of producing primary urine that is appreciably, hyperosmotic to the haemolymph. The proximal tubules may modify urine by reabsorption of some substances (*R. prolixus*), but many insects transfer the tubule fluid to the hindgut with few or no changes in its chemical composition or volume. The hindgut thenproceeds to concentrate waste products by reabsorbing water and useful substances (Nation, 2008).

6. THE BASIC TRANSEPITHELIAL TRANSPORT SYSTEM

Figure 6.4 illustrates the basic transepithelial transport system under control conditions in Malpighian tubules of the yellow fever mosquito. There are two pathways into the tubule lumen: a **transcellular pathway** through principal and stellate cells and a **paracellular pathway** between these cells. Transcellular transport involves solute entry from the peritubular medium into the cell across the basolateral membrane, movement through the cell interior and exit across the apical membrane to the tubule lumen. The paracellular pathway bypasses epithelial cells. It is a direct route from the haemolymph to the tubule lumen through septate junctions located between epithelial cells.Principal cells mediate the active transport for secreting $Na^+$and K^+ into the tubule lumen. The active transport step is located at the apical plasma membrane of the brush border, which is densely populated by an ATP-consuming proton pump, the V-type H+-ATPase (Beyenbach, 2001).

The pump consists of two major complexes, a cytoplasmic V1 complex capable of catalysing the hydrolysis of ATP, and a membrane spanningV0 complex with the properties of an H^+ channel (Muller andGruber, 2003). The reversible disassembly of the two complexes is thought of as one mechanism for regulating pump transport activity (Wieczorek *et al.*, 2000).

Protons secreted into the extracellular microenvironment of the brush border are thought to return to the cell in exchange for Na^+ and K^+, but it is unclear whether a single antiporter accepts both cations or whether separate Na^+/H^+ and K^+/H^+ antiporters are involved (Fig 6.4A). If antiport is electrically neutral, exchanging one H^+ion for one Na^+ or K^+ ion, voltage is not a driving force (Petzel, 2000). Therefore, only the net concentration difference of H^+ and Na^+ (or K^+) across the plasma membrane determines the direction and magnitude of the exchange transport. If the antiporter transports two H+ ions for each Na^+ (or K^+) ion, then voltage is an additional driving force (Petzel *et al.*, 1999). In this case, an apical membrane voltage of 120·mV (cell-negative) is able to drive Na^+ and K^+ into the tubule lumen against a 100-fold concentration difference.

Positive current passing through the septate junction from the tubule lumen tothe haemolymph is equivalent to that carried by Cl^- passing from haemolymph to lumen, which is the mechanism of transepithelial Cl^- secretion (Fig.6.4A,B). Positive current passing across the basolateral membrane is carried largely by K^+, which is the major mechanism for bringing K^+ intothe cell from the haemolymph (Fig 6.4A). One consequence of the intraepithelial current loopformed by active and passive transport pathways is that one Cl^- ion is secreted for every cation secreted into the tubule lumen. As a result, the sum of transepithelial Na^+ and K^+ secretion more orless equals the rate of transepithelial Cl^- secretion. Furthermore, the electrical coupling of active transcellular and passive paracellular transport pathways preserves electroneutrality of the solutions on both sides of the epithelium in spite of high rates of transepithelial salt and water flow.

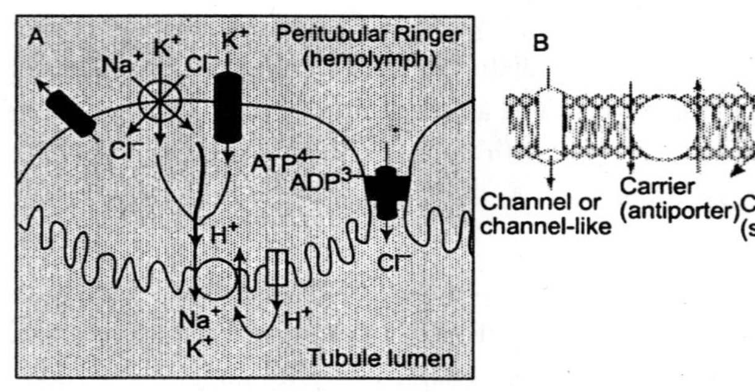

Fig. 6.4. Mechanism of transepithelial secretion in Malpighian tubules
A: Pehnotypic model B:Membrane complex
(Beyenbach(2003) J.ExptBiol, 206, 3485-3856)

7. PROTON PUMP IS THE DRIVING MECHANISM FOR URINE FORMATION AND HOMEOSTASIS

A proton pump is an integral membrane protein that is capable of moving protons across a biological membrane. Urine formation in Malpighian tubules relies upon a proton pump, which is located in the apical membrane (the side facing the lumen) of Malpighian tubule cells and in the midgut and rectum. Epithelial cells in Malpighian tubules are impermeable to sodium ions but contain a pump that transports potassium ions into the tubules.They actively secrete protons (H^+) into the tubule lumen against an electrochemical gradient. The pump consistsof a V_1 complex of proteins in the cytoplasm of principal cells, the primary cells in Malpighian tubules and anion channel formed by the transmembrane V_0 complex embedded in the lipid bilayer of the apicalmembrane (Box 6.4).

Box 6.4 Primary Urine formation

1. Urine formation in MT relies on a proton pump in the apical membrane of MT cells that actively secrete protons (H^+) into tubule lumen against electrochemical gradient.
2. Pump cause tubule lumen to become positive to hemolymph and creates highly variable gradient in pH across apical membrane of principal cells.
3. Proton gradient provides energy for antiporter mechanism that exchange K^+ for H^+ across the apical membrane.
4. Secretion of cations is coupled with Cl^- transport into the basolateral membrane of tubule cells.
5. Urine volume is dependent on K^+ concentration in haemolymph.

Malpighian tubules secrete a KCl- or NaCl rich solution that is essentially is osmotic to haemolymph and water and solutes follow by passive diffusion (Ramsay, 1954); however, some active transport is involved for, mainly, toxic substances (Maddrell 1981). The urine: haemolymph concentration ratio for many of the filtered substances approaches unity, indicating passive movement across the tubule cell membranes. But some components are actively secreted and their urine: haemolymph ratio is always greater than one. The driving force for secretion is a vacuolar-type proton pump, a V-H^+-ATPase, which is located in the apical membrane of MT principal cells in parallel with secondary activated antiporters for Na^+/H^+ and K^+/H^+ exchange.

The pump causes the tubule lumen to become positive (as much as +30 mV or more insome insects) to the haemolymph, and creates highly variable gradients in pH across the apical membrane of principal cells. The proton gradient provides the energy for an antiporter mechanism that exchanges K^+ for H^+ across the apical membrane (Wieczorek *et al.*, 2000). The net result is that K+ is secreted into the tubule lumen and concentrated against an electrochemical gradient (Nation, 2008). Through such mechanisms, non-blood-feeding insects secrete KCl into the tubule lumen, and bloodsucking insects, such as the tsetse fly *Glossina morsitans*, secrete NaCl. The formation of urine volume is highly dependent upon K^+ concentration in the bathing haemolymph or saline, but fluid formation stops even when K^+ concentration in the bathing saline ishigh if the H^+ pump is inhibited (Bertram *et al.*, 1991; Weltens *et al.*, 1992).The process driven by the proton pump has been called **standing gradient process** (Berridgeand Oschman, 1969). Cation pump plays a pivotal role in transcellular cation transport. For haemolymph homeostasis, insects must regulate their water and ion balance.

8. SELECTIVE REABSORPTION IN THE HIND GUT

The reabsorptive regions of the insect excretory system consist of an anterior ileum and the posterior rectum. If an insect is osmotically stressed due to a shortage of electrolytes and water, electrolytes and water from the filtrate are reabsorbed in the hind gut and returned to the haemolymph. Hindgut's ability to recover water from urine depends on ion movement. Reabsorption is highly selective and tightly regulated. Selective reabsorption of essential metabolites, including ions and water, occurs in the ileum and rectum and this largely determines the composition of the final excreta (Phillips, 1981). Selective reabsorption is responsible for regulatory changes in hindguts of most terrestrial insects; salt-water species experience the different problem of eliminating excess ingested salt rather than retaining water. The very large regulatory changes in the composition of the final excreta are generally achieved by selective reabsorption in the insect hindgut, particularly the rectum. This often leads to the production of either very hyposmotic or hyperosmotic urine or powder dry excreta.

8.1 Reabsorption in the Ileum

The principal function of the ileum is to act as a tubular epithelium that serves to transport to the rectum the undigested remains of the food from the midgut and fluid from the Malpighian tubules. This transport occurs by peristaltic movements of the circular and longitudinal muscles surrounding the ileum. The ileum also engages in important transepithelial transport functions. This has been investigated in considerable detail in the locust *Schistocerc agregaria* in which the ileum is a major site for isosmotic fluid reabsorption, active reabsorption of Na^+ and Cl^- and active secretion of proline as an energy source. In this species, potassium and chloride ions are transported from the lumen of the ileum into the haemolymph. This transport is iso-osmotic. It therefore does not contribute directly to osmotic regulation but serves instead to reduce the volume of the urine and to retain valuable ions and water in the haemolymph. This transport is under hormonal control, presumably to allow the insect to modulate the return of water to the haemolymph depending on whether osmotic condition of the animal dictates a diuretic or an antidiuretic response.

The driving mechanism for ion and water reabsorption in the ileum is an electrogenic Cl- pump (Phillips *et al.*, 1986,1988) (Box 6.5). A neuropeptide, the ion transport peptide (ITP), stimulates Na^+, Cl^-, and water reabsorption, and promotes passive reabsorption of $K^+$by electrical coupling (Meredith *et al.,* 1996). The movement of the excretory contents through the ileum of the desert locust reduced the volume of fluid as Na^+, Cl^-, K^+ by active mechanisms and reabsorbed the fluid following the osmotic gradients. The ileum plays a major role in acid-base balance by secretion of H^+ into the lumen, formation of NH_{4+}, and reabsorption of HCO_3^-. Metabolism of reabsorbed nonessential amino acids (alanine, asparagine, glutamine, serine, and proline) present in the urine releases energy for ATP synthesis, an essential power source needed to drive the active reabsorption of Na^+ and Cl^-.

About 80% of the ammonia produced in the epithelial cells from metabolism of theseamino acids is transported into the lumen (Phillips *et al.,* 1994) where itis excreted as NH4+ (Nation, 2008).

Box 6.5 Reabsorption in the ileum

1. The driving mechanism for ion and water reabsorption in the ileum is an electrogenic Cl^- pump.
2. The ion transport peptide (ITP) stimulates, Na^+, Cl^- and water reabsorption and promotes passive reabsorption of K^+ by electrical coupling.
3. Ileum plays a major role in acid-base balance by secretion of H+ into the lumen, formation of NH_4^+ and reabsorption of $HCO3^-$.

8.2 Reabsorption in the Rectum

The rectum is the final and major site for reabsorption of ions, water, and nutrients. All fluids and solids before being excreted enter the rectum. The rectum is therefore the last location in the gut in which the ionic and osmotic concentration of the excreta can be modified to meet the regulatory needs of the insect. It is capable of reabsorbing fluid against strong osmotic gradients, ultimately producing in many insects veryconcentrated, hyperosmotic excreta (Box 6.6). The driving mechanism for cation and water reabsorption, asin the ileum, is an electrogenic Cl- pump under the influence of a neuropeptide hormone, chloride transport stimulating hormone (CTSH), from the corpora cardiaca. It acts on the rectal epithelium to promote active Cl- absorption (Phillips, 1964; Phillips *et al.*, 1986, 1994; Harrison, 1995).

The pump provides the energy for K+ reabsorption. Water from the rectal lumen and dissolved solutes follows the osmotic gradient created by ion absorption. The result is that water is reabsorbed into the epithelial cells against an increasingly strong concentration gradient in the rectal lumen. The excreta in the rectum become very pasty or even dry in many insects as water is reabsorbed. The rectal epithelial cells actively reabsorb amino acids from the lumen, and metabolise them (primarily proline) to produce ATP needed to energise the pump. Thus, active ion secretion and electrogenic pumps play a major role in the formation of primary urine in the Malpighian tubules and then in reclaiming water from the hindgut. Several different hormones are responsible for regulating the different functions (Nation, 2008).

The role of the rectum in terrestrial insects has been most intensively studied in *S. gregaria*, in which the cells in the rectal pads serve to transport a hypo-osmotic fluid from the lumen into the haemolymph. This serves to produce excreta with a very high osmotic concentration and, in the process, conserve water in the haemolymph. Rectal function has also been investigated in aquatic insects residing in hyperosmotic media (coastal and desert saline waters). In species of *Aedes* inhabiting these waters, the rectum is differentiated into two segments. The anterior rectal segment is identical in function to the rectum of freshwater species and serves to remove ions from the urine under conditions in which the larvae find themselves in hypo-osmotic media (i.e., fresh water). When the larvae hatch in saltwater, or when the medium becomes concentrated because of evaporation, the posterior rectal segment becomes active. This segment has a single cell type, which is characterised by deep apical and basal infolds associated with numerous mitochondria. The cells actively transport ions from the haemolymph into the rectal lumen. Because the epithelium has a low osmotic permeability, ions are transported faster than water can follow. As a result, concentrated urine is produced by secretion in this segment, which has been called the **salt gland**.

Box 6.6 Reabsorption in the Rectum

1. Rectum is major site for reabsorption of ions, water and nutrients.
2. Capable of reabsorbing fluids against high osmotic gradients ultimately producing concentrated hyperosmotic excreta.
3. Driving mechanism for cation and water reabsorption is an electrogenic Cl^- pump under the influence of CTSH.

9. ROLE OF THE EXCRETORY SYSTEM IN MAINTAINING HOMEOSTASIS

Homeostasis is the dynamic tendency of a biological system to maintain constant internal equilibrium and relative stability. This involves the maintenance of a constant level of salts and water and osmotic pressure in the haemolymph (Osmoregulation), the elimination of toxic nitrogenous wastes derived from protein and purine metabolism and the elimination of other toxic compounds which may be absorbed from the environment. The excretory system plays a major role and primarily responsible for homeostasis. A key role of an excretory system is to eliminate automatically toxins not encountered before and to maintain chemical homeostasis. The alimentary canal and associated Malpighian tubules are the major organs in handling water and electrolyte balance in insects. The Malpighian tubules play a major role in eliminating metabolic wastes and toxins acquired with or from the food, typically by increasing the basal rate of fluid secretion, and transport of the toxin, into the lumen of tubules (Rheault *et al.*, 2006; Ruiz-Sanchez and O'Donnell, 2006, 2007a, 2007b; Ruiz-Sanchez *et al.*, 2007). The Malpighian tubules are an efficient system that maintains water and salt balance by producing primary urine and its subsequent elimination into gut. For haemolymph homeostasis, insects must regulate their water and ion balance.

9.1 Electrolyte Homeostasis

The fluid surrounding the cells in the body must maintain a specific concentration of electrolytes for the cells to function properly and electrolytes are a major component of body fluids. Electrolyte homeostasis is crucial for insect survival and excess electrolytes must be excreted before they lead to osmotic stress. In mosquito *Aedes aegypti*, feeding on a blood meal stimulates the release of mosquito natriuretic peptide (MNP) from the corpora cardiaca (Wheelock *et al.*, 1988; Beyenbach, 1995, 2003), and cAMP (which acts selectively to open Na+ channels in the basolateral membrane), is produced as a second messenger at the inner surface of the basolateral membrane of Malpighian tubule cells. A proton pump coupled with a H^+–Na^+ exchange mechanism secretes sodium into the tubule lumen.

The ion flex generated by MNP and cAMP is specifically an increase in secretion of Na^+-K^+ movement is not influenced. The large chloride load from the blood meal also must be excreted, and chloride (Cl^-) moves from haemolymph to tubule lumen in a passive transport pathway between the cells (paracellular pathway) (Pennabecker *et al.,* 1993) and the permeability of the paracellular pathway is increased by leucokinin-VIII, a neuropeptide (Wang *et al.*, 1996). Larval *A. aegypti* live in fresh water and, in response to an increase in salinity, they secrete 5-hydroxytryptamine (serotonin) into the haemolymph, leading to an increase in cAMP formation in the Malpighian tubules (Clark and Bradley, 1993). Serotonin and cAMP stimulate fluid and ion (Na^+ and K^+) secretion rates in isolated larval tubules, but the urine is not concentrated with respect to the ions (Clark and Bradley, 1996). Beyenbach (1995) reviews three potential physiological processes through which *A. Aegypti* may regulate (control) rates of ion and fluid excretion. These processes are (1) the proton pump that supplies energy for Na^+ and K^+ secretion to the tubule lumen, (2) the resistance Rc across the tubule cells that control ion channels in the basolateral membrane, and (3) the resistance of the passive transport pathway for chloride movement (Nation, 2008).

9.2 Water Homeostasis

In insects, the regulation of water balance is of critical importance in maintaining homeostasis (Phillips *et al* 1986). Water homeostasis delineates active physiological process that sustains and preserves an internal steady state of water within cells (cytoplasmic fluid) and two other compartments (extracellular spaces and circulatory fluids) of organisms. Water homeostasis is very important to insects because they face formidable problems of water and ion balance by virtue of their small size and high surface area:volume ratios. Water homeostasis is maintained by regulation of the osmotic pressure (Osmoregulation) of the organism's fluid and electrolyte within certain physiological range limits and tolerable boundaries. Water loss is minimised by a number of physical and behavioural mechanisms, but ultimately the homeostatic maintenance of the internal milieu is dependent upon the control of excretion. Water balance is maintained through excretion and retention which are regulated by a combination of two hormones (Box 6.7). These hormones, usually low molecular weight peptides, may enhance fluid loss (diuretic hormone) or retard it (antidiuretic hormone).

Diuretic hormones promote fluid formation and rapid excretion by the Malpighian tubules, while the currently known **antidiuretic hormones** act upon the hindgut and promote water reabsorption. Diuresis in insects is controlled by two antagonistic hormone groups: diuretic hormones, which promote water loss, and antidiuretic hormones, which inhibit it.

Box 6.7 Hormones in Water Homeostasis

1. **Diuretic hormones**- promote fluid formation and rapid excretion by MT.
   - Corticotropin-releasing factor (CRF)-related peptides: 30-46 amino acids.
   - Insect Kinins: small peptides of 6-15 amino acids.
2. **Antidiuretic hormones**- act on hindgut and promote water reabsorption.
   - Chloride transport stimulating hormone (CTSH).
   - Ion transport peptide (ITP).

9.3 Acid-Base Homeostasis

The excretory system is important in maintaining acid-base balance of body fluids and tissues (Harrison, 2001). Acid-base status in the haemolymph and gut lumen of insects is generally well regulated but varies within individuals owing to effects of temperature, activity, discontinuous ventilation, and diet. The epithelia of the renal system and gut participate in haemolymph acid-base regulation and it is one of the most critical aspects of homeostasis, as pH affects a wide variety of cellular functions. The respiratory (tracheal) system regulates haemolymph PCO_2 and the excretory system regulates haemolymph bicarbonate levels (Phillips *et al.*, 1986). As a result, haemolymph pH is maintained within narrow limits as reported for locusts during rest, temperature change and activity (Harrison, 1988; 1989; Harrison *et al.*, 1991).

Acid-base status influences many aspects of insect biology, including insect distributions in aquatic systems, insect-plant and insect-pathogen interactions, membrane transport phenomena, and the mode of action of pesticides. Acidosis or alkalosis may be experienced by an insect depending on various foods, the presence of certain types of chemical compounds in plants eaten, type of proteins metabolised (whether proteins yield a high proportion of acidic, basic, or neutral amino acids), and metabolic conditions, such as exercise (e.g., flight) that produce acids in the tissues (Harrison and Kennedy, 1994). Acid-base regulation in desert locust, *S. gregaria* injected with HCl excrete most of the acid equivalents by secretion of H+ and formation of ammonium ions ($NH4^+$) in the ileum, which is the principal mechanism for excreting excess acid equivalents (Harrison and Kennedy, 1994; Phillips *et al.*, 1994). The ileum is a major site of ammoniagenesis in locusts in which hindgut cells specifically metabolise amino acids and glucose for energy. Because of ileal and rectal secretion of ammonia/ammonium ions, about half of the total nitrogen excreted by desert locusts is ammonia nitrogen (Harrison and Phillips, 1992). Thus, excretion of total ammonia nitrogen serves several functions in locusts (Harrison and Phillips, 1992; Phillips *et al.*, 1994; Harrison, 1995), including:

1. Ammonium urate allows the insects to conserve Na^+.
2. Conversion of ammonia (NH_3) to ammonium (NH_4^+) in the ileal cells is equivalent to removal of protons (H^+).
3. Excretion of ammonia by locusts conserves water (because of precipitation of not very soluble ammonium urate salt) and recovers haemolymph pH.
4. Increases nitrogen excretion by 25% more than excretion of only sodium or potassium urate (Nation, 2008).

9.4 Nitrogen Homeostasis

Insects eliminate the excess quantities of nitrogen in one form or the other toxicity. Nitrogenous waste that accumulates in cells will alter the pH of the cell fluids and will affect the function of enzymes, thereby affecting the normal metabolic activity of the cell. Hence removal of nitrogenous waste from the cell is of greater importance for the normal metabolic functions of the cell to continue. Excess nitrogen that must be excreted may come from ingestion of proteins and nucleic acids of their food. Excess protein nitrogen is excreted as uric acid (a purine), as ammonia or ammonium salts and in several other (usually minor) forms and nitrogen from nucleic acids is excreted as uric acid.

9.4.1 *Ammonia Excretion*

Ammonia is a product of protein and amino acid metabolism. Free ammonia cannot be stored in tissues or cells because it is a very strong base influencing pH and, in its free form, it is very toxic to all cells. It must be rapidly excreted or transformed into a less toxic compound. If water is readily available for dilution, ammonia can be excreted as the free base or as an ammonium salt. Animals that excrete ammonia as their primary nitrogenous waste product are described as **ammoneoteleic** (Nation, 2008). Ammonia is a major excretory product for many aquatic insects, flesh feeding flies and insects that live in very wet environments and ammonia excretion is simple and inexpensive for them.

Ammonia is relatively toxic and usually excreted as dilute solution or else volatilised from the faeces (Cockroaches). Although some aquatic insects excrete substantial amounts of ammonia, many synthesise and excrete uric acid (or a further metabolic derivative of uric acid). Some terrestrial insects excrete the majority of their excretory nitrogen as ammonia or ammonium salts. The American cockroach, *P. americana*, excretes ammonia as a major excretory product. Enzymes involved in amino acid metabolism and ammonia production and which are detected in number of insects include amino transaminases (or transferases), glutamic acid and alanine dehydrogenases, L- and D-amino acid oxidases, adenosine deaminase, and monoamine oxidase. The amino transaminases are widely distributed in insect tissues, and enable the amino group of one amino acid to be transferred to a ketoacid, thereby forming a new amino acid. For example,

$$\text{L-aspartic acid} + \alpha\text{-ketoglutaric acid} \rightleftarrows \text{L-glutamic acid} + \text{oxaloacetate}$$

Although ammonia is not directly released in transaminase reactions, the reactions provide a way to interconvert nonessential amino acids and to make amino acids available to enzymes that deaminate with release of ammonia, such as glutamic dehydrogenase and alanine dehydrogenase in the following reactions:

$$\text{L-glutamic acid} + \text{NAD+} + H_2O \rightarrow \alpha\text{-ketoglutarate} + \text{NADH} + NH_3$$

$$\text{Alanine} + \text{NAD+} + H_2O \leftarrow \text{pyruvate} + \text{NADH} + NH_3$$

Ammonia formed in these reactions is rapidly excreted or converted into less toxic compounds. The α-ketoglutaric acid and pyruvate formed can be metabolised through the Krebs cycle as an energy source (Cochran, 1975).

Ammonia production also may come from turnover and replacement of an insect's own nucleic acids as well as from the metabolism of nucleic acids ingested with food. The enzymes adenosine deaminase, guanine deaminase, and adenine deaminase, all of which give rise to ammonia from metabolism of nucleic acids or their derived products, have been reported from various insects (Nation, 2008).

9.4.2 Uric Acid Synthesis and Excretion

Uric acid is synthesised in insects from protein nitrogen as well as from nucleic acid nitrogen (Fig 5). The major portion is synthesised from protein nitrogen simply because insects ingest relatively much more protein nitrogen (or amino acid nitrogen) in their diet than nucleic acid nitrogen (Cochran, 1975). Synthesis of one mol of uric acid from NH_3 requires the expenditure of 8 mols of ATP and four nitrogen atoms make up 33.3% of the molecular weight of uric acid (MW of uric acid = 168.11, 4 N = 56). Thus, excretion of uric acid as the main product of protein metabolism is energetically costly (Cochran, 1975). The fat body is the primary site for uric acid synthesis. Uric acid and related uricotelic compounds are admirably adaptive excretory products for organisms that live with water stress and that need to conserve water. Terrestrial insects excrete uric acid as nitrogenous waste, which serves to eliminate nitrogen in the precipitate form without losing water. Uric acid and urates require less water for synthesis and are non toxic and having low solubility in water and can be excreted essentially dry, without causing osmotic problem. Water concentration is essential for terrestrial insects and uric acid excretion (Uricotelism) is highly advantageous. Although the origin of the nitrogen atoms in uric acid from *Rhodnius* was not determined, it seems reasonably certain that the nitrogen atoms are derived from NH_3 resulting from metabolism of proteins, amino acids, and nucleic acids (Barrett and Friend, 1970).

Storage excretion of uric acid, or deposition in various parts of the body, is common in cockroaches and some other insects. Since uric acid is so insoluble, it can be easily stored without interacting with other physiological processes. Some cockroaches accumulate up to 10% of their dry weight in uric acid stored in specialised urate cells in the fat body, which can be utilised during periods of dietary stress and in some lepidoptera, the fat body shifts from excretion of uric acid to its storage during the last larval instar. Crystals of uric acid may occur in the haemolymph, and in other tissues, especially in fat body.

Many insects excrete small to large percentages of their excretory nitrogen as **allantoin**, a breakdown product of uric acid catalysed by the enzyme uricase (widespread in insects and in their tissues) and a few insects excrete some allantoic acid, an oxidation product of allantoin, but the enzyme allantoicase is not widespread in insects Razet (1966). Animals that excrete most or all of their excretory nitrogen as uric acid are described as uricotelic and insects should still be considered uricotelic in excretion even though some excrete a variety of nitrogenous compounds and a few excrete relatively little or no uric acid (Bursell, 1970 and Cochran, 1975).

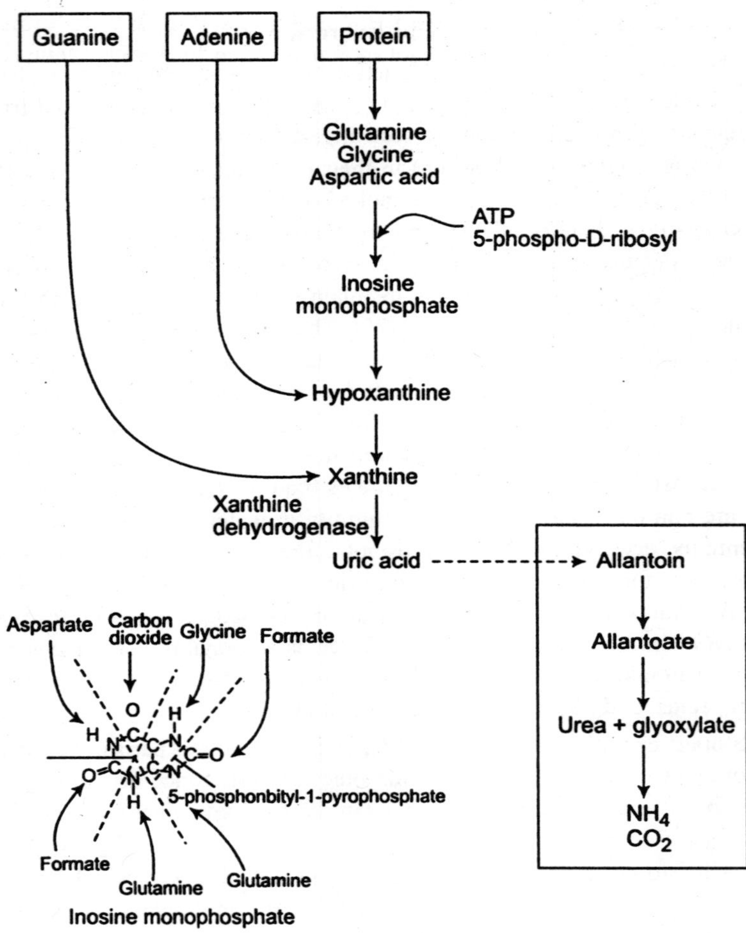

Fig 6.5. Pathway of synthesis of uric acid from nucleic acid and protein

10. HORMONAL CONTROL OF URINE FORMATION

The process of urine formation is under hormone control. The rates of urine formation and ion secretion are controlled by diuretic hormones and certain non-peptide compounds such as 5-hydroxytryptamine (5-HT or serotonin) (Box6.8). The diuretic hormone is capable of making Malpighian tubules actively produce urine. The level of hormone in the haemolymph varies with the state of hydration of the insect and the hormone is inactivated by the Malpighian tubules. The diuretic hormone by stimulating active

K^+ transport, changes only the rate of urine production (Pilecher, 1970). 5- hydroxytryptamine (serotonin) stimulates a rapid flow of urine in Rhodnius (Maddrell, 1969) but not in *Schistocerca* for which cyclic AMP was stimulatory.

Box 6.8 Hormonal control of urine formation

1. The rates of urine formation and ion secretion are controlled by diuretic hormones and certain non-peptide compounds such as 5-hydroxytryptamine (serotonin).
2. Diuretic neuropeptides isolated from insects are:
   - Corticotropin-releasing factor related peptides
   - Kinins

10.1 Diuretic hormones

Insect diuretic hormones (DH) play key roles in the regulation of water balance. They are generally released in response to feeding, which involves a large uptake of water. Water balance is brought about through elimination of excess water in the form of urine. Excretion and water balance are under the control of neuroendocrine, which are responsible for stimulating the secretion of urine by the Malpighian tubules. This is particularly important for blood-eating insects, such as *Rhodnius prolixus* (blood-sucking bug) and *A. aegypti* (yellow fever mosquito), which can consume up to 10-20 times their body weight during a single meal (Coast 2001).

There are three primary diuretic hormones found in insects: two are classified as **neuropeptides** (CRF-like DH and kinins) and one as a **neurotransmitter** (serotonin, 5-HT). Each diuretic hormone has its own way of mediating excretion, when presented together can act in synergism to further enhance the process of dieresis.

One important aspect regarding the insect excretory system is that it is coupled with the movement of ions. Tubule secretion of urine is driven by the active and/or passive transport of Na^+, K^+, and Cl^- ions from the haemolymph into the tubule lumen. Diuretic hormones increase the rate of tubule secretion by stimulating the ions (Na^+, K^+, and Cl^-) movement. CRF-like DH does this via cAMP as a secondary messenger. Through cAMP, CRF-like DH opens Na^+ channels on the tubules, thereby increasing the Na^+ concentration in urine. This pathway is especially significant in blood-feeding insects because blood is rich in NaCl (Williams and Beyenbach 1983). Although CRF-like DH is present in the majority of insects, each species of insects may have its own specific CRF-like DH such as *Locusta*-DH in locusts, *Acheta*-DH in crickets, *Stomoxys*-DH and *Musca*-DH in flies, *Periplaneta*-DH in cockroaches, and *Manduca*-DH in moths. In addition, some species of insects possess two CRF-like DH, such as *Manduca sexta*, which has *Manduca*-DH I and *Manduca*-DH II.

In contrast, Kinin causes the urine concentration of Na^+ to fall and the K^+ concentration to increase (Pannabecker *et al* 1992). This is particularly important in plant-feeding insects such as locusts and moths, which consume a diet rich in K^+ with relatively little Na^+ (Williams and Beyenbach 1983). Some examples of insect kinins include: leucokinins in *Leucophae amaderae* (cockroach), achetakinins in *Acheta domestics* (cricket), and locustakinins in *Locusta migratoria* (locust) (Veenstra 1994). The difference between CRF-like DH and kinins is that CRF-like DH stimulates the active transport of cations (Na^+; K^+) while kinins stimulate the passive transport of anions (Cl^-). Both responses have different effects on the ion transport processes and the composition of the excreted urine (Beyenbach 1995).

Unlike the other two diuretic peptides, serotonin (5-HT or 5-Hydroxytryptamine) is a neurotransmitter that functions as a hormone and is found in only some species of insects. One of these species is the blood-sucking bug, *Rhodnius prolixus*, which has both a CRF-like neuropeptide and serotonin. Although serotonin can stimulate urine secretion by the Malpighian tubules, it still requires a diuretic peptide (CRF-like DH or kinins) for maximal secretion (Maddrell *et al* 1991). When two diuretic hormones act together (kinin and a CRF-like DH, or a CRF-like DH and serotonin) the synergistic control of Malpighian tubule secretion results in the greater increase in tubule secretion than the sum of their separate responses. Advantage of synergism is the speed with which tubule secretion can be turned on and off and it reduces the amount of diuretic hormones needed to induce the desired urine secretion. In locusts, 50% of the insect's stored CRF-like DH (*Locusta*-DH) is required for maximal tubule secretion, whereas in the presence of a kinin (locustakinin), only 2.5% of the locust's stored *Locusta*-DH (Patel *et al* 1995) is required for maximum tubule excretion.

10.2 Antidiuretic Hormones

The antidiuretic hormones, in contrast with the diuretic hormones are those that inhibit the loss of water in the insect from inhibition of the secretions of MTs and/or by reabsorption of ions and water in the hindgut (rectal sacs). Terrestrial insects and insects that feed on dry solid food probably need to conserve water rather than excrete it. Antidiuretic hormones in insects may decrease ion transport and associated fluid secretion by the Malpighian tubules, or they may stimulate reabsorption of ions and water across the hindgut (Quinlan *et al.*, 1997). In the house cricket, *Acheta domesticus*, an antidiuretic hormone, detectable in the haemolymph of water-deprived crickets, was found to inhibit fluid secretion by the Malpighian tubules but has no effect on the hindgut. The hormone is released from specific axons and its release is triggered by dehydration. These results imply that an unusual mechanism regulates water balance in these insects (Spring *et al* 1988). For eg., *Rhodnius prolixus*, eliminates 40-50% of the fluid mass after ingestion of a blood meal (equals to 10-12 times their initial body mass) due to rapid activation of excretory system. Secretion of ions and water is stimulated 1000-fold by serotonin and diuretic hormone. These hormones cooperate synergistically to activate adenylate cyclase activity from MTs cells, which increase the level of intracellular cAMP.

The anti-diuretic hormones have also an important role in the fluid maintenance of *R. prolixus*. Reduction in urine flow may be due to decreased diuretic hormone release or from a novel mechanism of anti-diuresis involving insect cardioacceleratory peptide 2b (CAP2b) and cyclic GMP. The anti-diuretic hormones have also an important role in the fluid maintenance of *Rhodnius prolixus* (Sabrina, *et al* 2007). All insects possess a diuretic hormone, while only some of them also possess an anti-diuretic hormone. At the present time only a few antidiuretic hormones are known. Two anti-diuretic polypeptides, the **neuroparsins** (act at rectum) (Girardie*et al.*, 2001) and the **ion transport peptide (ITP),** which stimulate ions and water reabsorption on the ileum (King *et al.* 1999) have been isolated from corpora cardica (CC) of the desert locust, *Schistocerca gregaria* (Gäde, 2004).

10.3 Hormonal Regulation of Malpighian Tubule Transport

Both diuretic and anti-diuretic responses of Malpighian tubules to hormones have been reported. Endocrine factors involved in the regulation of Malpighian tubule function include several distinct structurally related groups of peptides, and the biogenic amine serotonin (5-hydroxytryptamine). Diuretic peptides identified to date include corticotropin-releasing factor (CRF)-related peptides, calcitonin-like peptides, kinins, tachykinins, and type 2b cardioacceleratory peptides (CAP2b) (Fig.6.5). Serotonin stimulates Malpighian tubule secretion in a number of insect species and is known to hormonally regulate excretory function in the blood sucking bug *Rhodnius prolixus* and in larval mosquitoes (*Aedes aegypti*). Short term hormonal adjustments in Malpighian tubule transport rates in response to immediate challenges may be superimposed upon longer term increases or decreases in basal and maximal transport capacities. These long-term responses may be mediated at least in part by prostaglandins.

Kinins

Kinins are small (6- to 15-amino-acid) peptides with a C-terminal amide. Those identified to date have similar C-terminal sequences of Phe-Xxx-Yyy-Trp-Gly-amide, where Xxx can be Asn, Ser, His, Phe, or Tyr and Yyy can be Pro, Ser, or Ala. The kinins have been isolated from several insect species but have not been detected in other animals. A single species may contain several kinins.

Serotonin or 5-hydroxytryptamine (5-HT)

Since nerves are not connected to MTs, the regulation of this epithelial transport is mediated via intrinsic mechanisms and via messengers that circulates in the haemolymph. After *Rhodnius prolixus* blood meal, distensionof the abdomen is observed, which trigger the release of serotonin and diuretic peptides from a variety of potential sources, including neurons located in the central nervous system, mesothoracic ganglia and the corpus cardiacum (Chiang and Davey 1988). The increased haemolymph serotonin concentration is correlated with an elevated rate of dieresis. The serotonin activates adenylate cyclase in cell membrane isolated from MTs of *Rhodnius prolixus*, increasing cyclic AMP second messenger. Stimulation of MTs with serotonin produces a characteristic triphasic change in transepithelial potential (TEP) (Ianowski and O'Donnell 2001).

The Na+/K+-ATPase participates in transepithelial transport and cell volume regulation in MTs of *Rhodnius prolixus* (Caruso-Neves *et al.* 2001). Serotonin, the primary diuretic agent in *Rhodnius prolixus* released during meal, inhibits the Na+/K+ pump, thereby bringing about the stimulation of transepithelial Na^+ secretion (Grieco and Lopes 1997). The inhibition is thought to increase intracellular Na^+ concentration, causing Na^+ and water movement to the lumen of the tubule and thus increasing the rate of fluid secretion. This hypothesis is supported by observation that ouabain increases the fluid secretion in isolated and unstimulated MTs (Nicolson 1993).

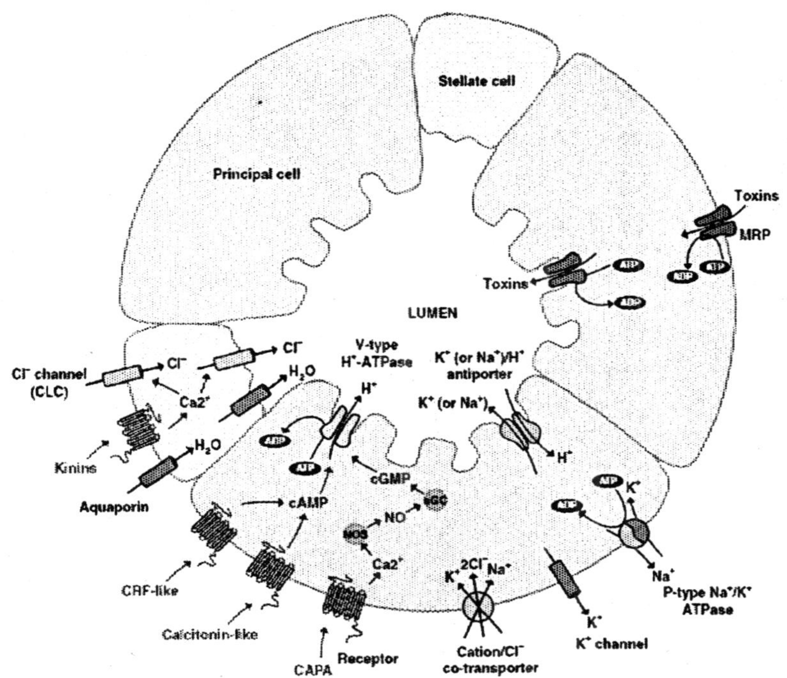

Fig 6.5 Some major ion transporters in the basolateral and apical membranes of Malpighian tubules of *Drosophila*

(Adopted from Esau Ruiz-Sanchez and m.J.O'Donnell (2015), Current Opinion in Insect Science 11: 14-20)

That transepithelial secretion continues in the presence of ouabain confirms the central role of the Vtype H^+ ATPase in powering transepithelial transport (Ianowski and O'Donnell 2001, Beyenbach *et al.*, 2000). Serotonin has also an important function in water transport in *Rhodnius prolixus* MTs. Cells membranes are exquisitely selective barriers that control the solute composition of enclosed compartments by regulating the entry of ions, solutes and water into cells. In general, osmotic forces drive water through the plasma membrane via selective channels called **aquaporins** (AQP) that are responsible for the rapid water fluxes through cell membranes (Agre *et al.*, 1993, Borgnia *et al.* 1999).

The Corticotropin-releasing factor

The corticotropin-releasing factor related diuretic hormone family (CFR-related DH) is also able to increase cyclic AMP content, transepithelial potential and rate of secretion in insect MTs (O'Donnell *et al.*, 1996, Coast *et al.*, 2002). The CRF-related diuretic peptides range from 30 to 46 amino acids in length and show structural homology to a family of peptides found in vertebrates that includes CRF (hence the name), sauvagine, and urotensin. Available evidence suggests that a single species produces more than one CRF-related diuretic peptide.

Cardio Acceleratory Peptides (CAPS)

Cardio acceleratory peptides (CAPs) are neuropeptides originally isolated from CNS of tobacco hawk moth *Manduca sexta*. The CAPs have also been implicated in the control of fluid secretion by the Malpighian tubules of *Drosophila melanogaster*.

Early biochemical studies using low-pressure column chromatography fractionated the CAPs into two distinct groups, CAP1 and CAP2 which are subdivided into at least two CAP1s (1a and 1b) and three CAP2s (2a, 2b and 2c) (Tublitz and Truman 1985a, b, c, Cheung *et al.*, 1992). The peptide CAP2b stimulates tubule fluid secretion via the cGMP second messenger system in *Drosophila* (Davies *et al.*, 1995). It has been suggested that modulate a vacuolar-type H+-ATPase (O'Donnell *et al.*, 1996).

10.4 Hormonal Regulation of Midgut Transport

Exchanges of materials between the midgut lumen and the haemolymph have important consequences for haemolymph composition, yet few studies have addressed the role of endocrine regulation of midgut transport in haemolymph homeostasis. In *Rhodnius*, serotonin stimulates both uptake of fluid across the midgut into the haemolymph and removal of that fluid from the haemolymph by the Malpighian tubules. Evidence suggests that a similar process may occur in larval mosquitoes. Acid/base exchanges between the haemolymph and the midgut lumen of larval Lepidoptera, creating conditions favourable to nutrient digestion and absorption, also appear to be regulated. Hormones inhibit ion transport across the lepidopteran midgut, and serotonin stimulates both acid/base transport and rhythmic muscular contractions of the larval mosquito midgut. Evidence for regulation of more than one midgut transport mechanism by multiple chemical signals suggests that the regulation of midgut transport may play an important role in haemolymph water and ion homeostasis.

10.5 Hormonal Regulation of Hindgut Transport

Hormones acting on the hindgut appear to stimulate recovery of fluid and ions from the primary urine, leading to cycling of materials through the excretory system and haemolymph, and clearance of wastes. The majority of our information about regulation of hindgut transport comes from studies on the locust. The locust hindgut consists of two distinct regions, a more anterior ileum and a more posterior rectum. Two peptide factors isolated from the corpus cardiacum are ion transport peptide (ITP), acting on the ileum, and chloride transport stimulating hormone (CTSH), acting on the rectum. The amino acid sequence of these peptides has been determined. Other factors that stimulate water reabsorption by the hindgut include neuroparsins A and B. In addition, a number of factors modulate contractions by the hindgut musculature.

II. CRYPTONEPHRIDIAL SYSTEMS

An arrangement of the excretory system in certain beetles (Coleoptera), Lepidoptera, and some sawfly larvae (Hymenoptera) in which the distal blind endings of the Malpighian tubules are held close to the surface of the rectum enveloped within a perinephric membrane. This arrangement is known as a **cryptosolenic** or **cryptonephridic tubule system**. Cryptonephric system is present in most of the larvae of Lepidoptera, Neuroptera, Diptera, Symphita and adults of Coleoptera, however, only a few Coleoptera and Diptera have been well documented (Bradley *et al.*, 1982). Cryptonephridic tubules do not penetrate the lumen of the rectum, but lie on the outer surface of the rectum, encased within a perinephric chamber bounded by the perinephric membrane. This arrangement is found in most lepidopteran larvae and many coleopterans in which the ends of the Malpighian tubules do not lie free in the hemocoel. The cryptonephridial complex performs two functions, reabsorbs water from the hindgut very efficiently and in some insects is able to absorb atmospheric water from the humidity in the hindgut. A high concentration of ions is established in the tubules, which draws water from the gut by osmosis. This creates a highly efficient mechanism for water conservation in species, such as mealworms (e.g. *Tenebrio* spp.), that feed on very dry materials or live in a relatively dry environment and therefore need to retain water (Chapman, 1998) (Fig. 6.6). In most of the Coleoptera, a cryptonephridic tubule in the outermost layer makes contact with the outer perinephric membrane through a single, highly modified cell of the tubule wall called the leptophragma cell. These points of contact when a tubule is separated from the haemolymph only by the thin cell membrane of the leptophragma cell and the very thin (at this particular point) perinephric membrane are called **leptophragmata** (Nation, 2008). Only two families of beetles, Ptinidae and Anobiidae, do not have leptophragmata, but none of the Lepidoptera studied by Saini (1964) has them.

The food of insects is variable in water content and some Coleoptera are phytophagous, eating food with relatively high water content, while others feed upon dry stored grain products or similar dry food materials. In phytophagous Coleoptera, the posterior rectal region containing the faecal pellet just prior to its being expelled is smaller and has fewer convolutions of the tubules than in Coleoptera that feed upon dry food (Saini, 1964). Coleoptera living in very dry environments, such as *Tenebrio molitor* and other grain-infesting beetles and weevils, have a large cryptonephridial system and an extraordinary ability to extract more water from the excrement faecal pellet (Meyran, 1982). While the majority of larvae of Lepidoptera, which have a diet rich in succulent leaves, this system appears to be related to remove excess ions from the haemolymph (Pacheco *et al.*, 2014). *Palpares annulatus* larvae (Neuroptera, Myrmeleontidae) living in arid and semi-arid savannah of the Kalahari in southern Africa, and are exposed to long periods of food deprivation and short periods of food abundance. The larvae of insects of this family are voracious predators, hence the form of digestion is very specialised, that is the extra-intestinal digestion takes place. The ingested food is trapped in a structure shaped bag in a blind bottom, slowly being absorbed by highly specialised internal epithelium (Van Zyl and Van Der Linde, 2000).

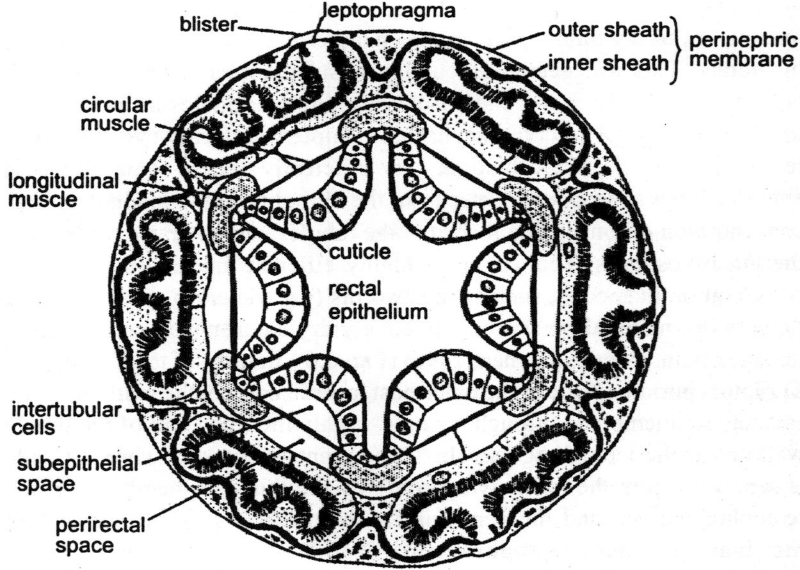

Fig. 6.6 Cross section of rectum showing cryptonephridian tubules in *Tenebrio molior* (Adapted from http://slideplayer.com/slide/1665123/)

12. THE INSECT EXCRETORY SYSTEM AS A TARGET FOR NOVEL PEST CONTROL STRATEGIES

The mechanism of fluid and ion transport by the epithelial cells of the excretory system provides a foundation for development of novel pest management strategies. Two important strategies are gaining importance- (1) impairment of osmoregulation by manipulation of diuretic or antidiuretic signaling pathways and (2) interference with toxin elimination by inhibition of toxin transport systems.

12.1 Peptide analogs of diuretic or antidiuretic factors

Insect pests may be controlled by using peptide analogs of diuretic or antidiuretic factors because severe disruption of ionic and osmotic homeostasis may lead directly to insect mortality. Analogs which act as antagonists of MT diuretic hormones may also act indirectly, through their effects on elimination of toxins, which are actively transported into the lumen of the MTs. Concentrations of toxins in the lumen of the tubule can be increased by reducing the rate of fluid secretion and thereby permit the toxins to diffuse passively into the haemolymph. Antagonists of diuretic factors can thus increase the retention time of toxins in the haemolymph, thereby enhancing their toxicity.

In *Drosophila*, the kinins act as diuretic factors through enhancement of Cl^- secretion through *ClC-a* channels. Native kinins are unsuitable for use as pest control agents because they are degraded, and thus inactivated by peptidases in the haemolymph and gut. Development of kinin analogs that may be used as pest control agents has been focused on increasing biostability through changes that impede peptidase action, and also maintaining or increasing efficacy.

Kinin analogs containing an aminoiso-butyric acid (Aib) as the 3rd (Xaa^2) residue of the pentapeptide are completely resistant to hydrolysis by angiotensin converting enzyme (ACE) and retain potent diuretic activity in the cricket *Acheta domesticus*. It is proposed that the Aibanalogs can interact with kinin receptors but that the branched chain at the alpa-carbon of Aib prevents effective binding with ACE. Analogs which are blocked also at the N-terminus with pyroglutamate (pGlu) residues are also resistant to aminopeptideases as well as to ACE and are also 8-fold more potent than native kinins in stimulating fluid secretion by cricket tubules set up in the Ramsay assay (Cabrero *et al* 2014). Kinin analogs incorporating one or two β amino acids (β-Pro, β-Phe, β-Trp) also show increased resistance to hydrolysis whilst retaining potency as diuretics (Zubrzak *et al.*, 2007).

β-Amino acids have their amino group bonded to the β carbon rather than the α carbon as in the 20 standard biological amino acids. The extended main chain of β-amino acid residues is thought to interfere with successful binding of endopeptidases.

The capa peptides have been shown to play important roles in mediating survival in response to both desiccation and cold stress through their effects on the MTs. In dipteran insects, the capa peptides promote diuresis and act through a G protein-coupled receptor (CapaR) in the basolateral membrane of the tubule principal cells to stimulate ion and fluid secretion. Targeted knockdown of capaR reduces capa-stimulated diuresis and thereby increases survival of flies exposed to desiccating conditions, whereas enhancement of capa-signaling renders flies more susceptible to desiccation.

12.2 Interference with membrane transporters for inorganic ions or toxins

Recent approaches with the potential for greater specificity for pest targets include RNAi-based strategies to silence the expression of essential ATP-binding cassette (ABC) transporter genes. RNA interference (RNAi) involves use of RNA molecules to inhibit gene expression, by causing degradation of specific mRNA molecules. The sequence specificity of insecticides based on RNA interference (RNAi) approaches is an important advantage with regards to limiting the mortality of non-target species during insecticide application. Selective killing of fruit flies (*D. melanogaster*), flour beetles (*Tribolium castaneum*), pea aphids (*Acyrtho siphonpisum*), and tobacco hornworms (*Manduca sexta*) has been achieved by feeding insects species-specific dsRNA targeting V-ATPase transcripts (Whyard *et al.*, 2009). RNAi has also been used to knock-down the expression of the gene encoding P-glycoprotein in mosquito larvae. In the absence of RNAi, Pgp gene expression increases 8-fold in temephos-treated larvae of *Aedes aegypti*. Toxicity of the temephos is increased approximately 2-fold in larvae after P-glycoprotein gene silencing (Figueira-Mansur *et al.*, 2013). ABC transporters in non-excretory tissues may also be targeted. In T. castaneum, RNAi knockdown of two ABC transporters involved in the transport of cuticular lipids was associated with reduced levels of epicuticular lipids and consequent mortality through desiccation (Broehan *et al.*, 2013). ABCB transporters (P-glycoproteins) in the MTs have been implicated inexcretion of nicotine and other alkaloids. Future studies of RNAi's targeting ABC transporters or other toxin extrusion mechanisms in lepidopteran pests will thus be of great interest. A major challenge of all approaches targeting toxin excretion mechanisms, including the use of RNAi, is the inherently broad substrate specificity of organic anion (OA) and organic cation (OC) transporters, including ABC transporters such as the Pgps and MRPs. The overlap in the substrate affinities of multiple transporters means that even effective and selective knockdown of a single OA or OC transporter may not be sufficient to induce lethality.

An alternative approach is the development of insect specific blockersfor inorganic ion channels. High-throughput screening of~30,000 small-moleculeslead to discovery of a potent and selective inhibitor ofthe inwardly rectifying K^+ channels in the MTs of *A. aegypti* (Raphemot *et al.*, 2014).

12.3 Microbial and plant-derived natural compounds

Various natural compounds inhibit the V-ATPase, which affects multiple transport processes in the MTs and midgut. Destruxins, produced by a wide variety of fungi, strongly inhibit V-ATPase [44] and have been shown to inhibit fluid secretion by isolated MTs of *Schistocerca gregaria, D. melanogaster* and *Rhodnius prolixus*. In tubules of *R. prolixus*, inhibition of fluid secretion by destruxin A, presumably through the inhibition of the V-ATPase, alters the transepithelial electrical potential and the pH of the secreted fluid. Similarly, the bacterial metabolite bafilomycin has been used for more than two decades as a potent inhibitor of the V-ATPase in MTs. More recently, PA1b (pea albumin 1, subunit b) isolated from *Pisum sativum* has been identified as the first peptide inhibitor of the V-ATPase and has proven effective as a bioinsecticide for multiple pest species.

Natural compounds can also potentially target second messenger signalling pathway. The urease-derived peptide jaburetox 2Ec and the alkaloid chelerythrine are both plant-derived compounds which inhibit fluid secretion by MTs. Both compounds alter second messenger signaling pathways that mediate the actions of diuretic and antidiuretic factors.

REFERENCES

Agree, P., Preston, G.M., Smith, B.L., Jung, J.S., Raina, S., Moon, C., Guggino, W.B. and Nielsen, S. (1993) Aquaporin CHIP: the archetypal molecular water channel. Am J Physiol 265: F463– 476

Barrett, F. M., and Friend, W. G., (1970). Uric acid synthesis in ***Rhodnius prolixus***, ***J. Insect Physiol.*16**:121–129

Berridge, M. J. and Oschman, J. L. (1969). A structural basis for fluid secretion by Malpighian tubules. *Tissue Cell* **1**,247 -27

Bertram, G., Schleithoff, L., Zimmermann, P. and Wessing, A. (1991). Bafilomycin A_1 is a potent inhibitor of urine formation by Malpighian tubules of ***Drosophila hydei***: is a vacuolar-type ATPase involved in ion and fluid secretion? J. Insect Physiol. 37, 201-209.

Beyenbach K W· (2003) Transport mechanisms of diuresis in Malpighian tubules of insects. J.Expt Biol, 206, 3485-3856

Beyenbach, K. W. (1995). Mechanism and regulation of electrolyte transport in Malpighian tubules. *J. Insect Physiol.* 41, 197-207.

Beyenbach, K. W. (2001). Energizing epithelial transport with the vacuolar H+-ATPase. *News Physiol. Sci.* **16**, 145-151.

Beyenbach KW, Pannabecker TL, Nagel W.(2000) Central role of the apical membrane H+-ATPase in electrogenesis and epithelial transport in Malpighian tubules. *J Exp Biol.*; 203:1459–1468.

Borgnia M, Nielsen S, Engel A And Age P. (1999). Cellular and molecular biology of the aquaporin water channels. *Annu Rev Biochem* 68: 425–458.

Bradley TJ, Stuart AM, Satir P(1982). The ultrastructure of the larval malpighian tubules of a saline-water mosquito. *Tissue Cell.* 14:759–773.

Broehan,G., T. Kroeger, M. Lorenzen, H. Merzendorfer (2013). Functional analysis of the ATP-binding cassette (ABC) transporter gene family of *Tribolium castaneum* BMC *Genomics,* 14

Bursell, E., (1970). An Introduction to Insect Physiology. Academic Press, London, New York

Cabrero P, Pollock VP, Davies SA, Dow JA.(2004). A conserved domain of alkaline phosphatase expression in the Malpighian tubules of dipteran insects. *J Exp Biol.* 207:3299–3305

Caruso-Neves C, Silva IV, Morales MM and Lopes AG. (2001). Cytoskeleton elements mediate the inhibition of the Na+/K+ ATPase activity by PKC in *Rhodnius prolixus* Malpighian tubules during hyperosmotic shock. *Arch Insect Biochem Physiol* 48: 81– 88.

Chapman, R. F. (1998). The insects:structure and function, 4th ed. Cambridge University Press, New York.

Cheung, C.C., Loi PK, Sylwester AW, Lee TD and Tublitz NJ. (1992). Primary structure of a cardioactive neuropeptide from the tobacco hawkmoth, *Manduca sexta. FEBS Lett* 313: 165– 168.

Chiang, R.G., and Davey, K.G. (1988). A novel receptor capable of monitoring applied pressure in the abdomen of an insect. Science 241: 1665– 1667.

Clark T.M. and Bradley, T. J. (1993). Short term changes in hemolymph properties of larval ***Aedes aegypti*** in response to physiological challenges affect Malpighian tubule secretion rates ***in vitro. Am. Zool.*** 33(5): 43A

Clark T.M., and Bradley T..J. (1996). Stimulation of Malpighian tubules from larval ***Aedes aegypti*** by secretagogues. ***J. Insect Physiol*** **42:** 593-602.

Coast, G. M. (2001). The neuroendocrine regulation of salt and water balance in insects. *Zoology* 103, 179-188.

Coast, G.M, Orchard I, Phillips JE and Schooley DA. 2002. Insect diuretic and antidiuretic hormones. Adv Insect Physiol 29: 279– 409.

Cochran, D. G., (1975). Excretion in insects, in: ***Insect Biochemistry and Function*** (D. J. Candy and B. A. Kilby, eds.), Chapman and Hall, London.

Davies, S.A, Huesmann GR, Maddrell SH, O'Donnel MJ, Skaer NJ, Dow JA and Tublitz NJ. (1995). CAP2b, a cardioacceleratory peptide, is present in Drosophila and stimulates tubule fluid secretion via cGMP. *Am J Physiol* 269 (6 Pt 2): R1321– 1326.

Esau Ruiz-Sanchez and m.J.O'Donnell (2015). The insect excretory system as a target for novel pest control strategiesCurrent Opinion in Insect Science 11: 14-20

Figueira-Mansur,J., A. Ferreira-Pereira, J.F. Mansur, T.A. Franco, E.S.L. Alvarenga, M.H.F. Sorgine, B.C. Neves, A.C.A. Melo, W.S. Leal, H. Masuda, M.F. Moreira (2013). Silencing of P-glycoprotein increases mortality in temephos-treated *Aedes aegypti* **larvae.** *Insect Mol. Biol.*22(6) 648-658

Gade G. (2004). Regulation of intermediary metabolism and water balance of insects by neuropeptides. *Annu Rev Entomol* 49: 93– 113.

Girardie J, Chaabihi H, Fournier B, Lagueux M and Girardie A. (2001). Expression of neuroparsin cDNA in insect cells using baculovirus vectors. *Arch Insect Biochem Physiol* 46: 26– 35.

Harrison, J. M. (1988). Temperature effects on haemolymph acid-base status in vivo and in vitro in the two-striped grasshopper Melanoplus bivittatus. *J. Exp. Biol.* 140, 421-435.

Harrison, J. F. (1989). Ventilatory frequency and haemolymph acid-base status during short-term hypercapnia in the locust, Schistocerca nitens. *J. Insect Physiol.* 35, 809- 814.

Harrison, J. F. and Kennedy, M. J. (1994). In vivo studies of the acid-base physiology of grasshoppers: the effect of feeding state on acid-base and Nitrogen excretion. *Physiol. Zool.* 67, 120-141

Harrison J F and Phillips, J.E 1992 Recovery from acute haemolymph acidosis in unfed locusts. II Role of ammonium and titratable acid excretion. *J. exp. Biol. 165, 97-110*

Harrison, J. F., Phillips, J. E. and Gleeson, T. T. (1991). Activity physiology of the two-striped grasshopper, *Melanoplus bivittatus*: Gas exchange, haemolymph acid–base status, lactate production and the effect of temperature. *Physiol. Zool.* 64, 451–472.

Harrison, J.F., 1995. Nitrogen metabolism and excretion in locusts. In: Walsh, P.J., Wright, P.A. (Eds.), Nitrogen Metabolism and Excretion. CRC Press, Boca Raton, pp. 119–131

Harrison JF (2001) Insect acid-base physiology. *Annu Rev Entomol* 46: 221–250.

Ianowski J P and O'Donnell M J 2001 Transepithelial potential in Malpighian tubules of *Rhodnius prolixus*: lumen-negative voltages and the triphasic response to serotonin. *J Insect Physiol* 46: 107– 117.

King, D.S., Meredith,J., Wang, Y.J., and Philips, J.E. 1999. Biological actions of synthetic locust ion transport peptide (ITP). *Insect Biochem Mol Biol* 30: 489– 498.

Maddrell, S. H. P. (1969). Secretion by the Malpighian tubules of *Rhodnius.* The movements of ions and water. *J. exp. Biol.* 52, 71-97.

Maddrell, S.H.P.(1971) Fluid secretion by the Malpighian tubules of insects. *Philos Trans R Soc Lond* B Biol Sci. 262:197–207.

Maddrell, S. H. P. (1981). Functional design of the insect excretory system. *J. Exp. Biol.* 90, 1-15

Maddrell, S. H. P., Herman, W. S., Mooney, R. L. and Overton, J. A (1991). 5-hydroxytryptamine: a second diuretic hormone in*Rhodnius prolixus*. *J. Exp. Biol.* 156, 557-566.

Meredith J., M. Ring., A. Macins, J. Marschall, N.N Cheng, D. Theilmann, H.W Brock and J.E. Philips (1996) Locust ion transport peptide (ITP): primary structure, cDNA and expression in a baculovirus system. *J.Expt. Biol.* 199: 1053-1061

Muller, V. and Gruber, G. (2003). ATP synthases: structure, function and evolution of unique energy converters. *Cell. Mol. Life Sci.* 60, 474-494.

Nation, J L (2008). Insect Physiology and Biochemistry, Second Edition, Published by CRC Press, 560 pp

Nicolson, S.W. (1993). The ionic basis of fluid secretion in insect Malpighian tubules: advances in the last ten years. *J Insect Physiol* 39: 451– 453.

O'Donnell,M.J., Dow, J.A., Huesmann, G.R., Tublitz, N.J., and Maddrell, S.H. (1996). Separate control of anion and cation transport in Malpighian tubules of *Drosophila melanogaster. J Exp Biol* 199 (Pt 5): 1163– 1175.

Pannabecker, T. L., Andrews, F., & Beyenbach, K. W. (1992). A quantitative analysis of the osmolytes in the hemolymph of the larval gypsy moth, Lymantria dispar. ***Journal of Insect Physiology*, *38***(11), 823-830.

Patel, M., Hayes, T. K. and Coast, G. M.(1995). Evidence for the hormonal function of a CRF-related diuretic peptide (*Locusta*-DP) in *Locusta migratoria*. J. Exp. Biol. **198**, 793-804

Pannabecker, T.L, Hayes,T.K, and Beyenbach, K.W.(1993) Regulation of epithelial shunt conductance by the peptide leucokinin. *J. Membr. Biol.* 132:63–76.

Petzel, D. H., Pirotte, P. T. and Van Kerkhove, E. (1999). Intracellular and luminal pH measurements of Malpighian tubules of the mosquito *Aedes aegypti*: the effects of cAMP. *J. Insect Physiol.* 45, 973-982.

Petzel, D. H. (2000). Na/H exchange in mosquito Malpighian tubules. *Am. J. Physiol.* 279, R1996-R2003

Phillips, J. E., (1964) Rectal absorption in the desert locust, ***Schistocerca gregaria*** Forskill. I. Water, ***J. Exp. Biol.*** 41:14–38

Phillips, J. E., (1981). Comparative physiology of insect renal function, ***Am. J. Physiol.*** 241:R241–R257.

Phillips, J. E., Hanrahan, J., Chamberlin, M., & Thomson, B. (1986). Mechanisms and control of reabsorption in insect hindgut. *Adv. Insect Physiol.*, 19, 329–422.

Philllips, J. E., Hanrahan, J., Chamberlin, M. and Thomson, B. (1986). Mechanisms and control of reabsorption in insect hindgut. *Adv. Insect Physiol.* 19, 329-421.

Phillips J.E., Thomson R.B., Audsley, N., Peach J.L., Stagg A.P. (1994) Mechanism of acid-base transport and control of locust excretory system. *Physiol. Zool.* 67(1) 95-119

Pilcher, D. E. M., 1970, The influence of the diuretic hormone on the process of urine secretion by the Malpighian tubules of ***Carausius morosus*, *J. Exp. Biol.* 53**:465–484

Quinlan, M.C., Tublitz, N.J., and O'Donnell, M.J. (1997). Anti-diuresis in the blood-feeding insect *Rhodnius prolixus* Stal: the peptide CAP2b and cyclic GMP inhibit Malpighian tubule fluid secretion. J Exp Biol 200 (Pt 17): 2363– 2367.

Ramsay, J.A. (1951). Osmotic regulation in mosquito larvae; the role of the Malpighian tubules. J Exp Biol. 28:62–73

Ramsay, J. A. (1954) Active transport of water by the Malpighian tubules of the stick insects, *Dixippus morosus* (Orthoptera: Phasmidae). J.Exp. Bil. 31: 104-113.

Raphemot,R, Matthew F. Rouhier,, Daniel R. Swale, Emily Days, C. David Weaver, Kimberly M. Lovell, Leah C. Konkel, Darren W. Engers, Sean F. Bollinger, Corey Hopkins,Peter M. Piermarini , Jerod S. Denton (2014). Discovery and Characterization of a Potent and Selective Inhibitor of ***Aedes aegypti*** Inward Rectifier Potassium Channels. PLoS ONE 9(11): e110772.

Razet, P. (1966). Les Elements terminaux du catabolisme axote chez les insects. *Anne Biol.* 5: 43-73.

Rheault, M., J.S. Plaumann and M.J.O'Donnel (2006) Tetraethylammonium and nicotine transport by the Malpighian tubules of insects. *J.Insect Physiol. 52:* 487-498.

Ruiz-Sanchez, E. and O'Donnell, M. J. (2006). Characterization of salicylate uptake across the basolateral membrane of the Malpighian tubules of ***Drosophila melanogaster.*** *J. Insect Physiol.* **52**,920 -928.

Ruiz-Sanchez, E. and O'Donnell, M. J. (2007a). Characterization of transepithelial transport of salicylate by the Malpighian tubules of ***Drosophila melanogaster*** and the effects of changes in fluid secretion rate. *Physiol. Entomol.* 32,157 -166.

Ruiz-Sanchez, E. and O'Donnell, M. J. (2007b). Effects of chronic exposure to dietary salicylate on elimination and renal excretion of salicylate by ***Drosophila melanogaster*** larvae. *J. Exp. Biol.***210**,2464 -2471.

Ruiz-Sanchez, E., Van Walderveen, M. C., Livingston, A. and O'Donnell, M. J. (2007). Transepithelial transport of salicylate by the Malpighian tubules of insects from different orders. *J. Insect Physiol.* **53**,1034 -1045.

Sabrina V., V. Martini, Suzete B. Nascimento and Marcelo M. Morales 2007 *Rhodnius prolixus* Malpighian tubules and control of diuresis by neurohormones. *Annals of the Brazilian Academy of Sciences* 79(1): 87-95

Saini R.S. (1964) Histology and physiology of the cryptonephridial system of insects. *Ecol.Ent.*Vol 116(14): 347-392

Spring, J.H., Morgan, A.M., Hazelton, S.R., 1988. A novel target forantidiuretic hormone in insects. Science 241, 1096–1098

Tublitz,N.J and Truman, J.W.(1985a). Intracellular stimulation of an identified neuron evokes cardioacceleratory peptide release. Science 228(4702): 1013– 1015.

Tublitz, N.J and Truman, J.W. (1985b). Insect cardioactive peptides. I. Distribution and molecular characteristics of two cardioacceleratory peptides in the tobacco hawk moth, *Manduca sexta*. J Exp Biol 114: 365– 379.

Tublitz, N.J and Truman, J.W. (1985c). Insect cardioactive peptides. II. Neurohormonal control of heart activity by two cardioacceleratory peptides in the tobacco hawk moth, *Manduca sexta*. J Exp Biol 114: 381– 395.

Van Zyl A and Van Der Linde, T.C.(2000) Anatomy and histology of the alimentary canals of the antlion larvae *Furgella intermedia* Markl and *Palpares annulatus* Stitz (Neuroptera: Myrmeleontidae), with reference to their feeding physiology. African Entomology2000 Vol.8 No.2 pp.179-188

Veenstra, J.A. (1999). Isolation and structure of the Drosophila corazonin gene. ***Biochem. Biophys. Res. Commun***. 204(1): 292--296

Wang, S., Rubenfeld, A.B., Hayes, T.K. and Beyenbach, K.W. (1996) Leucokinin increases paracellular permeability in insect Malpighian tubules. *J Exp Biol* 199: 2537–2542.

Weltens, R., Peferoen, M., Steels, P. And Van Kerkhove, E (1992) Electrophysiological measurements and cable analysis of coleopteran midgut epithelium: effect of ionic changes and of an insecticidal product. *Arch.Int. Phy*siol. *Biochem*. 100, 6

Wheelock, G. D., Petzel, D. H., Gillett, J. D., Beyenbach, K. B. And Hagedorn, H. H. (1988). Evidence for hormonal control of diuresis after a blood meal in the mosquito *Aedes aegypti*. *Arch. Insect Biochem. Physiol*. **7**, 75-90.

Whyard,, S., Singh A.D and Wong, S. (2009). Ingested double stranded RNAs can act as species- specific insecticides. Insect Biochem.Mol. Biol. 39: 824-832.

Wieczorek, H., Grueber, G., Harvey, W. R., Huss, M., Merzendorfer, H. and Zeiske, W. (2000). Structure and regulation of insect plasma membrane H+ V-ATPase. *J. Exp. Biol.* **203**, 127-135

Williams, J.C., Beyenbach, K.W.(1983) Differential effects of secretagogues on Na and K secretion in the Malpighian tubules of *Aedes aegypti* (L.). J. Comp. Physiol. 1983; 149:511–517.

Zubrzak P, Williams H, Coast G.M, Isaac R.E, Reyes-Rangel G, Juaristi E, Zabrocki J, Nachman R.J (2007) Beta-amino acid analogs of an insect neuropeptide feature potent bioactivity and resistance to peptidase hydrolysis. Biopolymers. *Peptide Science* 88(1): 76–82

Insect Nervous Coordination and Integration 7

Synopsis

1. Introduction
2. Basic plan of the insect nervous system
3. The central nervous system
   3.1 Brain structure and function
      3.1.1 Protocerebrum
         A. The mushroom bodies
         B. The central complex
         C. Optic lobes
         D. Ocelli
      3.1.2 Deutocerebrum
         A. Antennal lobes
         B. Dorsal lobe
      3.1.3 Tritocerebrum
   3.2 The suboesophageal ganglion and its function
   3.3 Thoracic and abdominal ganglia
4. Visceral nervous system/ Sympathetic nervous system
   4.1 The stomodaeal nervous system
5. The peripheral nervous system
6. Basic structure of ganglia
7. Morphology of neurons in the insect brain
8. Organisation of Glial cells
   8.1 Role of the Glial cells
9. Synapses
10. Neuromuscular junction
11. Physiology of impulse transmission
   11.1 Generation and conduction of nerve impulse
   11.2 Transmission of impulse
12. Neuropeptides and neurotransmitters
   12.1 Classical neurotransmitters
      A. Acetylcholine
      B. γ-aminobutyric acid
      C. Glutamate
   12.2 Biogenic amine neurotransmitters
      A. Serotonin
      B. Dopamine
      C. Tyramine and octopamine
13. Neurotransmitter receptors
14. Learning, memory and intelligence in insects
15. Hormonal control on the development of the nervous system
16. Nerve repair and regeneration in insects
   16.1 Degenerative responses

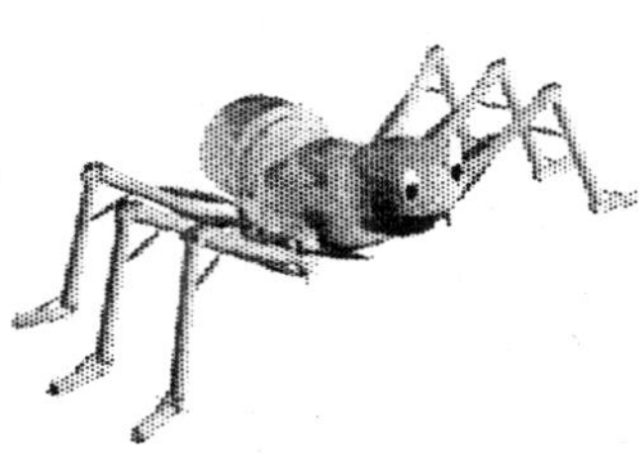

1. INTRODUCTION

Insects show co-ordination in behaviour, memory and possess intelligence due to well distributed nervous system. The nervous system functions as a link between the sense organs which respond to various external and internal stimuli and the effector organs such as muscles, glands etc. The sense organs include structures with various sensilla that respond to sounds, weather factors, smell etc. Nervous system consists of elongated cells, the neurons, which form the physiologically functional elements. These neurons carry the information in the form of electrical impulses.

A nervous system can be thought of as an organised collection of neurons. Neurons are electrically excitable cells whose membranes can generate and transmit changes of voltage that form a signal. These changes in voltage can be graded in amplitude or all-or-nothing events of fixed amplitude called **action potentials**. Synapses are points where the membranes of two or more neurons come very close together, associated with specialised features of the membranes. Electrical signals in one cell may be passed directly to other cells at electrical synapses, or more often are converted into chemical signals that diffuse across a tiny synaptic gap that remains between the membranes at chemical synapses. Communication between neurons may also take place without synapses, through long distance chemical transmission, but no known nervous system functions in the absence of the ubiquitous synapse.

The cellular components of an invertebrate nervous system include:

1. **Sensory (afferent) neurons**, which convert physical variables (e.g. light level or muscle force) into electrical signals.
2. **Motor (efferent) neurons,** which make synapses with muscles or other effector organs (e.g. light-producing organs, glands).
3. **Interneurons,** which transmit information between other neurons.
4. **Glia,** which are electrically inexcitable cells that influence the ionic environment surrounding neurons and the transmission of signals between them.

Among these neuron types, only sensory neurons have somata localised outside the rind cortex (outer soma layer) of the central nervous system. This placement of somata is a prominent feature of the insect brain (and other arthropod brains).

In invertebrates, unlike vertebrates, the somata of motor neurons and interneurons are not important sites of synaptic input. Instead they give rise to a single primary neurite that enters the core of the ganglion and divides into dendritic and axonal regions. The distinction between input and output regions is often quite blurred, and in most cases it is possible to find input and output synapses adjacent to one another. Nevertheless, the ratio of input to output synapses differs on different branches of a neuron, so it is usually possible to distinguish input and output regions. The ability of neurons to generate and transmit electrical signals, and for these signals to pass between cells, be summed, multiplied and transformed by different types of cells and synapses, enables nervous systems to process information.

Changes in the environment are detected by sensory structures, encoded in the activity of neurons, analysed for relevance against the backdrop of previous experience, transmitted through the nervous system, and used to drive appropriate actions. The actions may be compensatory (e.g. postural) or active (e.g. generating locomotion, or producing a sound). Many of these behavioural actions are completed in fractions of a second, limited only by the speed of neuronal activity; whereas others take place over hours, days or even the lifetime of the animal. These long-term changes in neuronal function can depend on neurohormonal chemical signals, or on morphological changes in the neurons and their interconnections.

2. BASIC PLAN OF THE INSECT NERVOUS SYSTEM

The insect nervous system comprises the central nervous system and the peripheral nervous system. The central nervous system (CNS) consists of brain and ventral nerve cord. The peripheral nervous system includes a stomatogastric ganglion and sensory and motor nerves not bundled in ganglia. In addition, insect nervous system includes pars intercerebralis-corpus cardiacum-corpus allatum and neuroendcrine system that consists of store and release organ.

The structure of the insect central nervous system resembles that of a rope ladder, therefore, it is also called **ladder-like nervous system**. The stiles of a ladder correspond to the connectives, running aligned to the body axis (except in the head, where the neuraxis is often bent). A ladder rung corresponds to a ganglion with commissures. (The CNS can be considered to be an ectodermal rudiment).

The most anterior ganglion is the cerebral ganglion (**supraoesophageal ganglion** or **brain**). It is a structure that evolved from the fusion of the three preoral neuromeres (anterior with respect to the oesophagus). The suboesophageal ganglion consists of the fusion of three postoral neuromeres.

The oesophagus is surrounded by the circumoesophageal connectives that link supra- and suboesophaegeal ganglia. More posterior to them are thoracic ganglia and abdominal ganglia. Except the brain, all the other ganglia lie ventral to the digestive system. The thoracic and abdominal ganglia connect to form the ventral nerve cord (Fig 7.1). In comparison, the vertebrate spinal cord lies dorsal with respect to the digestive tract. This difference is derived from their skeleton, each of which is external skeleton and internal skeleton. Vertebrate has a body plan that vertebral column puts up visceral organ whereas invertebrate has a body plan that ventral plate puts up nerve cord and visceral organ.

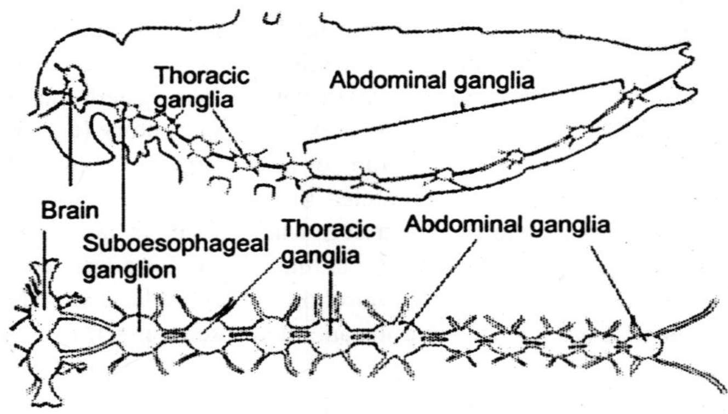

Fig. 7.1. Schematic sketch of the nervous system of a generalised insect

The variety of fusion patterns of ganglia reflects the phylogeny of taxonomic groups at the level of the central nervous system (CNS). For example, honey bee, fly, and moth have a fused form of brain and suboesophageal ganglion with an oesophageal foramen, through which the oesophagus extends towards the mouthparts. The locust embryo has eleven pairs of ganglion rudiments. During ontogeny, the first to third abdominal ganglia fuse with the metathoracic ganglion and the eighth to eleventh ganglia fuse to form a large terminal abdominal ganglion. In Diptera, fusion of thoracic and abdominal ganglia occurs resulting in a large segmented agglomerate ganglion.

3. THE CENTRAL NERVOUS SYSTEM

Insects have a relatively simple central nervous system consisting of a dorsal brain linked to a ventral nerve cord that consists of paired segmental ganglia running along the ventral midline of the thorax and abdomen.

Ganglia within each segment are linked to one another by short medial nerve (commissure) and also joined by intersegmental connectives to ganglia in adjacent body segments. In general, the central nervous system is rather ladder-like. In advanced insect orders, there is a tendency of individual ganglia to combine into larger ganglia that serve multiple body segments.

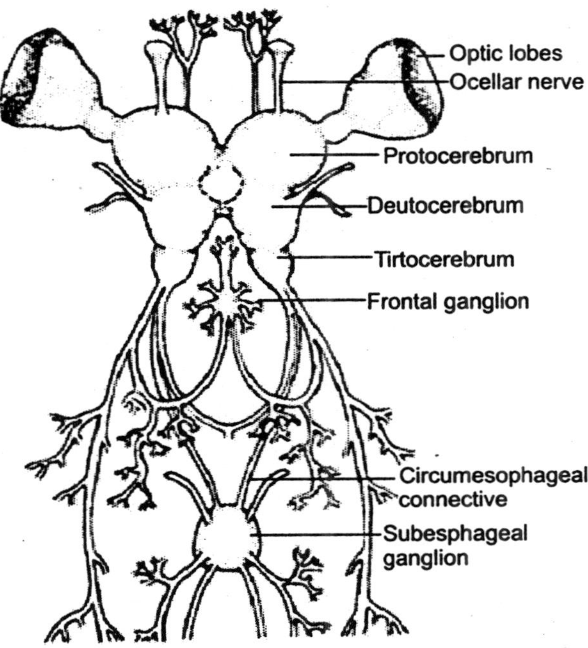

Fig 7.2. Insect brain, frontal ganglion and suboesophageal ganglion in a generalised insect

The brain is formed by the fusion of six ganglia (three pairs) located dorsally within the head capsule. It is composed of three parts (Fig 7.2):

1. **Protocerebrum:** The first pair of ganglia is largely associated with vision. They innervate the compound eyes and ocelli. This region contains the mushroom bodies, the pars intercerebralis, the central complex and the lateral accessory lobes.
2. **Deutocerebrum:** The second pair of ganglia process sensory information collected by the antennae.
3. **Tritocerebrum:** The third pair of ganglia innervates the labrum and integrates sensory inputs from proto- and deutocerebrums. They

also link the brain with the rest of the ventral nerve cord and the stomodaeal nervous system that controls the internal organs. The commissure for the tritocerebrum loops around the digestive system, suggesting that these ganglia were originally behind the mouth and migrated forward around the oesophagus during evolution.

Located ventrally in the head capsule, just below the brain and oesophagus, is another complex of fused ganglia, jointly called the **suboesophageal ganglion**. This ganglion innervates the mandibles, maxillae, labium, hypopharynx, salivary glands and neck muscles. A pair of circumoesophageal connectives loops around the digestive system to link the brain and suboesophageal complex together.

In the thorax, three pairs of thoracic ganglia control locomotion by innervating the legs and wings. The thoracic muscles and sensory receptors are also associated with these ganglia. Similarly, abdominal ganglia control movement of abdominal muscles. Spiracles in both the thorax and abdomen are controlled by a pair of lateral nerves that arise from each segmental ganglion. The terminal pair of abdominal ganglions is usually fused to form a large caudal ganglion that innervates the anus, internal and external genitalia, and sensory receptors located at the hind end of the insect (Fig. 7.1).

3.1 Brain structure and function

The brain is the principal association centre of the body, receiving sensory input from the sense organs of the head and from the more posterior ganglia. It is the seat of many long-term organised behaviour patterns and governs their modification by learning. The insect brain can be divided into three large regions, **protocerebrum, deutocerebrum**, and **tritocerebrum** (Box 7.1).

Box 7.1 General part of an insect brain

1. **Protocerebrum:** Connected to Optic lobes
   A. Mushroom bodies
      a: Calyx (input); b: Peduncle (transfer) c: αβγ lobes (output)
   B. Central complex
      a: Protocerebral bridge; b: central body (lower and upper);
      - Nodule
2. **Deutocerebrum:** Connected to antennal lobe and dorsal lobe
3. **Tritocerebrum:** Connected to suboesophageal ganglion

3.1.1 *Protocerebrum*

The protocerebrum bears two lateral protrusions, the optic lobes. Depending on the insect species, the optic lobe can be further subdivided into three or four neuropils (bunch of neurons). The optic lobes of the fly (an insect with particularly good vision) contain about 76% of the brain's neurons. The optic lobe connects directly to the sensory cells (retinula cells) in the retina of the compound eye. It contains three distinct regions (neuropils): the **lamina**, **medulla** and **lobula**, where processing of visual signals begins. The protocerebrum also receives inputs via the ocelli, when present, via the ocellar nerves.

Medially, in between the optic lobes are located another conspicuous pair of bilaterally symmetrical neuropils, the **mushroom bodies** (MB, corpora pedunculata, 'stalked bodies'). The mushroom bodies are made up of three regions- the **calices**, the **peduncle** and the **alpha, beta and gamma lobes** (Fig 7.3). The mushroom bodies are composed of thousands of small-diameter interneurons that share a common basic morphology. Each intrinsic neuron has dendrites in the calycal neuropil and a single projection fibre that runs through the stalk and bifurcates passing into the lobular divisions of the pedunculus. The structural organisation of the neuronal cells in the mushroom bodies is unique and therefore called **Kenyon cells**. The calices serve as the input areas, the peduncle is the transfer region and the lobes are the output region. The mushroom bodies are best developed in social insects, making up 20% of the brain of the bee and 50% of the brain of worker ants (Formica). These are thought to function as higher centres responsible for the most sophisticated computations occurring in the insect brain. The mushroom bodies receive sensory inputs from the lobula of the optic lobe and from the antennal lobes of the deutocerebrum. Most sensory inputs enter the mushroom bodies through the calyx. There are about 1000 to 100,000 Kenyon cells, in each mushroom body. These neurons have tree-like branching dendrites which receive inputs in the calyces of the MB. A single axon extends down the stalk of the MB and then gives of branches to two lobes of the MB. Dragonfly mushroom bodies have no calyces and no Kenyon cells. The mushroom bodies are also involved in learning, and in the honeybee have been shown to process memories, transferring data from short-term memory (STM) into long-term memory (LTM).

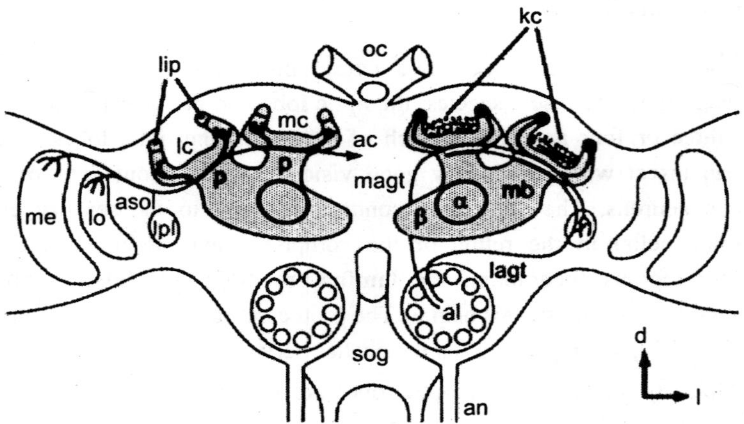

7.3 Schematic diagram of the frontal view of an adult honey bee brain (Taken from Dagmur Malun (1998) Learning & Memory 5: 90-101)

Shaded area- mushroom bodies; mc- medial calyx; lc-lateral calyx; p-peduncle; α- α lobe; β- β lobe; kc- Kenyon cells; al- antennal lobe; magt-medial antenna-glomerular tract; lip- lip region of calycal neuropil; lpl-lateral protocerebrum; lagt- lateral antennoglomerular tract; me- medulla; lo-lobula; ac- anterior commissure; an- antennal nerve; sog- suboesophageal ganglion; oc- ocelli; d-dorsal; l- lateral

The median protocerebrum also contains unpaired neuropils, in particular the protocerebral bridge and the central body, which together form the **central complex**. The shape, size, and disposition of these dense neuropils differ somewhat between insect groups but the basic morphology is essentially the same. The optic lobes process visual information, the mushroom bodies are important for learning and memory related in particular to olfactory stimuli, and the central complex is thought to play a crucial role in generating behavioural output.

The central body receives inputs from the mushroom bodies and integrates sensory inputs from different sensory modalities (such as small and vision) - so-called **multimodal sensory perception**. It functions as an activating centre, switching on appropriate locomotor activity patterns which are central programs located in the thoracic ganglia. That is, it instructs the thoracic ganglia which programs to run - programs that control the legs and wings. These hard-wired programs are sometimes called **central pattern generators** and require no sensory input for their execution, though sensory inputs may start and stop these programs or modify them slightly.

The pars-intercerebralis is a mass of cell bodies, including neurosecretory cells which send their axons to the pair of corpora cardiaca. The corpora cardiaca are sometimes fused into a single medial ganglion. They send out nerves to innervate the dorsal blood vessel, forming a cardio-aortic system, which controls the rate of heart beat, as well as having a secretory hormonal function

Another function associated with the protocerebrum is time-keeping. Insect activity is timed with the daily light/dark cycle - the circadian cycle. This timing is due to internal clocks within the insect, which update themselves according to external cues from the environment (zeitgeibers or time-givers) such as the number of hours of light and dark. Many body parts and organs have their own circadian clocks, indeed each cell appears capable of keeping time, but these appear to be set and synchronised by a central master clock, which resides in the protocerebrum and is both neural and hormonal. In some insects, a master clock is found in each optic lobe, which makes sense as these process light signals. There is also a daily movement of screening pigments in the ommatidia of the compound eye, as the insect adjusts to night-time darkness by increasing the sensitivity of its retina. Severing of the optic lobes prevents these clocks from synchronising bodily activities. In other species, however, the clock is only abolished if the brain is cut in two, which suggests that it may reside in the central body.

A. The Mushroom bodies

In all insects the mushroom bodies consists of three major divisions- an upper cup shaped part called the calyx, a stalk or pedunculus and the lobes. The lobes represent a bipartite or tripartite division of pedunculus and are referred to as **alpha, beta and gamma lobes**. In some insects such as bees and cockroaches there are a pair of calyces on either side of the brain, while in other species the calyces are single. In most groups of insects, the calyx is subdivided into a number of neuropil regions. Three concentric circular neuropil zones called the **lip, collar** and **basal ring** make up each of the bee's paired calyces. The lip and collar calycal neuropils are microgromerular while the basal ring appears to consist of rather looser and larger gromerulli. Microglomeruli are the sites of contact between pre-synaptic blebs representing the calycal inputs and the spiny dendrites of the K cells. The pedunculus and its lobular divisions are generally considered to represent the sites of output from the mushroom bodies.

Two different kinds of K cells have been identified in the mushroom bodies. One kind has very spiny dendrites while the other has peculiar bunched or clawed specialisations. Both cell types have lambda-shaped projections within the pedunculus and the lobes. In insects with gamma lobes such as fly and Sphinx moth, K cells have been described that, instead of bifurcation, terminate in a single lobe. The spiny and clawed cell groups each have further different morphological classes. For example in housefly there are twelve different geometries, in ant's three types and in bees five groups.

In most insect species, the mushroom bodies primarily receive olfactory inputs (Box 7.2). Odorant information is relayed from the olfactory receptors, through the antennal lobes and into the calices. In the case of cockroaches and locusts, tactile (touch) and gustatory (taste) inputs are also found. Bees and wasps, which are predominantly visual insects, have two exceptionally large optical lobes connected to the calyces.

The diversity of insect brains is most apparent in the mushroom bodies, since they are often compartmentalised, with each compartment dedicated to each input. The functional output of mushroom bodies is still a matter of debate. They are mostly involved in sorting out information from the different sensory inputs, since there is little evidence of any direct downstream connection to other brain areas. They may be involved in pattern recognition, as in locusts, where specific Kenyon cells activate depending on specific smells.

The mushroom bodies function as a centre for learning. Insects are smart and have considerable memorisation ability. Manipulating the mushroom bodies (not allowing them to develop properly) has always resulted in loss of short- and long-term memory. There is a strong correlation between mushroom body size and memory in hymenopterans (bees, wasps, ants) as well as between size of the mushroom bodies and behavioural complexity. The reason for this is the Kenyon cells' remarkable plasticity, in that they will readily rebuild the neural fibers, acting as a sort of neural substrate on which new memories can grow.

Box 7.2 Mushroom body

1. Center for higher-order sensory integration and learning.
2. Receive information mainly from olfactory (antennal lobe). In Hymenoptera also from optic lobe.
3. Olfactory learning and memory, place memory, associative memory, and roles in motor control.

B. The central complex

In contrast to the mushroom bodies, we don't know that much about the central complex. It forms a line in the middle of the brain, and is made up of four interconnected regions: the **protocerebral bridge, the center body** (which has a lower and upper part), and the **noduli**. It's connected to almost all other protocerebrum regions, with the notable exception of the mushroom bodies.

In locusts and bees, the compound eyes directly connect to the protocerebral bridge. The ocelli are also innervated by the central complex, and there seems to be a strong connection to the pathway responsible for polarised vision. Blind insects also have strongly developed central complexes indicating that they must have some other role.

In locusts, it is connected to the ventral cord, suggesting a role in movement. This is an interesting proposition; given that it is between the left and right hemispheres of the brain, and it is further supported by the fact that those insects that can perform complex, asymmetrical leg movements (like weaving silk threads) have highly-developed central complexes.

There is also the possibility of the central complex playing a vital role in behaviour, since neural activity at this region is always correlated with the expression of specific behaviours. For example, changes in activity levels are seen at the start and end of flight in locusts. This is not limited to just walking and flying, but also includes visual searches.

The most definite proof comes when central complex development is suppressed. In *Drosophila*, this leads to inability to fly and strange walking, indicating that while it is not necessary for initiating leg movement, it is needed for control. If only the central part of the central complex is disrupted, the fly has difficulty orienting itself and cannot deal with asymmetries. They are unable to fly in a straight line (Box 7.3).

Box 7.3 Roles of central complex

A. Control of locomotor activity, particularly flight and walking.
B. Center for direction perception and spatial navigation.
C. Possibly also coordinates L and R brain.

Combined with their likely role in the visual system, there is also a chance that they are integral for insects to calculate their travel paths, especially in migrating insects like the locust. Locusts have a very well-developed polarised light detection system. They see polarised light in the sky and use it as a compass to guide them on their journey.

C. *Optic lobes*

The optic lobes are lateral extensions of the protocerebrum to the compound eye. Each consists of three neuropile masses, known as **lamina ganglionaris, medulla externa** and **medulla interna**, the lobula and their associated perikarya and connections.

The lamina cartridge is composed of the projections of receptor cell terminals, monopolar neurons, centripetal climbing fibres, and centrifugal, tangential and amacrine cells. In the case of fly lamina, there are 19 cells contributing to each cartridge in a regular and predictable fashion.

The medulla consists of a series of columnar fibres that run through different levels of anaxonal neurons and the dendritic strata of tangential cells. In fly more than 34 axons contribute to each column. These perpendicular elements can be divided into a number of groups that connect the medulla to the lobula and the lobula plate. There are as many medulla columns as there are lamina columns, and each medulla column receives first-order inputs from the long visual fibres and the second-order inputs from lamina monopolars and tangential cells. The output of the medulla to lobula represents an abrupt change in periodicity of the retinal mosaic as the number of columns reduces to a third. The medulla has a number of outputs to the midbrain, the neurons of which bypass the lobula neuropils.

In fly the lobula plate is linked to the medulla by uncrossed connections and the lobula is similarly linked to the lobula plate. Each sheet of the second optic chiasma comprises of three rows of medulla-derived columns. The columnar mosaics precisely retained into the outer layers of the medulla become represented with less structural precision in the distal lobula.

Axons of the retinula cells pass through the basement membrane at the back of the eye and into the lamina ganglionaris where they synapse with large monopolar neurons, the axons of which extend to the medulla externa.

Axons of the mechanoreceptors on the surface of eye pass directly from the lamina ganglionaris to the protocerebrum, but the majority pass to the medulla externa, crossing over each other to form the external chiasma between two neuropile layers. A second cross over, the internal chiasma, lies between the medulla externa and the medulla interna. Cell bodies, supplying axons to these systems, lie peripherally and in a group on the outside of the medulla externa.

In fly, each ommatidia provides six receptor axons (from the R1-R6 photoreceptors) that terminate in six separate columns (optic cartridges) of the lamina. The R1-R6 photoreceptors in an ommatidium have different optical alignments. However, six R1-R6 photoreceptors distributed amongst six different ommatidia share the same optical alignment. The axons of six optically coherent photorecpetors converge to a single optic cartridge. Each optic cartridge represents a single visual sampling point and gives rise to a set of 5 monopolar cells and the T1 basket cell. The axons of these six neurons from a cartridge project coherently across the first optic chiasma accompanied by the axons of R7 and R8 photoreceptors that share the same optical alignment.

Each group of 8 axons from the lamina and retina comprises the inputs to a medulla column. There are as many columns as there are optic cartridges or ommatidia. Successive chiasmata link the lamina to the outer medulla, and the inner medulla to the lobula, lobula plate. Each neuropil of the lobula complex receives an isomorphic arrangement of retinotopic input neurons. Thus, throughout the optic lobes the arrangement of columns is representative of the arrangement of ommatidia and, hence, visual sampling points. Dendritic trees of individual neurons can be precisely mapped within these columnar neuropils and related out to specific areas of the eye's visual field.

The lobula plate has provided a rich source of knowledge about motion detecting mechanisms in large flies (e.g. *Calliphora* and *Phaenicia*) because of its large tangential neurons that receive inputs from chromatic motion-sensitive interneurons and which are easily accessed by microelectrodes.

The lobula consists of mainly palisades of columnar neurons which, in larger flies, have small receptive fields and are orientation-selective, rather like pyramidal neurons in the cat's visual cortex. These smaller neurons are arranged as retinotopic arrays, each made up of an isomorphic ensemble of one morphological species of columnar cell. In larger flies, about 30 different types of columnar cells have been recognised and thus 30 ensembles. Each cell type targets an optic glomerulus in the ventro-lateral deutocerebrum. Thus, each glomerulus receives all the axons of a specific ensemble. The dendrites of descending neurons embrace and invade optic glomeruli in a manner reminiscent of output neurons from olfactory glomeruli.

D. *Ocelli*

Many insects possess ocelli in addition to compound eyes. Ocelli are cluster of retinular cells beneath a poor condenser lens. It is generally thought that the ocelli are responsible for flight attitude correction and maintenance. Each ocellus is a collection of 1000 receptor cells that converge on to a dozen or so large-diameter fibres (L) and upto 100 smaller cells (S). The receptor cells are not generally organised into discrete layers but are aimed at collecting the maximum possible amount of light at the expense of image resolution. Other specialisation in the ocellar design includes palisade devices and movable horizontal and vertical pigment irises.

In all insects the L neurons run from the ocelli, as tracts through the superficial dorsal protocerbrum, to the dorsal deuto/tritocerebral neuropil. The ribbons of the ocellar fibres pass behind the protocerebral bridge. Within the deutocerebrum the L neurons terminate amongst the dendrites of the large descending neurons that form the median dorsal longitudinal tracts of the thoracic ganglia. The S neuron population consists of a large number of neurons of heterogeneous morphology projecting to several different regions of the brain.

3.1.2 *Deutocerebrum*

The deutocerebrum is a bilaterally symmetrical brain region that consists of the antennal lobe (AL) and the dorsal lobe. The antennal lobe receives the axons of olfactory receptor neurons from the antenna. In at least some insects such as flies and moths, the antennal lobe also receives olfactory receptor axons from the labium. The AL receives inputs from the third (terminal) antennal segment (the flagellum, which is made-up of sub-segments called **flagellomeres**) via the antennal nerves. It contains from less than 10 to more than 200 sub-centres called **glomeruli**. Inputs to the AL appear to be mainly or exclusively from chemoreceptors (i.e. chemical sensors - olfactory and gustatory, smell and taste) on the flagellum. Each antenna sends signals to the AL on the same side of the head (ipsilateral pathways) although some may also send signals to the AL on the opposite side (contra lateral pathways).

Each glomerulus is a region of neuropil where computations occur. It is thought that each glomerulus may, in some species at least, receive inputs from a specific class of receptor (sensor) on the antenna. For example, in the males of some species there is a specially large glomerulus, called the**macroglomerular complex** (MGC) which receives inputs from pheromone olfactory sensors on the antenna.The AL does not receive one input line from each chemoreceptor, as sensors of the same type converge - their axons fuse into a smaller number of axons in the antennal nerve (typically inputs from 15 sensors are combined, a 15:1 ratio). These sensory input axons, and also input axons from the central body of the protocerebrum, synapse with local interneurones within the AL (amacrine cells). Outputs from the AL are carried along the axons of output neurons to the mushroom body of the protocerebrum.

A. *Antennal lobes*

Antennal lobes are the first order neuropils of the olfactory chemosensory pathway. They comprise glomeruli- discrete glial bound partitions of neuropil- which are invaded by olfactory receptor terminal.

Glomeruli provide projection neurons. **Uniglomerular projections** neurons (having dendrites in a single glomerulus) send their axons to the superior lateral protocerebrum on the same side of the brain *via* the inner antennocerebral tract. Most types of **multiglomerula projections** (those whose dendrites invade more than one glomerulus) reach the superior lateral protocerebrum *via* the outer and medial antennocerebral tracts and do not send collaterals to the mushroom body calyces.

Although the projections from the antennal lobes to the brain are similar between different groups of insects, the organisation and neuro-architecture of their antennal lobes can differ markedly. In honeybees, their glomeruli are uniquely identifiable on the basis of their shape and location within the antennal lobe. Each glomerulus has a complex architecture. Each supplies at least three and some as many as seven uniglomerular projection neurons, the axons of which send collaterals to relatively simple and small mushroom bodies.

In Hymenoptera, ball-like glomeruli are clustered into two to four distinct populations. The internal organisation of each glomerulus, regarding the number of its uniglomerular projections neuron dendrites, is simpler than in moths and flies. Each glomerulus provides a pair of projections neurons, the axons of which provide collaterals to complex, large mushroom bodies.

B. Dorsal lobe

The dorsal lobe receives mechanosensory and gustatory receptor neurons from the antenna and also contains the motor neurons controlling the antennal muscles. The dorsal lobe is called **antennal mechanosensory** and **motor center** (AMMC) in some insects. For flies, this name may be retained because they differ from almost all other insects in not having an **antennal gustatory system**. The anntennal nerve carries all sensory axons from the antenna to the deutocerebrum and also contains efferent neurons, in particular the motor neurons of the pedicellus muscles. Generally, a second nerve emanates from beneath the antennal lobe that contains the axons of the motor neurons of the scapus muscles, which are located in the head capsule.

The AMMC receives mechanosensory inputs from mechanosensors (mechanoreceptors) on the first two antennal segments (scape and pedicel) *via* the antennal nerves. It also sends motor outputs to the muscles of the scape. It also receives inputs from mechanosensors on the labial palps, some tegument (body wall) mechanosensors, and some inputs from the flagellum (possibly from the mechanosensors found on the flagellum). The antennal nerve is therefore a mixed nerve - containing both sensory and motor axons. Some of the antennal mechanoreceptors also send outputs to the SOG, the protocerebrum and the thoracic ganglia.

3.1.3 Tritocerebrum

Tritocerebrum is the smallest region and little studied. In the primitive insect ground plan, the tritocerebrum is connected to the suboesophageal ganglian by a pair of circumoesophageal connectives (for example in locusts, crickets, cockroaches). In more derived groups, the tritocerebrum, the posterior part of the deutocerebrum, and the suboesophageal ganglion are fused such that it is difficult to determine the borders between them. The tritocerebrum contains at least two commissures, one of which is running above the oesophagus in the primitive ground plan. The tritocerebrum receives sensory inputs from the labrum through the labral nerve, from the tegument through the tegumentary nerve, and at least in some insects also direct sensory input from the mouthparts (mandible, maxilla, and labium). The frontal nerves of the tritocerebra on each side run to the frontal ganglion that connects the brain and the stomatogastic system, lying dorsal with respect to the oesophagus and innervating cardiac and gastrointestinal muscles. The frontal ganglion is connected to the hypocerebral ganglion through the recurrent nerve. In some insects (honeybee), the frontal and labral nerves are fused at the root to form a labro-frontal nerve. Tritocerebrum is connected to suboesophageal ganglion through a pair of connective. Frontal ganglion is center of stomatogastric ganglion innervating cardiac and gastrointestinal muscle and is connected to suboesophageal ganglion *via* recurrent nerve.

The higher order neuropils in protocerebrum, such as mushroom body, central complex, and lateral protocerebrum, receive sensory information from optic lobe, deutocerebrum, tritocerebrum, suboesophageal ganglion, and ventral nerve cord through several layers of filters in a preprocessed form. The higher neuropils integrate information from various modalities and relay outputs that can eventually results in activating output neurons such as motor neurons of the brain or descending interneurons transmitting command information to the ventral nerve cord, which is important in controlling a wide range of behaviours. Thus, overall control mechanisms such as for triggering particular behaviours are localized in the brain and are separated from local control mechanisms that are responsible for aspects like the maintenance of motor patters for wing flapping or walking are localised in the thoracic ganglia. This is not to say that the ventral nerve cord is a fully hardwired system: In seminal experiments, it was demonstrated that cockroaches and locusts are capable of leg position learning, which is even improved by removal of the head. Thus, plasticity is possible at short time scales even in thoracic circuits controlling very basic behaviours.

3.2 The suboesophageal ganglion and its function

The suboesophageal ganglion (SOG) and the segmental ganglia of the double ventral nerve-cord each send out pairs of nerves, one of which innervate the pair of spiracles on that segment and so help regulate breathing. (In some insects the segmental ganglia are absent, e.g. in *Dytiscus*, in which case the lateral abdominal nerves send out nerves to innervate the spiracles). The SOG is a composite ganglion, formed by fusion of the ganglia from the mandibular, maxillary and labial segments of the head and the SOG also sends out nerves to the mouthparts (mandibles, palps, etc.) and so controls feeding behaviours.

The SOG is largely concerned with the processing of sensory information from and the motor control of the mouth parts. The mouth parts consist of three appendages, the mandible, the maxilla, and the labium. These appendages can be modified in various ways in different insect groups. For instance, many insects have paired maxillary and labial palps. In the bee, a part of the labium forms the unpaired glossae Axons of the receptor cells of the sensilla distributed on the mouth-parts project to the SOG and in at least some insects also to higher areas. The suboesophageal ganglion also receives direct sensory input from the labrum and through non-olfactory sensory cells of the antenna. Generally, there are separate mandibular, maxillary, and labial nerves innervating the appendages (for example in the bee), but they can also be fused (such as in Dipera, in which there is a fused maxillary-labial nerve). At least in the bee, a little-known fourth nerve emanates from the labial neuromere: the labial gland nerve.

3.3 Thoracic and abdominal ganglia

From the SOG two connectives or nerve cords run back along the ventral side of the insect. These connect to the thoracic ganglion of the first thoracic segment, T1, which is actually a pair of ganglia, more-or-less fused into a single structure. T1 then gives off two connectives to the second thoracic ganglion, T2 and the sequence continues with a chain of connected ganglia running throughout the length of the insect, in the basic plan. Thus, we say that insects have a double ganglionated ventral nerve cord (VNC). Each ganglion functions as a local processor, regulating the functions of its body segment. The thoracic ganglia are especially well-developed as they have to carry out complex computations to generate patterns of movement in the legs and wings. These output patterns or central programs are contained in the ganglia, but the brain is normally required to switch them on and off. Sensory inputs have little effect on the basic patterns, but do modify them. For example, stress sensors in the wings feedback information to allow fine-adjustments to the wings and control of the angle of attack and wing-twisting. Typically, however, the basic pattern of movement is pre-coded.

The thoracic ganglia possess motor centres that control important behaviours, such as flight, walking, and vocalisation. The thoracic ganglia are not only responsible for controlling the appendages (wings, legs, etc.) but also collect information from the environment (wind strength, sound), and relay these back to the brain.

The abdominal ganglia possess motor centres related to posture control, rhythmic behaviour, such as respiration, and circulation and the control of copulation and egg laying.

The motor centres residing in thoracic or abdominal ganglia autonomously express and correct motor patterns in response to their own local sensory inputs (trigger inputs) largely without depending on inputs from the brain. However, the initiation and maintenance of their motor patterns are generally controlled by descending interneurons from the brain.

Within the ventral nerve cord of cockroach, various Orthoptera, dragonfly larvae and other insects, there are a number of axons which are much bigger in diameter (60 μ in comparison to the normal 5μ) and called the **giant fibres**. In the cockroach there are six to eight giant fibres, 20-60μ in diameter, in each connective as well as 10-12 medium sized axons between 5 and 20 μ diameters. Giant fibres are interneurons, running for considerable lengths of nerve cord without synapses. They are believed to be concerned with the co-ordination of rapid evasive movements.

4. VISCERAL NERVOUS SYSTEM/ SYMPATHETIC NERVOUS SYSTEM

This consists of three separate subsystems (1) the stomodeal (stomatogastric) which consists of frontal ganglion and connected with aorta, foregut and midgut, (2) ventral visceral nervous system associated with the ventral nerve cord and (3) caudal visceral nervous system associated with the posterior segments of abdomen and reproductive organs. Together, the nerves and ganglia of these subsystems innervate the anterior and posterior gut, several endocrine organs (corpora cardiaca and corpora allata), the reproductive organs, and the tracheal system including the spiracles.

4.1 The Stomodaeal Nervous system

An insect's internal organs are largely innervated by a stomodaeal or stomatogastric nervous system (Fig 7.4). The stomodaeal nervous system consists of sensory and motor neurones having their centres in small ganglia developed from the dorsal, or dorsal and lateral walls of the stomodaeum. The nerve fibres are distributed to all the stomodaeal parts of the alimentary canal and in certain insects they are continued over the entire length of the mesenteron. The labral muscles also, in some cases, appear to receive at least a part of their innervation from stomodaeal nerves, as do likewise the salivary ducts, the aorta, the corpora allata, and some of the mandibular muscles.

A pair of frontal nerves arising near the base of the tritocerebrum links the brain with an unpaired frontal ganglion on the anterior wall of the oesophagus. This ganglion innervates the pharynx and muscles associated with swallowing. A recurrent nerve along the anterio-dorsal surface of the foregut connects the frontal ganglion with a hypocerebral ganglion that innervates the heart, corpora cardiac, and portions of the foregut. Gastric nerves arising from the hypocerebral ganglion run posteriorly to ingluvial ganglia in the abdomen that innervate the hind gut.

In the back of the head there is usually a second nerve centre of the stomodaeal system, which consists typically of a pair of ganglia lying just behind the brain. These ganglia may be termed the **occipital ganglia**, though they are variously called also **pharyngeal**, **oesophageal**, or **hypocerebral** ganglia. Each occipital ganglion is connected with the back of the brain by a short **occipital ganglion nerve** and communicates with the frontal ganglion by a branch from the recurrent nerve. In some insects the paired occipital ganglia are united in a single median occipital ganglion situated beneath the aorta. In such cases the recurrent nerve ends in this ganglion, and the latter has two connectives with the back of the brain. Several nerves are given off from the occipital ganglia, or ganglion, but the pattern of the postcerebral stomodaeal innervation varies much in different insects.

The frontal ganglion is encased by a neural lamella. It is further characterised by a central neuropil surrounded dorsally and laterally by neurons. In Lepidoptera two neurosecretory cells are clearly identified in the frontal ganglion. In other cases, neurosecretory material has been found in the central neuropil of the ganglion as well as in other parts of the stomatogastric nervous system.

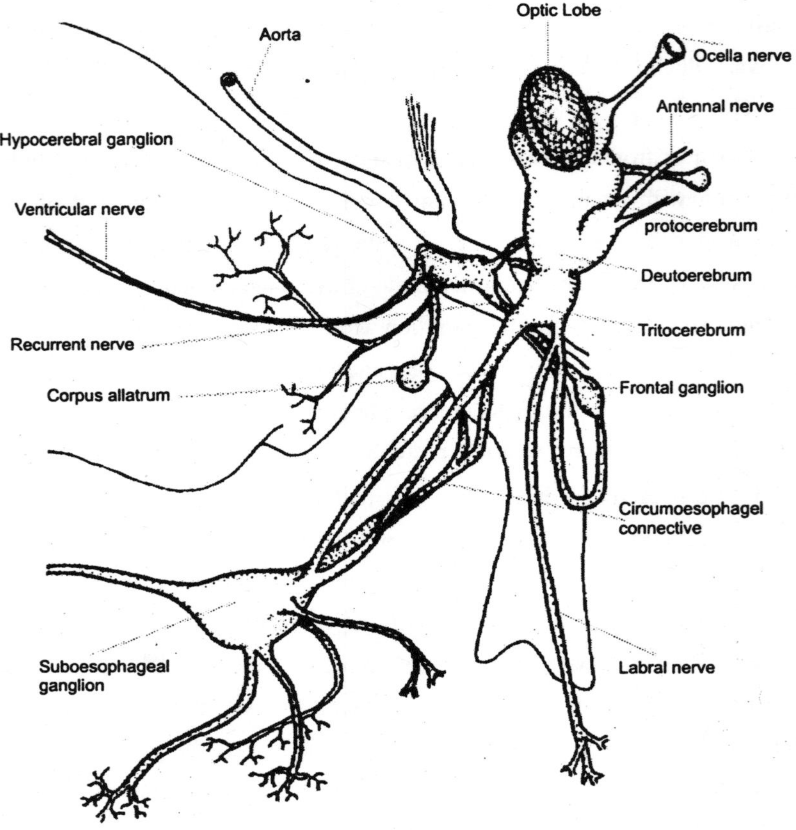

(Modified from http://biodidac.bio.uottawa.ca/thumbnails/filedet.htm/File_name/inse017b)

Fig. 7.4. Stomatogastric nervous system in a generalised insect

A heavy spontaneous electrical activity is produced in the frontal ganglion and leaves it *via* the recurrent nerve and *via* the frontal connectives to the tritocerebrum. A similar heavy activity reaches the frontal ganglion through the frontal connectives from the tritocerebrum.

The most accepted function of the frontal ganglion is the control of crop-emptying. After ganglionectomy, a massive accumulation of undigested food in the foregut, in consequence of a restricted passage of food from the foregut into the mid- and hind-gut has been observed.

5. THE PERIPHERAL NERVOUS SYSTEM

The peripheral nervous system includes the nerve trunks radiating from the ganglia, and the distal branches and terminal organs of the motor and sensory fibers contained in the nerve trunks. When the sensory fibers of a nerve trunk are traced outward from the central ganglia, they are found to end in cells lying either within the epidermis of the body wall, or immediately beneath it, or on the somatic muscles or the wall of the alimentary canal. These cells appear to be the true neurocytes of the sensory neurones. The peripheral cells of the sensory nerves are either bipolar or multipolar. The distal processes of most of them go direct to specific ectodermal sense organs. Cells of this kind, which are always bipolar, are those designated **sensory cells of Type I**. The others, which may be either bipolar or multipolar, but which are typically multipolar, are provided with one or more distal processes that branch elaborately and end with fine varicose fibrils, which terminate on the inner surface of the body wall, on the somatic muscles and connective tissue, and on the muscles and wall of the alimentary canal. Cells of this kind are those distinguished as **sensory cells of Type II**. Sensory neurones of Type I are most numerous in arthropods having sclerotic plates in the body wall and are evidently developed asa means of circumventing the loss of sensitivity to external conditions, which would otherwise result from the hardening of the cuticula.

Sensory Neurones of Type I

The neurocytes of the first type of sensory neurones are the so-called sense cells of the specific ectodermal sense organs. They are always bipolar, and their distal processes are immediately connected with the cuticular parts of the receptors. The proximal processes are the centripetal axons which end in terminal arborisation's within the central ganglia. In most cases the sense cell (or cells) of a sense organ lies within the epidermis in close association with the constructive cells of the receptor; but in various larvae the sense cells of the tactile setae distributed over the body lie beneath the epidermis, and their long distal processes penetrate the basement membrane to enter the receptor.

Sensory Neurones of Type II

Sensory neurones of this type are particularly abundant in the Annelida and in soft skinned arthropods, such as the larvae of holometabolous insects, but they also occur in arthropods, including crustacea and insects, that have a sclerotised integument. Neurones of type II are never connected with specific sense organs. Their neurocytes lie on the inner face of the body wall, on the muscles, or on the wall of the alimentary canal, and their finely branching distal processes end in free terminals which innervate the epidermis, the somatic muscles, connective tissue, and the muscles and epithelium of the alimentary canal. The centripetal axons go to the ganglia of the central nervous system.

6. BASIC STRUCTURE OF GANGLIA

Ganglia are externally delimited by an insulating cell layer, the **neurolemma**, forming a chemical and mechanical barrier. Outer layer of the neurolemma is called **neural lamella** (or sheath) that is composed of extracellular matrix produced by an inner layer, the perineurial sheath cell layer consisting of glial cells (Box 7.4). The lamella protects the neuronal tissue mostly mechanically while the sheath cell layer acts as a blood-brain barrier, being relatively impermeable.

Box 7.4 Inside ganglion: glial cells

Surround ganglion

1. Neural lamella: mechanical support for NS, secreted by perineurium
2. Perineurium = brain-blood barrier: a layer of glial cells that maintain stable ionic environment
3. Nerve sheath = lamella + perineurium

Surround individual nerve (axon)

1. Protect, insulateand repair neuron
2. Pass nutrients to nerve and control ionic environment
3. More glial cells than neurons

Neuronal tissue within a ganglion is grouped into two layers, an **outer soma layer** and an inner neuropil core. Insect central neurons are usually monopolar cells. A single neurite extends from the soma and branches into several neurites, forming synapses with adjacent neurites of other neurons. Generally, the majority of synapses are localised in the neuropil.

The neuropil is classified into glomerular neuropil, non-glomerular neuropil, and neuronal tracts. **Glomerular neuropil** is a region where finely branched neurites form dense tangles and numerous synapses. The border of dense neuropil is clearly discernable and often delineated by a glia cell layer. **Non-glomerular neuropil** has a low synapse density and poorly defined borders. Its internal structure cannot be revealed with general staining methods. A neuronal tract ·is a bundle of axons that connects neuropil regions. Tracts formed by a large number of axons, in particular large diameter ones, are easily delineated, but this may not be so when the number of axons is small.

Dense neuropils are further grouped into small modular structures of different shapes such as spherical, columnar, layered. The word, module reminds a column in cerebral cortex of vertebrate. Although the vertebrate column can be distinguished functionally they are not compartmented morphologically. On the other hand the insect module is functionally and morphologically compartmented so that it can be used in the study easily.

Modules in the insect nervous system are local microcircuits with similar properties supposedly arranged to process information in a parallel and distributed manner. For example a single lamina cartridge in the peripheral optic lobe of a fly processes information received by six photoreceptors looking at the same point in space whereas a glomerulus in the antennal lobe generally processes information from a population of olfactory receptor cells that express the same odorant receptor protein. Neuropil regions specifically concerned with a sensory modality (i.e. lamina of the optic lobe or antennal lobe) are built from such modules that are concerned with more fine-grained aspects within this modality (spatial distribution of light for the lamina, specific odorants for the antennal lobe). In other words, one densed neuropile takes charge of one modality and each module in the neuropile takes charge of a certain quality of modality.

7. MORPHOLOGY OF NEURONS IN THE INSECT BRAIN

Figure 7.5 shows examples of neurons in the insect brain. The neuron is a cell containing the nucleus and is called **soma** (cell body/perikaryon). The size of somata is about 10 μm in diameter in many neurons (but varies considerably depending on the cell type within a species and also across species). From the soma protrusions extend: the neurites. Neurons with a single neurite are called **monopolar cells** (A). Those with two neurites are bipolar cells (B). Those with more than two neurites are **multipolar cells** (C). Based on their function they may be sensory, motor or association neurons (Box 7.5). Most insect neurons are monopolar cells. Their primary neurite runs into the neuropil and forms complex branching patterns. In general, the longest neurite within a neuron is the axon and other neurites are dendrites. The axon is representing the output channel of a neuron. On the axon, the output is relayed to other neurons through synapses, the axon being the output element. On the other side of a synapse, a dendrite, the input region of another neuron, receives this output. The patterns of connectivity of synapses between neurons results in a complex network for information processing.

Box 7.5 Types of neuron: two ways of classification

By the number of extensions

1. Unipolar neurons have one projection extending from the soma.
2. Bipolar neurons have two projections extending from the soma.
3. Multipolar neurons have many projections extending from the soma. However, each has only one axon.

By the direction of information that they send (function)

1. **Afferent (sensory) neurons**-bipolar or multipolar cells have dendrites that are associated with sense organs. They carry information *towards* the central nervous system (CNS).
2. **Efferent (motor) neurons**-unipolar cells that conduct signals *away* from CNS and stimulate responses in muscles and glands.
3. **Interneuron (association neuron)**-unipolar cells that form connections between afferent and efferent neurons and conduct signals within CNS.

8. ORGANISATION OF GLIAL CELLS

Glial cells constitute the second major cell type of the nervous system. They wrap the neuronal somata and axons with their cytoplasmic processes and play a number of roles in neuronal development, activity and during regeneration. Even though glial cells have been identified in higher invertebrates such as Annelids, Arthropods and Molluscs; they have been studied for a long time in vertebrate model systems. This is due to the fact that invertebrate glial cells are highly heterogeneous, which hampers an easy classification. Recently, however, enormous progress has been made in insect neurobiology and developmental biology due to the availability of genetic and molecular tools developed in *Drosophila melanogaster*.

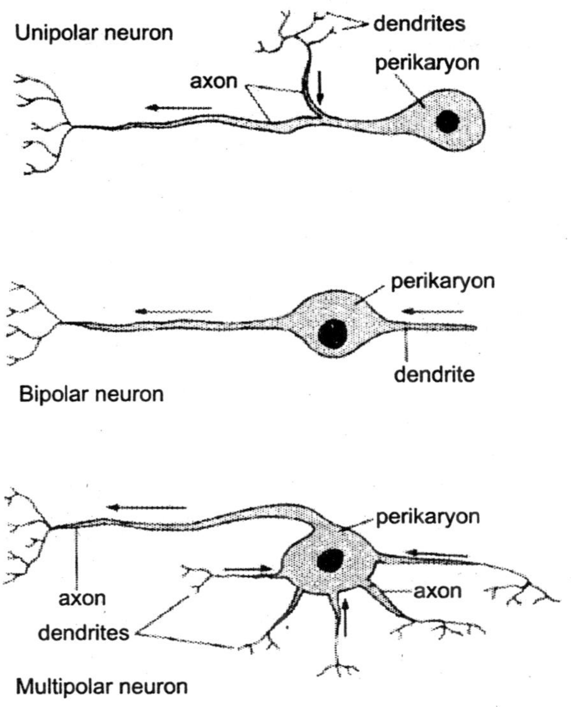

Fig. 7.5 Structural types of neurons

Glial cells have been subdivided into three major classes depending on their position within the ventral nerve cord. The **first class** is constituted by surface associated glial cells which include the outermost glial cells, perineural and subperineural cells. These cells form a sheath around the ventral cord functioning as a blood-brain barrier. Surface associated glial cells also include channel glial cells at the segment border. The **second class** includes the cortex glial cells, or cell body glial cells, found underneath the surface glial cells and associated with neuronal cell bodies. The **third class** of glial cells, neuropile associated glial cells, are found along the axonal structures and provide a scaffold for the axon bundles. To this class belong midline glial cells, wrapping the anterior and posterior commissures, longitudinal glial cells, lining the longitudinal connectives, and nerve root glial cells, which lie along the peripheral nerve root.

Glial cells are also present in the embryonic as well as the adult peripheral nervous system where they wrap sensory and motor fibers. Peripheral glial cells ensheath multiple axons and are devoid of myelin. Glial cells which originate from the optic lobe display a rather complex organization, which reflects the layer structure of axonal fibers and neuronal somata within the optical ganglia. Depending on the position in the ganglia, glial cells have been subdivided into different sub-types, most of which have been characterised in detail in large insects. A very specialised form of glial cell in the peripheral nervous system is the thecogen cell, one of the accessory cells of the external sensory organ. This cell is tightly associated with the sensory organ. The external sensory organ originates from a sensory organ precursor cell, a committed cell that divides twice to give the neuron, the thecogen cell and two other cells, tormogen and trichogen, which secrete the cuticular structures of the sensory organ, the socket and the shaft. In insects, sensory neurons are located at the periphery, beneath the epidermis. These bipolar neurons send a basal process, the axon, towards the central nervous system, and a short apical process, the dendrite, towards the shaft. The axon fasciculates with other fibers to form a nerve wrapped by glial cells whereas the dendrite is wrapped by an electron-dense sheath secreted by the thecogen, which also ensheathes the neuronal cell body.

8.1 Role of the Glial cells

The name **glia** has a Greek origin and means glue or cement. Glial cells originally were thought to play a passive role by providing a compact structure to the nervous system. Considerable evidence now points to glial cells playing a number of active roles. Invertebrates, the insulating sheath provided by glial cells ensures a correct transmission of the electrical signal along neurons. Some central glial cells are required during development for neuronal migration while Schwann cells are necessary for axonal regeneration upon wounding. Furthermore, the presence of different typesof ion channels on glial cell membranes might also suggest additional regulatory roles. As in vertebrates, fly glial cells provide electrical insulation. **Gliotactin**, a protein expressed in peripheral glial cells, belongs to a family of transmembrane proteins that mediate cell-cell or cell-matrix interactions. Embryos mutant for the gliotactin product are paralysed due to the presence of gaps and openings in the glial wrappings. Glial cells form an incomplete sheath around the axons, which are thus exposed to high K^+ extracellular environment. As a result, action potentials cannot propagate. Glial cells are necessary for several aspects of nervous system organisation such as axonal fasciculation and nerve cord condensation. Glial cells are also required for neuronal survival. Finally, it has been proposed that anachronism, a glycoprotein produced by central glial cells alter embryonic neuroblast proliferation, regulates postembryonic development of neuroblasts in the brain and in optic lobes.

9. SYNAPSES

A synapse is a junction formed between two neurons which allow the cells to communicate. The neuron approaching the synapse is called **presynaptic neuron**. On the other side of the synapse is the postsynaptic neuron. At the synapse the presynaptic neuron sends chemical or electrical signals to the post-synaptic neurons. The signal received by the post-synaptic neuron produce either an inhibitory or excitatory response.

Neurons at **electrical synapses** are joined by **gap junctions**. In these areas, the intercellular space narrows to 2-4 nm and adjacent neurons communicate through interconnecting channel proteins. The channels are called **connexions** and each is composed of six connexion protein subunits. Pairs of opposing connexions form water-filled pores that bridge the gap between neuron plasma membranes. The channels allow ionic current or action potentials to move quickly from one neuron to the next.

The most common type of neuron synapse is the **chemical synapse.** At chemical synapses, the presynaptic neuron is separated from the postsynaptic neuron by a narrow (20nm) water-filled space called the **synaptic cleft**. Neurotransmitter molecules are used by the presynaptic neuron to send a message across the cleft to the postsynaptic neuron. The neurotransmitter molecules are stored in **synaptic vesicles** located in the expanded tips (synaptic knobs) of the presynaptic neuron. Action potential at the synaptic **knob signal** the synaptic vesicle to fuse with the presynaptic membrane and release their contents into the synaptic cleft. Some of the transmitter molecules diffuse across the synaptic cleft and bind with receptors in the postsynaptic membrane. The bound transmitter causes channels in the postsynaptic membrane to open. Ions diffuse through the open channels and the exchange either excites or inhibits the post synaptic neuron.

Cholinergic synapses are chemical synapses that use acetylcholine molecules as the neurotransmitter. Embedded in the membrane of the postsynaptic neuron are acetylcholine (ACh) receptors. Two forms of ACh receptors occur, the **nicotine type** and **muscarine type**. Nicotinic receptors are composed of five polypeptide subunits, which together form a water-filled ion channel in the lipid bilayer. Normally, the gates are closed and ions are prevented from passing through the channel. Two of the receptor's subunits have binding sites for ACh molecules. When ACh occupies both sites, the gate subunits change shape and the channel opens. When it is open sodium and potassium ions diffuse through the channel. The exchange of ions changes the membrane potential of postsynaptic neuron.

Types of synapses

In the central nervous system, where most of the synapses occur, presynaptic neurons can approach and interact with postsynaptic neurons at three locations. Most presynaptic neurons synapse along the dendritic spines of the postsynaptic neuron. Because of the location, this type of contact is called **axodendritic synapse.** Presynaptic neurons also commonly terminate along the cell body of the postsynaptic neuron. These junctions are termed as **axosomatic synapse.** Synapses can form between the axon of a presynaptic neuron and axon of a postsynaptic neuron. This type is called **axoaxonic synapse.**

10. NEUROMUSCULAR JUNCTION

Neuromuscular junction is the point of contact between the nerve and muscle. A motor nerve supplies to an insect muscle. It makes contact with each muscle fibre at a number of points. The nature of the ending varies considerably. In the simplest forms as in the flight muscles of Diptera, the fine nerve branches pass longitudinally over the surface of the muscle or sometimes, as in *Tenebrio* (Coleoptera), the terminal axon is completely invaginated into the muscle fibre so that it makes contact with the muscle all round its circumference. In Orthoptera, the axon divides at the surface of the muscle and the branches with their sheaths form a claw like structure. Each junction may contain only one axon as in *Tenebrio* or more than one as in *Blatta* (Dictyoptera).

The fine structure appears to be similar in all these forms. Glial cells are lacking along the nerve/muscle interface so that the axon and muscle fibre plasma membrane are close together, separated only by a synaptic gap of about 100 Å. The terminal axoplasm contains synaptic vesicles some 250-45 Å across.

11. PHYSIOLOGY OF IMPULSE TRANSMISSION

All the nerve fibres carry information in the form of nerve impulse. Nerve impulse is the sum total of physical and chemical disturbances created by a stimulus (electrical, chemical or mechanical) in a neuron or nerve fibre which result in the movement of a wave along the nerve fibre.

The nerve fibre or axon is like a cylinder, the interior of which is filled with axoplasm (i.e., the cytoplasm of the nerve cell) and is covered with a thin membrane, the axon membrane or axolemma. The axon is immersed in the extracellular fluid (ECF). Through axolemma movement of solute takes place between the axoplasm and ECF. Generally the solutes in ECF and axoplasm are in ionic form. In the axoplasm negatively charged protein molecules are present which are neutralized due to the presence of large amount of K^+ ions. In the ECF (outside the axon) the -vely charged Cl^- ions are neutralised by the presence of +vely charged Na^+ ions.

Conduction of nerve impulse is an electrochemical process. Membrane of a non-conducting nerve cell or neuron is positive on the outside and negative inside. The difference in charge is about 70 to 90 millivolts which is called **resting potential** and the membrane is said to be **polarised.** To maintain resting potential, sodium potassium metabolic pump operates. This pump which is located on the axon membrane, pump Na^+ from axoplasm to ECF and K^+ from ECF to axoplasm. It pumps more positive charges (3 Na^+) from axoplasm to ECF than in the reverse direction ($2K^+$), and is run by an enzyme called **sodium potassium-ATPase**. The concentration of sodium ions will be about 14 times more in ECF (outside) and concentration of potassium ions will be about 28-30 times more in axoplasm (inside). When a stimulus (may be mechanical, electrical or chemical) is applied to the membrane of the nerve fibre, its permeability changes and sodium potassium pump stop operating. Sodium ions rush inside and potassium ions rush outside. This results in the positive charge inside and negative charge outside the membrane. The nerve fibre is said to be in **action potential** or depolarised. The resting potential inside the membrane is about -70 mV and the action potential inside the membrane is about +30 mV. The travelling of action potential along the membrane is a nerve impulse. After a period of action potential, again sodium pump operates and axon membrane will get resting potential by **repolarisation**. During this process the sodium ions will rush outside and potassium ions will move inside (reversal of the process taken place during action potential). Refractory period is the period of complete inexcitibility (restoration of nerve fibre) between deplorisation and repolarisation (1-6 milli seconds in mammals). During refractory period nerve fibre never transmits impulse.

11.1 Generation and Conduction of Nerve Impulse

Neurons are excitable cells because their membranes are in a polarised state. Different types of ion channels are present on the neural membrane. These ion channels are selectively permeable to different ions. When a neuron is not conducting any impulse, i.e., resting, the axonal membrane is comparatively more permeable to potassium ions (K^+) and nearly impermeable to sodium ions (Na^+). Similarly, the membrane is impermeable to negatively charged proteins present in the axoplasm. Consequently, the axoplasm inside the axon contains high concentration of K^+ and negatively charged proteins and low concentration of Na^+. In contrast, the fluid outside the axon contains a low concentration of K^+, a high concentration of Na^+ and thus forms a concentration gradient. These ionic gradients across the resting membrane are maintained by the active transport of ions by the sodium-potassium pump which transports 3 Na^+ outwards for 2 K^+ into the cell. As a result, the outer surface of the axonal membrane possesses a positive charge while its inner surface becomes negatively charged and therefore is polarised. The electrical potential difference across the resting plasma membrane is called **resting potential.**

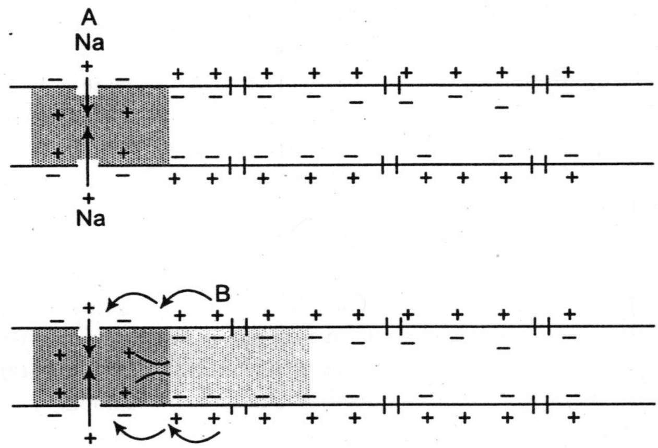

Fig. 7.6. Diagrammatic representation of impulse conduction through an axon (at points A and B)

When a stimulus is applied at a site (Fig 7.6 e.g., point A) on the polarised membrane, the membrane at the site A becomes freely permeable to Na^+. This leads to a rapid influx of Na^+ followed by the reversal of the polarity at that site, i.e., the outer surface of the membrane becomes negatively charged and the inner side becomes positively charged. The polarity of the membrane at the site A is thus reversed and hence depolarised. The electrical potential difference across the plasma membrane at the site A is called the **action potential,** which is in fact termed as a **nerve impulse.** At sites immediately ahead, the axon (e.g., site B) membrane has a positive charge on the outer surface and a negative charge on its inner surface. As a result, a current flows on the inner surface from site A to site B. On the outer surface current flows from site B to site A (Fig 7.6) to complete the circuit of current flow. Hence, the polarity at the site is reversed, and an action potential is generated at site B. Thus, the **impulse** (action potential) generated at site A arrives at site B. The sequence is repeated along the length of the axon and consequently the impulse is conducted. The rise in the stimulus-induced permeability to

Na^+ is extremely short lived. It is quickly followed by a rise in permeability to K^+. Within a fraction of a second, K^+ diffuses outside the membrane and restores the resting potential of the membrane at the site of excitation and the fibre becomes once more responsive to further stimulation.

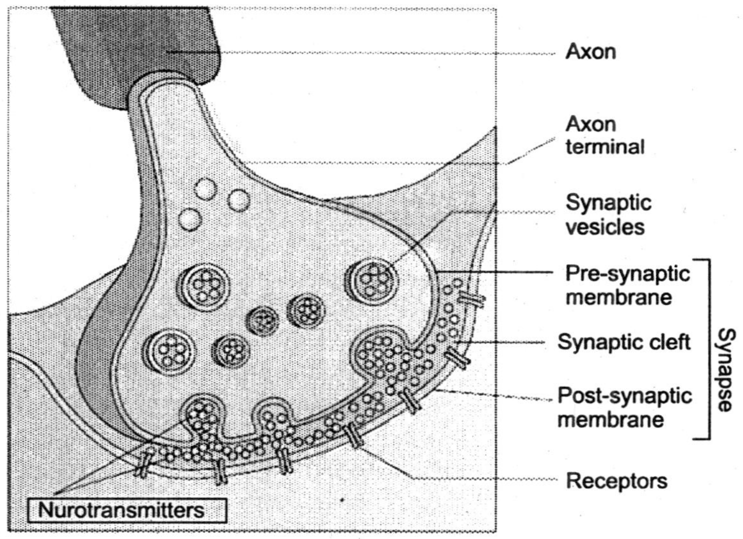

Fig. 7.7. Schematic diagram showing axon terminal and synapse

11.2 Transmission of Impulse

A nerve impulse is transmitted from one neuron to another through junctions called **synapses.** A synapse is formed by the membranes of a presynaptic neuron and a post-synaptic neuron, which may or may not be separated by a gap called **synaptic cleft.** There are two types of synapses, namely, **electrical synapses** and **chemical synapses**. At electrical synapses, the membranes of pre- and postsynaptic neurons are in very close proximity. Electrical current can flow directly from one neuron into the other across these synapses. Transmission of an impulse across electrical synapses is very similar to impulse conduction along a single axon. Impulse transmission across an electrical synapse is always faster than that across a chemical synapse.

At a chemical synapse, the membranes of the pre and postsynaptic neurons are separated by a fluid-filled space called **synaptic cleft** (Fig 7.7). Chemicals called **neurotransmitters** are involved in the transmission of impulses at these synapses. The axon terminals contain vesicles filled with these neurotransmitters. When an impulse (action potential) arrives at the axon terminal, it stimulates the movement of the synaptic vesicles towards the membrane where they fuse with the plasma membrane and release their neurotransmitters in the synaptic cleft. The released neurotransmitters bind to their **specific receptors,** present on the post-synaptic membrane. This binding opens ion channels allowing the entry of ions which can generate a new potential in the post-synaptic neuron. The new potential developed may be either excitatory or inhibitory.

12. NEUROPEPTIDES AND NEUROTRANSMITTERS

Signal transmission between neurons in the nervous system is mainly performed by chemical substances such as neuropeptides and various types of neurotransmitters. These signalling substances regulate different physiological processes such as locomotion, metabolism, homeostasis, reproduction and growth in multicellular organisms.

Synthesis of neuropeptides occurs within the ribosomes in the neuronal cell body. Gene transcription in the nucleus leads to the synthesis of a precursor protein, a pre-propeptide, in the rough endoplasmic reticulum. The pre-propeptide is then transported to the Golgi apparatus where it is further modified and packaged into large dense-core vesicles (100- 200 nm in diameter). The mature peptide in vesicles is then transported along the microtubules by motor proteins towards the pre-synaptic terminal and vesicles are stored in preparation for release. If an action potential reaches the pre-synaptic terminal, voltage gated calcium channels are opened resulting in an influx of calcium ions. The rise of cytosolic calcium triggers fusion between the vesicle membrane and the plasma membrane of the pre-synaptic neuron and causes release of neuropeptides. The neuropeptide then binds to a receptor on the post-synaptic neuron and causes a post-synaptic effect which can either be an **excitatory** or **inhibitory effect**, but more common a modulatory effect in the CNS (Nässel, 2002). Neuropeptides can function as neuromodulators when released at a short distance, synaptically or non-synaptically, or they can also act as neurohormones when released at a longer distance into the blood stream (Nässel, 2009; O'Shea and Schaffer, 1985). Neuropeptides in insects are found expressed in interneurons, neurosecretory neurons, and in motor and sensory neurons, suggesting multifunctional roles of peptides.

Neuropeptides are sometimes found co-localised with classical neurotransmitters within the same neuron termination. **Classical neurotransmitters** include acetylcholine, glutamate, γ-aminobutyric acid (GABA) and biogenic amines (such as serotonin, dopamine, octopamine, tyramine and histamine). In contrast to neuropeptides that are synthesised in the cell bodies, neurotransmitters are synthesised and packaged into small synaptic vesicles (50 nm in diameter) within the pre-synaptic terminal. One or more biosynthetic enzymes are required for production of the mature neurotransmitter. Moreover, removal of a neurotransmitter from the synaptic cleft (to terminate a signal) is accomplished by uptake into the pre-synaptic terminal or into glial cells *via* vesicular transporters whereas neuropeptides are removed from the synaptic cleft by enzymatic degradation or diffusion (Nässel, 2002).

12.1 Classical neurotransmitters

Acetylcholine

Acetylcholine is the principal excitatory neurotransmitters in the insect CNS. It is synthesised from the compounds choline and acetyl-CoA by the enzyme choline acetyl transferase (ChAT). Immunoreactivity against ChAT in *Drosophila* has revealed an abundant and widespread distribution in the nervous system including the calyces of the mushroom bodies, antennal lobes, optic lobes, suboesophageal ganglia and in the central complex (Buchner *et al.*, 1986; Yasuyama and Salvaterra, 1999). The importance of acetylcholine in numerous behaviours, such as optomotor behaviour and courtship, was shown (Greenspan *et al.*, 1980). Acetylcholine receptors are divided into **ionotropic nicotinic acetylcholine receptors** (nAChRs) and **metabotropic muscarinic acetylcholine receptors** (mAChRs). nAChRs are activated by nicotine and blocked by α-bungarotoxin, a snake poison, whereas mAChR is activated by the mushroom alkaloid muscarine and blocked by atropin. Five mAChR subtypes (M1-M5) has been identified in mammals (Eglen *et al.*, 2001). M1, M3 and M5 are coupled to the stimulatory G protein (Gq) thus, stimulating phospholipase C whereas M2 and M4 receptors are coupled to the inhibitory G protein, (Gi), inhibiting adenylyl cyclase.

In *Drosophila*, only one mAchR gene (Dm1) with similar pharmacological profile to the mammalian M3 and M1 subtypes has been identified (Shapiro *et al.*, 1989). mAchR in *Drosophila* is expressed in the whole nervous system with strong expression in the antennal lobe glomeruli (Blake *et al.*, 1993). The *Drosophila* genome predicts 10 genes that express nAChRs (Sattelle *et al.*, 2005). Of these seven genes are encoding α subunits; Dα1 or ALS (alpha-like subunit), Dα2 or SAD (second alpha-like subunit), Dα3-Dα7 and the remaining three genes are coding for non α subunits, Dβ1-Dβ3. Using immunocytochemistry and in situ hybridisation Dα1- Dα4 and Dβ1 and Dβ2 have been shown to be expressed only in the nervous system.

B. γ-aminobutyric acid

γ-Aminobutyric acid (GABA) is synthesised from glutamate through the biosynthetic enzyme glutamic acid decarboxyalse (GAD). GABA is the major inhibitory neurotransmitter in both vertebrates and insects. Immunocytochemical studies using antibodies against GABA and GABA signalling components, such as GAD and vesicular GABA transporter (vGAT), have shown GABA producing neurons in the nervous system. GABA is expressed in major neuropil regions such as the anntenal lobe, the optic lobe, the mushroom body calyces and in some substructures in the central complex. (Enell *et al.*, 2007; Fei *et al.*, 2010; Hamasaka *et al.*, 2005; Kolodziejczyk *et al.*, 2008; Okada *et al.*, 2009). In *Drosophila,* GABA activates two types of GABA receptors known as ionotropic GABAA receptors and metabotropic GABAB receptors. The **GABAA receptors** are ligand gated chloride channels that are blocked by picrotoxin. The GABAA receptors can be formed by three subunits, the RDL (for resistance to dieldrin) (Buckingham *et al.*, 2005; French-Constant *et al.*, 1993), LCCH3 (ligand-gated chloride channel homologue 3) (Zhang *et al.*, 1995) and Grd (GABA and glycine-like receptor of *Drosophila*) (Harvey *et al.*, 1994). Immunocytochemical studies have localised RDL mostly in neuropil structures such as the optic lobes, the antennal lobes, the mushroom bodies and in the central complex in the adult *Drosophila* brain (Enell *et al.*, 2007). LCCH3 was found mostly distributed in neuronal cell bodies in the adult nervous system (Harrison *et al.*, 1996). The distribution of Grd has not yet been mapped in *Drosophila*. RDL has been shown to be significant for olfactory learning (Liu *et al.*, 2009; Liu *et al.*, 2007). Three types of GABAB receptors exist in *Drosophila,* GABABR1-3 (Mezler *et al.*, 2001). GABABR1 and GABABR-2 heterodimerise to mediate their synaptic transmission *via* Go/Gi protein (Kaupmann *et al.*, 1998; Mezler *et al.*, 2001). Based on in situ hybridisation in the *Drosophila* embryo, GABABR1 and GABABR2 have been shown to have similar expression patterns whereas the GABABR3 displayed a unique distribution pattern and different pharmacology (Mezler *et al.*, 2001). The GABABR protein is found distributed in the adult nervous system, most abundantly in the mushroom body calyces, central complex, optic lobe and antennal lobe (Enell *et al.*, 2007; Hamasaka *et al.*, 2005).

Glutamate

Glutamate is known as one of the principal excitatory neurotransmitters in the vertebrate brain and in insect motoneurons. Several studies have concentrated on its role in the insect neuromuscular junction (Jan and

Jan, 1976). However, more recent studies using the vesicular glutamate transporter (vGluT) as a marker in *Drosophila* have localised glutamate in sensory neurons, interneurons as well as in motoneurons in *Drosophila* (Daniels *et al.*, 2008; Mahr and Aberle, 2006). *Drosophila* express both ionotropic and metabotropic glutamate receptors. Until now, out of the 30 predicted ionotropic receptor types encoded in the *Drosophila* genome only 9 have been cloned (Littleton and Ganetzky, 2000). Six of the nine cloned receptors are expressed in NMJ (Qin *et al.*, 2005) and the rest of the NMDA-like receptors are found distributed in the *Drosophila* brain (Xia *et al.*, 2005). The only functional metabotropic receptor known in *Drosophila* is DmGluRA (*Drosophila* melanogaster glutamate receptor A) (Parmentier *et al.*, 1996). DmGluRA is expressed in most neuropil regions except the lobes of the mushroom bodies (Devaud *et al.*, 2008). In addition, it is expressed in the neuromuscular junction where it modulates excitatory neurotransmission (Bogdanik *et al.*, 2004) and in certain clock circuits regulating circadian locomotor behaviour (Hamasaka *et al.*, 2007).

12.2 Biogenic amine neurotransmitters

Serotonin

Serotonin (5-HT) is synthesised from the amino acid tryptophan by two enzymes: **tryptophan hydroxylase** (TRH) and **dopa decarboxylase** (Ddc). 5-HT in *Drosophila* has been reported to play an important role in aggression (Alekseyenko *et al.*, 2010) and in learning and memory (Sitaraman *et al.*, 2008). Both ionotropic and metabotropic 5-HT receptors exist in mammals. However in insects, only metabotropic 5-HT receptors have been identified (Tierney, 2001). In *Drosophila* four types of 5-HT receptors that exhibit sequence similarity to the mammalian 5-HT receptors are known; these are 5-HT1A, 5-HT1B, 5-HT7 and 5-HT2 receptors. 5-HT1A and 5-HT1B belong to the same family of receptors and inhibit adenylate cyclase while stimulating phospholipase C whereas 5-HT7 activates adenylate cyclase. The transduction mechanism of the fourth receptor, 5-HT2, is not clearly investigated, however, it is proposed that it is similar to the mammalian 5-HT2 receptor that activates phospholipase C. 5-HT1A is expressed in the mushroom body where it was shown to have a role in sleep (Yuan *et al.*, 2006). 5-HT1B is also expressed in the mushroom body and is additionally found in certain clock neurons (Yuan *et al.*, 2005). 5- HT2 is expressed mainly in the EB of the central complex and in certain neurons in the protocerebrum, and its role in circadian behaviour has been shown (Nichols, 2007). In addition 5-HT1A and 5-HT2 has been implicated in modulating aggressive behaviour in *Drosophila* (Johnson *et al.*, 2009).

Dopamine

Dopamine is synthesised from the amino acid tyrosine into its precursor molecule LDOPA by the rate limiting enzyme tyrosine hydroxylase (TH), there after L-DOPA is converted into dopamine by the enzyme dopadecarboxyalse (Ddc). Based on immunocytochemical studies in the nervous system of *Drosophila* and in two other flies, the distribution pattern of dopamine was found to match with that of TH (Friggi-Grelin *et al.*, 2003; Nässel and Elekes, 1992). In the *Drosophila* brain, using antibodies to TH and GAL4 expression analysis, about 600 dopamine expressing neurons, organised in 15 cell cluster have been reported to innervate all major neuropil structures (Friggi-Grelin *et al.*, 2003; Mao and Davis, 2009). Dopamine is involved in behaviours such as locomotion, courtship, sleep and arousal. Three types of metabotropic dopamine receptors are known in *Drosophila*. Two of these, DopR (also known as dDA1) and DopR2 (also known as DAMB) are orthologs of the mammalian D1 receptors. The third one, D2R (also known as Dop2R) is the ortholog of the mammalian D2 receptors. The D1-like receptors are coupled to the stimulatory G protein (Gαs) and the D2 like receptors are coupled to the inhibitory G protein (Gαi/o) thereby activating and inhibiting adenylate cyclase, respectively. In addition a novel GPCR that is activated by both dopamine and ecdysteroids has also been proposed as a fourth likely dopamine receptor (Srivastava *et al.*, 2005).

Tyramine and octopamine

Tyramine, a precursor to octopamine is synthesised by the enzyme tyrosine decarboxyalse (TDC) from the amino acid tyrosine. Octopamine, an end product in the synthesis of tyramine is synthesised from tyramine by the enzyme tyramine β- hydroxylase (TBH). Immunocytochemical data has revealed about 100 octopamine expressing neurons in the *Drosophila* brain (Busch *et al.*, 2009; Monastirioti, 1999; Sinakevitch and Strausfeld, 2006). Subsets of these neurons were shown to innervate the central complex (Busch *et al.*, 2009; Sinakevitch and Strausfeld, 2006). Octopamine in *Drosophila* has been shown to be involved in several behaviours such as locomotion, ovulation and egg laying, ethanol tolerance, appetitive reinforcement, cocaine sensitisation and aggression. Tyramine has recently been implicated in locomotor activity (Saraswati *et al.*, 2004) and cocaine sensitisation (Hardie *et al.*, 2007). Two GPCRs that are activated by both octopamine and tyramine has been cloned in *Drosophila*. One receptor that has a preference to tyramine over octopamine is known as **tyramine/ octopamine receptor (Octyr).**

The second receptor known as OAMB (octopamine receptor in the mushroom body) is activated by both octopamine and tyramine but preferentially binds octopamine (Han *et al.*, 1998). Both receptors stimulate the production of intracellular calcium signal; however Octyr and OAMB seem to exhibit opposite action towards adenylate cyclase, OAMB being the activator and Octyr the inhibitor (Han *et al.*, 1998; Saudou *et al.*, 1990). OAMB is found widely distributed in the mushroom body and in the central complex in *Drosophila*. Reports have also revealed the expression of OAMB in the thoracic and abdominal ganglion and in the reproductive regions.

I. Neuropeptides and neurotransmitter receptors

Signal transduction most commonly occurs through two types of receptors; G-protein coupled receptors (GPCRs) also known as **metabotropic receptors** and **ligand-gated ion channel receptors** (LGICs). GPCRs are characterised by seven transmembrane (7TM) protein domains that are connected by three extracellular and three intracellular loops (Fig 7.8). The extracellular N-terminus domain can be glycosylated whereas the intracellular C terminus domain can be phosphorylated. GPCRs are directly coupled to heterotrimeric GTP-binding (G) proteins that are composed of three subunits α, β, and γ. G proteins exist in two functional states, in an inactive state when the α subunit is bound to GDP and an active state when the α subunit is bound to GTP. Activation of a GPCR leads to binding of the α subunit to GTP, the α-GTP then disassociates from the βγ subunits. Depending on the signal transduction pathway, the Gα-GTP complex activates or inhibits one of the effector enzymes; adenylyl cyclase, guanylyl cyclase, phospholipace C or calcium channels; to generate second messenger such as cAMP, cGMP inositol 1,4,5- triphosphate (IP3) and diacylglycerol and calcium respectively (Lodish *et al.*, 1999). LGICs are transmembrane proteins usually made of five subunits with an extracellular ligand-binding site and an ion channel that selectively transmits small ions such as sodium, potassium, calcium or chlorine. LGICs mediate fast post-synaptic effects, usually lasting less than a millisecond. GPCRs on the other hand signal slower post-synaptic transmission lasting from seconds to hours (Lodish *et al.*, 1999). If a receptor is continually stimulated by a signalling substance, the post-synaptic neuron can decrease its responsiveness for that ligand; this reaction is known as **receptor desensitisation**.

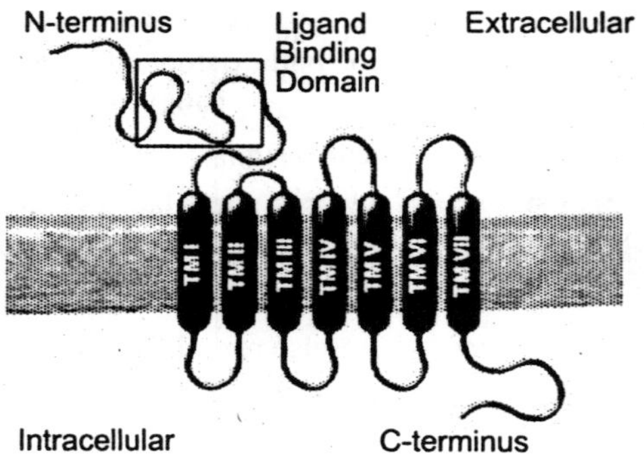

Fig. 7.8. Metabotropic receptor
(Adapted from http://www.bristol.ac.uk/synaptic/receptors/mglur/)

14. LEARNING, MEMORY AND INTELLIGENCE IN INSECTS

Learning may be defined as an adaptive change in individual behaviour as a result of experience. Although the mushroom bodies of the brain have been shown to be involved in learning, ganglia other than the brain are also capable of learning. Learning has been demonstrated in decapitated cockroaches. In a classical experiment it was shown that if a headless cockroach is wired so that one of its legs receives an electric shock when lowered, then it will learn to avoid the shocks by keeping the leg raised. The thoracic ganglia are responsible for controlling leg movement, and it is these ganglia that learn the new behaviour. Through another experiment it was shown that intact cockroaches have a preference for darkness if given a choice between the illuminated half of a chamber and the darkened half. However, if they receive electric shocks in the darkened half, then they will learn to avoid the dark-half and remain in the light.

Habituation

This term is employed for learning not to respond to stimuli which tend to be without significance in life of the insect. *Nemeritis* (Hymenoptera) for instance move away from the smell of cedar wood oil, but if it is continuously exposed to the smell it learns to tolerate it. An isolated cockroach leg exhibits another phenomenon related to learning - **sensory habituation**. If a touch-sensitive bristle is stimulated on an isolated leg and the activity in the leg nerve recorded, then it will be found that the strength of the stimulus diminishes if the stimulus is repeated rapidly, or sustained.

This is due to fatigue in the periphery nerve and sensor, and this ensures that insects respond to changes in the environment, and learn to ignore persistent stimuli that are of no relevance. For example, body lice prefer rough fabrics, like wool, to smooth fabrics like silk. When crossing from wool to silk they will keep turning as they try to find their way back onto the wool. However, if they fail to find the wool, then this behaviour stops as the lice learn to make do - they habituate. Habituated insects are still responsive to changes in the stimulus, however. Thus, if the texture of the fabric for our louse on silk changes again, then we expect it will respond again, and habituate again if necessary.

Learning one's way about - route learning

Some insects, such as ants and cockroaches, are capable of learning the route of a maze, if the maze exit leads to reward, or if escaping from the maze avoids punishment. A cockroach will navigate a maze to it home-pot, so long as the home-pot contains recognisable cockroach odours. On subsequent trials it will reach its home-pot with increasing ease as it gradually learns the route. When compared to rats, ants learn mazes at half the speed: a rat will master a maze after about 15 runs, an ant after about 30, though the ants still make a few errors. However, if a change is made to the maze, such as reversing the pattern, then rats learn the new maze more rapidly than before, recognising the similarity in the patterns and transferring their previous learning to the new situation (**transfer learning**). In contrast, the ant starts all-over again, treating the maze as entirely new. Thus ants have little or no capacity for transfer learning, that is they seem unable to apply what they have learnt to a novel situation. One of the most impressive feats of insect learning is **locality learning**. In addition to learning routes, insects can recognise the locality in which their nest is situated, or in which food is found. This involves exploratory learning- the insect will typically fly around a bit after leaving the nest, learning the position of many landmarks very rapidly and then leave. This is latent learning, meaning that a period of time elapses between learning and reward. The reward occurs when the insect returns home and locates its nest. This can be demonstrated by experiment. For example, in classic experiments on the beewolf, *Philanthus triangulum* a type of digger wasp, which brings back food to its developing young in the nest, the wasp will learn to recognise a ring of pine cones placed around the nest entrance and that the nest hole is the centre of this ring. If the pine cones are displaced a few centimetres when the wasp leaves, however, it will return to the centre of the ring of pine. Generally, however, these insects are only temporarily confused by changes to one or two landmarks, falling back on other landmarks further from the nest.

Associative learning

This occurs when a stimulus, irrelevant by itself, is made relevant by pairing it with something meaningful, like a food reward and the animal learns to identify the hitherto unconnected stimulus with food. For example, a cockroach can be rewarded by being given food when presented with a particular, but unrelated, odour. It will then learn to associate that odour with food and be attracted to it when foraging. Bees can be trained to associate a particular colour or pattern, though their perception of shape is limited with food.

Short-term and long-term memories

When an insect has just completed a task, learning is abolished if a new task is undertaken immediately after. Learning requires a latent period of rest in-between activities. To some extent the same is true of humans - learning becomes greatly enhanced if breaks are taken every 20 minutes or so during study. It is during these rest periods that the brain processes the information (often subconsciously in humans) and the appropriate neural pathways are reinforced. This can be explained by the existence of short-term memory (STM) and long-term memory (LTM). The nervous system requires time to process information in short-term memory, making sense of it and discarding information deemed irrelevant and transferring more useful information into long-term memory. In insects LTM typically retains information for several days.

Do Insects Sleep?

Insects certainly exhibit circadian cycles. Insects enter a dormant or semi-dormant state when chilled. This may occur seasonally, during winter, or daily at nightfall. At night time insects may assume a particular posture in which to rest, bees may hold onto vegetation with their tightly clamped jaws, for example. They may also return to the same resting spot each night. During these quiescent states the body temperature drops and energy is conserved. Interestingly, learning increases when insects are given rest between tasks or training sessions, as already mentioned, but also if they are chilled during these rest periods. Could it be that when chilled into torpor at night time insects are forming memories? It has been suggested that two chief functions of sleep are: energy conservation and memory formation. In this case the insect night time torpor is not so very different. Many insects are, of course, nocturnal. Cockroaches are nocturnal and learn better at night than during the day. It has been suggested that this encourages them to remember useful things like lessons learned during night time foraging.

Temporal learning

This occurs in insects. As we have seen insects can measure time, by the use of internal body-clocks, and they can also compensate for the changing position of the Sun in the sky when they use the Sun to navigate. To navigate successfully by the Sun they must know what time it is, in the sense of what portion of daylight has passed. Observations have shown that this may require learning, with young insects making mistakes by assuming that the Sun stays fixed in the sky.

Insight learning

This is a higher form of learning, similar to transfer learning, in that it takes prior learning and applies that to a new problem. One example is tool use. Some insects use tools, for example, some digger wasps fill in their nest burrow once the young is mature and ready to pupate (and so requires protection but no more food). The female may then hold a small pebble in her mandibles and use it to pound-down the earth, and then discard the pebble. However, this behaviour is not intelligent in the sense that the insect reasoned a solution; rather it is inherited genetically and is instinctive and so not true insight learning.

15. HORMONAL CONTROL ON THE DEVELOPMENT OF THE NERVOUS SYSTEM

The action of juvenile hormone (JH) and ecdysteroid hormone on the CNS is in line with their action on other tissues. In immature stages, JH appears to have no action on its own, but only when paired with the ecdysteroids. Changes in the form of sensory neurons and motorneurons, maturation of immature neurons, and some aspects of neuronal death are prevented if JH is given along with ecdysteroids. JH is most likely involved in early determinative events that occur during the cell's response to ecdysteroids. This early action is evident from the example of neurite outgrowth which occurs over a span of 8-10 days but which can only be blocked by JH at its onset. Thus, once a development programme has been initiated, JH appears to have no role in the execution of the programme. Not all ecdysteroid related processes are modulated by JH. The degeneration of motoneurons after adult ecdysis cannot be prevented by JH at the time of ecdysteroid withdrawal. Both JH and ecdysteroids appear to be acting directly on the cells which show the respective changes.

In terms of its action on the CNS, the primary effect of ecdysteroids seems to be in permitting neurons to undergo any sort of change. This is best seen in the case of topical application of JH treatment to the sensory neurons. It is highly likely that local applications in the ganglion are responsible for shaping the axon arbour, but ecdysteroids are required in order for the cell to undergo the growth which would make such an interaction possible. Likewise, preganglion cells in the abdominal lineages of *Manduca* are presumably poised for either maturation or death but this decision is not made until the ecdysteroid signal is given. There are undoubtedly local cues and interactions that are important but the neurons must be exposed to ecdysteroids before they can take advantage of these cues or signals.

An intriguing aspect of the effect of ecdysteroids on the insect CNS is that the same hormone can have so many different effects on the same cell. At one time a neuron loses dendrites when confronted with ecdysteroids, the next exposure might cause it to grow dendrites, and the next may cause it to die. How this comes about, whether it is intrinsic to the cells or related to changing environments as growth and metamorphosis proceeds, is not known.

A key to understanding how ecdysteroids promote change within the CNS undoubtedly lies in the pattern of gene expression evoked by these hormones. These include not only the structured genes through which changing phenotypes are established, but also the regulatory genes that presumably orchestrate the patterns of structural gene expression.

16. NERVE REPAIR AND REGENERATION IN INSECTS

Nerve repair and regeneration is an intricate process involving a variety of specialised cell types and often producing widespread cellular responses beyond the lesions. The impressive regenerative power of insects was first recognised by Bodenstein (1955,1957). Insects are capable of extensive axonal re-growth and accurate re-formation of synaptic connections which contrasts with the more limited power of the brains of vertebrates. Insects are used as a model system of understanding the regenerative processes in more complex mammalian central and peripheral nervous systems.

16.1 Degenerative responses

Surgical cutting of the insect axons causes degenerative changes. Severance of the central nervous connectives induces rapid ultrastructural changes at the cut ends of both proximal and distal axonal segments. These include disruption of neurotubules, rapid appearance of amorphous material in the axoplasm and a slower accumulation of large numbers of mitochondria, various sized vesicles and quantities of smooth endoplasmic reticulum. In the giant axons of cockroach, these changes are similar at the severed ends of both the distal segment and the proximal one suggesting that they do not depend on the presence of the soma and thus are locally mediated.

Extensive changes also occur in the surrounding tissues of connectives or nerves when they are cut or crushed. These include damage to the superficial neuroglial layer, the perineurium. A notable feature at the damaged site is the swift accumulation of haemocytes and cells containing electron-dense granules which together contribute to the formation of scar tissue.

The neuroglia also shows early response to surgical lesions. Within 24 hrs, the glial processes become associated with the invagination of the axolemma, which form at both proximal and distal axonal tips. By 4 days many of the glial cells in the lesion zone are clearly disorganized and there is also enlargement of extracellular spaces and marked changes in the appearance of the intercellular matrix, which is now highly fragmented and separated from the glial membranes.

The degenerative response of the peripheral nerves share similar features of those following axotomy of central neurons. The sensory axons show regressive changes in the proximal stump and the motor neurons in the distal stump. There is swift accumulation of haemocytes and transient appearance of granule-containing cells among the damaged perineural and underlying glia. Changes also occur in the neural lamella which becomes darker and thinner. There is considerable glial disruption.

16.2 Regenerative responses

A notable feature of the central nervous regeneration in insects is the vigorous axonal re-growth which can occur following neuronal lesioning. In cockroach CNS, regenerating giant interneurons can traverse extensively along the abdominal ganglionic chain. Insect neurons have the capacity to re-establish precise functional connectives within the CNS. Reconnection of the severed motor axons with their target sites is achieved by re-growth from the proximal stump. Where lesioning is by nerve crush, guidance for axonal re-growth can be provided by residual basal laminae, extracellular matrices and/or surviving neuroglia. It has been reported that specific neuromuscular reconnection can also be achieved in insects following complete nerve severance. This precludes the possibility of immediate contact guidance of regenerating axons by pre-existing structures associated with the distal stump and suggests an additional involvement of long-range factors in guidance to the appropriate target muscle.

The regeneration of insect motor axons takes place in a complex miro-environment consisting of freshly secreted matrix with closely associated haemocytes, neuroglia, fat body cells, and scar tissue. All these elements are important in axonal regeneration. Besides stimulating axonal extension, the extracellular environment also provides directional guidance. There is a possibility of long distance signals for directing growth of severed axons to the denervated muscle target.

17. NERVOUS CONTROL OF LOCOMOTION

Insects are masters of movement. The cockroaches run, bees swarm, moths fly, mantids strike, diving beetles swim, caterpillars crawl, dragonflies dart, maggots squirm, water boatmen paddle, mole crickets burrow, mosquito larvae wriggle, fleas jump, water striders skate, army ants march, and backswimmers dive. Indeed, the capacity for independent, goal-directed movement is one of the distinguishing characteristics that set animals apart from most other forms of life on this planet. The nervous system of insects is capable of producing the proper sequence of nerve impulse to generate co-ordinated locomotion.

17.1 Walking and Running

Insects have three pairs of legs positioned laterally in a wide stance. The body's center of mass is low and well within the perimeter of support for optimal stability. Each leg serves both as a strut to support the body's weight and as a lever to facilitate movement.

At very slow walking speeds an insect moves only one leg at a time, keeping the other five in contact with the ground. At intermediate speeds, two legs may be lifted simultaneously, but to maintain balance, at least one leg of each body segment always remains stationary. This results in a wave-like pattern of leg movements known as the **metachronal gait**. When running, an insect moves three legs simultaneously. This is the tripod gait, so called because the insect always has three legs in contact with the ground: front and hind legs on one side of the body and middle leg on the opposite side. Clearly, it is no coincidence that insects have exactly six legs, the minimum needed for alternating tripods of support.

Coordination of leg movements is regulated by networks of neurons that can produce rhythmic output without needing any external timing signals. Such networks are called **central pattern generators** (CPGs). There is at least one CPG per leg. Individual networks are linked together *via* interneurons and output from each CPG is modified as needed by sensory feedback from the legs. It is interesting that these interneurons are non-spiking neurons. That is, they exert their effects on the post-synaptic neurons in the absence of action potential. It is based on the release of transmitter substance by small fluctuations in the cell's membrane potential. Thus, even in the absence of spikes these interneurons can have strong excitatory or inhibitory influence on post-synaptic neurons. Because of this feature of their physiology, non-spiking neurons can act to bias follower cells towards either increased or decreased probabilities of action.

Only animals with a rigid body frame can use the tripod gait for movement. Soft-bodied insects, like caterpillars, have a hydrostatic skeleton. They move with peristaltic contractions of the body, pulling the hind prolegs forward to grab the substrate, and then pushing the front of the body forward segment by segment. This type of movement is exaggerated in larvae of Geometrid moths. While grasping the substrate with their six thoracic legs, they hunch the abdomen up toward the thorax, grasp the substrate with their prolegs, and then extend the anterior end as far as possible. This distinctive pattern of locomotion has earned those nicknames like inchworms, spanworms, and measuring worms.

17.2 Flight

In all flying insects, the base of each wing is embedded in an elastic membrane that surrounds two (or three) axillary sclerites. One of these sclerites articulates with the pleural wing process, a finger-like sclerite that acts as a fulcrum or pivot point for the wing; a second sclerite articulates with the lateral margin of the mesonotum (or metanotum). Together, these elements form a complex hinge joint that gives the wing freedom to move up and down through an arc of more than 120 degrees. The hinge is a **bi-stable oscillator**. In other words, it stops moving only when the wing is completely up or completely down. During flight, the wing literally "snaps" from one position to the other.

Power for the wing's upstroke is generated by contraction of dorsal-ventral muscles (also called **tergosternal muscles**). These are called **indirect flight muscles** because they have no direct contact with the wings. They stretch from the notum to the sternum. When they contract, they pull the notum downward relative to the fulcrum point and force the wing tips up. Elasticity of the thoracic sclerites and hinge mechanism allows as much as 85% of the energy involved in the upstroke to be stored as potential energy and released during the downstroke.

In the more primitive insect orders (*e.g.* Odonata and Blattodea), the downstroke is initiated by basalar muscles that attach through ligaments directly to the wing's axillary sclerites. Contraction of these "direct flight muscles" literally pulls the wings into their "down" position. Most other insects have dorsal-longitudinal muscles attached like bow strings to apodemes at the front and back of each thoracic segment. These are "indirect flight muscles". When they contract, they cause the edges of the notum to flex upward (relative to the fulcrum point) causing the wings to snap down.

During flight, upstroke and downstroke muscles must contract in alternating sequence. There are two different mechanisms for controlling this muscle action, **synchronous** (neurogenic) and **asynchronous** (myogenic). Insects with synchronous control have neurogenic flight muscles, meaning that each contraction is triggered by a separate nerve impulse. Central pattern generators in the thoracic ganglia coordinate the rate and timing of these contractions. Since nerve cells have a refractory period that limits how often they can fire, insects with neurogenic flight muscles have relatively slow wing beat frequencies (typically 10-50 beats per second).

Insects with asynchronous control depend almost entirely on indirect flight muscles for upstroke (dorsal-ventrals) and downstroke (dorsal-longitudinals). These muscles have developed myogenic properties, that is, they contract spontaneously if stretched beyond a certain threshold. When the nervous system sends a **start** signal, the dorsal-longitudinal and dorsal-ventral muscles begin contracting autonomously, each in response to stretching by the other. Contractions continue until the muscles receive a **stop** signal from the nervous system. Asynchronous control is not limited by the nerves' refractory period, so wing beat frequency in some of these insects (notably flies and bees) may be as high as 500-1000 beats per second. Such high frequencies produce greater lift with smaller surface area and also improve manoeuvrability (*e.g.* hovering, flying backwards).

As an insect's wing moves up and down during flight, it also twists about the vertical axis so that its tip follows an ellipse or a figure eight. This sculling motion maximizes lift on the downstroke and minimises drag on the upstroke. Turning, hovering, and other acrobatic maneuvers are controlled by small muscles attached to the axillary sclerites. These muscles adjust the tilt and twist of the wing in response to feedback from the central nervous system and sensory receptors that monitor lift and thrust.

Controlling the rhythmic contractions of thoracic muscles to produce the coordinated movements of flight depends on a large number of interactions, both among neurons in the CNS and between central neurons and receptors. The CNS in the absence of phasic excitation from peripheral receptors can produce a version of the flight motor pattern. Reflex actions could also be sufficient to generate a flight motor pattern. In actuality both mechanism contribute to flight behaviour.

Among insect species, similarities exist in the receptors that provide sensory information important in flight. These include wind-sensitive hairs on the head, antennal receptors that monitor antennal position and air-speed, wing mechanoreceptors that contribute to the synchronisation of motor outputs, and photoreceptors that are important in orientation.

With respect to the production of motor output, there appear to be differences between neurogenic and myogenic fliers. The motor neurons controlling synchronous muscles appear to be driven by interneurons, whereas the motor neurons controlling asynchronous muscle appear to set their own rhythm of firing.,

Although the evidence is limited, it is likely that each wing is provided with its own rhythm generating mechanism with both central and peripheral components. In addition some interneurons and some sensory cells have processes and synapses widely distributed in the thoracic ganglia and beyond. They function to coordinate the activity of individual rhythm generators and the motor neurons.

17.3 Swimming and skating

Many aquatic beetles (Coleoptera) and bugs (Hemiptera) use their middle and/or hind legs as oars for swimming or diving. These legs are usually flattened or equipped with a fringe of long, stiff hairs to improve their performance and efficiency in the water. Legless larvae and pupae of mosquitoes, midges, and other flies (Diptera) manage to swim by twisting, contorting, or undulating their bodies. Dragonfly naiads (Odonata) have a jet propulsion system: they can propel themselves forward by contracting abdominal muscles and forcing a jet of water out of the rectal chamber that houses their respiratory gills.

A few aquatic insects, such as water striders, have a whorl of hydrophobic hairs on the tips of their feet. These hairs prevent the insect's legs from breaking the surface tension of the water and allow them to skate on the surface.

REFERENCES

Alekseyenko, O.V., Lee, C., and Kravitz, E.A. (2010). Targeted manipulation of serotonergic neurotransmission affects the escalation of aggression in adult male *Drosophila melanogaster*. *PLoS One* 5(5):e10806.

Blake, A.D., Anthony, N.M., Chen, H.H., Harrison, J.B., Nathanson, N.M., Sattelle, D.B. (1993). *Drosophila* nervous system muscarinic acetylcholine receptor: transient functional expression and localization by immunocytochemistry. *Mol Pharmacol* 44(4):716-724.

Bogdanik L, Mohrmann R, Ramaekers A, Bockaert J, Grau Y, Broadie K, Parmentier ML. (2004). The *Drosophila* metabotropic glutamate receptor DmGluRA regulates activity-dependent synaptic facilitation and fine synaptic morphology. *J Neurosci* 24(41):9105-9116

Buchner E, Buchner S, Crawford G, Mason T, Salvaterra, P.M., Satelle, D.B. (1986). Choline acetyltransferase-like immunoreactivity in the brain of *Drosophila melanogaster*.Cell Tissue Res 246:57-62.

Buckingham, S.D., Biggin, P.C., Sattelle, B.M., Brown, L.A., Sattelle, D.B. (2005). Insect GABA Receptors: Splicing, Editing, and Targeting by Antiparasitics and Insecticides. *Molecular Pharmacology* 68(4): 942-951

Busch S, Selcho M, Ito K, Tanimoto H. (2009). A map of octopaminergic neurons in the*Drosophila* brain. *J Comp Neurol* 513(6):643-667.

Dagmur Malun (1998) Early development of mushroom bodies of the honeybee *Apis mellifera* as revealed by BrdU incorporation and ablaltion experiments. *Learning & Memory* 5: 90-101

Daniels RW, Gelfand MV, Collins CA, DiAntonio A. (2008). Visualizing glutamatergic cell bodies and synapses in *Drosophila* larval and adult CNS. *J Comp Neurol* 508(1):131-152.

Devaud, J.M., Clouet-Redt C, Bockaert J, Grau Y, and Parmentier M.L. (2008). Widespread braindistribution of the *Drosophila* metabotropic glutamate receptor. *Neuroreport* 19(3):367-371.

Eglen RM, Choppin A, Watson N. (2001). Therapeutic opportunities from muscarinic receptor research. *Trends Pharmacol Sci* 22(8):409-414.

Enell L, Hamasaka Y, Kolodziejczyk A, Nässel, D.R. (2007). gamma-Aminobutyric acid (GABA) signaling components in *Drosophila*: immunocytochemical localization of GABA(B) receptors in relation to the GABA(A) receptor subunit RDL and a vesicular GABA transporter. *J Comp Neurol* 505(1):18-31.

Fahrbach, S.E. (2006). Structure of the mushroom bodies of the insect brain. *Annu. Rev. Entomol.* 51:209–32.

Fei H, Chow DM, Chen A, Romero-Calderon R, Ong WS, Ackerson LC, Maidment NT,Simpson JH, Frye MA, Krantz DE. (2010). Mutation of the *Drosophila* vesicular GABA transporter disrupts visual figure detection. *J Exp Biol* 213(Pt 10):1717- 1730.

French-Constant RH, Rocheleau TA, Steichen JC, Chalmers AE. (1993). A point mutation in a *Drosophila* GABA receptor confers insecticide resistance. *Nature* 363(6428):449-451.

Friggi-Grelin F, Coulom H, Meller M, Gomez D, Hirsh J, Birman S. (2003). Targeted gene expression in *Drosophila* dopaminergic cells using regulatory sequences from tyrosine hydroxylase. *J Neurobiol* 54(4):618-627.

Greenspan RJ, Finn JA, Jr., Hall JC. (1980). Acetylcholinesterase mutants in *Drosophila* and their effects on the structure and function of the central nervous system. J Comp Neurol 189(4):741-774.

Hamasaka Y, Rieger D, Parmentier ML, Grau Y, Helfrich-Forster C, and Nassel, D.R. (2007).Glutamate and its metabotropic receptor in *Drosophila* clock neuron circuits. J Comp Neurol 505(1):32-45.

Hamasaka Y, Wegener C, and Nässel, D.R. (2005). GABA modulates *Drosophila* circadian clock neurons via GABAB receptors and decreases in calcium. *J Neurobiol* 65(3):225-240.

Han, K.A, Millar N.S, Davis R.L. (1998). A novel octopamine receptor with preferential expression in *Drosophila* mushroom bodies. *J Neurosci* 18(10):3650-3658.

Hardie, SL, Zhang JX, Hirsh J. 2007. Trace amines differentially regulate adult locomotoractivity, cocaine sensitivity, and female fertility in *Drosophila melanogaster*. DevNeurobiol 67(10):1396-1405.

Harvey R.J, Schmitt, B, Hermans-Borgmeyer I, Gundelfinger, E.D, Betz H, Darlison MG.(1994). Sequence of a *Drosophila* ligand-gated ion-channel polypeptide with an unusual amino-terminal extracellular domain. J Neurochem 62(6):2480-2483.

Homberg, U. Christensen, T.A. and J.G. Hildebrand (1989). The structure and function of the deuterocerebrum in insects. *Annu. Rev. Entomol.* 34: 477-501.

Howse, P.E. (1975). Brain structure and behaviour in insects. *Annu. Rev. Entomol.* 20: 359-379.

Jan, L.Y, and Jan, Y.N. (1976). L-glutamate as an excitatory transmitter at the *Drosophila* larval neuromuscular junction. *J Physiol* 262(1):215-236.

Johnson O, Becnel J, Nichols, C.D. (2009). Serotonin 5-HT(2) and 5-HT(1A)-like receptors differentially modulate aggressive behaviors in *Drosophila melanogaster*. Neuroscience 158(4):1292-1300.

Kolodziejczyk A, Sun X, Meinertzhagen, I.A, and Nässel, D.R. (2008). Glutamate, GABA and acetylcholine signaling components in the lamina of the *Drosophila* visual system. *PLoS One* 3(5):e2110

Littleton, J.T., and Ganetzky, B. (2000). Ion channels and synaptic organization: analysis of the *Drosophila* genome. *Neuron* 26(1):35-43.

Liu. X., Krause, W.C, and Davis, R.L. (2007). GABAA Receptor RDL Inhibits *Drosophila* Olfactory Associative Learning. Neuron 56(6):1090-1102.

Liu X, Buchanan M.E, Han K.A, and Davis, R.L. (2009). The GABAA receptor RDL suppresses the conditioned stimulus pathway for olfactory learning. *J Neurosci* 29(5):1573-1579.

Lodish H, Berk A, Zipursky SL, Matsudaira P, Baltimore D, and Darnell, J.E. (1999). Molecular Cell Biology

Mahr A, and Aberle H. (2006). The expression pattern of the *Drosophila* vesicular glutamatetransporter: a marker protein for motoneurons and glutamatergic centers in the brain. *Gene Expr Patterns* 6(3):299-309.

Mao Z, and Davis, R.L. (2009). Eight different types of dopaminergic neurons innervate the *Drosophila* mushroom body neuropil: anatomical and physiological heterogeneity. *Front Neural Circuits* 3:5.

Monastirioti M. (1999). Biogenic amine systems in the fruit fly *Drosophila melanogaster*. *Microscopy Research and Technique* 45(2):106-121.

Nässel D. (2009). Neuropeptide signaling near and far: how localized and timed is the action of neuropeptides in brain circuits? *Invertebrate Neuroscience* 9(2):57-75.

Nässel DR, Elekes K. (1992). Aminergic neurons in the brain of blowflies and *Drosophila*:dopamine- and tyrosine hydroxylase-immunoreactive neurons and their relationship with putative histaminergic neurons. *Cell Tissue Res* 267(1):147-167.

Nichols, C.D. (2007). 5-HT2 receptors in *Drosophila* are expressed in the brain and modulate aspects of circadian behaviors. Dev Neurobiol 67(6):752-763.

Okada R, Awasaki T, Ito K. (2009). Gamma-aminobutyric acid (GABA)-mediated neuralconnections in the *Drosophila* antennal lobe. *J Comp Neurol* 514(1):74-91.

O'Shea M, Schaffer M. 1985. Neuropeptide function: the invertebrate contribution. *Annu Rev Neurosci* 8:171-198.

Parmentier M.L, Pin J.P, Bockaert J, and Grau Y. (1996). Cloning and functional expression of a *Drosophila* metabotropic glutamate receptor expressed in the embryonic CNS. *J Neurosci* 16(21):6687-6694.

Qin G, Schwarz T, Kittel R.J, Schmid A, Rasse T.M, Kappei D, Ponimaskin E, Heckmann M, and Sigrist SJ. (2005). Four different subunits are essential for expressing the synaptic glutamate receptor at neuromuscular junctions of *Drosophila*. *J Neurosci* 25(12):3209-3218.

Saraswati S, Fox L.E, Soll D.R, Wu C.F. (2004). Tyramine and octopamine have opposite effects on the locomotion of *Drosophila* larvae. *J Neurobiol* 58(4):425-441.

Sattelle D.B, Jones A.K, Sattelle B.M, Matsuda K, Reenan R, and Biggin P.C. (2005). Edit, cut and paste in the nicotinic acetylcholine receptor gene family of *Drosophila melanogaster*. *Bioessays* 27(4):366-376.

Saudou F, Amlaiky N, Plassat J.L, Borrelli E, Hen R. 1990. Cloning and characterization of a *Drosophila* tyramine receptor. *EMBO J* 9(11):3611-3617.

Shapiro R.A, Wakimoto B.T, Subers E.M, and Nathanson, N.M. (1989). Characterization and functional expression in mammalian cells of genomic and cDNA clones encoding a *Drosophila* muscarinic acetylcholine receptor. *Proc Natl Acad Sci U S A* 86(22):9039-9043.

Sinakevitch I, and Strausfeld, N.J. (2006). Comparison of octopamine-like immunoreactivity in the brains of the fruit fly and blow fly. *J Comp Neurol* 494(3):460-475.

Sitaraman D, Zars M, Laferriere H, Chen Y.C, Sable-Smith A, Kitamoto T, Rottinghaus G.E, and Zars T. (2008). Serotonin is necessary for place memory in *Drosophila*. *Proc Natl Acad Sci U S A* 105(14):5579-5584

Srivastava D.P, Yu E.J, Kennedy K, Chatwin H, Reale V, Hamon M, Smith T, Evans P.D. (2005). Rapid, nongenomic responses to ecdysteroids and catecholamines mediated by a novel *Drosophila* G-protein-coupled receptor. *J Neurosci* 25(26):6145-6155.

Tierney, A.J. (2001). Structure and function of invertebrate 5-HT receptors: a review. *Comp Biochem Physiol A Mol Integr Physiol* 128(4):791-804.

Yasuyama K, and Salvaterra P,M. (1999). Localization of choline acetyltransferase-expressing neurons in *Drosophila* nervous system. *Microsc Res Tech* 45(2):65-79.

Yuan Q, Joiner W.J, Sehgal A. (2006). A sleep-promoting role for the *Drosophila* serotonin receptor 1A. *Curr Biol* 16(11):1051-1062.

Physiology of Insect Reproduction 8

Synopsis

1. The male reproductive system
   - 1.1 Structure of testis follicle
   - 1.2 Accessory glands of the male tract
   - 1.3 Structure of insect spermatozoa
   - 1.4 Production of spermatozoa
   - 1.5 Spermatophores
   - 1.6 Seminal fluid
   - 1.7 The Male external genitalia
   - 1.8 Male courtship behaviour and insemination
   - 1.9 Endocrine control of the male reproductive system and spermatogenesis

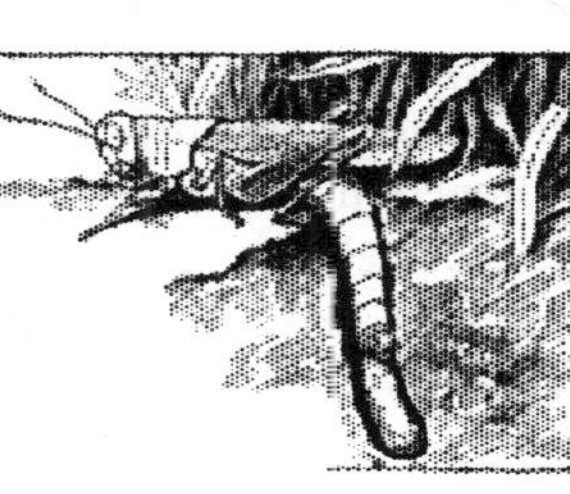

2. The female reproductive system
   - 2.1 Structure of ovariole
   - 2.2 The spermatheca
   - 2.3 Other accessory glands
   - 2.4 The female external genitalia
3. Oogenesis
   - 3.1 Formation of oocytes and nurse cells
   - 3.2 The follicle cells
   - 3.3 Alimentary aspects of oogenesis
   - 3.4 Vitellogenesis
   - 3.5 Formation of egg membranes and specialised chorionic structures
   - 3.6 Oosorption
   - 3.7 Development of ovarian structure in Drosophila
4. Oviposition
5. Structure and physiology of eggs and eggshells in insects
   - 5.1 Eggshell layers
   - 5.2 Physiological function of eggshell
6. Modes of Reproduction
   - 6.1 Oviparity
   - 6.2 Viviparity
   - 6.3 Parthenogenesis
   - 6.4 Paedogenesis and Neoteny
   - 6.5 Polyembryony
   - 6.6 Hermophroditism
   - 6.7 Castration
   - 6.8 Alternation of generation (Heterogony)
   - 6.9 Sex determination
7. Endocrinology of female reproduction
8. Physiological and reproductive effects of Wolbachia

Parasitoid wasp laying eggs in an insect

INTRODUCTION

Reproduction is a function of the organism in the maintenance of the species. It involves the re-shuffling and storage of developmental information in such a way that natural selection can play its role in fitting the organism into its environment. Several environmental factors play a role in the regulation of reproduction. Food, temperature, humidity and photoperiod are important besides social conditions. Several internal factors such as nutritional state, activity of fat bodies and the state of development of oocytes are also important in the regulation of reproduction. The evolutionary success of insects is mainly due to their remarkable reproductive capacity. Most insects are bisexual; however there are species that are hermaphrodites, gynandromorphs or intersexes. The most complex of the situation is the alternation between a one sex generation and a bisexual generation as in aphids.

The reproductive system is divided into two parts namely internal genitalia and external genitalia. The internal genitalia serve to the development of germ cells and external genitalia accomplish the union of two sexes and enable the female to deposit eggs.

1. THE MALE REPRODUCTIVE SYSTEM

The internal male reproductive organ consists of a pair of testis, a pair of vas deferens, seminal vesicle, ejaculatory duct, accessory glands and genitalia. This system is mainly concerned with the manufacturing of spermatozoa and delivering them to the female. It also provides them with an appropriate environment in the form of seminal fluid to nourish them until they reach the spermatheca of the female. The reproductive system also delivers to the female certain physical or chemical stimuli which signal the neuroendocrine system of the female that mating has occurred.

Testis

The testes lie in visceral cavity above the alimentary canal and are connected to the body wall through translucent ducts. They are well supplied with trachea and fat body tissues. Each testis consists of number of oval shaped structures known as **follicles** or **testicular tubes**. Each follicle has a layer of epithelial cells. The entire follicle is covered by a peritoneal membrane where as the testis is completely enveloped within a coat known as **scrotum.** Each testicular tubule opens *via* a vas efferens into the mesodermal sperm duct, the vas deferens. The paired vasa deferentia unite to connect with the ectodermally derived ejaculatory duct which terminates in the gonopore (Fig 8.1). Only in the Ephemeroptera are the original paired gonopores still present. The vas deferens is usually expanded over part of its length to form a seminal vesicle or sperm storage organ. In most insects, the accessory glands are formed as diverticula from the vas deferens. When the mesodermally derived vas deferens is itself glandular they are called **mesadenia**. The mesadenia commonly supply the secretions which form the spermatophore. Similarly there may be glandular diverticula of the ejaculatory duct as in *Drosophila* or *Rhodnius*. Such ectodermally derived glands are called **ectadenia**.

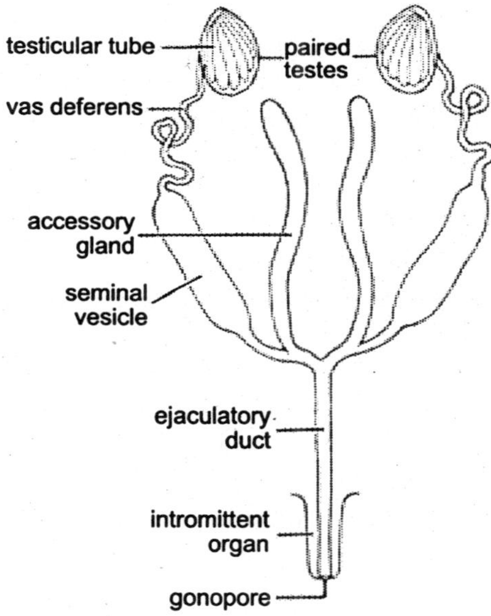

Fig. 8.1 Reproductive system of a typical male insect

Vas deferens

These are the long tubes formed by the union of vas efferens which receives the sperms from testis and allow their transport to the ejaculatory duct.

Seminal vesicles

Each vas deferens becomes enlarged posteriorly to form a sac like structure called **seminal vesicle** for storage of spermatozoa for some time.

Ejaculatory duct

The vas deferens of the two testes unites posteriorly to form a common median ejaculatory duct. The terminal section of ejaculatory duct is enclosed in a finger like evagination of body wall and represents the male copulatory organ or aedeagus or penis.

Accessory glands

These are 1-3 pairs of glands which open in to the ejaculatory duct. In most cases their secretion mix with spermatozoa. These glands are called **mushroom glands** in cockroaches and mantids because of their appearance as mushrooms. Their secretions facilitate sperm transmission from male to female.

1.1 Structure of testis follicle

Each individual follicle is divided into a series of zones (or) areas characterised by the presence of the sex cells i.e. sperms in different stages of development (Fig 8. 2). These zones are zone of germarium, zone of growth, zone of division and reduction and zone of transformation.

1. **Germarium:** It is the region containing primordial germ cells or spermatogonia which undergo multiplication (Zone of spermatogonia).
2. **Zone of growth**: It is the area where spermatogonia increase in size, undergo repeated mitosis and develop in to spermatocytes (Zone of spermatocytes).
3. **Zone of division and reduction**: It is the area where spermatocytes undergo meiosis and give rise to spermatids (Zone of spermatids).
4. **Zone of transformation**: It is the area where spermatids get transformed in to spermatozoa (Zone of spermatozoa).

The follicle number varies considerably; there is only 1 follicle in some coleopterans and over 100 in some orthopterans. The follicle is formed when a stem cell divides mitotically at the anterior region of the follicle to produce a gonialblast and is surrounded by cyst cells in the stem cell niche (Fig 8.2). Mitotic division of the gonialblast results in spermatogonia that are enclosed by the somatic tissue that form cysts. The development of the spermatozoa occurs within these cysts, which elongate as development proceeds. As more cysts are synthesised, they move the more mature ones posteriorly so that a range of cysts that contains gametes in progressive developmental stages is produced.

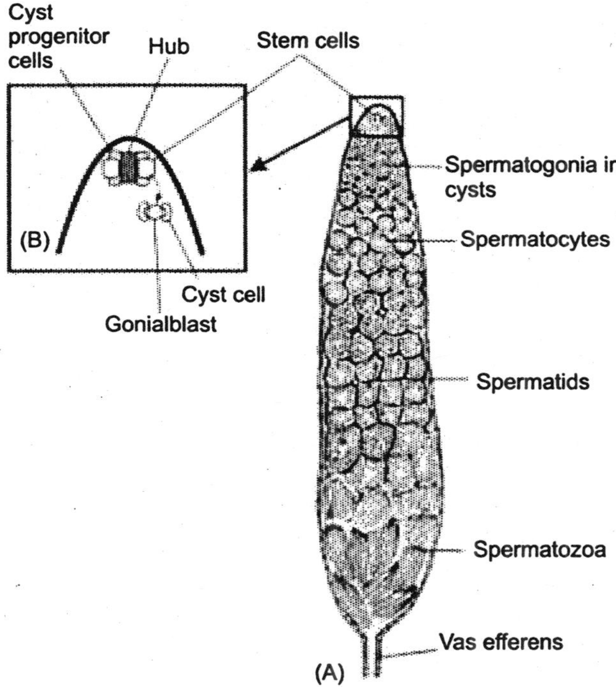

Fig. 8.2 (A) Longitudinal section of testis follice. (B)The stem cell niche at the anterior of the follicle

1.2 Accessory glands of the male tract

In most insects the accessory glands are **mesadenes** and only a small number of **ectadenes**, or ectodermally derived glands have been described. Those arising from the vas deferens during development are mesodermal in origin, whereas in some insects in which they originate from the ejaculatory duct they have an ectodermal origin. The male accessory glands serve a variety of functions, including producing the seminal fluid that serves as a transport, activation, and nourishing medium for sperm; forms a vaginal mating plug that temporarily blocks sperm from another male from entering; and forms spermatophores, the proteinaceous secretions of the glands that enclose the sperm and protect them while they are being transferred to the female.

The male accessory glands also produce substances that affect the behaviour and physiology of the female (Box 8.1). During insemination, peptides produced by the accessory glands that are contained in the semen are transported to various sites within the female. They are frequently involved in triggering her reproductive processes, including oogenesis and oviposition, or terminating subsequent female mating behaviour so as to prevent her from acquiring the sperm of another male. Male accessory gland substances are frequently responsible for monogamy in females.

Box 8.1 Male accessory glands – influence of secretions on females

1. Stimulate muscle contraction of genital tract to facilitate sperm movement
2. Stimulate oviposition
3. Accelarate oocyte maturation
4. Inhibit subsequent mating by forming vaginal plugs or effecting behaviour

1.3 Structure of Insect Spermatozoa

Insect spermatozoa have elongated heads and especially long tails (Fig 8.3). Within the head are the haploid nucleus and the acrosomal complex at the tip that arises from the Golgi apparatus during differentiation. The acrosome contains the enzyme acrosin, a trypsin-like enzyme and is associated with the penetration of the plasma membrane of the egg for fertilisation. Insect sperm are unique in that they are the only cells of these animals that bear a flagellum. Within the flagellum is an axial filament that develops from the centriole and consists of an array of microtubules in a 9+9+2 arrangement, with nine outer accessory tubules, nine inner doublets, and a pair of central tubules comprised of the dimeric protein tubulin. One member of each inner doublet also bears the protein dynein, which breaks down ATP and converts the energy released into flagellar movement.

The tail length of flagella may be variable even within a species; in *D. melanogaster*, the sperm range in length from 1.6 to 2 mm, but *D. bifurca* males produce sperm with a total length of over 58 mm, the longest described for any animal and many times the length of the adult fly that produces them. Males of all species within the *Drosophila obscura* group produce at least two different sperm lengths, and four different sperm lengths ranging from 69 to 430 μm are produced by *Drosophila subobscura*. Individual males of some anopheline mosquito species produce polymorphic sperm that vary widely in length from 50 to over 500 μm. *Drosophila* that produces longer sperm tends to also take longer to become reproductively mature.

Lepidopteran males are unusual as they produce two very different types of sperm in their fused testes, namely the *eupyrene sperm* and the *apyrene sperm*. Both types are transferred to the spermatheca of the female, but only the conventional eupyrene sperm are involved in normal fertilization of the oocyte. The other apyrene sperm have no nuclei, losing them at a late stage of spermatogenesis, and thus cannot have any genetic function. Actin filaments on the sperm bundles produce peristaltic contractions that squeeze both types out of their cysts. Eupyrene sperm differentiate during the larval stage, but apyrene do so later after the larvae initiate spinning. In the absence of any genetic role, the function of apyrene sperm may be to assist in moving the eupyrene sperm through the female reproductive tract or provide them with nutrients. Apyrene spermatozoa are generally shorter than eupyrene sperm, are initially more motile, and are present singly in the seminal vesicles when the eupyrene sperm are still in bundles, gaining their motility when they are placed into the spermatophore. As many as 70% to 90% of the spermatozoa transferred may be apyrene, and they may displace the eupyrene sperm that are already present in the spermatheca from the previous matings of the female. Both types are necessary for fertilization to occur; artificial insemination of female *Bombyx mori* with only eupyrene sperm fails to fertilise eggs unless apyrene sperm are added.

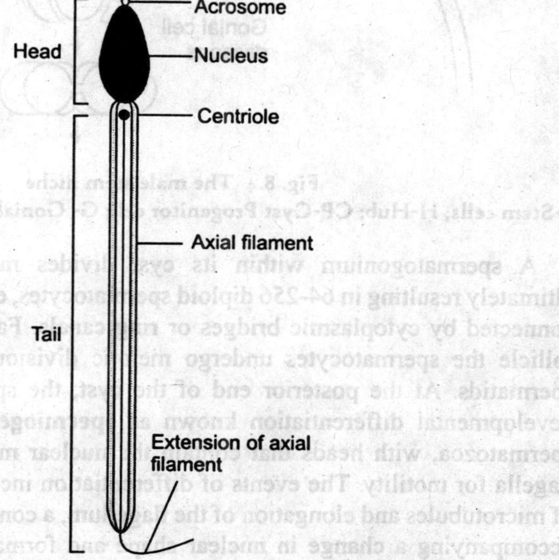

Fig. 8.3 A generalised insect Spermatozoa

1.4 Production of Spermatozoa (Spermatogenesis)

The male gametes are produced in the follicles of the paired testes. The physiology of production of spermatozoa has been thoroughly investigated in *Drosophila*. The development of the testes during metamorphosis is under the control of the insect hormone 20-hydroxyecdysone (20HE). In the reproductively mature male, apical stem cells at the anterior end of the testes are surrounded by cyst progenitor cells and specialised non-dividing somatic cells, known the **hub**, that together form the stem cell niche (Fig. 8.2B and Fig 4). Signalling from the niche regulates the renewal, maintenance, and survival of the stem cells. The stem cells surrounding the hub divide mitotically to produce a gonial blast. The cyst progenitor cells give rise to the cyst cells. These cyst cells enclose the gonial blast to form the follicle (Fig 8.4). The development and maturation of spermatozoa occurs within these cysts.

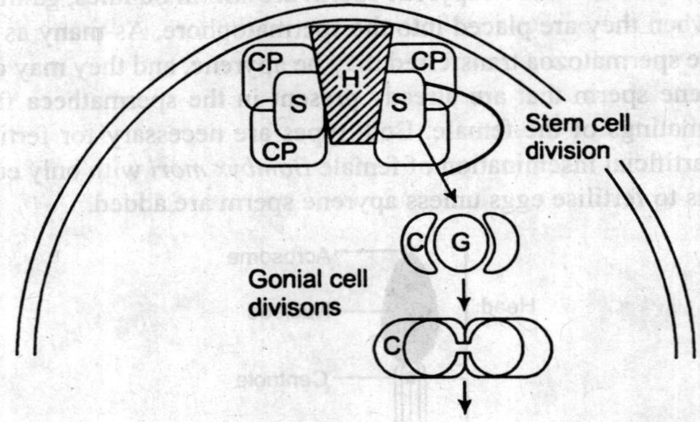

Fig. 8.4 The male stem niche

S-Stem cells; H-Hub; CP-Cyst Progenitor cell; G- Gonialblast; C- New cyst cell

A spermatogonium within its cyst divides mitotically 6-8 times, ultimately resulting in 64-256 diploid spermatocytes, each of which remains connected by cytoplasmic bridges or ring canals. Farther down the testis follicle the spermatocytes undergo meiotic division to become haploid spermatids. At the posterior end of the cyst, the spermatids complete a developmental differentiation known as **spermiogenesis** to become the spermatozoa, with heads that contain the nuclear material and elongated flagella for motility. The events of differentiation include the development of microtubules and elongation of the flagellum, a condensation of the DNA accompanying a change in nuclear shape and formation of a specialized mitochondrial derived structure, the *Nebenkern*, that is also associated with the flagellum. The acrosomal complex develops at the sperm head as a skeletal glycoprotein associated with the enzymes acrosin and hyaluronidase. Finally, the spermatozoa break through the cyst and are released into the vas efferens. These spermatozoa move to the seminal vesicles, affected by the contraction of the walls of the reproductive tract, where they remain as a reservoir of sperm until the insect copulates with a female. In Lepidoptera, this migration occurs before adult emergence and depends on a circadian clock located in the reproductive tissues that produces a two-step release, first to the vas deferens and then to the seminal vesicles. The circadian clock gene, *period*, is expressed in the epithelium of the vas deferens. In the gypsy moth, *Lymantria dispar*, the sperm descend from the testes into the reproductive tract 4 days before the adult encloses.

1.4.1 Spermatogenesis in Postembryonic Development

In most cases, spermatogenesis is completed at the time of the adult moult. In the early days of adult life, the spermatozoa migrate to the vesiculae séminales in full number and the testes gradually atrophy. In species in which copulations are frequently repeated throughout adult life and in which the males are even active in successive years, e.g., in many Chrysomelids, the testes remain active until the end of the adult life span.

In *Pediculus corporis,* throughout the first larval instar, the testes merely contain spermatogonia. In the second instar, the formation of spermatocytes begins. At the end of the third instar, all stages of spermatogenesis are present in the follicle.

In *Melanoplus differentialis* (Acridiidae), secondary spermatogonia are formed in the beginning of the second larval instar. Gonial divisions take place throughout the second, third, and fourth instars. Maturation divisions begin in the sixth (last) instar, and spermatids complete their metamorphosis shortly after the adult moult.

In the honey bee drone, spermatid formation is completed during the pupal stage. Sperm differentiation begins 4 days before emergence. Sperm begins to descend to the vesiculae séminales 2 days before emergence, and completes migration during the first 3 days of adult life. By this time, the testes start degenerating.

1.5 Spermatophores

Spermatophores are structures produced by the male which more or less enclose the semen. They provide an appropriate medium for the semen while it is transferred to the female. Structurally they vary widely in form. Among the lower orders they are usually flask-shaped, with the bulb of the flask partially or wholly protruding from the terminalia of the female and the neck penetrating to a greater or lesser distance into the duct of the female.

In lower neopteroid orders the spermatophore consists of a membranous sac enclosing the spermatozoa embedded in a gelatinous matrix which may in turn be enclosed in a fibrous or membranous covering (Fig. 8.5). In more advanced insects, the spermatophore is usually reduced. In *Rhodnius* it is pear-shaped. Since the spermatophore is formed within the male tract its shape is determined by the ducts within which it is formed.

In many groups such as crickets, mantids and phasmids, the prominent bulb of the spermatophore protrudes from the genitalia and the neck penetrates into the female ducts. The spermatophore is thus exposed and may be eaten by the female after copulation. In these insects the spermatophore represents considerable investment on the part of the male and in crickets it accounts for as much as 40% of the weight of the male. The eating of the spermatophore by the female is thus a means of ensuring that the investment is not lost to the species.

In acridids, the gelatinous portion of the spermatophore has disappeared and only the membranous sperm sac remains. It is reduced to a long tubular structure which in some species may function as an extension of the intromittent organ. In Heteroptera, on the other hand, the sperm sac is lost, and only the gelatinous matrix is retained to function as a sperm plug to hold the semen in place while contractions of the muscles of the female tract transport semen to the spermatheca.

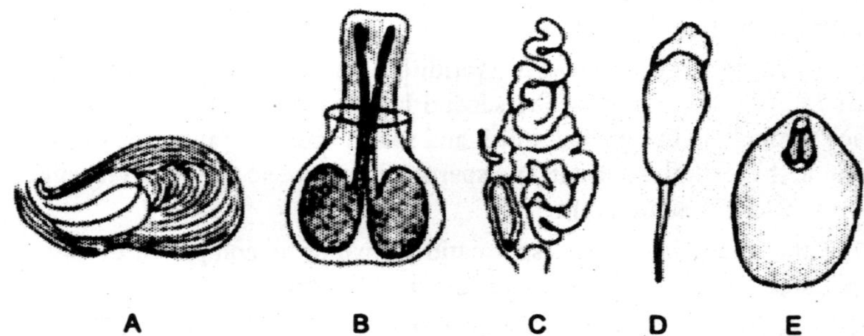

Fig. 8.5 Different types of spermatophores
A, *Blatella germanica;* B, *Sialis lutaria* (Neuroptera); C, *Anabolia nervosa* (Trichoptera); D, *Galleria mellonella* (Lepidoptera); E, *Pimpla instigator* (Hymenoptera) (after Khalifa, 1949, 1950)

1.6 Seminal fluid

Honey bees transfer relatively large amounts of semen. An individual ejaculate amount to a little more than 1 μl, with seminal plasma accounting for about 50% of the volume. The pH of ejaculated semen is 6.8-7.0. The spermatozoa in insects occurs in three places: the seminal vesicles, where they are stored in an inactive state for a long time; in the spermatophore where they are mixed with other constituents of the semen; and in the spermatheca of the female where they are nourished by the secretions of the spermathecal gland. The composition of semen in these three places is relatively less known. The seminal vesicle does not appear to produce any secretion. The seminal fluid here in honey bees has glucose, fructose and trehalose and it is thought that they enter by diffusion from the haemolymph. The accessory glands contribute to the seminal fluid. In *Drosophila* the ejactulatory bulb is a source of lipid, primarily an acetate ester of *cis*-11-octadecanoic acid. A non-specific carboxylesterase known as **esterase 6** has also been identified in the anterior ejaculatory duct of *Drosophila*. Relatively high levels of cyclic GMP have been observed in the accessory glands of crickets suggesting that nucleotide is transferred during copulation.

Insect semen may also contain a wide variety of materials which have a function in the female related to events such as semen transport, egg development, ovulation and oviposition. Among these chemical signals are prostaglandins which are also transferred in mammalian semen.

1.7 The Male external genitalia

The external genitalia of a male are generally formed from the ninth sternite from an originally unpaired invagination, the lobi phallici. During further development, this anläge becomes paired and gives rise to paired mesomeri and parameri. The mesomeri generally fuse to form an unpaired intromittent organ, or phallus, which may bear an aedeagus; however, depending on the species, the parameres become phallomeres, claspers, or harpagones which are functional during copula. Parameres are considered by some investigators to be homologous to abdominal appendages similar to the gonapophyses of the female. Some recent studies on the musculature of the genitalia in a few species, however, do not support this view; these muscles are of sternal origin and are homologous to those muscles of the pregenital segments; if the external genitalia were appendages, they would have no metameric muscles.

The primary function of the male genitalia in insects is insemination of the female. Methods of achieving insemination that involve special functions of the external genitalia include clasping and holding the female, retaining the connection with the female gonopore, the construction of spermatophores, and the deposition of spermatophores or semen into the female genital tract. In some insects the injection of semen takes place directly into the female body (traumatic insemination of some Hemiptera). Other functions of the male genitalia include excretion and various sensory functions.

The genitalia of male insects exhibit such an enormous variety of shapes and constituent parts, often further complicated by structural rotation or inversion of all or some of the parts, that determination of a ground plan is virtually impossible. Examination of ancient orders shows highly variable and specialised conditions. In general, the coxites of the eighth segment in most apterygotes are reduced and without gonapophyses, and they are absent altogether in the Pterygota. Thus, the male external genitalia are derived from the ninth abdominal coxites.

Again, the Thysanura have genitalia that closely resemble that of the pterygote orders: a median intromittent organ or phallus, and paired lateral accessories (the periphallus of Snodgrass). The phallus is a conical, tubular structure of variable complexity (Fig 8.6). Primitive insects may not display differentiated parts, and the entire structure may be long, sclerotised, and tapering apicad. In a ground plan condition for pterygote insects, there is a sclerotised basal portion termed the phallobase and a distal sclerotised portion called the aedeagus (Fig 8.6). The phallobase in insects is characterised by highly variable development: sometimes sclerotised and supporting the aedeagus, sometimes forming a sheath for the aedeagus. The phallobase often contains an apodeme, which may provide support or a point for muscle attachment. The phallobase and aedeagus are joined by a membranous phallotheca (Fig 8.6). The external walls of the phallobase and aedeagus are called the **ectophallus**. The gonopore is positioned at the apex of the ejaculatory duct and is concealed within the phallobase. The gonopore is connected to the apex of the aedeagus via a membranous tube called the **endophallus**. In some insects the endophallus may be everted through the aedeagus. The circular aperture at the apex of the aedeagus is called the **phallotreme**. In some insects the endophallus and the gonopore may be everted through the phallotreme and into the female's bursa copulatrix. Genital lobes referred to as **phallomeres** form at the sides of the gonopore in the ontogeny of some insects. Usually the phallomeres unite to form the phallus.

The external genitalia of males show an extreme morphological divergence. Indeed, their shapes often serve as the only reliable basis for species identification in taxonomic keys. In species in which the females mate often, there may be selective pressure for males to develop genitalia that compete for more effective ways to inseminate the females. There is a selective pressure for the evolution of male genitalia that offer males the greatest success; males with genitalia that more effectively transfer sperm are more likely to have their genetic contributions remain in the gene pool. Because copulation may last for a long period, males must have a way of avoiding detachment by other males that are also attempting to mate. Male insects have thus evolved a diverse array of grasping structures that allow the male to remain attached to the female; these devices include claspers, spines on the aedeagus, and even the use of mandibles to grasp a portion of the female's body. Structures have also evolved for the removal of existing sperm from a prior mating. For example, the aedeagus of the damselfly male scoops out the sperm present in the female before he adds his own. He thus eliminates any competition among sperm that may already be present in the female. Some male bedbugs inseminate the female by piercing her cuticle and depositing sperm directly into the body cavity.

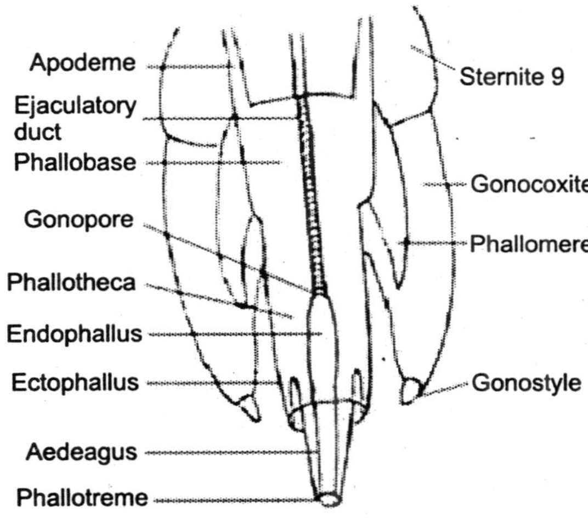

Fig. 8.6 Diagrammatic sketch of male genitalia

1.8 Male courtship behaviour and insemination

Male reproductive behaviour typically involves first searching for a female and, once she is found, precopulatory behaviours follow that increase her receptivity for copulation. This may include the presentation of a nuptial gift that occupies the female and increases her nutritional state. Once copulation is initiated, the actual transfer of sperm may require from a few minutes to a few hours. To protect their genetic investment, some males remain with the female for a variable interval after insemination has taken place. Male termites guard their mates for life. Some male scarab beetles gather dung for the female to lay her eggs on or help her to build a burrow and even remain inside the burrow with her for months.

Male paternal investment in offspring is highly expressed in the giant water bugs of the family Belostomatidae. Females lay their eggs on the backs of partially submerged males, interrupted only to mate repeatedly with the male. The males carry the eggs on their backs until the young hatch, actively maintaining their own position in the water in order to properly aerate the eggs.

1.8.1 Role of pheromones

In long-distance attraction between males and females, as well as in aggregation of individuals of both sexes, chemical messengers is often involved namely **sex pheromones** and **aggregation pheromones**. Pheromones are often secreted in mixtures of two or more compounds. One compound might possibly be involved in the orientation of the insect, another compound in courtship behaviour, synchronising the partners physiologically. Sometimes, as in the case of *Bombyx mori,* one single compound: 10-trans,12-cis-hexadeca-dienol, secreted by the females, apparently triggers both orientation and mating behaviour in the male. In Lepidoptera, sex pheromones are often organic acids and their esters, saturated or unsaturated alcohols, with 12-16 C atoms. Most frequent are acetates and C12.i4 unsaturated alcohols. In Tortricidae, the combination between *9-cis*- and 1 l-trans-tetradecenyl acetate is known to occur.

Male pheromones which have a role in the interplay of stimuli and responses involved in courtship are probably found in many orders, but are especially known to occur in the Rhopalocera. In *Eumenis semele,* the male acquiesces the female by bringing two brushlike structures on the forewing in contact with her antennae. In *Lycorea,* **hair pencils,** situated on the male abdomen, are brushed against the female antennae. A similar use of hair pencils is reported from *Danais plexippus.*

1.8.2 Insemination

The accessory structures promote the attachment of the partners during copulation and the deposition of sperm at the critical sites in the female gonoducts. During insemination, the sperm is deposited in the bursa copulatrix, and sometimes in the common oviduct and even the lateral oviducts (Apis) (Box 8.2). In extreme cases, as in several Hemiptera-Heteroptera, the sperm is placed directly in the female spermatheca.

Box 8.2 Most Spectacular Mating

Apis mellifera is judged to have the most spectacular mating because a **comet** of drones pursues the female with the winner forfeiting a portion of his phallus at the end of coitus and dying soon thereafter (Woyke and Tuttner 1958; Winston 1987)

Erection of the phallus may be brought about by special hydraulic mechanisms. In the milkweed bug *Oncopeltus fasciatus,* the ejaculatory duct is a blind tube folded back in itself to form a crosier-shaped organ. The anterior end of the duct is modified into an erection fluid reservoir and its posterior end into an erection fluid pump provided with dilator muscles. Erection of the phallus is accomplished by the displacement of the fluid into the phallus by internal compression of the abdomen and by pulsations of the erection fluid pump. As erection proceeds during copulation, the aedeagus gradually penetrates into the spermathecal duct and spermatheca. The same mechanism probably holds for *Lygaeus* (Bonhag and Wick, 1953). In the honey bee drone, the phallus takes an inverted position in the male bursa genitalis. The innermost part is enlarged into a bulbus. The basic part forms a thin walled chamber from which two large membranous pneumophyses project. Contraction of the abdominal wall results in eversion of the endophallus together with its lobes. As this takes place while the phallus is being introduced, the pneumophyses attach themselves to the wall of the female vaginal pouch. The semen is released from the top of the bulb after it fills the vaginal pouch, and is immediately followed by the mucus, which in its turn is pushed forward by an epithelial plug torn off the mucous glands. The abdominal wall contracts with such vigour that the mucous glands, the seminal vesicles, and even a large part of the intestine slide into the endophallus (Woyke, 1955). The pressure thus exerted causes the sperm to fill the median and lateral oviducts, and, as multiple matings take place, the latter may be distended considerably. In the Lepidoptera (Norris, 1932) contractions of the bursa copulatrix, peristaltic movements of the seminal ducts, and compression of the abdomen have been thought to aid in the filling of the spermatheca. In *Rhodnius,* the seminal fluid associated with the spermatophore, deposited in the bursa causes peristaltic contraction in the oviduct (Davey, 1958). In the honey bee queen, the "sperm pump" of the spermathecal duct may either act as a sphincter muscle, or may aid in sperm transport and, thereby, in the mechanical filling of the spermatheca. It is known that the vaginal pouch of the honey bee queen is provided with a **valve fold**. When the proximal part of the vaginal pounch and the oviducts are filled with semen, compression of the abdominal wall causes the valve to close and to occlude and compress the genital chamber.

1.8.3 Traumatic insemination

In four families of Cimicoidea, i.e., Cimicidae, Polyctenidae, Anthocoridae, and Nabidae, peculiar conditions of mating and insemination are found. Courtship behaviour is very rudimentary; larvae, males and females, even of other species, may be subject to copulation attempts. Traumatic insemination is observed wherein sperm is injected outside the female genital apparatus, and often in the haemocoel. In the resulting condition of spermathemia, the spermatozoids find their way to the ovarioles which are penetrated from the outside. In *Alloeorhynchus, Prostemma,* and *Pagasa* (Nabidae), injection takes place through the female genital wall, which may be provided with a diverticulum in which sperm is received (Spermalege) (Carayon, 1952a,b). In the primitive *Primicimex cavernius,* and in some Anthocoridae, the abdominal integument is perforated in predestined regions. In some Anthocoridae, the underlying region is provided with a mesodermal mesospermalege (Carayon, 1964) composed of prohemocytes. In Cimicidae, an ectodermal spermalege is formed at the adult moult, and situated at the dorsal or ventral side of the abdomen, according to the species. This ectospermalege is developed in both males and females and homosexual inseminations apparently occur, e.g., in *Afrocimex* (Carayon, 1959). Only in females a mesospermalege is formed underlying the ectospermalege. The resulting complex spermalege varies greatly between species. In *Cimex lectularius,* sperm is injected through the ectospermalege into the mesospermalege, the cells of which phagocyte the seminal fluid and part of the spermatozoids. The common oviduct is provided with two seminal conceptacles, in which the remaining sperm is collected within a few hours. It subsequently ascends into the ovarioles. In many Cimicidae, a strand of tissue develops between the mesospermalege and the ovary and the spermatozoids, into which the spermatozoa take their course to the ovary (Carayon, 1970). In Polyctenidae, females may be fertilised when still in the last larval instar (Carayon, 1964).

1.9 Endocrine Control of the Male Reproductive System and Spermatogenesis

Relatively little is known about the endocrine regulation of male reproduction compared to what is known about the female. In short-lived insects that may not feed as adults, spermatogenesis occurs early during the larval and pupal stages. In longer-lived males, spermatogenesis continues throughout adult life. With the vastly different hormonal conditions that exist during the immature and adult periods, a unifying scheme for the control of spermatogenesis has not been possible. Indeed, it has been suggested that insect spermatogenesis may simply be a sequential process of differentiation that is completely independent of hormones. The testes of several lepidopterans produce ecdysteroids at the end of the last larval instar and during pupal development and also contain ecdysteroid receptors that regulate the transcription of target genes. The testes are induced to synthesize the ecdysteroids when stimulated by an ecdysiotropin produced by the medial neurosecretory cells of the brain. This ecdysiotropin is different than PTTH and fails to act on the prothoracic glands; its activity appears to be limited to the testes. A testis ecdysiotropin is also produced by the brain, subesophageal ganglion, and the testes sheath of male *Rhodnius*.

Both 20HE and juvenile hormone (JH) have been implicated in the regulation of spermatozoa within the testes. In the absence of the prothoracic glands of the adult male, the 20HE is produced by the testes themselves in response to an ecdysiotropin produced by the neurosecretory cells of the brain. High levels of 20HE increase the rate at which the spermatogonia undergo mitotic divisions to form spermatocytes, but this increase is abolished by high levels of JH.

When spermatogenesis begins during the larval period, as in some Lepidoptera, the spermatocytes initiate their meiotic divisions but are arrested at prophase until the end of the larval period is reached. In many holometabolous insects, a peak of 20HE occurs before the end of the larval period that induces the wandering behaviour that allows the larva to find a secluded spot in which to pupate. A post wandering peak of 20HE unblocks the meiotic division of the spermatocytes and allows the cells to proceed to metaphase. In some insects, JH has also been found to accelerate spermatogenesis. The release of mature spermatozoa from the cysts has a circadian rhythmicity in some insects that is inhibited by 20HE and occurs when levels of 20HE decline.

The developmental sequence in which spermatozoa develop from spermatids involves an elongation of both the nucleus and the flagellum. In Lepidoptera, the nuclear elongation is triggered by the declining concentrations of 20HE, but the elongation of the flagellum appears to be independent of any hormone.

Spermatogenesis is interrupted in those lepidopterans that undergo a larval or pupal diapause but resumes once diapause has been completed. It is not any developmental activity that causes this interruption but rather the lysis of developing gametes before they become mature. The renewal of spermatogenesis occurs with increasing titers of 20-hydroxyecdysone that occur when diapause is terminated. In lepidopterans, the differentiation of apyrene sperm from eupyrene sperm occurs with the exposure to a haemolymph-borne apyrene spermatogenesis- inducing factor.

Male accessory gland function is regulated by several hormones, but JH has the most prominent role in stimulating protein synthesis in the gland. Stimulation of protein synthesis has also been reported by 20-hydroxyecdysone in some lepidopterans and by brain neuropeptides in *Rhodnius*. The glands are also a source of JH.

2. THE FEMALE REPRODUCTIVE SYSTEM

The internal organs of reproduction in female insects vary from species to species. But in general they have a basic plan (Fig 8.7). It consists of a pair of ovaries which possess number of ovarioles, a pair of oviducts, common oviduct / Median oviduct, spermatheca, a pair of accessory glands and Bursa copulatrix or copulatory pouch or genital chamber or vagina. Typically the common oviduct is heavily invested with muscles than the lateral oviducts.

Ovaries

These are the prominent visceral organs present on the either side of alimentary canal. Anteriorly the ovaries get connected with the body wall by means of thread like suspensory ligaments. The ovaries are covered with fat body and are also richly covered with trachea. Each ovary consists of a number of ovarioles or egg tubes.

Ovarioles

Each ovariole is enveloped by a double layered cellular wall. The outer wall is called **ovarial sheath** which has an abundant supply of tracheae. The inner layer called **tunica propria** is elastic in nature. Each ovariole at its terminal has a filament which unites with other filaments to form a suspensary ligament. The ligament is attached to the body wall or dorsal diaphragm and thus helps the ovaries to remain suspended at a proper place. The eggs are discharged in to the lateral oviducts.

Lateral oviducts

Proximal end of the ovarioles of each ovary join to form a lateral oviduct on each side. The wall of oviduct is glandular and muscular. The lateral oviducts may be either mesodermal as extensions of the calyx of the ovary into which the ovarioles opens, or ectodermal as outpocketings of the common oviduct.

Median Oviduct

Two lateral oviducts combine to form a median oviduct. The common oviduct is clearly ectodermal in origin and possesses a cuticular intima.

Vagina

In most of the insects, median oviduct does not open directly to outside. It opens in to a tubular genital chamber or vagina formed by invagination of bodywall from VIII segment. The vagina opens outside by an opening called **vulva**. Vulva serves both purposes of receiving the sperms and discharging the eggs.

Bursa Copulatrix

In some insects the genital chamber or vagina develops a separate pouch called Bursa Copulatrix in to which insects have two reproductive openings. One is vulva for receiving the sperms open on VIII sternum and another one is ovipore or gonopore on IX segment for discharging eggs. For example Lepidoptera and water beetles.

Spermathea

It is a sac like structure consisting of a spermathecal gland and opens in to vagina through spermathecal duct. This is mainly used for storing the sperms. It also produces some fluids responsible for longevity of cells for several hours.

Accessory glands

These are a pair of collateral glands which open in to the distal portion of vagina and secrete the substance responsible for the formation of ootheca of cockroach, preying mantid and poisonous secretions in case of Hymenoptera. These sticky substances are useful for attachment of egg to the substrate on which they are laid.

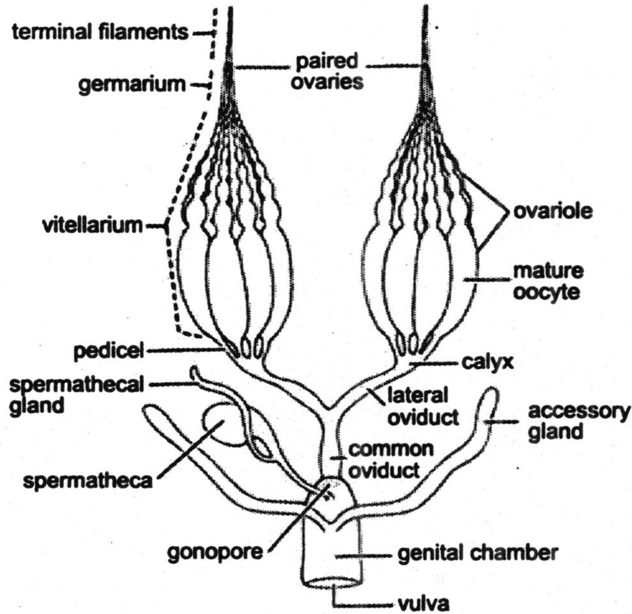

Fig. 8.7 Reproductive system of a typical female insect

2.1 Structure of ovariole

Each ovary in insects consists of a group of tapering units called **ovarioles**. The number of ovarioles in an ovary varies greatly in different insects, usually 4 to 8. In Isoptera more than 2000 ovarioles are present. Typical ovariole consists of 3 parts namely: 1. Terminal filament, 2. Egg tube and 3. Supporting stalk or pedicel

Terminal filament

All the ovarioles in an ovary unite distally with one another in a suspensory ligament. The ligaments from the two ovaries are combined in a single median ligament which is attached to the tergal plate of the thorax.

The Egg tube is divided into two parts: (1). Egg chamber or Germarium and (2). Zone of growth or vitellarium

Germarium

The germarium is also called **egg chamber**. It contains the primordial germ cells or undifferentiated cells. These cells give rise to three types of cells namely: (1).Germ cells developing into oogonia and finally oocytes, (2). Nutritive cells or nurse cells, the trophocytes and (3). Follicle cells.

Vitellarium or Zone of growth

It occupies the major part of the ovariole and contains large number of oocytes and eggs in different stages of development. The egg cells grow and attain their mature stage. In the anterior region of the vitellarium, the nurse cells and oocytes remain mixed up and assume the central position while follicle cells take peripheral position. In the posterior end, oocytes are enclosed by follicle cells to form follicular layer. The nurse cells absorb nutrients from haemolymph through follicular cells and transmit them to oocytes. In some cases follicle cells provide nutrients to the oocytes where nurse cells are absent.

2.1.1 *Types of ovarioles*

Based on the presence or absence of nutritive cells and their location, ovarioles are categorised in to two types (Fig 8.8):

1. **Panoistic ovarioles**: In these, the nutritive cells are absent and the development of oocytes takes place with the help of follicular epithelial cells e.g.: Odonata, Dictyoptera, Orthoptera and Ephemeroptera. In panoistic ovarioles, all germ line stem cells become functional oocytes, which are surrounded by a monolayer of follicular cells and thus constitute an ovarian follicle; all RNA types that accumulate in the oocyte cytoplasm are expressed in

the oocyte nucleus. Each ovariole originates as a germarial region which usually persists into the adult. However, the cystoblasts either develop directly into oocytes or the clusters of cystocytes separate into single cells each of which develops into an oocyte.

2. **Meriostic ovarioles**: They contain trophocytes / nutritive cells which vary in their position. Based on the position of trophocytes meriostic ovarioles are classified into:

   A. **Polytrophic ovarioles:** Where developing oocyte and trophocytes are arranged alternatively within the vitellarium e.g.: Mecoptera, Dermaptera, Psocoptera. The polytrophic ovarioles of *Drosophila* has been well studied. In *Drosophila*, the **germarium** produces growing entities, the **follicles**, which undergo previtellogenic and vitellogenic growth and chorionogenesis in the **vitellarium**. Each follicle consists of a clone (= cluster) of germ cells: one oocyte syncytially connected to a set of 15 sister cells, which differentiate to nurse cells and supply the oocyte with euplasmatic elements needed by the oocyte and by early stages of embryogenesis. The cluster of germ cells is surrounded by a monolayer of follicular cells that function in vitellogenesis and synthesize and deposit the egg envelopes at the end of oogenesis. The cluster of germ cells develops in a stereotypic manner from stem cells, which reside in a **niche** at the apex of the ovariole. The niche consists of a few somatic cells, named cap cells near the anterior tip of each ovariole, next to the terminal filament. Permanent cell to cell communication between cap cells and the germ line stem cells is necessary to maintain the function of stem cells and to keep them in the anterior most position. As a rule, stem cells divide regularly only in an anterior-posterior direction, giving birth, with each division, to a **cystoblast**, which undergoes a strict program of four synchronous mitotic cycles, each followed by incomplete cytokinesis. The result is a cluster of 16 **cystocytes**. Cystocyte divisions and differentiation into nurse cells and oocytes take place subsequently in four compartments of the germarium. Cystoblasts and clusters are guided by specialised somatic cells. Only one of the two innermost cells of each cluster becomes the oocyte while all other cystocytes develop into nurse cells. This final differentiation is regulated by intrinsic factors.

   Subsequent growth of oocytes proceeds in the vitellarium in so-called **follicles**. In polytrophic ovaries, each follicle consists of one oocyte and its sister cells, which develop as nurse cells surrounded by follicular cells. Follicular cells deposit the vitelline membrane and egg shell as a final act of oogenesis.

B. **Acrotrophic ovarioles:** This also called **telotrophic ovarioles** where the trophocytes are present in the germarium (apex) and are connected with the growing or developing oocytes by cytoplasmic strands. e.g.: Hemiptera and Coleoptera.

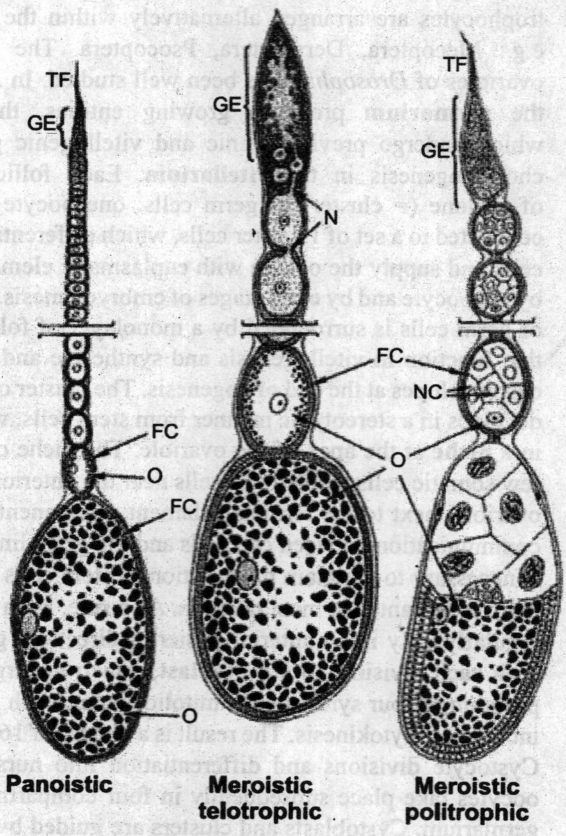

Fig. 8.8 Types of ovarioles in insects
FC: follicle cells; GE: germarium; NC: nurse cells; O: oocyte; TF: terminal filament;
N: nutritive cord. (Taken and modified from Mahowald 1972)

Instead of an anterior germarial region in which the proliferation of germ cells takes place, each telotrophic ovariole has an anterior **tropharium** in which nurse cells synthesize all cytoplasmic products stored in the oocyte. Thus, in contrast to polytrophic and panoistic ovarioles, all germ cells maintain their anterior position, except for growing oocytes. In other words, the function of the cap cells to maintain the anterior position of stem cells has expanded to all inner sheath cells surrounding the germarium/ tropharium. Thus the functional ovarioles of telotrophic ovaries lack a germarium. Consequently, the number of presumptive oocytes is fixed and the proliferation of germ cells is shifted into the larval stages. The transport of products from nurse cells to oocyte is maintained through **nutritive (trophic) cords**, which are anterior extensions of the oocytes. A similar, but shorter and, in most cases, preliminary, extension of oocytes occurs also in clusters of many polytrophic females where it is called the **nutritive appendix**. Therefore, the existence of nutritive cords is not unique to telotrophic ovarioles but the character belongs to all meroistic ovaries.

2.2 The Spermatheca

The spermatheca functions as a storage organ of the spermatozoa until they are used to fertilise the egg. In honeybees, the storage period may be for 2 to 3 years. The spermatheca also has glandular elements. In many insects, the glandular portion is distinct from the storage portion as in honeybees and other hymenopterans. In these insects the distal end is specialised for secretion and is referred to as the **spermathecal accessory gland**. In other insects such as the cockroaches, the storage portions contain the glandular cells. Morphologically, the secretory cells form a cup shaped cavity at the apical surface, and the membrane in this region is thrown into a number of microvilli. This cavity is connected to a pore in the cuticular intima of the spermatheca by a ductule lined with cuticle. The junction between the ductile and the cup-shaped reservoir beneath it is occluded by fibres. In its simplest form, there is a ductule and secretory pore associated with each secretory cell.

The secretions in the spermatheca consist of mucoproteins and mucopolysacharides. It is presumed that the spermatozoa utilise this as a source of energy. The epithelium of the spermatheca, particularly in the storage portion, exhibits many characteristics of ion-transporting epithelia. Thus it is thought that the spermatheca provides for the ionic milieu required for the spermatozoa. Many insects show some sort of control of the secretions of the spermatheca. When semen reaches the preapical diverticulum, there is a massive release of secretions. The corpus allatum is also thought to control secretory activities of the spermatheca. The neurosecretory myotropin also acts on the muscles of the spermatheca to squeeze the secretions from the reservoirs beneath the intima into the lumen.

2.3 Other accessory glands

Collaterial gland

In addition to the spermatheca, several other glands are seen in the female reproductive tract. In most orthopteroid insects, group of eggs are encased in an ootheca to protect the developing embryos from desiccation. In cockroach the ootheca is a purse-like structure, in mantids it is a hardened mass of frothy secretions and in acridids it is a foamy secretion surrounding the eggs which seals them in the underground chamber. These secretions are produced by the **collaterial gland**. In *Periplaneta americana* the collaterial gland consists of two parts. The left collaterial gland is larger and more prominent. It secretes the proteins which make up the structure of the ootheca, together with polyphenol oxidase and 4-o-β glucoside of protocatechuic acid. It also secretes calcium oxalate. The left collaterial gland branches into tubules. Each tubule contains four types of cells:

1. Basal, presecretory, epidermal
2. Glandular, secreting structural protein
3. Glandular, secreting oxidase
4. Glandular, secreting structural protein + calcium oxalate

The right collaterial gland contains a more transparent secretion, and is composed of three cell types:

1. Presecretory, epidermal
2. Glandular, secreting structural protein
3. Glandular, secreting protocatechuic acid

The right collaterial gland secretes a β-glucosidase. When the secretions of the right and left are mixed, the β-glucosidase liberates protocatechuic acid from the glucoside which is then oxidised by phenol oxidase to a quinine which forms cross-links in the structural protein, yielding the hard and tanned material of ootheca.

In Orthoptera these are mesodermal glands which consist of convoluted blind tubules opening into the proximal end of the lateral oviducts. The ootheca is a product of not only these cells, but the entire lateral oviduct which are mesodermal in origin.

Milk glands

Adenotrophic viviparity refers to the process by which embryos develop within the mother and receive nourishment for development from the mother. In such cases, the accessory glands of the female supply a secretion, usually known as **milk** upon which the embryos feed. The best known example is the *tse-tse* fly *Glossina.* The milk gland in *Glossina* is a branched tubular structure ramifying throughout the abdomen and emptying into the uterus via a single muscular collecting duct which contains two channels. Each chitinous channel connects with a proximal tubule which branches once to produce four proximal tubules. Each of the proximal tubule opens into a distal tubule which ramifies repeatedly. The distal tubules consist of a single layer of prominent secretory cells. The secretions of each cell accumulate in a single large reservoir which empties into the lumen of the gland via a short ductule which penetrates the intima. The secretory reservoir is an extracellular structure, formed by a cup-shaped invagination of the apical membrane of the secretory cell. The milk consists of equal quantities of lipid and proteins or amino acids. The source of these secretions is the blood ingested by the mother. The secretions are cyclic in nature indicating that some sort of control mechanism exists. The corpus allatum is said to be involved in the control.

Conglobate gland

The conglobate (phallic) gland of male cockroaches is a reproductive organ anatomically close to the accessory glands. It is a club-shaped ectodermal structure which opens separately to the outside by a duct. The lumen of the gland is lined with cuticle containing epi-, exo-, and endocuticles and layer of epidermal cells. Ductules penetrate the cuticular layer and establish continuity with intracellular ductules in gland cells. The epicuticular layer of the ductule becomes highly reticulated on entering the gland cell, possibly for the passage of secretion. The microvilli abut on this reticulate layer. In the luminar border of the gland cells, rows of small vesicles occur rather frequently in association with the microvilli. These small vesicles constitute units of a transport system.

From a functional point of view it has been postulated that the conglobate gland is involved in the formation of the spermatophore, although no clear demonstration of this role has been published so far. In addition, endocrinological studies have suggested that juvenile hormone might influence conglobate gland maturation. It is suggested that juvenile hormone stimulates the accumulation of proteins in the conglobate gland and that these proteins contribute to the formation of the spermatophore.

2.4 The female external genitalia

The female genitalia originate from abdominal segments 7 through 9. During development ectodermal invaginations occur in all three sterna, but later these can no longer be recognised as separate parts.

Morphologists often use the Thysanura as a starting point for developing a generalized model to explain the evolution of the external reproductive system of pterygote insects. The thysanuran abdomen has basal sclerotised plates called **coxopodites** on which styli are attached. These plates are serially homologous along the abdomen, and the pregenital plates are regarded as identical with the genital plates. The plates located on segments 8 and 9 are considered to be genital plates. The styli associated with these segments are called **gonapophyses**. There are four gonapophyses on segments 8 and 9 (i.e., a pair of styli on each segment). The gonapophyses are medially concave and directed rearward. The basal sclerite is called a gono-coxa, and in some Thysanura it may be fused with the style.

The primitive pterygote with a gonopore on segment 8 has an appendicular ovipositor that consists of three components. A basal apparatus corresponds to the basal plate or primitive gonocoxite of the thysanuran abdominal appendage. The second part is the first valvifers (on the eighth sternum), and second valvifers (on the ninth sternum) are responsible for providing support and points of articulation for the tube through which the egg passes (Fig. 7). Interpolated between the first and second valvifers is a small sclerite called a **gonangulum**, which articulates with the second gonocoxite and tergum 9. The gonangulum is present in Odonata and Grylloblatoidea. It apparently is fused with the first valvifer in Dictyoptera and Orthoptera. In the remaining orders these structures are highly variable.

The shaft of the ovipositor consists of two pairs of elongate, closely appressed sclerites called the **first** and **second valvulae** (Fig.8.9). The first pair of valvulae is positioned on the eighth abdominal sternum. The second pair of valvulae is located on the ninth abdominal sternum and is dorsal in position. **Third valvulae** are positioned on the posterior end of the second valvifers. These valvulae usually serve as a sheath for the shaft of the ovipositor.

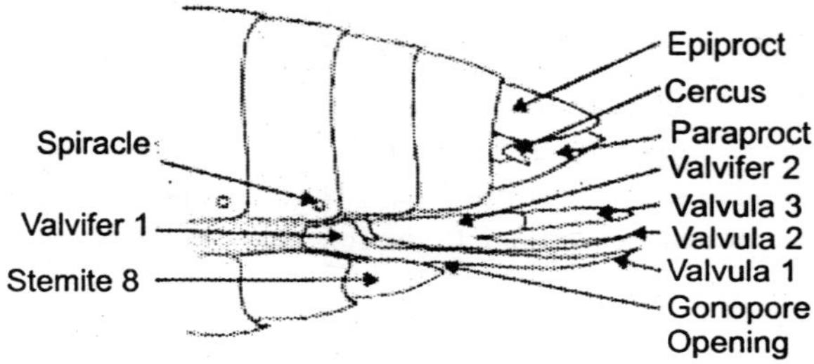

Fig. 8.9 Diagrammatic sketch of an orthopteran female genitalia

In Ephemeroptera paired gonopores are found on the seventh sternum and it is believed that phylogenetically this is the original type. In the most common type, the genital opening is on the eighth sternum and joins the common oviduct (arising from the anläge of the seventh sternum) leading anteriad. The opening may be located more anteriorly, due to secondary modifications of the terminal segments. Accessory sex glands (arising from the anläge of the ninth sternum) open into the genital atrium or vagina. This type is found in the Orthoptera, Hymenoptera, Diptera, Neuroptera, and others. While the common oviduct has an ectodermal lining, the paired oviducts often are composed of mesodermal and ectodermal parts. Ovipositors, which are of great taxonomic value, are formed by the eighth and ninth sternum and probably in the female can be considered as abdominal appendages.

A further modification of the female genitalia is found in the Lepidoptera. With the exception of the Microlepidoptera, all species have two genital openings. The posterior opening constitutes the actual gonopore, while the anterior one (eighth segment) forms a bursa copulatrix. The two openings are apparently homologous to the corresponding embryonic anlagen. The bursa has a connection to the common oviduct via the ductus seminalis through which the sperm migrate to reach the receptaculum seminis. All these organs are lined with a cuticular intima since they are of ectodermal origin.

3. OOGENESIS

3.1 Formation of oocytes and nurse cells

Panoistic Ovarioles

Oogonial differentiation takes place in the germarium. All germ cells differentiate into oocytes which are subsequently enveloped by epithelial cells derived from the actively dividing prefollicular tissue. It is not possible to differentiate between primary and secondary oogonia.

Polytrophic Ovarioles

Polytrophic ovarioles are characterised by follicles consisting of one oocyte together with a well-defined number of trophocytes or nurse cells, the whole complex being surrounded by a sheath of follicle cells. Oocyte differentiation in this type of ovariole has been extensively studied in *Drosophila*. In this insect, a primary oogonium stem cell occurs in the apical region of each germarium, close to the cells of the terminal filaments. The primary oogonium divides into a cystoblast and another primary oogonium. As the cystoblast moves down the germarium it undergoes four mitoses giving rise to a cyst of sixteen cells (cystocytes) which form an egg-nurse-cell complex. One of the cystocytes will be the oocyte, the others the trophocytes. Subsequently, this complex is enveloped within a sheath of follicle cells, originating from the mesodermal prefollicular tissue. As the oocyte grows, the follicle cells undergo divisions. They will not only alter in shape and size but their function will also change. Finally the cysts pinch off from the germarium and thus become a follicle. The oocyte and trophocytes are interconnected by fifteen canals, each surrounded by a protein-rich ring, which will allow the transport of RNA (ribosomes) and mitochondria to the oocyte during vitellogenesis. Trophocyte nuclei usually become polyploid through endomitosis, but the oocyte nucleus remains diploid. This mechanism of oocyte differentiation is probably applicable to the greater part of insects with polytrophic ovarioles. The number of trophocytes is variable from species to species but is always constant in the same species. The numerical relation between trophocytes and oocytes is 1:1 in Dermaptera and *Chironomus,* 5:1 in Mallophaga and Siphunculata, 7:1 in *Anopheles* and *Gyrinus,* 15:1 in *Drosophila* and *Musca,* and 48:1 in *Apis.* Descending into the germarium, each oocyte is followed by its trophocytes.

Telotrophic Ovarioles

In telotrophic ovarioles different conditions may prevail. In *Oncopeltus,* and probably other Hemiptera, Heteroptera, the trophocyte nuclei partly fuse to form giant nuclei arranged in the periphery of the germarium. The central part is occupied by the cytoplasmic nutritive chamber or trophic core, into which the nuclei eventually release their contents. The oocyte remains connected to this trophic core by means of a cytoplasmic strand serving as a **feeding channel**. Alimentary egg formation is realized both by way of this **trophic strand** and by the follicle. In *Tenebrio* and probably other Coleoptera Polyphaga, the situation is different. The trophic core is absent, nutritive cells being scattered throughout the germarium. They break down individually into trophic strands, but these may often disappear as the follicle is completed.

3.2 The follicle cells

As the oocyte, moving down the germarium, passes through the mesodermal prefollicular tissue, it is enveloped by a layer of follicle cells, which still have a primitive subcellular cyto-architecture at this time. Upon descending into the vitellarium, the follicle cells increase in number by mitotic divisions and form a monolayer. Soon their number remains stationary and the steady increase in volume of the oocyte merely results in a tangential stretch of the follicle cells. Consequently, these cells are first columnar, then cuboidal, and at last flattened. They often become binucleate by amitotic divisions. The movement of the follicles in the ovarian envelopes may be effected by crowding, but may be assisted by peristaltic contraction waves of the myoepithelial cells of the outer envelope.

At the onset of proteid yolk formation, both the follicle and the terminal oocyte may develop microvilli. During this stage of vitellogenesis, the follicle cells play a role in the selective absorption of vitellogenic blood proteins by the oocyte. After vitellogenesis is complete, the follicle cells secrete the vitelline membrane and the chorion, which in many cases may show impressions of follicle folds and villi. Finally, the egg leaves the vitellarium by rupturing the follicle and the epithelial plug separating it from the ovarial stalk. The follicle cells, remaining in the ovariole, degenerate. In some cases they form a corpus luteum, which, however, does not show progressive development and secretory activity.

3.3 Alimentary aspects of oogenesis

Yolk is deposited in the ooplasm in the form of globules or platelets, consisting of proteins or lipids. Glycogen occurs in a typical particulate form. The yolk-precursor substances are provided by the trophyocytes, the follicle cells, the oocyte itself, or are absorbed from the haemolymph. Proteid bodies account for most of the oocyte volume. Lipid yolk globules are distributed in between while free glycogen, when present, is either dispersed throughout the ooplasm or in spaces between the proteid yolk bodies.

In most insects, the major protein component of the yolk is vitellin (Vn), the crystallised storage form of Vg. Insect Vgs are large, conjugated proteins. They are translated as primary products with masses of approximately 200 kDa, which are usually cleaved into subunits (apoproteins) ranging in size from 50 to 180 kDa. In Drosophila, the yolk proteins are designated as **yolk polypeptides** or **YPs**. They are relatively small proteins (approximately 45 kDa) that are glycosylated, phosphorylated, and sulphated. A number of supplemental proteins have been described that accumulate selectively in developing oocytes. Some of them originate from the fat body and some from the cells of the follicular epithelium. These proteins include both female-specific and bisexually distributed members and, normally, they represent minor components of the yolk. A 30 kDa YPP, called **microvitellogenin,** is produced by the fat body in *Hyalophora cecropia* and *M. sexta*. The fat body of the domesticated silkworm *B. mori* also produces multiple 30 kDa proteins homologous to microvitellogenin.

Lipids are a critical source of energy during insect embryogenesis. They represent 30–40% of the insect egg's dry weight. Lipophorin (Lp), the insect lipid transport protein, have a major role in accumulation of lipids in oocytes. In addition to its role as a lipid carrier, Lp plays a role as a YPP. In several studied insects, many other proteins are incorporated into the yolk. For example, the iron transport protein, transferrin (~65 kDa) is selectively deposited in the yolk of the flesh fly, *Sarcophaga peregrina*, and a similar protein is deposited in the bean bug *Riptortus clavatus*. The haemolymph and oocytes of *R. prolixus* also contain a 15 kDa heme-binding protein. In Lepidoptera, a blue biliprotein called insecticyanin, a component of larval haemolymph, is deposited in developing oocytes and is present in mature eggs (Kang *et al*., 1995). Biliverdin is a chromophore covalently bound to Vg in *Spodoptera litura*, colouring the eggs green or blue, presumably providing camouflage against green vegetation. Finally, an arylphorin-like cyanoprotein, which is a haemolymph storage hexamer, also is deposited in eggs of the bean bug, *R. clavatus*, as a YPP (Chinzei *et al*., 1990; Miura *et al*., 1994).

The Origin of Proteid Yolk

It is evident that the relative importance of proteid yolk synthesis by the oocyte, the trophocytes, or follicle cells, on one hand, and the uptake by pinocytosis of proteid yolk precursors from the haemolymph, on the other hand, may vary from one species to another, depending upon the peculiarities of its life cycle and habitat.

Exogenous supply by the haemolymph is the most important source of proteid yolk precursors in *Hyalophora.* Proteins enter into the vitellogenic oocyte from four sources: (1) the oocyte itself synthesises proteins, presumably in the oocyte cortex, which are deposited in the yolk spheres along with the blood proteins. (2) The nurse cells or trophocytes deliver part of their cytoplasm, including proteins, to the oocyte (3) The follicle cells release a nondialyzable product, believed to be at least partly protein in nature, into the intercellular space between the follicle cells and the oocyte. This material seems to be necessary to enhance pinocytotic activity in the oolemma. (4) By far the most important source of vitellogenic proteins is the haemolymph.

In *Drosophila,* unlike *H. cecropia,* the exogenous supply of yolk proteins is only of minor importance, but that the nurse cells, and presumably the oocyte itself, synthesise these proteins. In *Leptinotarsa decemlineata,* oocytes selectively absorb and accumulate a hemolymph-borne female protein which is a glycoprotein with a molecular weight of about 250,000. At the onset of proteid yolk deposition, an intercellular space arises between follicle cells and oocyte. Haemolymph proteins are taken up in the oolemma by pinocytosis. The follicle cells, being highly specialized for protein synthesis, release a mucopolysaccharide-protein complex. A selective reaction between this product and the female protein, leading to the precipitation or crystallisation of the female protein, is supposed.

In *Periplaneta americana* (Panoistic Ovariole), trophocytes being absent, the possible sources of vitellogenic proteins are the oocyte, the follicle cells, and the haemolymph. A striking feature of the cytoplasm of all oocytes, regardless of their position within the ovariole, is the absence of an organised endoplasmic reticulum. Instead, the cytoplasmic matrix is filled with a host of free ribosomes. During active yolk deposition, the follicle cells are even highly specialised for protein synthesis as can be deduced from the very extensive endoplasmic reticulum and the numerous Golgi systems.

RNA Supply

The oocyte nucleus of polytrophic and telotrophic ovarioles is inactive and does not participate in the supply of ribonucleic acids to the oocyte. The trophocytes deliver RNA to the oocyte. In telotrophic ovarioles, a stream of RNA material is transported from the trophic tissues to the oocyte via the trophic core. A directed flow of RNA proceeds from the more anterior nurse cells through the posterior ones into the oocyte. RNA is delivered from the trophocytes to the oocyte as ribosomes.

The Occurrence of DNA in the Ooplasm

The presence of DNA in the cytoplasm of the oocyte has been reported in several insects. In *Rhodnius,* there appeared to be a transfer of some of the DNA in a partially depolymerised form from the trophic tissues of the ovary to the growing oocytes. A reservoir of DNA seems to exist in the cytoplasm of the oocyte of *Drosophila.* An acceptable explanation could be that the DNA originates from the disintegrating endopolyploid nurse cells.

Origin of Carbohydrate Yolk

Carbohydrate yolk occurs in the oocyte as glycogen and mucopolysaccharide-protein complexes in the proteid yolk bodies. Glycogen is synthesised in the ovary itself. Glycogen synthetases are formed early in vitellogenesis and are multiplied during the growth of the oocyte (Engels and Bier, 1967). Only at the end of vitellogenesis is glycogen very quickly condensed from glucose. In *Hyalophora cecropia* and in the Colorado beetle the follicle cells release a product, probably of a glycoproteid nature, which is necessary to induce pinocytotic activity in the oocyte surface and which is absorbed by the oocyte. The vitellogenic female protein, the most important proteid yolk precursor in the haemolymph, is also a glycoproteid in the Colorado beetle and in *Gryllus*.

The Origin of Lipid Yolk

In *Oncopeltus,* lipid yolk bodies are first formed in the oocyte from precursor bodies near the entrance of the trophic strand. In later stages of oogenesis they occur near the follicle epithelium, suggesting their origin from this source. The precursor bodies are Golgi vacuoles or lipochondria.

3.4 Vitellogenesis

Vitellogenesis is the formation and incorporation of yolk into the growing oocytes. As might be expected from the structural differences of ovarioles among the different species, several sites and mechanisms of yolk synthesis and incorporation are found: oocyte, follicular epithelium, nurse cells and extraovarian tissues participate. In any one species all of these tissues may be functional in yolk deposition. In some of the species one or the other may be dominant, but in none of the species do we reliably know about the relative importance of the various routes in yolk deposition.

Insect vitellogenesis can be accomplished by two general mechanisms. The first, in which the follicle produces its own yolk, is called **autosynthetic vitellogenesis** and has been reported in only a few species of the most primitive insects, the Apterygota (Bilinski, 1979). In the second, called **heterosynthetic vitellogenesis** and used by almost all advanced species, the follicle obtains the bulk of its yolk from an extra-ovarian source (the fat body) via the maternal blood stream through a hormonally (juvenile hormone and or 20- hydroxyecdysone) regulated process (Postlethwait and Giorgi, 1985; Bownes, 1994). The major yolk components in insect eggs are protein granules containing mainly yolk proteins, lipid droplets whose contents (neutral lipids) are probably synthesised by the nurse cells and glycogen rich particles synthesized by the oocyte itself (Postlethwait and Giorgi, 1985).

Since insect vitellogenesis is almost exclusively a heterosynthetic process, the major-female specific protein sequestered from the blood, modified co- and post translationally by the addition of glycosyl, phosphate and sulfate residues (Izumi *et al.*, 1994) and stored in yolk granules, is defined as vitelline and its soluble counterpart in the haemolymph as vitellogenin. Insect yolk proteins may be divided into five categories.

1. **Vitellines**, found in most insect species, such as moths, bees, wasps, locusts and lower flies such as the mosquitoes. These proteins derive from a vitellogenin precursor molecule that is synthesized

in the fat body in a female specific manner, cleaved in a large and small subunit, secreted into the haemolymph and sequestered in the developing oocytes. Reported cases where no vitellogenin precursor cleavage occurs refer to higher Hymenoptera (suborder Apocrita) (Nose *et al.*, 1997) and two species of Homoptera (Tu *et al.*, 1997), while vitellogenin can be also cleaved into three or four subunits, as occurs in *Locusta migrutoria* and in sawflies (Izumi *et al.*, 1994; Takadera *et al.*, 1996).

2. Yolk proteins of the higher Diptera, including fruit flies, flesh fly and house fly, which are synthesised in both the female fat body and ovarian follicle cells and are incorporated into developing oocytes; there are no differences between the yolk proteins originating in the fat body and those originating in the follicle cells. In *D. melanogaster* three major yolk proteins have been isolated with molecular weight of 4446 kDa, encoded by single copy genes (Bownes, 1994), while in *Dacus oleae* and *Ceratitis capitata* yolk proteins are represented by only two polypeptides with molecular weight ranging between 45 and 49 kDa (Rina and Mintzas, 1988; Trougakos *et al.*, 1999). Among Diptera, vitellogenin synthesis in the ovary has been documented for *D. melanogaster* (Brennan *et al.*, 1982), Ducus oleue (Trougakos *et al.*, 1999), *Calliphora erythrocephala* (Martinez and Bownes, 1994), *Musca domestica* (White and Bownes, 1997) and *Ceratitis capitata* (Rina and Mintzas, 1988).
3. Proteins such as the egg-specific protein of *Bombyx mori* (Ire and Yamashita, 1983), the paravitellogenin of *Hyalophora cecropia* (Telfer *et al.*, 1982), the two proteins isolated from *Manduca sexta* eggs (Tsuchida *et al.*, 1992) and a follicle cell-specific protein recently isolated from *Rhodnius prolixus* (Melo *et al.*, 2000). These proteins are synthesised only in the ovarian follicle cells and sequestered into the oocytes, but appear to be distinct from the vitellogenins produced in the fat body.
4. Proteins such as the 30 kDa protein of *Bombyx mori*, the microvitellogenin found in *Manducu sexta* and *Hyalophora cecropia*, the 21 kDa protein of *Locusta migratoria* and the female specific 53 (a proenzyme of vitellogenic carboxypeptidase) and 44 kDa (40KP) proteins of *Aedes aegypti* (Raikhel and Snigirevskaya, 1998). These proteins are synthesized in a non sex-specific manner in the fat body cells and appear to be distinct from the major vitellogenins. They are however secreted into the haemolymph and sequestered in the developing oocytes.

5. The cochineal yolk protein in the homopteran *Dactylopius confusus* (Ziegler *et al.*, 1996).

The yolk proteins secretory pathway in the fat body of the mosquitoes (Raikhel and Snigirevskaya, 1998) and the ovary of fruit flies (Trougakos *et al.*, 1999; Trougakos *et al.*, 2001) has only recently been analysed by means of transmission electron microscope (TEM) immunolocalisation and appear to be quite similar. In the fruit fly follicle cells, however, which at the same time are engaged in the vitelline membrane proteins secretion, the yolk proteins are differentially sorted from the co-secreted vitelline membrane proteins either in distinct parts of biphasic vesicles or in entirely distinct vesicles. Following secretion, yolk proteins, along with their haemolymph originating counterparts, diffuse through permeable channels in the vitelline membrane maintained by the extended follicle cell microvilli and finally reach the plasma membrane of the oocyte. Yolk sequestration in the oocyte follows a similar pathway in all insects studied so far and is accompanied by two major changes in the egg chambers (Raikhel and Dhadialla, 1992), namely **follicle cell patency** (referring to formation of large intercellular spaces between the adjacent follicle cells) and the appearance of coated pits on the surface of the oocyte that is related to receptor mediated endocytosis (Sappington and Raikhel, 1998; Schonbaum *et al.*, 2000).

Several observations have suggested the possibility that the oocyte and the germ nucleus participate in yolk formation in some species. For example, follicles of panoistic ovarioles do not obtain nutrients from special nutritive tissues except, perhaps, from the follicular epithelium. It is interesting to note that oocyte nuclei of this type of ovariole contain giant lampbrush chromosomes, which are never found in either telotrophic or polytrophic ovarioles. These oocytes also contain several nucleoli. The lampbrush chromosomes and the nucleoli appear to be involved in a high rate of RNA synthesis—as is exemplified by *Locusta migratoria*—thus suggesting their involvement in yolk protein synthesis.

Further evidence for protein synthesis in the oocyte comes from studies in *Drosophila.* Here, as in some other species of Diptera, most of the ribosomal RNA in the oocytes is derived from the nurse cells. During the major growth phase of the oocyte, yolk spheres which are not membrane bound grow in close association with the rough surfaced endoplasmic reticulum (Cummings, 1968). Since these yolk spheres are not membrane bound in contrast to those of the preceding growth phase, one can conclude that they are not derivatives of pinocytotic vesicles. This finding probably indicates that proteins are synthesized on RNA from the nurse cells since no evidence has been obtained for nucleolar RNA synthesis in the germ nucleus.

Two lines of information point to the fact that, in Blattaria, yolk is not exclusively synthesised by the oocyte. **First** of all, immunological techniques show that in *Leucophaea maderae* several of the blood serum proteins are incorporated into the maturing oocytes. These proteins are synthesised elsewhere and pass through the ovarian sheaths and follicular epithelium as intact molecules. **Secondly**, electron micrographs of the yolk-depositing oocyte in *Periplaneta* show an excessive development of oolemma foldings which seem to be pinched off later, giving rise to the growing yolk spheres. Evidence from other species also supports the view that the small vesicles pinched off from the oolemma contain blood proteins.

The first demonstration of the presence of extraovarian proteins in the oocytes of insects was given by Wigglesworth (1943), using a number of blood-sucking Hemiptera. The eggs of *Rhodnius prolixus, Triatoma infestans, T. brasiliensis,* and *Cimex lecturlarius* contained haemoglobin, which could only be derived from the host blood. During oocyte maturation the blood protein concentration, particularly that of the specific female protein, drops. It is interesting to note that this specific female protein is selectively accumulated in the oocytes and reaches a concentration 20 times higher than in the haemolymph: it is transferred to the oocytes against a concentration gradient.

Scientists have tried to unravel the exact route of entry for blood proteins into the growing oocytes. Histologically, one finds that during oocyte growth in *H. cecropia* (King and Aggarwal, 1965), intercellular spaces open up between the follicular epithelium. Sections treated with fluorescein- labelled antibodies clearly and elegantly show how blood proteins are first taken up through these gaps; they are next to be found in the brush border, and then later in the yolk globules of the oocytes (Telfer, 1961; Telfer and Melius, 1963). Thus, it appears that a major portion of the proteins does not pass through the follicular epithelium. In the cortex of the oocytes of *H. cecropia,* the oolemma is thrown into innumerable folds and pits which may contain ferritin after injection of the animal with these molecules. Small vesicles in the cortex have the same content as the pits, indicating that these spheres are pinched-off oolemma pits (Stay, 1965; King and Aggarwal, 1965). It is thus obvious that the oocytes obtain extraovarian proteins which may be both secreted elsewhere by the animal and derived from foods via this process of pinocytosis.

Participation of the epithelium in yolk formation has been suggested. Follicular epithelial cells of a variety of species are rich in RNA. This is demonstrated, for example, in the orthopteran *Gryllus bimaculatus*, the dipterans *Calliphora* and *Musca*, and the mecopteran *Panorpa communis*. It is possible that the high rate of RNA synthesis in these cells is associated with yolk protein synthesis, particularly when this can be shown during the early phases of egg maturation. On the other hand, it has also been shown on the basis of electron micrographs that follicle cells, at least in several Diptera (King and Koch, 1963; Cummings, 1968), are associated with the formation of the vitelline membrane. This membrane is later thrown into innumerable folds.

The role of the nutritive tissues in polytrophic and telotrophic ovarioles during egg maturation undoubtedly is complex and may differ in different species. In telotrophic ovarioles of *Oncopeltus* and *Rhodnius*, for example, a flow of RNA and other materials is observed in the nutritive cord, particularly during the early phases of egg maturation. This nutritive cord breaks down in the later stages of maturation; therefore, at this point the germinal nutritive tissues can no longer play a role. The bulk of the oocyte proteins is apparently derived from both follicular epithelium and extraovarian tissues in these species. As is apparent from these and other reports, nutritive tissues seem to supply mostly RNA to the oocytes both in telotrophic and in polytrophic ovarioles. For example, in the ovariole of *Panorpa,* the nurse cells contain large amounts of RNA which reach the oocyte via the ring canal; but, as yolk deposition begin, the nurse cells break down; they do not supply yolk proteins. In several Carabidae and Hymenoptera likewise, the nurse cells do not seem to supply protein yolk to the oocytes; rather a flow of RNA is seen going into the oocytes from the nurse cells throughout the growth period of the egg. On the other hand, dipteran oocytes apparently derive most of their RNA as well as some proteins and lipids from their nurse cells. In *Drosophila,* for example, precursors of both RNA and proteins are derived from the nurse Cells.

Once the oocyte is fully grown, the follicular epithelium secretes the egg chorion. It is generally agreed that, in all the species studied, the chorion which can be a complex array of layers, is formed by this epithelium. Another egg membrane, the vitelline membrane, is secreted during the early phase of the oocyte growth period. Early studies in *Culex pipiens* and several grasshoppers caused belief that this membrane is secreted by the oocyte itself. However, it has now become apparent in several dipteran species that it is actually secreted by the follicle cells.

After the mature egg is ovulated, the follicular epithelium breaks down and degenerates. The degenerating tissues form a plug which stays in the ovariole for some time. This plug usually attains a reddish or yellowish colour and was, therefore, termed the corpus luteum. In some species of the Acrididae, a variable number of eggs which have grown to various sub mature sizes degenerate while eggs in the neighbouring ovarioles continue to mature. These eggs also form so-called corpora lutea. Oosorption in *Nasonia vitripennis* is associated with aminopeptidase and esterase production in the follicular epithelial cells. Depending on the species, the corpora lutea disappear from the ovarioles more or less rapidly, although, in some species, one can recognise these plugs even after several eggs have been ovulated.

3.5 Formation of Egg Polarity

In all insect eggs (with only a few exceptions) the embryonic anterior-posterior and dorso-ventral axes are already established before fertilisation. These axes are apparent in the external morphology of the eggshell (e.g. localisation of the micropyle apparatus) and the location of the nurse cells or the trophic cords. The bulk of our current knowledge concerning the genetics and the molecular basis of pattern formation during follicle maturation in the ovariole refers to the model insect *Drosophila melanogaster*. In this species once the oocyte is determined it must be positioned at the extreme posterior of the cyst as this determines the anterior-posterior polarity for the rest of the development (Gonzalez-Reyes and St Johnston, 1994). Polar follicle cells are determined in the germarium by the hedgehog gene (Dobens and Raftery, 2000), while establishment of terminal cells is Notch dependent and occurs before oogenic stage 7 (Larkin *et al.*, 1999). The follicle cell layer becomes polarised along both the anterior-posterior and dorso-ventral axis through two rounds of gurken signalling from the oocyte. At stages *4-6,* when oocyte nucleus is still localised posteriorly, the early gurken signals to the follicle cells adjacent to the oocyte to induce them to adopt a posterior fate (Gonzalez-Reyes *et al.*, 1995).

Dorso-ventral follicle cells and embryonic polarity arise at stage 8 following a microtubule directed migration of the nucleus to an asymmetric position in the anterior of the oocyte where it determines where gurken mRNA is translated and localised (late gurken signaling), resulting in the induction of two prominent dorsal anterior fates, the operculum and the dorsal appendages (Ray and Schupbach, 1996). The polarised microtubule cytoskeleton is also required for creating an oocyte anterior-posterior polarity (that is eventually transmitted to the embryo) by directing the localisation and localised translation of bicoid mRNA anteriorly and oskar mRNA-staufen protein posteriorly (Gonzalez-Reyes and St Johnston, 1998). Gurken signaling by activating the epidermal growth factor receptor (EGFR) signalling pathway in the dorsal follicle cells (rhomboid, vein, kekkon and sprouQ), restricts pipe expression to the ventral follicle cells, defining thus the ventral side of the embryo. High level EGFR activation induces the dorsal expression of the EGFR inhibitor urgos and this may split the single peak of EGFR activation into two lateral peaks (van Eeden and St Johnston, 1999). The establishment of follicle polarity is also accompanied by five impressive migration events of small groups (or subpopulations) of follicle cells (Dobens and Raftery, 2000). The first major follicle cell migration is that from the wall of the germarium to surround the 16-cell germinal cysts in germarium region two. During stage 8 a group of anterior follicle cells (border cells) separate from the follicular epithelium, migrate posteriorly through the nurse cell cluster (second migration) and eventually locate at the nurse cell-oocyte border at stage 10. During late stage 9 most follicle cells retract from the nurse cell cluster to cover only the oocyte (third migration), leaving only a few stretched cells to be associated with the nurse cells. At stage 10b concurrently with nurse cell **dumping** the anterior most columnar cells migrate centripetally along the nurse cell-oocyte border to cover the anterior end of the oocyte (fourth migration). Finally, two groups of columnar follicle cells in the dorsal-anterior region of the follicle migrate anteriorly to produce the respiratory appendages (Deng and Bownes, 1998).

3.6 Formation of egg membranes and specialised chorionic structures

After termination of yolk deposition, the oocyte becomes surrounded by two protective layers, the vitelline membrane and the chorion. The structure of these layers is such that respiration remains possible while water evaporation is kept to a minimum. The follicle cells synthesise the precursor material for the vitelline membrane. The precursor material is synthesized in the Golgi complexes and is a polysaccharide-protein complex. The chorionic layers are synthesised by the follicle cells. This activity of the follicle cells is undoubtedly one of the most remarkable and complicated features of oogenesis. If we consider the vast variety of forms and structures of insect eggs, adapted to a great many oviposition media and substrates, their micropylar apparatus, ridges and caps, their finely sculptured outer surface, it is difficult to realize that this is accomplished by one layer of cells. These cells have the ability to secrete subsequently and synchronously a variety of substances, together forming the several layers of the chorion.

Most authors, following Korschelt (1887), recognise two main groups of layers of the chorion, secreted during two distinct phases: the more or less rigid chorionic outer layers and the finely structured chorionic inner layers. The composition of these layers in *Rhodnius prolixus,* according to Beament (1946), is as follows:

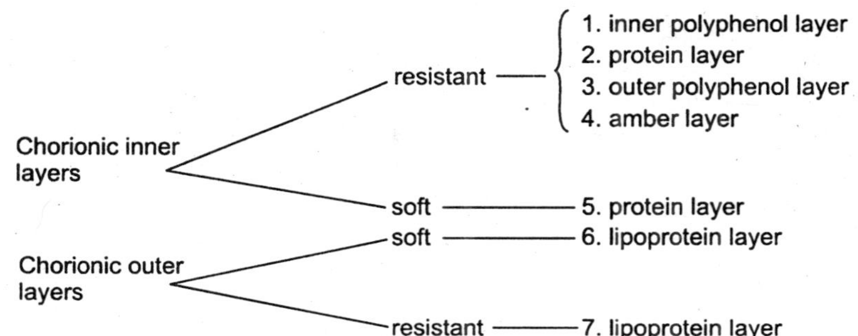

In the **first secretory phase**, the follicle cells produce the inner layers and the soft outer layer. In the **second secretory phase**, they produce the resistant outer layers not very permeable to water, and possessing mechanical rigidity.

The crucial problem for eggs laid in dry environments is how to minimize water evaporation and at the same time allow the free exchange of 0_2 and $C0_2$. Specialised areas of the chorion are aeropylar, hydropylar, and micropylar structures. The unspecialised region of the shell may be thin and entirely solid in eggs regularly deposited within plants or animals. In most cases the shell is made up of one or more layers of fine meshwork, filled with gas. This gas store is in exchange either with the ambient air or, when the egg is immersed in water, with the gases dissolved in this medium. In the latter case, the respiratory system functions as a plastron, a gas layer of constant volume and extensive water-air interface. Such layers are held in position by a system of hydrofuge structures that resist the entry of water under pressure (Hinton, 1962). In dipteran eggs, the whole surface or a more restricted area may function as a plastron. In *Calliphora* the vacuolar shell of the egg has a plastron at its waist, the functioning of which is vital to the egg. In other cases, respiratory horns with aeropylar filaments are present. When the eggs are not immersed, respiratory horns with a plastron merely function as aeropyles.

In Microphysidae (Heteroptera) the aeropylar filaments alternately radiate out and close again with humidity changes, undoubtedly affecting the rate of gas exchange. Hydropyles are chorionic structures with a coarse meshwork conducting fluid to the embryonic system. In *Nepa,* the inner surface of the hydropyle has openings large enough to permit the passage of water, although these ducts could not be distinguished optically (Hinton, 1961). The micropylar apparatus, situated at the anterior pole of the egg, is a system of canaliculae serving the entrance of sperm. At these places, the chorion is secreted around long protoplasmic projections of the follicle cells. In *Rhodnius,* all seven layers of the chorion are represented in the micropylar complex. In this region, the cells start to secrete at various stages in the seven phase cycle. The rim of the characteristic cap of the anterior pole is produced by four rings of follicle cells. Thickness is achieved by increase of exochorion layers, secreted around long follicular villi, and forming pits. The sealing bar, produced by one ring of follicle cells, is composed of the inner four layers of the chorion only. Three rings of pits become closed, the fourth row remains. Its pits become micropylae and associated structures.

The micropyle-forming cells are determined in the earliest stages of secretion, being pressed out of their ring. There are 200 pits, about 10-20 of which become micropylae. The remainder, closed at both ends, are pseudomicropylae (Beament, 1947). The number of micropylae may vary greatly, within a species as well as between species. In *Locusta migratoria* the number is 35-43 (Roonwal, 1936). Tettigoniids have from 1-23 micropylae.

3.6 Oosorption

Under the influence of a variety of factors, vitellogenesis may be interrupted and the oocyte, wholly enveloped in its follicle, may die. Resorption of oocytes can be triggered by starvation, short photoperiod, failure to mate, absence of adult males, unavailability of a parasitic host, or abnormally long retention of a spermatophore. Oosorption may also be induced by allatectomy. In *Labidura riparia* (Dermaptera), the penultimate follicle degenerates just before oviposition and in the period during which the female tenders the egg. In non re-productive female castes of social insects, oosorption is also observed. In normal females of *Schistocerca,* there is always a limited number of degenerating ovarioles. Partial ovariectomy results in less resorbed oocytes in the remaining ovary, indicating that the concentration of yolk precursors in the haemolymph may be an important factor. The follicle cells cease to participate in alimentary egg formation. They sometimes divide amitotically and absorb the dead oocyte. Their nuclei subsequently become pycnotic and the cells break down and are absorbed through the ovarian sheaths. After this phenomenon was first observed by Wigglesworth (1936) in *Rhodnius,* it has been described in many other insects, e.g., in *Oncopeltus*, *Calliphora*, Hymenoptera, and *Schistocerca*. In *Rhodnius* and *Oncopeltus,* both with telotrophic ovarioles in which the oocyte remains connected to **trophic strands** for some distance along its passage through the vitellarium, oosorption only occurs after this phase of oogenesis has finished (Wigglesworth, 1936; Johansson, 1958). As a consequence, one may observe alimentary egg formation as well as oosorption occurring in one ovariole.

Following allatectomy, the follicle cells of the Colorado beetle ovary do not divide amitotically and do not take part in the résorption of the oocyte. The proteid yolk bodies are dissolved by lysosome like bodies, which are numerous in the degenerating oocyte. All products which resist this enzymic digestion, such as the crystals in the protein yolk bodies, degenerating nuclei of the follicle cells, lipid yolk droplets and numerous myelin figures, will form the plug with autolysing residues (corpus luteum). As oosorption advances, the volume of the oocyte gradually diminishes so that the follicular epithelium becomes too large. As a result the follicle cells glide along each other, giving the faulty impression of multiplication. Absence of juvenile hormone not only induces lysosomal breakdown of the oocyte, but also arrests the uptake of proteid yolk precursors from the haemolymph, which are nevertheless available in large quantities, and the synthesis of mucopolysaccharide protein complexes by the follicle cells.

3.7 Development of ovarian structure in *Drosophila*

The organogenesis of the ovary and the production of eggs in *Drosophila* is one of the most completely examined developmental systems. Throughout most of the larval stages, the germ cells divide mitotically to produce an unorganised mass of primordial germ cells mixed with somatic cells. Midway through the third instar, however, the metamorphosis of the ovary is initiated, leading to the organisation of the somatic and germ cells into the adult organ. The morphogenetic changes in the undifferentiated ovary begin approximately 12 h after the moult to the third (last) instar, when the anterior somatic cells of the ovary express Ultraspiracle (USP) and the A isoform of the ecdysteroid (or ecdysone) receptor (EcR-A). In the presence of low levels of ecdysteroids, these two proteins interact to activate downstream gene expression including that of the bric a` brac (bab) gene. Under the influence of the BAB protein, approximately 20 terminal filament stacks begin to differentiate from the somatic cells.

After initiating differentiation, the terminal filament cells maintain high levels of USP but no detectable EcR. The first two terminal filament stacks form on the medial side of the ovary, with new stacks forming more laterally, as progressively more are added until 32 h after the moult. By the time of pupariation, the 13–20 terminal filaments, which consist of stacks of 8–9 disk-shaped cells, have been formed in the ovary. These terminal filament stacks establish the positions for the formation of each of the ovarioles within the ovary.

Within 48 h from pupariation, the morphogenesis of the ovary is completed. A small cluster of cells forms at the base of the terminal filaments, in direct contact with the 2–3 germ cells localised within each ovariole. The somatic cap cells and inner sheath cells surround the germ cells and form a niche that provides the signalling that controls germ cell divisions during the adult reproductive period. The primary germ cells function as stem cells and divide asymmetrically to produce one daughter cell that becomes a cystoblast, which enters follicle differentiation, and another daughter cell that retains a stem cell function.

Each ovariole in a mature ovary is divided into three regions, the terminal filament, the germarium, and the vitellarium. Within the germarium, four functional regions, 1, 2A, 2B, and 3, have been described where the previtellogenic follicle forms (Fig 8.10). A cluster of 2–3 germ stem cells is found within a niche in the anterior-most section of the germarium (region 1), immediately adjacent to the terminal filaments. During egg production by the adult female, germ stem cells divide asymmetrically to produce cystoblasts and new stem cells. The asymmetry of the cell division is necessary for maintenance of the stem cells, so that eggs can be produced throughout the adult stage.

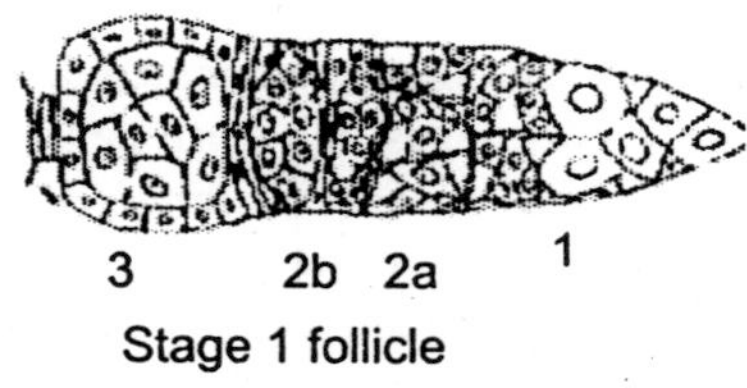

Fig. 8.10 Functional regions of the germarium in *Drosophila*

Following its separation from the germ stem cell, a cystoblast divides four times, with incomplete cytokinesis, within region 2a. This produces the cystocyte complex. The complex is composed of the oocyte that is connected by cytoplasmic bridges to the 15 nurse cells. The asymmetric division of the cystocytes leads to the assignment of one cell as the oocyte, with the remaining daughters of the cystocyte progeny becoming nurse cells. After the formation of the oocyte–nurse cell complex, the dividing mesodermal cells, associated with the tunica propria in region 2b, separate from this membrane and attach to the oocyte–nurse cell complex as follicle cells. These eventually intercalate between the germ cells. The formation of the differentiated previtellogenic follicle (oocyte–nurse cell complex surrounded by the follicular epithelium cell monolayer) is completed within region 3, and the assembled follicle is expelled into the vitellarium. There it undergoes the transition from previtellogenesis to vitellogenesis and, subsequently, the transition to choriogenesis and ovulation.

During vitellogenesis, the oocyte receives material from the nurse and follicle cells and from the haemolymph. Some of these components represent differentially distributed determinants that establish pattern formation during early embryonic development, while others are nutrients necessary for the completion of embryogenesis. Upon completion of provisioning the oocyte and before the follicle is expelled into the oviduct, the follicle undergoes the transition into the choriogenic period. During this period, the oocyte is first sealed within a vitelline membrane before being covered by the chorion.

3.7.1 Egg chamber development in *Drosophila*

Egg-chamber development in the *Drosophila* ovary begins in the germarium. At the anterior tip, specialised somatic- follicle cells make a niche that supports the germline stem cells, which divide asymmetrically to produce two daughters — a cystoblast and a daughter stem cell that remains in the niche. The cystoblast goes through four subsequent rounds of cell division that give rise to a cluster of 16 germline cells, which remain connected to one another as a result of incomplete cleavage. Of the 16 germline cells, one differentiates into the oocyte and enters meiosis, while the other 15 become polyploid nurse cells. Follicle cells then surround the 16-cell cyst and the oocyte moves into the most posterior position. At this point, the nascent egg chamber buds off the germarium to form stage 1 (Fig 8.11). Stages 2–7 are characterised by polyploidisation of the nurse cells, mitotic division of the follicle cells and increases in the size of the egg chamber. As the egg chamber continues to grow, it is pushed further posterior within the ovary as successive egg chambers bud from the germarium. Egg chambers remain connected to each other by chains of stalk cells. In addition, a specialised pair of follicle cells, polar cells, differentiates at each end of the egg chamber. They stop dividing soon after they become specified, whereas the rest of the follicle cells continue to divide until stage 6. The oocyte begins to accumulate yolk at stage 8. At stage 9, most of the follicle cells change from their original cuboidal shape to a columnar shape and stack up in the posterior half of the egg chamber, directly contacting the oocyte. A few cells remain in the anterior and stretch out into a thin layer that covers the nurse cells. At the same time, the anterior pair of polar cells recruits 4–8 neighbouring follicle cells to form the border-cell cluster. The outer border cells detach from the epithelium, become invasive and carry the polar cells down the centre of the egg chamber to the oocyte–nurse-cell border. By stage 10, these rearrangements are complete and the ooctye occupies half the total volume of the egg chamber. The nurse cells nourish the oocyte by transferring cytoplasm. In the final few hours of oogenesis, the nurse cells undergo apoptosis and the follicle cells produce the eggshell.

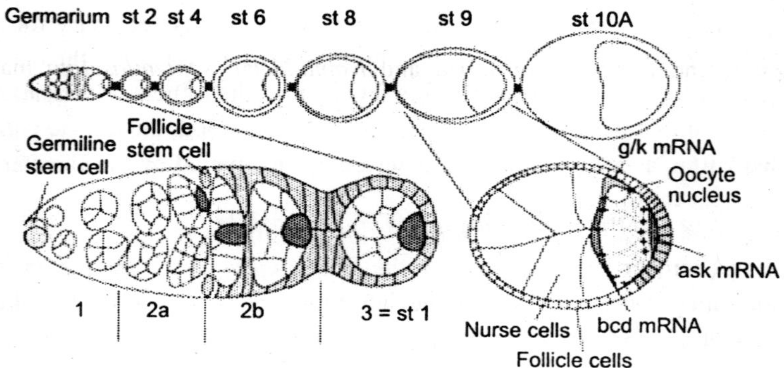

Fig 8.11 Schematic diagram of germarium and egg chamber of increasing age in *Drosophila*
(http://*www.cshperspectives.net/*)

4. OVIPOSITION

Oviposition behaviour in insects is adjusted mostly too specific requirements and living conditions of the larvae. Oviposition occurs in a wide range of cryptozoic (in-litter), endogean (inground), subcortical (under-bark), endophytic (within plant), endozoic (within animal), exophytic (on plant), exozoic (on-animal), aquatic (in-water), semi-aquatic (periodically submerged) and other exposed habitats.

Ovulation and oviposition may not take place simultaneously, i.e. some species store a considerable number of mature eggs in the outer genital ducts until suitable conditions for oviposition arise. Many species of insects show behaviour which encompasses various levels of parental care. For example, Phasmida apparently do not exhibit any kind of adaptive oviposition behaviour. The females simply drop their eggs, which then fall among the dead leaves below the food plant. Eggs of the phasmids have a hard shell and do not require additional protection for survival. Similar situations have been observed in many species of several orders and are usually related to the species' particular habitat. The other extreme in oviposition behaviour is to be found in the digger wasps, *Ammophila campestris*. This species digs a burrow, provisions it with a paralysed caterpillar onto which it lays one egg, and then closes the burrow. Later, the burrow is inspected and, after the larva has hatched, more prey is provided as needed; several inspections take place. After the larva has begun to pupate, no more prey is provided and the burrow is finally closed. A female may work on several burrows simultaneously and always provisions them as required.

Grasshoppers and locusts, generally deposit their egg pods into loose and moist soil. Some such species are *Orthacanthacris aegyptia*, *Schistocerca gregaria*, and *Locusta migratoria* and *Nomadacriti septemfasciata* may probe in dry or sodden soil but then will abandon the site without depositing any eggs. Relatively high salinity (0.05 M NaCl or higher) causes the females of *Locusta* not to lay their eggs. Chemosensors seem to be located on the tip of the abdomen.

Several factors, such as salinity, pollution, reflectance, and texture of the surface, influence oviposition in mosquitoes. Generally, waters of above 0.5- 1.0 per cent NaCl are avoided by *Culex fatigans* and *Aedes aegypti*. Salt receptors have been found on the tarsi.

Some insects chose specific plants to lay her eggs. *Manduca sexta,* the tobacco hornworm, oviposits preferably onto tomato rather than other solanaceous plants. The female is attracted to tomato olfactorily, since, after amputation of her antennae, she oviposits indiscriminately. The leaves of the tomato plant contain an ethanol and water-extractable compound which is the attractant. *Leptinotarsa decemlineata,* a phytophagous beetle, normally oviposits on potato, which is the normal food plant of the larvae. However, if another solanaceous plant, *Solarium nigrum,* is available, the females will select this plant over the potato, even though *S. nigrurn* is not a food plant.

Oviposition behaviour may depend on a complex array of information that the species obtains from its environment. This may apply even more so in certain species of Odonata. Species of this order may simply drop their eggs into the water during flight, or deposit them into submerged plants, even going under water for up to 30 min for oviposition. Species of the genus *Lestes* and *Sympecma*, as well as others, oviposit while in tandem with the males. The male of the species apparently guides the female to the oviposition site.

The oviposition behaviour of insects can be peculiarly suited to the conditions of the species habitat or the substrate that they oviposition. For example, species of the tettigoniid *Scudderia* insert their flat eggs into leaves of various plants in an interesting manner. The female first chews away the edge of the leaf and then parts the upper and lower surface of the leaf with her ovipositor. This peculiar mode of oviposition certainly protects the eggs from desiccation, since the leaves stay fresh throughout the embryonic development of the species. Another curiosity found in the dipteran parasite *Villa quinquefasciata* may be worth mentioning, since it is a case where a parasite oviposits independently of the presence of the host (Biliotti *et al.*, 1965). The female of this species possesses a large ventral abdominal pouch which she fills with fine sand prior to oviposition. Then, while in flight, the female shoots her eggs—together with the sand particles—underneath stones, leaves, or debri; this behaviour continues until there is no more sand in the pouch.

Mating is important in the stimulation of oviposition but it is also important in the inducement of the regular oviposition behaviour. For example, virgins of *Porthetria dispar* lay irregularly scattered masses of eggs which are not covered with the glandular secretion and scales (so-called "Afterwolle") that are observed in the mated females (Schedle, 1936). Virgins of *Coccinella undecimpunctata aegyptica* scatter their eggs irregularly, whereas mated females lay theirs in an orderly fashion.

In *Locusta migratoria manilensis,* the entire process of oviposition appears to be controlled by the last abdominal ganglion since no oviposition took place after its removal whereas cutting the ventral nerve cord anteriorly did not affect oviposition behaviour (Quo, 1959). However, the control of oviposition may be different in different species, as may be illustrated by *Polistes gallicus* which generally does not oviposit after ventral nerve cord severance. Also, other nervous centers may be superimposed on the last abdominal ganglion, as is suggested by the observation that decapitation often induces oviposition. Nervous control of oviposition most likely is only part of the entire control mechanism which governs egg laying in any given species. Possibly the actions of humoral or endocrine agents are superimposed. This is suggested by the following reports. In *Drosophila melanogaster,* implantation of paragonia (male accessory sex glands) into virgins stimulated egg laying (Leahy, 1966), as did mating with sterile males which had functional paragonia (David, 1963). It appears that the paragonia of *Drosophila* and the accessory glands of *Aedes aegypti* (Leahy and Craig, 1965) contain a principle that causes oviposition. Other hormonal principles are possibly involved in other species. For example, in *Bombyx mori,* injections of blood from females or males or of a 2 per cent saline solution stimulated oviposition.

Oviposition behaviour often consists of a sequence of stimulus-response interactions. This is found particularly often in the more sophisticated cases, such as grasshoppers, mosquitoes, or parasites. In all of these instances, the animal receives chemical and tactile stimuli via antennae, tarsi, and ovipositor which make her decide to lay or not to lay the eggs. The parasite *Mormoniella vitripennis* illustrates this particularly well (Edwards, 1954). First the female is attracted by the odour of the host or the area in which the host lives, then she recognizes the shape of the host; together these stimuli cause her to drill. If the host is alive—which is realised after insertion of the ovipositor—the number of the eggs deposited can vary according to the size of the host. Quite similar sequences of events can be established for many species.

Entomologists and physiologists are attracted by yet another aspect of insect oviposition behaviour—the often rhythmic circadian peak of egg laying. For example, many noctuids, such as *Porthetria,* oviposit at night (Klatt, 1920). Circadian oviposition behaviour also appears in two moths which are injurious to European vineyards, *Polychrosis botrana* and *Clysia ambiguella.*

5. STRUCTURE AND PHYSIOLOGY OF EGGS AND EGGSHELLS IN INSECTS

Insect egg shapes vary significantly ranging from cylindrical with rounded or pointed poles to spherical or hemispherical, or irregular. They also exhibit surface projections and a variable number of appendages originating from the anterior, the posterior or both poles. Their size also varies considerably (Box 8.3) , ranging from the tiny eggs of the dipteran *Acnemia amoena* (Mazzini and Santini, 1983) with about 140 mm diameter to the more than 10 mm long eggs of the orthopteran *Saga pedo* (Mazzini, 1976).

5.1 Eggshell layers

In *Drosophila,* the main body of the mature egg were divided into two major overlays, which from the oocyte outwards are the vitelline membrane (otherwise termed the vitelline envelope) and the chorion. Furthermore, in the chorion four successive layers were recognized, namely the plaques of the wax layer, the crystalline innermost chorionic layer, the amorphous tripartite endochorion and the fibrous exochorion.

Box 8.3 The largest and smallest insect egg

The largest insect eggs are those of carpenter bees (Xylocopini). The eggs of many species remains to be measured but *X.auripennis* produced the largest egg reported so far: 16.5mm in length and 3.0 mm in diameter.

Taking into account female size, the smallest insect eggs are the microtype eggs of Tachinidae which are usually 0.02 to 0.2mm long but very rarely as long as 0.4mm. The eggs of *Zenillia pullata* are exceptionally minute, only 0.027 by 0.02mm. The volume of the macrotype egg of *Gymnosoma* sp., is approximately 2000 times the volume of the microtype egg of *Z.pullata.*

The vitelline membrane represents a continuous amorphous, porous, granular or perforated layer (Chauvin and Barbier, 1979), with regular or irregular shape covering the entire oocyte. Similarly, the wax layer that according to permeability studies should exist in most insect eggshells is organized in 4 to 7 planes of overlapping hexagonally shaped plaques that are squeezed one upon the other following completion of oogenesis due to oocyte expansion against the rigid endochorion. No wax layers have been observed in the parasitic hymenopteran species, presumably because these eggs never develop in a dry substrate (Margaritis and Mazzini, 1998).

The endochorion can be further subdivided into the perforated inner endochorion, the pillars creating a network of cavities (otherwise termed the **trabecular layer**) and the continuous layer of the outer endochorion. The described configuration of the eggshell layers seen in the main body o f the egg is significantly altered across the extremely variable specialized structures seen across the anterio-posterior axis of the insect egg. In Drosophila species these regions constitute the so-called regional complexity and include the posterior pole, the main body, the operculum, the collar, the micropyle and the respiratory appendages.

The endochorionic layers vary significantly even in very closely related insect species and show hardly any specific features within a larger taxon. This variability could be attributed to the fact that the endochorion is directly affected by the micro-environment conditions of the ovipositional site and thus should exhibit unique features to accommodate specific physiological functions during embryogenesis.

5.2 Physiological function of eggshell

The eggshell accomplishes several functions of the reproductive machinery; it protects the embryo from environmental hazards such as temperature fluctuations, humidity, dryness and bacterial attack, provides elasticity to accommodate oviposition through a narrow ovipositor, allows and facilitates sperm entry through the micropyle apparatus, ensures an adequate oxygen supply for proper embryo development and finally liberates the larva.

Occasionally, several other species-specific functions are attributed to the eggshell such as the one achieved by the anterior cup in mosquito eggs which serves as a hydrating device to hold the eggs upright on the water surface (Hinton, 1981) or the bristles and projections in the eggshells of mallophagans and anoplurans (Phthiraptera) that facilitate attachment to feathers of the host.

5.2.1 *Spermatozoon Entry: Micropyle Apparatus*

Spermatozoon entry is facilitated by the micropyle apparatus. The structural organisation, the number of micropyles and the micropyle apparatus location on the egg varies significantly among insect species. When only one micropyle apparatus is present it is usually located at the anterior pole of the egg, as seen in dipteran Drosophilidae and Tephritidae. It appears to be a protrusion that occasionally can be very long, as in the case of the hymenopteran Eurytomu umygduli. Micropyle apparatus have similar architecture in all Drosophilidae studied so far, consisting of an external orifice (the micropyle) and a micropylar canal that connects the micropyle with the oocyte (Zarani and Margaritis, 1991). The micropyle apparatus is believed to block polyspermy by restricting passage to a single sperm.

In contrast, in Lepidoptera, the micropyle apparatus forms a depression and is usually surrounded by a rosette consisting of numerous petal-shaped imprints. In the lepidopteran *Bombyx mori* and *Antheraea polyphemus* the number of canals range from 2 to 6 per egg, radiating asymmetrically from a common external orifice (Margaritis, 1985). The micropyle apparatus of the stick insect *Carausius morosus* has two micropylar canals (Godecke and Pijnacker, 1984) whereas in the anopluran *Pediculus humanus cupitis* numerous micropyles are located at the anterior pole (Mazzini and Ermini, 1978). The micropyle apparatus is located at the dorsal side of the egg in stick insects, while in the orthopteran *Eyprepocnemis plorans* about 40 micropyles located posteriorly can be seen arranged in a regular ring-like manner (Viscuso *et al.*, 1990). It is worth mentioning that in some Apterygota no micropyle apparatus has been reported (Larink and Bilinski, 1989). Nevertheless, despite the vast micropyle apparatus heterogeneity, it seems that this structure has been acquired early in insect evolution since it is present even in the most primitive insect orders.

5.2.2 *Oxygen Entry: Aeropyles and Plastron*

One of the most important physiological functions of the eggshell is to allow respiratory gas molecules to diffuse freely in and out during a passive process based on diffusion and governed by the co-efficient and the relative partial pressure of oxygen on either side of the eggshell (Hinton, 1981). To meet embryonic oxygen the chorion in some species is perforated by aeropyles, ranging in diameter from 0.2 to 2.5 μm, an order of magnitude greater than the mean free path of oxygen in air. Aeropyles connect to a gas-filled meshwork in the endochorion. Oxygen uptake occurs by diffusion down a partial pressure gradient generated by embryonic metabolic activity, although passive gas uptake in eggs with thin chorion could also directly occur from the egg surface that often appears rough, revealing in most cases imprints of the follicle cells that participate in eggshell production and secretion. Prolonged submersion on the other hand, requires more elaborate eggshell modifications to enable oxygen absorption through plastron respiration (i.e. retention of a film of gas around a submerged body by means of hydrofuge structures) like the respiratory appendages seen in Drosophilidae. These structures are tubular chorionic extensions serving as physical gills when the egg is submerged under water, where an extensive water-air interface is established. They are seen in many insect orders and their number could be as many as 26 in the *Borborphilus primitiva* (Hemiptera), whereas their length in some *Nepidae* can be as much as 10 mm (Hinton, 1981).

Drosophila respiratory appendages are impermeable to water due to their hydrophobic properties and surface texture and they extract oxygen from the ambient water, transporting it by means of diffusion via partial pressure gradients towards the oocyte. In Drosophilidae the specialised regions of the operculum and the posterior pole may also function as plastrons. In insect eggs bearing aeropyles the water-air interface across the aeropyles themselves acts as a plastron. In these species the efficiency of the plastron can be enhanced by increasing the number and the width of the aeropyles. Plastron-bearing chorions are also prevalent among species ovipositing in decaying animal matter (such as carrion and dung) where water surface tension is reduced by the organic solutes.

5.2.3 *Eggshell Structures for larval Hatching*

Hatching is accomplished through regions of weakness that differ substantially in structure from the rest of the eggshell. In *Drosophila* species larva hatching at the end of embryogenesis is facilitated through the collar region that consists of two discontinuous endochorionic halves held together by a common layer of fragile inner endochorion (Margaritis *et al.*, 1980). On the other hand, eggs of species like the fruit flies *Dacus oleae* or *Ceratitis capitata* feature no visible structural specialization involved in hatching and in these species hatching is accomplished by random tearing of the eggshell at the main body. The lack of weakness lines is apparently related to the extreme stretching that these eggs receive during their passage through the very narrow ovipositor. Some of these eggshell related physiological properties, however, like elasticity, might represent evolutionary remnants since eggshell elasticity in Drosophilidae is rather useless as compared to species like *Dacus oleae*.

6. MODES OF REPRODUCTION

Insects are bisexual; they can undergo sexual reproduction for producing either the eggs (or) the young ones. However they also reproduce by other means. Different types of reproduction in insects are (Box 8.3):

6.1 Oviparity

Insects reproduce by laying eggs by the female on any substrate either singly (or) in mass (or) in groups which later hatch and produce the young ones. e.g.: moths and butterflies.

6.2 Viviparity

It is the phenomenon of reproduction where the female gives birth to the young ones instead of laying eggs. Embryonic development is completed within the female and the embryo get nourishment from the mother and produces the larva/young ones within the haemocoel. These young ones after some time emerges out from the mother, based on the source of nourishment. Viviparity is classified into four types.

1. **Ovo-viviparity:** Insects retain the eggs within the genital track until the eggs are ready to hatch (or) giving birth to young ones. However immediately after hatching , the young ones will be released outside. Hence, no special structures are developed for nourishment. e.g.: Thysanoptera
2. **Adenoparous viviparity:** It is a type of viviparity where the eggs have sufficient yolk, complete their embryonic development and retain in the uterus. Eggs hatch and the young ones get nourishment from special nourishment glands called **milk glands** which contains milk that have lipids and proteins. When young ones are fully developed, it emerges from the mother and forms in to a pupa within short time and no feeding phase is seen. e.g.: *Glossina pupipara* of Diptera.
3. **Pseudoplacental viviparity:** It is a phenomenon where insect have eggs with little (or) no yolk. Hatching takes place within the mother and the nourishment for the young one is received through embryonic maternal structure called **pseudoplacenta** e.g. Psocoptera, Dermaptera, aphids etc.
4. **Haemocoelous viviparity:** It is a type of reproduction where the eggs are retained within the haemocoel and the embryonic development as well as the nourishment of young one takes place through the transfer of nutrients from the haemolymph of mother. After development, young one comes out either through genital canal or by the rupture in the walls of the parent. Eggs have no chorion but become surrounded from an early stage by a trophic membrane through which nutrient material are supplied from the maternal tissues. e.g: strepsipterans & some larvae of cecidomyids (Diptera).

Box 8.3 Means of Producing offsprings

1. Oviparity: egg laying initiates development
2. Viviparity
   A. Ovoviviparity: allows offsprings to take advantage of highly ephemeral resource
   B. Pseudoplacental: egg nourished in reproductive tract
   C. Hemocoelous: larvae free in hemocoel
   D. Adenoparous: milk glands, larva pupates immediately on deposition

6.3. Parthenogenesis

It is the ability of the females to reproduce without fertilisation / copulation with males. This usually occurs due to the genetic characters, due to heredity, failure in finding a mate, hormonal changes within the body and weather factors. Parthenogenesis is classified a;s

1. **Sporadic parthenogenesis**: Occurs occasionally e.g.: silkworm.
2. **Constant parthenogenesis**: Occurs regularly. e.g.: thrips
3. **Cyclic parthenogenesis**: it is nothing but the alternation of generations where parthenogenesis occurs in alternation with the sexual reproduction. e.g.: aphids.

Based on the **sexes of the off springs produced**, parthenogenesis can be (Box 8.4)

1. **Arrhenotoky:** Only males are produced e.g. Hymenoptera.
2. **Thelytoky:** Only females are produced e.g. acridids.
3. **Amphytoky:** Both females and males are produced e.g. hymenopterans.

Arrhenotoky would seem to have certain disadvantages with respect to the propagation of the species since no females are produced; in certain cases, however, it may even save the population. In long-lived species, the sons can mate the females of the earlier generation which then produce daughters. An example of this is the chalcid wasp *Melittobia acasta*. Thelytoky can be obligatory; even if males occasionally occur, they may have no function in reproduction, they may be an atavism. This is found, for example, in *Culicoides bermudensis*, *Pycnoscelus surinamensis*, *Pristiphora erichsonii*, *Carausim morosus*, and in the coccid *Pulvinaria mesenbryanthemi*, as well as in some others. Thelytokous parthenogenesis apparently can occur in many species in which it is often not the regular mode of reproduction. Embryos and larvae of these facultative parthenogenetic species frequently have a very low viability and generally only a few animals reach a reproductive age. This is, for instance, known to be true for **parthenogenetic** *Culex fatigans* (Kitzmiller, 1959) and several species of grasshoppers and cockroaches.

In many species of the Aphidoidea, one to several thelytokous generations alternate with a sexual generation, i.e. these species have annual or biannual cycles known as **heterogony** Fundatrices, virgo, emigrantes, exules and sexuparae reproduce parthenogenetically; the former three forms produce exclusively females, the latter one either male or female sexuals. It appears that environmental factors, such as photoperiod and temperature, influence the succession of generations, i.e. control to some extent the number of thelytokous generations. This is dramatically illustrated by the observation that *Aphis fabae* could be reared for 258 generations exclusively by thelytoky under long photoperiods (16-18 hr light) and elevated temperatures (Müller, 1954). In nature, some species may have only one thelytokous generation, while others are solely parthenogenetic, as for example, some species of the Pemphigidae. In some cynipids, a parthenogenetic thelytokous generation alternates with a bisexual one. For example, in *Neuroterus lenticular Is,* the thelytokous females arise from fertilized eggs and lay either female or male eggs.

Parthenogenetic species have a certain advantage over bisexual ones in that their reproductive capacity is theoretically doubled. If a parthenogenetic strain arises within a population and these females are as fecund as those of the bisexual strain, the entire population may become parthenogenetic in time (Stalker, 1956). The distribution and occurrence of parthenogenetic races and species of the Curculionidae in the northern parts of Scandinavia or in the isolated Alpine valleys indicate their success in finding suitable ecological niches.

Box 8.4 Types of parthenogenesis

1. **Thelytokous parthenogenesis:** Production of females from unfertilized eggs:
   A. Can occur via **automixis**: meiosis occurs, two products of meiosis refuse to form diploid female.
   B. Can occur via **apomixis**: no meiosis, diploid egg is produced via mitosis.
2. **Arrhenotokous parthenogensis:** Production of males from unfertilized eggs:
   A. Always produced via automixis
3. **Amphytoky parthenogensis**: both females and males are produced

A distinct disadvantage of parthenogenetic species is their lack of adaptability to changes in environmental conditions. In these species there is no exchange of genetic information between individuals, and new forms can only arise through mutations. The lack of adaptability must eventually lead to extinction of the species, as is perhaps exemplified by the sporadic occurrence of *Saga pedo;* this species is found today only in small isolated populations, and it is assumed that the species became extinct in intervening regions.

6.4 Paedogenesis and Neoteny

It is a phenomenon where the immature insects or stages give birth to young ones. This usually occurs due to the hormonal imbalance. Most of the insects which reproduce by paedogenesis also reproduce by parthenogenesis. e.g. cecidomyids.

Some insects abbreviate their life cycle by loss of the adult stage, or even loss of both adult and pupal stages. In this precocious stage, reproduction is almost exclusively by parthenogenesis. Larval paedogenesis, the production of young by the larval insect, has arisen at least three times in the gall midges and once in Coleoptera. In some gall midges, in an extreme case of haemocoelous viviparity, the precocially developed eggs hatch internally and the larvae may consume the body of the mother-larva before leaving to feed on the surrounding fungal medium. In the gall midge *Heteropeza pygmaea*, eggs develop into female larvae, which may metamorphose into female adults or produce more larvae paedogenetically. These larvae may be males, females or a mixture of both sexes. Female larvae may become adult females or as repeat the larval paedogenetic cycle, whereas male larvae must develop to adulthood.

In pupal paedogenesis, which sporadically occurs in gall midges, embryos are formed in the haemcoel of a paedogenetic mother-pupa, termed a **hemipupa** as it differs morphologically from the normal pupa. This production of live young in pupal paedogenetic insects also destroys the mother-pupa from within, either by larval perforation of the cuticle or by the eating of the mother by the offspring. Paedogenesis appears to have evolved to allow maximum use of locally abundant but ephemeral larval habitats, such as a mushroom fruiting body. When a gravid female detects an oviposition site, eggs are deposited, and larval population builds up rapidly through paedogenetic development. Adults are developed only in response to conditions adverse to larvae, such as food depletion and overcrowding. Adults may be female only, or males.

In contrast, in neoteny, non-terminal instars develop reproductive features of the adult, including ability to locate a mate, copulate and deposit eggs.

For example, scale insects have neotenous females. In this insect, moult to the winged adult males follows the final immature instar, whereas in females, one or more instars relative to the males is omitted. In appearance, the female is a sedentary nymph-like or larviform instar, resembling a larger version of the previous instar. Neoteny also occurs in all female members of order Strepsiptera. In these insects, female development ceases at the puparium stage.

6.5 Polyembryony

It is a type of reproduction where insects reproduce by giving birth to two or more young ones instead of a single one, as two or more embryos are produced from a single egg. e.g.: endo parasitic Hymenoptera like *platygaster.*

6.6 Hermaphroditism

It is a type of reproduction where both male and female gonads are present in the same individual. It may be a functional hermaphroditism as in case of *Icerya purchsi* (or) non functional as in case of stonefly, *Perla marginata.* Hermaphroditic species may be both male and female simultaneously or in succession. Offspring having the same mode of reproduction are produced, often by self-fertilisation.

In *Perla marginata* the testes are fused anteriorly where they seem to form 70 to 100 ovarioles. However, as soon as the testes follicles have differentiated, oogenesis proceeds no further. Oogenesis is not completed and the partially grown eggs degenerate; apparently no vitellogenesis takes place beyond an initial stage. No report is available whether any egg has ever matured in the functional male. The female is normal and has no identifiably testicular tissues.

In *Icerya purchase* both male and female gonads are seen in the same animal. Both gonads differentiate in the larval stages in this species, as in other insects. Definitely, spermatogenesis, as well as oogenesis, does occur in the hermaphroditic gonad. All females have ovo-testes in which spermatogenesis usually precedes oogenesis (proterandry). No pure females were found so far. Self-fertilization is the common pattern of reproduction. Among 2500 individuals, only nine males were found and it is assumed that they arose from occasionally unfertilized eggs. That the males are haploid can be verified cytologically. Males can be produced by either mated or unmated hermaphrodites.

6.7 Castration

It is a type of reproduction where the separation of the individuals occurs mainly due to the development of the reproductive organs. The insects with well developed ovaries develop in to females (queens), the insects with well developed testis develop in to males (drones) and insects with underdeveloped ovaries develop in to workers. e.g.: social insects such as honey bees.

6.8 Alternation of generation (Heterogony)

Insects reproduce by parthenogenesis and bisexual reproduction by alteration of generations. This phenomenon is also known as **heterogony** e.g.: aphids which reproduce by parthenogenesis in summer and undergo sexual reproduction in winter. Heterogony may be obligatory, i.e. intrinsically controlled, or it may be facultative; in the latter case environmental factors, such as day length, temperature, or crowding, influence the appearance of either sexual or parthenogenetic individuals.

A very simple case of heterogony is exhibited by *Acanthochermes quercus,* which lives on *Quercus petraea* or *Q. robur* in Italy. The fundatrix (arising from a fertilized egg) appears in April or early May; she parthenogenetically lays eggs which become either males or females (the sexuales). These females lay the fertilised eggs in May from which the fundatrix will hatch the following April. In this species, parthenogenesis strictly alternates with bisexual reproduction. Here, the fundatrix is at the same time the sexupara, known as a separate generation in most aphid species. The reproductive period in this species of aphid is restricted to only 2 months, but the egg lasts for the rest of the annual cycle.

A primitive heterogonic cycle has also been found in *Mindarus abietinus,* in which species a separate sexupara generation gives rise to the sexuales. As in *Acantho chermes,* the reproductive period is condensed into a short portion of the year, from April to June. Further complications in the life cycle may be seen in the European monoecious species *Dysaphis devecta* which has four generations: **fundatrix, virgo, sexupara,** and **sexualis** in succession. It appears that the sequence of generations is as obligatory in this species as it is in those mentioned above: there is no duplication of any of the generations. Eggs which overwinter are laid in June or July by the females of the sexualis generation. The virgo (fundatrigenia of some authors) can be either apterous or alate, whereas the male sexualis is alate, the female sexualis is apterous. It ought to be mentioned here that the sexupara of this species is either an andropara or gynopara: any given animal gives rise either to the male or to the female sexualis but never to both. This is in contrast to *Acantho chermes* and many other species in which the sexupara produces both males and females.

6.9 Sex determination

Sex determination depends on molecular switches that signal whether the male or the female sex-differentiating pathway will be followed during development; it can be triggered by genetic, epigenetic, or environmental cues.

The most common pattern of chromosomal sex determination is for females to be the homogametic sex, with two copies of one sex chromosome (XX), whereas males are the heterogametic sex, with two different sex chromosomes (XY). Some insect groups (e.g., Lepidoptera, Trichoptera) have the opposite pattern, in which males are the homogametic sex (ZZ) and females are heterogametic (ZW). Y chromosomes are usually much smaller than X and do not successfully recombine with the X. Unlike mammals, few sex-determining genes are located on Y chromosomes. Thus, the Y chromosome has been lost entirely in numerous orders of insects, leading to a system in which females are XX but males are haploid for sex chromosomes (denoted XO). A molecular genetics mechanism interacting with chromosomal sex determination has been described in *Drosophila*, and it entails a **molecular counting** mechanism that assesses the ratio of X chromosomes to autosomes early in development. Thus, chromosomal sex determination may generally depend on this type of genetic balance, but whether this mechanism is a general feature of insects or is unique to *Drosophila* is unknown.

Sex chromosomes vary greatly in number and size in insects. Translocations and duplications of sex chromosomes are common and fall into two general categories. First, in groups with XO males, small parts of the X chromosomes may be duplicated and form what are called **neo-Y** chromosomes. Second, entire X and Y chromosomes may be duplicated, so that some species have sexes that are homogametic or heterogametic, but contain multiple copies of the sex chromosomes. For instance, in the oriental rat flea, *Xenopsylla cheopis*, males are X1X2Y, whereas females are X1X1X2X2, although in other insect groups it is not uncommon for there to be up to five copies of the X chromosome. Multiple sex chromosomes like these would change the balance between sex chromosomes and autosomes and thus bring into doubt the generality of genetic balance as a sex-determining signal.

Many insects have no identifiable sex chromosomes, but still have specific genetic loci encoding genes that act as key sex-determining signals. A common pattern is to have a dominant sex-determining factor that specifies male development for individuals with an Ml+ genotype, whereas females have a +l+ genotype. In this regard, M represents a key gene in a system similar to heterogametic sex determination, but lacking sex chromosomes that differ in size. House flies (*Musca domestica*) demonstrate further variations of the same theme, in which intermediate genes change the sexual phenotype. Another dominant genetic factor, F, interacts with rare MlM genotypes to produce females with an MlM Fl+ genotype. Individuals with an MlM + l+ genotype develop as males. Finally, some house fly populations have epigenetic sex determination via a maternally expressed gene. Here, a gene found in females determines whether they will produce only female or only male progeny, regardless of the genotype of their male mates.

Haplodiploid sex determination (sometimes called arrhenotoky—the production of males) is found in all Hymenoptera and Thysanoptera and in some Homoptera and Coleoptera. It provides an interesting study of combinations of epigenetic and genetic sex determination. In haplodiploidy, females arise from fertilised eggs and therefore expression during development, with numerous switches in pathways that eventually lead to somatic sexual differentiation and germ-line differentiation. The sex-determining trigger for male development (XY) is an imbalance between genes on the X chromosome and one of the autosomes (X:A = 0.5), whereas females (XX) have an equal ratio of genes on the X chromosome and autosomes (X:A = 1.0). A molecular counting mechanism controls early transcription of sexlethal (Sxl), the key gene in the sex-determining cascade in *Drosophila* that controls sex determination and dosage compensation. Intermediate genes, especially transformer (tra), interact with one another to control expression of the double-switch gene double sex (dsx). This pathway controls somatic sexual differentiation of male and female morphology and behaviour, in addition to germ-line differentiation leading to mature ovaries and testes.

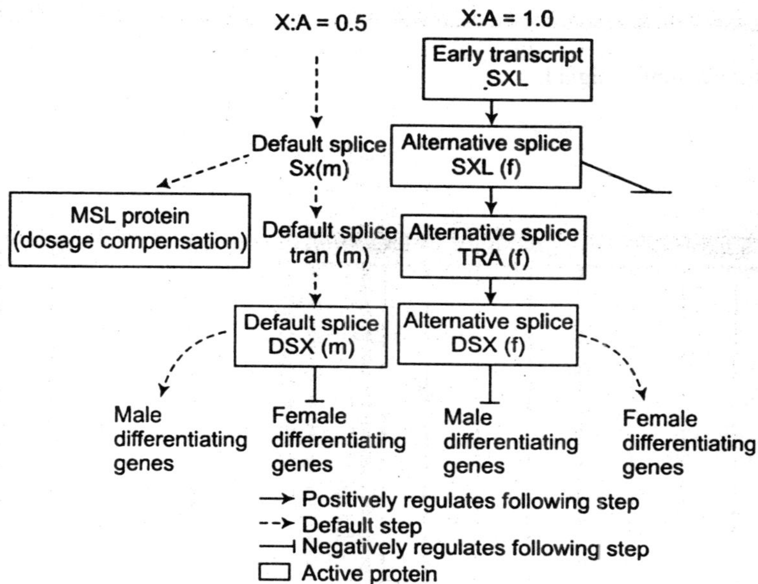

Fig. 8.12 A simplified diagram of heterogametic (XX:XY) sex determination in *D. melanogaster*

In brief, the cascades work as follows (Fig 8.12). In females (X: A = 1.0), the molecular counting mechanism activates the early transcription of Sxl mRNA. SXL protein produced by this mRNA regulates the splicing of subsequent Sxl mRNAs to produce the active female-specific form of SXL (f) protein. Active SXL (f) protein regulates splicing of tra mRNA to produce female-specific TRA (f) protein. In turn, TRA (f) protein interacts with other proteins to regulate the splicing of dsx mRNA, generating the DSX (f) form. DSX (f) protein negatively regulates the genes responsible for male somatic sexual differentiation, but allows genes responsible for somatic sexual differentiation to produce a morphologically female fly.

In male *Drosophila* (X:A = 0.5), the molecular counting mechanism prevents transcription of the Sxl gene early in development. In the next step, the absence of SXL protein results in unspliced Sxl (m) mRNA. This form includes an exon with stop codons embedded in its sequence, so that Sxl (m) transcripts make no functional protein. In turn, transcripts of the tra gene also are not spliced, again resulting in non-functional protein because of internal stop codons. Without TRA protein, there is no alterative splicing of dsx, and DSX (m) protein is made. This pathway is the default, because of the absence of functional proteins from key and intermediate genes early in the pathway. Without these proteins, transcription of the double-switch gene dsx results in male-specific DSX (m) protein. DSX (m) negatively regulates the genes responsible for female somatic sexual differentiation, but allows genes responsible for somatic sexual differentiation to produce a morphologically male fly.

Because male and female *D. melanogaster* have different numbers of X chromosomes (males have one X and females have two), a mechanism is required to compensate for the difference in the number of genes present in the two genotypes. This mechanism is known as dosage compensation. In male *Drosophila*, a group of genes collectively known as **male-specific lethal** (msl) genes are active and increase transcriptional translation of genes on the single X chromosome. Thus, genes on the single X chromosome in males generate approximately twice the gene products, which are approximately equal to those from genes on the two X chromosomes in the female fly. In female *Drosophila* the SXL (f) protein prevents msl genes from acting, and there is no subsequent increase in transcription and translation of genes on the X chromosome in females.

A fairly large number of genes also control germ-line sex differentiation and the production of egg and sperm. The regulation of the genes responsible for germ-line sex differentiation is poorly understood. Although different forms of DSX protein appear to affect the morphology of the sex cells in *Drosophila*, which gene products comprise the switch that determines the fate of germ cells as either egg or sperm remain to be identified.

Insects other than *Drosophila* do not use exactly the same genetic pathways for sex determination. Not only are sex-determining triggers different, but for many insects that have been studied some or all of the genes described above either do not exist at all or do not perform the same function. In addition to *Drosophila*, the gene Sxl has been isolated from four other Diptera (the fruit fly *Bactrocera*, Mediterranean fruit fly *Ceratitis*, phorid fly *Megaselia*, house fly *Musca*), but in none of these does SXL protein differ between males and females. Thus Sxl is not involved in sex determination in these insects, and since no homolog of tra has been found in any other insects outside of Drosophila, it is unlikely that the same genetic pathway for sex determination exists in other insects. On the other hand, the gene dsx appears to be conserved in structure and function in *Bactrocera, Ceratitis, Megaselia,* and the silkworm *Bombyx mori* (Lepidoptera). Apparent homologs of dsx can also be found in the nematode *Caenorhabditis* (mab-3) and in mammals (DMT1), and in each they have sex-specific functions in somatic sexual differentiation andlor germ-line differentiation. The control of sex determination is highly variable in insects, but control of sex differentiation leading to somatic sexual differentiation and germ-line differentiation in males and females appears to be conserved, with the possibility that genes within pathways of sex differentiation of all metazoans have common origins.

7. ENDOCRINOLOGY OF FEMALE REPRODUCTION

In long lived insects that undergo multiple cycles of reproduction, vitellogenesis occurs when the fat body is activated by hormones. In most insects, the nutrients that are available for oogenesis are regulated by insulin-like signalling. The specific hormones that are involved in cyclic vitellogenesis vary among insects and there is no uniform mechanism of hormonal control. In some insects, the production of vitellogenin is dependent on JH alone as in earwigs, grasshoppers, cockroaches and *Rhodnius*. JH actually acts directly on the fat body causing them to initiate the translation and secretion of vitellogenin and its uptake by oocytes.

In tenebrionid beetles, however, JH has no role in ovarian growth and oocyte maturation. Instead, 20E is required for the development of the follicle cells and oocytes. As the oocyte matures, the ovary signals either directly or indirectly to the fat body, with JH initiating the expression of the vitellogenin gene and the synthesis of vitellogenin that is taken up by the oocytes. This system is reverse from what is seen in other insects where JH is involved in oocyte maturation and 20E in vitellogenin synthesis.

In most dipteran, both JH and 20E are involved. JH regulates the formation of new endoplasmic reticulum in the fat body and the sequestration of the vitellogenin produced, while 20E regulates the rate of its production.

8. PHYSIOLOGICAL AND REPRODUCTIVE EFFECTS OF *WOLBACHIA*

The parasitic endosymbiont, *Wolbachia pipientis,* induces a variety of reproductive disorders in their host organisms. These disorders are often species specific and include: parthenogenesis in Hymenoptera and Thysanoptera; feminization in isopod crustaceans; male-killing in Lepidoptera, Drosophila and Coleoptera; and sperm competition and cytoplasmic incompatibility in a diverse array of arthropods. *Wolbachia*, also has many physiological and reproductive effects on arthropods, such as mosquitoes. The deleterious effects of *Wolbachia* on mosquitoes are very important as mosquitoes have several species that act as vectors for deadly infectious diseases.

Thelytokous parthenogenesis is a form of asexual reproduction, in which, females develop from eggs that are unfertilised. While most microbe-associated thelytokous are reversible, *Wolbachia* induced thelytokous parthenogenesis, however, is irreversible. *Wolbachia* bacteria induce irreversible parthenogenesis by causing the reduction and loss of genes: for sperm maturation; for mating behaviour of females; for a major muscle in female sperm reservoir; and the spermatheca. If females infected by parthenogenesis-inducing *Wolbachia* are successfully cured by using antibiotics to remove the bacteria, they can then produce male offspring. Most often, however, the male offspring are not sexually functional, and the formerly infected females become unable or unwilling to mate.

Male-killing strains of *Wolbachia* are elicited in organisms of the order Lepidoptera, Diptera and Coleoptera. The reduction of the male population in a species by *Wolbachia* is to increase the chances of the bacterial cells being transmitted down the generational line. To *Wolbachia* the males in a population are unimportant because the male sperm cells are too small to harbour the bacterial cells for propagation. In contrast the female egg cells are important, therefore, females are favoured. *Wolbachia* infected males in some host species experience decreased sperm competition in comparison to uninfected males of the same species. Significantly fewer sperm is produced by males that are infected with *Wolbachia*, due to the reduction in sperm competitive ability.

The endosymbiont, *Wolbachia*, imposes cytoplasmic incompatibility in many arthropods, resulting in embryonic mortality. When an infected male mates with an uninfected female, unidirectional cytoplasmic incompatibility arises; however, uninfected males can successfully mate with infected females, as can two infected parents. At the other end of the spectrum, bidirectional cytoplasmic incompatibility occurs between individuals infected with different *Wolbachia* strains, when they mate. *Wolbachia* affects mitotic division as a result of spermatogenesis paternal chromosome modification. This modification thus causes a loss in mitotic synchrony. The importance of cytoplamsic incompatibility induced by *Wolbachia* is for use in stopping mosquito vector disease transmission. *Wolbachia* can manipulate reproduction by causing a form of sterility known as cytoplasmic incompatibility (CI), which prematurely stops the development of early embryos.

When a *Wolbachia* infected male mates with an uninfected female or a female infected by a different strain mates, then cytoplasmic incompatibility occurs. Some *Wolbachia* strains stimulate an extremely high cytoplasmic incompatibility (CI) which leads to increased population invasion and also increased mortality and suppression in the mosquito population. Strains such as *wRi Wolbachia* strain spread at a rate of 100km/yr as a consequence of cytoplasmic incompatibility. The increased mortality of vector mosquitoes has no impact on the *Wolbachia* strains because the host organism reproduces prior to death. The life-shortening strain spreads, due to the benefit of higher reproductive success that arises from cytoplasmic incompatibility of individuals with *Wolbachia* bacterial infection.

The most virulent *Wolbachia* strain, popcorn *(wMelpop)* associated with brain degeneration in its host fly, was isolated from *D. melanogaster*. This strain, *wMelpop*, reduces the life-span of the infected flies to about 50% compared to uninfected flies. Density of *wMelpop* strain in different tissues, at the adult stage, is drastically higher than during the earlier stages of life, such as the larval and pupal stages. Bacterial cells accumulate at such high densities, in different tissues, resulting in early death because of the overwhelming host cell pathology. There are several physiological factors that influence how effective *wMelpop* is at shortening the life-span of infected organisms. Temperature is a very important physiological factor. At higher temperatures, the virulence of *wMelpop* increases while the life-span of the bacterial host organism drastically decreased. For instance in *D. melanogaster*, *wMelpop* had little to no effect on the life-span at 19 °C, however, at 25 °C there was a considerable decrease in life-span.

REFERENCES

Beament, J. W. L. (1947). The formation and structure of the micropylar complex in the egg-shell of *Rhodnius prolixus* Stahl. (Heteroptera Reduviidae),. *Exp. Biol.* 23, 213-233

Bilinski, S. (1979). Oogenesis in *Campodea sp.* (Diplura). The ultrastructure of the egg chamber during vitellogenesis. *Cell Tiss. Res.* 202: 133-143.

Bonhag, P.F. and Wick, J.R. (1953) The functional anatomy of the male and female reproductive

systems of the milkweed bug, *Oncopeltus fasciatus* (Dallas) (Heteroptera: Lygaeidae). *Journal of Morphology* **93**, 177–283

Bownes, M. (1994) The regulation of yolk protein genes, a family of sex differentiation genes in *Drosophila melanogaster. Bioessay.* 16: 745-752

Brennan, M.D, Weiner A.J, Goralski, T.J, and Mahowald A.P. (1982) The follicle cells are a major site of vitellogenin synthesis in ***Drosophila melanogaster*. *Developmental Biology***. 89:225–236

Carayon, J. (1952a) Les fecundations hemocoeliennes chez les Hemipteres Nabides du genre *Alloeorhynchus. C.R. Acad. Sci. Paris 234' 751-754*

Carayon, J. (1952b). Les fecundations hemocoeliennes chez les Hemipteres Nabides du genre *Prostemma. C. R. Acad. Set. Paris* 234, 1220-1222

Carayon J. (1959) Inséminatiôn par 'spermalège' et cordon conducteur de spermatozoids chez *Stricticimex brevispinosus* Usinger (Heteroptera, Cimicidae) *Rev. Zool. Bot. Afr.*;60:81–104.

Carayon, J. (1964) Un cas d'offrande nuptial chez les Heteropteres. *C.R. Acad*.Sci. Paris. 259' 4815-4818.

Chauvin, G and Barbier,R. (1979). Morphogenese de l'enveloppe vitelline, ultrastructure du chorion et de la cuticle serosale chez *Korscheltellus lupulinus* L. (Leipidoptera, Hepialidae). *Int.J.Insect Morph. Embryol.* 8: 375-385

Chinzei, Y., Haruna, T., Miura, K., Numata, H., and Nakayama, S. (1990). Purification and characterization of biliverdin-associated cyanoprotein from eggs and hemolymph of the bean bug, Riptortus clavatus (Heteroptera: Alydidae). *Insect Biochem.* 20, 545–555

Cummings, M R and King, R.C. (1969) The cytology of the vitellogenic stages of oogenesis in ***Drosophila melanogaster***. I. General staging characteristics. *Morphology* 128 (4); 427–441

Davey, K. G. (1958) The migration of spermatozoa in the female of Rhodnins prolixus Stal. *J.* Exp.Biol. 35: 694-701.

David, J. (1963) Influence de la fecundation de la femelle sur le bobre et al taille des oeufs pondus. *J.Insect Physiol.* 9: 13-24

Deng W.M. Bownes, M. (1998) Patterning and morphogenesis of the follicle cell epithelium during Drosophila oogenesis. *Int. J. Dev. Biol.* 42: 541 - 552

Dobens, L.L.,and Raftery, L.A., (2000). Integration of epithelial patterning and morphogenesis in Drosophila ovarian follicle cells. *Devel. Dyn.* 218, 80–93.

Edwards, R. L. (1954). The effect of diet on egg maturation and resorption in Mormoniella vitripennis (Hymenoptera, Pteromalidae). *Quarterly Review of Microbiological Science*, 95,459–468

Engels, W., and Bier, K. (1967). *Wilhelm Roux' Arch. Entwicklungsmech. Organismen* 158, 64-88.

Godekeke, J. and Pijnacker, P (1984) Slruclure of the micropyle in Ihe eggs of the parlhenogenetic stick insect Carausius morosus Br. (Phasmalodea, Phasmatidea). - *Neth J Zool* 34: 407-4 13; Bri ll/ Nederland.

Gonzalez-Reyes A and St Johnston D, (1994) Role of oocyte position in establishment of anterior-posterior polarity in Drosophila. *Science*. 266(5185):639-42.

Gonzalez-Reyes A and St Johnston D, 1998 Patterning of the follicle cell epithelium along the anterior-posterior axis during Drosophila oogenesis. *Development*125(15):2837-46.

González-Reyes A, Elliott H, **and** St Johnston D.(1995) Polarization of both major body axes in Drosophila by gurken-torpedo signalling *Nature*.;375(6533):654-8.

Hinton, H. E. (1962) A key to the eggs of the Nepidae (Hemiptera) Phy. Ent. Volume 37, Issue 4-6; 65–68

Hinton, H. E. (1981). Biology of Insect Eggs. Vol. 1-3. Pergamon Press, Oxford.

Ire, K and Yamashita,O (1983) Egg-specific protein in the silkworm, *Bombyx mori:* Purification, properties, localization and titre changes during oogenesis and embryogenesis. *Insect Biochem.* 13: 71-80

Izumi, S., Yano, K., Yamamoto, Y., Takahashi, S.Y., (1994). Yolk proteins from insect eggs: structure, biosynthesis and programmed degradation during embryogenesis. *J. Insect Physiol.* 40, 735–746.

Johansson, A. S. (1958). Relation of nutrition to endocrine reproductive functions in the Milkweed-bug, *Oncopeltus fasciatus* (Dallas). *Nytt. Mag. Zool.*, 7, 1-132.

Khalifa, A. (1949). The mechanism of insemination and the mode of action of the spermatophore in Gryllus domesticus. *Quart. J.Microscop.Sci. 90: 281-292*

Khalifa, A. 1950. Spertmatophore production and egg laying behaviour in *Rhodnius prolixus* Stal.(Hemiptera: Reduviidae) (*Parasitology* 40, 283-289.

King, R.C. & S.K. Aggarwal. 1965. Oogenesis in *Hyalophora cecropia. Growth* 29: 17-83.

King, R. C. and Koch, E. A. (1963). Studies on the ovarian follicle cells of *Drosophila. Q. Jl microsc. Sci.* **104,** 297-320.

Klatt, B. (1920). Beitrage zur sexual physiologic des schewammspinners. *Biol, zbl.* 40: 539-58.

Korscheit, E. (1887). *Nova Ada Leopold. Carol.* 51, 181-252.

Larink, O. and Bilinski S. (1989) Fine structure of the egg envelopes of one Proturan and two collembolan species (Apterygota). *Int. J. Insect Morphol. Embryol.* 18: 1731-1737

Larkin, M., Deng, W M., Holder K., Tworogerm M., Clegg, N., Ruohola-Baker, H.(1999) Role of Notch pathway in terminal follicle cell differentiation during *Drosophila* oogenesis. *Dev.Genes Evol.* 209: 302-311

Leahy, M.G, and Craig, G.B., Jr (1965) Accessory gland substance as a stimulant for oviposition in Aedes *aegypti* and *Ae. albopictus*. *Mosq News.* 1965;25:448–452.

Leahy, M.G. (1966) Egg deposition in *Drosophila melanogaster* increased by transplant of male pargonia. *Drosophila Inform. Serv.* 41: 145-146

Mahowald, A.P. (1972) Ultrastructural observations on oogenesis in Drosophila. *J.Morphol.* 137: 29-48

Margaritis, L.H. and Mazzini, M. (1998) Structure of the egg. In: Harrison, F.W, Locke, M.(eds). *Microscopic anatomy of Invertebrates.* Vol 11C "Insecta". Pp 995-1037 .Wiley-Liss Publ., New York

Margaritis L.H, Kafatos F.C, and Petri W.H. (1980) The eggshell of ***Drosophila melanogaster***. I. Fine structure of the layers and regions of the wild-type eggshell. *J. Cell Sci.* ;43:1–35

Margaritis, L.H. (1985). Structure and physiology of the eggshell. In: Kerkut, G.A., Gilbert, L.I. (Eds.), *Comprehensive Insect Physiology, Biochemistry and Pharmacology*, vol. 1. Pergamon Press, NewYork, pp. 153–173.

Martinez, A., and Bownes, M. (1994) The sequence and expression pattern of the *Calliphora erythrocephala* yolk protein A and B genes. *J.Mol. Evol.*38: 336-351

Mazzini, M and Ermini,G. (1978) Sulla fine struttura del micropilo negli insetti. IX. Morfologia dell'uovo di *Pediculus humanus capitis* De Geer (Anoplura: Pediculidae). Redia LXI: 29-267

Mazzini,M., and Santini,L. (1983) Sulla fine struttura del micropilo negli insetti. XVII. L'uovo di *Acnemia amoena* Winnertz (Diptera Mycetophilidae, Sciophilinae). *Frustula Entomol.* N.S VI: 1-12

Mazzini, M. (1976) Sulla fine structura del micropilo negli insetti. IV. Le sculture corionidee come mezzo di identificazione della uova degli Ortotteri Tettigoniodei. Redia LIX: 109-134

Melo, A.C., Valle, D., Machado. E.A., Salerno, A.P., Paiva-Silva, G.O., Cunha, E., Silva, N.L., de Souza, W., and Masuda. H. (2000). Synthesis of vitellogenin by the follicle cells of *Rhodnius prolizus*. *Insect Biochem.Mol.Biol.* 30: 549-557

Miura, K.,Nakagawa, M., Chinzei,Y., Shinoda, T.,Nagao, E., et al., (1994). Structural studies of biliverdin-associated cyanoprotein from the bean bug, Riptortus clavatus. *Zool. Sci.* 11, 537–545.

Mu¨ ller K. (1954) Investigations on the organic drift of the north Swedish streams. *Report of the Institute for Freshwater Research, Drottningholm*, **35**, 133–148.

Norris, M. J. (1932). Contribution towards the study of Insect fertility. I. The structure and operation of the reproductive organs of the genera Ephestla and Plodla (Lepidoptera : Phycltldae).*Proc. Zool. Soc. London* 102, 595-611.

Nose, Y., Lee, J.M., Takatoshi, U., Hatakeyama. M. and Oishi. K. (1997) Cloning of cDNA for vitellogenin of the parasitoid wasp, *Pimpla nipponica* (Hymenoptera: Apocrita: Ichneumonidae): Vitellogenin primary structure and evolutionary consideration. *Insect Bioche. Mol.Niol.* 27: 1047-1056.

Postlethwait, J.H. and Giorgi, F. 1985. Vitellogenesis in insects. In: Browder, L.W (ed): *Developmental biology: A comprehensive synthesis.* Pp. 85-126. Plenum Press. New York.

Raikhel, A.S., and Dhadialla, T.S., (1992). Accumulation of yolk proteins in insect oocytes. *Annu. Rev. Entomol.*37, 217–251.

Raikhel, A.S., Snigirevskaya, E.S. (1998). Vitellogenesis. In: Harrison, F.W., Locke, M. (Eds.*), Microscopic Anatomy of Invertebrates,* Vol. 11C: Insecta. Wiley- Liss, New York, pp. 933–955.

Ray, R.P., and Schupbach, T., (1996). Intercellular signalling and the polarization of body axes during Drosophila oogenesis. *Genes Devel.* 10, 1711–1722.

Rina, M.D., and Mintzas,A.C. (1988) Biosynthesis and regulation of two vitellogenins in the fat body and ovaries of *Ceratitis capitata* (Diptera). *Roux's Arch.Dev.Biol.* 197: 167-174

Roonwal, M.L. (1936). The growth changes and structure of the egg of African migratory locust, *Locusta migratoria migratoiodes* R&F. (Orthoptera : Acrididae). *Bull. ent. Res.*, London, **27** (1) : 1-24.

Sappington, T.W., and Raikhel, A.S., (1998). Ligand-binding domains in vitellogenin receptors and other LDL-receptor family members share a common ancestral ordering of cysteine-rich repeats. J. Mol. Evol. 46, 476–487.

Schonbaum, C.P., Perrino, J.J., and Mahowald, A.P. (2000). Regulation of the vitellogenin receptor during *Drosophila melanogaster* oogenesis. *Mol. Biol. Cell* 11(2): 511--521

Stalker, H.D. (1956) Onthe evolution of parthenogenesis in Lonchoptera (Diptera). *Evolution* 10 : 345-359.

Stay, B. (1965). "Protein uptake in the oocytes of the cecropia moth," J. Cell Biol. , 26, 49-62.

Takadera, K., Yamashita, M., Hatakeyama, M., and Oishi, K. 1996 Similarities in vitellin antigenicity and vitellogenin mRNA nucleotide sequence among sawflies (Hymenoptera: Symphyta: Tenthredinoidae). *J.Insect Physiol.* 42: 417-422

Telfer, W.H. and Melius,M.E. (1963). The mechanism of blood protein uptake by insect oocytes. ***Am. Zool.***, 3: 185-191.

Telfer, W.H., Huebner, E., Smith, D.S., (1982). The cell biology of vitellogenic follicles in Hyalophora and Rhodnius. In: King, R.C., Akai, H. (Eds.), Insect Ultrastructure, vol. 1. Plenum, New York, pp. 118–149.

Telfer,W.H. 1961 The rout of entry and localization of blood proteins in the oocytes of Saturniid moths. *J.Biophys. Biochem. Cytol.* 9: 747-759

Trougakos I P., Lamnissou K, **and** Margaritis LH. (1999) Biochemical and immunocytochemical analysis of vitellogenesis in the olive fruit fly Dacus (bactrocera) oleae (Diptera: Tephritidae). *Cell Biol Int.*;23(6):417-29.

Trougakos, I.P., Papassideri, I.S., Waring, G.L., Margaritis, L.H., (2001). Differential sorting of constitutively

co-secreted proteins in the ovarian follicle cells of Drosophila. *Eur. J. Cell Biol.* 80, 271–284.

Tsuchida, K., Kawooya, J K., Law, J H., and Wells, M.A. (1992) Isolation and characterization of two follicle-specific proteins from eggs of *Manduca sexta. Insect Biochem. Mol Biol.* 22: 89-98

Tu ,H.M., Kim,S.W., Salvatore, D., Bartha, t., Legradi, G., Larsen P.R and Lechan, R.M. (1997). Regional distribution of type 2 thyroxine deiodinase messenger ribonucleic acid in rat hypothalamus and pituitary and its regulation by thyroid hormone. *Endocrinology*, 138: 3359-3368

Van Eeden, F., St. Johnston, D., (1999). The polarization of the anterior–posterior and dorsal–ventral axes during Drosophila oogenesis. *Curr. Opin. Genet. Devel.* 9, 396–404.

Viscuso, R., G. Longo and A.Giuffrida (1990) Ultrastructural features of chorion and micropyles in eggs of Eyprepocnemis plorans (Orthoptera, Acrididae). Italian Journal of Zoology 57(4):303-308

White N M and Bownes, M. 1997 Cloning and characterization of three *Musca domestica* yolk protein genes. *Insect Mol.Biol.* 6: 329-386

Wigglesworth, V.3. 1936; Symbiotic bacteria in a blood sucking insect Rhodnius prolixus stal (Hemiptera?

Triatominae) Parasit», 28: 234-89.

Wigglesworth, V.B. 1943 The fate of haemoglobin in *Rhodnius prolixus* (Hemiptera) and other blood sucking arthropods. *Proc. Roy.Soc.* B131: 313-339

Winston, M.L. (1987) *The Biology of the Honey Bee.* Harvard University Press, Cambridge, MA

Woyke, J. and F. Ruttner. 1958. An anatomical study of the mating process in the honeybee. *Bee World* 39: 3-18.

Woyke, J. (1955). Multiple mating of the honeybee queen (*Apis mellifera* L) in one nuptial flight. *Bull. Acad. Pol. Sei.* 3,175-180.

Zarani F E, Marg and Aritis L H 1991 Ultrastructural fealUres and formation 01' the micropylar apparatus in the cherry fl y Rhago letis cerasi. - J Morphol 208: 205- 214; Philadelphia, USA.

Ziegler R, Engler DL, Bartnek F, Van Antwerpen R, Bluestein HA, Gilkey JC, Yepiz-Plascencia GM. 1996 A new type of highly polymerized yolk protein from the cochineal insect Dactylopius confusus. Arch Insect Biochem Physiol. ;31(3):273-87.

Insect Endocrine System and Regulatory Mechanisms 9

Estradiol

Ecdysone

Retinoic Acid

Juvenile Hormone

INTRODUCTION

For long it was thought that invertebrates lacked a hormone-like system similar to the one observed in vertebrates until Kopeć (1917, 1922) proposed that a factor secreted from the insect brain was responsible for metamorphosis. The idea of an endocrine function in neuronal cells was new and it took almost an additional two decades until Wigglesworth (1934, 1939) demonstrated that neuronal cells are able to synthesise hormones. On the assumption that a brain hormone (neurohormone) activates the release of a moulting hormone, which is involved in the control of the metamorphosis, neuroendocrinology was born. Nevertheless, it took nearly another two decades until the first neurohormone, the brain hormone of the silkworm *Bombyx mori*, was crystallised (Kobayashi and Kirimura, 1958; Kobayashi *et al*., 1962). It then became much more apparent that almost every aspect of life in invertebrates is under hormonal control. Insects and other invertebrates became the perfect organisms to study neuroendocrinology. The fact that insects can survive several brain operations and manipulations in the ventral nerve cord, and still stay alive for days after decapitation provided the possibility of several basic endocrine experimentations. Parabiosis, a common used method, where two or more individuals share the same blood circulation, established the evidence of the brain hormone in experiments with assassin bugs and giant silkworms (Wigglesworth, 1934; Williams, 1946, 1952). An additional fundamental experimental method, which was only possible to be performed in insects, was the ectomy of an endocrine gland or neurosecretory cell cluster to switch off the hormonal related effect. For the 'revitalization' of the endocrine switch, the re-implantation of the ectomized organ or tissue established the function. Ectomy and re-implantation were used to find the source of the brain hormone and the moulting hormone of *Rhodnius* spp. and other insects (Wigglesworth, 1934, 1939).

1. INSECT ENDOCRINE SYSTEM

In insects, two different endocrine systems can be distinguished: the **glandular** and the **neurosecretory system**. In the glandular system, there are just two true endocrine glands known- the **prothoracic glands** and the **corpora allata**. Both glands are specialised for the synthesis and secretion of hormones, i.e. ecdysteroids in the prothoracic glands and juvenile hormones in the corpora allata. In the adult insect, ovaries and testes can also function as an endocrine gland. The other endocrine machinery is the neurosecretory system, characterized by a group of specialised neurons, the **neurosecretory cells**. Typically found in the central nervous system (brain and ventral nervous system), they produce neurohormones (small polypeptides) and neurotransmitters. The neuropeptides are released at the end site of the axon, called **neurohaemal organ**. Neurohormones often control the release of the hormones and vice versa. However, the nervous system also has a regulatory function in the endocrine system. The nervous and endocrine systems are closely linked and are strongly dependent on the developmental and physiological processes of the insect life (Fig 9.1).

A **hormone** is a chemical signal sent from cells in one part of an organism to cells in another part (or parts) of the same individual. They are often regarded as **chemical messengers**. Although typically produced in very small quantities, hormones may cause profound changes in their target cells. Their effect may be stimulatory or inhibitory. In some cases, a single hormone may have multiple targets and cause different effects in each target. There are at least four categories of hormone-producing cells in an insect's body:

1. **Endocrine glands:** The secretory structures adapted exclusively for producing hormones and releasing them into the circulatory system.
2. **Neurohemal organs:** They are similar to glands, but they store their secretory product in a special chamber until stimulated to release it by a signal from the nervous system (or another hormone).
3. **Neurosecretory cells:** They are specialised nerve cells (neurons) that respond to stimulation by producing and secreting specific chemical messengers. Functionally, they serve as a link between the nervous system and the endocrine system
4. **Internal organs:** Thehormone-producing cells are associated with numerous organs of the body, including the ovaries and testes, the fat body, and parts of the digestive system.

Together, these hormone-secreting structures form an **endocrine system** that helps maintain homeostasis, coordinate behaviour, and regulate growth, development, and other physiological activities.

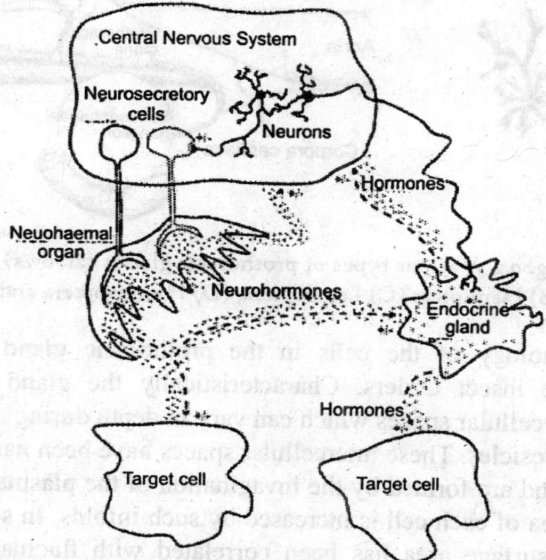

Fig. 9.1 General scheme of the endorine and nervous system in insects

2. ANATOMY OF THE ENDOCRINE SYSTEM

2.1 The Prothoracic glands

These endocrine glands, also called the **head gland, moulting gland, thoracic gland**, or **ventral gland**, probably developed from a maxillary gland, an exocrine gland that evolved into an endocrine gland. They are the largest and most obvious endocrine glands found in the prothorax, just behind the head. In a few insects, part of the head gland is still found in the head, but as its main name indicates, these glands are mainly situated in the pro-, but sometimes also in the mesothorax (Figure 9.2 A to D). In higher Diptera, the prothoracic gland fuses with the corpora cardiaca and corpora allata and form together the Weissmann's ring (later known as the ring gland, Figure 9.2 E).

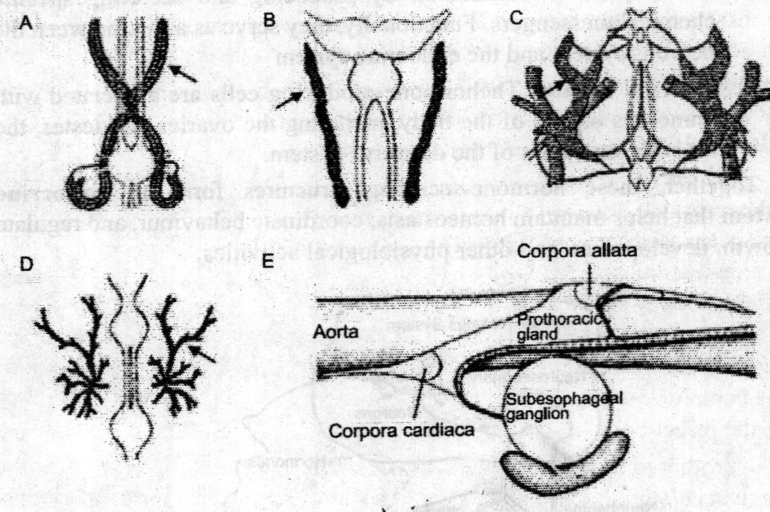

Fig. 9.2 Various types of prothoracic glands (arrows)
(A) Blattodea (B) Hemiptera (C) Lepidoptera, (D) Hymenoptera and (E) Diptera.

The morphology of the cells in the prothoracic gland is uniform throughout the insect Orders. Characteristically the gland consists of peripheral intercellular spaces which can vary in depth during development and has microvesicles. These intercellular spaces have been named lacunae and channels and are formed by the invagination of the plasma membrane. The surface area of each cell is increased by such infolds. In some insects, the increased surface area has been correlated with fluctuations in the moulting hormone. It has been observed that the increase in area occurs when there is a surge in the ecdysteroids.

Box 9.1 The Prothoracic gland (PTG)

1. PTG is destructed in adults of most insects in the absence of JH except in Apterygota which keep moulting during adulthood.
2. Produces moulting hormone (MH)- Ecdysteroids, initiates moulting (larvae) or egg maturation (adults).
3. Insects must obtain sterols (cholesterol or phytosterol), the precursor of ecdysteroids.
4. Ecdysone is converted to active 20-hydroxyecdysone by P450 in fat body.
5. In larvae, PTG synthesize and release MH which targets epidermis.
6. Synthesis of MH is stimulated by prothoracicotropic hormone (PTTH).

Generally the prothoracic gland is not present in the adult insect, except in insects that do not go through metamorphosis and instead still moult as an adult, e.g. the firebrat, *Thermobia domestica*, member of the wingless subclass Apterygota. Almost all other insects start with the apoptosis of these glands during or after the incomplete or complete metamorphosis (Box 9.1).

The prothoracic glands manufacture **ecdysteroids**, a group of closely-related steroid hormones (including ecdysone) that stimulate synthesis of chitin and protein in epidermal cells and trigger a cascade of physiological events that culminates in moulting. For this reason, the ecdysteroids are often called **moulting hormones**. Once an insect reaches the adult stage, its prothoracic glands atrophy and it will never moult again. Ecdysone triggers the transition from larval to pupal moults. It is also responsible for the complex metamorphic remodeling processes that shape the adult organs of the insect body.

Prothoracic glands produce and release ecdysteroids only after they have been stimulated by another chemical messenger, **prothoracicotropic hormone** (PTTH). This compound is a peptide hormone secreted by the Corpora cardiac, a pair of neurohemal organs located on the walls of the aorta just behind the brain. The corpora cardiaca release their store of PTTH only after they receive a signal from neurosecretory cells in the brain. In a sense, they act as signal amplifiers, sending out a big pulse of hormone to the body in response to a small message from the brain.

The prothoracic glands synthesise the ecdysteroids, which are not stored, but immediately secreted in the inactive α-form (α-ecdysone or ecdysterone). The hydroxylation to the active β-form (β-ecdysone or 20 hydroxy-(OH)-ecdysterone) takes place in the fat bodies. In the imagos of

Pterygota, where the prothoracic glands are missing, the ovaries or testes are the sources of the ecdysteroids.

The exact mechanism of release of the hormone from the prothoracic gland is not known, but the microvesicles present in the intercellular spaces of several species may be involved in the transport of material between the gland and the hemolymph.

2.2 The Corpora Allata

The second true endocrine gland pair, the corpora allata, located in the neck region of insects, synthesises and releases the juvenile hormones (JH), a compound that inhibits development of adult characteristics during the immature stages and promotes sexual maturity during the adult stage. Normally it is a pair of small glands (Figure 9.3 A to D), but in larvae of higher Diptera the corpora allata (CA) is incorporated into the ring gland (Fig 9.2 E and 9.3 E).

The CA are made up of closely interdigitated cells, and in many species the gland region is surrounded by a stromal sheath of variable diameter ranging from a thin basal lamina to a lamellated layer within which may be strewn tracheoles and neurosensory cell (NSC) axons. These NSC axons ending within the CA have their cell bodies residing within the brain and the axons traverse the CC before reaching the CA. Corpora allata is usually rich in mitochondria and have a predominance of smooth endoplasmic reticulum and rough endoplasmic reticulum or both. Other organelles present include nucleus, ribosomes, Golgi complex, lysosomes and various inclusions, all appear to vary in their presence and individual morphology for each species.

Neurosecretory cells in the brain regulate activity of the CA, stimulating them to produce JH during larval or nymphal instars, inhibiting them during the transition to adulthood, and reactivating them once the adult is ready for reproduction (Box 9.2). The chemical structure of JH is rather unusual; it is a sesquiterpene compound more similar to defensive chemicals found in pine trees than to any other animal hormone. JH has profound effects on larval growth, metamorphosis, egg development and sexual behaviour. The pars lateralis in the brain via its projections to the CA is the source of positive and negative control over JH production. One of the neuropeptides reaching the Corpora Allata was identified as allatostatin, which inhibits JH release.

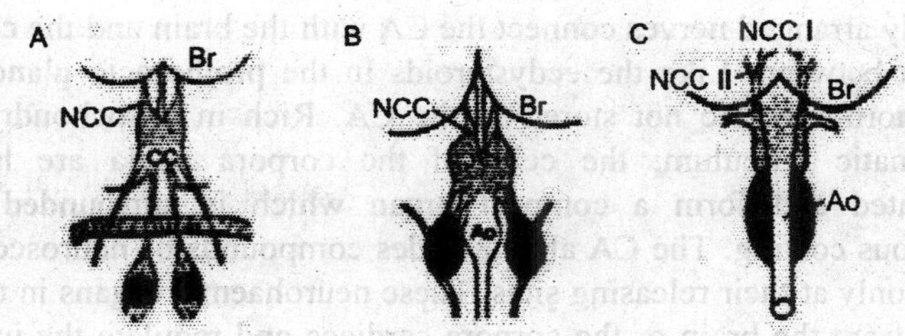

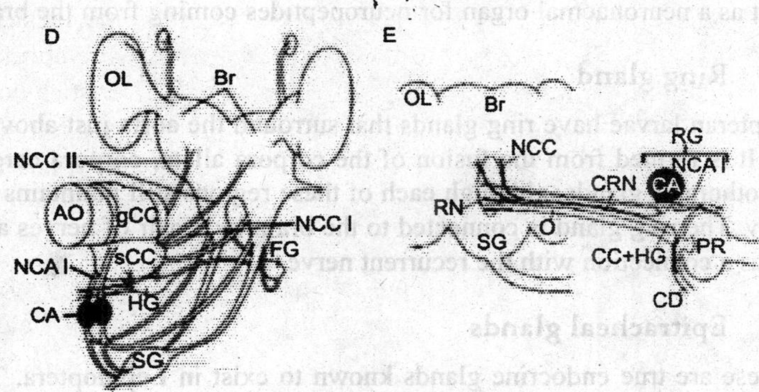

Fig 9.3 Location of the corpora allata (CA), and the corpora cardiaca (CC) in adult (A) Diptera (B), Blattoidea (C) Hemipteran member (D) brain anatomy of *Locusta migratoria* (E) *Phormia terraenovae* (E). (Illustrations taken from Cymborowski (1992), A-C;Diederen et al. (2002), D; and Shiga (2003), E).

Abbreviations: aorta (Ao), brain (Br) corpora allata (CA), glandular (gCC) and storage (sCC) part of corpora cardiaca (CC), crop duct (CD), cardiac recurrent nerve CRN, frontal ganglion (FG), hypocerebral ganglion (HG), nervi corporis allati (NCA-I), nervi corporis cardiaci (NCC, NCC I, or -II), oesophagus (O), optical lobe (OL), proventriculus (PR), ring gland (RG), cardiac recurrent nerve (RN), and suboesophageal ganglion (SG).

Box 9.2 Hormone produced by Corpora Allata

- Juvenile hormone

  JH-I : Mainly in Lepidoptera

  JH-II : Mainly in Lepidoptera

  JH-III : Most insects

  JH-O : Only from Lepidoptera egg

  4-Methyl JH-O
- Chemical : Sesquiterpenoids
- Function : Regulation of moulting, metamorphosis, reproduction

Serially arranged nerves connect the CA with the brain and the corpora cardiac in between. Like the ecdysteroids in the prothoracic glands, the juvenile hormones are not stored in the CA. Rich in mitochondria and endoplasmatic reticulum; the cells of the corpora allata are heavily interdigitated and form a compact organ which is surrounded by a membranous coating. The CA also includes compounds of neurosecretory cells, but only at their releasing sites. These neurohaemal organs in the CA originate from the brain or the corpora cardiaca and regulate the juvenile hormone secretion activity. In a few species of Lepidoptera, the CA also can act as a neurohaemal organ for neuropeptides coming from the brain.

2.3 Ring gland

Dipteran larvae have ring glands that surround the aorta just above the brain. It is formed from the fusion of the corpora allata, corpora cardiaca and prothoracic glands, although each of these regions still maintains their identity. The ring gland is connected to the brain by a pair of nerves and it also has a connection with the recurrent nerve.

2.4 Epitracheal glands

These are true endocrine glands known to exist in Lepidoptera. Their presence in other insect orders is not sure. They are a three celled gland complex found near each of the larval and pupal spiracles. They produce the ecdysis triggering hormone (ETH), a peptide hormone that plays a role in ecdysis behaviour.

2.5 The Neurosecretory System

The neurosecretory system is located in the brain and ventral nervous system. The neurosecretory cells are found in clusters, both medially and laterally in the insect's brain. Axons from these cells can be traced along tiny nerves that run to the corpora cardiaca and corpora allata. The cells produce and secrete **brain hormone**, a low-molecular-weight peptide that appears to be the same as (or very similar to) prothoracicotropic hormone (PTTH) manufactured by the corpora cardiaca. Insect physiologists suspect that brain hormone is bound to a larger carrier protein while it is inside the neurosecretory cell, and some believe that each cluster of cells may produce as many as three different brain hormones (or hormone-carrier combinations). Large numbers of neurosecretory cells also occur in the ventral ganglia of the nerve cord, but their function is unknown.

Neurosecretory cells communicate with organs, different glands and the insect's periphery, or in some cases, they are self-regulating. Neurosecretory cells are different from neurons, because they do not release neurotransmitters from synaptic ends, but secrete neurohormones from the neurchaemal organ. The secretion is granular with characteristic staining properties and the granules pass down the axons. It is probable that the visible secretion is only a carrier molecule, possibly a large protein, to which the smaller hormone molecule is attached. When the hormone is finally released it becomes separated from the carrier molecule and is then free to enter the blood. The general organization of a neurosecretory cell is illustrated in Figure 9.4.

The amount of granular material in a cell at any one time is the result of a balance between its rate of production and the rate of dissipation. Some scientists believe that the cell full of granular material indicate active secreting, while others regard such a cell as inactive one accumulating material.

The endocrine activity of the neurosecretory cells may take one of the two forms. Either the cells may produce hormones which act directly on effector organs or they may act on other endocrine organs, which in turn are stimulated to produce hormones. In this case the neurosecretory cells act as intermediaries between the nervous system and the endocrine glands, responding to the overall situation perceived and analysed by the nervous system. Large number of neurosecretory cells may be present in the nervous system, but it is possible that they do not all produce hormones.

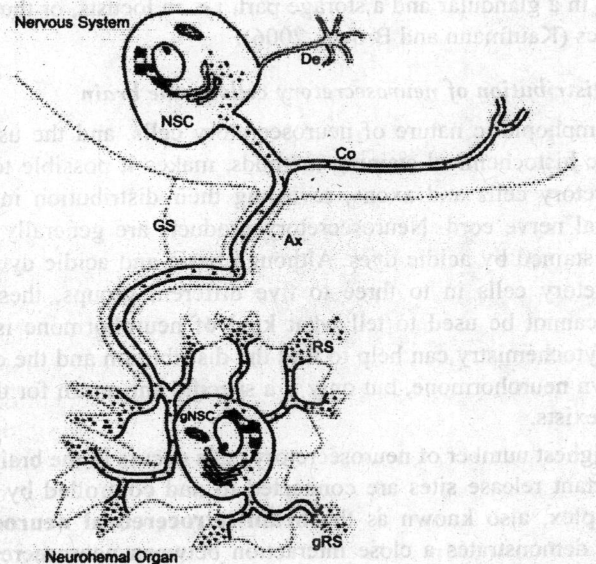

Fig 9.4 General structure of a neurosecretory cell (NSC) and the neurohaemal organ with a glandular neurosecretory cell (gNSC)
glialsheath (GS); axon (Ax), release site of the neurohormone/ hormone (RS/ gRS); collaterals (Co); dendrites (De). (Illustration modified from Raabe,1982)

The synthesis of the neurohormones takes place in the perikaryon, the main cell body of the neurosecretory cell that contains the nucleus and organelles. The dendrites and collaterals are like those in the insect neuron, thus the neuropil, the grey matter of the brain, contains many axo-dendritic and axo-axonal contacts. The often very long axons branch in a pre-terminal bulge at their distal ending, the neurohaemal organ. The glial sheath around the neurosecretory cell and its axon supports their structure and provides insulation. There is less glial sheath found at the terminal end of the axon, where the neurohormones are released from the neurohaemal organ. The peptide hormones are synthesized in the rough endoplasmatic reticulum of the perikaryon, and are later further processed in the Golgi bodies. Granules (carrier proteins) transport the active neuropeptides along the axon to its release site. In case of neurohormone overproduction, lysosomes reabsorb the abundance. The neurohormones are released by exocytosis into the haemocoel. The vesicles burst and release the neurohormones; the perikaryon then reabsorbs and reutilizes the used empty vesicles (Raabe, 1982; Nijhout, 1994). The neurohaemal organ can also include hormone producing cells or glandular neurosecretory cells, like the adipokinetic hormone producing cells in the corpora cardiaca. Some insects even have the corpora cardiaca separated in a glandular and a storage part, i.e. in locusts, or the X-cells in mosquitoes (Kaufmann and Brown, 2006).

2.5.1 *Distribution of neurosecretory cells in the brain*

The amphophilic nature of neurosecretory cells, and the use of basic and acidic histochemical staining methods, makes it possible to stain the neurosecretory cells and axons, revealing their distribution in the brain and ventral nerve cord. Neurosecretory products are generally basic and therefore stained by acidic dyes. Although basic and acidic dyes separate neurosecretory cells in to three to five different groups, these staining methods cannot be used to tell what kind of neurohormone is secreted. Immunocytochemistry can help to find the distribution and the cell source of a known neurohormone, but only if a specific antiserum for the peptide hormone exists.

The highest number of neurosecretory cells occurs in the brain and thus the important release sites are connected to and controlled by the brain. This complex, also known as the **brain-retrocerebral neuroendocrine complex**, demonstrates a close interaction between neurosecretory cells, neurohaemal organs, the endocrine glands, and the central nervous system. Together they manage the regulation of the synthesis, release, and inhibition of the hormones.

Generally, the insect brain is divided into three regions called the **protocerebrum**, the **deutocerebrum**, and the **tritocerebrum** (Fig 9.5). Collaterals, and possible synaptoid contacts, maintain coordination between different neurosecretory centres in the brain. The protocerebrum has two lobes and contains the pars intercerebralis in the middle. Each lobe is divided into the main and lateral protocerebrum. The vision (optic lobes and ocelli, if available) is also associated with the protocerebrum. Most neurosecretory cells are found in this part of the brain. The deutocerebrum contains the antennal lobes and processes the sensory information collected by the antennae. The tritocerebrum, together with the stomodaeal nervous system, regulates the function of the mouthparts, and links the brain with the rest of the ventral nerve cord. The stomodaeal nervous system, which is also referred to as the stomatogastric nervous system, contains the frontal, hypocerebral and the suboesophageal ganglion, and is, linked to the tritocerebrum by the frontal commisure, the recurrent nerve, and the circumoesophageal connectives.

Some of the neurosecretory cell clusters in the brain (Fig 9.5) are named due to their location. The **median neurosecretory cells** are found in the protocerebrum and normally have one or two paired clusters of cells on each side of the brain lobes. Their axons cross over to the other brain lobe through the deutocerebrum and the tritocerebrum, into the collateral located internal nervus corporis cardiaci or NCC-I, which runs to the corpora cardiaca and the thoracic endocrine organs. The **anterior neurosecretory cells** found in the pars intercerebralis also contain two clusters of cells. Their axons do not cross over, but connect ipsilateral into the NCC-I. The other axonal bifurcation runs into the circumoesophageal connective and into the ventral nerve cord. One cluster of the **intermediate neurosecretory cells** is found between the median neurosecretory cells. Their axons run exclusively contralateral in the circumoesophageal connective posterior to the ventral nerve cord. The **lateral neurosecretory cells** show one or two clusters of cells and are situated dorsally and laterally of the protocerebrum in each brain lobe. Their axons do not cross over to the other lobe, but run together into the ipsilateral located external nervus corporis cardiaci or NCC-II, which runs towards the corpora cardiaca, like NCC-I. A third nervus corporis cardiaci, the NCC-III connects to neurosecretory cells found in the tritocerebrum. All three NCCs are bundled into the nervus corporis cardiaci (NCC), which runs in the direction of the corpora cardiaca. The corpora cardiaca act as an enormous neurohaemal organ, but not exclusively, it also contains neurosecretory cells and neurons. A last mentioned group of neurosecretory cells are the **ventral neurosecretory cells** found in the ventral and dorsal part posterior of the tritocerebrum. These neurosecretory cells are not found in all species, but in larvae of *Manduca sexta,* their axons run to the ventral nerve cord and have their release site at the terminal ganglion.

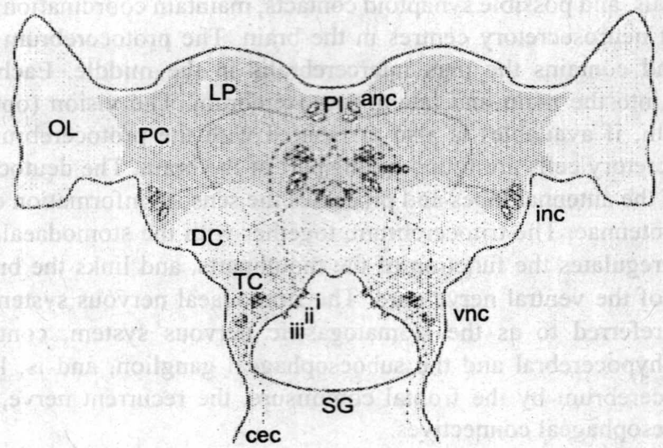

Fig 9.5 Distribution of the anterior (anc), median (mnc), intermediate (inc), lateral (lnc), and ventral (vnc) neurosecretory cells in an insect brain.

Abbreviations :protocerebrum (PC), deutocerebrum (DC), and tritocerebrum (TC); Çec, circumoesophageal connective; i, ii, and iii, nervous corpọri cardiaci I, II, and III; OL, optical lobe; LP, lateral protocerebrum; PI, pars intercerebralis; and SG, sub-oesophageal ganglion. Ipsilateral axons are illustrated as dotted lines and contralateral as dashed lines (After Raabe, 1982; Clements, 1992; Nijhout, 1994).

The architecture of the neurosecretory cells has been the object of many studies, describing them as monopolar neurons with dendrites spreading in the hemispheres and axons joining the first nerve to the corpora cardiac. Neurosecretory cells of the pars intercerebralis secrete insulin-like peptides, FMRFamide-like peptides, Locusta-diuretic hormone, pigment-dispersing hormone, *Manduca sexta*-allatostatin, ovary ecdysteroidogenic hormone, and myomodulin (Box 9.3). The NSCs forming the pars lateralis of the brain produce FMRFamide-like peptides, pigment-dispersing hormone, corazonin, and *M. sexta*-allatostatin (Box 9.4). Their axons form the second nerve of the corpora cardiaca (nccII), which in most insects travels alongside the ncсI; in *Drosophila*, both nerves are enclosed by a single perineurial (glial) sheath.

The hormones secreted by the NSC in ganglia of ventral nerve cord are the Crustacean cardioactive peptide (CCAP) which is involved in ecdysis behaviour and heart rate modulaton; Bursicon, a peptide hormone necessary for tanning of cuticle; Pheromone biosynthesis activating neuropeptide (PBAN) which stimulates pheromone biosynthesis.

Box 9.3 Hormones produced by neurosecretory cells (NSC): Pars intercerebralis

1. Prothoracicotropic hormone (PTTH): Peptide; stimulate PTG to produce moulting hormone (Ecdysteroids)
2. Ovarian ecdysteroidogenic hormone (OEH) : Peptide, stimulate ovary to produce MH
3. Allatostatin : Peptide; inhibit JH production by CA
4. Insulin-like peptides, FMRFamide-like peptides, Locusta-diuretic hormone, pigment-dispersing hormone
5. Myomodulin

Box 9.4 Hormones produced by neurosecretory cells (NSC): Pars lateralis

- FMRFamide-like peptides
- Pigment-dispersing hormone
- Corazonin
- *Manduca sexta*-allatostatin

The tritocerebrum and the ventral nerve cord, as well as the ganglia of the stomatogastric nervous system (SNS) contain neurosecretory cells. NSC axons of the tritocerebrum and subesophageal ganglion project toward the corpora cardiaca form the nccIII nerve. Neurosecretory axons from the SNS also form a compact axon bundle connecting the hypocerebral ganglion with the corpora cardiaca. NSCs of the ventral nerve cord have release site associated with the dorsal glial sheath of the cord and the segmental peripheral nerves.

2.5.2 *Corpora Cardiaca*

The corpora cardiac (CC) is named after the location of its releasing site, which is associated with the insect hearts. The CC is a conglomeration of several neurohaemal organs coming mainly from the brain, but also from the stomodaeal nervous system. The nervous corpori cardiaci (bundle of NCC-I, -II, and -III) runs from the brain to the corpora cardiaca. This large neurosecretory 'cable' ends in the form of the main insect neurohaemal organ, the corporus cardiacum. The release of the neurohormones from the protocerebrum (anterior, median, and lateral neurosecretory cells) and the tritocerébrum takes place mainly from this organ. The CC has two main cellular components which contain neurosecretory cell granules-the extrinsic cells storing brain hormones and the intrinsic cells storing products of their own making. The neurotransmitters released from synaptic connections, and the neurohormones released from intrinsic neurosecretory cells of the corpora cardiaca itself, can both have regulatory function on the corpora cardiaca. The hypocerebral ganglion or the recurrent nerve connects the stomodaeal nervous system and corpora cardiaca, and in some cases the hypocerebral ganglion and the corpora cardiaca are fused together. As mentioned above, the corpora cardiaca also is the source of intrinsic neurosecretory cells, which synthesise and release hormones locally (Box 9.5). Some members of Orthoptera (and some Diptera) have the corpora cardiaca split into two different parts, a glandular and a storage lobe (gCC and sCC in Fig. 9.3D). The storage part contains two compartments, a neuronal one located to the haemocoel site and a neurohaemal site connected to the aorta. The glandular lobe has neurohaemal organs, but also intrinsic neurosecretory cells, in which the main source of adipokinetic hormones synthesis is found (Diederen *et al.*, 2002). In addition, the corpora cardiac produces certain glycemic factors, cardiac-accelerating factors, and melanin-inducing factor. These and other hormones are released from the corpora cardiaca into the aorta directly in the haemolymph. The AKH hormones are the major products of the corpora cardiaca, which are secreted into the haemolymph to mobilize lipids and carbohydrates during flight.

Other intrinsic neurosecretory cells of the corpora cardiaca have axons running towards the corpora allata. The connection between these two organs is called the **nervi corporis allati I** (NCA-I, Fig. 9.3D and E). The secretions from the corpora cardiaca may have regulatory function, but transsection of the nervi corporis allati-I did not affect the function of the corpora allata. Some members of Lepidoptera and Orthoptera contain a second nervi corpori allati, the NCA-II, which does not connect to the corpora cardiaca, but to the suboesophageal ganglion. So far, there has been no evidence found to show that there are any neurohaemal organs coming from the ventral nervous system, connecting to, and releasing from the corpora cardiaca. However, in the ventral nerve cord, most neurosecretory cells either have a neurohaemal organ locally located around the ganglia, or extend their axonal connection that ends into a perisympathetic organ at the periphery.

Box 9.5 The Corpora Cardiaca (CC)

1. CC stores and releases brain hormones
2. Produces its own peptide hormones
   A. Adipokinetic hormone (AKH): Fat (energy) metabolism
   B. Hypertrehalosemic hormone (HTH): control of sugar metabolism, resulting in high concentration of trehalose in blood
   C. Diuretic hormone (DH): Also produced by the NSC in brain and ventral ganglia; water balance and urine production

The mode of release of the NSC material has grabbed the attention of many workers. The two most commonly recorded features associated with the release at the cellular level are exocytosis figures and clusters of electron-lucent synaptic-like vesicles which are called **synaptoids**. In exocytosis, areas of the plasma membrane in CC cells invaginate in a characteristic omega shape representing a recently exocytosed NSC granule. Synaptoids are presumed to be the release site within NSC axons. Whether or not these microvesicles represent membrane retrieval following release of the granule has not been determined with certainty, but it can be stated that the synaptoids do represent release sites of the hormones

2.6 Other endocrine organs

Many other tissues and organs of the body also produce hormones. Ovaries and testes, for example, produce gonadal hormones that have been shown to coordinate courtship and mating behaviours. In adult Diptera the ovary produce moulting hormone that stimulates the fat body to produce vitellogenin and egg maturation. The synthesis is regulated by ovary ecdysteroidogenic hormone (OEH). Ventral ganglia in the nervous system produce one compound (eclosion hormone) that helps an insect shed its old exoskeleton and another compound (bursicon) that causes hardening and tanning of the new one. There are still other hormones that control the level of sugar dissolved in the blood, adjust salt and water balance, and regulate protein metabolism.

2.7 Neuroendocrine System in the Ventral Nervous System

The anatomy of the rope ladder like ventral nerve cord starts with the suboesophageal ganglion, then the thoracic ganglia, followed by the abdominal ganglia, and at the end the terminal abdominal ganglion. The connection between the brain and the ventral nerve cord is through the suboesophageal ganglion, which is often closely combined with the brain. In the thorax, there are three ganglia, one for each pair of legs. Noticeably, they are not only responsible for a functional leg apparatus, but they also regulate the flight muscles and like all ganglia, they contain neurosecretory cells. Higher developed species show a fusion of the thoracic and/or abdominal ganglia. For example in most Culicidae, the thoracic ganglia are fused together, but each abdominal segment still contains one abdominal ganglion, except in the terminal ganglia where fusion is also found. Contrarily, in the housefly, *Musca domestica* all thoracic and abdominal ganglia are fused together. As mentioned above, all ganglia are sources of various neurosecretory cells types. Several hormones are produced by the neurosecretory cells in ganglia of ventral nerve cord (Box 9.6). Most neurosecretory cells end in the segmental analogues of the corpora cardiaca, the perivisceral organs, also called the **perisympathetic organs**, but some act locally on neighbouring cells or are of self-regulating nature. The axons of the perisympathetic organs normally start on the posterior or anterior site of the ganglion and run out to the lateral periphery.

Box 9.6 Hormones produced by NSC in ganglia of ventral nerve cord

1. Crustacean cardioactive peptide (CCAP): peptide; ecdysis behaviour and heart rate modulation
2. Bursicon: peptide; tanning of cuticle
3. Pheromone biosynthesis activating neuropeptide (PBAN): peptide; stimulation of pheromone biosynthesis.

2.8 The Nervous and Endocrine System

The endocrine and nervous systems interact closely to regulate various animal life processes. These two systems have a regulatory nature and function with a messenger system that interacts with a specific target location. The nervous system is mainly responsible for the **short-term regulation** and action, whereas the endocrine system functions as a **long-term regulator**. Influences from the outside (i.e. temperature, humidity, water, characters of colours and surfaces, food sources, or light and darkness) generally activate the insects short-term system first. Insects have to act quickly, if a predator is approaching, or if the wind is changing during flight. Several neurons link receptors on the outside of the insect to the inside; e.g. mechanoreceptors detect wind changes by contact receptors, and insects are able to smell with olfactory chemoreceptors, located in the antennae and palpi. Thermo and hygroreceptors, and naturally the complete visual apparatus, with the compound eyes and the ocelli, recognise other kinds of messages from the outside. However, often after a short-term reaction, the endocrine response follows, showing that the nervous and endocrine systems are dependent on each other. During and after an escape of a predator, hormones regulate the metabolism, e.g. by mobilising metabolites for the leg or flight muscles and they are also responsible for resorting the storage nutrients after the succeeded escape, or initiate wound healing after a fight.

Hormones can also cause a hunger response due to the lack of reserves. There are many behaviour that are under hormonal control, including the endocrine interactions that take place after mating, feeding, or during the gonadotrophic cycle in haematophagous insects like mosquitoes.

2.9 The Midgut Endocrine System

In insects, the midgut is the primary tissue for digestion and absorption of nutrients. The stomatogastric nervous system, which consists of a nerve trunk along the dorsal surface of the foregut, and the proctodeal nervous system, which originates from the terminal ganglion of the ventral nerve cord, innervates the midgut. These nervous systems innervate only the midgut musculature, indicating that the digestive and absorptive cells in the midgut epithelium are not regulated by nerve tissues. Thus, nonneural regulatory mechanisms are expected to regulate digestion and absorption in insect midgut. Apart from the visceral nervous system, the insect midgut epithelium contains secretory cells called **endocrine cells** or **midgut paraneurons**. In insects, the distribution of these endocrine cells was demonstrated in larvae, pupae, and imago of a diverse number of insects. Antibodies, which were used in vertebrates, showed staining for several mammalian-like hormones, as shown for insulin, glucagons, somatostatin, β-endorphins, leucomyosuppressin and more. It was demonstrated that neuropeptide F is found to be expressed in the brain and midgut. Neuropeptide F is involved in food intake and digestion. Another study showed that **cardio accelerating hormone** (CCAP) is secreted from midgut cells and is most probably involved in the release of α-amylase a digestive enzyme (Sakai *et al.*, 2006). It is well reported that tachykinin, AKHs and other neuropeptides can induce midgut muscle contraction (Klowden, 2002), but no evidence has yet been found as to whether these neurohormones also activate the release of hormones that originate from the digestive tract.

Ultrastructural studies have detected exocytosis of secretory granules from insect midgut endocrine cells that have morphological similarity to vertebrate gut endocrine cells. Immuno-histochemical observations using antibodies against mammalian neuropeptides also support the presence of diverse neuropeptide-like compounds in insect midgut endocrine cells. To date, gene expression of insect neuropeptides such as allatostatin, leucomyosuppressin, and neuropeptide F have been detected in the midgut endocrine cells of insects, which is consistent with the physiological role of midgut endocrine cells as a major source of gut neuropeptide.

The midgut endocrine cells probably originate in the midgut nidi and show a variable lifespan. Peptidergic axons also extend along the length of the gut and originate from either the proctodeal or stomatogastric nervous system. Their presence, in concert with the endocrine cells, suggests that multiple modes of regulation of physiological processes by neuropeptides are possible. Examples of peptides present in the midgut include members of the allatostatin, tachykinin, and FMRFamide-related peptide (FaRP) families.

The allatostatins (ASTs) are a family of peptides originally isolated on the basis of their ability to inhibit the production of juvenile hormone in cockroaches. They are characterised by a common pentapeptide C terminus (Y/F)XFG(L/I) amide and a variable N terminus of 1 to 13 residues. In addition to their ability to inhibit juvenile hormone production, the ASTs are able to inhibit both proctolin-induced and spontaneous contraction of muscle of the hindgut and the antennal pulsatile organ. These peptides are present in putative endocrine cells, particularly in the anterior portion of the midgut of the cockroach, *Diploptera punctata,* as well as in nerves running much of the length of the gut. The proctodeal nerves seem to terminate in the densest area of AST IR (immune reactive) endocrine cells and do not extend anterior beyond this region. Cells that show immunoreactivity to other peptide families, including some FaRPs, are abundant in the posterior region of the midgut whereas still other members of this family [e.g. leucomyosuppressin (LMS)], are concentrated in the anterior portion.

Leucomyosuppressin (LMS) was first isolated from the cockroach, *Leucophaea maderae*, based on its ability to inhibit hindgut muscle contractions and has since been shown to inhibit proctolin- induced midgut contractions as well. The myosuppressin subfamily comprises a group of decapeptides that share the common sequence XDVXHXFLRFamide, and that seem to be encoded on a gene separate from the other FaRPs. The actions of myosuppressins on visceral muscle contractions of insects have been well documented. However, members of this family can also stimulate contractions of flight muscle and cardiac muscle, as well as a-amylase secretion from weevil midgut and scallop digestive gland. Studies on ASTs and LMS localisation suggest that there is a differential spatial distribution, as well as a difference in function between LMS and other midgut FaRPs.

3. INSECT HORMONES

Two different types of hormones can be distinguished, the non-polar, lipid hormones and the polar proteinaceous, peptide hormones. They have different surface and solubility conditions, and can connect at different places to the target receptor. Lipid hormones penetrate through cell membranes and act with receptors in the nucleus at the genome, whereas the peptide hormones cannot penetrate membranes and only interact with receptors on the cell membrane. A list of some of the known insect hormones, their functions and chemical nature is provided in table 9.1

3.1 Lipid Hormones

The non-polar and lipophilic nature of the lipid hormones makes it possible for them to pass through the cell membranes into the nucleus, where they often act directly at the chromosomal level. When bound to a nuclear receptor, lipid hormones activate or inhibit the expression of a specific gene. The regulation on the transcriptional level results in a physiological feedback with the consequences of a *de novo* synthesis of regulatory proteins, enzymes, or the disappearance of previously active proteins. The two main lipid hormones in insects are the ecdysteroids and the juvenile hormones.

Table 9.1 Some of the known insect hormones and their functions

Hormone	Source	Function
Ecdysteroid *	Prothoracic gland	Initiation of moulting and egg maturation
Prothoracico tropic hormone (PTTH)	Brain	Control of ecdysteroid production
Bombyxin #	Brain	Ecdysteroid production, ?carbohydrate metabolism
Ecdysis triggering hormone (ETH)	Epitracheal gland	Ecdysis behaviour
Eclosion hormone (EH)	Brain	Ecdysis behaviour
Crustacean cardioactive peptide (CCAP)	Ventral Nerve ganglia	Ecdysis behaviour, heart rate modulation
Bursicon	Brain and ventral nerve ganglia	Tanning of cuticle
Juvenile hormone (JH) **	Corpora allata	Regulation of metamorphosis and reproduction
Allatostatin	Brain	Inhibition of JH production
Allatotropin	Brain	Stimulation of JH production
Adipokinetic hormone (AKH)	Corpora cardiaca	Fat (energy) metabolism
Hypertrehalosemic hormone	Corpora cardiaca	Control of sugar metabolism
Diuretic hormone (DH)	Brain, corpora cardiac, ventral ganglia	Water balance and urine production
FMRFamides	Central nervous system	Neuromodulators, myosuppression
Diapause hormone	Brain, sub-oesophageal ganglion	Initiation of diapauses in embryos
Pheromone biosynthesis activatin neuropeptide	Suboesophageal ganglion	Initiates pheromone production

* Steroid; ** Terpenoid; # Peptide (insulin-like); all remaining are peptides

3.1.1 *Ecdysteroid*

The moulting hormone, today known as **ecdysone**, was the first structurally identified insect hormone (Butenandt and Karlson, 1954). Out of around 500 kg of *Bombyx mori* pupae, 25 mg ecdysone was purified. α-ecdysone and β-ecdysone were isolated, of which the latter represents the active form. Today over 60 different ecdysteroid analogues are characterized in insects. The synthesis of the inactive α-ecdysone takes place in the prothoracic gland. The precursor is a sterol, like cholesterol. Surprisingly, unlike most animals, insects cannot synthesise cholesterol from acetate precursors and have to get the cholesterol they require from their diet. Cholesterol is easily accessible for zoophagous insects, and especially for those of a haematophagous nature, whereas phytophagous insects have to convert campesterol, β-sitosterol, stigmasterol, or ergosterol into cholesterol. Phytophagous insects are also confronted with phytoecdysteroids, which probably work as a natural insect repellent for the plants. The inactive α-ecdysone penetrates through the cell membrane into the haemocoel. In the haemolymph, α-ecdysone binds rapidly to the ecdysone carrier protein that transports it to the fat body. The hydroxylation of α-ecdysone in the fat bodies results in the active β-ecdysone (Fig 9.6) Experiments with *D. melanogaster* showed that ecdysteroids act directly on the transcriptional level. They cause two sequences of specific puffs in the giant polytene chromosomes. Puffs manifest the diffuse uncoiled regions of the polytene chromosome that are the sites of the RNA transcription. The first specific puffs induced by the ecdysteroids are responsible for the translation of the proteins that cause the sec ond sequence of puffs. The translational products of the second puffs shut down the larval activities and make the insect ready for the pupal instar and the metamorphosis (Ashburner *et al.*, 1974; Baehrecke, 1996). An abbreviated scheme of ecdysteroidogenesis is provided in Fig 9.7 and a detailed description is given under the section on ecdysteroid biosynthesis.

In most Pterygota, the prothoracic gland degrades after adult life begins, and ecdysteroid production shifts into the ovaries or testes, but other sources also are described, like for example oenocytes. In the ovaries, the ecdysteroids are synthesised in the follicle cells and are incorporated into the eggs for the use during embryogenesis, or act on the fat body to activate the synthesis of yolk protein. In males, the steroid hormone is involved in spermatogenesis, and acts like a sex hormone, as in mammals. However, the overall discussion on whether insects do have sex hormones, or not, is stills an area of great controversy (de Loof and Huybrechts, 1998).

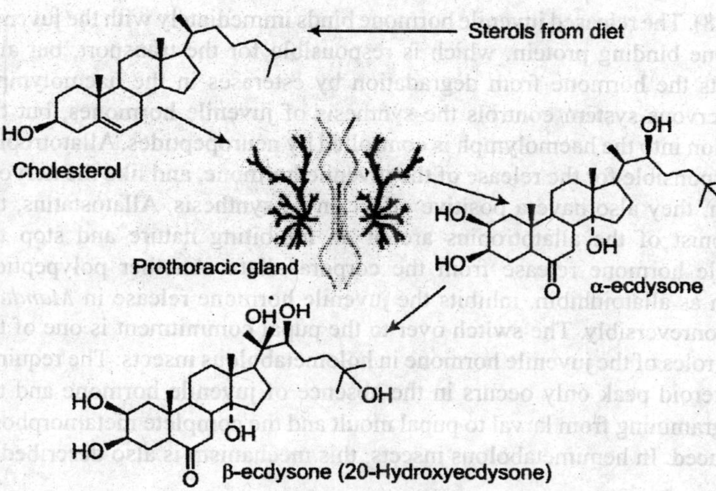

Fig. 9.6. Structure of cholesterol, α- and β-ecdysone

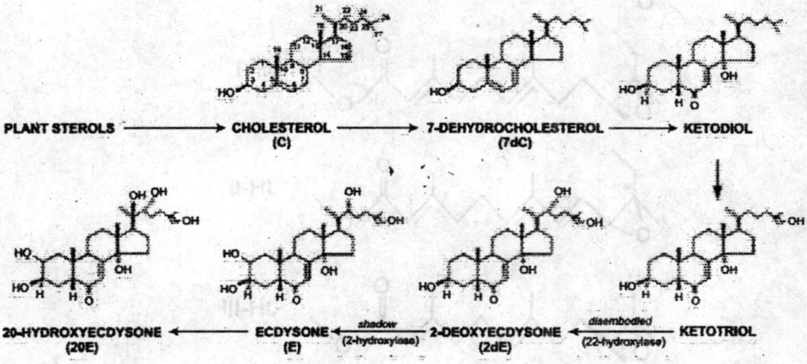

Fig. 9.7. Abbreviated scheme of ecdysteroidogenesis

3.1.2 *Juvenile Hormones*

Juvenile hormones (JH) are probably the most adaptable hormones in insects. It influences development, reproduction, diapause, polyphenism, and behaviour. The only source of this lipid hormone is in the corpora allata. As a non-polar hormone, it can diffuse through cell membranes and acts directly at the transcriptional level like ecdysterone, but juvenile hormones can as well act through receptors on the outer cell membrane, like a peptide hormone. Juvenile hormones are methyl esters with a terminal or subterminal epoxide group and today five structurally related forms are known in insects

(Fig 9.8). The released juvenile hormone binds immediately with the juvenile hormone binding protein, which is responsible for the transport, but also protects the hormone from degradation by esterases in the haemolymph. The nervous system controls the synthesis of juvenile hormones, but the secretion into the haemolymph is controlled by neuropeptides. Allatotropins are responsible for the release of the juvenile hormone, and like the nervous system, they also have a positive effect on its synthesis. Allatostatins, the antagonist of the allatotropins are of an inhibiting nature and stop the juvenile hormone release from the corpora allata. Another polypeptide, known as allatoinhibin, inhibits the juvenile hormone release in *Manduca* spp. nonreversibly. The switch over to the pupal commitment is one of the major roles of the juvenile hormone in holometabolous insects. The required ecdysteroid peak only occurs in the absence of juvenile hormone and the reprogramming from larval to pupal moult and the complete metamorphosis is induced. In hemimetabolous insects, this mechanism is also described.

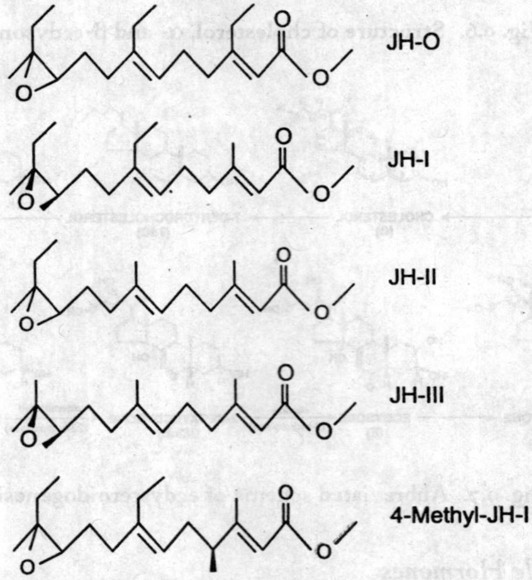

JH-O

JH-I

JH-II

JH-III

4-Methyl-JH-I

Fig. 9.8. Structure of five common juvenile hormones in insects

3.2 Peptide Hormones

Peptide hormones are small polypeptides and they cannot pass through cell membranes. Normally they act as a ligand on membrane receptors, which send the endocrine message further into the interior of a cell. The action of binding to a receptor instigates a conformational change that initiates a cascade, which ends in activation or inhibition of an enzyme, or in the change of the ionic level in the cytoplasm. Peptide hormones often bind to the G-protein coupled receptors (GPCR), which are widely dispersed in the animal kingdom. These GPCR's will send a signal transducer (second messenger) into the cell's interior, which activates pre-existing enzymes and proteins. Neuropeptides activate enzymes or proteins, which normally are already stored, whereas the lipid hormones on the contrary are responsible for the new production of activators or inhibitors. Most known hormones in insects are peptide hormones, the following examples describe a list of important peptide hormones.

3.2.1 *Prothoracicotropic Hormone*

This peptide hormone was the first described hormone in insects (Kopeć, 1917, 1922) and was named the brain hormone, but today more than one hormone source is known in the brain and it was renamed prothoracicotropic hormone (PTTH). This polypeptide hormone interacts with the prothoracic glands and regulates the synthesis of ecdysteroids.

PTTH is synthesised in the lateral neurosecretory cells of the brain and is released from the corpora cardiaca, or in some species from the corpora allata into the aorta. In brain ectomized pupae, the lack of PTTH keeps them in the pupal diapause. Such pupae can survive for up to 3 years until all nutrients are depleted. Reactivation of the prothoracic gland is only possible by injecting active PTTH or by connecting the debrained and intact pupa. After stimulation of the prothoracic gland, ecdysteroid synthesis starts again, the pupal diapause ends, and the pupa undergoes the 'normal' adult development. *In vitro* experiments showed that PTTH is directly responsible for the ecdysteroid synthesis in the prothoracic gland (Bollenbacher *et al.*, 1979).

Not many PTTHs were identified and characterised. In *B. mori* a small and big form was isolated. The small form showed a predicted effect in species like *Rh. prolixus*, but not in *B. mori*. Later it was renamed to bombyxin, which showed homology with the vertebrate insulin, but its effect in *B. mori* is not hypo-, but hypertrehalosaemic (Satake *et al.*, 1997). Today bombyxin is a member of the insulin-like peptides, which are characterised in species of Orthoptera, Lepidoptera, and Diptera. The insulin-like peptides can affect developmental regulation, metabolism, longevity, and female reproduction.

3.2.2 *Diuretic and anti-diuretic hormones*

Most insects are terrestrial animals, which make them particularly susceptible to dehydration; therefore, they are more concerned in maintaining water rather than expelling it. On the contrary, aquatic, herbivorous, and haematophagous insects are adapted to a water rich environment and are concerned with efficient water elimination. In general, diuretic hormones are responsible for the increased water loss, due to the enhanced secretion through the Malpighian tubes, and by inhibition of the reabsorption of water. The antagonists of the diuretic hormones, the anti-diuretic hormones inhibit the water excretion through the Malpighian tubes, and increase the reabsorption of water in the hindgut, especially in the rectum. This water balance is closely linked with the insect's ion balance. Ion receptors in the Malpighian tubes pump ions from the haemolymph into their lumen and water is absorbed for excretion. Once the water is in the Malpighian tubes, it passes through the hindgut and the rectum, where it is finally excreted. To regain the water, insects reabsorb the water at the hindgut and rectum (Gäde, 2004; Schooley *et al.*, 2005).

By selective use of proteolytic enzymes it has been established that the insect diuretic hormone (DH) is a peptide or polypeptide. Trypsin cleaves specifically peptide bonds in which the carbonyl group is donated either by lysine or arginine. Chymotrypsin hydrolysis peptide bonds in which phenylalanine, tryptophan or tyrosine donate the carbonyl group. A variety of other enzymes, including neuraminidae, α-amylase, DNase, RNase and lipase have effects on the diuretic activity of DH. As a result of immunochemical and physiological studies, it has been suggested in *Locusta migratoria* that DH is a vasopressin-like molecule.

3.2.3 *Muscle stimulating neuropeptides*

Proctolin, a pentapeptide, which functions as a neuromodulator and neurohormone, is involved in the contraction of the longitudinal muscles in the hindgut of the American cockroach, *Periplaneta americana* and also affects the visceral and skeletal muscle in a range of insect species (Isaac *et al.*, 2004). This neuromodulator is found in the central and ventral nervous system Proctolin is a pentapeptide, Arg-Tyr-Leu-Pro-Thr. It is suggested that proctolin is either an excitatory neurotransmitter or acts as a neurohormone to modulate muscle excitability.

Another well-studied muscle stimulating neuropeptide group is the cardioaccelerating peptides. The CAPs (**cardioactive peptides**) are found in the abdominal ganglia and are released at the perisympathetic organs. These peptides not only accelerate heart muscle rate, but also activate the contraction of the hindgut (Tublitz *et al.*, 1991) and play a crucial part in the ecdysis. In addition, a cardioaccelerating effect is reported for members of the AKH-family. It was also found that other neuropeptides, neuromodulators, or neurotransmitters have a cardioaccelerating effect, i.e. acetylcholine, 5-hydroxytryptamine, octopamine, neurohormone D, periplanetins, corazonin, and members of the leukosulfakinin family, as well as the above mentioned proctolin. However, this effect could occur as a secondary result, perhaps to enhance the transport in the haemocoel (Mordue and Morgan, 1985).

3.2.4 *Adipokinetic hormones*

Adipokinetic hormones (AKHs) also known as **hyperprolinaemic**, or **hypertrehalosaemic** peptides, are one of the most extensively characterised peptide hormone families in insects with more than 40 isoforms identified representing almost all insect orders. The classical AKHs are generally synthesised and secreted from a distinct region of the corpora cardiaca, found in all life stages. Prior to its secretion, they are modified at the amino and the carboxy termini to a pyroglutamic acid and amide, respectively; the two aromatic amino acid residues at position four (Phe/Try) and eight (Trp) typically are conserved. Cells in the brain and ganglia may also synthesise and secrete such peptides as shown in the mosquitoes *Aedes aegypti* and *A. gambiae* and in other insect species, but none of those AKHs were proven to have any metabolic function. Since they are also not synthesised in the corpora cardiaca, they are also referred to as non-classical AKHs.

Most insects synthesise AKHs in the corpora cardiaca, but it is also reported in the brain and the ventral nerve system. AKH synthesis is a continuous process and is independent of life stages. It has been shown that newly synthesised AKH is preferably secreted into the haemolymph and as a result only a small amount of the stored AKHs will ever be used; consequently, the older an insect, the more stored AKH.

The identity of the releasing factor for AKH is still a matter of discussion. Neurons from the lateral part of the protocerebrum transport messengers through the NCC-II to a synaptic ending, which is connected to the glandular part of the corpora cardiaca (Rademakers, 1977) and electrical stimulation of the NCC-II resulted in AKH release (Orchard and Loughton, 1981b; Orchard and Lange, 1983b). In studies with *L. migratoria* the neurohormones tachykinin-I and –II (Locmi-TK-I, and -II) induced AKH release. *In vitro* studies with crustacean cardioactive peptides (CCAP), high potassium concentration, or the 3-isobutyl-l-methylxanthine (IBMX) also activated the release of AKHs, which supports a hormonal release of AKHs. On the other hand, it was found that high trehalose levels in the haemolymph inhibit AKH release (Cheeseman *et al.*, 1976; Passier *et al.*, 1997). Since AKHs act as multifunctional hormones, their release may be controlled differently depending on its need; especially for AKH cells that originate in the brain or ventral nerve cord.

The members of the AKH family are typical examples of multifunctional neurohormones. The main function of the AKHs is linked to the insect's nutrient metabolism, but not all members of the AKH-family are involved in hyperlipaemia, some induce hypertrehalosaemia (hypertrehalosaemic hormone, HrTH) or hyperprolinaemia (hyperprolinaemic hormone, HrPH). It is also demonstrated that AKH members can inhibit lipid (Gokuldas *et al.*, 1988), protein, and RNA synthesis (Kodrik and Goldsworthy, 1995). Other functions induced by AKHs are reported, in *M. sexta*, where similar to octopamine, species-specific AKH injection increased motor activity of the mesothoracic muscle, whereas in different cockroaches, acceleration of the heart rate is reported. In *P. americana* AKH-family members induce an intracellular Ca^{2+}-influx and consequently amplify the spontaneous spike frequency in the dorsal unpaired median (DUM) neurons of the terminal ganglion. An increase of the cytochrome P450 (P4504C1) was reported for the discoid cockroach, *Blaberus discoidalis* after HrTH injection. Cytochrome P450 was also enhanced in starved animals, which is probably derived from an up-regulation of AKHs during starvation (Bradfield *et al.*, 1991). In locusts, there is even a possible immune function attributed to AKH enhancing indirectly the phenoloxidase and the nodule formation (Mullen and Goldsworthy, 2003, 2006; Mullen *et al.*,2004).

3.2.5 *Chloride transport stimulating hormone (CTSH)*

The rectum of insects is an important organ involved in the regulation of both ion and water balance. It has been shown that the transport of Cl^- from the lumen of the rectum to the hemolymph is under the control of hormone from the corpus cardiacum. Since this hormone stimulates increased transport of chloride ions, it has been named chloride transport stimulating hormone (CTSH). CTSH molecule contains peptide linkage in its structure. The hormone can be extracted using both aqueous and 80% methanol solvents. It has a net negative charge at neutral to alkaline pH.

3.2.6 *Neurohormone D*

Neurohormone D was first isolated from the corpus cardiacum of *Carausius morosus* by paper chromatography. It is reported to have an effect on a variety of physiological functions including increasing the frequency and amplitude of heart beat, stimulating colour change in *Carausius morosus,* induction of nervous activity in the phallus nerve of *P.americana*, and increasing the rate of dye excretion by the Malpighian tubules. Neurohormone D isolated from *P.americana* corpora cardiac is able to increase flux of water across the toad bladder, but has an opposite diuretic effect in insects. Although several functions have been attributed to this hormone, it is not clear if it is due to a single hormone or different compounds.

Chemically, the peptide nature of this hormone has been confirmed and it was also suggested that one of the aminoacids contains a heterocyclic ring.

3.2.7 *Neurohormones that activate ecdysis and sclerotisation*

Ecdysis is a periodic shedding of the exoskeleton by arthropods, which allows growth. Juvenile hormone, prothoracicotropic hormone, and ecdysteroids are the regulators for keeping the larval status quo or initiating metamorphosis. In respect to maintaining the larval condition, ecdysis is the last crucial part for the next larval instar and is controlled by the eclosion hormone, the ecdysis-triggering hormone, and the crustacean cardioactive peptide. The eclosion hormone is synthesised in the ventral median neurosecretory cells (VM cells) of the protocerebrum and is released from the corpora cardiaca. The ecdysis-triggering hormone is synthesised and released from the endocrine cells located in the epitracheal glands, the inka cells. Both peptide hormones activate the release of each other. Before the actual ecdysis occurs, the release of ecdysis triggering hormone starts the pre-ecdysis, induced by ecdysteroid and other cuticular related cues (Fig 9.9). Once the ecdysis-triggering hormone is released, a hormonal positive feedback mechanism occurs until the right proportion of ecdysis triggering hormone and eclosion hormone is reached, then the crustacean cardioactive peptide is secreted, thus activating the ecdysis program. After the successful ecdysis, the hardening of the cuticle follows. This tanning process, called **sclerotisation**, is activated by bursicon. This neuropeptide is synthesised in the same neurosecretory cells (located in the ganglia) as the crustacean cardioactive peptide, and is secreted into the haemocoel from the perisympathetic organs. It is not yet known what activates bursicon and how it is activated, but it is vital that the tanning takes place after the old cuticle is shed (Mesce and Fahrbach, 2002).

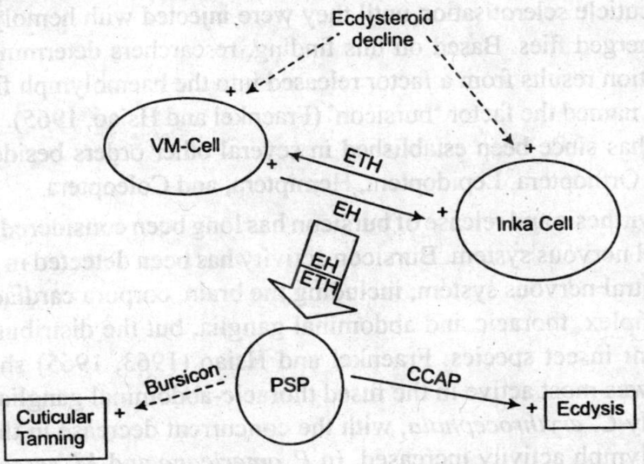

Fig. 9.9. Graphic illustration of the hormonal positive feedback mechanism of the eclosion hormone (EH) and the ecdysis-triggering hormone (ETH).

VM-Ventral median neurosecretory; CCAP- crustacean cardioactive peptide PSP- perisympathetic organ.

3.2.8 *Neurohormones acting as neurotransmitters*

The neurogenic amines are synthesised in neurosecretory cells and released at the neurohaemal organ and they can act as a neurohormone, -transmitter, or -modulator. Two well-identified neurogenic amines are octopamine and tyramine. Both are involved in metabolic and behavioural functions and are the only known non-peptide transmitter/hormone in invertebrates (Roeder, 2005). Octopamine, which may be analogue to the vertebrate hormone adrenalin, has been proposed to be a modulator in the releasing cascade of the neuropeptides of the AKH-family.

3.2.9 *Bursicon - the insect cuticle sclerotisation hormone*

Bursicon is a neurohormone that regulates cuticle sclerotisation (tanning and hardening) and the wing expansion processes in insects. Bursicon was discovered over forty five years ago from the blowfly *Calliphora erythrocephala* in a neck-ligated fly assay. However due to the difficulties in hormone protein purification and the lack of molecular techniques, the genes encoding bursicon were not identified until in the years of 2004-2005. Studies in *Drosophila melanogaster* have indicated that bursicon is actually a heterodimeric cystine knot family protein containing two subunits, bursicon α (bur α or burs) and bursicon β (bur β or pburs), which are encoded by two individual genes *burs* α and *burs* β (or *burs* and *pburs*).

The initial discoveries in the physiology of cuticle sclerotisation, or 'tanning' using, blowflies, *Calliphora erythrocephala,* established that some hormonal control factor was involved in tanning because decapitated flies would not undergo cuticle sclerotisation until they were injected with hemolymph from newly emerged flies. Based on this finding, researchers determined that the sclerotisation results from a factor released into the haemolymph from the fly head, and named the factor 'bursicon' (Fraenkel and Hsiao, 1965). Bursicon's presence has since been established in several other orders besides Diptera, including Orthoptera, Lepidoptera, Hemiptera, and Coleoptera.

The synthesis and release of bursicon has long been considered to be from the central nervous system. Bursicon activity has been detected in most parts of the central nervous system, including the brain, corpora cardiaca/corpora allata complex, thoracic and abdominal ganglia, but the distribution varies in different insect species. Fraenkel and Hsiao (1963, 1965) showed that bursicon was most active in the fused thoracic-abdominal ganglia (TAG) of the blowfly *C. erythrocephala,* with the concurrent decrease in this activity as haemolymph activity increased. In *P. americana* and *M. sexta,* bursicon was also acknowledged to be released from the abdominal ganglia.

Results from several studies using different insect species have established the polypeptide nature of bursicon, including evidence indicating that the tanning factor is non-dialysable, relatively insoluble in organic solvents, able to be precipitated with loss of activity by reagents such as alcohol and acetone, and able to be inactivated by trypsin or pronase treatments. The variability in the reported size and release sites suggests that bursicon may exist in different forms between various insect species, but its activity, with an initial post-emergence surge and subsequent steady decline, was common in most species studied.

The partially purified bursicon from nerve cords of the cockroach *P. americana* also made it possible for exploring the genomic information of this hormone. By comparing the partial sequences with available genomic data, Dewey *et al.* (2004) identified the *D. melanogaster* gene *CG13419* (*burs*, or *burs α*) as a candidate bursicon gene.

The *burs α* gene encode a peptide with a predicted final molecular weight of 15 kDa. This predicted bursicon protein belongs to the cystine knot family, which includes vertebrate transforming growth factor-β (TGF-β) and various glycoprotein hormones. The link between this *Drosophila* gene and the bursicon hormone was established from the defective post-ecdysial behaviour of flies, such as the cuticle sclerotisation and wing expansion, with loss-of-function mutations in this gene, and the corresponding decrease in bursicon bioactivity in mutant extracts. Accordingly based on the previous finding of the putative molecular mass of the mature bursicon hormone in several insect species, it was proposed that bursicon would be a homodimeric 30 kDa protein made of two 15 kDa burs *α* subunits (Dewey *et al.*, 2004). However, later results showed that the recombinant burs protein had no bioactivity, and it was then hypothesised that the *burs α* gene product need to heterodimerise with another cystine knot protein encoded by the *D. melanogaster* gene *CG15284* (*pburs* or *burs β*) (Luo *et al.*, 2005; Mendive *et al.*, 2005), which encodes a protein of approximately 14 kDa, for a valid bursicon activity. R-bursicon α+β heterodimer was shown to be capable of inducing tanning in the neck-ligated fly bioassay *in vivo*.

Bursicon is the first heterodimer cystine knot hormone protein found in insects (Mendive *et al.*, 2005). Based on the available cystine knot protein structures, the structure of bursicon can be predicted. Both burs α and burs β subunits belong to the cystine knot protein family, with an uneven number of cystine residues containing five intramolecular disulfide bonds between C2-C8, C3-C9, C4-C10, and C5-C11, and a possible formation of an intermolecular bridge by C6, as shown in Fig. 9.10 (Luo *et al.*, 2005). This structure ensures a covalent association between the burs α and burs β subunits and possible ligand binding ability. However it is not known why these two subunits with similar structure have to bind together for a valid bursicon hormone function.

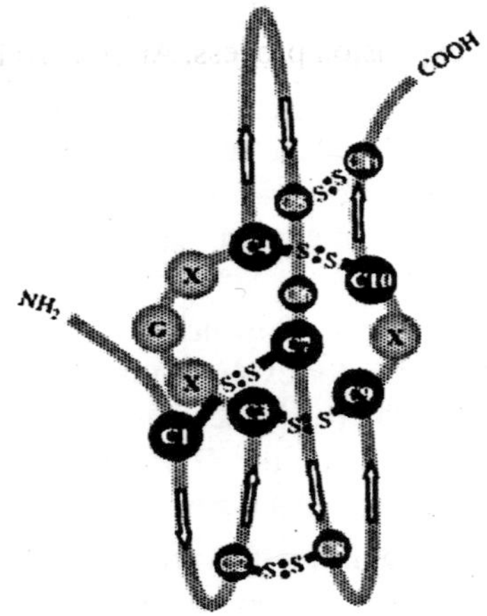

Fig. 9.10. Predicted cystine knot structure of burs α or burs β subunit. (Luo *et al.*, 2005)

Bursicon signal transduction pathway

Although bursicon's discovery was made a long time ago, it has not been elucidated as to how it plays its role in the insect cuticle sclerotisation process. Currently great efforts are being made to study the signal transduction pathway(s) that bursicon regulates. In recent years, a subfamily of G protein-coupled receptors, named LGRs, was identified, which have a large ecto-domain containing leucine-rich repeats. Genes in this family are conserved in invertebrates and vertebrates. LGRs can be divided into three groups: group A, vertebrate glycoprotein hormone receptors; group B, *D. melanogaster* LGR2 (DLGR2) (Eriksen *et al.*, 2000) and vertebrate orphan receptors LGR4, 5, 6 (Hsu, 2003); and group C, mammalian relaxin_INSL3 receptors (Hsu *et al.*, 2002; Kumagai *et al.*, 2002).

Based on molecular genetic analyses in *D. melanogaster*, bursicon was proposed to act through the G protein-coupled receptor DLGR2, which is encoded by the gene *rickets*. Baker and Truman (2002) reported that a loss-of-function mutation in the gene *rickets* resulted in failure in initiating the behavioural and tanning process following ecdysis, and this gene was confirmed to encode the glycoprotein hormone receptor DLGR2, which belongs to the group DLGRs. In mutant flies, although the hormone bursicon was produced and released, insects were not able to perform cuticle sclerotisation. Mutation in the *rickets* gene also blocked the behavioural program for wing expansion, a process that occurs after the eclosion and accompanies cuticle sclerotisation. So DLGR2 is likely one of the downstream elements for bursicon function in the cuticle sclerotisation and wing expansion process. An *in vitro* binding assays using cells over-expressing DLGR2 and r-bursicon heterodimer also indicated that r-bursicon is capable of high-affinity binding to the DLGR2 (Luo *et al.*, 2005; Mendive *et al.*, 2005). Although it has not yet been identified which tissue of the insect express DLGR2, it has been reported that this receptor is only expressed in embryo and pupal stages of *Drosophila*, but not in the larval and adult stages (Eriksen *et al.*, 2000).

Another component that is possibly involved in the bursicon-mediated cuticle sclerotisation process is cyclic AMP (cAMP). cAMP is an important secondary messenger for many signal transduction pathways. Earlier studies on blowflies have implicated an increase in cAMP upon the action of bursicon. Recent studies using an *in vitro* binding assay also indicated that r-bursicon stimulated cAMP production in a dose-dependent manner. Injection of the *rickets* mutated flies with a cAMP analog resulted in melanization of the abdominal tergites, showing that at least this part of the response machinery is functional in the mutants and cAMP might be a critical element downstream of the receptor DLGR2 (Baker *et al.*, 2002). Although these findings imply the role of cAMP in mediating the bursicon-mediated signal transduction pathway, so far little is known about what kind of downstream elements are regulated by cAMP and how cAMP is involved in this process. Also it should be noted that cAMP injections do not have any effects on the components of the wing expansion program in either *rickets* mutated or wild-type flies. So if cAMP is involved in the bursicon- mediated signal pathway, it is probably only associated with cuticle sclerotisation (Baker *et al.*, 2002).

Using RNAi knock down of bursicon expression in *B. mori*, Huang *et al.* (2007) reported three genes that are regulated by bursicon. One gene shows a similarity with a thioesterase superfamily member 2 in *D. melanogaster*, and the second gene is the silkworm trehalase gene, while the third gene is an unknown gene. Since it has long been proposed that several enzymes are possibly involved in the final synthesis of the sclerotisation agents, Davis *et al.* (2007) further investigated two enzymes, tyrosine hydroxylase (TH, encoded by *ple*) and dopa decarboxylase (DDC, encoded by *Ddc*), which are responsible for the synthesis of dopamine (DA), in *D. melanogaster*. Data showed that both *ple* and *Ddc* transcripts begin to accumulate before eclosion, and the activity of DDC is high before eclosion, indicating DDC is not regulated in the sclerotisation process. However, TH activity is present before eclosion, disappears at eclosion, then re-accumulates rapidly within one hour of eclosion, and is transiently activated by phosphorylation during sclerotisation, indicating TH might be the final regulatory switch to control cuticle sclerotisation.

Bursicon- wing expansion and programmed cell death

In addition to inducing the insect cuticle sclerotisation process, bursicon is also involved in the insect wing expansion. The insect wing expansion process is another physiological event occurring after insect ecdysis and accompanying cuticle sclerotisation. These processes occur at the same time, but the mechanisms under which they are controlled and regulated are still not clear. Recent research showed that mutation in the bursicon genes causes the failure of initiation of the behavioural program for wing expansion and results in defects in wing expansion in both the fruit fly *D. melanogaster* and the silkworm *B. mori* (Dewey *et al.*, 2004; Huang *et al.*, 2007), which indicates that bursicon is the hormone required to initiate both of these processes in insects. It was also shown that mutation of the *rickets* gene, which encodes the DLGR2 receptor as a putative bursicon receptor, inhibited wing expansion in *Drosophila* (Kimura *et al.*, 2004).

Although it is not clear how wing expansion in insects is controlled, it has been demonstrated that wing development and expansion is strongly associated with programmed cell death and removal of cell debris from the wing tissue (Kiger *et al.*, 2007; Natzle *et al.*, 2008). Shortly after eclosion, haemolymph pressure forces the expansion of the folded wing blade. At the same time, the dorsal and ventral epithelial cell layers delaminate from the outside cuticle, undergo a process called epithelial-mesenchymal transition (EMT), and then exit the wing accompanied by the initiation of a programmed cell death under the control of a yet to be identified signaling pathway (Kiger *et al.*, 2007; Natzle *et al.*, 2008).

Programmed cell death (apoptosis) is crucial for the development and maintenance of multicellular organisms, to control cell number, to remove infected, mutated or damaged cells, and to eliminate cells that are no longer required at a given stage of development. During development, apoptotic, autophagic and nonlysosomal cell death have been widely reported. The components and mechanisms for apoptosis are conserved in a wide variety of organisms, from worms and flies to humans. They involve the activity of cysteine proteases (caspases) and their negative regulators, such as Inhibitor of Apoptosis Proteins (IAPs), and the pro-apoptotic activators, such as *reaper* (*rpr*), *head involution defective* (*hid*; *Wrinkled*, *W* - FlyBase), *grim* and *sickle* (*skl*), which promote cell death by inhibiting the function of the IAPs. There should be multiple cell death and survival signals that regulate the turning on and shutting off of the apoptotic genes. However the nature of these signal transduction pathways is nearly unknown. At the last step of metamorphosis, newly emerged adults undergo extensive cell death.

Many of the abdominal muscles and associated neurons die within a day of adult emergence. In addition, the wing epidermis is also removed by cell death at the time of wing spreading in the large fly *Lucilia cuprina* (Seligman *et al.*, 1975) and in *D. melanogaster* (Johnson and Milner, 1987). At the last step of metamorphosis in *Drosophila*, the wing epidermal cells are removed by programmed cell death during the wing spreading behaviour after eclosion. Kimura *et al.* (2004) demonstrated that cell death was accompanied by DNA fragmentation and that this cell death exhibited extensive vacuoles, implying possible autophagy. Ectopic expression of an anti-apoptotic gene, *p35*, inhibited the cell death, indicating the involvement of caspases.

It was reported that both cell death and the removal process is controlled by changes in the levels of the steroid hormone 20-hydroxyecdysone (20E) through regulation of a series of early response and late response genes, including the expression of the cell death activator genes, *rpr* and *hid*, which then repress the *Drosophila* inhibitor of apoptosis proteins (IAPs) *Diap2*. Changes in 20-E hormone, in combination with the fine-scale expression of the ecdysone receptor, control the timing and spacing of cell death during metamorphosis. Though ecdysone-dependent regulation is the possible pathway controlling cell death, it does not exclude the possibility of the involvement of other hormones in wing expansion and programmed cell death. It is very likely that bursicon co-regulates the programmed cell death with other factors during the post-eclosion wing maturation process. It has been shown that stimulation of components downstream of bursicon, such as a membrane permeant analog of cAMP, or ectopic expression of constitutively active forms of G proteins or PKA, induced precocious cell death; and conversely, cell death was inhibited in wing clones lacking G protein or PKA function. Therefore, activation of the cAMP/PKA signaling pathway is likely required for transduction of the hormonal signal that induces wing epidermal cell death after eclosion (Kimura *et al.*, 2004).

Role of Bursicon in Integumentary Structure Development and Expansion

Recent evidence indicated that bursicon also plays important roles other than the classic functions attributed to it, such as cuticle tanning and wing expansion. Injection of the bursicon receptor Tcrk dsRNA into early stage last instar *T. castaneum* larvae inhibits development and expansion of integumentary structures and adult eclosion, in addition to cuticle tanning (Bai and Palli, 2010). These results indicate that bursicon could also be released before eclosion to regulate development and expansion of integumentary structures. PCR analysis of burs and pburs transcripts in *Drosophila* and *Musca* clearly shows that the burs and pburs transcripts are highly expressed in larval and pupal stages (Luo *et al.*, 2005; Wang *et al.*, 2008), suggesting that the products of *burs* and *pburs* may be released into haemolymph during larval and pupal development. Loveall and Deitcher

(2010) reported that the release of bursicon in *Drosophila* surprisingly occurs in two waves: preceding and following ecdysis, indicating that the release preceding ecdysis may play a role in development and expansion of integumentary structures as shown in the red flour beetle (Bai and Palli, 2010) and the release following ecdysis is responsible for cuticle tanning and wing expansion.

Role of Bursicon in immune response

A DNA microarray analysis in the neck-ligated flies using r-bursicon as a probe identified a set of 87 genes whose expression were up- or down-regulated by r-bursicon in *D. melanogaster*. However, not all identified genes could be related to the cuticle sclerotisation and wing expansion processes, suggesting that bursicon might have unidentified functions. From the microarray identified genes, seven immune-related genes were up regulated by r-bursicon. These genes include three from the turandot gene family (*turandot* X, *turandot* F, and *turandot* C), two from the attacin gene family and one each from the cecropin and immunoglobulin families (An *et al.*, 2008). Since bursicon is a heterodimer, it is quite possible that in addition to its classic roles in cuticle tanning and wing expansion, each subunit (burs and pburs) may have its own physiological function. Neck-ligated flies when injected with r-bursicon, r-burs, and r-pburs separately, each could induce a set of immune response genes in the final moulting cycle of *D. melanogaster.* Such induction was both time- and dose-dependent. When the homogenates from the flies injected with r-burs or r-pburs were incubated with the gram negative bacterium *Escherichia coli*, the homogenate from each fly equivalent was able to completely clear bacterial growth while the control that received the cell medium transfected with blank vector could not. Recent studies suggest that bursicon mediates the immune response via the IMD (Immune Deficiency) pathway by activation of the transcription factor relish. However, it is not clear whether burs or pburs forms a homodimer or heterdimerizes with a yet unidentified factor to mediate the immune response. It is also not clear if burs or burs bind to DLGR2 or to an as yet unidentified receptor leading to the regulation of antibacterial peptide expression.

3.2.10 Hormones and carbohydrate metabolism

Presence of hyperglycaemic and hypoglycaemic factors in insects has been reported. The source of hyperglycaemic factor in the locust is the glandular lobe of the corpus cardiacum, whereas it is speculated that a hypoglycaemic factor is synthesised in the brain neurosecretory cells and stored and released from the storage lobes of the corpus cardiacum. There are numerous reports confirming the presence of hyperglycaemic factor in

Periplanta Americana. The physiological function of the hyperglycaemic hormone is to stimulate the breakdown of glycogen in the fat body, which results in an elevation of haemolymph trehalose. This action is mediated by the activation of the fat body phosphorylase system. Chemically it is a glucagon-like molecule.

Similarly, insulin-like hormone has also been isolated from the brain of blowfly *Calliphora* and the haemolymph of *Manduca* and its peptide nature has been reported. Insulin-like peptides (ILPs) are encoded by multiple, distinct genes within each species and, upon secretion, serve as hormones, neurotransmitters, and growth factors during post-embryonic life stages. These diverse messages are transduced in target cells through an insulin receptor and a signalling network of activated proteins that directly affect biochemical pathways and gene expression. ILP genes are expressed in specific cells in both the nervous system and midgut of insects. As peptide hormones, ILPs can be stored or secreted from a cellular source in a bioactive form. ILPs function to activate the insulin receptor and a series of intracellular proteins, comprising the insulin signalling pathway. This pathway is responsive to nutrient intake and signalling through the "target of rapamycin" (TOR) pathway, and together regulates growth, development, and reproduction in insects, as best characterised for *Drosophila melanogaster*. Coordination of growth and development are regulated jointly by the Insulin signalling pathway (ISP) via ILP activation and by the TOR pathway through nutrient sensing.

Several studies demonstrate that ILP treatment directly stimulates the growth or proliferation of cells in different organs. Imaginal discs of lepidopteran species are model organs for *in vitro* studies of the hormonal regulation of growth. Wing discs from last instar *Precis coenia* or *Manduca sexta* are synergistically stimulated to grow *in vitro* by 20-hydroxyecdysone (20-ECD) and an ILP (bombyxin-II or partially purified brain extract) (Nijhout and Grunert, 2002; Nijhout *et al.*, 2007). Insulin signalling is activated in these discs by ILP treatment and not by 20-ECD. *In vivo* studies showed that commitment to pupal development in wing discs was stimulated by the injection of mammalian insulin or purified *M. sexta* ILP into newly moulted, starved last instar *M. sexta* larvae, and further indicated that these treatments suppress JH action later in the instar. The ISP (Insulin signalling pathway) has also been implicated in the regulation of JH synthesis in *Drosophila*.

Synthetic ILPs directly affect metabolic processes in insects. A silkmoth ILP (bombyxin-II), when injected into neck ligated larvae, was observed to lower trehalose and glycogen levels, apparently promoting consumption of carbohydrate reserves and not the accumulation of reserves, as for insulin in mammals (Satake *et al.*, 1997, 1999). Synthetic *Ae. Aegypti* ILP3 and ILP4 were injected into decapitated, sugar-fed female *Ae. aegypti* to resolve whether different ILPs redundantly act as insulins. Only ILP3 stimulated trehalose depletion and glycogen and lipid storage in a dose-responsive manner (Brown *et al.*, 2008; Wen *et al.*, 2010). This result shows that ILPs differ in their ability to activate anabolic processes in the fat body. However, the two ILPs activated ovarian ECD production *in vitro* at similar doses. This differential action likely reflects the binding of ILP3 to the IR and ILP4 to another membrane protein, as determined for ovary membranes. ILPs are also implicated in the regulation of insect reproduction, diapauses, immunity and aging.

4. ECDYSTEROID BIOSYNTHESIS

As in vertebrate steroidogenesis, cholesterol is the precursor of arthropod ecdysteroid biosynthesis but, unlike vertebrates, arthropods cannot synthesise this molecule de novo, and must acquire it from their diet. Insects can obtain sterols from plant or animal sources, which may require modification to cholesterol before it can be used for ecdysteroid biosynthesis. The extra- or intracellular source of cholesterol used in arthropod ecdysteroid biosynthesis is not known, but cholesterol used in mammalian steroidogenesis can be derived from three separate sources: (1) exogenous cholesterol transported to the cell via lipoproteins, (2) the plasma membrane, or (3) intracellular de novo synthesis. Given the lack of de novo cholesterol synthesis in arthropods, its source for ecdysteroid biosynthesis must either come from haemolymph lipoproteins or the plasma membrane. It has been shown in lepidopteran insects that cholesterol absorbed by the midgut is loaded into lipoproteins and transported to peripheral tissues for internalization (Chino and Gilbert, 1971; Jouni *et al.*, 2002).

The prothoracic glands are considered the major site of ecdysteroid synthesis in immature insects. The detailed ecdysteroid biosynthetic pathway is not known, but it is believed to mimic vertebrate steroidogenesis in that precursor steroid molecules shuttle between the endoplasmic reticulum and the inner mitochondrial membrane during enzymatic processing (Rees, 1995; Gilbert *et al.*, 2002). In contrast to the ability of vertebrate animals to synthesise the steroid precursor cholesterol de novo, insects must obtain it from their diet. Cholesterol absorbed by the midgut is transported to peripheral tissues via the haemolymph by a transport protein, lipophorin (Soulages and Wells, 1994). Once inside the cell, various cholesterol "transporters" have been proposed to shuttle the ecdysteroid precursor to the organelles for modification.

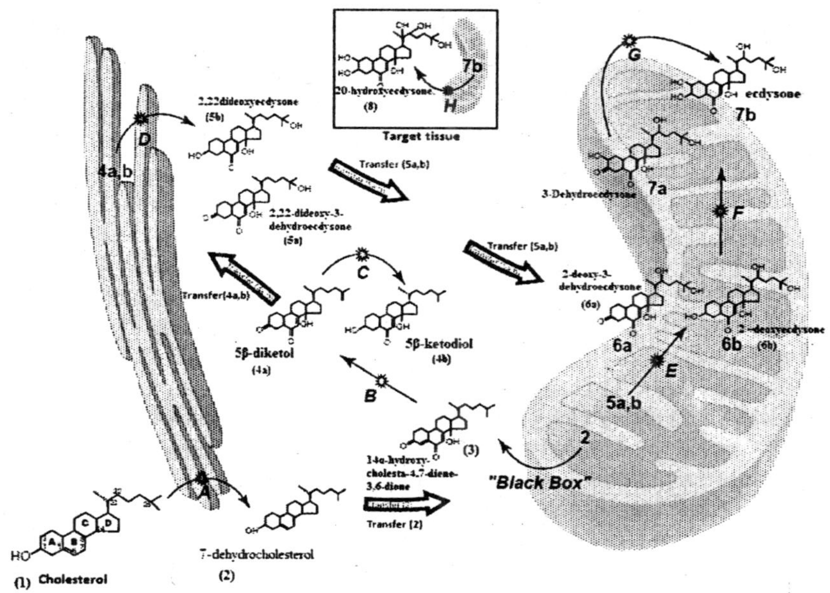

Fig. 9.11. A generalised pathway for ecdysteroid biosynthesis (Modified from Brown et al 2009)

(1) Cholesterol; **(2)** 7-dehydrocholesterol; **(3)** 14α-hydroxy-cholesta-4,7-diene-3,6-dione; **(4a)** 5β-diketol; **(4b)** 5β-ketodiol; **(5a)** 2,22-dideoxy-3-dehydroecdysone; **(5b)** 2,22dideoxyecdysone; **(6a)** 2-deoxy-3-dehydroecdysone; **(6b)** 2 –deoxyecdysone; **(7a)** 3-Dehydroecdysone; **(7b)** ecdysone; (**8**) 20-hydroxyecdysone. **(A)** 7,8-dehydrogenase; **(B)** 5β-reductase; **(C)** 3β-reductase; **(D)** 25-hydroxylase; **(E)** 22- hydroxylase; **(F)** 2-hydroxylase; **(G)** 3-dehydroecdysone 3β-reductase; **(H)** 20-hydroxylase

Figure 9.11 provides a generalised scheme for the ecdysteroid biosynthesis. The numbers in this scheme represent the substrate and the capital letters in bold represent the enzymes. The first step of ecdysteroid biosynthesis is the delivery of cholesterol (1) to the ER where the first enzymatic modification is conducted by the cytochrome P450 enzyme, 7,8-dehydrogenase (A), producing 7-dehydrocholesterol (2). This step is not a rate limiting step because of the abundance of 7-dehydrocholesterol in ecdysteroidogenic glands and the lack of trophic hormonal stimulation of its synthesis. Various sterol transfer protein have been hypothesised to be responsible for the delivery of cholesterol to the ER or mitochondria, such as the diazepam-binding inhibitor, start1 and sterol-carrier protein 2, but none have been shown conclusively to mediate this process. Considerable evidence supports the role for the mammalian homologs of the three sterol transfer proteins in intracellular cholesterol transporter or promoters of steroidogenesis (Uribe *et al.*, 2003; Tuckey *et al.*, 2004; Alpy *et al.*, 2005). The three putative sterol transfer proteins identified in insects are expressed in insect ovaries, but their expression in insect testes has not been addressed (Krebs and Lan, 2003; Roth *et al.*, 2004).

The next step is the transfer of 7-dehydrocholesterol to the mitochondria, and this is considered the rate limiting step of insect ecdysteroid biosynthesis. At the mitochondria, 7- dehydrocholesterol is transported into the inner mitochondrial membrane (IMM), where a series of reactions termed the **black box** take place, and various steps and precursor members within this **black box** have been proposed. The modifications that occur within this black box involve A/B ring modification, introduction of the 6-keto group and 14-hydroxylation, but the order of such modifications is not clear (Rees, 1989; Grieneisen, 1994). The ecdysteroid precursor following the black box modifications has alluded characterisation, and the identity of the functional group on 3-C has been of much speculation. The possibility of a 3-oxo-Δ4 intermediate (3) following black box modifications has been addressed through both the metabolism of radiolabeled precursors by *S. gregaria* ovaries (Davies *et al.*, 1981; Rees, 1985; Fujimoto *et al.*, 1989), and the metabolism of such a 3-oxo-Δ4 intermediate by *C. maenas* Y-organ preparations (Blais *et al.*, 1996). When provided the 3-oxo-Δ4 intermediate, *C. maenas* Y-organ preparations conducted a 5β-reduction, and the 5β-reductase (B) responsible was cytosolic requiring NADPH as a cofactor (Blais *et al.*, 1996). No such 5β-reductase has yet been characterised from insects, but is a probable step following the black box (Gilbert *et al.*, 2002). The retention of the 3-oxo is a feature of ecdysteroids secreted by lepidopteran prothoracic glands (Kiriishi *et al.*, 1990; Grieneisen, 1994) and the Y-organ of many crustaceans. The enzyme that conducts the 3β- reduction of secreted 3-oxo ecdysteroids has been characterised in numerous lepidopterans and later identified as 3-dehydroecdysone-3β-reductase (G) in *S. frugiperta* (Kiriishi *et al.*, 1990; Chen *et al.*, 1999). The fact that both 5β-diketol (4a) and its 3β-reduced product 5β-ketodiol (4b) have been isolated and metabolically converted by insect gonads, and the lack of 3β-reductase (C) activity in the haemolymph of non-lepidopteran insects (Kiriishi *et al.*, 1990) strongly suggests that 3β-reduction may occur within gonads of various insect orders. The subsequent hydroxylation reactions made to either 5β-diketol (4a) or 5β-ketodiol (4b) have been extensively studied. The identified cytochrome P450 enzymes conduct their hydroxylation reactions in a preferred sequence C25 → C22 → C2 (Warren *et al.*, 2002; Warren *et al.*, 2004). Again, the sequence of "terminal hydroxylation" reactions may not be steadfast. An example of such an "alternative" biosynthetic pathway is observed in *B. mori* in which ovarian ecdysteroids possess a hydroxyl group at C-20 but lack hydroxyl groups at C-2 or C-22 (Mizuno *et al.*, 1981; Ohnishi *et al.*, 1981). This observation has leaded some investigators to propose an alternative biosynthetic pathway that would result in a different sequence of "terminal hydroxylations" (Ohnishi *et al.*, 1981; Sonobe and Yamada, 2004). The cytochrome P450 enzymes conducting the **terminal hydroxylation** reactions of ecdysteroid biosynthesis have been localized to the ER (25-hydroxylase, CYP306a1) (D) or mitochondria (22-hydroxylase, CYP302a1) (E) (2-hydoxylase, CYP315a1) (F) (Petryk *et al.*, 2003; Warren *et al.*, 2004), further corroborating that ecdysteroid biosynthesis requires the transfer of ecdysteroid precursors between these organelles during processing (Kappler *et al.* 1989; Rees, 1995; Gilbert *et al.*, 2002).

The final hydroxylation reaction at C-20 is the most studied reaction of ecdysteroid biosynthesis, and generally occurs in tissues peripheral to the original source of ecdysteroid biosynthesis (Smith, 1985). It is widely accepted that this reaction creates the most bioactive form of ecdysteroid, 20-hydroxyecdysone (8). This reaction is common to all arthropods, because 20-hydroxylated ecdysteroids are observed in species throughout Arthropoda (Gupta, 1990; Lachaise, 1990). 20-hydroxylase activity has been localised to both ER and mitochondrial preparations, depending on the species or tissue under study (Smith, 1985; Rees, 1995). The cytochrome P450 enzyme CYP314a1 of *D. melanogaster* was recently identified as a 20- hydroxylase (H), and its transcripts were observed primarily within tissues peripheral to the original source of ecdysteroids, the ring glands (e.g., Malpighian tubules, fat body) (Petryk *et al.*, 2003).

5. MODE OF ACTION OF ECDYSTEROIDS AT THE GENE LEVEL

In recent years, significant progress has been made in understandin[illegible]
the molecular mechanisms of ecdysone action in the fruit fly, *Dr*[illegible]
melanogaster (Box 9.7). Ecdysone, bound to its receptor p[illegible]
activates primary-response genes. A subset of [illegible]
transcription factors that induce large batterie[illegible]
genes. Dynamic changes in ecdysone tit[illegible]
interactions among these tra[illegible]
order of regulatory gen[illegible]
Furtherm[illegible]

Box 9.7 Molecular mechanism of ecdysone action

1. Ecdysone binds to receptor protein and activates primary-response genes
2. Transcription factors induce secondary-response genes
3. Cross-regulation among transcription factors determine order of gene expression
4. Gene regulation hierarchies and biological responses.

5.1 Ecdysone Pulses During Drosophila Development

At specific times during development, and in response to environmental cues, neurons in the insect central nervous system release prothoracicotropic hormone (PTTH). This neuropeptide signals the adjacent ring gland to secrete the steroid hormone ecdysone into the haemolymph of the circulatory system. Some peripheral tissues, most notably the larval fat body, modify ecdysone into its biologically active form, 20-hydroxyecdysone. This hormone can then directly regulate gene expression through its interaction with the ecdysone receptor. Although insect endocrinologists use the term **ecdysone** to refer to the inactive form of the hormone, this word is synonymous with 20-hydroxyecdysone in the molecular biology literature. Ashbuner proposed a model for ecdysone action. In this model the hormone combined with its receptor binds to the target DNA to induce directly the **early puff genes** and to repress the **late puff genes**. The protein products encoded by the early genes then activate the late genes while inhibiting their own synthesis. In support of this model, three genes from early puffs have been cloned and shown to encode multiple proteins that contain DNA-binding motifs of various types and therefore are likely transcription factors.

Growth of insect larvae is punctuated by a series of moults to allow expansion of the hardened cuticle as the larva becomes larger. These moults
initiated by surges of the steroid hormone ecdysone, which is converted
ive form 20-hydroxyecdysone by the fat body and some other
r moult period of each larval instar usually begin with low
transiently elevates towards the end of the instar.
etarmorphosis will take place is determined
the time of ecdysone release. In the
inor ecdysone pulses in the
are important for

and the onset of prepupal development. This is followed, 10-12 hours after puparium formation, by another ecdysone pulse that initiates the 3.5 days of pupal development, followed by eclosion of the adult fly. Most larval tissues are destroyed during the early stages of metamorphosis and are replaced by adult tissues that develop from clusters of imaginal progenitor cells. The crawling larva thus undergoes a complete transformation, resulting in the formation of a highly motile, reproductively active adult fly. Remarkably, this transformation occurs in apparent response to a single steroid hormone, ecdysone.

5.2 Ecdysone Regulation of Primary-Response Genes

Ecdysone manifests its effects on development by activating its receptor, a heterodimer of two members of the nuclear receptor superfamily: EcR and USP (Table 9.2). Interestingly, at least two of the three protein isoforms encoded by the EcR gene are expressed in a tissue-restricted manner that can contribute to the specificity of developmental responses to ecdysone. EcR-B1 is expressed primarily in larval tissues that are fated to die, whereas EcR-A is expressed in developing adult structures and tissues. Both high-affinity DNA binding and ecdysone binding require EcR heterodimerization with USP, a homologue of vertebrate retinoid X receptor. The ecdysone/EcR/USP complex can then binds DNA and activate the transcription of target genes. It is interesting to note that a Drosophila orphan receptor, DHR38, a homologue of the vertebrate NGFI-B receptor, can also heterodimerize with USP. This raises the possibility that ecdysone signalling may be modified through distinct heterodimer combinations of receptors, similar to the interactions that have been demonstrated in vertebrate organisms.

Many genes are induced directly by the ecdysone-receptor complex, presumably reflecting their immediate requirement following ecdysone pulses. Some of these genes, like IMP-E1 and Fbp-1, are induced by ecdysone in a restricted tissue-specific manner, whereas others, like hsp23 and Eip28 / 29, are expressed in multiple ecdysone target tissues.

A subset of primary-response genes encode transcription factors that function at the top of ecdysone-triggered genetic regulatory hierarchies. Most of these genes were defined as puffs that are induced directly by ecdysone in the larval salivary gland polytene chromosomes. Detailed studies by Michael Ashburner and co-workers demonstrated that these so-called early puffs appeared to encode regulatory proteins that both repress their own activity and induce large sets of secondary-response late puffs. The protein products of the late puffs were, in turn, thought to play a more direct role in specifying appropriate biological responses to the hormone.

Four early puff genes have been described at the molecular level. One of these, E63-1, encodes a calcium binding protein related to calmodulin, providing the possibility of cross-regulation between hormone and calcium signalling pathways. The remaining three genes each encode multiple transcription factor isoforms. The broad-complex (BR-C) encodes more than a dozen proteins, each of which carries one of four possible pairs of zinc fingers. The E74 early gene contains nested promoters that direct the synthesis of two related proteins, E74A and E74B, that share an identical C-terminal ETS domain that binds DNA. Similarly, the E75 early gene contains nested promoters that direct the synthesis of three orphan members of the nuclear receptor superfamily, designated E75A, E75B, and E75C. Curiously, three other orphan receptor genes, DHR3, E78, and bFTZ-F1, correspond to ecdysone-regulated puffs in the polytene chromosomes, suggesting that this family of transcription factors may play an important role in ecdysone signalling during development.

Table 9.2. Members of the steroid hormone receptor superfamily

Receptor	Function or properties	P-box
EcR	Ecdysone receptor, expressed in most tissues. Directly induced by 20E	EGCKG
USP	Formation of functional 20E receptor heterodimer with EcR. Abdominal segmentation, larval development, eye morphogenesis	EGCKG
E75	75B early Puff product. Directly induced by 20E in most tissues	EGCKG
DHR3	Product of the 46F puff. Directly induced by 20E?	EGCKG
E78	78C early late puff product. Full induction by 20E requires protein synthesis.	EGCKG
FTZ-F1	Embryonic pattern formation. Activator of the ftz segmentation gene and Adh gene	ESCKG
DHR 39	Expressed mainly in embryonic brain and ventral CNS. Coregulator of Adh and ftz?	ESCKG
SVP	Embryonic CNS development. Cell fate determination in compound eye morphogenesis	EGCKS
TLL	Embryonic pattern formation of terminal structures	DGCAG
HNF-4D	Embryonic midgut, fat body and malpighian tubule organogenesis	DGCKG

Ecdysone dose-response studies have revealed that each early promoter is activated at a specific critical threshold hormone concentration. Combined with ecdysone titre measurements, these dose-response studies have led to a model for the timing of early gene expression during the onset of metamorphosis. The EcR, BR-C, and E74B promoters are most sensitive to ecdysone, and are induced in mid-third instar larvae in apparent response to a low- hormone pulse. The high- late larval ecdysone pulse then represses some of these messenger RNAs, such as EcR and E74B, as it induces the less-sensitive promoters, including E74A, E75A, E75B, and DHR3 (Fig. 9.11). The ecdysone drops during mid-prepupal development, allowing bFTZ-F1 to be expressed. This gene is repressed by ecdysone and thus depends on the decrease in hormone titre for its induction (Fig 9.12). The rapid rise in ecdysone titre in late prepupae then recapitulates the larval pattern of early gene expression, with a burst of EcR and E74B expression preceding that of E74A and E75A. In addition to these dose-dependent responses to ecdysone pulses, many early genes also cross-regulate their activity, ensuring that they will be expressed at the correct time and for the appropriate duration.

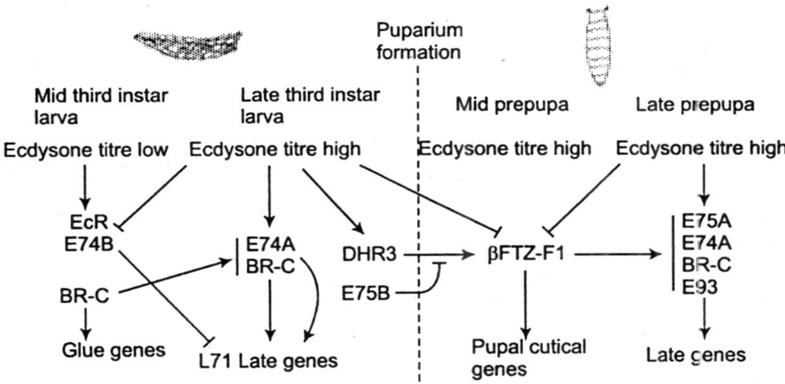

Fig. 9.12. Ecdysone triggered regulatory hierarchies in Drosophila metamorphosis

5.3 Regulation of Ecdysone Secondary-Response Genes

Molecular and genetic studies have demonstrated that these stage-specific patterns of primary-response regulatory gene expression are transduced into corresponding patterns of secondary-response gene activity. Two classes of secondary-response genes have been studied in the larval salivary glands: the glue genes and the L71 genes from the 71E late puff. The glue genes are induced in mid-third instar salivary glands and encode polypeptide glue that is used by the animal to affix itself to a solid surface for pupariation. In contrast, the L71 genes are of unknown function and are induced at puparium formation, as the glue genes are repressed. Both BR-C and E74B are required for proper glue gene induction, defining a mid-third instar regulatory hierarchy in the salivary glands (Fig. 9.12). BR-C binding sites in the glue gene enhancers are required for proper induction, indicating that this is a direct regulatory link. The tissue specificity of this response is directed by the Fork head homeodomain transcription factor that binds adjacent to the BR-C in glue gene enhancers.

The L71 late genes also depend on BR-C and E74 for their proper induction in late third instar larval salivary glands (Fig 9.12). E74B maintains the L71 late genes in a repressed configuration during third instar larval development. The high-titre late larval ecdysone pulse then shuts off the E74B repressor and induces the E74A activator, which, together with the BR-C, directly induces L71 transcription (Fig 9.12). These studies provide molecular support for Ashburner's proposal that early puff proteins direct the induction of late puffs genes.

5.4 Cross-Regulation among Ecdysone-Induced Transcription Factors

In addition to their roles in regulating ecdysone secondary-response genes, the ecdysone-induced transcription factors also regulate one another. BR-C is required for maximal ecdysone-induced early gene transcription in late third instar larvae (Fig 9.12). In this sense, BR-C is functioning as a competence factor that facilitates the subsequent regulatory response to ecdysone. Similarly, the early induction of EcR by a mid-third instar hormone pulse allows for increased levels of ecdysone receptor in preparation for the high hormone titre at the end of larval development.

Evidence has also been obtained for cross-regulatory interactions among the ecdysone-regulated orphan receptor genes. DHR3, which is induced by ecdysone in newly formed prepupae, is an inducer of bFTZ-F1 expression in mid-prepupae (Fig 9.12). Furthermore, this induction can be inhibited by heterodimerisation of DHR3 with E75B. E75B protein levels normally decay during early prepupal development, thus determining the time at which DHR3 can activate bFTZ-F1, in turn, appears to be a critical competence factor for maximal reinduction of the early genes (Fig 9.12). Furthermore, bFTZ-F1 is sufficient to induce the stage-specific E93 early gene in the salivary glands of late prepupae. The DHR3, E75B, and bFTZ-F1 orphan receptors thus provide a regulatory link between the late larval and prepupal responses to ecdysone. Their stage-specific expression confirms that a critical developmental stage has been achieved and that the next response to the hormone should be distinct from the last response. These receptors thus provide a mechanism by which a hormonal signal can be refined into stage-specific developmental pathways.

5.5 From Gene Regulation to Biological Responses

Studies are currently under way that attempt to link ecdysone regulatory hierarchies with their appropriate biological responses during metamorphosis. Most larval tissues are destroyed during prepupal and early pupal development. Recent studies have revealed that the destruction of the larval midgut and salivary gland occurs as stage-specific programmed cell death responses that display many of the hallmark features of apoptosis. Furthermore, the *Drosophila* death inducer genes reaper and head involution defective are coordinately induced in these tissues immediately preceding their destruction. A current focus of research is aimed at understanding how these death inducers are regulated by ecdysone at the appropriate time and place during metamorphosis. Specific neurons in the central nervous system also undergo programmed cell death in newly eclosed adults, in response to a decrease in ecdysone titre. Interestingly, these neurons can be distinguished by their high expression of the EcR-A isoform, suggesting that this receptor may play a key role in directing the death of these neurons.

As the larval tissues are being destroyed, the adult fly is being constructed. During prepupal development, the imaginal discs evert and elongate to form rudiments of the adult appendages. BR-C and E74 both play a critical role in this process. Studies have also shown that bFTZ-F1 is sufficient to induce the pupal cuticle genes in mid-prepupal imaginal discs, providing a mechanism for the stage specificity of pupal cuticle deposition (Fig 9.12). It seems likely that future studies will provide more links between the ecdysone regulatory hierarchies and the biological responses to ecdysone during the onset of metamorphosis.

6. BIOSYNTHESIS OF JH

The JH biosynthetic pathway may be viewed as two distinct biosynthetic units or "arms" (Fig 9.13). In the first arm, biosynthesis proceeds through the mevalonate pathway (MP), which insects share with other organisms and for which cholesterol is the best known end product. The MP uses acetyl-CoA as starting material to generate mevalonate, which is then converted to isopentenyl disphosphate (IPP). Biogenesis of the sesquiterpene (i.e., C-15) precursor of JHs, farnesyl diphosphate (FPP), is effected by an FPP synthase (FPPS, a prenyltransferase), which catalyses the head-to-tail condensation of three isoprene units.

Insects and other arthropods have lost the ability to synthesise cholesterol and depend on dietary sources for acquisition of this essential compound.

Insects lack the genes encoding the enzymes required for the production of cholesterol from FPP, including squalene synthase, which catalyzes the reductive condensation of two FPP molecules (Clark and Bloch, 1959). Although insects utilise FPP in other pathways (e.g., ubiquinone synthesis, dolichol synthesis, protein prenylation, etc.), its use as a JH precursor is not only well documented, but is a hallmark feature of insect biochemistry.

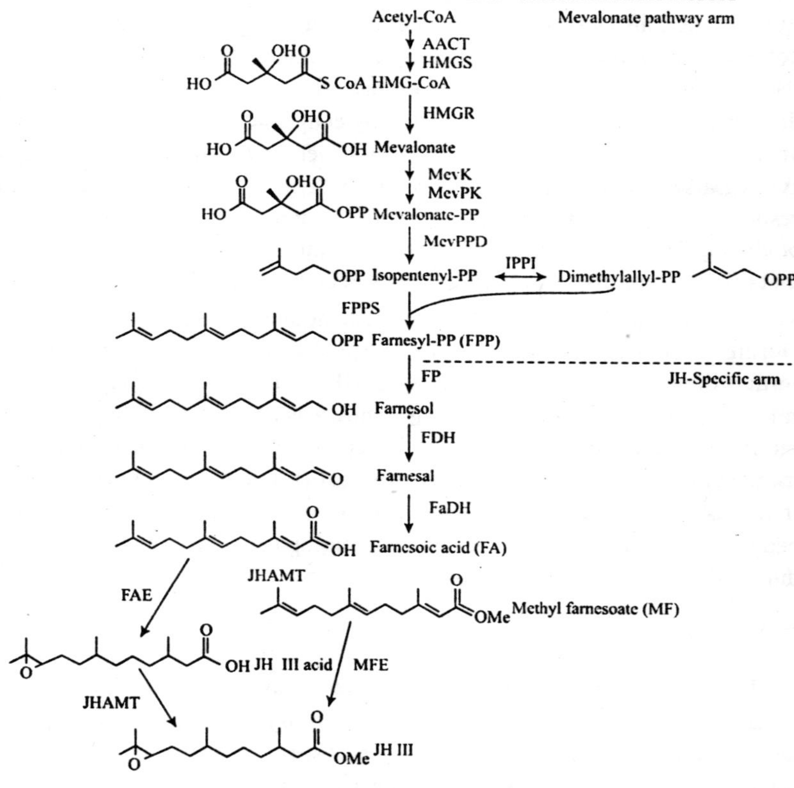

Fig. 9.13. The JH biosynthetic pathway

AACT, acetoacetyl-CoA thiolase; HMGS, HMG-CoA synthase; HMGR, HMG-CoA reductase; MevK, mevalonate kinase; MevPK, phosphomevalonate kinase; MevPPD, diphosphomevalonate decarboxylase; IPPI, isopentenyl diphosphate isomerase; FPPS,farnesyl diphosphate synthase; FP, farnesyl phosphatase; FDH, farnesol dehydrogenase; FaDH, farnesal dehydrogenase;JHAMT, JH acid methyltransferase; FAE, farnesoic acid epoxidase; and MFE, methyl farnesoate epoxidase

The second arm of JH biosynthesis, the **JH-specific arm**, comprises enzymatic steps unique to JH-producing organisms. FPP is first converted to farnesol by a farnesyl phosphatise (FP), while farnesol undergoes two sequential oxidation reactions that generate farnesal and farnesoic acid (FA). The order of the two last steps, epoxidation and methyl esterification, catalysed by an epoxidase and a JH acid methyltransferase (JHAMT), may vary between species.

Although all insects produce JHs using the same general biosynthetic scheme, variations are evident for taxa known to produce JHs other than JH III. For example, production of ethyl-branched JHs by Lepidoptera involves the substitution of propionyl-CoA for acetyl-CoA in an early step of JH synthesis. This substitution leads to the production of the C-6 homologue of IPP (homo-IPP), which may then be used as substrate, with its allylic isomer homo- DMAPP, by FPPS to generate the various FPP homologue precursors of ethyl-branched JHs.

In the higher Diptera and the Hemiptera that produce two different bisepoxy forms of JH III, catalytic features of the epoxidases are expected to differ from those found in other insects (Helvig *et al.*, 2004). Finally, as noted earlier, the order in which methylation and epoxidation take place, in the last two steps of JH biosynthesis, varies among different groups of insects. In Lepidoptera, in which JH III acid (i.e., epoxyfarnesoic acid) is secreted by larval CA during the second half of the last stadium (Bhaskaran *et al.*, 1986) and is the only secreted product of adult male CA. The same appears to be true for the production of JHB3 by the higher Diptera. In cockroaches, however, esterification precedes epoxidation because the epoxidase only accepts MF as substrate.

7. PHEROMONES

Pheromones are volatile chemical compounds secreted by insects and animals. They act as chemical signals between individuals influencing physiology and behaviour in a manner similar to hormones. Pheromones are important to a variety of behaviours including mate attraction, territorality, trail marking, danger alarms, and social recognition and regulation.

The term pheromone is derived from the Greek words *pheran* (to transfer) and *horman* (to excite). Pheromones are found throughout the insect world. They are active in minute amounts. In fact, the pheromones released by some female insects (e.g., Silkworm Moth) are recognized by the male of the species as far as a mile away. The pheromone secreted by the female gypsy moth can be detected by the male in concentrations as low as one molecule of pheromone in 1×10^{17} molecules of air. Insects detect pheromones with specialised chemosensory organs. Based on their effect, pheromones can be divided into at least the following categories:

1. **Aggregation pheromones:** These compounds that increase the concentration of insects at the pheromone source.
2. **Alarm pheromones:** The compounds that stimulate insects' escape or defence behaviour.
3. **Sex pheromones:** The compounds that help individuals of the opposite sex to find each other.
4. **Trail pheromones:** Among social insects, compounds used by workers to mark the way to a food source, for example.
5. **Marking pheromones**: These compounds are used by insects to mark the boundaries of a territory.

The structure of the first insect sex pheromone bombykol (*E,Z*)-10,12-hexadecadien-1-ol from the silkworm moth, *Bombyx mori* (L.) was elucidated after utilising a half million female abdomens. Rapid improvements in analytical instrumentation and techniques reduced the number of insects needed for pheromone extracts from a half million or more to where now individual insects can sometimes provide sufficient material for chemical analysis. Most pheromones are biogenetically derived blends of two or more chemicals that must be emitted in exact proportions to be biologically active.

Most insect pheromone components are synthesised *de novo* by insect tissue. The complexity of the pheromone producing tissue varies from simple unicellular glands distributed throughout the integument to elaborate internal cellular aggregates connected to a reservoir. The most common location for pheromone production is the abdomen. The oxygenated lepidopteran pheromone components are usually produced and released from extrudable glands located between the 8th and 9th abdominal segments. In contrast, the site of synthesis of the hydrocarbon and hydrocarbon-derived pheromone components is more complicated. 2-Methylheptadecane is not synthesised in the pheromone gland of *Holomelina aurantiaca,* but it is synthesised by epidermal tissue, presumably oenocytes, transported by lipophorin, and then sequestered into and released from the pheromone gland. Male Lepidoptera produce aphrodisiac scent from glands which are commonly associated with scales. These scales are known as androconia and occur on the wings as in Pieridae, and they may be either scattered or grouped together. Comparable scales may also occur on the legs or abdomen. Honey bees have two important glands producing pheromones: the mandibular glands in the head and Nassanoff's glands in the abdomen.

The products of normal metabolism, particularly those of the fatty acid and isoprenoid pathways, were modified by a few pheromone- tissue specific enzymes to produce many of the myriad of pheromone molecules. Many of the lepidopteran pheromones could be formed by the appropriate interplay of highly selective chain-shortening and unique Δ11 and other desaturases, followed by modification of the carboxyl carbon. The honeybee also uses highly specific chain shortening of fatty acids to produce the major component of the queen pheromone. In some insects, fatty acid elongation followed by oxidative decarbonylation produces the hydrocarbon and hydrocarbon-derived pheromones (Tillman *et al.*, 1999; Blomquist, 2010; Millar, 2010). Recent work in bark beetles has shown that *Ips* and *Dendroctonus* spp. produce their monoterpenoid derived pheromones ipsenol, ipsdienol, and frontalin by modification of isoprenoid pathway products (Blomquist *et al.*, 2010).

It is worth noting that the production of pheromones is usually affected, either directly or indirectly, by feeding or starvation, even in insects that do not sequester pheromone precursors from the diet. Indirect effects may be through feeding-induced stimulation of endocrine factors. Foster (2009) showed in *Heliothis virescens* moths that *de novo* production of the fatty-acid-derived pheromone components is strongly influenced by sugar feeding. The pheromone titre rapidly declined in mated females, as did haemolymph trehalose, but ingestion of sugar restored the pheromone to near virgin levels. He suggested that redirection of carbon to the developing oocytes lowers blood sugar, and females can elevate both trehalose concentration and their pheromone with sugar feeding.

Most female moths produce their pheromone components through modifications of fatty acid synthesis pathways. The pheromone components are produced through a fatty acid biosynthesis pathway involving a Δ11 desaturase and limited chain-shortening enzymes. The appropriate chain length fatty acid is then reduced and acetylated to form the acetate ester. For example, it was determined that Z11-16:Ald is produced by Δ11 desaturation of 16:CoA followed by reduction in the moths *Helicoverpa zea* and *H. assulta* (Choi *et al.*, 2002). Whereas Z9-16:Ald, the major pheromone component in *H. assulta*, is produced by a Δ9 desaturase using 16:CoA as a substrate, in *H. zea* Z9-16:Ald is produced by Δ11 desaturation of 18:CoA to produce Z11-18:CoA that is then chain-shortened to Z9-16:CoA (Choi *et al.*, 2002).

Most female moths utilise a blend of components produced in a specific ratio for pheromone attraction of conspecific males. These ratios are produced by the inherent specificity of certain enzymes present in the biosynthetic pathways. The combination of these enzymes acting in concert produces the species-specific pheromone blend. The cabbage looper, *T. ni*, uses *Z*7-12:OAc as the major sex pheromone component, which is produced by Δ11 desaturation of 16:CoA followed by two rounds of chain shortening, reduction, and acetylation. The next step in the pathway is limited chain shortening, and it was shown that *Z*11-16:CoA is the preferred substrate for these enzymes (Jurenka *et al.*, 1994). After chain shortening the 14- and 12-carbon intermediates are reduced to an alcohol and acetylated. The acetyltransferase enzyme is not specific and will accept a variety of substrates (Jurenka and Roelofs, 1989). From these observations, it can be inferred that the final ratio of pheromone components is produced by the specificity found within the 11 desaturase and chain-shortening enzymes.

A blend of seven acetate esters used by *A. velutinana* is produced in a biosynthetic pathway similar to the one just described, except that the Δ11 desaturase starts with 14:CoA as the substrate and produces both *Z* and *E* isomers of 11-14:CoA in about a 60/40 ratio (Wolf and Roelofs, 1987). However, the final ratio of *Z*11- to *E*11 14:OAc is 92/8 (Miller and Roelofs, 1980). A selective increase in the *Z* isomer occurs within the biosynthetic pathway and it was determined that acetyl:CoA fatty alcohol acetyltransferase shows specificity for the *Z* isomer (Jurenka and Roelofs, 1989). Therefore, selective acetylation of *Z*11-14:OH and production of >60% *Z*11-14:CoA indicates that these enzymes have the inherent specificity to produce the 92:8 ratio of the major pheromone components *Z*11- and *E*11-14:OAc. Two minor pheromone components are produced by chain-shortening *Z*11- and *E*11-14:OAc. The ratio of Z9- to *E*9-12:OAc is about 1 to 2. This indicates that the chain-shortening enzymes may prefer *E*11-14:CoA or that very little Z11 14:CoA is available to chain-shorten. This combined information indicates that in *A. velutinana* pheromone glands the final ratio of pheromone components can be produced through the concerted action of a Δ11 desaturase that produces at least a 60:40 ratio of *Z/E* intermediate isomers. The final ratio of acetate esters (92/8) is produced through the specificity for the *Z* isomer by the acetyltransferase. The minor components are produced by specificity in chain shortening.

Moths in the families Geometridae, Arctiidae, Amatidae, Lymantriidae, Lyonetiidae, and some Noctuidae utilize hydrocarbons or epoxides of hydrocarbons as their sex pheromones (Millar, 2010). Biosynthesis of hydrocarbons occurs in oenocyte cells associated with either epidermal cells or fat body cells. Once the hydrocarbons are biosynthesised they are transported to the sex pheromone gland by lipophorin. Most moth sex pheromones that are straight chain hydrocarbons usually have an odd number of carbons. Most of these are polyunsaturated with double bonds in the 3,6,9- or 6,9-positions, indicating that they are derived from linolenic or linoleic acid, respectively. Linolenic and linoleic acid cannot be biosynthesised by moths so they must be obtained from the diet. A few even-chain-length hydrocarbon sex pheromones have been identified that also have 3,6,9- or 6,9-double bond configurations (Millar, 2000, 2010), indicating they too are derived from linolenic or linoleic acids. A few moths utilize a combination of hydrocarbon and fatty-acid-derived pheromone components (Ando *et al.*, 2004). The navel orangeworm moth is one of these and labelling studies supported different pathways producing the hydrocarbons and the aldehydes. The aldehydes were produced through a fatty acid biosynthetic route in pheromone glands while the hydrocarbons were produced by oenocytes and transported to the pheromone gland for release (Wang *et al.*, 2010).

8. HORMONAL CONTROL OF DIAPAUSE

Diapause is an arrest in development accompanied by major shutdown in metabolic activity usually when environments are seasonally hostile. Unlike a simple quiescence that is an immediate response to an unfavourable environmental condition, diapause is a genetically programmed response that occurs at a specific stage for each species. Sometimes the insect will enter diapause at this particular stage in each generation regardless of the environmental conditions and this is referred to as **obligatory diapause.** Much more frequently, the decision to enter diapause is determined by environmental factors, usually day length, received by that individual or its mother at an earlier developmental stage. This is referred to as **facultative diapause** (Box 9.8). Diapause can take place at any stage in the development of the insect. Embryonic diapauses have been documented for nearly all possible stages of embryonic development, ranging from early blastoderm formation to pharate larvae. **Larval diapauses** are most common in late instars, but early instar diapauses are not uncommon. **Pupal diapause** is well documented, and a few species diapause as fully developed pharate adults. The most prevalent cases of adult diapause involve newly emerged adults that have not yet reproduced, but there are also examples of diapause interrupting periods of reproduction. Diapausing insects usually have a prolonged pre-diapausing period when they attain a larger size, accumulate additional lipid reserves, and add additional waterproofing to their cuticle. The longest diapause among insects is seen in the yucca moth *Prodoxus y-inversus* (Box 9.9).

Box 9.8 Diapause

Diapause is a neurohomonally mediated, dynamic state of low metabolic activity. Associated with this is reduced morphogenesis, increased resistance to environmental extremes, and altered or reduced behavioural activity.

1. **Obligate:** Every individual in every generation enters diapause. These insects usually have 1 generation/yr, thus a univoltine life cycle. It is genetically controlled.
2. **Facultative:** Some generations may be completely free of diapauses while some or all of the individuals in a generation may enter diapause. These insects usually have several generations/yr and are known as multivoltine. It is environmentally controlled.

It is evident that diapause is a dynamic process, as evidenced by characteristic changes in the utilisation of energy reserves, systematic changes in patterns of oxygen consumption, changes in responsiveness to environmental stress and hormones, and distinct patterns and changes in gene expression. A progression of events must transpire before diapause can be terminated. In some cases, development is not completely halted. There are several examples in which developmental processes continue during diapause, albeit at much slower rates; for example, ovarian maturation during adult diapauses in the mosquito *Culex pipiens.*

Box 9.9 Longest Diapause among insects

The yucca moth *Prodoxus y-inversus* Riley (Lpidoptera: Prodoxidae) has the longest reported diapause. Structures of *Yucca baccata* (Agavaceae) containing prepupae of this species were collected in Nevada, and successful emergence of adults was observed 19 years later, under artificial condition.

The events in a diapausing insect can be grouped into three periods namely the pre-diapause period, diapause period and post diapauses period (Box 9.10). The pre-diapause period is characterised by the preparatory phase when accumulation of food reserves takes place. The diapause period has four phases – 1) Induction phase characterised by gradual cessation of DNA, RNA synthesis and lowering of metabolism, 2) Refractory phase characterised by activation process and supercooling reactions, 3) Activated phase when individuals show potency to develop general physiology as before and 4) Termination phase when endocrine reactivation, and biochemical readjustment for DNA, RNA synthesis takes place. The post diapauses period is characterised by resumption of mitosis, growth and development.

Box 9.10. Periods in Diapause

1. Prediapause period

   A. Preparatory phase : Accumulation of food reserves

2. Diapause period

   A. ***Induction phase***: Gradual cessation of DNA,RNA synthesis, lowering of metabolism

   B. ***Refractory phase***: Activation process, super cooling reactions

   C. ***Activated phase***: Individuals with potency to develop general physiology as before

   D. ***Termination phase***: Endocrine reactivation, biochemical readjustments for DNA,RNA synthesis

3. Post diapause period

   Resumption of mitosis, growth and development.

During the preparative phase, the insect must make the decision to enter diapause (and sometimes how long to remain in diapause) and make the accompanying alterations that will result in the sequestration of additional energy reserves and waterproofing agents as well as initiate specific behavioural programs needed to find protected sites appropriate for overwintering. Some species, such as the Monarch butterfly, *Danaus plexippus*, engage in long-distance flights to reach their overwintering site, but more commonly, insects programmed for diapause migrate to closer, buffered sites.

Short day length is the most common cue used for the programming of overwintering diapauses. Insects that enter a summer diapauses may use long day lengths instead of short day lengths for the programming of diapauses. With a few exceptions, temperature also influences the decision to enter diapause. While it may act alone as the environmental cue used for diapause induction in some tropical species, temperature usually acts in concert with the photoperiod. Other factors contributing to the diapause decision include food presence, food quality, plant volatiles, pheromones, and crowding.

8.1 Hormonal basis for embryonic diapauses

Hormonal regulation has been examined only in a few species of Lepidoptera, and it is evident that no single regulatory scheme prevails. The most intensely studied embryonic diapause is that of *B. Mori.* Diapause in *B. mori* differs from that of most other species. While both temperature and daylength are key components contributing to the diapauses decision, the specifics are reversed as compared to other diapause models. High temperature and long daylength prompt diapause induction, while low temperature and short daylength lead to non-diapause development. Diapause induction in this species is under strict maternal control. The photosensitive stage occurs far in advance of the actual diapauses stage: exposure of developing female embryos and larvae to diapause-inducing conditions cause the adult female to produce eggs that are destined to enter diapause. This effect is mediated by diapause hormone (DH), a neuropeptide released by the female adult during her period of egg maturation, which acts upon the ovarioles and leads to production of diapause-destined eggs. After being oviposited, the silkworm embryos enter a diapause characterised by G2 cell cycle arrest (Nakagaki *et al.*, 1991). Once diapauses has been established, this arrest can be broken after the embryos have been chilled for 1–3 months at 5°C, after which development can be reinitiated by exposure to higher temperatures. It has been shown that the brain and subesophageal ganglion controlled the release of DH.

The neural control of DH expression and release has also been characterised. Both *in vivo* and *in vitro* studies show that DH transcription is under control of the neurotransmitter dopamine. Dopamine levels in both the haemolymph and brain–subesophageal ganglion complexes of diapause-type individuals are significantly higher than in their non-diapause type counterparts during the late larval and early pupal stages, correlating with an increase in dopamine decarboxylase Expression. Neuronal control of DH is not restricted to the transcriptional level. The release of DH from the subesophageal ganglion is under control of GABAergic (Gamma aminobutyric Acid) neurons.

Downstream of DH, regulation of diapause appears to involve accumulation of sorbitol. A key indicator of DH activity has long been a metabolic switch, resulting in the accumulation of glycogen in diapause-destined eggs. This accumulation is due to the enzyme trehalase, which is transcriptionally upregulated after DH injection. Accumulated glycogen is then converted to sorbitol, which acts as a classic cryoprotectant, lowering the supercooling point of the diapausing eggs. Upon diapauses termination, an ERK/MAPK-mediated pathway signals the reconversion of sorbitol to glycogen (Fujiwara *et al.*, 2006), which is subsequently used as an energy reserve.

8.2 Hormonal basis for larval and pupal diapauses

Larval insects that diapause mostly do so in the last larval instar, but some larvae diapauses in earlier instars or as pre-pupae. Pupal diapause has been examined extensively in Lepidoptera and the higher Diptera, but appears to be rare in other taxa. It is the true pupal stage that is usually involved in diapause, but occasional reports indicate a diapause late in pharate adult development. Larval and pupal diapauses share much in common. Central to both is a failure to progress to the next metamorphic stage. This suggests that one of the key endocrine events regulating such a diapause is likely to be a failure of the prothoracic gland to produce the ecdysteroids needed to initiate the next moult. Invariably, this has proven to be a key element of larval and pupal diapauses. One might also predict a shutdown in the production of PTTH until the end of diapauses. JH may or may not play a role in diapauses depending on the species. In *Diatraea grandiosella*, a species that diapauses late in the final larval instar, PTTH activity is far lower in brains and brain extracts from diapauses destined larvae and diapausing larvae than in non-diapausing larvae (Yin *et al.*, 1985). Likewise, in the European corn borer, *Ostrinia nubilalis*, PTTH activity is lower in brains of diapausing larvae than in non- diapausing brains (Gelman *et al.*, 1992). Similar low levels of PTTH activity are noted in pupal diapauses of the cabbage armyworm, *Mamestra brassicae.* Yet, brains from all of these species clearly show some PTTH activity, and pupal brain extracts from *M. sexta* (Bowen *et al.*, 1984) and *Sarcophagia argyrostoma* (Richard and Saunders, 1987) actually have as much or more PTTH activity than brain extracts from nondiapausing pupae. Likewise, PTTH activity is higher in than in larvae not destined for diapause (Moribayashi *et al.*, 1992). These observations suggest that the hormone may be present during diapause but is simply not released.

Several experiments suggest that high dopamine levels in the brain may be critical for diapause induction, possibly by preventing release of PTTH. High levels of dopamine in brains of diapause-destined larvae of *C. costata* (Kostal *et al.*, 1998), diapause-destined pupae of *M. brassicae* (Noguchi and

Hayakawa, 1997), and pupae of *P. brassicae* suggest a causal relationship. In diapausing pupae of *A. pernyi*, a drop in brain dopamine is noted shortly before the surge of ecdysteroids that elicits diapause termination (Matsumoto and Takeda, 2002). But the most compelling evidence suggesting a role for dopamine is the fascinating observation that feeding L DOPA to last instar larvae of *M. brassicae* reared under long daylengths (not diapauses inducing) elicits a diapause-like state in the pupae (Noguchi and Hayakawa, 1997). Elevation of brain serotonin (5-hydroxytryptamine) has also been linked to diapause in several cases.

8.3 Hormones in adult diapauses

Diapuse in adults is related to cessation of reproduction. In females there is arrest in occyte development and in males it is failure to mate with receptive females. The accessory reproductive glands are usually reduced in size during diapauses and flight muscles may degenerate. Mating may occur either before the entry into diapauses or after diapause has been completed. When mating occurs prior to diapause, the males usually die without entering diapause and only the female's bridge the diapauses period. When both sexes diapause, mating usually occurs only after diapause has been terminated.

JH has been implicated to play a major role in diapause. JH titre in diapause-programmed *Leptinotarsa decemlineata* decreases after adult emergence, remains low throughout diapause, and then rises after termination of diapause. Earlier results arguing a role for JHs in regulating the adult diapause of *C. pipiens* were based on the ability of JH applications to terminate diapause; more recently this evidence was supplemented by demonstrating that cultured CA dissected from diapausing females secrete very little JH, whereas those dissected from non- diapausing females are quite active, especially during the first week after adult eclosion. Gene silencing experiments (Sim and Denlinger, 2008), as well as with experiments demonstrating that a knockdown of two ribosomal genes, S2 and S3a, causes a diapause-like arrest in *C. Pipiens* that can be reversed with application of JH III (Kim and Denlinger, 2010; Kim *et al.*, 2010) strengthens the role of JH in diapause.

The JH haemolymph is a function of both JH synthesis and its degradation. There is an inverse relationship between the JH haemolymph and JH esterases (JHE) activity in pre-diapausing *L. decemlineata* adults, with peak esterase activity occurring just prior to diapause initiation. This peak in JHE activity contributes to the low JH titre noted during the first two days of adult life (Kramer, 1978). The effect of JH on JHE activity is indirect and appears to require a factor from the brain.

Juvenile hormone is not the exclusive cause of diapauses induction and maintenance. Results, even in the Colorado potato beetle, suggest that lack of JH cannot account for all observations. Juvenile hormone application, at best, can stimulate only a short burst of oviposition in diapausing females, suggesting that the actual termination of diapause requires more than a single boost in the JH titre. This observation has usually been interpreted as the need for sustained stimulation of the CA by the brain. At the end of diapause the brain must become reactivated to promote JH synthesis by the CA. Brain activation clearly needs to precede activation of the CA and it is possible that the brain independently directs specific events associated with diapause termination.

Another intriguing connection between JH and adult diapause has been proposed by Hunt *et al.* (2007) for *Polistes* wasps. In this insect accumulation of a storage protein, hexamerin 1, may dictate the JH titre, determining the diapause fate of the emerging females. Low levels of hexamerin are suggested to elevate JH, which in turn promotes the non-diapause phenotype that is primed to reproduce, while high levels of hexamerin generate a JH deficient diapause phenotype that fails to reproduce until the following year

The role of ecdysteroids in the regulation of adult diapause has been investigated most thoroughly in *D. melanogaster*, although this arrest appears to be a low-temperature induced halt, similar to a quiescence than a photoperiodically induced diapause (Emerson *et al.*, 2009). The arrest occurs at the pre-vitellogenic stage of ovarian follicular development. Yolk proteins (YPs) are present in the haemolymph of these females, but the oocytes fail to incorporate the YPs. Ecdysteroids are proposed to activate endocytosis at termination of the arrest allowing initiation of vitellogenesis (Richard *et al.*, 1998). Significantly lower ecdysteroid titres are reported for diapausing females of *L. migratoria* than for nondiapausing females (Tawfik *et al.*, 2002), a result that is consistent with *D. melanogaster*. The ecdysteroid titre in *L. migratoria* fluctuates between 22 and 40 ng ml−1 in diapausing females, whereas the titre reaches a peak of 354 ng ml−1 in 18- to 22-day old non-diapausing females. Treating diapausing females with three consecutive applications of JH III induces ovarian development and increases ovarian and haemolymph titres of ecdysteroids to levels similar to those noted in sexually active females.

9. HORMONAL REGULATION OF MOULTING

Ecdysis is the process of an arthropod moulting its exoskeleton. Moulting is necessary as the arthropod exoskeleton is inflexible and so, to grow larger, arthropods must moult. During this intricate process, insects shed the old cuticle surrounding the external surface, and the lining of the foregut, hindgut, and tracheal tubes of the respiratory system. Moulting is a critical but vulnerable time for arthropods. Their existing cuticle weakens at specific points and by taking in water or air the animal splits its old exoskeleton. The animal then extracts itself from the old skin and inflates its new skin. The new exoskeleton is soft and the arthropod can't move because its musculature is attached to this exoskeleton. Once the new exoskeleton has dried and hardened the animal can move again. While ecdysis is the gateway to the next developmental stage, it is also effectively a life-threatening process, since failure to execute the sequence on schedule can result in death of the organism. Its success depends on coordinated gene expression controlled by ecdysteroids and JH responsible for the formation of new cuticle and other tissues appropriate for the next stage. Ecdysteroids also participate in activation of a peptide signalling cascade that regulates the ecdysis sequence at the end of the moult.

As soon as the egg hatches, the young larvae feed, grows and undergo a series of successive immature stages that culminates in the reproductive adults. At the beginning of each larval instar, the animal feeds and grows until it acquires the proper size and nutritional state. The feeding stage or intermoult proceeds until integrative processes in the brain produce a decision to move on to the next stage, whereupon production and release of ecdysteroids occurs. The appearance of elevated ecdysteroid levels brings an end to the intermolt and onset of the moult associated with production of new cuticle (Box 9.11). At the end of last larval instar, holometabolous insects empty the gut and engage in wandering behaviour aimed at finding an optimal location for pupation. Metamorphosis is initiated in the pharate pupa (pupal development) and continues after pupation (adult development). This complex process involves remodelling or programmed death of larval organs and proliferation of new cells and tissues specific for pupal and adult stages.

Box 9.11 Sequential steps in the moult cycle

1. Feeding and growth shortly after ecdysis of previous stage.
2. Initiation of moult by increased ecdysteroid levels plus JH to regulate gene expression for next stage phenotype.
3. Termination of moult through decline in steroid levels and secretion of peptide hormones.
4. Post-ecdysial process- expansion, sclerotization and melanisation of new cuticle.

The critical endocrine signal that switches the animal from feeding to moulting is elevation of ecdysteroids, including ecdysone, 20-hydroxyecdysone (20E), and other hydroxylated analogs of ecdysone, depending on the insect. Epidermal cells respond by changing shape and expressing a set of genes that prepare the animal for passage to the next stage. An early step of this process is detachment of the epidermal cell layer from the existing cuticle, or apolysis, whereupon synthesis of a new cuticle layer appropriate for the next stage begins. Secretion of enzymatic moulting fluid by the epidermal cells dissolves unsclerotized old cuticle so that its chitin and protein components are recovered and re-cycled into new cuticle or other tissues. Similarly, the old cuticular lining of tracheal tubes detaches from underlying epidermal cells and the gap between these layers is filled with moulting fluid. This fluid is later reabsorbed from the old head capsule and space between the old and new cuticles, while the new tracheal tubes become inflated with air. Apolysis begins the pharate stage, defined as the incipient new animal covered by its old "coat." Thus, a pharate fifth instar larva has already produced a new cuticle, but is still wearing its old coat from the fourth instar. A pharate pupa is a developing pupa still encased in its old larval cuticle. A pharate adult has developed all adult organs, but is still covered by its old pupal cuticle. The order of these events and their precise timing are determined by a peptide and neurotransmitter signalling cascade.

Neuroendocrine signals are critical for the initiating and scheduling of the ecdysis sequence. Current evidence shows that corazonin and (eclosion hormone) EH act sequentially to regulate release of ecdysis triggering hormones (ETHs) from endocrine Inka cells of the epitracheal endocrine system. These peptides, called PETH (pre-ecdysis triggering hormone) and ETH in moths and ETH1 and ETH2 in other insects, act directly on the central nervous system (CNS) to initiate and schedule preecdysis, ecdysis, and post-ecdysis behaviours. The scheduling of behavioural steps occurs through release of numerous neuropeptides within the CNS to coordinate the ecdysis sequence and associated processes.

9.1 Peptides of the epitracheal endocrine system implicated in ecdysis

Each epitracheal gland of *Manduca sexta* is composed of one prominent peptidergic cell and three smaller cells. The smaller cells are present in all larval stages, but degenerate after pupation, leaving only nine pairs of large peptidergic cells in developing adults. These smaller remarkable peripheral endocrine cells were named Inka cells. Inka cells produce pre-ecdysis triggering hormone (PETH) and ETH, which are released into the haemolymph to activate pre-ecdysis and ecdysis motor programs in the CNS. Inka cells are wide spread in various insects and they occur in various sizes and shapes. Inka cells of hemimetabolous insects appear to degenerate after adult ecdysis as indicated by permanent disappearance of PETH-IR on tracheal surfaces, whereas Inka cells in some representative holometabolous insects (beetles, moths, and dipterans) persist in adults and continue to produce ETH. Physiological roles for these peptides in adults have not been assigned.

9.2 Roles of rising and falling ecdysteroids in signalling

The transition from feeding to pharate stage during each stadium of development is initiated by a rise in ecdysteroid levels. This event triggers successive bouts of gene expression that have been referred to as early, early-late, and late phases, according to the appearance of sequential puffs at various positions on polytene chromosomes of *Drosophila.* Early events that occur around the peak of ecdysteroid levels include acquisition of CNS sensitivity to ETH and increased production of ETH in Inka cells. Although Inka cells accumulate large amounts of peptides during this phase, they are incapable of releasing their contents until ecdysteroids decline to low levels. Acquisition of competence to release ETH is associated with late gene expression events that occur once ecdysteroids decline.

In *Manduca,* only pharate stages are sensitive to PETH and ETH. Pharate larvae become responsive to PETH or ETH injections at the time of peak ecdysteroid levels, which correspond to initiation of head capsule slip (approximately 30 h prior to ecdysis). These animals respond to ETH injection by performing pre-ecdysis behaviour only, but after the head capsule is completely slipped 2 h later, ETH injection induces the entire sequence in most animals. ETH treatment of pharate pupae elicits ecdysis behaviour about 48 h prior to expected ecdysis. These animals do not show pre-ecdysis contractions. The latency from ETH injection to the onset of ecdysis behaviour is progressively shorter as animals approach the time of normal ecdysis. These studies showed that sensitivity to PETH and ETH is correlated with appearance of the highest ecdysteroid titers in the haemolymph.

9.3 Phases of peptide release from Inka cells

Inka cells trigger pre-ecdysis and ecdysis through two modes of peptide release (Kim *et al.*, 2004b). The early phase of Inka cell release is initiated by corazonin release into the haemolymph by neurosecretory cells of the brain that project to the CA. This early, low-level phase of release leads to appearance of PETH and ETH in the haemolymph. Corazonin association with ecdysis is indicated by *in vivo* and *in vitro* experiments in *Manduca* and *Bombyx* pharate larvae and pupae. These experiments showed that low concentrations of corazonin (25–50 pM) induce the release of PETH and ETH to initiate the ecdysis sequence.

EH elicits a second phase of massive peptide release from Inka cells. Blood-borne EH acts on a receptor guanylyl cyclase in Inka cells to induce the release of PETH and ETH. Studies have established that EH actions on the Inka cell provide a mechanism for the complete depletion of Inka cells necessary for the transition from pre-ecdysis to ecdysis behaviour.

Pre-ecdysis is initiated by low levels of PETH/ETH induced by corazonin. ETH then acts on its receptor (ETHR-A) [ETH receptor subtype A] in VM neurons to elicit the EH release (Gammie and Truman, 1999; Kim *et al.*, 2006). Blood-borne EH acts back on Inka cells to cause cGMP production and PETH/ ETH depletion in a positive feedback loop (Ewer *et al.*, 1997). The appearance of cGMP in Inka cells approximately 30 min into pre-ecdysis corresponds to the steep elevation of PETH/ETH in the haemolymph, known to occur upon exposure of Inka cells to EH *in vitro* (Kingan *et al.*, 1997). Furthermore loss of EH-IR in larval proctodeal nerves, which are neurohemal organs for EH, suggests that this peptide is released at this time (Novicki and Weeks, 1996).

9.4 Ecdysis Sequence by Peptide Signalling Cascade

During normal development, *Manduca* larvae undergo one embryonic and four post embryonic ecdyses. PETH and ETH production in Inka cells is detectable in embryos at about 45% of embryonic development. Peptide levels in these cells reach a peak at about 65% of development. Embryonic ecdysis at 65–70% coincides with depletion of PETH and ETH in Inka cells indicating that these peptides have been released into the haemolymph.

The natural ecdysis behavioural sequence in *Manduca* larvae is composed of three distinct patterns initiated in the following order: pre-ecdysis I, pre-ecdysis II and ecdysis. Pre-ecdysis I is characterised by synchronous dorsoventral contractions of abdominal segments, and lasts for 60–70 min. The larvae show occasional weak, intermittent pre-ecdysis I contractions for 2–5 min interrupted by approximately 5 min quiet periods for about 30 min before launching into robust, continuous behaviour. Intermittent pre-ecdysis behaviour coincides with the first appearance of corazonin in the haemolymph and low levels of circulating PETH and ETH. About 20 min after onset of pre-ecdysis I, larvae initiate pre ecdysis II (posteroventral and proleg contractions) and continue performing both behaviours simultaneously for about 40 min.

During this time, posteroventral and proleg contractions alternate with dorsoventral contractions. About 1 h after initiation of preecdysis I, a switch from pre-ecdysis to ecdysis behaviour occurs, characterised by successive anteriorly directed peristaltic movements of abdominal segments and proleg retractions. Ecdysis movements last for about 10 to 12 min, serving first to rupture the old cuticle along the anterior midline of the thorax and subsequently to shed the cuticle with attached old tracheae.

Pre-ecdysis of pharate pupae always occurs with the animal positioned ventral side down, and is visible as weak rhythmic contractions of the abdomen. Pre-ecdysis of pharate pupae always occurs with the animal positioned ventral side down, and is visible as weak rhythmic contractions of the abdomen. Robust ecdysis peristaltic contractions begin approximately 1h later in the most posterior abdominal segment and move anteriorly. After rupture of the old cuticle along the dorsal thoracic midline, animals turn ventral side up and with the aid of strong peristaltic movements the cuticle is shed. During post-ecdysis, the proboscis, wing pads, and legs of the fresh pupa are then expanded and sclerotized to their typical shapes.

The adult eclosion sequence is quite different from larval and pupal ecdysis sequences and is composed of pre-eclosion and eclosion behaviours, as well as digging from the underground pupation site. Pre-eclosion behaviour is initiated by occasional rotations of the abdomen, which may be missing in some animals. The onset of eclosion behaviour is characterised by strong movements of wing bases that break anterior parts of the pupal cuticle while abdominal retractions and extensions help the adult to emerge from the pupal case and to escape to the surface.

Initiation of pre-ecdysis I and II involves direct actions of blood-borne PETH and ETH on abdominal ganglia. Pre-ecdysis I behaviour is driven by a single pair of ascending interneurons whose cell bodies are located in the posterior region of the TAG (terminal abdominal ganglion), the IN-402 (interneurons 402) cells. They make monosynaptic connections with MN-2,3 (motoneurons 2, 3) in each abdominal ganglion to synchronise bursting of these motor neurons in all abdominal segments. IN-402 cells probably produce acetylcholine to provide the necessary input for normal synchronised rhythmic contractions. It seems likely that centrally released kinins and DHs (diuretic hormone) are proximal signalling molecules for recruitment of a specific network of inter- and motoneurons that give rise to pre-ecdysis I behaviour.

In contrast to pre-ecdysis I, where motor neurons in each abdominal ganglion (AG) are slaves to ascending interneurons from the TAG, the pacemakers for pre-ecdysis II reside in each ganglion, such that motor bursts occur independent of inputs from other ganglia. Neurons and neurotransmitters controlling pre-ecdysis II have not been identified, but it seems likely that the principal and accessory planta retractor motoneurons (PPR, APR) involved in proleg retractions during ecdysis also participate in similar contractions during pre-ecdysis II (Weeks and Truman, 1984b). The ability of ETH to elicit pre-ecdysis II contractions in individually isolated abdominal ganglia indicates the presence of command ETHR neurons in each ganglion. Unidentified dorsolateral neurons expressing ETHR-B (ETH receptor subtype B) in AG1-7 (abdominal ganglion) are prime candidates for modulation of the pre-ecdysis II circuit

9.5 Mechanistic model for endocrine control of the ecdysis

Successful ecdysis depends on a signalling cascade consisting of ecdysteroid and peptide hormones. In *Manduca* fourth instar larvae, a surge in ecdysteroid levels initiates the moult about 48 h prior to initiation of the ecdysis sequence. Elevated ecdysteroids exert both positive and negative effects on a variety of cellular targets. One target is the Inka cell, where steroids upregulate expression of their own receptors (EcR) and the ETH gene; the latter effect results in increased production of the peptides PETH and ETH. Elevated ecdysteroid levels also target specific neurons in the CNS, inducing sensitivity to Inka cell peptides through expression of ETH receptors (ETHR). A negative regulatory effect of elevated steroids in the Inka cell is to suppress peptide secretion. Upon decline of ecdysteroid levels, a wave of late gene expression confers secretory competence in Inka cells and peptidergic neurons in the CNS. Expression of the transcription factor β-FTZ-F1 coincides with the appearance of secretory competence in Inka cells, suggesting that this factor may regulate key event(s) necessary acquisition of secretory competence. In *Manduca*, initial secretion of PETH and ETH from Inka cells is regulated by the brain neuropeptide corazonin. It is not known what factors control corazonin release, it is speculated that sensory input to the larval brain or the circadian clock in pharate adults may determine when the animal is ready to initiate the ecdysis sequence. Corazonin action on its receptor, which is highly expressed in Inka cells, causes secretion of PETH and ETH into the haemolymph. Low levels of these peptides then act on abdominal ganglia to activate motor programs for pre-ecdysis I and II. Pre-ecdysis I is activated by PETH and ETH via ETHR-A neurons L3,4, which co-release kinins and DHs (Kim *et al.*, 2006). These neuropeptides modulate activity of interneurons IN-402 in the posterior TAG region to control synchronous bursts of paired motoneurons MN-2,3 in abdominal ganglia and corresponding dorsoventral contractions (Novicki and Weeks, 1995). Approximately 15–20 min later, ETH acts on additional neurons expressing ETHR-B in each abdominal ganglion to induce posteroventral and proleg contractions (pre-ecdysis II). The neuronal network and neurotransmitters that control pre-ecdysis II remain to be identified and thus the roles of ETHR-B are currently hypothetical.

The ecdysis/post-ecdysis neuronal network is activated through parallel pathways mediated (1) via ETH and EH jointly or (2) via ETH only.

1. **Circulating ETH** acts through ETHR-A on brain VM neurons to elicit central and peripheral release of EH. Blood-borne EH acts back on its receptor guanylyl cyclase in Inka cells to induce cGMP production and depletion of PETH and ETH. The resulting surge of ETH causes EH depletion from VM axons in ventral ganglia to activate EHR in the entire network of 27/704 cells to increase cGMP production. This results in activation of the ecdysis and post-ecdysis circuitry producing CCAP, MIPs, bursicon, and other neuropeptides. This activation step precedes behaviour initiation due to descending inhibition from anterior ganglia.
2. **High ETH** levels also activate the 27/704 network via ETHR-A expressed in these neurons. After some delay, this leads to co-release of CCAP (crustacean cardioactive peptide), MIPs (myoinhibitory peptide), bursicon, and other neuropeptide(s), which control ecdysis and post-ecdysis processes. ETH can activate 27/704 in the absence of EH, but in this case the timing of ecdysis onset may be altered and post-ecdysis processes may be compromised.

Although cGMP elevation and activation of the ecdysis/ post-ecdysis network in abdominal ganglia are apparent already 30 min into natural pre-ecdysis, the switch to ecdysis behaviour occurs only after an additional 30 min delay. This delay from activation of the CCAP/MIP network to initiation of ecdysis behaviour may be controlled by putative descending inhibitory ETHR-B neurons in the SOG (subesophageal ganglion) and TG1-3 (thoracic ganglion) and abdominal neurons expressing both ETHR-A and ETHR-B (ETHR-A/B) and FGLa (cockroach allatostatintype A). Segmental interneurons expressing ETHR-A,B and NPF (neuropeptide F) may be responsible for disinhibition of IN-704 and consequent central co-release of CCAP and MIPs. CCAP and MIPs initiate the ecdysis motor circuit, while MIPs suppress activities of pre-ecdysis neurons. CCAP probably modulates activities of several identified motoneurons (MN2,3, ISMM, PPR, and

APR) controlling ecdysis movements. After the old cuticle is shed, the ecdysis motor program and hindgut contractions are terminated irreversibly by visceral modulatory neurons expressing ETHR-A,B and inhibitory neuropeptides — MIPs and sNPF (short neuropeptide F) in the TAG. Post-ecdysis is mediated by cells 27 expressing both ETHR-A and EHR. These cells release bursicon and CCAP after ecdysis to regulate plasticization, expansion, and hardening of the cuticle. Activation of ETHR-A in CA and ETHR-B in MLb cells in the SG results in the release of JH and PBAN/MRCH, (pheromone biosynthesis activating hormone/melanisation and reddish coloration hormone) respectively, which participate in regulation of cuticle pigmentation. Peripheral release of kinins and DHs from L3,4 cells and visceral modulatory neurons may be important for regulation of tracheal inflation and water balance during and after ecdysis.

REFERENCES

Alpy, F., Latchumanan, V. K., Kedinger, V., Janoshazi, A., Thiele, C., Wendling, C., Rio, M. C. and Tomasetto, C., (2005). Functional characterization of the MENTAL domain. J Biol Chem

An, S., Wang, S., Gilbert, L., Beerntsen, B., Ellersieck, M., and Song, Q. (2008). Global identification of bursicon-regulated genes in *Drosophila melanogaster*. *BMC Genomics*, *9*, 424

Ando, T., Inomata, S., and Yamamoto, M. (2004). Lepidopteran Sex Pheromones. In S. Schulz (Ed.), *Topics in Current Chemistry Vol. 239*. (pp. 51–96). Heidelberg: Springer-Verlag

Ashburner, M., Chihara, C., Meltzer, P., & Richards, G. (1974). Temporal control of puffing activity in polytene chromosomes. *Cold Spring Harbor Symposia on Quantitative Biology*, *38*, 655–662.

Baehrecke E. H. (1996). Ecdysone signaling cascade and regulation of *Drosophila* metamorphosis. *Arch Insect Biochem Physiol* 33, 231-244.

Bai, H., & Palli, S. R. (2010). Functional characterization of bursicon receptor and genome-wide analysis for identification of genes affected by bursicon receptor RNAi. *Dev. Biol.*, *344*, 248–258.

Baker, J. D., & Truman, J. W. (2002). Mutations in the *Drosophila* glycoprotein hormone receptor, rickets, eliminate neuropeptide-induced tanning and selectively block a stereotyped behavioral program. *J. Exp. Biol.*, *205*, 2555– 2565.

Baker, J., & Truman, J. W. (2002). Mutations in the *Drosophila* glycoprotein hormone receptor, rickets, eliminate neuropeptide- induced tanning and selectively block a stereotyped behavioral program. *J. Exp. Biol., 205*, 2555–2565

Bhaskaran, G., Sparagana, S. P., Barrera, P., and Dahm, K. H. (1986). Change in corpus allatum function during metamorphosis of the tobacco hornworm *Manduca sexta*: regulation at the terminal step in juvenile hormone biosynthesis. *Arch. Insect Biochem. Physiol., 3*, 321–338.

Blais, C., Dauphin-Villemant, C., Kovganko, N., Girault, J. P., Descoins, C., Jr. and Lafont, R.,1996. Evidence for the involvement of 3-oxo-delta 4 intermediates in ecdysteroid biosynthesis. Biochem J 320 (Pt 2), 413-419.

Blomquist, G. J. (2010). Biosynthesis of cuticular hydrocarbons. In G. J. Blomquist, & A.-G. Bagneres (Eds.), *Insect Hydrocarbons: Biology, Biochemistry and Chemical Ecology* (pp. 35–52). Cambridge: Cambridge Press.

Blomquist, G. J., Figueroa-Teran, R., Aw, M., Song, M., Gorzalski, A., Abbott, N., Chang, E., and Tittiger, C. (2010). Pheromone production in bark beetles. *Insect Biochemistry and Molecular Biology, 40*, 699–712.

Bollenbacher W. E., Agui N., Granger N. A. and Gilbert L. I. (1979). *In vitro* activation of insect prothoracic glands by the prothoracicotropic hormone. *Proc Natl Acad Sci U S A* 76, 5148-5152.

Bowen, M. F., Bollenbacher, W. E., & Gilbert, L. I. (1984). in vitro studies on the role of the brain and prothoracic glands in the pupal diapause of *Manduca sexta. J. Exp. Biol., 108*, 9–24.

Bradfield J. Y., Lee Y. H. and Keeley L. L. (1991). Cytochrome-P450 family-4 in a cockroach - molecular-cloning and regulation by hypertrehalosemic hormone. *Proceedings of the National Academy of Sciences of the United States of America* 88, 4558-4562.

Brown, M. R., Sieglaff, D. H., & Rees, H. H. (2009). Gonadal ecdysteroidogenesis in Arthropoda: Occurrence and regulation. *Annu. Rev. Entomol., 54*, 105–125.

Brown, M. R., Clark, K. D., Gulia, M., Zhao, Z., Garczynski, S. F., et al. (2008). An insulin-like peptide regulates egg maturation and metabolism in the mosquito Aedes aegypti. *Proc. Natl. Acad. Sci. USA, 105*, 5716–5721.

Butenandt A. and Karlson P. (1954). Über die Isolierung eines Metamorphose-Hormone der Insekten in kristallierter Form. *Z Naturforsch* 9b, 389-391.

Cheeseman P., Jutsum A. R. and Goldsworthy G. J. (1976). Quantitative studies on the release of locust adipokinetic hormone. *Physiological Entomology* 1, 115-122.

Chen, J. H., Turner, P. C. and Rees, H. H., (1999). Molecular cloning and characterization of hemolymph 3-dehydroecdysone 3beta-reductase from the cotton leafworm, *Spodoptera littoralis*. A new member of the third superfamily of oxidoreductases. J Biol Chem 274, 10551-10556.

Chino, H. and Gilbert, L. I., (1971). The uptake and transport of cholesterol by haemolymph lipoproteins. Insect Biochem 1, 337-347.

Choi, M.-Y., Han, K. S., Boo, K. S., and Jurenka, R. (2002). Pheromone biosynthetic pathway in *Helicoverpa zea* and *Helicoverpa assulta*. *Insect Biochemistry and Molecular Biology*, *32*, 1353–1359.

Clark, A. J., & Bloch, K. (1959). Conversion of ergosterol to 22-dehydrocholesterol in *Blatella germanica*. *J. Biol. Chem.*, *234*, 2589–2593.

Clements A. N. (1992). *The Biology of Mosquitoes*, vol. 1. Chapman & Hall, London.

Cymborowski ,B. (1992). *Insect Endocrinology*. Elsevier Press, New York.

Davies, T. G., Dinan, L. N., Lockley, W. J., Rees, H. H. and Goodwin, T. W., 1981. Formation of the A/B cis ring junction of ecdysteroids in the locust, *Schistocerca gregaria*. Biochem J 194, 53-62.

Davis, M. M., O'Keefe, S. L., Primrose, D. A., and Hodgetts, R. B. (2007). A neuropeptide hormone cascade controls the precise onset of post-eclosion cuticular tanning in *Drosophila melanogaster*. *Development*, *134*, 4395–4404.

de Loof A. and Huybrechts R. (1998). "Insects do not have sex hormones": a myth? *Gen Comp Endocrinol* 111, 245-260.

Dewey, E. M., McNabb, S. L., Ewer, J., Kuo, G. R., Takanishi, C. L., et al. (2004). Identification of the gene encoding bursicon, an insect neuropeptide responsible for cuticle sclerotization and wing spreading. *Curr. Biol.*, *14*, 1208–1213.

Diederen J. H., Oudejans R. C., Harthoorn L. F. and van der Horst D. J. (2002). Cell biology of the adipokinetic hormone-producing neurosecretory cells in the locust corpus cardiacum. *Microsc Res Tech* 56, 227-236.

Emerson, K. J., Uyemura, A. M., McDaniel, K. L., Schmidt, P. S., Bradshaw, W. E., and Holzapfel, C. M. (2009). Environmental control of ovarian dormancy in natural populations of *Drosophila melanogaster. J. Comp. Physiol. A, 195*, 825–829.

Eriksen, K. K., Hauser, F., Schiott, M., Pedersen, K. M., Sondergaard, L., and Grimmelikhuijzen, C. J. (2000). Molecular cloning, genomic organization, developmental regulation, and a knock-out mutant of a novel leu-rich repeats-containing G protein-coupled receptor (DLGR-2) from *Drosophila melanogaster. Genome Res*, *10*, 924–938.

Foster, S. P. (2009). Sugar feeding via trehalose haemolymph concentration affects sex pheromone production in mated *Heliothis virescens* moths. *Journal of Experimental Biology*, *212*, 2789–2794

Fraenkel, G., and Hsiao, C. (1963). Tanning in the adult fly: A new function of neurosecretion in the brain. *Science*, *141*, 1057–1058

Fraenkel, G., and Hsiao, C. (1965). Bursicon, a hormone which mediates tanning of the cuticle in the adult fly and other insects. *J. Insect Physiol.*, *11*, 513–556.

Fujimoto, Y., Hiramoto, M., Kakinuma, K. and Ikekawa, N., (1989). Elimination of C-6-hydrogen during the formation of ecdysteroids from cholesterol in *Locusta migratoria* ovaries. Steroids 53, 477-485

Fujiwara, Y., Tanaka, Y., Iwata, K. I., Rubio, R. O., Yaginuma, T., Yamashita, O., and Shiomi, K. (2006). ERK/MAPK regulates ecdysteroid and sorbitol metabolism for embryonic diapause termination in the silkworm, *Bombyx mori. J. Insect Physiol.*, *52*, 569–575.

Gäde G. (2004). Regulation of intermediary metabolism and water balance of insects by neuropeptides. *Annu Rev Entomol* 49, 93-113

Gelman, D. B., Thyagaraja, B. S., Kelly, T. J., Masler, E. P., Bell, R. A., et al. (1992). Prothoracicotropic hormone levels in brains of the European corn borer, *Ostrinia nubilalis:* diapause vs. the non-diapause state. *J. Insect Physiol.*, *38*, 383–395.

Gilbert, L. I., Rybczynski, R. and Warren, J. T. (2002). Control and biochemical nature of the ecdysteroidogenic pathway. *Annu Rev Entomol* 47, 883-916.

Gokuldas M., Hunt P. A. and Candy D. J. (1988). The inhibition of lipid synthesis in-vitro in the locust *Schistocerca gregaria* by factors from the corpora cardiaca. *Physiological Entomology* 13, 43-48.

Grieneisen, M. L., 1994. Recent advances in our knowledge of ecdysteroid biosynthesis in insects and crustaceans. *Insect Biochem Mol Biol* 24, 115-132

Gupta, A. P., 1990. Morphogenetic hormones and their glands in arthropods: evolutionary aspects. In: Gupta, A. P. (Ed.), Morphogenetic Hormones of Arthropods: Discoveries,Syntheses, Metabolism, Evolution, Modes of Action, and Techniques. Rutgers University Press, New Brunswick, pp. 1-32

Helvig, C., Koener, J. F., Unnithan, G. C., & Feyereisen, R. (2004). CYP15A1, the cytochrome P450 that catalyzes epoxidation of methyl farnesoate to juvenile hormone III in cockroach corpora allata. *Proc. Natl. Acad. Sci. USA*, *101*, 4024–4029.

Hsu, S. Y. (2003). New insights into the evolution of the relaxin-LGR signaling system. *Trends Endocrinol. Metab.*, *7*, 303–309.

Hsu, S. Y., Nakabayashi, K., Nishi, S., Kumagai, J., Kudo, M., Sherwood, O. D., and Hsueh, A. J. (2002). Activation of orphan receptors by the hormone relaxin. *Science*, *295*, 671–674

Huang, J., Zhang, Y., Li, M., Wang, S., Liu, W., et al. (2007). RNA interference-mediated silencing of the bursicon gene induces defects in wing expansion of silkworm. *FEBS Letters*, *581*, 697–701, Epub 2007 Jan 24.

Isaac R. E., Taylor C. A., Hamasaka Y., Nassel D. R. and Shirras A. D. (2004). Proctolin in the post-genomic era: new insights and challenges. *Invert Neurosci* 5, 51-64.

Johnson, S. A., & Milner, M. J. (1987). The final stages of wing development in *Drosophila melanogaster*. *Tissue Cell*, *19*, 505–513

Jouni, Z. E., Zamora, J. and Wells, M. A., (2002). Absorption and tissue distribution of cholesterol in *Manduca sexta*. Arch Insect Biochem Physiol 49, 167-175.

Jurenka, R. A., Schal, C., Burns, E., Chase, J., & Blomquist, G. J. (1989). Structural correlation between cuticular hydrocarbons and female contact sex pheromone of German cockroach *Blattella germanica* (L.). *Journal of Chemical Ecology*, *15*, 939–949.

Jurenka, R. A., Haynes, K. F., Adlof, R. O., Bengtsson, M., and Roelofs, W. L. (1994). Sex pheromone component ratio in the cabbage looper moth altered by a mutation affecting the fatty acid chain shortening reactions in the pheromone biosynthetic pathway. *Insect Biochemistry Molecular Biology*, *24*, 373–381.

Kappler, C., Hetru, C., Durst, F. and Hoffmann, J. A., 1989. Enzymes involved in ecdysone biosynthesis. In: Koolman, J. (Ed.), Ecdysone: From Chemistry to Mode of Action. Thieme Medical Publishers, New York, pp. 161-166

Kaufmann C. and Brown M. R. (2006). Adipokinetic hormones in the African malaria mosquito, *Anopheles gambiae*: identification and expression of genes for two peptides and a putative receptor. *Insect Biochem Mol Biol* 36, 466-481.

Kiger, J. A., Natzle, J. E., Kimbrell, D. A., Paddy, M. R., Kleinhesselink, K., and Green, M. M. (2007). Tissue remodelling during maturation of the *Drosophila* wing. *Dev. Biol.*, *301*, 178–191.

Kim, M., & Denlinger, D. L. (2010). A potential role for ribosomal protein S2 in the gene network regulating reproductive diapause in the mosquito *Culex pipiens*. *J. Comp. Physiol. B*, *180*, 171–178.

Kim, M., Sim, C., and Denlinger, D. L. (2010). RNA interference directed against ribosomal protein S3a suggests a link between this gene and arrested ovarian development during adult diapause in *Culex pipiens*. *Insect Mol. Biol.*, *19*, 27–33.

Kimura, K., Kodama, A., Hayasaka, Y., and Ohta, T. (2004).Activation of the cAMP/PKA signaling pathway is required for post-ecdysial cell death in wing epidermal cells of *Drosophila melanogaster*. *Development*, *131*, 1597–1606.

Kiriishi, S., Rountree, D. B., Sakurai, S., and Gilbert, L. I. (1990). Prothoracic gland synthesis of 3-dehydroecdysone and its hemolymph 3β-reductase mediated conversion to ecdysone in representative insects. *Experientia*, *46*, 716–721.

Klowden M. J. (2002). *Physiological Systems in Insects*. Academic Press, London.

Kobayashi M. and Kirimura J. (1958). The brain hormone in the silkworm, *Bombyx mori*. *Nature* 181.

Kobayashi M., Kirimura J. and Saito M. (1962). The brain hormone in an insect. *Mushi* 36,85

Kodrik D. and Goldsworthy G. J. (1995). Inhibition of RNA synthesis by adipokinetic hormones and brain factor(s) in adult fat body of *Locusta migratoria. Journal of Insect Physiology* 41, 127-133.

Kopec, S. (1917). Experiments on metamorphosis of insects. *Bull. Int. Acad. Cracovie B*, 57–60.

Kopec, S. (1922). Studies on the necessity of the brain for the inception of insect metamorphosis. *Biol. Bull., 42*, 322–342.

Kostal, V., Noguchi, H., Shimada, K., and Hayakawa, Y. (1998). Developmental changes in dopamine levels in larvae of the fly *Chymomyza costata:* comparison between wild-type and mutant-nondiapause strains. *J. Insect Physiol., 44*, 605–614.

Kramer, S. J. (1978). Regulation of the activity of JH specific esterases in the Colorado potato beetle, *Leptinotarsa decemlineata. J. Insect Physiol., 24*, 743–747.

Krebs, K. C. and Lan, Q., (2003). Isolation and expression of a sterol carrier protein-2 gene from the yellow fever mosquito, *Aedes aegypti.* Insect Mol Biol 12, 51-60.

Kumagai, J., Hsu, S. Y., Matsumi, H., Roh, J. S., Fu, P., Wade, J. D., Bathgate, R. A., & Hsueh, A. J. (2002). INSL3/Leydig insulin-like peptide activates the LGR8 receptor important in testis descent. *J. Biol. Chem., 277*, 31283–31286.

Lachaise, F., 1990. Synthesis, metabolism, and effects on molting ecdysteroids in Crustacea, Chelicerta, and Myriapoda. In: Gupta, A. P. (Ed.), Morphogenetic Hormones of Arthropods: Discoveries, Syntheses, Metabolism, Evolution, Modes of Action, and Techniques. Rutgers University Press, New Brunswick, pp. 275-323.

Loveal, B. J., & Deitcher, D. L. (2010). The essential role of bursicon during *Drosophila* development. *BMC Dev. Biol., 10*, 92.

Luo, C. W., Dewey, E. M., Sudo, S., Ewer, J., Hsu, S. Y., et al. (2005). Bursicon, the insect cuticle-hardening hormone, is a heterodimeric cystine knot protein that activates G proteincoupled receptor LGR2. *Proc. Natl. Acad. Sci. U S A., 102*, 2820–2825, Epub 2005 Feb 9.

Matsumoto, M., & Takeda, M. (2002). Changes in brain monoamine contents in diapausing pupae of *Antheraea pernyi* when activated under long-day conditions. *J. Insect Physiol., 48*, 765–771.

Mendive, F. M., Van Loy, T., Claeysen, S., Poels, J., Williamson, M., et al. (2005). *Drosophila* molting neurohormone bursicon is a heterodimer and the natural agonist of the orphan receptor DLGR2. *FEBS Lett.*, *579*, 2171–2176.

Mesce K. A. and Fahrbach S. E. (2002). Integration of endocrine signals that regulate insect ecdysis. *Front Neuroendocrinol* 23, 179-199.

Millar, J. G. (2000). Polyene hydrocarbons and epoxides: a second major class of lepidopteran sex attractant pheromones. *Annual Review of Entomology*, *45*, 575–604.

Millar, J. G. (2010). Polyene hydrocarbons, epoxides, and related compounds as components of lepidopteran pheromone blends. In G. J. Blomquist, & A.-G. Bagneres (Eds.), *Insect Hydrocarbons: Biology, Biochemistry and Chemical Ecology* (pp. 390–447). Cambridge: Cambridge University Press.

Miller, J. R., & Roelofs, W. L. (1980). Individual variation in sex pheromone component ratios in two populations of the redbanded leafroller moth, *Argyrotaenia velutinana*. *Environmental Entomology*, *9*, 359–363.

Moribayashi, A., Kurahashi, H., & Ohtaki, T. (1992). Physiological differentiation of the ring glands in mature larvae of the flesh fly, *Boettcherisca peregrina,* programmed for diapause or non-diapause. *J. Insect Physiol.*, *38*, 177–183.

Mullen L. and Goldsworthy G. (2003). Changes in lipophorins are related to the activation of phenoloxidase in the haemolymph of *Locusta migratoria* in response to injection of immunogens. *Insect Biochem Mol Biol* 33, 661-670.

Mullen L. M. and Goldsworthy G. J. (2006). Immune responses of locusts to challenge with the pathogenic fungus *Metarhizium* or high doses of laminarin. *J Insect Physiol* 52, 389-398.

Mullen L. M., Lightfoot M. E. and Goldsworthy G. J. (2004). Induced hyperlipaemia and immune challenge in locusts. *J Insect Physiol* 50, 409-41

Nakagaki, M., Takei, R., Nagashima, E., and Yaginuma, T. (1991). Cell-cycles in embryos of the silkworm, *Bombyx mori:* G2 arrest at diapause stage. *Roux's Arch. Devel. Biol.*, *200*, 223–229. 1991

Natzle, J. E., Kiger, J. A., and Green, M. M. (2008). Bursicon signaling mutations separate the epithelial-mesenchymal transition from programmed cell death during *Drosophila melanogaster* wing maturation. *Genetics*, *180*, 885–893.

Nijhout, H. F., and Grunert, L. W. (2002). Bombyxin is a growth factor for wing imaginal disks in Lepidoptera. *Proc. Natl. Acad. Sci. USA*, *99*, 15446–15450.

Nijhout, H. F., Smith, W. A., Schachar, I., Subramanian, S., Tobler, A., et al. (2007). The control of growth and differentiation of the wing imaginal disks of *Manduca sexta*. *Dev. Biol.*, *302*, 569–576.

Nijhout H. F. (1994). *Insect Hormones*. Princeton University Press, Princeton.

Noguchi, H., & Hayakawa, Y. (1997). Role of dopamine at the onset of pupal diapause in the cabbage armyworm, *Mamestra brassicae*. *FEBS Lett.*, *413*, 157–161.

Passier P. C., Vullings H. G., Diederen J. H. and van der Horst D. J. (1997). Trehalose inhibits the release of adipokinetic hormones from the corpus cardiacum in the African migratory locust, *Locusta migratoria*, at the level of the adipokinetic cells. *J Endocrinol* 153, 299-305

Petryk, A., Warren, J. T., Marques, G., Jarcho, M. P., Gilbert, L. I., Kahler, J., Parvy, J. P., Li, Y., Dauphin-Villemant, C. and O'Connor, M. B., (2003). Shade is the *Drosophila* P450 enzyme that mediates the hydroxylation of ecdysone to the steroid insect molting hormone 20-hydroxyecdysone. Proc Natl Acad Sci U S A 100, 13773-13778.

Raabe M. (1982). *Insect Neurohormones*. Plenum Press, New York

Rademakers L. H. (1977). Identification of a secretomotor centre in the brain of *Locusta migratoria*, controlling the secretory activity of the adipokinetic hormone producing cells of the corpus cardiacum. *Cell Tissue Res* 184, 381-395.

Rees, H. H. (1985). Biosynthesis of ecdysones. In G. A. Kerkut, & L. I. Gilbert (Eds.), *Comprehensive Insect Physiology Biochemistry and Pharmacology* (Vol. 7), (pp. 249–293). Oxford: Pergamon Press

Rees, H. H. (1989). Pathways of biosynthesis of ecdysone. In: Koolman, J. (Ed.), Ecdysone: From Chemistry to Mode of Action. Thieme Medical Publishers, Inc., New York, pp. 152-160.

Rees, H. H. (1995). Ecdysteroid biosynthesis and inactivation in relation to function. Eur. J.

Entomol. 92, 9-39.

Richard, D. S., & Saunders, D. S. (1987). Prothoracic gland function in diapause and non-diapause *Sarcophaga argyrostoma* and *Calliphora vicina*. *J. Insect Physiol.*, *33*, 385–392.

Roeder T. (2005). Tyramine and octopamine: ruling behavior and metabolism. *Annu Rev Entomol* 50, 447-477.

Roth, G. E., Gierl, M. S., Vollborn, L., Meise, M., Lintermann, R. and Korge, G. (2004). The *Drosophila* gene Start1: A putative cholesterol transporter and key regulator of ecdysteroid synthesis. Proc Natl Acad S i U S A 101, 1601-1606.

Sakai T., Satake H. and Takeda M. (2006). Nutrient-induced [alpha]-amylase and proteaseactivity is regulated by crustacean cardioactive peptide (CCAP) in the cockroachmidgut. *Peptides* 27, 2157

Satake S., Masumura M., Ishizaki H., Nagata K., Kataoka H., Suzuki A. and Mizoguchi A. (1997). Bombyxin, an insulin-related peptide of insects, reduces the major storagecarbohydrates in the silkworm *Bombyx mori. Comp Biochem Physiol B Biochem MolBiol* 118, 349-357.

Satake, S., Nagata, K., Kataoka, H., & Mizoguchi, A. (1999). Bombyxin secretion in the adult silkmoth *Bombyx mori*: Sex-specificity and its correlation with metabolism. *J. Insect Physiol., 45*, 939–945.

Schooley D. A., Horodyski F. M. and Coast G. M. (2005). Hormones Controlling Homeostasis in Insects. In *Comprehensive Molecular Insect Science*, pp. 493-550. Elsevier Pergamon, Amsterdam.

Seligman, L. M., Doy, E. A., & Crossley, A. C. (1975). Hormonal control of morphogenetic cell death of the wing hypodermis in *Lucilia cuprina. Tissue Cell, 7*, 281–296.

Shiga, S. (2003). Anatomy and functions of brain neurosecretory cells in Diptera. *Microsc. Res. Tech., 62*, 114–131

Sim, C., and Denlinger, D. L. (2008). Insulin signaling and FOXO regulate eh overwintering diapause of the mosquito *Culex pipiens. Proc. Natl. Acad. Sci., USA, 105*, 6777–6781

Smith, S. L., 1985. Regulation of ecydsteroid titer: synthesis. In: Kerkut, G. A. and L. I. Gilbert (Ed.), Comprehensive Insect Physiology, Biochemistry, and Pharmacology. Pergamon Press, New York, pp. 295-341

Soulages, J. L., and Wells, M. A. (1994). Metabolic fate and turnover rate of hemolymph free fatty acids in adult *Manduca sexta. Insect Biochem. Mol. Biol., 24*, 79–86.

Tawfik, A. I., Tanaka, Y., and Tanaka, S. (2002). Possible involvement of ecdysteroids in photoperiodically induced suppression of ovarian development in a Japanese strain of the migratory locust, *Locusta migratoria*. *J. Insect Physiol.*, *48*, 411–418.

Tillman, J. A., Seybold, S. J., Jurenka, R. A., and Blomquist, G. J. (1999). Insect Pheromones – an overview of biosynthesis and endocrine regulation. *Insect Biochemistry and Molecular Biology*, *29*, 481 514.

Tublitz N., Brink D., Broadie K. S., Loi P. K. and Sylwester A. W. (1991). From behavior to molecules: an integrated approach to the study of neuropeptides. *Trends Neurosci* 14, 254-259.

Tuckey, R. C., Bose, H. S., Czerwionka, I. and Miller, W. L., 2004. Molten globule structure and steroidogenic activity of N-218 MLN64 in human placental mitochondria. Endocrinology

Uribe, A., Strauss III, J. F. and Martinez, F., 2003. Contact sites from human placental mitochondria: characterization and role in progesterone synthesis. Arch Biochem Biophysics 413, 172-181.

Wang, S. J., An, S., & Song, Q. (2008). Transcriptional expression of bursicon and novel bursicon regulated genes in the house fly *Musca domestica*. *Arch. Insect Biochem. Physiol.*, *68*, 100–112.

Wang, Y., Mutti, N. S., Ihle, K. E., Siegel, A., Dolezal, A. G., et al. (2010). Down-regulation of honey bee IRS gene biases behavior toward food rich in protein. *PLoS Genet.*, *6*, e1000896

Warren, J. T., Petryk, A., Marques, G., Jarcho, M., Parvy, J. P., Dauphin-Villemant, C., O'Connor, M. B. and Gilbert, L. I., (2002). Molecular and biochemical characterization of two P450 enzymes in the ecdysteroidogenic pathway of *Drosophila melanogaster*. Proc Natl Acad Sci U S A 99, 11043-11048.

Warren, J. T., Petryk, A., Marques, G., Parvy, J. P., Shinoda, T., Itoyama, K., Kobayashi, J., Jarcho, M., Li, Y., O'Connor, M. B., Dauphin-Villemant, C. and Gilbert, L. I., (2004). Phantom encodes the 25-hydroxylase of *Drosophila melanogaster* and *Bombyx mori*: a P450 enzyme critical in ecdysone biosynthesis. Insect Biochem Mol Biol 34, 991-1010.

Wen, Z., Gulia, M., Clark, K. D., Dhara, A., Crim, J. W., et al. (2010). Two insulin-like peptide family members from the mosquito Aedes aegypti exhibit differential biological and receptor binding activities. *Mol. Cell Endocrinol.*, *328*, 47–55

Wigglesworth V. B. (1934). The physiology of ecdysis in *Rhodnius prolixus* (Hemiptera). II.Factors controlling moulting and "metamorphosis". *Q J Microsc Sci* 77, 191-222.

Wigglesworth V. B. (1939). *The principles of insect physiology*. London.

Williams C. M. (1946). Physiology of insect diapause: I. The role of the brain in the production and termination of pupal dormancy in the giant silkworm, *Platysamia cecropia*. *Biol Bull* 93, 234-243.

Williams C. M. (1952). Physiology of insect diapause: I. The brain and prothoracic glands as an endocrine system in the cecropia silkworm. *Biol Bull* 103, 120-138.

Wolf, W. A., and Roelofs, W. L. (1987). Reinvestigation confirms action of Δ11-desaturase in spruce budworm moth sex pheromone biosynthesis. *Journal of Chemical Ecology*, *13*, 1019–1027

Yin, C.-M., Wang, Z. S., and Chaw, W. D. (1985). Brain neurosecretory cell and ecdysiotropin activity of the nondiapausing, pre-diapausing and diapausing southwestern corn borer, *Diatraea grandiosella* Dyar. *J. Insect Physiol*, *31*, 659–667.

Metabolism in Insects 10

Introduction

1. Carbohydrate metabolism
   - 1.1 Glycogen
   - 1.2 Trehalose
   - 1.3 Chtin metabolism
2. Lipid metabolism
   - 2.1 Lipids for storage and liberation of metabolic energy
   - 2.2 Structural lipids
   - 2.3 β Oxidation of fatty acids
   - 2.4 Ketone bodies
   - 2.5 Biosynthesis of fatty acids
3. Amino acid and protein metabolism
   - 3.1 Steps in protein synthesis
   - 3.2 Aminoacid metabolism in general
   - 3.3 Formation of ammonia from glutamate
   - 3.4 Storage proteins
   - 3.5 Glycoprotein and lipoprotein
4. Energy storage and release- Role of fat body
   - 4.1 Fat body cell types
   - 4.2 Accumulation of energy reserves
   - 4.3 Mobilisation of carbohydrate reserves
   - 4.4 Mobilisation of lipid reserves
   - 4.5 Hormonal regulation of fat body energy metabolism
5. Metabolic regulation in insect cold hardiness
   - 5.1 Control of polyol metabolism
   - 5.2 Metabolic rate depression
   - 5.3 AMPK and metabolic poise over the winter
   - 5.4 Mitochondrial metabolism
6. Fuels for flight and control of flight metabolism
   - 6.1 Energy substrates for flight
   - 6.2 Metabolic pathways of proline during insect flight
   - 6.3 Regulation of flight metabolism

INTRODUCTION

The enzyme catalysed chemical reactions of cells linked together in series to form pathways, is collectively referred to as **metabolism**. Metabolic pathways are interdependent and exquisitely regulated for efficient extraction of energy from fuels and for synthesis of biological macro-molecules. Cellular processes produce unique chemical fingerprints or metabolite profiles, and a complete quantitative set of metabolic intermediates associated with a cellular process is referred to as **metabolome. Metabolomics** is the study of changes in the metabolome that may arise from metabolic regulation or alteration in gene expression, or a combination of both mechanisms. Evolutionary adaptation of insects to different environments and life styles has led to considerable specialisation at the morphological, physiological and biochemical levels. Examples of insect specialisation and the metabolic implications include ecdysis, metamorphosis, flight, nutritional specialisation, cold hardiness, silk synthesis, to name a few. Studies of metabolism and metabolomics are subject areas of biochemistry, which also includes the structural chemistry of biological molecules and the chemistry of molecular genetics.

The chemical changes involved in the synthesis of complex molecules are called anabolim. These complex molecules are then utilized to form cellular structures that are formed from small and simple precursors that act as building blocks. Catabolism is the set of metabolic pathways that breaks down molecules into smaller units that are either oxidized to release energy, or used in other anabolic reactions. Whenever molecules are degraded or synthesised, there is a change in the energy states between the starting substrates and the end products, and living systems can capture this change in energy and store it in molecules that conserve the energy as phosphate bonds. Adenosine triphosphate (ATP) is the universal energy currency of cells, serving as a transfer medium for energy and its temporary storage. The ATP molecule consists of adenine, ribose and phosphate units. The phosphate bonds in ATP give it its ability to store energy, which is liberated when it is hydrolyzed to adenosine diphosphate (ADP). The energy locked up in organic molecules is released primarily by oxidation, the removal of electrons from the molecules and the transfer of those electrons to other molecules within the cell that undergo reduction when they acquire the electrons. Electrons are shuttled within the cell and are ultimately passed to oxygen, resulting in the formation of water and the synthesis of ATP. The major electron acceptor molecules during food oxidation are nicotinamide adenine dinucleotide (NAD+) and flavin adenine dinucelotide (FAD). Successive decarboxylation of food substrates releases carbon dioxide into the environment while degradation of proteins releases nitrogen molecules as ammonia. The net result of these biochemical transformations is the production of useful energy from the degradation of organic molecules. Energy that is not required immediately is stored as trehalose, glycogen or fat.

The fat body plays a central role in insect metabolism and is commonly compared to the liver of vertebrates, performing multiple functions that satisfy the changing metabolic needs of the insect throughout its development.

It originates from embryonic mesoderm and is distributed throughout the insect body, mostly surrounding the gut and reproductive tissues. **Trophocytes** are the principal cells of the fat body and may change their form according to the developmental and physiological state of the insect. They may show regional differentiation for performing different functions. They contain numerous vacuoles of lipid droplets, glycogen that is free or bound to lipids and proteins, and granules of proteins. Trophocytes may be modified as **urate cells** that sequester and detoxify urates, or mycetocytes that harbour microbial symbionts.

1. CARBOHYDRATE METABOLISM

Carbohydrates are aldehyde or ketone derivatives of polyhydric alcohols. They are involved in energy storage and structure and the more complex and conjugate forms participate in a number of physiological activities. The major storage form of carbohydrates in insects is glycogen and trehalose, while chitin is a major structural component. Glycoproteins and glycolipids have mixed functions such as in cell-cell recognition and protein translocation.

1.1 Glycogen

Glycogen consists of long polymer chains of glucose units which are bonded by an **alpha acetal linkage**. An acetal linkage forms by the combination of carbonyl group and alcoholic group. If the carbonyl group is an aldehyde group (-CHO), it is termed as **hemiacetal** and if there is ketonic group, it forms **hemiketal** bond. If two alkoxy groups are bonded on same carbon atom, it called **acetal group** and bond termed as **acetal linkage**. In case of glycogen, all alpha-D-glucose bonded to each other by alpha acetal linkage between C1 of one monomer unit and C4 of other monomer unit, hence called α,α-1,4-glycosidic linkage (Fig. 10.1). Since glycogen is a branched polymer, branching occurs at intervals of 8-10 glucose units (Fig 10.2). These branches are formed by acetal linkage between C1 and C6. The purpose of the high branched structure is to have many non-reducing ends so that glucose can be rapidly mobilised in times of metabolic needs.

Glycogen is stored in large amount in the flight muscles of insects which use it as an immediate source of energy. Glycogen is also stored in the fat body from where glucose is provided to other tissues after transformation into trehalose, the circulating blood sugar.

Fig. 10.1. Chemical bonds between monomers (glucose) glycogen

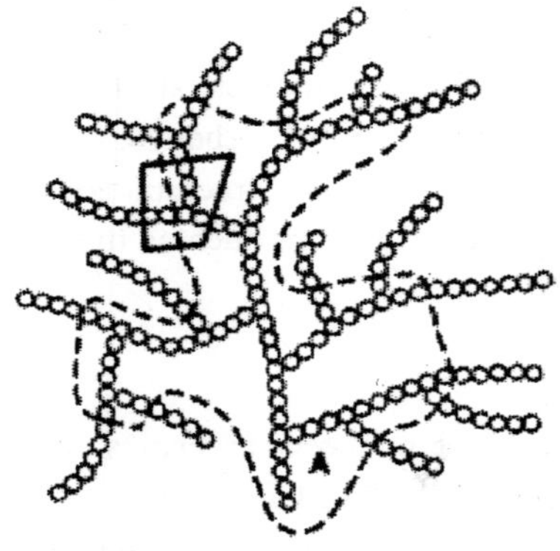

Fig.10.2. Diagram of a molecule of glycogen

(A is the "aldehyde" origin of the chain; the small circles are glucose radicals. The boundaries of β-dextrin are shown by the dotted line, and the quadrangle is the part of the molecule whose formula is given in Figure 10.1)

1.1.1 Glycogen metabolism pathway

Glycogen metabolism pathway includes its anabolic and catabolic pathways. The **anabolic** (synthetic pathways) includes formation from pyruvate to glucose (gluconeogenesis) and from glucose to glycogen (glycogenesis).The **catabolic** pathway includes conversion of glycogen into glucose 6-phosphate (glycogenolysis) followed by digestion of glucose 6-phosphate into pyruvate (glycolysis).

Anabolic pathways

The glycogen synthesis is called **glycogenesis** and is an endergonic process, hence require some input of energy which is provided by UTP (Uridine triphosphate). Glycogen synthesised from non-carbohydrate source is called **glyconeogenesis**.

Glycogenesis is a four steps process, each step catalysed by certain enzymes (Fig. 10.3).

1. In first step, glucose is converted into glucose-6-phosphate in the presence of glucokinase or hexokinase enzymes.
2. This glucose-6-phosphate forms glucose-1-phosphate by the action of phosphoglucomutase.
3. UTP reacts with glucose-1-phosphate to form UDP-glucose and pyrophosphate in the presence of UDP-glucose pyrophosphorylase or Uridyl Transferase.
4. Pyrophosphate further hydrolysed into 2 molecules of Pi in the presence of pyrophosphatase.
5. The UDP-glucose acts as monomer units progressively lengthen the glycogen chain with bonded glucose in the presence of enzyme glycogen synthase.
6. This step is needed to initiate protein glycogenin.

The branching process is catalysed by branching enzyme (amylo ($\alpha1\rightarrow4$) to ($\alpha1\rightarrow6$) transglycosylase) which can act upon only a branch having at least 11 residues. It catalyses the transfer of a terminal fragment of 6-7 glucose residues from a non-reducing end to the C_6 hydroxyl group of a glucose residue deeper into the interior of the glycogen molecule. The reactions of glycogen biosynthesis are shown below.

	Reaction	Enzyme
1.	Glucose-6-phosphate → Glucose -1-phosphate	Phosphoglucomutase
2.	Glucose-1-phosphate + UTP → UDP-Glucose + PPi	UDP-glucose pyrophosphorylase
3.	PPi + H_2O → 2Pi	Inorganic pyrophosphatase
4.	UDP-Glucose + Glycogenin UDP + $Glycogen_{n+1}$	Glycogen synthase
5.	UDP + ATP → UTP + ADP	Nucleotide diphosphokinase

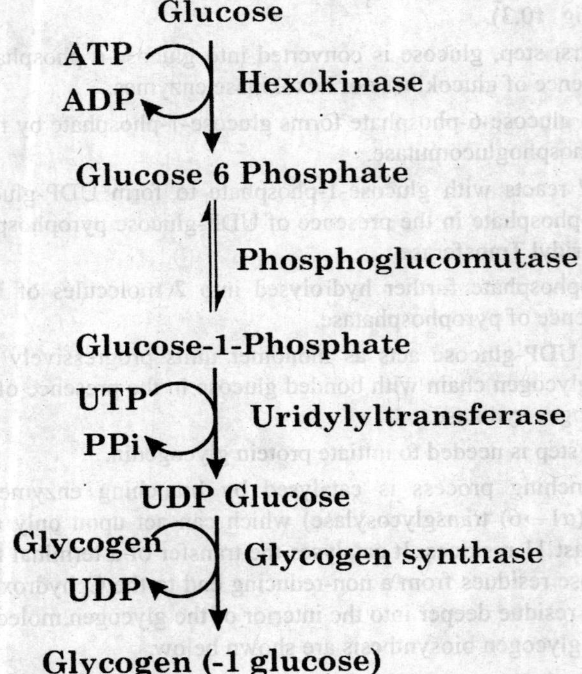

Fig. 10.3. Glycogenesis

Gluconeogenesis is synthesis of glucose from nonhexose precursors. Once glucose is formed, the glucose 6-phosphate is converted to glucose 1-phosphate through phosphoglucomutase reaction. The glucose 1-phosphate is then converted to UDP glucose under enzymatic reaction of UDP-glucose pyrophosphorylase. This is the key step of glycogen biosynthesis. UDP-glucose serves as immediate donor of glucose residues during the reaction catalyzed by glycogen synthase. This results in transfer of the glucose residue from UDP-glucose to a nonreducing end of a branched glycogen molecule.

Branching points of glycogen are formed by the glycogen-branching enzyme which is also called **amylo (1 → 4) to (1 → 6)** transglycosylase. The glycogen-branching enzyme serves in transfer of a terminal fragment of 6 or 7 glucose residues from the nonreducing end of a glycogen branch having at least 11 residues to the C-6 hydroxyl group of a glucose residue at a more interior position of the same or another glycogen chain. The overall result is formation of a new branch.

There are nine steps and one sub-step in gluconeogenesis:

1. Pyruvate gets converted into phosphoenolpyruvate. This is the step that requires a sub step in order for it to occur. When phosphoenolpyruvate is converted into pyruvate in glycolysis, a lot of energy is released. So doing the reverse is not energy-favourable. This is why two steps are needed. The sub-step makes it so that less energy needs to be used. The first step adds a carbon dioxide into the pyruvate-forming oxaloacetate. By then removing the carbon dioxide, the energy is created to add the phosphate into the pyruvate and rearrange the double bond to form phosphoenolpyruvate. After the phosphoenolpyruvate is formed, the steps are similar to glycolysis, but in the reverse. Most of these steps are just rearranging the previous compound:
2. Phosphoenolpyruvate rearranges into 2-phosphoglycerate.
3. 2-phosphoglycerate rearranges into 3-phosphoglycerate.
4. 3-phosphoglycerate gets another phosphate added, forming 1,3-bisphosphoglycerate.
5. 1,3-bisphosphoglycerate rearranges into glyceraldehyde.
6. Glyceraldehyde combines with another 1,3-bisphosphoglycerate to form fructose 1,6- bisphosphate.
7. Fructose 1,6-bisphosphate loses a phosphate and becomes fructose 6-phosphate. This reaction is irreversible because it is catalysed by a different enzyme, Mg^{2+}-dependent fructose 1,6 bisphosphatase.
8. Fructose 6-phosphate rearranges to form glucose 6-phosphate.
9. Glucose 6-phosphate loses the phosphate to form the final product, glucose. This reaction is also not reversible because it is catalysed by another enzyme- the Mg^{2+}-dependent glucose-6-phosphatase.

For most of these steps, the same enzymes are used to go either direction in glycolysis and gluconeogenesis. Step 7, fructose 1,6-bisphosphate to fructose 6-phosphate, and step 9, glucose 6-phosphate to glucose, are the exceptions. Both of these steps require different enzymes than used for glycolysis. This is because in glycolysis, these steps actually require energy to occur, and so the same enzyme is not needed when doing the reverse.

Catabolic pathways

Glycogen catabolism pathways are glycogenolysis and glycolysis.

1. Glycogenolysis: The phosphorolytic reaction catalysed by glycogen phosphorylase degrades glycogen. In presence of this enzyme, inorganic phosphate attack on 14 glycosidic linkage between the last two glucose residues at a nonreducing end. This reaction results in generating glucose 1-phosphate and a polymer one glucose unit shorter. Repetitive reaction of glycogen phosphorylase on last 1-4 glycosidic linkage continues until its arrival on 1-6 branch point. The branch point inhibits its reaction. Debranching enzymes come in action to remove the branching points. The debranching enzyme has two active sites, each with its own unique activity. One active site, the transferase, catalyses the transfer of blocks of 3 glycosyl residues from one outer branch to another. The other enzyme active site, the α-1,6- glucosidase, cleaves the α(1→6) glycosidic linkage.

The glucose-1-phosphate formed by glycogen phosphorylase is isomerized into glucose-6-phosphate by phosphoglucoisomerase. The enzyme has an active site serine residue that must be phosphorylated for the enzyme to be active. The mechanism of the isomerisation is very similar to that of phosphoglycerate mutase. The phosphoryl group attached to the serine is transferred to the C6 hydroxyl group to form a glucose-1-6-bisphosphate intermediate followed by the transfer of the phosphoryl group attached to the C1 position back to the serine to produce glucose-6-phosphate.

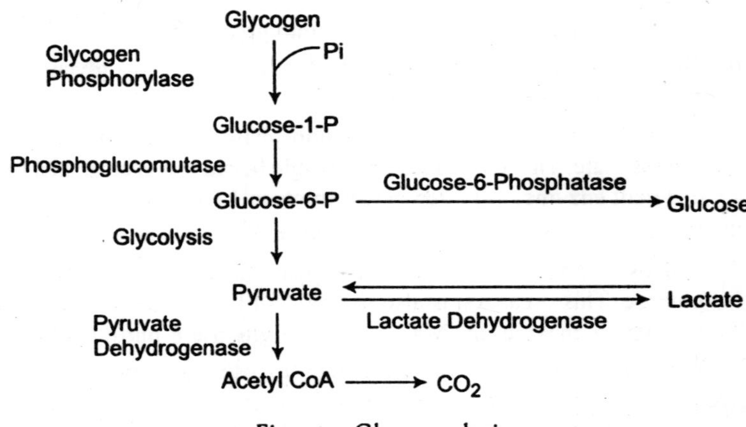

Fig. 10.4. Glycogenolysis

The glucose-1,6-bisphosphate formed can dissociate out of the active site before transferring the C1 phosphoryl group. When this happens, the active site serine must be rephosphorylated. There is an enzyme called **phosphoglucokinase** which phosphorylates glucose-1-phosphate to form glucose-1,6-bisphosphate which can bind to the dephosphorylated enzyme and transfer the C1-phosphoryl group to reactivate the enzyme and produce glucose-6-phosphate.

Phosphoglucomutase converts the glucose 1-phosphate produced by glycogen phosphorylase into glucose 6-phosphate. This glucose 6-phosphate then enters glycolysis or pentose phosphate pathway to be degraded into pyruvate. The pyruvate formed is then converted into acetyl CoA by pyruvate dehydrogenase.

2. Glycolysis: Glycolysis is the metabolic process that serves as the foundation for both aerobic and anaerobic cellular respiration. In glycolysis, glucose is converted into pyruvate. Glucose is a six- memebered ring molecule found in the blood and is usually a result of the breakdown of carbohydrates into sugars. It enters cells through specific transporter proteins that move it from outside the cell into the cell's cytosol. All of the glycolytic enzymes are found in the cytosol (Fig 10.5).

The overall reaction of glycolysis which occurs in the cytoplasm is represented simply as:

$C_6H_{12}O_6 + 2\ NAD^+ + 2\ ADP + 2\ P \rightarrow 2$ pyruvic acid, $(CH_3(C{=}O)COOH + 2\ ATP + 2\ NADH + 2\ H^-$

Step 1. ***Conversion of D-glucose into glucose-6-phosphate. The enzyme that catalyses this reaction is hexokinase.***

Here, the glucose ring is phosphorylated. Phosphorylation is the process of adding a phosphate group to a molecule derived from ATP. As a result, at this point in glycolysis, one molecule of ATP has been consumed. The reaction occurs with the help of the enzyme hexokinase, an enzyme that catalyses the phosphorylation of many six-membered glucose-like ring structures. Atomic magnesium (Mg) is also involved to help shield the negative charges from the phosphate groups on the ATP molecule. The result of this phosphorylation is a molecule called glucose-6-phosphate (G6P), thusly called because the 6′ carbon of the glucose acquires the phosphate group.

Step 2. ***Rearrangement of glucose 6-phosphate (G6P) into fructose 6-phosphate (F6P) by glucose phosphate isomerase (Phosphoglucose Isomerase).***

The second step of glycolysis involves the conversion of glucose-6-phosphate to fructose-6-phosphate (F6P). This reaction occurs with the help of the enzyme phosphoglucose isomerase (PI). As the name of the enzyme suggests, this reaction involves an isomerisation reaction. The reaction involves the rearrangement of the carbon-oxygen bond to transform the six-membered ring into a five-membered ring. To rearrangement takes place when the six-membered ring opens and then closes in such a way that the first carbon becomes now external to the ring.

Step 3. ***Phosphofructokinase, with magnesium as a cofactor, changes fructose 6-phosphate into fructose 1,6-bisphosphate.***

In the third step of glycolysis, fructose-6-phosphate is converted to fructose- 1,6-bisphosphate (FBP). Similar to the reaction that occurs in step 1 of glycolysis, a second molecule of ATP provides the phosphate group that is added on to the F6P molecule. The enzyme that catalyzes this reaction is phosphofructokinase (PFK). As in step 1, a magnesium atom is involved to help shield negative charges.

Step 4. ***Splitting of fructose 1, 6-bisphosphate into two sugars that are isomers***

This step utilises the enzyme aldolase, which catalyzes the cleavage of FBP to yield two 3-carbon molecules. One of these molecules is called glyceraldehyde-3-phosphate (GAP) and the other is called dihydroxyacetone phosphate (DHAP).

Step 5. ***Triophosphate isomerise***

The enzyme triophosphate isomerase rapidly inter- converts the molecules dihydroxyacetone phosphate (DHAP) and glyceraldehyde 3-phosphate (GAP). Glyceraldehyde phosphate is removed / used in next step of glycolysis. GAP is the only molecule that continues in the glycolytic pathway. As a result, all of the DHAP molecules produced are further acted on by the enzyme triphoshpate isomerase (TIM), which reorganizes the DHAP into GAP so it can continue in glycolysis. At this point in the glycolytic pathway, there are two 3-carbon molecules, but have not yet fully converted glucose into pyruvate.

Step 6. ***Glyceraldehyde-3-phosphate dehydrogenase***

Glyceraldehyde-3-phosphate dehydrogenase (GAPDH) dehydrogenates and adds an inorganic phosphate to glyceraldehyde 3-phosphate, producing 1,3-bisphosphoglycerate. In this step, two main events take place: 1) glyceraldehyde-3-phosphate is oxidised by the coenzyme nicotinamide adenine dinucleotide (NAD); 2) the molecule is phosphorylated by the addition of a free phosphate group. The enzyme that catalyzes this reaction is glyceraldehyde-3-phosphate dehydrogenase (GAPDH). The enzyme GAPDH contains appropriate structures and holds the molecule in a conformation such that it allows the NAD molecule to pull hydrogen off the

GAP, converting the NAD to NADH. The phosphate group then attacks the GAP molecule and releases it from the enzyme to yield 1,3 bisphoglycerate, NADH, and a hydrogen atom.

Step 7. *Phosphoglycerate kinase*

Phosphoglycerate kinase transfers a phosphate group from 1,3-bisphosphoglycerate to ADP to form ATP and 3-phosphoglycerate. In this step, 1,3 bisphoglycerate is converted to 3-phosphoglycerate by the enzyme phosphoglycerate kinase (PGK). This reaction involves the loss of a phosphate group from the starting material. The phosphate is transferred to a molecule of ADP that yields our first molecule of ATP. Since we actually have two molecules of 1,3 bisphoglycerate (because there were two 3-carbon products from stage 1 of glycolysis), we actually synthesize two molecules of ATP at this step. With this synthesis of ATP, we have cancelled the first two molecules of ATP that we used, leaving us with a net of 0 ATP molecules up to this stage of glycolysis. Again, an atom of magnesium is involved to shield the negative charges on the phosphate groups of the ATP molecule.

Step 8. *Phosphoglycero mutase*

The enzyme phosphoglycero mutase relocates the P from 3-phosphoglycerate from the 3rd carbon to the 2nd carbon to form 2-phosphoglycerate. This step involves a simple rearrangement of the position of the phosphate group on the 3 phosphoglycerate molecule, making it 2 phosphoglycerate. The molecule responsible for catalyzing this reaction is called phosphoglycerate mutase (PGM). A mutase is an enzyme that catalyses the transfer of a functional group from one position on a molecule to another. The reaction mechanism proceeds by first adding an additional phosphate group to the 2′ position of the 3 phosphoglycerate. The enzyme then removes the phosphate from the 3′ position leaving just the 2′ phosphate, and thus yielding 2 phosphoglycerate. In this way, the enzyme is also restored to its original, phosphorylated state.

Step 9. *Enolase*

The enzyme enolase removes a molecule of water from 2-phosphoglycerate to form phosphoenolpyruvic acid (PEP). This step involves the conversion of 2 phosphoglycerate to phosphoenolpyruvate (PEP). The reaction is catalyzed by the enzyme enolase. Enolase works by removing a water group, or dehydrating the 2 phosphoglycerate.

Step 10. *Pyruvate kinase*

The enzyme pyruvate kinase transfers a P from phosphoenolpyruvate (PEP) to ADP to form pyruvic acid and ATP. The final step of glycolysis converts phosphoenolpyruvate into pyruvate with the help of the enzyme pyruvate kinase. As the enzyme's name suggests, this reaction involves the transfer of a phosphate group. The phosphate group attached to the 2′ carbon of the PEP is transferred to a molecule of ADP, yielding ATP. Again, since there are two molecules of PEP, here we actually generate 2 ATP molecules.

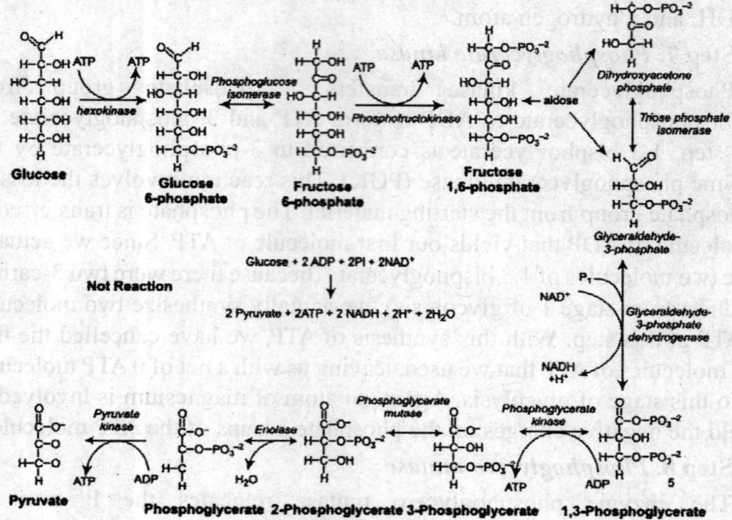

Fig. 10.5. Steps in glycolysis

3. The TCA Cycle: The citric cycle is a central driver of cellular respiration (Fig 10.6). It takes acetyl CoA —produced by the oxidation of pyruvates and originally derived from glucose—as its starting material and, in a series of redox reactions, harvests much of its bond energy in the form of NADH, FADH2, and ATP molecules. The reduced electron carriers-NADH, and FADH2- generated in the TCA cycle will pass their electrons into the electron transport chain and, through oxidative phosphorylation, will generate most of the ATP produced in cellular respiration.

In eukaryotes, the citric acid cycle takes place in the matrix of the mitochondria, just like the conversion of pyruvate to acetyl CoA.

Step 1. In the first step of the citric acid cycle, acetyl CoA joins with a four-carbon molecule, oxaloacetate, releasing the CoA group and forming a six-carbon molecule called citrate.

Step 2. In the second step, citrate is converted into its isomer, isocitrate. This is actually a two-step process, involving first the removal and then the addition of a water molecule.

Step 3. In the third step, isocitrate is oxidized and releases a molecule of carbon dioxide, leaving behind a five-carbon molecule—α-ketoglutarate. During this step, NAD^+ is reduced to form NADH. The enzyme catalysing this step, isocitrate dehydrogenase, is important in regulating the speed of the citric acid cycle.

Step 4. The fourth step is similar to the third. In this case, it's α-ketoglutarate that's oxidised, reducing NAD^+ to NADH and releasing a molecule of carbon dioxide in the process. The remaining four-carbon molecule picks up Coenzyme A, forming the unstable compound succinyl CoA. The enzyme catalysing this step, **α-ketoglutarate dehydrogenase**, is also important in regulation of the citric acid cycle.

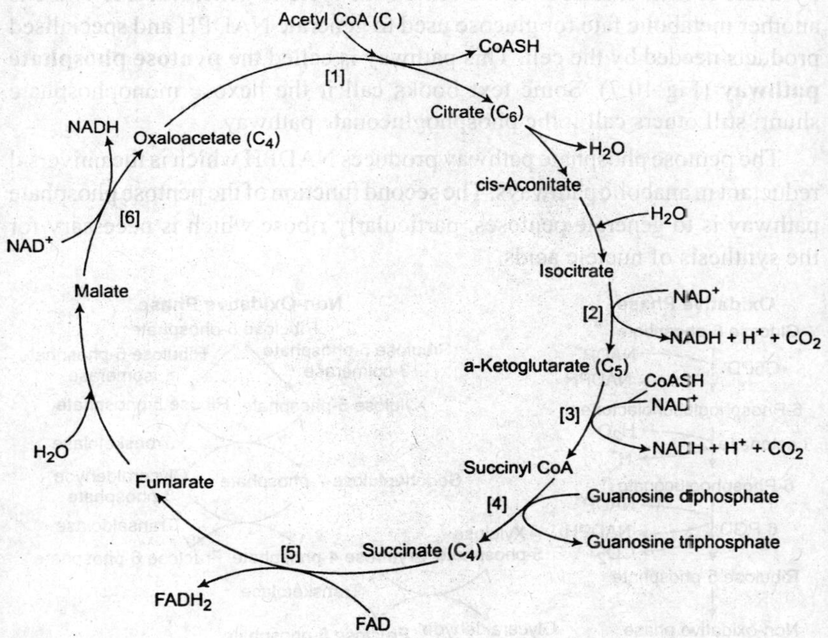

Fig. 10.6. The TCA cycle

Step 5. In step five, the CoA of succinyl CoA is replaced by a phosphate group, which is then transferred to ADP to ATP. In some cells, GDP—guanine diphosphate—is used instead ADP forming GTP—guanine triphosphate—as a product. The four-carbon molecule produced in this step is called **succinate.**

Step 6. In step six, succinate is oxidised, forming another four-carbon molecule called **fumarate**. In this reaction, two hydrogen atoms—with their electrons—are transferred FAD, producing $FADH_2$. The enzyme that carries out this step is embedded in the inner membrane of the mitochondrion, so $FADH_2$ can transfer its electrons directly into the electron transport chain.

Step 7. In step seven, water is added to the four-carbon molecule fumarate, converting it into another four-carbon molecule called malate.

Step 8. In the last step of the citric acid cycle, oxaloacetate—the starting four-carbon compound—is regenerated by oxidation of malate. Another molecule of NAD^+ is reduced to NADH in the process.

4. Pentose phosphate pathway: In most animal tissues, glucose is catabolised via the glycolytic pathway into two molecules of pyruvate. Pyruvate is then oxidised via the citric acid cycle to generate ATP. There is another metabolic fate for glucose used to generate NADPH and specialised products needed by the cell. This pathway is called the **pentose phosphate pathway** (Fig 10.7). Some text books call it the hexose monophosphate shunt; still others call it the phosphogluconate pathway.

The pentose phosphate pathway produces NADPH which is the universal reductant in anabolic pathways. The second function of the pentose phosphate pathway is to generate pentoses, particularly ribose which is necessary for the synthesis of nucleic acids.

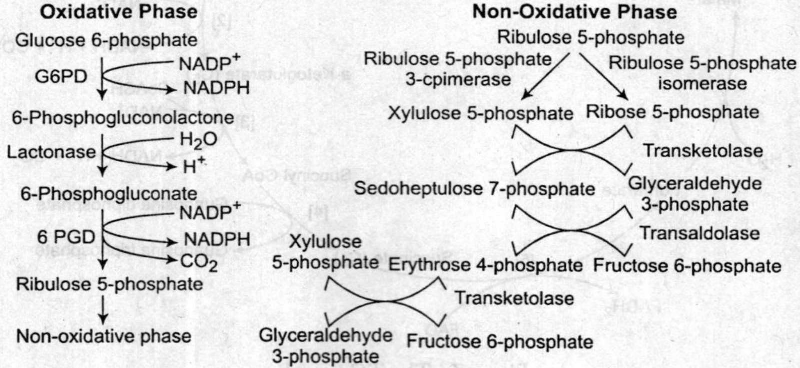

Fig. 10.7. Pentose phosphate pathway

It is convenient to think of the pentose phosphate pathway as operating in two phases. The **first phase** is the oxidative phase. Two of the first three reactions of the first phase generate NADPH. The second phase is the non-oxidative phase. In the first step glucose-6-phosphate is oxidised into ribulose-5-phosphate, CO_2. During the oxidation of glucose-6-phosphate NADP+ is reduced into NADPH. The **second step** of the pathway coverts the ribulose 5-phosphate into other pentose-5- phosphates including ribose-5-phosphate used to synthesise nucleic acids. The **third step** includes a series of reactions that convert three of the pentose-5-phosphates into two molecules of hexoses and one triose. In the fourth step, some of these sugars are converted into glucose-6-phosphate so the cycle can be repeated.

The first step of the phosphopentose pathway is the irreversible committed step. This reaction is catalysed by glucose-6-phosphate dehydrogenase. This step is of course allosterically regulated. The product of this reaction NADPH is a strong inhibitor. So when the cytosol concentration of NADPH is high, the enzyme's activity is low. It is also allosterically regulated by fatty acid acyl esters of coenzyme A. The transcription of the gene for this enzyme is under hormonal regulation.

1.1.2 *Conversion of glycogen to glycerol and sorbitol*

Diapause is an alternative developmental pathway with its own metabolic demands. On initiation of diapause, metabolic activities are suppressed and the suppression facilitates the insect to extend its food reserves to bridge the hostile conditions. During diapause, survival is also enhanced by the synthesis of polyols and other cryoprotectants and several longevity classes of heat shock proteins. On termination of diapause metabolic rate rapidly increases, initiating the development.

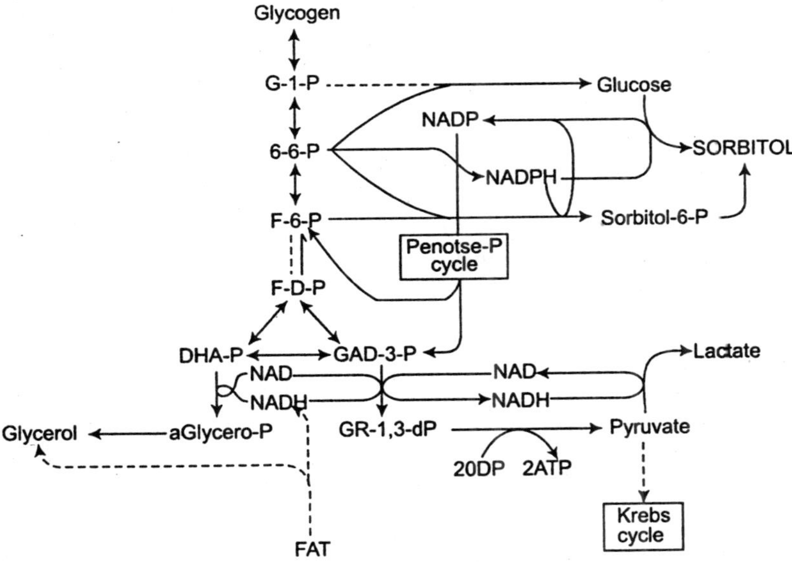

Fig. 10.8. Metabolic pathway for synthesis of sorbitol and glycerol in diapausing silkmoth egg (Kageyama and Ohnishi 1973)

In the silkworm *Bombyx mori* (L.), diapause is triggered by a neuropeptide diapause hormone (DH), released from the neurosecretory cells of the subesophageal ganglion (SG), in the mother *via* corpora cardia - corpora allata into the haemolymph, which then targets the developing ovaries (Yamashita 1996; Mitsumasu *et al.* 2008). Only eggs originating from the oocytes receiving the DH signal stage enter diapause at gastrula (Homma *et al.* 2006). Glycogen is stored in the eggs of the silkworm during oogenesis and is utilised as and when energy is required during the course of diapause as well as embryonic development.

In the diapausing eggs, bulk of the glycogen is converted into two polyols, namely **sorbitol** and **glycerol**, soon after oviposition and the glycogen is resynthesised from the polyols at the termination of the diapause. The main pathway of carbohydrate metabolism leading to the polyol formation is pentosephosphate cycle. Sorbitol or sorbitol-phosphate is formed by way of reduction of hexose or hexose-phosphate. The reduction requires NADPH. Glycerol is formed by way of reduction and hydrolysis of triose-phosphate. The reduction requires NADH. Lactate is formed by reduction of pyruvate. This reaction requires NADH. In order to meet the demand for the reduced co-enzymes for these polyols and lactate formation, NADPH is supplied by pentose phosphate cycle at the reactions of glucose-6- phosphate dehydrogenase and 6-phosphogluconate dehydrogenase and NADH would be formed by the reaction of glyceraldehyde-3-phosphate dehydrogenase. In this way, oxidation-reduction cycles of NADP-NADPH and NAD-NADH are formed.

In the anerobic conditions, ATP formation is restricted to glycolysis. Enzymes concerned for the ATP generation are 3-phosphoglycerate kinase and pyruvate kinase. Although these enzymes have not yet been demonstrated in the silkworm eggs, they must be operative. Phosphofructokinase was not demonstrated in the diapausing eggs or in the early stage of non diapausing eggs (Kageyama and Ohnishi 1971). Therefore, glyceraldehyde-3-phosphate would be formed only by way of the pentose-phosphate cycle. When glycogen is degraded to lactate, 2 moles of ATP would be formed for 1 mole of glucose equivalent. In this scheme (Fig 10.8) concerns, lactate formation is inevitable in anoxia as long as ATP is needed in the eggs. The driving force which shifts the total metabolic flow to the side of the polyols or lactate formation is considered to be generated by ATP consumption. In *Bombyx* eggs, glycerol may be accumulated as a side reaction and the reaction would compete with pyruvate reduction for NADH. The relative rates of the accumulation of lactate and glycerol would be determined by the relative activities of lactate dehydrogenase and glycerophosphate dehydrogenase. Sorbitol will be accumulated in conjugation with NADPH consumption, the latter being generated in pentose-phosphate cycle.

The onset of embryonal diapause in silkworm, is accompanied by the conversion of glycogen into glycerol and sorbitol. The phosphofructokinase (PFK) acts as a primary enzyme playing a key role in the conversion. Diapause termination is followed by glycogen synthesis from these sugar alcohols, i.e., glycerol and sorbitol. The sorbitol is then converted to glycogen, which is a source of energy for embryogenesis. The conversion is controlled by the key enzyme NAD dependent sorbitol dehydrogenase-2(SDH-2) (Chino 1958).

1.2 Trehalose

Trehalose is the unique blood sugar of insects. Trehalose serves multiple functions, as a storage carbohydrate that serves as a fuel for flight and as a cryoprotectant, protecting insects from damage during overwintering in cold climes. The haemolymph level of trehalose plays an important role in regulating carbohydrate intake and maintaining nutritional homeostasis. Levels of trehalose in the haemolymph are maintained by a complex interaction of nutrient intake and metabolism.

Trehalose is synthesised in the fat body from two glycolytic intermediates, glucose-1-phosphate and glucose-6-phosphate. The reactions are catalysed by trehalose-6-phosphate synthase and trehalose-6-phosphate phosphatase. Sources of glucose for trehalose synthesis include dietary sucrose, glycogen, and gluconeogenesis; dietary sugar being the sole source of glucose under fed conditions. Trehalose formation from glycogen has been described in several insects including the cockroach *Periplaneta americana* and moth *Manduca sexta* during starvation. The breakdown of glycogen to glucose is due to activation of the enzyme glycogen phosphorylase and is under endocrine control by a neurohormone released from the corpora cardiaca in the brain. Induction of a hypertrehalosemic hormone RNA transcript in the central nervous system of the cockroach, *Blaberus discoidalis*, in response to starvation was recently demonstrated. Glucose synthesis, followed by trehalose formation, via gluconeogenesis, has only been reported in *M. sexta* and was induced when larvae were maintained on low carbohydrate diets. Starvation did not induce gluconeogenesis.

Insects obtain energy principally from aerobic respiration, but many species have some capacity for anaerobic energy metabolism when exposed to hypoxic or anoxic conditions. This is best known in aquatic insects such as midge larvae where the fermentation products may include lactate, ethanol, and acetate. The midge *Chaoborus crystallinis* accumulates succinate, suggesting that this species is able to use anaerobic respiration for ATP production.

1.2.1 *Metabolic pathway of Insect trehalose*

In insects, trehalose forms the major haemolymph sugar and is synthesised in the fat body following a pathway (Fig 10.9) that involves two enzymes, namely, trehalose-6-phosphate synthase (TPS; EC 2.4.1.15) and trehalose 6-phosphate phosphatase (TPP; EC 3.1.3.12). TPS catalyse the transfer of glucose from UDP-glucose to glucose 6-phosphate forming trehalose 6-phosphate (T6P) and UDP, while TPP dephosphorylates T6P to trehalose and inorganic phosphate (Figure 1B). This stored energy reserve in the form of trehalose is hydrolyzed by trehalase (EC 3.2.1.28) to meet the energy demands for flight and development (Reyes-DelaTorre *et al.*, 2012).

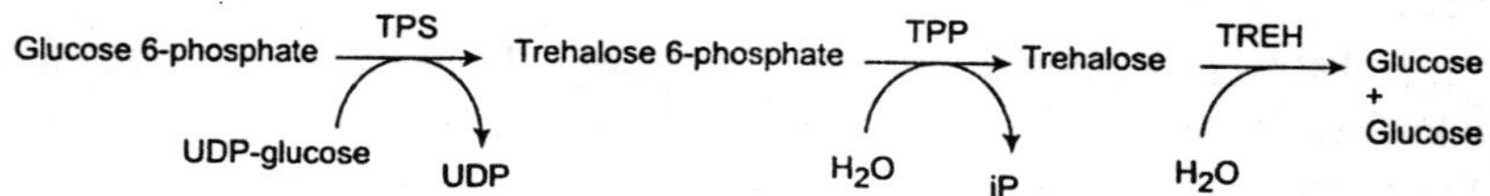

Fig 10.9 Metabolic pathway of trehalose
TPS- Trehalose – 6- phosphate synthase; TPP- Trehalose-6-phosphate phosphatase; TREH- Trehalase; UDP-Uridine diphosphate

Insect trehalase has been shown to exist in two distinct forms, Tre-1 as the soluble form, while Tre-2 being the membrane bound form. Tre-1 has been purified from haemolymph, cavity of goblet cells in midgut and from egg homogenates of some insects (Silva *et al.*, 2009). On the other hand, Tre-2 has been described in flight muscles, follicle cells, ovary cells, spermatophore, midgut, brain and thoracic ganglia (Wegener *et al.*, 2003). Tre-2 is believed to face the blood side, degrading the extracellular trehalose (Mitsumasu *et al.*, 2005). The glucose so generated then enters into the cell and is utilised for physiological activities.

The mechanism for regulation of trehalase is not fully understood; however, trehalase activity in various insects is believed to be partly under hormonal control. For instance, in certain diapausing insect larvae, juvenile hormone ensures complete growth during diapause by controlling trehalase activity. Further, enhanced trehalase activity has been found at the end of diapause following which pupation begins, an observation that has been experimentally tested by exposing diapausing larvae to juvenile hormone analog (JHA) (Singtripop *et al.*, 1999). In addition, ecdysone (20E) in insects functions similar to JHA leading to pupation (Mitsumasu *et al.*, 2008). Diapause hormone has been reported to induce embryonic diapause by modulating Tre-2 activity in the oocytes of *Bombyx mori* (Kamei *et al.*, 2011). The increase in trehalase activity at the end of diapause justifies the energy requirements before the beginning of pupation and further metabolic activities. Thus, these hormones play a role in controlling stage-specific insect development by regulating trehalase activity. In addition, certain other environmental stressors have indicated changes in trehalase expression in insects. Recently, the effects of pesticides and inhibitors have also been shown to modulate soluble trehalase activity (Ge *et al.* 2011; Zhao *et al.*, 2011).

Insect trehalase ranges from 60 to 140 kDa in molecular weight and the two forms, namely Tre-1 and Tre-2, basically differ in three aspects: localisation, size and kinetic properties. Tre-2 possesses a transmembrane domain and is usually larger while Tre-1 lacks a transmembrane domain. It is a glycoprotein. The optimum pH range for trehalase activity is 5–7. As revealed by isoelectric focusing, Tre-2 has a lower pI value when compared with Tre-1. The Km value of Tre-2 has also been shown to be lower than that of Tre-1.

1.2.2 *Biological role of insect trehalase*

Growth and development, energy supply for flight, regulation of chitin biosynthesis and recovery from abiotic stress are important physiological roles of trehalase in insects (Box 10.1). The enzyme has been functionally characterized using different inhibition approaches including RNA interference (RNAi) and potent inhibitors (Fig 10.10).

Box 10.1 Biological role of insect trehalase

1. Sugar metabolism and growth
2. Metamorphosis and reproduction
3. Flight
4. Chitin synthesis
5. Stress recovery

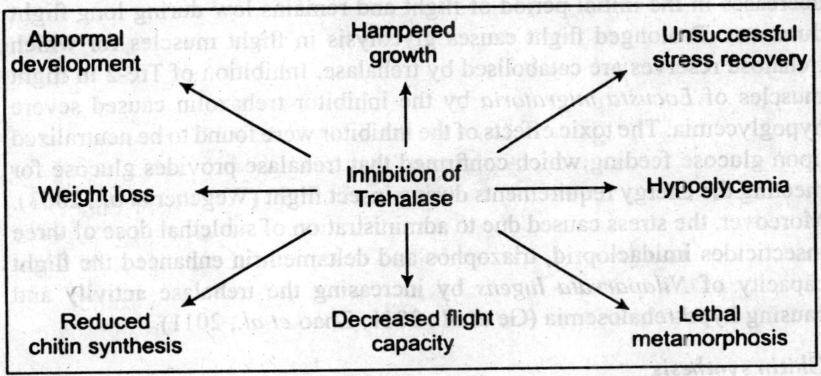

Fig. 10.10. Effect of trehalase inhibition on insects

Sugar metabolism and growth

Trehalase converts the insect sugar reserve trehalose into glucose which is then utilised to release energy. This stringent control of trehalase is responsible for the regulation of trehalose concentration vital for normal functioning by a phenomenon known as **enantiostasis** which ensures the maintenance and conservation of metabolism in response to environmental variations (Wharton 2009). Therefore, one can expect fatal consequences by inhibiting trehalase activity. RNAi studies reported weight loss and death in brown plant hopper (*Laodelphax striatellus*). Silencing of trehalase genes has not only been reported for its lethal effects by decreasing the mRNA expression and enzyme activity but has also shown to prevent normal growth and development (Zhang *et al.*, 2012).

Metamorphosis and reproduction

Injection of certain trehalase inhibitors into insect larvae have led to unsuccessful pupation and lethal metamorphosis into the adult (Wegener *et al.*, 2003). The larvae and pupae of *Spodoptera exigua* treated with double-stranded RNA of Tre-1 and Tre-2 produced unusual and anomalous phenotypes like severe-abnormal, abdomen-abnormal, misshapen-wings and half-eclosion (Chen *et al.*, 2010). Association of Tre-2 with oogenesis was also described in insects like *Rhodnius prolixus* and *Bombyx mori*. It was therefore hypothesized that this membrane bound trehalase contributes to sugar accumulation in the oocytes for the sustenance of development and viability of embryonic stages (Kamei *et al.*, 2011).

Flight

In flight muscles of many insects, the concentration of trehalose generally decreases in the initial period of flight and remains low during long flight durations. Prolonged flight causes glycolysis in flight muscles for which trehalose reserves are catabolised by trehalase. Inhibition of Tre-2 in flight muscles of *Locusta migratoria* by the inhibitor trehazolin caused severe hypoglycemia. The toxic effects of the inhibitor were found to be neutralized upon glucose feeding which confirmed that trehalase provides glucose for meeting the energy requirements during insect flight (Wegener *et al.*, 2003). Moreover, the stress caused due to administration of sublethal dose of three insecticides imidacloprid, triazophos and deltamethrin enhanced the flight capacity of *Nilaparvata lugens* by increasing the trehalase activity and causing hypotrehalosemia (Ge *et al.*, 2011; Zhao *et al.*, 2011).

Chitin synthesis

Insect cuticle is composed of chitin, a polymer of N-acetyl- β-D-glucosamine. During the process of ecdysis, the chitinous exoskeleton in insects is periodically replaced with fresh cuticle, a process essential for proper growth and development of insects. Both forms of insect trehalase, namely Tre-1 and Tre-2, are shown to participate, though differently, in the activation of chitin synthesis pathway in insects. Chitin biosynthesis utilises trehalose as the starting substrate and the pathway involves eight enzymes, trehalase being the first and chitin synthase being the last in the cascade. Chitin synthase A (CHSA) mainly exists in the cuticle while chitin synthase B (CHSB) is the major contributor in peritrophic matrix. RNAi studies in *Spodoptera exigua* demonstrated that interference with Tre-1 and Tre-2 adversely affected CHSA and CHSB, respectively (Chen *et al.*, 2010). In *Locusta migratoria*, the expression pattern of Tre-1 during moulting was also found to be very similar to that of CHSA (Liu *et al.*, 2012). Another experiment described that over-expression of trehalase genes increased the mRNA levels of four other genes which are involved in chitin biosynthesis, including CHSA and CHSB (Chen *et al.*, 2010). Inhibition of trehalase would simply indicate insufficient glucose content (which is required in large amounts for chitin synthesis) thereby affecting chitin formation. Thus, changes in the expression levels of the two trehalase genes have been shown to affect chitin biosynthesis in insects.

Stress recovery

Some organisms are able to survive extreme environmental stresses by entering into a latent state of suspended animation known as **cryptobiosis**. Depending upon the abiotic stressor in question, cryptobiosis can be further categorised as **anhydrobiosis** (desiccation stress; lack of water), **cryobiosis** (chill stress; cold temperature), **anoxybiosis** (aerobic stress; lack of oxygen) or **chemobiosis** (chemical stress; presence of toxin). Cryptobiotic organisms possess a remarkable ability to sustain the latent state from few days to several months until the return of favourable conditions upon which they resume active life. Survival of these organisms under such stressful conditions has often been correlated with the accumulation of compatible solutes, mainly sugars. Owing to its unique physicochemical properties, trehalose is the preferred solute found in several taxa of plants as well as animals. It is synthesised as a stress–responsive or an anti-stress metabolite when cells are exposed to environmental stresses. Being a non-reducing sugar, trehalose demonstrates exceptional stability and does not undergo the characteristic Maillard reaction that is known to cause cellular damage. It stabilises cellular membranes and protects proteins by replacing water molecules and facilitating cytosolar vitrification when cells are exposed to environmental stresses. In the context of desiccation stress, the chironomid midge *Polypedilum vanderplanki* is known to sustain the dry state by its ability to synthesize higher amounts of desiccation-induced trehalose (Cornette and Kikawada, 2011). Desiccation tolerance in other insects like *Drosophila melanogaster* and mosquito has also been attributed to the stress protectant ability of trehalose (Thorat *et al.*, 2012).

1.2.3 Insect trehalase: Target for insect pest control

Development of novel targets based on unexploited biochemical pathways is a promising area towards formulating strategies for insect pest control. Since trehalase plays an essential role in insect glycometabolism and homeostasis, it has been considered as a possible target for insect pest control. Exposure of insects to insecticides like fenitrothion was found to affect trehalose catabolism (Forcella *et al.*, 2007). Several trehalose mimetics and analogs have thus been proposed as potential insecticides due to the biological relevance of trehalose and trehalase (Asano *et al.*, 1987; Daviess and Henrissat, 1995; Asano, 2003).

Trehalase is highly specific to trehalose and hence is competitively inhibited by most glycosidase inhibitors. Most of the trehalase inhibitors reported to date are chemically iminosugars (nojirimycin and pyrrolidine derivatives) or pseudosaccharides. Iminosugars are polyhydroxylated secondary and tertiary amines which resemble mono or disaccharide sugars with the ring oxygen replaced by nitrogen atom. Iminosugars along with certain β-glucosides which imitate glucose units, fall under the catalysis-site targeting inhibitors. Pseudosaccharides are amide-linked sugars where the glycosidic oxygen atom is replaced by an amide group. These are structurally similar to trehalose and mimic its transition state during the reaction, synergistically interacting with both the subsites (nitrogen-containing heterocycle with the recognition site and sugar moiety with the catalysis site) (Bini *et al.*, 2012). These compounds act as competitive inhibitors targeting one or both the subsites in the active center of the enzyme and are able to specifically block trehalase activity in micromolar or submicromolar range.

Due to its crucial role in carbon metabolism in insects, inhibitors against trehalase have been proposed to form the future class of insecticides. Studies concerning alterations of trehalase activity in vivo are of great interest because of the involvement of glucosidases in a wide range of life processes. Inhibitors to trehalase have been shown to impart insecticidal or larvicidal action, by arresting insect growth, development and flight (Wegener *et al.*, 2003; Forcella *et al.*, 2010). Unfortunately, despite the remarkable efficacy, potency and specificity shown by some inhibitors, there are few difficulties with the practical use of these compounds as insecticides. Major reasons include their poor absorption through insect cuticle and poor stability (Asano, 2003; Gibson *et al*., 2007). Furthermore, there also occurs some discrepancy between their in vitro and in vivo potency (Knuesel *et al*., 1998). Thus, exploiting the available methods and improving the mode of targeting, potential trehalase inhibitors should pave way for development of novel insect pest control agents. However, comparative studies with trehalases from species belonging to different insect orders need to be explored to ascertain whether the inhibition is detrimental for the non-target/beneficial insects and crop plants and whether any shared features exist amongst the agricultural pests.

1.3 Chitin metabolism

Chitin, a polymer of *N*-acetyl-β-D-glucosamine, is a major component of the insect cuticle. Chitin is found in the exo- and endocuticle or in the newly secreted, unsclerotised procuticle but not in the epicuticle, the outermost part of the integument. It functions as light but mechanically strong scaffold material and is always associated with cuticle proteins that mainly determine the mechanical properties of the cuticle. Chitin is also an integral part of insect peritrophic matrices, which function as a permeability barrier between the food bolus and the midgut epithelium, enhance digestive processes and protect the brush border from mechanical disruption as well as from attack by toxins and pathogens. Variation of peritrophic matrix formation rate is observed frequently in insects, depending on the physiological condition (Locke, 1991). Some insects even completely cease peritrophic matrix production during periods of starvation or moult. The old peritrophic matrix then gets expelled or reabsorbed and regenerates when the animal starts feeding again. Thus, insect growth and development is strictly dependent on the capability to remodel chitinous structures. Therefore, insects consistently synthesise and degrade chitin in a highly controlled manner to allow ecdysis and regeneration of the peritrophic matrices.

1.3.1 Chitin structure

Chitin is composed largely of alternating *N*-acetylglucosamine residues, which are linked by β-(1-4)- glycosidic bonds (Fig 10.11). Chitin polymers tend to form microfibrils (also referred to as rods or crystallites) of ~3·nm in diameter that are stabilised by hydrogen bonds formed between the amine and carbonyl groups. Chitin microfibrils of peritrophic matrices may even exceed 0.5 mm in length and frequently associate in bundles containing parallel groups of 10 or more single microfibrils (Peters *et al.*, 1979; Lehane, 1997). X-ray diffraction analysis suggested that chitin is a polymorphic substance that occurs in three different crystalline modifications, termed α-, β- and γ- chitin. They mainly differ in the degree of hydration, in the size of the unit cell and in the number of chitin chains per unit cell (Rudall and Kenchington, 1973; Kramer and Koga, 1986). In the α form, all chains exhibit an anti-parallel orientation; in the β form the chains are arranged in a parallel manner; in the γ form sets of two parallel strands alternate with single antiparallel strands. The α form is most prevalent in chitinous cuticles, whereas the β and γ forms are frequently found in cocoons (Kenchington, 1976; Peters, 1992). Peritrophic matrices usually consist of α- and β-chitin.

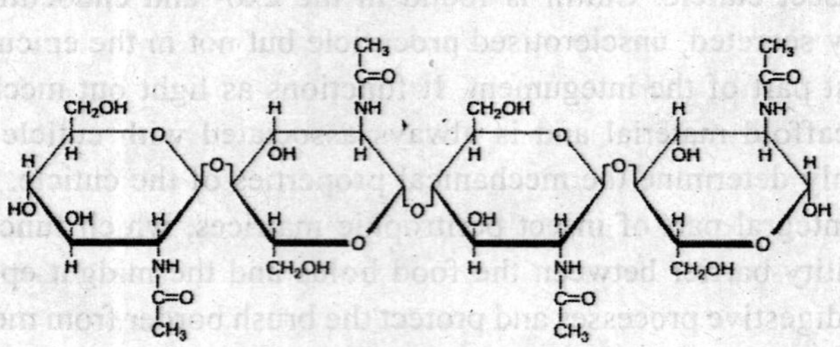

Fig. 10.11. Structure of chitin

1.3.2 *Chitin formation*

Formation of the different chitin forms is catalysed by chitin synthase (UDP-*N*-acetyl-D glucosamine: chitin 4-β *N*-acetyl glucosaminyl transferase; EC 2.4.1.16), a highly conserved enzyme found in every chitin synthesising organism. It utilises UDP-*N*-acetyl glucosamine (UDPGlcNAc) as the activated sugar donor to form the chitin polymer. Candy and Kilby (1962) were the first to propose a biosynthetic pathway for insect chitin synthesis starting with glucose and ending with UDP-GlcNAc. The pathway from UDP-GlcNAc to chitin was finally established in insects by Jaworski *et al.* (1963) using cell-free extracts from the southern armyworm *Spodoptera eridania* (Fig 10.12). Chitin synthase is definitely the key enzyme within the biosynthetic pathway, but detailed understanding of its mode of action seems to be a distant prospect, in particular because the enzyme has not yet been purified to homogeneity.

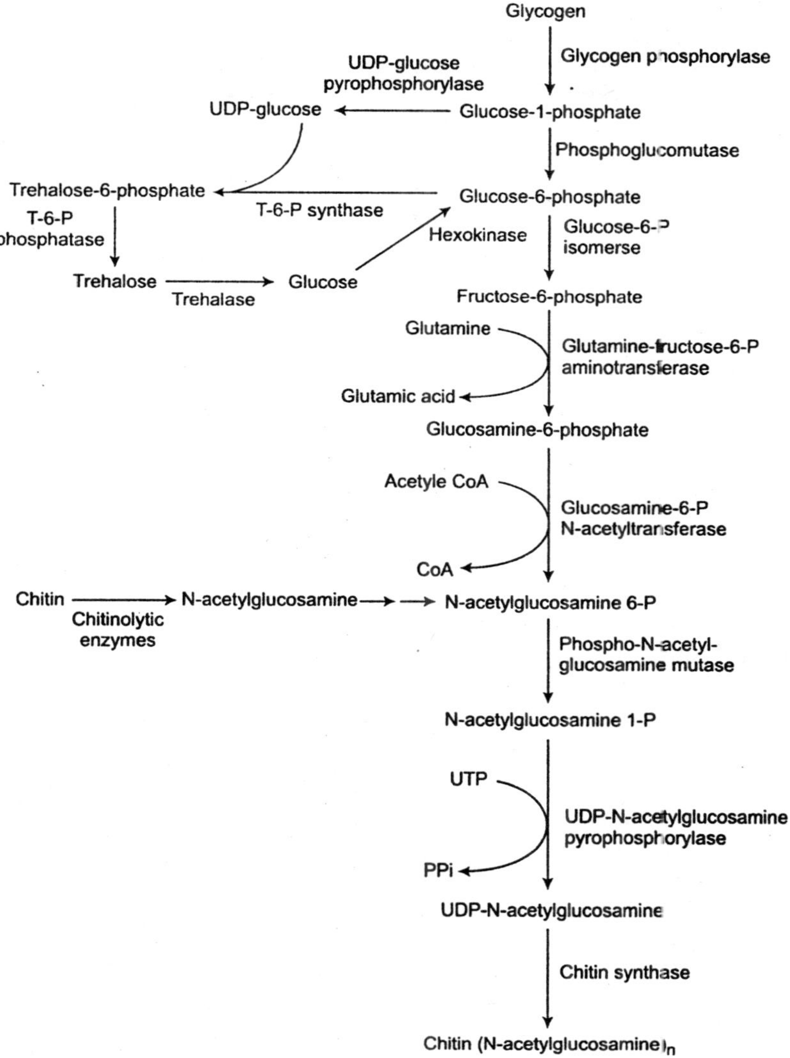

Fig. 10.12. Biosynthesis of chitin in insects

In accordance with the predicted site of chitin synthesis, immunohistochemistry using polyclonal antibodies raised against a conserved region of the chitin synthase showed strong labelling within the apical region of the epidermis from the epiproct of the American cockroach *Periplaneta americana.*

Similar results have been obtained for the chitin synthase found in insect intestinal systems to produce chitin for peritrophic membranes, which are thought to be secreted by the microvilli of gut epithelial cells. By secreting the peritrophic matrix, the microvilli act as a mold that causes microfibril spacing and, in doing so; contribute to the formation of regular patterns that are sometimes found in peritrophic matrices. Immunohistochemistry conducted with polyclonal antibodies raised against a conserved polypeptide of the *Manduca sexta* chitin synthase demonstrated that the enzyme is restricted to the apical tips of microvilli from columnar cells, one major cell type found in larval midgut (Zimoch and Merzendorfer, 2002)

1.3.1 Chitin degradation

The degrading enzymes include the chitinases {poly[1,4-(*N*acetyl-β-D-glucosaminide)] glycanohydrolase; EC 3.2.1.14} and β-*N*-acetyl glucosaminidases (β-*N*-acetyl-b-D hexosaminide *N*-acetyl hexosaminohydrolase; EC 3.2.1.52). All of them catalyse the hydrolysis of β-(1-4)-glycosidic bonds of chitin polymers and oligomers. Some of them, including one insect enzyme, additionally catalyze transglycosylation reactions (Kondo *et al.*, 2002). Since chitin-degrading enzymes can be used to convert chitin containing raw material into biotechnologically utilisable components, they are of significant interest for the chemical and pharmaceutical industry. Moreover, chitinases and their inhibitors may be adopted as insecticides to combat pests or as fungicides for the treatment of microbial infections.

In insects, chitin-degrading enzymes play a crucial role in post-embryonic development, especially during larval moult and pupation. During the moult, proteases and chitinases are synthesised by epidermal cells and accumulate in the moulting fluid between the epidermis and the old cuticle. Most of the digestion products are transported *via* the moulting fluid to the mouth and anal openings and are subsequently accumulated in the midgut (Yarema *et al.*, 2000). However, direct reabsorption by the epidermis may also occur. In any case, the reincorporated constituents seem to be recycled and used to produce the new procuticle (Reynolds and Samuels, 1996). In addition, some larvae ingest the shed exuvia to regain its constituents. This behaviour coincides with the period of chitinase expression in the gut. Moreover, the midgut chitinases seem to be involved in the formation, perforation and degradation of the midgut peritrophic matrix, which protects the gut epithelium from damaging factors.

Since chitin is hard to break due to its physicochemical properties, its degradation usually requires the action of more than one enzyme type. Endo-splitting chitinases produce chitooligomers that are subsequently converted to monomers by exo-splitting β-*N*-acetyl glucosaminidases. The latter enzyme cleaves off *N*-acetyl glucosamine units from non-reducing ends and prefers smaller substrates than chitinases. As a consequence of these properties, the overall rate of chitin hydrolysis is limited by the action of the chitooligomer- producing chitinase, which drastically increases the effective substrate concentration for the β-*N*-acetyl glucosaminidase.

1.3.4 Chitin synthase

Insect chitin synthases are large transmembrane proteins with theoretical molecular masses ranging from 160·kDa to 180·kDa and exhibit a slightly acidic isoelectric point between 6.1 and 6.7. Alignments of the amino acid sequences from *Lucilia*, *Drosophila* and *Caeno rhabditis* revealed a tripartite domain structure (Fig. 13) (Tellam *et al.*, 2000).

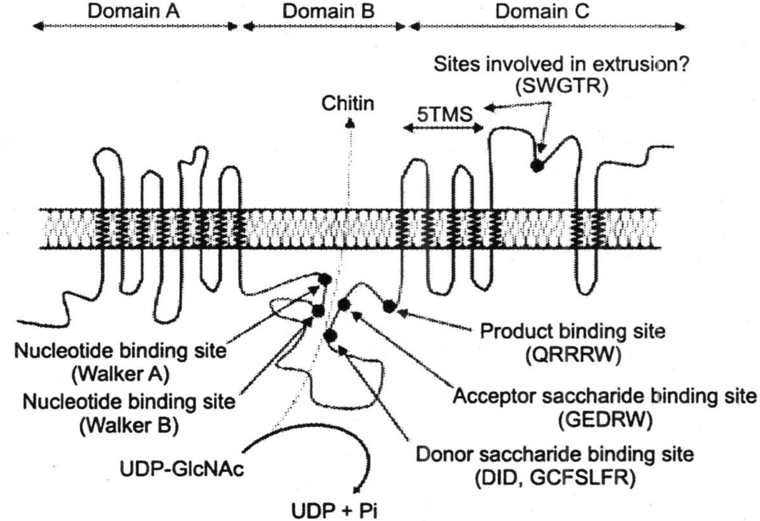

Fig. 10.13. Structural model of the tripartite domain organization of *Drosophila* chitinase

(Taken from Muthukrishnan, S., Merzendorfer, H., Arakane, Y., Kramer, K.J., 2012. Chitin metabolism in insects. In: Gilbert, L.I. (Ed.), Insect Molecular Biology and Biochemistry. Elsevier B.V., London, pp. 193–235).

Domain A, found in the N-terminal region, has varying numbers of transmembrane helices and shows the least sequence similarity among any of the species. Depending on the number of predicted transmembrane helices in the A domain, the N-terminus appears to be located at different sides of the membrane, facing either the extracellular environment or the cytoplasm.

However, this may also reflect shortcomings regarding the computer-based prediction of transmembrane helices.

Domain B, found in the center of chitin synthases, comprises ~400 amino acids and contains the catalytic center of the protein. The B domain is highly conserved and contains two unique motifs, EDR and QRRRW, that are present in all types of chitin synthases and therefore can be regarded as **signature sequences**.

Domain C comprises the C-terminal part of the enzyme and contains two amino acids that might also be involved in catalysis. Both residues are conserved in insects at positions comparable to those of the yeast enzyme, immediately following transmembrane helix five of the C domain. Although this domain is far less conserved than the catalytic domain, it exhibits seven transmembrane helices as a common feature.

Based on relative sequence differences, chitin synthases have been grouped into two classes, class CHS-A and class CHS-B enzymes. So far, most insects seem to have one gene copy for each enzyme. Since both genes are located at one chromosome in both *Drosophila* and *Anopheles*, it is likely that they have evolved from a common ancestor by gene duplication.

1.3.5 Hormonal regulation of chitin metabolism

The chitinolytic activity in the moulting fluid rises just prior to each moult and falls shortly thereafter. These changes are accompanied by the increasing and falling ecdysteroid titers prior to ecdysis. A direct role for ecdysteroids in inducing chitinase expression was demonstrated using *M. sexta* larval abdomens that were precluded from receiving hormonal signals from the brain by a ligature below the second thoracic segment. Injection of 20-hydroxyecdysone (20HE) into these ligated abdomens resulted in a sharp and rapid increase in transcripts for chitinase. This increase was abolished by a simultaneous injection of a juvenile hormone mimic (Fukamizo & Kramer, 1987). Koga *et al.*, (1992) reported a similar induction of chitinase by ecdysteroid, utilising isolated *Bombyx* abdomens. Zheng *et al.*, (2003) observed that injection of an ecdysteroid agonist resulted in induction of expression of a chitinase gene in epidermal tissue of *C. afumiferana*, and demonstrated the accumulation of chitinase in moulting fluid. It appears that hormonal regulation of chitinase genes occurs in a broad range of insect species. However, the presence of multiple genes encoding chitinases was not appreciated when these early studies were done, and it was not apparent which class of chitinases was induced by the ecdysteroid treatment. Based on our present knowledge about the tissue specificity of expression of different groups of chitinases, it is likely that these early studies were only focused on the expression of group I chitinases.

A group II chitinase gene with five catalytic domains from the beetle *Tenebrio molitor* has also been shown to be hormonally regulated (Royer *et al.*, 2002). During pupal-adult metamorphosis, the abundance of transcripts for this gene paralleled the changes in ecdysteroid (20HE) titers during metamorphosis. Interestingly, even topical application of the JH analog, methoprene, induced transcripts for this chitinase within 8 hours after treatment. These results are somewhat contradictory to the studies on *M. sexta* chitinase, in which JH had no inductive effect on chitinase transcript levels (Kramer *et al.*, 1993). In *B. mori*, another chitinase gene, BmChiRl, required 20HE for induction, and was suppressed by the simultaneous application of a JH analog (Takahashi *et al.*, 2002).

In general, chitin synthesis reaches peak levels in between moults when new cuticle is being synthesised at the maximal rate. In larval stages, this is also the period when PM-associated chitin is synthesised. Thus, both CHS-A and CHS-B (Chitin synthase) levels are high during feeding periods in larval stages. In the pupal stage CHS-B transcripts are undetectable, whereas levels of transcripts for CHS-A have multiple peaks roughly corresponding to periods of synthesis of pupal cuticle, adult epidermal cuticle, and tracheal chitin (Hogenkamp *et al.*, 2005; Arakane *et al.*, 2008).

1.3.6 *Inhibition of chitin metabolism as a strategy for insect control*

Since chitin metabolism is crucial for fungal and arthropod development, inhibition or deregulation of the key enzymes are important objectives for the development of fungicides and insecticides, including anti-malarial agents.

Inhibitors of chitin synthesis have been classified into three major groups: peptidyl nucleosides, acyl ureas and substances interfering with hormonal control. Peptidyl nucleosides isolated from diverse *Streptomyces* species act as substrate analogues and include polyoxins and nikkomycins (Zhang and Miller, 1999). They competitively inhibit both fungal and insect chitin synthases. It is believed that inhibition occurs *via* binding to the catalytic site (Ruiz-Herrera and San-Blas, 2003). Generally, the application of peptidyl nucleosides is complicated by low permeability, hydrolytic lability, varying susceptibility of fungal species and the multitude of responses found in animals.

In contrast to the peptidyl nucleosides, acyl ureas play an important role in integrated pest management. Although it is well established that acyl ureas such as diflubenzuron and teflubenzuron affect chitin synthesis (Post *et al.*, 1974; van Eck, 1979), their mode of action is still puzzling. However, several lines of experiment argue against a direct interaction of these inhibitors with the chitin synthase. For instance, in cellfree systems, acyl ureas do not inhibit chitin synthesis (Cohen, 1985). Instead of directly blocking chitin synthase activity, they may alter either vesicle transport or fusion, inhibit the translocation of chitin fibrils across the plasma membrane (Cohen, 2001) or interfere with the hormonal regulation of chitin synthesis by influencing ecdysteroid production (Fournet *et al.*, 1995).

The third group of inhibitors evidently affects hormonal regulation of insect growth and development. One of the manifold effects of these substances is certainly deregulation of chitin synthesis, probably by preventing the expression of the chitin synthase or regulating factors. Hormonal regulation can already be disturbed at the level of hormone biosynthesis. Some synthetic imidazole and cholesterol derivates have been shown to prevent ecdysteroid biosynthesis (Lorenz *et al.*, 1995). By contrast, the heterocyclic compound brevioxime and the alkaloid arborine show significant blocking of juvenile hormone synthesis (Moya *et al.*, 1997; Muthukrishnan *et al.*, 1999). The auxiliary application of isolated molting hormones or their synthetic agonists and antagonists leads to abnormalities in insect development as well. The ecdysteroid agonist tebufenozide manifests its effect by interacting with the ecdysone receptor, the juvenile hormone agonists fenoxycarb, methoprene and pyriproxyfen mimic the hormone action, and the juvenile hormone antagonists precocene I and II act *via* their cytotoxicities on the corpora allata (Kostyukovsky *et al.*, 2000; Retnakaran *et al.*, 2001).

Chitinase inhibitors can generally be grouped into two major classes: they mimic either carbohydrate substrates or the oxocarbonium reaction intermediate of family 18 chitinases. The most-studied chitinase inhibitor is allosamidin, a pseudotrisaccharide. It was isolated from the mycelium of *Streptomyces* sp. and exhibits a strong inhibitory activity against family 18 chitinases of insects and fungi with a *K*i in the nano- to micromolar range (Berecibar *et al.*, 1999). Most strikingly, it blocks malaria parasite transmission by inhibiting the chitinase of *Plasmodium* that is essential to penetrate the host's peritrophic matrix (Filho *et al.*, 2002). The structural basis of interactions between the inhibitor and several family 18 chitinases has been solved by X-ray crystallography (Terwisscha van Scheltinga *et al.*, 1995; van Aalten *et al.*, 2001). As a result of these studies, it was proposed that allosamidin mimics the catalytic transition state. Allosamidin consists of two *N*-acetylallosamine sugars linked to an allosamizoline that may resemble the catalytic intermediate and cannot be hydrolysed because it lacks the pyranose oxygen (Rao *et al.*, 2003). Allosamidin is a potent chitinase inhibitor; however, its production is expensive because it is difficult to synthesise. A new, alternative class of inhibitors includes the cyclopentapeptides argifin and argadin. These molecules are as potent inhibitors as allosamidin but synthesis by peptide chemistry is less expensive

(Arai *et al.*, 2000; Omura *et al.*, 2000). While the cyclopentapeptides are carbohydrate mimics, the small peptide CI-4, which was recently identified in the marine bacterium *Pseudomonas*, functions like allosamidin as a mimic of the family 18 chitinases' catalytic transient state (Izumida *et al.*, 1996).

Recombinant baculoviruses expressing mutant insect chitinase genes may themselves constitute potent biocides since they could perturb peritrophic matrix function of infected insects, leading to enhanced susceptibility to other harmful compounds. Insect chitinases have even been alienated to induce insect resistance in plants by expressing a truncated but active version of the enzyme in transgenic tobacco (Ding *et al.*, 1998). Plants expressing an insect chitinase gene may have agronomic potential for insect control.

2. LIPID METABOLISM

Lipids are heterogeneous compounds characterised by their solubility in solvent of low polarity. They include fatty acids, fatty acid esters of glycerol (acylglycerol) and sphingosine, waxes, terpenes, steroids etc. From a physiological point of view lipids can be categorized according to their functional criteria for storage and liberation of energy and into a structural criterion for a role in biological membranes. Besides, there are lipids which are regulatory such as pheromones and hormones and protective lipids such as waxes, paraffins, alcohols.

2.1 Lipids for storage and liberation of metabolic energy

Acylglycerols

Insect species mostly have triacylglycerol as a dominant source as reserve of metabolic energy. Triacylglycerol has advantages over glycogen in terms of having a higher caloric content/unit weight, the liberation of more than two times metabolic water than from carbohydrates and the capacity for storage in anhydrous form. For these reasons triacylglycerol is used during diapause, long migratory flights, and during non-feeding stages of development. Fat body is the major repository of triacylglycerol. The ovaries of mature females also accumulate them during vitellogenesis.

Fatty acids

Major fatty acid complement of insects is represented by eight fatty acids. They are the saturated fatty acids, myristic acid, palmitic acid, stearic acid, the mono-unsaturated fatty acids, myristoleic acid, palmitoleic acid and oleic acid and the poly-unsaturated fatty acid, linoleic acid and linolenic acid. The long chain poly-unsaturated fatty acids serve as a precursor for prostaglandin synthesis.

2.2 Structural lipids

Phospholipids

Phospholipids are lipids containing glycerol phosphate. Their primary function is to serve like building blocks of the most membrans. Diacylglycerol backbone is esterified by the phospho-group in *sn-3*-glycerol position and long chain fatty acids at position *sn-1* and *2* – positions are connected via ester bond. The diversity of PLs is dependent on a diversity of the long chain fatty acids occurring on a phospholipid headgroupThe diversity of PLs is dependent on a diversity of the long chain fatty acids occurring on a phospholipid headgroup.

Phosphatidylcholines (PCs) and phosphatidylethanolamines (PEs) are the most abundant phospholipids classes through insects which represent more than 70% of total lipid components (Downer, 1985). The proportion of PEs and PCs and also of particular FAs is species specific and also dependent on the physiological state of insect (overwintering, diapause etc.). Other phosphoglycerolipid classes are further important components of biological membrane and their presence and abundance is organism dependent.

Sterols

Sterols are another lipid class essential in insect living with several crucial functions: a fundamental component in subcellular membranes, a precursor of the moulting and vitellogenic hormone ecdysone and a constituent of surface wax of insect cuticle and lipoprotein carrier molecules (Downer, 1978; Canavoso *et al.*, 2001). All insects require sterol in their diets. This is a result of inability of insect tissues to synthesise squalene by reductive dimerisation of farnesyl pyrophosphate. Cholesterol usually earns in food of predators or blood sucking insects. Among phytophagous insects, phytosterols are often an adequate and, in some cases, better substitute for cholesterol in the diet. These insects have a metabolic capacity for conversion of phytosterols to cholesterol.

Sphingolipids

Sphingolipids are composed from a fatty acid moiety bound to sphingosine base. They are important in number of cellular processes and are involved in essentially all aspects of cellular regulation. Sphingolipids (SLs) serve as ligands for receptors and mediate change in cell behaviour in response to cells environment. SLs are also involved in membrane trafficking for example influence receptor internalisation, sorting and recycling

2.3 Oxidation of fatty acids

Fatty acid β-oxidation is a multistep process by which fatty acids are broken down by various tissues to produce energy. Fatty acids primarily enter a cell via fatty acid protein transporters on the cell surface. Fatty acid transporters include fatty acid translocase (FAT/CD36), tissue specific fatty acid transport proteins (FATP), and plasma membrane bound fatty acid binding protein ($FABP_{pm}$). Once inside the cell, a CoA group is added to the fatty acid by fatty acyl-CoA synthase (FACS), forming long-chain acyl-CoA. Carnitine palmitoyltransferase 1 (CPT1) conversion of the long-chain acyl-CoA to long-chain acylcarnitine allows the fatty acid moiety to be transported across the inner mitochondrial membrane via carnitine translocase (CAT), which exchanges long-chain acylcarnitines for carnitine. An inner mitochondrial membrane CPT2 then converts the long-chain acylcarnitine back to long-chain acyl-CoA. The long-chain acyl-CoA enters the fatty acid β-oxidation pathway, which results in the production of one acetyl-CoA from each cycle of fatty acid β-oxidation. This acetyl-CoA then enters the mitochondrial tricarboxylic acid (TCA) cycle. The NADH and FADH2 produced by both fatty acid β-oxidation and the TCA cycle are used by the electron transport chain to produce ATP (Fig 10.14).

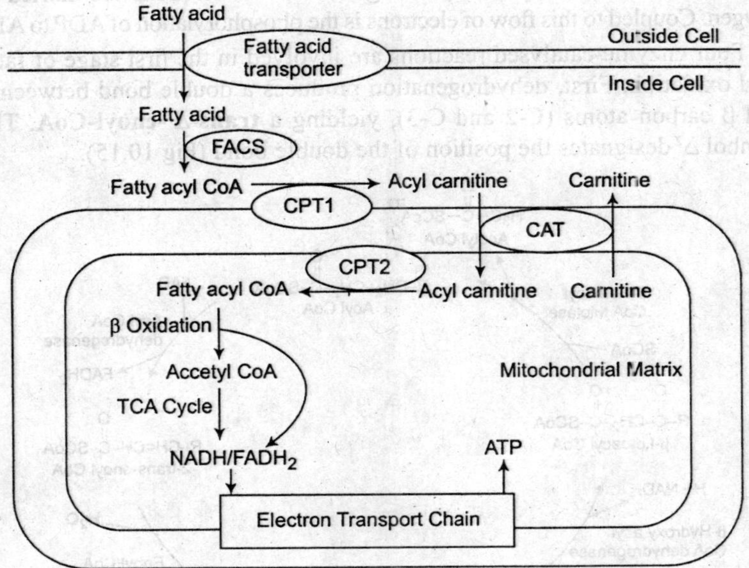

Fig. 10.14. Summary of fatty acid metabolism

Mitochondrial oxidation of fatty acids takes place in three stages. In the first stage-β oxidation-the fatty acids undergo oxidative removal of successive two-carbon units in the form of acetyl-CoA, starting from the carboxyl end of the fatty acyl chain. For example, the 16-carbon fatty acid palmitic acid (palmitate at pH 7) undergoes seven passes through this oxidative sequence, in each pass losing two carbons as acetyl-CoA. At the end of seven cycles the last two carbons of palmitate (originally C-15 and C-16) are left as acetyl-CoA. The overall result is the conversion of the 16-carbon chain of palmitate to eight two-carbon acetyl-CoA molecules. Formation of each molecule of acetyl-CoA requires removal of four hydrogen atoms (two pairs of electrons and four H^+) from the fatty acyl moiety by the action of dehydrogenases.

In the second stage of fatty acid oxidation the acetyl residues of acetyl-CoA are oxidised to CO_2 via the citric acid cycle, which also takes place in the mitochondrial matrix. Acetyl-CoA derived from fatty acid oxidation thus enters a final common pathway of oxidation along with acetyl-CoA derived from glucose via glycolysis and pyruvate oxidation.

The first two stages of fatty acid oxidation produce the reduced electron carriers NADH and $FADH_2$, which in the third stage donate electrons to the mitochondrial respiratory chain, through which the electrons are carried to oxygen. Coupled to this flow of electrons is the phosphorylation of ADP to ATP.

Four enzyme-catalysed reactions are involved in the first stage of fatty acid oxidation. First, dehydrogenation produces a double bond between α and β carbon atoms (C-2 and C-3), yielding a **trans-Δ^2-enoyl-CoA**. The symbol Δ^2 designates the position of the double bond (Fig 10.15).

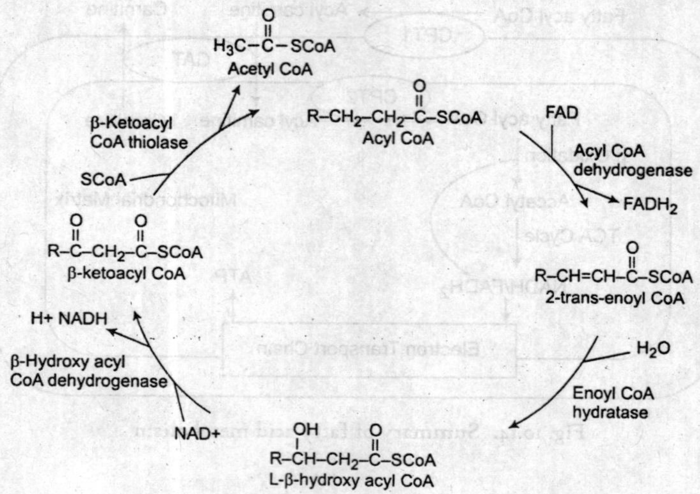

Fig. 10.15. β-oxidation of fatty acids

The enzyme responsible for this first step, **acyl-CoA dehydrogenase**, includes FAD as a prosthetic group. The electrons removed from the fatty acyl-CoA are transferred to the FAD, and the reduced form of the dehydrogenase then immediately donates its electrons to an electron carrier, the **electron-transferring** flavoprotein (ETFP). ETFP, an integral protein of the inner mitochondrial membrane, is one of the electron carriers of the mitochondrial respiratory chain. The transfer of a pair of electrons from the FADH2 of acyl-CoA dehydrogenase to O_2 via the respiratory chain provides the energy for the synthesis of two ATP molecules.

The oxidation catalysed by acyl-CoA dehydrogenase is analogous to succinate dehydrogenation in the citric acid cycle; in both reactions the enzyme is bound to the inner membrane, a double bond is introduced into a carboxylic acid between the α and β carbons, FAD is the electron acceptor, and electrons from the reaction ultimately enter the respiratory chain and are carried to O_2 with the concomitant synthesis of two ATP molecules per electron pair.

In the second step of the fatty acid oxidation cycle, water is added to the double bond of the trans-enoyl-CoA to form the L stereoisomer of **-hydroxyacyl-CoA** (also designated **-hydroxyacyl-CoA**). This reaction, catalysed by enoyl-CoA hydratase, is formally analogous to the fumarase reaction in the citric acid cycle, in which H_2O adds across an α-β double bond.

In the third step, the L-β-hydroxyacyl-CoA is dehydrogenated to form **β-ketoacyl-CoA** by the action of **β-hydroxyacyl-CoA** dehydrogenase; NAD^+ is the electron acceptor. This enzyme is absolutely specific for the r. stereoisomer. The NADH formed in this reaction donates its electrons to **NADH dehydrogenase (Complex I)**, an electron carrier of the respiratory chain. Three ATP molecules are generated from ADP per pair of electrons passing from NADH to O_2 via the respiratory chain. The reaction catalyzed by β-hydroxyacyl-CoA dehydrogenase is closely analogous to the malate dehydrogenase reaction of the citric acid cycle.

The fourth and last step of the fatty acid oxidation cycle is catalyzed by **acyl-CoA acetyltransferase** (more commonly called **thiolase**), which promotes reaction of β-ketoacyl-CoA with a molecule of free coenzyme A to split off the carboxyl-terminal two-carbon fragment of the original fatty acid as acetyl-CoA. The other product is the coenzyme A thioester of the original fatty acid, now shortened by two carbon atoms. This reaction is called **thiolysis**, by analogy with the process of hydrolysis, because the β-ketoacyl-CoA is cleaved by reaction with the thiol group of coenzyme A.

The carbon-carbon single bond that connects methylene ($-CH_2-$) groups in fatty acids is relatively stable. The β-oxidation sequence represents an elegant solution to the problem of breaking these bonds. The first three reactions of β oxidation have the effect of creating a much less stable C-C bond, in which one of the carbon atoms (the α carbon, C-2) is bonded to two carbonyl carbons. The ketone function on the β carbon (C-3) makes it a good point for nucleophilic attack by -SH of coenzyme A, catalysed by thiolase. The acidity of the a carbon makes the terminal $-CH_2$-CO-S-CoA a good cleaving group, facilitating breakage of the α-β bond.

2.4 Ketone Bodies

Acetyl-CoA formed during oxidation of fatty acids may enter the citric acid cycle or it may be converted to the ketone bodies- **acetoacetate, D-β-hydroxy-butyrate,** and **acetone** for export to other tissues. Acetone, produced in smaller quantities than the other ketone bodies, is exhaled. Acetoacetate and D-β-hydroxybutyrate are transported by the blood to the extrahepatic tissues, where they are oxidised via the citric acid cycle to provide much of the energy required by tissues.

A major determinant of the pathway taken by acetyl-CoA in mitochondria is the availability of oxaloacetate to initiate entry of acetyl-CoA into the citric acid cycle. Under some circumstances (such as starvation) oxaloacetate is drawn out of the citric acid cycle for use in synthesizing glucose. When the oxaloacetate concentration is very low, little acetyl-CoA enters the cycle, and ketone body formation is favoured (Box 10.2).

The first step in formation of acetoacetate is the enzymatic condensation of two molecules of acetyl-CoA, catalyzed by thiolase; this is simply the reversal of the last step of β oxidation. The acetoacetyl-CoA then condenses with acetyl-CoA to form **-hydroxy-β-methylglutaryl-CoA** (HMG-CoA), which is cleaved to free acetoacetate and acetyl-CoA.

The free acetoacetate so produced is reversibly reduced by **D-β-hydroxybutyrate dehydrogenase**, a mitochondrial enzyme, to D-β-hydroxybutyrate. This enzyme is specific for the D stereoisomer; it does not act on L-β-hydroxyacyl-CoAs and is not to be confused with L-β-hydroxyacyl-CoA dehydrogenase, which acts in the β-oxidation pathway The free acetoacetate so produced is reversibly reduced by**D-β-hydroxybutyrate dehydrogenase**, a mitochondrial enzyme, to D-β-hydroxybutyrate. This enzyme is specific for the D stereoisomer; it does not act on L-β-hydroxyacyl-CoAs and is not to be confused with L-β-hydroxyacyl-CoA dehydrogenase, which acts in the β-oxidation pathway. Acetoacetate is easily decarboxylated; the carboxyl group may be lost spontaneously or by the action of acetoacetate decarboxylase (Fig 10.16).

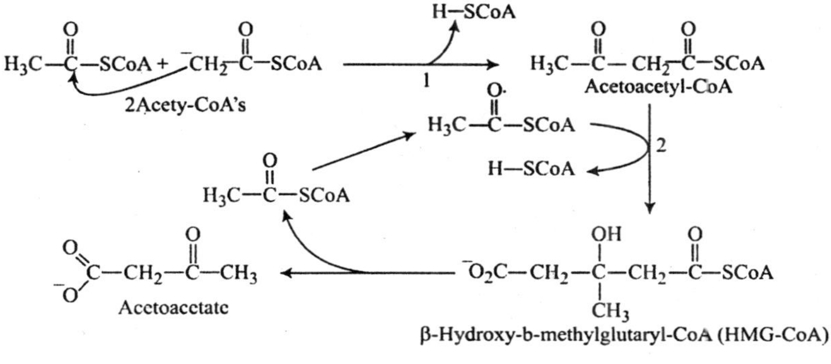

$H_3C-\overset{O}{\overset{\|}{C}}-SCoA + {}^{-}CH_2-\overset{O}{\overset{\|}{C}}-SCoA \xrightarrow{1} H_3C-\overset{O}{\overset{\|}{C}}-CH_2-\overset{O}{\overset{\|}{C}}-SCoA$ + H—SCoA

2Acety-CoA's

Acetoacetyl-CoA

$H_3C-\overset{O}{\overset{\|}{C}}-SCoA$ → (2) H—SCoA

$H_3C-\overset{O}{\overset{\|}{C}}-SCoA$

$^{-}O_2C-CH_2-\underset{CH_3}{\overset{OH}{C}}-CH_2-\overset{O}{\overset{\|}{C}}-SCoA$

β-Hydroxy-b-methylglutaryl-CoA (HMG-CoA)

$^{-}O(O=)C-CH_2-\overset{O}{\overset{\|}{C}}-CH_3$

Acetoacetate

Fig. 10.16. Ketogenesis

Box 10.2 Ketogenesis and ketone bodies

1. Large amounts of acetyl CoA accumulate.
2. Two acetyl CoA molecules combine to form acetoacetyl CoA.
3. Acetoacetyl CoA hydrolyses to acetoacetate- a ketone body.
4. Acetoacetate is reduced to β-hydroxybutyrate or loses CO2 to form acetone, both ketone bodies.

Ketone bodies have been found in the haemolymph, fat body and flight muscles of locusts with acetoacetate predominating in the haemolymph and flight muscles whereas D-β-hydroxy-butyrate is the major ketone body in the fat body. The ketone bodies may also provide an alternative fuel to glucose for musculature during starvation and for use by nervous tissue.

2.5 Biosynthesis of fatty acids

When fatty acid oxidation was found to occur by oxidative removal of successive two-carbon (acetyl-CoA) units, biochemists thought that the biosynthesis of fatty acid might proceed by simple reversal of the same enzymatic steps used in their oxidation. However, fatty acid synthesis and breakdown occur by different routes, are catalysed by different sets of enzymes and take place in different parts of the cell.

2.5.1 Acetyl-CoA Transport into the Cytosol

Acetyl-CoA serves as a key intermediate between lipid and carbohydrate metabolism. For the production of fatty acids, acetyl-CoA (which is produced in mitochondria) must first be transported across the organelle's membrane into the cytosol. Since acetyl-CoA itself cannot traverse the membrane, this transfer relies on the transport of the acetyl moiety as citrate (produced from actyl-CoA and oxaloacetate). After citrate is transferred *via* the tricarboxylate transport system from mitochondria into the cytosol, it is cleaved by **ATP-citrate lyase** to produce acetyl-CoA by the following reaction:

$$\text{Citrate} + \text{CoA} + \text{ATP} = \xrightarrow[\Delta G' = -3{,}400 \text{ cal/mol}]{\text{ATP-citrate lyase}} \text{Acetyl-CoA} + \text{Oxaloacetate} + \text{ADP} + \text{P}i$$

Although carnitine has been assigned the role as a carrier of acetyl groups, as well as of fatty acids, current evidence supports the contention that citrate and not acetylcarnitine is the principal source of cytosolic acetyl CoA. The acetyl-CoA is now ready to serve as a substrate with the required amounts of ATP and NADPH to form palmitate.

2.5.2 *Production of Malonyl-CoA: The Initiation Phase*

The production of malonyl-CoA is the initial committed step in the fatty acid synthesis. In 1961, Salih Vakil's observation that CO_2 greatly stimulates the incorporation of acetyl-CoA into fatty acid structures was an important finding in the elucidation of this process. In fact, his studies revealed that acetyl-CoA must be converted, rather carboxylated, into **malonyl-CoA** prior to its utilisation for fatty acid synthesis. This irreversible two-step reaction is the committed step in fatty acid synthesis and, as would be expected, it is also the primary rate-limiting reaction of the process. The reaction is catalysed by the enzyme, acetyl-CoA carboxylase.

Malonyl-CoA is synthesised in two steps by the action of two enzymes, each of which employs the biotin carrier protein as one substrate. The two steps are:

1. **First Step:** The biotin carboxylase (BC) catalyses carboxylation of biotin carboxyl carrier protein (BCP) to yield carboxybiotin carboxyl carrier protein (BCP–COO–); the carboxyl group being derived from bicarbonate (HCO_3^-). This is an ATP-dependent reaction.

$$\underset{\text{(Biotinyl enzyme)}}{\text{BCP} + \text{HCO}_3^- + \text{ATP}} \xrightarrow{\text{Biotin carboxylase}} \underset{\text{(Carboxybiotinyl enzyme)}}{\text{BCP-COO}^- + \text{ADP} + \text{P}i}$$

2. **Second Step:** The transcarboxylase transfers the "bound" CO_2 from BCP–COO– to acetyl- CoA, forming malonyl-CoA and regenerating BCP.

$$\text{BCP} - \text{COO}^- + \text{CH}_3\text{CO-SCoA} \xrightarrow{\text{Transcarboxylase}} \text{BCP} + \text{-OOC-CH}_2\text{CO-SCoA}$$

The overall result of these two reactions would be the production of a mole of malonyl-CoA by the addition of a mole of CO_2 (actually as HCO_3^-)

to a mole of acetyl-CoA; the ATP mole providing energy for driving the reaction. The net equation then would be:

$$CH_3CO - SCoA + HCO_3^- + ATP \xrightarrow{\text{Acetyl-CoA carboxylase}} {}^-OOC - CH_2CO - SCoA + ADP + Pi$$

(Acetyl CoA) (Malonyl-CoA)

The malonyl-CoA provides 14 out of 16 carbon atoms of palmitate.

2.5.3 *The Fatty Acid Synthase Complex*

All of the reactions in the biosynthesis of fatty acids are catalysed by a multienzyme complex, the fatty acid synthase. The detailed structure of this multienzyme complex and its location in the cell differ from one species to another, but the reaction sequence is identical in all organisms.

The **fatty acid synthase** system from *E.coli,* which is the best studied, consists of 7 separate polypeptides (and hence 7 different active sites) that are tightly associated in a single, organised complex (Table 1). The proteins act together to catalyse the formation of fatty acids from acetyl-CoA and malonyl- CoA. Throughout the process, the intermediates remain covalently attached to one of the two thiol (—SH) groups of the complex.

Table 1. Seven components of the fatty acid synthase complex from *Escherichia coli*

Component	*Abbreviation*	*E.C. No.*	*Role*
Acyl carrier protein	ACP	6.4.1.2	Carries acyl groups in thioester linkage
Acyl-CoA–ACP transacetylase CoA to cysteine	AT	2.3.1.9	Transfers acyl group from residue of KS
Malonyl-CoA–ACP transferase	MT	2.8.3.3	Transfers malonyl group from CoA to ACP
β-ketoacyl–ACP	KS	2.3.1.16	Condenses acyl and malonyl groups synthase
β-ketoacyl–ACP reductase	KR	1.1.1.36	Reduces β-keto group to β-hydroxy group
β-hydroxyacyl–ACP dehydrogenase	HD	4.2.1.17	Removes H2O from β-hydroxyacyl-ACP, creating double bond
Enoyl–ACP reductase	ER	1.3.99.2	Reduces double bond, forming saturated acyl-ACP

The growing fatty acid is shifted between these two —SH groups. One is relatively fixed in position because it is on a cysteine residue. It acts as a parking place for acyl groups that are to be lengthened. The **other** —SH group carries the extended chain while it undergoes the reactions necessary for reduction to a saturated acyl group, and it also accepts the acetyl and malonyl groups from which the fatty acid is built. This —SH group can swing across the 7 different catalytic sites because it is located in a residue of phosphopantetheine.

The two thiol groups are designated as central and peripheral. The **central** one is the — SH group of acyl carrier protein (ACP), with the intermediates of fatty acid synthesis form a thioester and the **peripheral** one is the —SH group of a cysteine residue in β-ketoacyl-ACP synthase, one of the 7 proteins of the multienzyme complex.

2.5.4 *Priming of the Fatty Acid Synthesis by Acetyl-CoA : The Priming Phase*

Before the condensation reactions that build up the fatty acid chain can begin, the two —SH groups on the enzyme complex must be charged with the correct acyl groups. The priming of the system, as it is called, takes place in 2 steps :

In the **first step**, the acetyl group of acetyl-CoA is tranferred to the cysteine—SH group of the β- ketoacyl-ACP synthase. This reaction is catalysed by **acetyl-CoA-ATP transacetylase.** In the **second step**, the malonyl group from malonyl-CoA is transferred to the —SH group of ACP by the enzyme **malonyl-CoA-ACP transferase**, also part of the complex.

2.5.5 *Growth of the Fatty Acyl Chain by Two Carbons : The Elongation Phase*

In the **first round,** charged synthase complex, the acetyl and malonyl groups are very close to each other and are activated for the chain-lengthening process, which consists of the following four steps :

1. **Condensation:** The first step in the formation of a fatty acid chain is condensation of the activated acetyl and malonyl groups to form an acetoacetyl group bound to ACP through the phosphopantetheine —SH group, thus producing **acetoacetyl-ACP** ; simultaneously, a mole of CO_2 is eliminated from the malonyl group. In this reaction, catalyzed by β-*ketoacyl-ACP synthase*, the acetyl group is transferred from the cysteine —SH group of this enzyme to the malonyl group on the —SH of ACP, becoming the methyl-terminal two-carbon unit of the new acetoacetyl group. The carbon atom in the CO_2 formed in this reaction is the same carbon atom that was originally introduced into malonyl-CoA from HCO_3^- by the acetyl-CoA carboxylase reaction. Thus, CO_2 is only transiently in covalent linkage during fatty acid biosynthesis; it is removed as such as each two-carbon unit is inserted. Thus, the net effect of condensation reaction is the extension of the acyl chain by 2 carbon atoms

By using activated malonyl groups in the synthesis of fatty acid and activated acetate in their degradation, the cell manages to make both processes favourable, although one is effectively the reversal of the other. The extra energy, needed to make fatty acid synthesis favourable, is provided by the ATP used to synthesize malonyl-CoA from acetyl-CoA and HCO_3^-. In effect, the condensation reaction is driven by ATP, although ATP does not directly participate in the condensation reaction. Rather, ATP is used to form an energy-rich substrate in the carboxylation of acetyl-CoA to malonyl-CoA. The free energy stored in malonyl-CoA in the carboxylation reaction is released in the decarboxylation accompanying the formation of acetoacetyl-ACP. Although HCO_3^- is required for fatty acid synthesis, its carbon does not appear in the product. Rather, all of the carbon atoms of even-chain fatty acids are derived from acetyl-CoA. The next 3 steps in fatty acid synthesis reduce the keto (> CO) group at C-3 to a methylene (—CH2—) group, the result being the conversion of acetoacetyl-ACP into butyryl-ACP.

2. **Reduction of the Carbonyl group:** The acetoacetyl group formed in the condensation step next undergoes reduction of the carbonyl group at C-3 to form **D-β-hydroxybutyryl-ACP.** This reaction is catalysed by β-ketoacyl-ACP reductase and the electron donor is NADPH. This reaction differs from the corresponding one in fatty acid degradation in two respects:

   A. The D- rather than the L-epimer is formed

   B. NADPH is the reducing agent, whereas NAD+ is the oxidising agent in βoxidation. This difference exemplifies the general principle that NADPH is consumed in biosynthetic reactions, whereas NADH is generated in energy-yielding reactions.

3. **Dehydration:** In the third step, the elements of water are removed from C-2 and C-3 of d-β-hydroxybutyryl-ACP to yield a double bond in the product, **trans-Δ^2-butenoyl-ACP** (also called **crotonyl-ACP**). The enzyme that catalyzes this dehydration is β-hydroxyacyl-ACP dehydratase.

4. **Reduction of the double bond:** Finally, the double bond of *trans*-Δ^2-butenoyl-ACP is reduced (or saturated) to form **butyryl-ACP** by the enzymatic action of *enoyl-ACP reductase*; again NADPH is the electron donor or the reductant. Note that FAD+ is the oxidant in the corresponding reaction in β-oxidation.

These 4 reactions, taken together, complete the first round of elongation cycle. Thus, after the first round of elongation, the C4 (butyryl) precursor of palmitate has been synthesised from a C2 (acetyl) and a C3 (malonyl) unit, with the acetyl group constituting the two terminal carbons of the growing fatty acid chain (C15 and C16 in palmitate, for example).

The general sequence of condensation and reduction by fatty acid synthase may be schematically represented as :

$$\text{Acyl group} \xrightarrow[-CO_2]{+\text{ malonyl group}} \text{3-ketoacyl group} \xrightarrow{+\text{NADPH}} \text{3-hydroxyacyl group} \xrightarrow[-H_2O]{} \text{Enoyl group} \xrightarrow{+\text{NADPH}} \text{Acyl group}$$

n carbons ←—————— (n+2) carbons ——————→

In **successive rounds,** the production of C-4 saturated fatty acyl-ACP (*i.e.*, a C4-butyryl-ACP) completes one round through the fatty acid synthase complex in fatty acid synthesis. During the second round of elongation phase, the butyryl group is now transferred from the phosphopantetheine —SH group of ACP to cysteine —SH group of β -ketoacyl-ACP synthase (KS). To start the next cycle of 4 reactions, that lengthens the chain by 2 more carbons, another malonyl group is linked to the now vacant phosphopantetheine —SH group of ACP. Condensation occurs as the butyryl group, acting exactly as did the acetyl group in the first round, is linked to two carbons of the malonyl-ACP with simultaneous release of a mole of CO_2. The product of this condensation is a C-6 acyl group, covalently bound to the phosphopantetheine —SH group of ACP (*i.e.*, a **C6-β- ketoacyl-ACP**). Its β-keto group is reduced in the next 3 reactions of the second round of synthesis cycle to yield the C-6 saturated fatty acyl-ACP (*i.e.*, a **C6-fatty acyl-ATP**), exactly as in the first round of reactions.

The C6-fatty acyl-ACP is now ready for a third round of elongation. Seven such cycles of condensation and reduction produce the **C-16 saturated palmitoyl group,** still bound to ACP. This intermediate is not a substrate for the condensing enzyme, β-ketoacyl-ACP synthase (KS) and the chain elongation generally stops at this point. Rather, it is hydrolysed to yield **palmitate** and ACP.

The overall reaction for the synthesis of palmitate from acetyl-CoA can be broken down into 2 parts :

A. *The formation of 7 malonyl-CoA molecules*

7 Acetyl-CoA + 7 CO_2 + 7ATP →7 Malonyl-CoA + 7 ADP + 7 P*i* ...(*i*)

B. *The 7 cycles of condensation and reduction*

Acetyl-CoA + 7 Malonyl–CoA + 14 NADPH + 14 H^+

Palmitate + 7 CO_2 + 8 CoA + 14 NADP+ + 6 H_2O ...(*ii*)

Hence, the overall process (the sum of Equations *i* and *ii*) for the synthesis of palmitate is :

8 Acetyl-CoA + 7 ATP + 14 NADPH + 14 H^+

Palmitate + 8 CoA + 7 ADP + 7 P*i* + 14 NADP+ + 6 H_2O ...(*iii*)

Note that the CO_2 utilised (formation of malonyl-CoA) and the CO_2 produced (condensation reaction) cancel each other when the overall stoichiometry is tabulated.

The biosynthesis of fatty acids such as palmitate, thus, requires acetyl-CoA and the input of chemical energy in 2 forms: the group transfer potential of ATP and the reducing power of NADPH. The ATP is required to attach CO_2 to acetyl-CoA to produce malonyl-CoA; the NADPH is required to reduce the double bonds to form the corresponding saturated fatty acyl group.

3. AMINO ACID AND PROTEIN METABOLISM

Studies on insect proteins are hampered by three inherent complications. Firstly, insects have various physiological adaptations in their organ system such that even proteins produced in homologous organs of closely related species differ strikingly in patterns and biochemical properties. Secondly, the response of hormones to the same tissue or organs in different insects can be totally different. Thirdly, in holometabolous insects, drastic morphological changes take place during transformation from larva to pupa and from pupa to adult.

During embryogenesis the amino acids required for protein synthesis are derived from degradation of yolk reserves. In the developing eggs of beetles and housefly, protein synthesis has been reported to begin at the cleavage stage and increase rapidly at the time of blastoderm formation. It seems that protein synthesis at the early cleavage stage use pre-existing messages and ribosomal RNAs that are transferred to the oocyte from the nurse cells. Growth is the predominant phenomenon during larval stages and there is an increase in total protein content. The protein concentration in the haemolymph also increases, the increase being rapid during the later half of larval development, falls at metamorphosis and declines to zero in early adult life.

Early in the last instar, the fat body synthesises and exports proteins into the blood, while later in the instar, the proteins are sequestered from the blood into the fat body. In diapausing pupae, protein synthesis remains at this low level. The similarity between the electrophoretic patterns of proteins from the fat body and those from the blood provides strong evidence that fat body is the site of synthesis of many of the blood proteins. Some of the blood proteins have been shown to possess enzymatic properties, while others are thought to play a role in the transportation of various types of compounds. Ecdysone and juvenile hormone both stimulate the rate of protein synthesis by larval fat body. Proteins synthesis in fat body from diapausing pupae is stimulated after injury to the pupae.

3.1 Steps in protein synthesis

Transcription is one of the first processes in the mechanism of protein synthesis (Box 10.3). In transcription, a complementary strand of mRNA is synthesised according to the nitrogenous base code of DNA. To begin, the enzyme RNA polymerase binds to an area of one of the DNA molecules in the double helix. (During transcription, only one DNA strand serves as a template for RNA synthesis. The other DNA strand remains dormant.) The enzyme moves along the DNA strand and reads the nucleotides one by one. Similar to the process of DNA replication, the new nucleic acid strand elongates in a 5'-3' direction. The enzyme selects complementary bases from available nucleotides and positions them in an mRNA molecule according to the principle of complementary base pairing. The chain of mRNA lengthens until a "stop" message is received.

The nucleotides of the DNA strands are read in groups of three. Each group is a **codon**. Thus, a codon may be CGA, or TTA, or GCT, or any other combination of the four bases, depending on the codon's complementary sequence in the DNA strand. Each codon will later serve as a **code word** for an amino acid. First, however, the codons are transcribed to the mRNA molecule. Thus, the mRNA molecule consists of nothing more than a series of codons received from the genetic message in the DNA. After the **stop** codon is reached, the synthesis of the mRNA comes to an end. The mRNA molecule leaves the DNA molecule, and the DNA molecule rewinds to form a double helix. Meanwhile, the mRNA molecule passes through a pore in the nucleus and proceeds into the cellular cytoplasm, where it moves toward the ribosomes located in the cytoplasm or on the rough endoplasmic reticulum.

The genetic code is transferred to an amino acid sequence in a protein through the translation process, which begins with the arrival of the mRNA molecule at the ribosome. While the mRNA is being synthesised, tRNA molecules are uniting with their specific amino acids according to the activity of specific enzymes. The tRNA molecules then begin transporting their amino acids to the ribosomes to meet the mRNA molecule. After it arrives at the ribosomes, the mRNA molecule exposes its bases in sets of three, the codons. Each codon has a complementary codon called an anticodon on a tRNA molecule. When the codon of the mRNA molecule complements the anticodon on the tRNA molecule, the latter places the particular amino acid in that position. Then the next codon of the mRNA is exposed, and the complementary anticodon of a tRNA molecule matches with it. The amino acid carried by the second tRNA molecule is positioned next to the first amino acid, and the two are linked. At this point, the tRNA molecules release their amino acids and return to the cytoplasm to link up with new molecules of amino acid. When it's time for the next amino acid to be positioned in the growing protein, a new codon on the mRNA molecule is exposed, and the complementary three-base anticodon of a tRNA molecule positions itself opposite the codon. This brings another amino acid into position, and that amino acid links to the previous amino acids. The ribosome moves farther down the mRNA molecule and exposes another codon, which attracts another tRNA molecule with its anticodon.

One by one, amino acids are added to the growing chain until the ribosome has moved down to the end of the mRNA molecule. Because of the specificity of tRNA molecules for their individual amino acids, and because of the base pairing between codons and anticodons, the sequence of codons on the mRNA molecule determines the sequence of amino acids in the protein being constructed. And because the codon sequence of the mRNA complements the codon sequence of the DNA, the DNA molecule ultimately directs the amino acid sequencing in proteins. The primary start codon on an mRNA molecule is AUG, which codes for the amino acid methionine. Therefore, each mRNA transcript begins with the AUG codon, and the resulting polypeptide begins with methionine.

Box 10.3 Steps in protein synthesis

1. mRNA transcribed from DNA
2. mRNA exits through nuclear pore to cytoplasm/ribosomes
3. tRNA binds to amino acids
4. Loaded tRNA bonds mRNA at ribosomes
5. Polypeptide chain grows in response to mRNA codons
6. Completed protein either used by cell or is packaged and exported.

After the protein has been synthesised completely, it is removed from the ribosome for further processing and to perform its function. For example, the protein may be stored in the Golgi apparatus before being released by the cell or it may be stored in the lysosome as a digestive enzyme. Also, a protein may be used in the cell as a structural component, or it may be released as a hormone, such as insulin. After synthesis, the mRNA molecule breaks up and the nucleotides return to the nucleus. The tRNA molecules return to the cytoplasm to unite with other molecules of amino acids, and the ribosome awaits the arrival of a new mRNA molecule.

3.2 Aminoacid metabolism in general

Amino acids, derived largely from protein in the diet or from degradation of intracellular proteins, are the final class of biomolecules whose oxidation makes a significant contribution to the generation of metabolic energy. The fraction of metabolic energy derived from amino acids varies greatly with the type of organism and with the metabolic situation in which an organism finds itself. Carnivores, immediately following a meal, may obtain up to 90% of their energy requirements from amino acid oxidation. Herbivores may obtain only a small fraction of their energy needs from this source.

The α-amino groups of the 20 L-amino acids commonly found in proteins are removed during the oxidative degradation of the amino acids. If not reused for synthesis of new amino acids or other nitrogenous products, these amino groups are channelled into a single excretory end product. Fig 10.17 provides details of the inter conversion of various amino acids.

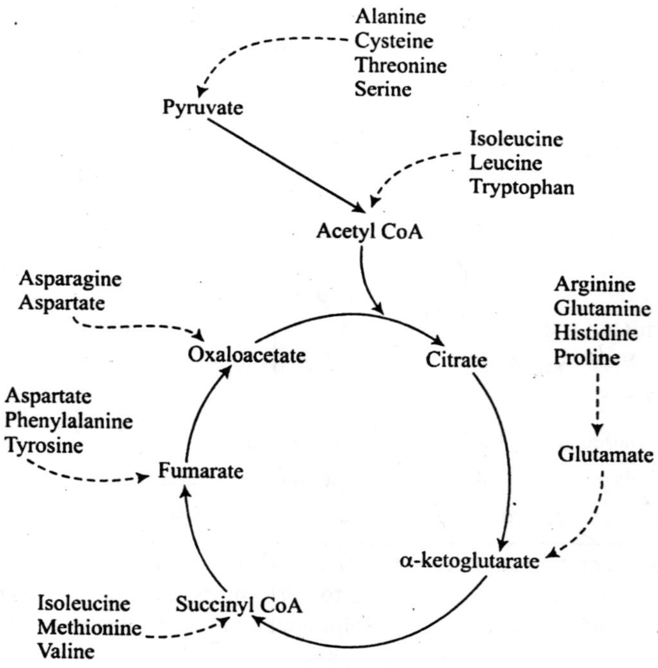

Fig. 10.17. Interconversion of amino acids

The removal of the α-amino groups, the first step in the catabolism of most of the L-amino acids, is promoted by enzymes called **aminotransferases** or **transaminases**. In these **transamination** reactions, the α-amino group is transferred to the α-carbon atom of α-ketoglutarate, leaving behind the corresponding α-keto acid analog of the amino acid. The effect of transamination reactions is to collect the amino groups from many different amino acids in the form of only one, namely, L-glutamate. The glutamate channels amino groups either into biosynthetic pathways or into a final sequence of reactions by which nitrogenous waste products are formed and then excreted.

All aminotransferases share a common prosthetic group and a common reaction mechanism. The prosthetic group is **pyridoxal phosphate** (PLP), the coenzyme form of pyridoxine or vitamin B_6. Pyridoxal phosphate functions as an intermediate carrier of amino groups at the active site of aminotransferases. It undergoes reversible transformations between its aldehyde form, pyridoxal phosphate, which can accept an amino group, and its aminated form, pyridoxamine phosphate, which can donate its amino group to an α-keto acid. Pyridoxal phosphate is generally bound covalently to the enzyme's active site through an imine (Schiff base) linkage to the ε-amino group of a Lys residue.

Amino acids are required for the formation of tissues and enzymes. Usually present as proteins, they form major part of the diet contributing about 30-40% in most synthetic diets. Amino acids have various metabolic functions (Box 10.4). There are ten essential amino acids and the absence of any one of them usually prevents growth. They are arginine, lysine, leucine, isoleucine, tryptophan, histidine, phenylalanine, methionine, valine and threonine. However, there are some variations in their requirement in different insects. For instance, glycine is essential for several dipterous species, alanine for *Blatella* and proline for *Phormia*. It is probable that the balance between different amino acids is particularly important.

Box 10.4 Functions of amino acids

1. Responsible for nearly 30% of total osmotic activity
2. Prominent in many biochemical pathways
   A. Proline: as a metabolic substrate for flight energy
   B. Tyrosine: necessary for cuticular sclerotization
   C. Glutamate: neurotransmitter of motor and muscle
3. Building blocks for protein structure
4. Consists of the vitellogenins
5. Source of storage protein

3.3 Formation of Ammonia from Glutamate

Glutamate is transported from the cytosol to the mitochondria, where it undergoes **oxidative deamination** catalysed by **L-glutamate dehydrogenase** releasing ammonia (Fig10.18). This enzyme, which is present only in the mitochondrial matrix, requires NAD^+ (or $NADP^+$) as the acceptor of the reducing equivalents. The combined action of the aminotransferases and glutamate dehydrogenase is referred to as **transdeamination**. A few amino acids bypass the transdeamination pathway and undergo direct oxidative deamination. Ammonia is quite toxic to animal tissues. In most animals excess ammonia is converted into a nontoxic compounds, urea and uric acid.

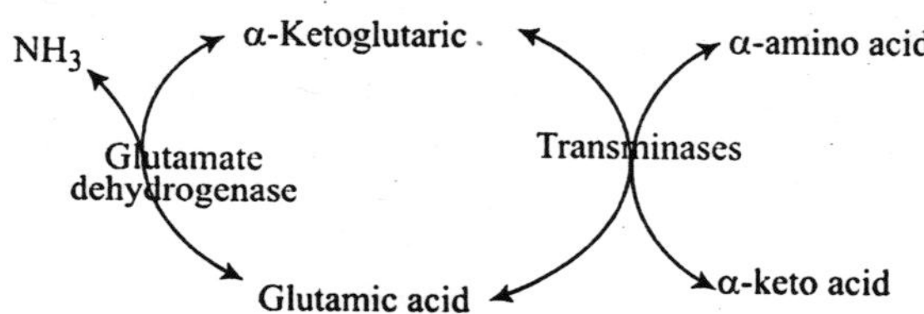

Fig. 10.18. Formation of ammonia from glutamate

3.4 Storage proteins

The term **storage protein** implies uptake from the haemolymph and storage in fat body tissues. They serve as a storage pool for amino acid resources needed later in development. Because of their high content of aromatic amino acids these storage proteins are classified as arylphorins. They are large hexameric proteins composed of 80kDa subunits (Box 10.5).

Box 10.5 Storage Proteins

1. Essential for metamorphosis, reproduction, general body maintenance
2. Storage Hexamerins – MW 500,000
   A. Consists of 6 similar subunits; MW 70,000 – 90,000
   B. Provides the amino acid source for protein synthesis in the development phase that do not feed
   C. As hemocyanins in crustaceans and celicerate arthropods
   D. Synthesised by fat body
   E. Released into hemolymph during the later larval stage
   F. May be recaptured by fat body after metamorphosis.

Different types of hexameric proteins have been described in insects. Most dipteran species possess a protein rich in aromatic amino acids (>15%) and methionine (>4%) and are called as arylphorin. A second storage hexamer common to Diptera has a much lower content of aromatic amino acids and methionine, called LSP-2. Two common lepidopteran storage proteins are rich in either aromatic amino acids or methionine, but not both. The second type of protein is called **methionine rich storage protein**. Based on sequence identities, four distinct groups of storage proteins are known, namely (1) lepidopteran methionine rich proteins, (2) lepidopteran arylphorins, (3) certain lepidopteran juvenile hormone-suppressible proteins and (4) the dipteran storage proteins.

Arylphorins

In Diptera, this protein is synthesised in the fat body of larvae in the penultimate and ultimate larval stage, and secreted into the haemolymph where it accumulates in high concentrations. When the insect begins wandering, arylphorins biosynthesis stops; instead the protein is gradually taken up into pupal fat body, where it is stored in protein granules. The protein granules are most abundant in the posterior region of the fat body. During the development of the pharate adult, much of the arylphorin is proteolytically destroyed and its constituent amino acids can be found in the cuticle.

Lepidopteran arylphorins resemble dipterans. They are synthesised in high rates in the fat body of last instar larvae and released into the haemolymph. Reabsorption into the fat body begins during the prepupal stage.

LSP-2 from Diptera

This is a second type of storage protein known in Diptera. This protein is also taken up into the fat body and is thought to play an important role in adult insects as suggested by the fact that its expression resumes after adult eclosion.

Methionine rich storage proteins from Lepidoptera

These proteins contain more than 4% methionine. Their biosynthesis in the fat body seems to commence later than the arylphorins, towards the end of the last larval stage. Then methionine-rich protein disappears completely from the haemolymph and is taken up by pupal fat body. Its deposition in granules has not been studied. These proteins are much more abundant in female *M.sexta*, than the males.

Juvenile hormone-suppressible proteins

The expression of storage proteins is mostly confined to the last larval instar, a period where juvenile hormone titres are low. It has been suggested that the hormone prevents production of storage proteins, but this is restricted to some, not all storage proteins. These proteins are acid hexamerins as characterised in *T.ni.*

Riboflavin-binding proteins

Riboflavin-binding hexamerins found in *H. cecropia, H. virescens* and *G. mellonella* are not true storage proteins. Although they are expressed only in the last larval stage, they are not actively sequestered by the fat body before pupation and do not accumulate in pupal fat body. Their haemolymph composition diminishes rapidly during the pupal-adult eclosion.

Very high density lipoprotein

A non-hexameric storage protein has been characterised in Helicoverpa zea and other Noctuid species. This protein, a dimeric or tetrameric complex of 150kDa subunits is characterised by two unusual properties: it is coloured blue, due to non-covalently bound biliverdin and it is a very high density lipoprotein (VHDL) with approximately 10% lipid. It is synthesised in the peripheral fat body of last instar larvae, and accumulates in high concentrations in haemolymph, before it is specifically taken up by perivisceral fat body prior to pupation. Within the fat body it can be found in protein granules.

Tyrosine-rich proteins

These proteins are rich with upto 27% tyrosine and found in coleopteran fat body. Their synthesis and accumulation in the fat body occurs shortly before pupation, and these proteins are deposited in the protein granules of pupal fat body. Later in development these granules are partially broken down and are used in the biosynthesis of cuticular structures of pharate adult.

3.5 Glycoprotein and lipoprotein

Glycoproteins are glycosylated molecules. Glycosylation is the covalent attachment of an oligosaccharide chain to a protein backbone and is considered to be a very common protein modification. The structure and size of the carbohydrate chain can be very diverse and can alter the physico-chemical characteristics of a protein. Two major types of glycosylation, referred to as ***N*- and *O*-linked glycosylation**, can be distinguished. *N*-glycans are attached to Asn residues of the peptide backbone while *O*-glycans are connected to Ser or Thr residues. Only in recent years, it has been acknowledged that glycosylation of proteins modulates various processes such as subcellular localisation, protein quality control, cell-cell recognition and cell-matrix binding events. In turn, these important functions control developmental processes such as embryogenesis or organogenesis. However, for specific biological processes relative differences can be found between insects belonging to different orders. For example, the relative amount of glycoproteins associated with transport was 11%, 16%, 15%, 6% and 5% for the glycoproteins derived from *T. castaneum, B. mori, A. mellifera, D. melanogaster and A. pisum*, respectively. Between the highest and the lowest relative amount of glycoproteins for the category transport, a three-fold difference was observed (*A. mellifera* versus *B. mori*). This illustrates a potential differential importance of glycosylation for a particular biological process between insect species belonging to different orders. In addition, it is striking that a large part of the glycoproteins was associated with several metabolic processes (Vandenborre *et al*., 2011).

Lipoproteins mediate most of the lipid transport in the circulation of animals. In mammals, a single protein component, the nonexchangeable apolipoprotein B (apoB), provides the structural basis for the biosynthesis of neutral fat transporting lipoproteins. Interestingly, the major lipoprotein of insects, lipophorin, contains two structural apolipoproteins, because the insect apoB homolog, the precursor apolipophorin-II/I (apoLp-II/I), is cleaved during lipoprotein biosynthesis by the fat body.

In insects, lipophorins (Lps) are the major haemolymphatic lipoproteins, which carry and distribute lipids of many classes among the tissues involved in lipid absorption, storage and utilisation. Lps contain two apolipoproteins, apolipophorin I (apoLp-I, ~250 kDa) and apolipophorin II (apoLp-II, ~80 kDa), which comprise about 60% of the mass of the particle. A third apoprotein (apolipophorin III, apoLp-III), of molecular mass about 17 kDa, is present in adults of some species in a molecular ratio that depends on physiological state of the insects.

Lps have been suggested to function as a shuttle for the transport of most of the lipid classes in the haemolymph. The best-known transport system is that of diacylglycerol. Insects capable of sustained flight use lipids as their main energy source. Lipophorins particles transfer diacylglycerols to flight muscles and can then return to the fat body to be reloaded with diacylglycerol, thus acting as a reusable shuttle.

Insect oogenesis is another physiological process that also requires lipids. This process is characterised by a rapid accumulation of proteins and lipids by the growing oocytes and the production of a great number of eggs in a relatively short period of time. This process is very complex and the involvement of several organs coordinated by hormones has been recognized. In this process, Lp is the main lipoprotein responsible for lipid delivery to oocytes. Evidence for the role of Lp as the major vehicle for lipid transport to the oocytes was first provided in *Philosamia cynthia* (Lepidoptera). In *Manduca sexta* (Lepidoptera), Kawooya and Law (1988) showed that low-density lipophorin unloads its lipids to the ovaries and then returns to the hemolymph. In addition, high-density lipophorin was also found to transfer lipids, resulting in its conversion to very high-density lipophorin, but without a recycling of apoproteins to the haemolymph.

4. ENERGY STORAGE AND RELEASE- ROLE OF FAT BODY

Fat body is a dynamic tissue involved in multiple metabolic functions. One of these functions is to store and release energy in response to the energy demands of the insect. Insects store energy reserves in the form of glycogen and triglycerides in the adipocytes, the main fat body cell. In addition, it also synthesise most of the haemolymph proteins and circulating metabolites. Large amounts of relevant proteins, such as storage proteins used as an amino acid reservoir for morphogenesis, lipophorins responsible for the lipid transport in circulation, or vitellogenins for egg maturation, are secreted by the fat body. Most of the insect's intermediary metabolism takes place in this organ, including lipid and carbohydrate metabolism, protein synthesis, and amino acid and nitrogen metabolism.

The fat body is a relatively large organ distributed throughout the insect body, preferentially underneath the integument and surrounding the gut and reproductive organ. Unlike the solid structure of the liver, the insect fat body is a loose tissue. Generally, the organ is arranged in thin lobes that are bathed by the haemolymph. This type of organization provides maximal exposure to the haemolymph. Ready access to haemolymph is vital for the organism to adjust appropriately to the changes in the concentration of energy precursors in circulation. This is especially critical under conditions of extreme energy demand such as insect flight, in which the metabolic rate increases 50-to 100-fold. Because only small amounts of energy precursors are present in the flight muscle, the energy consumed during flight is provided by circulating energy substrates (trehalose, lipids, and proline), which are replenished by the fat body.

4.1 Fat body cell types

The basic cell of the fat body is the adipocyte, characterised by the presence of numerous lipid droplets. Triglycerides are the major component of the lipid droplets, and at the end of the feeding period lipid droplets occupy most of the intracellular space along with glycogen and protein granules. A less common fat body cell is the urocyte, which is specialized for urate storage. Urocytes have been described in the fat bodies of cockroaches and locusts but are not present in Lepidoptera. In these insects adipocytes themselves acquire the ability to accumulate urate at the end of the larval stage, and the urate contents disappear during development of the adult. A third type of fat body cell is the mycetocyte, found in cockroaches, aphids, and some Hemiptera. These cells contain microorganisms living inside vacuoles in permanent symbiosis with the insect. Symbionts have been speculated to produce essential components that are not incorporated with the diet. In some insects oenocytes can be found distributed among adipocytes. Unlike the fat body, which has a mesodermal origin, oenocytes are derived from the ectoderm. These are specialised cells that can be associated with either epidermal cells or fat body cells whose function is linked with the synthesis of cuticular lipids, proteins, and hydrocarbons.

4.2 Accumulation of energy reserves

Glycogen and triglyceride are the energy reserves in animal cells. Glucose is stored in a polymeric form, glycogen, and can be readily degraded on demand to be used as a glycolytic fuel. Fatty acids stored as triglyceride can be used for energy production through β-oxidation. Triglyceride is stored in an anhydrous form, whereas glycogen is stored in a bulky hydrated form. Triglycerides also have a higher caloric content per unit of weight than glycogen, and provide a useful source of water upon oxidation, yielding almost two times more metabolic water than glycogen. These considerations have direct implications on energy metabolism of insects. Fat reserves are the most important reserve used by insects to meet their energy demand during diapause, to provide energy for the developing embryo, and to fuel prolonged periods of flight. Fat body adipocytes are cells with great plasticity, able to store large amounts of triglycerides. Diapausing insects accumulate lipids. Extreme lipid deposition can cause fat body hypertrophy, and this is a mechanism that enhances overwintering survival in *Culex pipiens* females. Males do not enter diapause and do not survive the winter. Prior to diapause adult females accumulate twice the lipid reserves of their non-diapausing counterparts. Fat body lipid accumulation is built entirely by feeding on plant juices rich in carbohydrates. These individuals feed more frequently prior to diapause than nondiapausing females. Diapausing females do not take blood meals, and the molecular machinery to digest the blood meal is not present (Arrese and Soulages, 2010).

4.3 Mobilisation of carbohydrate reserves

Glycogen is mobilised for use by other tissues, mostly in the form of trehalose. The utilisation of glycogen depends on the activity of glycogen phosphorylase, which provides glucosyl residues for trehalose synthesis (Thompson, 2003). Fat body phosphorylase activity increases during larval development. An increase in activity prior to pupation is linked to the need for energy and glucose for chitin synthesis. Likewise, pupal-adult development is marked by high phosphorylase activity. Different insects exhibit different patterns of phosphorylase activity during the pupal period. In some insects the increase in phosphorylase activity is observed at the beginning of the period, whereas in others phosphorylase activity increases toward the end of the pupal stage. Apart from the use of trehalose for maintenance of energy metabolism during fasting or nonfeeding periods, trehalose is a substrate for insect flight in general. Long-term flyers, such as locusts and mosquitoes, subjected to several hours of flight, start flying using trehalose and after some time switch to lipids. Short-term flyers such as the cockroach *Periplaneta americana* use mostly trehalose. On the other hand, the Colorado potato beetle (*Leptinotarsa decemlineata*), which oxidizes proline to fuel flight, also needs a concomitant utilization of glucose to produce pyruvate and to allow the proline-alanine cycle to occur (Gade and Auerswald, 2002). A particular group of insects is represented by the bees, which exclusively utilise carbohydrates to power flight (Suarez *et al.*, 2005). These insects support flight by using sugars stored in the crop rather than reserves from the fat body.

4.4 Mobilisation of lipid reserves

The first required step of fatty acid mobilisation consists in the action of fat body triglyceride lipases to catalyse the hydrolysis of the triglyceride molecules contained in the lipid droplets. Two lipases that are expressed in the fat body have been identified so far: insect adipose triglyceride lipase (ATGL), or Brummer lipase (Gronke *et al.*, 2007), and triglyceride lipase (TGL) (Arrese *et al.*, 2006). TGL is the main lipase in *M. sexta*.

4.4.1 Utilisation of Lipids

Lipids are used for direct support of flight and for the synthesis of trehalose and proline, as well as during starvation, embryogenesis, and the immune response.

Flight

Direct utilisation of fat reserves to support flight is in general required in long-term flying insects. Studies using half-thorax preparations of *Locusta migratoria* showed the ability of electrically stimulated flight muscles to metabolise lipids and carbohydrates. Fat body lipids are commonly secreted into the haemolymph as diacylglycerol, which is transported to the tissues by the insect lipoprotein, lipophorin.

Synthesis of trehalose and proline

The synthesis of trehalose in the fat body is an energy-dependent process. Studies in two cockroach species have shown that synthesis and release of trehalose is coupled to fatty acid oxidation. In fact, inhibition of β- oxidation prevents AKH-induced release of trehalose. These studies suggest that at least in some insect's trehalose synthesis is dependent on fatty acid mobilisation and oxidation.

Proline is an abundant substrate found in the haemolymph of most insect species. It is synthesised in the fat body from acetyl-CoA and alanine and then released into the hemolymph. Studies based on, the tsetse fly, *Glossina morsitans*, and the Colorado potato beetle, *Leptinotarsa decemlineata*, that use proline as the main fuel to power flight indicate that the synthesis of proline requires fatty acid mobilisation. The exceptional ability of these insects to use proline resides in the properties of their muscle mitochondria, which have the enzymes required to oxidise proline as well as a low capability to oxidize fatty acids and pyruvate. Utilisation of proline during flight is accompanied by an increase in the concentration of alanine in haemolymph. Alanine is reused for the synthesis of proline in the fat body and is considered a shuttle for the transport of acetate units from the fat body to muscles. Several studies have shown that fatty acids stored in the fat body provide the acetate units required for proline synthesis (Bursell, 1981). Furthermore, flight and AKH promote a parallel increase in lipase activity and proline synthesis in beetles, and extracts of corpora cardiaca induce both lipolysis and proline synthesis in tsetse fly fat body. The haemolymph lipids also increase during flight in the Colorado potato beetle. Overall, the studies of insects whose flight muscles have a high selectivity for proline oxidation indicate that the mobilisation of triglyceride depots is central and tightly linked to the use of proline.

Starvation

Starvation is another condition in which lipid mobilisation is observed. For instance, starvation of *L. migratoria* is characterised by a threefold increase in haemolymph lipid concentration due to a prominent elevation of diglycerides (Beenakkers *et al*., 1985). A similar situation is observed in *M. sexta*, in which starvation induces a twofold increase in haemolymph lipid concentration (Ziegler, 1991). Unlike flight-induced lipid mobilisation, the increase of haemolymph diglyceride during starvation is independent of

AKH. The signal that promotes activation of lipolysis during starvation remains unknown. However, in both insects, injection of trehalose reduces the lipid concentration in haemolymph, implying that an inverse relationship exists between the haemolymph concentration of lipid and trehalose. Trehalose has an inhibitory effect on the lipolytic activity of the fat body as judged by the decrease of *M. sexta* fat body lipase activity and the production of diglyceride by the locust fat body

Embryogenesis

Lipids comprise 30–40% of the dry weight of insect oocytes, and they are the main source of energy for the developing embryo. During their maturation, insect oocytes increase their lipid content several fold in a short period, usually one or two days. Although oocytes are able to synthesise fatty acids de novo, this contribution does not account for more than 1% of the egg lipid content. The vast majority of the lipid accumulated in oocytes originates in the fat body and is transported to the ovaries by lipophorin. The accumulation of lipid in the ovaries is concomitant to a massive reduction in the lipid content of the fat body. Consistent with the significance of lipophorin in oogenesis, expression of apolipoprotein genes and lipoprotein receptors increases during vitellogenesis. Ovaries take up some of the lipids by receptor-mediated endocytosis of the lipoprotein particle. This mechanism is evident in the appearance of the protein components of lipophorin inside the growing oocytes.

Immune response

Lipids are also mobilised to the haemolymph in response to immune challenge. A study of mosquito-parasite interactions (*Plasmodium falciparum* and *Anopheles gambiae*) confirmed that infected field mosquitoes present an increased transcriptional level of lipophorin proteins.

4.5 Hormonal Regulation of Fat body energy metabolism

The mobilisation of fat body energy reserves is essentially regulated by adipokinetic hormone (AKH), which is produced by the corpora cardiaca. The most widely recognised action of the AKH family of peptides is their stimulation of the fat body to convert stored glycogen and triglyceride into haemolymph trehalose and diglyceride, respectively. In some insects AKH also stimulates the synthesis of proline.

AKH mobilises glycogen through the activation of glycogen phosphorylase during the larval stages and promotes a lipolytic response in the adult. The lipolytic response induced by AKH in *M. sexta* involves a rapid increase in Ca^{2+} influx and an increase in the intracellular concentration of cAMP, which leads to the activation of cAMP-dependent protein kinase A (PKA).

Mobilisation of energy reserves is also involved in the acute stress response of insects. This response is controlled by the neurohormone octopamine. The action of octopamine is mediated by G-protein-coupled receptors that are coupled to either an increase or a decrease of the intracellular level of cAMP, or to the generation of intracellular calcium signals. Octopamine stimulates lipid mobilisation in several insects. In *M. sexta* it stimulates glycogen phosphorylase in the larval stage, but the response is moderate compared with that elicited by AKH.

Large amounts of energy are needed to support flight. The release of hormones during flight has been studied best in locusts and an integrated relationship between octopamine and AKH during flight has been shown. The increase of haemolymph diglyceride has a biphasic pattern, with aplateau between 10 and 20 min of flight. During the first few minutes of flight octopamine is released, inducing the first release of diglyceride from the fat body. The subsequent, more prolonged phase of lipid elevation is produced by the release of AKHs, whose titers increase sharply after 15 min of flight. Some studies suggest that octopamine elicits the secretion of AKH by the corpora cardiaca during flight (Orchard *et al.*, 1993). However, incubation of corpora cardiaca with octopamine failed to induce AKH secretion. The release of AKH could be triggered by the decrease in haemolymph trehalose that occurs soon after the initiation of flight. In Drosophila a decrease in extracellular trehalose concentration stimulated the secretion of AKH from the corpora cardiaca (Kim and Rulifson, 2004). The available information suggests that the levels of haemolymph trehalose and/or glucose play a direct role in the secretion of AKH by the corpora cardiaca.

5. METABOLIC REGULATION IN INSECT COLD HARDINESS

Numerous insect species have strategies that allow them to avoid exposure to winter temperatures below 0 °C (Box 10.6); for example, honey bees thermoregulate to keep their hives warm, various butterflies and dragonflies migrate south, some insect species spend the winter as aquatic larvae in thermally buffered water (often 0–4 °C), and others burrow underground below the frost line (Storey and Storey, 2010). For those that need to endure subzero temperatures in terrestrial environments, two main forms of cold hardiness have traditionally been defined: **freeze avoidance** and **freeze tolerance. Freeze avoidance** involves deep supercooling of body fluids to temperatures well below those normally encountered in the microenvironment of overwintering insects. This is potentiated by primary adaptations including the synthesis of antifreeze proteins (AFPs) and the accumulation of extremely high levels of carbohydrate cryoprotectants (most often glycerol), as well as other features such as partial dehydration of body fluids and the use of water impermeable cocoons. **Freeze tolerance** involves the regulated freezing of up to about 65% of total body water in extracellular spaces. Ice formation is often triggered by the action of specific ice nucleating agents or proteins (INAs, INPs), whereas low molecular weight cryoprotectants are used to maintain a liquid intracellular space and protect membrane bilayer structure in shrunken cells. More recently, a third category of cold hardiness— **cryoprotective dehydration**—was defined for a number of soil invertebrates including a collembolan. This strategy combines extreme dehydration (losing virtually all freezable water) with high cryoprotectant levels that stabilise macromolecules. The cryoprotective dehydration strategy has the advantage of bringing remaining body fluids into vapour pressure equilibrium with ice (Elnitsky *et al*., 2008). Finally, a fourth option for winter survival by an insect has recently been reported for the Alaskan bark beetle (*Cucujus clavipes puniceus* Mannerheim) (Sformo *et al.* 2010). Although previously known to supercool to at least –40 °C, Sformo *et al.* (2010) used differential scanning calorimetry to show that at lower temperatures the larvae do not freeze but transition into a **vitrified state** and can survive temperatures as low as –100 °C. Vitrification is favoured by extensive dehydration (to about 20% of summer values) and accumulation of AFPs and high concentrations of polyols. Briefly, both freeze-avoiding and freeze-tolerant species accumulate polyol cryoprotectants; in freeze-avoiding species polyols permit colligative suppression of supercooling point to prevent body freezing, whereas in freeze-tolerant species polyols offer protection against intracellular freezing when ice accumulates in extracellular compartments. In cryoprotective dehydration, the disaccharide trehalose plays the major role in macromolecular protection. AFPs are critical to supercooling point suppression in freeze-avoiding species but can also be found in freeze-tolerant species where they appear to function in inhibition of ice recrystallisation.

Box 10.6 Most tolerant of Cold

In laboratory test, Hinton (1960) found that dehydrated larvae of the African chironomid *Polypedilum vanderplanki* Hinton (Diptera) could survive submersion in liquid helium (-270 °C). This phenomenon seems related to its ability to tolerate extreme desiccation. The larvae feed in small pools on unshaded rocks in northern Nigeria and Uganda where they withstand an environment which is alternately dry and flooded. It is the only insect definitely known to endure cryptobiosis and survive drying to <3% moisture.

5.5 Control of Polyol Metabolism

Glycerol is by far the most common protectant but pairing of two polyols (often glycerol and sorbitol) occurs frequently and some species have multiple components in their protectant mix (glucose, trehalose, sorbitol, mannitol, erythritol) and several minor components. Trehalose as a minor component of the cryoprotectant mix is widespread and contributes to stabilisation of the lipid bilayer of membranes particularly during cell volume shrinkage in freeze-tolerant animals (Crowe *et al*., 1992). During cryoprotective dehydration, extremely high levels of trehalose effectively replace water in the protection of cellular macromolecules. Cryoprotectant synthesis is supported by the accumulation of high concentrations of glycogen in cells (mainly fat body) during summer–autumn feeding and enhanced activities of enzymes in the biosynthetic pathways, particularly the sugar phosphatases and polyol dehydrogenase that are needed for the two-step conversion of intermediates of glycolysis or the pentose phosphate cycle into polyols. Similarly, polyol catabolism at winter's end begins when the catabolic enzymes appear, particularly glycerol or glyceraldehyde kinase and sorbitol dehydrogenase.

Cryoprotectant synthesis involves a massive conversion of glycogen to polyols or sugars. Such a massive undertaking needs close regulation both during autumn synthesis and spring degradation. Cold activation of glycogen phosphorylase (GP) is well known to initiate cryoprotectant synthesis. The underlying mechanism is phosphorylation to convert inactive *GP b* to the active *GP a* form. This was one of the earliest demonstrations of the use of reversible protein phosphorylation (RPP) to mediate a stress response (Ziegler *et al*., 1979). However, GP activation cannot determine the fate of the glucose-1-phosphate units that are cleaved off glycogen and so controls on other enzymes are needed.

Because of the high demand for NADPH-reducing power for the synthesis of polyols, almost all carbon must be pushed through the pentose phosphate pathway during the production of polyols resulting in a close match between the outputs of the cycle: NADPH and various sugar phosphate precursors—glyceraldehyde-3-phosphate (GAP) in the case of glycerol synthesis (Storey and Storey, 1991). Regulation is needed to control glycolytic vs. pentose phosphate cycle flux, prevent GAP (or other intermediates) recycling back to glycogen, and promote production of the polyol product.

Previous studies have shown that dephosphorylation mediated inhibition of the gluconeogenic enzyme fructose- 1,6-bisphosphatase (FBPase) plus phosphorylation inhibition of glycogen synthase (GS) are key to preventing GAP recycling back into glycogen. Temperature-dependent control of phosphofructokinase (PFK) that determines the synthesis of glycerol vs. Sorbitol in *E. solidaginis* (Storey, 1982) is probably also due to RPP control, as also occurs for this enzyme in *B. mori*. These enzymes (GP, GS, FBPase, PFK) have long been known to be regulated by RPP. There is also good evidence that glucose-6 phosphate dehydrogenase (G6PDH), the initial enzyme of the pentose phosphate pathway, is also regulated by RPP in *E. scudderiana*.

5.6 Metabolic rate depression

To survive in severe environmental stress such as extreme heat or cold, oxygen deprivation, water limitation, food scarcity, organisms often resort to hypometabolic states. Strong metabolic rate depression underlies hypometabolism in multiple forms including diapause, dauer, hibernation, estivation, anaerobiosis, and anhydrobiosis. Diapause is widely used among insects and may occur at one or more developmental stage in each species. They may be facultative or obligate depending on the species and its poly, bi-, or uni-voltine status. The metabolic rate of diapausing *E. solidaginis* larvae in midwinter was just 35%–40% of the early-autumn value at the same temperature, confirming the occurrence of metabolic rate suppression (Irwin and Lee, 2002; Levin *et al.*, 2003).

Metabolic suppression in overwintering insects could come from several sources. Cessation of feeding saves energy expenditures on digestion, may also cause gut regression, and would reduce or halt biosynthesis of storage fuels (e.g., glycogen, triglycerides). The diapause program of developmental arrest coordinates the shutdown of major energy expenditures associated with cell division and tissue differentiation and growth. Low environmental temperatures directly suppress the rates of all metabolic reactions but, as a regulatory mechanism, temperature itself is likely not very effective for accomplishing differential regulation of metabolic functions. However, temperature (and (or) thermoperiod) is clearly a triggering mechanism for various features related to cold hardiness including stimulating cryoprotectant synthesis and degradation. Cold exposure and diapause also interact in upregulating aquaporins to permit water and glycerol transport. Central to this signaling is RPP (reversible protein phosphorylation) that manipulates responses by protein kinases, protein phosphatases, target proteins and transcription factors to control and coordinate both intermediary metabolism and the expression of genes. Although coarse controls on protein synthesis or degradation can be associated with hypometabolism, for example, leading to altered patterns of enzymes of polyol metabolism, or cold induction of desaturase enzymes to restructure lipids for function in the cold (Kayukawa *et al.*, 2007), the transition into a hypometabolic state primarily involves reversible controls that can provide both global suppression of metabolic functions and differential regulation of cell processes.

A prime example of RPP-mediated metabolic suppression in cold-hardy insects is the control of ion-motive ATPases. The plasma membrane Na^+/K^+-ATPase is often the greatest single consumer of ATP in cells and hence, a major reduction in cellular ATP turnover can be achieved by the coordinated suppression of Na^+/K^+ ATPase ion pumps and the oppositely directed ion channels that allow facilitated diffusion of Na^+ and K^+ ions down their concentration gradients.

5.7 AMPK and metabolic poise over the winter

Cold-hardy species go through major metabolic transitions as they acquire cold hardiness. Metabolism shifts from an anabolic focus over the summer and early autumn (eating, growing, storing fuels) to both a catabolic focus (cryoprotectant synthesis, consumption of stored fuel reserves for energy and a hypometabolic focus (energy-saving mode associated with diapause entry or triggered by freezing, hypoxia, etc.) over the late autumn and winter. Multiple signalling pathways are undoubtedly involved. In freeze-tolerant *E. solidaginis* larvae, for example, signalling via the hypoxiainducible transcription factor (HIF-1) is clearly one of the mechanisms involved, responding to oxygen restriction in the frozen state (Morin *et al.* 2005). The alpha subunit of HIF-1 is stabilised under low oxygen conditions, whereas under normoxic conditions HIF-1a is rapidly destroyed after oxygen mediated hydroxylation of key proline residues (Semenza, 2007).

Eurosta solidagin is larvae responded to anoxia exposure, as expected, with a strong 2.7-fold increase in hif- 1a transcript levels. Cold exposure also had the same effect, transcript levels increased by 1.6-fold in response to an acute temperature decrease from 15 to 3 °C and remained high when larvae were frozen at –16 °C. HIF-1a protein level also increased sharply by 2.3-fold when larvae were transferred from 15° to 3 °C. In addition, analysis of HIF-1a protein seasonal patterns showed a progressive increase over the winter months to peak in February. Hence, the data suggest a significant role for this transcription factor in mediating metabolic responses to freeze-induced ischemia or hypoxia. HIF- 1 mediates responses to oxygen restriction, whereas other protein kinases and transcription factors mediate responses to other limiting factors including low energy and low nutritents. The AMP-activated protein kinase (AMPK) is proving to be a major regulator of catabolic vs. anabolic poise in cells, its actions favouring the former and inhibiting the latter. AMPK was first discovered as a protein kinase that was allosterically activated by AMP accumulation under low energy conditions (e.g., hypoxia) and it is often called the **energy sensor** or the **fuel gauge** of the cell (Hardie, 2007; Hue and Rider, 2007). In addition, AMPK is now known to respond to various upstream kinases that allow other signals to have input on AMPK targets. The best known action of AMPK is phosphorylation and inactivation of acetyl-CoA carboxylase (ACC), which inhibits lipogenesis and promotes fatty acid oxidation under energy-limiting conditions. AMPK activation also exerts inhibitory control over carbohydrate storage (by inhibiting glycogen synthase) and protein synthesis (by activating the protein kinase that inactivates the ribosomal eukaryotic elongation factor-2 (eEF2).

5.8 Mitochondrial metabolism

Cold temperatures, cessation of growth, and entry into diapause or quiescence by cold-hardy insects would all reduce the demand on mitochondrial oxygen-based ATP production over the winter months. Furthermore, as evidenced by glycolytic end product (lactate, alanine) accumulation during freezing, freeze-tolerant species rely on anaerobic metabolism while frozen and so have a reduced need for mitochondria over much of the winter (Storey and Storey 1985). By contrast, there is no evidence of a disruption of aerobic metabolism in freeze-avoiding species. Hence, mitochondrial capacity could be reduced in winter and this could have at least two beneficial functions: (i) reduced production of potentially damaging reactive oxygen species (ROS) because superoxide radicals made by complex I of the electron transport chain are the major source of ROS in most cells, and (ii) in freeze-tolerant species, reduced potential damage to mitochondrial membranes and macromolecules during freezing episodes as the result of freeze concentration of intracellular and intra mitochondrial fluids. Indeed, activities of multiple mitochondrial enzymes are suppressed over the winter months in both freeze avoiding *E. scudderiana* and freeze-tolerant *E. solidaginis*, except for increased activities of fatty acid oxidizing enzymes in *E. scudderiana*. Activities of citrate synthase, NAD-isocitrate dehydrogenase and glutamate dehydrogenase were all reduced by about 50% between about November and March (compared with September) in both species and rose again in the spring (Joanisse and Storey, 1994).

Lowered mitochondrial enzyme activities could come from metabolic regulation or from a decrease in total mitochondrial numbers over the winter months. Accumulating evidence suggests that it is the former for freeze-avoiding and the latter for freeze-tolerant insects.

6. FUELS FOR FLIGHT AND CONTROL OF FLIGHT METABOLISM

Many insects have developed the ability to fly over various distances and for different durations. In some insects, flight plays an important role in their life cycle. Locusts, for example, perform long-distance migratory flights to locate food sources and breeding sites. During these migrations they form large swarms which cause enormous damage to crops. Other insects migrate on a smaller scale. One example is the African fruit beetle, *Pachnoda sinuata*. Its larvae live underground and after the adult moult, the beetles have to fly to their food plants. Although these beetles do not migrate in the true sense, their ability to fly has aided their distribution. Regardless of its duration, lift generating flight is very energy demanding. The high energy demands during flight are met through oxidation of lipids, carbohydrates and the amino acid proline. While carbohydrates can be used in combination with lipids or with proline, no case has yet been reported for a oxidation of lipids in combination with proline. The insect with the highest reliably measured airspeeds are desert locusts (*Schistocerca gregaria*) and corn earworm moths (*Helicoverpa zea*) (Box 10.7).

Box 10.7 Fastest Flyer

The insect with the highest reliably measured airspeeds are desert locusts (*Schistocerca gregaria*) and corn earworm moths (*Helicoverpa zea*). These fly at average airspeeds of 33 and 28 km/h respectively (about 21 and 17 mph). Many insects surely fly faster, but their airspeeds have yet to be studied with modern methods. The highest sustained ground speed recorded is that of the black cutworm (*Agrotis ipsilon*) which flies at speeds of between 97 and 113 km/h (60-70 mph) (Showers & Sappington 1992). Insect air speed is affected by mass, size, age, gender, feeding, water content, activity type, temperature, humidity, solar radiation, wind, oxygen level, ascent angle and even habitat isolation.

6.1 Energy substrates for flight

Lipids are the major fuel for long-term flights in insects such as *Locusta migratoria* (Jutsum and Goldsworthy, 1976) and the moth *Manduca sexta* (Ziegler and Schulz, 1986). They are also the energy substrate in insects which do not feed as adults, such as the moth *Philosamia cynthia* (Beenakkers, 1969). In both long-term flight and nonfeeding, the storage and use of lipids is advantageous, because lipids are the least oxidised of all possible energy substrates and, in contrast to glycogen, they can be stored without being hydrated. Lipids are mainly stored in the fat body as triglycerides (Wheeler, 1989). In a multi-step procedure, these triglycerides are converted into diglycerides during flight and released into the haemolymph where they are loaded onto a lipoprotein carrier. When arriving at the flight muscles, the diglycerides are released from the carrier and hydrolysed. The resulting fatty acids can be oxidised after a carrier (carnitine) has transported them across the mitochondrial membrane.

The use of carbohydrates as a fuel for flight is widespread amongst the insects. Many hymenopterans and dipterans are well known for their exclusive use of carbohydrates to power flight (Sacktor, 1975). In locusts (Jutsum and Goldsworthy, 1976) and certain moths, for example *Manduca sexta* (Ziegler and Schulz, 1986a, b), carbohydrates are oxidised during short-term flight and in the initial stages of long-term flight. Later, however, these insects switch to the use of lipids, and the energy yield from carbohydrate oxidation becomes minor. Carbohydrates are stored as glycogen in the flight muscles and the fat body. The breakdown of fat body glycogen during flight is catalysed by glycogen phosphorylase, which shortens the glycogen chains by single glucose units, resulting in the formation of glucose-I-phosphate. The latter is converted via several steps to trehalose (Candy and Kilby, 1961) and released into the haemolymph. Trehalose is subsequently converted into two glucose units by the action of the enzyme trehalase (Sacktor, 1975). Thereafter, the glucose enters the glycolysis and its carbon units are oxidised in the Krebs cycle.

In many insects proline is present in haemolymph and flight muscles (Wheeler, 1989). As early as 1963 it was shown by Bursell that proline serves as an energy substrate during flight in the tsetse fly, *Glossina morsitans*. Since then, several investigations have led to the conclusion that proline is partially oxidised to supply the flight muscles with energy and that alanine is the end product of this partial oxidation. However, in some insects there are indications that a portion of proline is completely oxidised: in the Japanese beetle, *Popillia japonica* (Hansford and Johnson, 1975) and some dragonfly species (Janssens, 1995) the consumption of proline does not result in an equimolar appearance of alanine. The degree of proline utilisation for flight ranges from the sparker function proposed for the blowfly, *Phormia regina* (Sacktor and Childress, 1967), via the combined use of proline and carbohydrates in the Colorado potato beetle, *Leptinotarsa decemlineata* (Weeda *et al.*, 1979), and the African fruit beetle, *Pachnoda sinuata* (Zebe and Glide, 1993), to the exclusive breakdown of proline in the tsetse fly and possibly some scarab beetles (Gade, 1997a,b). Although high concentrations of proline can be found in haemolymph, flight muscles and fat body of insects, there are no known specialised storage organs for proline and it must therefore be provided through synthesis during flight. The fat body was found to be the site where this re-synthesis of proline takes place in the tsetse fly (McCabe and Bursell, 1977) and the Colorado potato beetle (Weeda *et al.*, 1980). The alanine generated during breakdown of proline, is in turn the precursor for its re-synthesis.

6.2 Metabolic pathways of proline during insect flight

Results of several investigations suggest that proline serves as flight substrate in a number of insects. However, biochemical pathways of proline oxidation and its resynthesis have only been fully described for the tsetse fly (Bursell, 1981) and partially described for the Colorado potato beetle (Weed *et al.*, 1980a,b). The whole process is depicted in Figure 10.19. In brief, during flight proline is degraded in the flight muscles in two steps to glutamate, the first step being catalysed by proline dehydrogenase. Glutamate subsequently serves as the substrate for alanine aminotransferase. This enzyme makes the carbon skeleton of glutamate available for oxidation in the Krebs cycle in the form of α-ketoglutarate. The 5-carbon moiety is only partially oxidised. Malic enzyme decarboxylates malate from the cycle and supplies a 3-carbon unit in the form of pyruvate. Pyruvate subsequently acts as the acceptor for the ammonia released from glutamate by the action of alanine aminotransferase, resulting in the formation of alanine. The latter is released into the haemolymph and travels to the fat body where it serves as the precursor for the re-synthesis of proline. The additional two carbon units necessary for this re-synthesis are supplied by the lipid stores in the fat body of the tsetse fly and most likely in the Colorado potato beetle (Weeda *et al.*, 1980b). The two overall reactions which describe the oxidation (1.) and re-synthesis (2.) of proline are:

1. Pro + 2.5 O_2 + 14 ADP + 14 Pi -+ Ala + H_2O + 2 CO_2 + 14 ATP
2. Ala + acetyl-CoA + NADH + rr -+ Pro + H_2O + NAD+ + CoAH

The tsetse fly and the Colorado potato beetle belong to different orders within the Insecta and the question arises whether these are special metabolic cases within their respective groups. Because dipterans usually power their flight by the breakdown of carbohydrates (Sacktor, 1975) this seems likely for the tsetse fly, However results of various investigations on beetles indicate that this feature may be more common in the order Coleoptera.

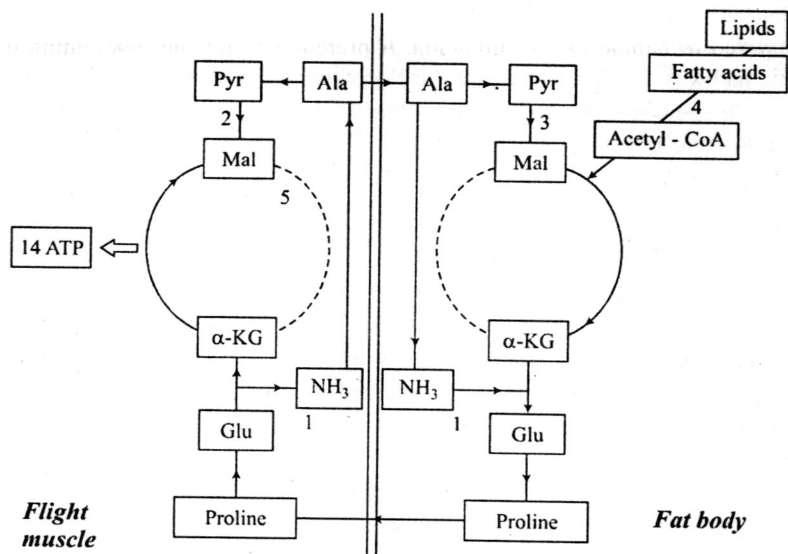

Fig. 10.19. Biochemical pathways during proline oxidation in the flight muscles and resynthesis in the fat body (adapted from Weeda et al., 1980b).

1 = alanine aminotransferase, 2 = NAD-dependent malic enzyme, 3 = NADP-dependent malic enzyme, 4 3-hydroxyacyl-CoA dehydrogenase, 5 malate dehydrogenase.

The first step of proline breakdown is mediated by the proline dehydrogenase (ProDH) which catalyses the conversion of proline to pyrroline-5-carboxylate which then converts spontaneously to glutamate-5-semialdehyde. The latter intermediate is subject to oxidation by the glutamate semialdehyde dehydrogenase. The carbon moiety of the formed glutamate can theoretically be introduced into the Krebs cycle as a-ketoglutarate by the action of glutamate dehydrogenase (GluDH). This reaction, however, would lead to the release of ammonia which is thought to inhibit the GluDH in insect flight muscles (Mordue *et al.*, 1980). This was shown for the GluDH of tsetse fly flight muscles (Bursell, 1975). In addition, the carbon skeleton would be completely oxidised when this pathway was used. Although there were some indications that at least a portion of the proline was broken down via this pathway in the tsetse fly (Hargrove, 1976) and the Japanese beetle (Hansford and Johnson, 1975), the major route in these insects was shown to be the transamination catalysed by the action of alanine aminotranferase (AlaT) (Hansford and Johnson, 1975). Because there is no release of ammonia during the course of this pathway, it has a clear advantage over the GluDH pathway. No additional mechanisms are required to handle excess ammonia. A prerequisite for the functioning of this pathway is high activity of AlaT in the flight muscles. However, the activity of the GluDH is also high. In the blister beetle, both enzymes have similar levels of activity. However, the accumulation of alanine in the body during flight confirmed that the AlaT pathway was most likely the major route of proline catabolism.

When the pathway via the AlaT is used, pyruvate is necessary as an acceptor for the ammonia of glutamate. A pre-requisite of that pathway is therefore an enzyme which removes a carbon moiety from the Krebs cycle and provides pyruvate for the transamination (Bursell, 1963). This function is fulfilled by the NAD-dependent malic enzyme (ME), found to have high activities in flight muscles of insects which use the partial oxidation of proline. The activity of the ME must be sufficiently high to ensure the removal of malate from the Krebs cycle despite the presence of a strong MDH which uses the same substrate as the ME (Bursell, 1981). Therefore, Bursell (1981) proposed the MDH:ME ratio to describe the importance of the partial proline oxidation in the flight muscles of an insect. According to this measure, the pathway is best developed in the tsetse fly, where a ratio of 1 was found (Hoek *et al.,* 1976). A ratio of 8 was found in *L. decemlineata* (Weeda *et al.,* 1980a).

6.3 Regulation of flight metabolism

The transition from rest to flight in an insect requires the establishment of a completely different level of metabolism with the metabolic rate increasing dramatically. During rest-flight transition and during the early phase of flight, regulation of the metabolic flux is necessary at various levels to organise the oxidation of substrates as well as the supply of additional substrates from the fat body.

6.3.1 *Regulation of flight metabolism at physiological level*

Flight preparation (importance of thoracic temperature)

In many insects, flight is only possible when certain flight muscle temperatures are reached (Heinrich, 1993). This was shown to be the case in *P. sinuate.* The beetle needs a thoracic temperature of 28 *°C* to fly effectively. Before flight, *P. sinuata* warms up endothermically to about 33 *°C,* if passive elevation by basking in the sun is not possible. This is similar to the closely related Protea beetle, *Trichostetha fascicularis* (Nicolson and Louw, 1980) but different from certain dung beetles, which maintain a high body temperature during terrestrial activity (Bartholomew and

Heinrich, 1978; Heinrich and Bartholomew, 1979). Endothermic pre-flight warm-up requires large amounts of energy, as can be seen from the oxygen consumption rate during this period. Unequivocal evidence was provided that proline supplies most of the energy required when elevation of body temperature has to be achieved endothermically: while no changes occurred in haemolymph lipids and carbohydrates, proline concentration decreased steeply and alanine concentration rose concomitantly. In addition, glycogen stores in the flight muscles seem unaffected by warm-up, while a dramatic drop of proline concentration occurred, accompanied by the concomitant elevation of alanine levels.

Some insects, such as locusts, use only small amounts of proline shortly after the commencement of flight (Worm and Beenakkers, 1980) which probably serves a sparker function in order to provide the Krebs cycle with intermediates as proposed for the blowfly. The use of proline as a major or even the sole flight substrate in certain scarab beetles such as *P. sinuata, Onitis* and *Scarabaeus* (Gade, 1997a,b) might have derived from its original function as a sparker. Its high resting concentration in flight muscles and haemolymph of the above beetles and the immediate availability of its carbon skeleton for the Krebs cycle also makes it an ideal fuel for warm-up.

Influence of metabolic neuropeptides

Insect flight is very energy demanding and the reserves immediately available from stores in the haemolymph and flight muscles are rapidly depleted. Haemolymph proline and flight muscle proline and glycogen are diminished to about 50 % of the resting levels within the first 30 s of flight. Fuels from other sources, namely the fat body, have to be mobilised to meet the demand during long-term flight. In locusts lipid stores of the fat body are made available (Goldsworthy, 1983) while in cockroaches the glycogen of the fat body is broken down and the resulting trehalose released into the haemolymph (Beenakkers *et al.,* 1984). From locusts it is well known that this process is under the control of neuropeptides of the *AKH/RPCH* family, which are synthesised and stored in the intrinsic cells of the corpora cardiaca and released from there into the haemolymph during flight.

Proline is an important flight substrate in some insects and there are some indications that AKH peptides are involved in the regulation of its metabolism during flight. An investigation *in vitro* with fat body preparations of *L. decemlineata* demonstrated that crude extract of corpora cardiaca is capable of increasing the release of proline from the fat body, while it also leads to a quicker disappearance of alanine from the medium (Weeda, 1981 a). In the same study it was shown that synthetic Lom-AKH-I had the same effect when added to the incubation. However, because the beetle's own AKH peptides, Pea-CAH-I and II, were only discovered and fully characterised later (Gade and Kellner, 1989), the stimulation of proline synthesis could not be linked conclusively to the endogenous AKH peptides of *L. decemlineata.*

Although alanine concentration in the haemolymph of the Colorado potato beetles was shown to decrease upon injection of corpora cardiaca extract of the beetles, locusts and cockroaches as well as Lom AKH-I, no hyperprolinaemic activity *in vivo* with either substance was demonstrated (Weeda, 1981a). An increased proline release by isolated fat body cells, incubated with conspecific corpora cardiac a material, was also observed in the *tse-tse* fly, G. *morsitans* (Pimley and Langley, 1982; Pimley, 1984). It shown by Gade (1997a,b) that con specific injection of corpora cardiac extract into onitine and scarabaeine dung beetles caused hyperprolinaemia *in vivo.* In contrast to findings in the tsetse fly, it was able to demonstrate that this effect is caused by AKHIRPCH peptides separated by HPLC from the beetles' corpora cardiaca as well as by synthetic peptides.

6.3.2 *Regulation of flight metabolism at the cellular and sub-cellular level*

Co-oxidation of proline and pyruvate in mitochondria

Respiration rates of isolated flight muscle mitochondria with proline and pyruvate as substrates are additive only when proline is present in low concentration. At high concentrations of proline the oxidation of pyruvate seems to become negligible. That would be the case *in vivo* when the animal is at rest. As soon as the pre-flight warm-up or flight starts, proline concentration declines dramatically. Because the proline has to pass at least the cell membrane and the inner mitochondrial membrane (the outer membrane is permeable), it is most likely that a steep gradient is created immediately after any flight muscle activity starts. The proline concentration should therefore drop rapidly to a level which enables pyruvate to participate in the energy supply for flight. Such a situation should be reached within seconds of flight with lift or, if the beetle warmed up endothermically, at the end of the flight preparation period. The increase of haemolymph carbohydrate concentration shortly after the onset of lift generating flight might be a result not only of the release of substrates into the haemolymph by the fat body, but also of the limited ability of the flight muscle mitochondria to oxidise pyruvate at this time of flight. This suppression of pyruvate oxidation indicates the exclusive oxidation of proline. A decrease of the RQ because of the participation of fatty acid oxidation is unlikely in light of the poor capability of the flight muscle mitochondria to oxidise palmitoyl-carnitine.

Weeda *et al.*, (1980a) demonstrated an almost complete suppression of the pyruvate respiration by proline in mitochondria from flight muscles of *L. decemlineata.* The authors assumed that co-oxidation does not occur *in vivo.* They proposed that the pyruvate deriving from haemolymph carbohydrates and flight muscle glycogen was used as amino acceptor *in vivo.* According to this theory, the pyruvate would never be oxidised and would therefore accumulate. This seems unlikely. In contrast to the inhibitory role of proline on pyruvate respiration in the beetles are the results of Sacktor and Childress (1967) using the blowfly, *P. regina.* They observed that proline stimulates pyruvate oxidation. However, in these insects proline is not a major flight substrate.

Feedback inhibition of proline synthesis

Results of bioassays with injection of AKH peptides and alanine as well as the observations concerning the circadian pattern of the proline concentration in the haemolymph of *P. sinuate* demonstrated that the maximum level of proline at 120 µmol/ml. Also, the pre-injection level of alanine seemed to be less important than that of proline for the maximum response of proline concentration upon injection of active material. Such results can be best interpreted by a feedback inhibition of proline synthesis. Even when alanine is injected into the beetles in addition to bioactive material, the maximum proline concentration cannot be elevated. The presence of a feedback inhibition seems also likely when one compares the maximum increase in proline levels by about 20 % of the pre – injection level with the increase of haemolymph lipids in locusts by more than 300 % (Gade, 1990) and haemolymph carbohydrates in cockroaches by almost 200 %. Even carbohydrate mobilisation by Mem-CC in *P. sinuata* amounted to an increase of about 100 % (Lopata and Gade, 1994). When an incubation experiment was conducted in order to investigate the influence of proline on its own synthesis, such a feedback inhibition could be confirmed *in vitro.*

A similar feedback inhibition was found during experiments *in vitro* by Bursell (1977) in the fat body of *tsetse* flies and by Weeda (1981a) with fat body of the Colorado potato beetle. The site of inhibition of proline synthesis is probably the transamination system: Bursell (1981) showed that the transamination of alanine by AlaT of *tsetse* fly fatbody was inhibited by proline. Feedback control of proline synthesis was also found in bacteria (Adams, 1970). The site of inhibition is probably the glutamyl kinase, which catalyses the reaction from glutamate to y-glutamyl phosphate, a step during proline synthesis. This enzyme is known to be sensitive to low concentrations of proline in *Escherichia coli* (Baich, 1969) and in *Pseudomonas aeruginosa* (Krishna and Leisinger, 1979).

REFERENCES

Adams E 1970 Metabolism of proline and hydroxyproline. *Int. Rev. Connect. Tissue* 5:1-91.

Arai, N., Shiomi, K., Yamaguchi, Y., Masuma, R., Iwai, Y., Turberg, A., Kolbl, H. and Omura, S. (2000). Argadin, a new chitinase inhibitor, produced by *Clonostachys* sp. FO-7314. *Chem. Pharm. Bull.* 48, 1442-1446.

Arakane, Y., Specht, C. A., Kramer, K. J., Muthukrishnan, S., and Beeman, R. W. (2008). Chitin synthases are required for survival, fecundity and egg hatch in the red flour beetle, *Tribolium castaneum*. *Insect Biochem. Mol. Biol., 38*, 959–962.

Arrese E L and Soulages, J L 2010). : Insect fat body: energy, metabolism, and regulation. *Annu. Rev. Entomol.* 55: 207–225

Arrese, E.L, Patel, R.T, Soulages, J.L.(2006) The main triglyceride-lipase from the insect fat body is an active phospholipase A(1): identification and characterization. *J Lipid Res;* 47:2656–67.

Asano A. (2003). Glycosidase inhibitors: Update and perspectives on practical use. Glycobiology. 13:93R–104R.

Asano N, Yamaguchi T, Kameda Y, and Matsui K. (1987). Effect of validamycins on glycohydrolases of Rhizoctonia solani. *J Antibiot.* 4:526–532.

Baich A (1969) Proline synthesis in *Escherichia coli.* A proline inhibitable glutamic acid kinase. *Biochim. Biophys. Acta* 192, 462-467.

Bartholojew, G A and Heinrich B (1978) Endothermy in African dung beetles during flight, ball making, and ball rolling. 1. *Exp. Bioi.* 73, 65-83.

Beenakkers A M T, Van der Horst D J, Van Marrewijk W J A (1985) Insect lipids and lipoproteins, and their role in physiological processes. Prog. Lipid Res, 24, 19

Beenakkers, A.M.TH, (1969) Carbohydrate and fat as a fuel for insect flight. A comparative study. 1. *Insect Physiol.* 15, 353-361.

Berecibar, A., Grandjean, C. and Siriwardena, A. (1999). Synthesis and biological activity of natural aminocyclopentitol glycosidase inhibitors: mannostatins, trehazolin, allosamidins, and their analogues. *Chem. Rev.* 99, 779-844

Bini D, Cardona F, Forcella M, Parmeggiani C, Parenti P, Nicotra F, Cipolla L. (2012). Synthesis and biological evaluation of nojirimycin and pyrrolidinebased trehalase inhibitors. Beilstein J Org Chem. 8:514–

521.

Bursell, E. 1963 Aspects of metabolism of amino acids in the tsetse fly, *Glossina* (Diptera).1. *Insect Physiol.* 9, 439-452.

Bursell, E. (1975) Glutamic dehydrogenase from sarcosomes ofthe tsetse fly *(Glossina morsitans)* and the blowfly *(Sarcophaga nodosa). Insect Biochem. 5,* 289-297.

Bursell, E. (1977) Synthesis of proline by fat body of the tsetse fly *(Glossina morsitans):* metabolic pathways. *Insect Biochem.* 7,427-434..

Bursell, E. (1981) The role of proline in energy metabolism. In *Energy Metabolism of in Insects (R. G. H. Downer, ed.), pp. 135-154, Plenum Press, New York.*

Canavoso L E, Jouni Z E, Karnas K J, Pennington J E and Wells M A 2001: Fat metabolism in insects. *Annu. Rev. Nutr*. 21: 23–46.

Candy D I and Kilby B A (1961) The biosynthesis of trehalose in the locust fat body. *Biochem.* 1. 78, 531-536.

Candy, D.J.,and Kilby, B.A. (1962). Studies on chitin synthesis in the desert locust. J. Exp. Biol. 39, 129–140.

Chen J, Tang B, Chen H, Yao Q, Huang X, Chen J, Zhang D, ZhangW. 2010. Different functions of the insect soluble and membrane-bound trehalase genes in chitin biosynthesis revealed by RNA interference. *PLoS ONE*. 5: e10133.

Chino, H (1958) Carbohydrate metabolism in the diapause egg of the silkworm,Bombyx *mori.* Conversion of glycogen into sorbitol and glycerol during diapause. *J. Insect Physiol*., 2, 1-12.

Cohen, E. (1985). Chitin synthetase activity and inhibition in different insect microsomal preparations. *EXS* 41, 470-472.

Cohen, E. (2001). Chitin synthesis and inhibition: A revisit.*Pest Manag. Sci., 57,* 946–950.

Cornette R, Kikawada T. (2011). The induction of anhydrobiosis in the sleeping chironomid: Current status of our knowledge. IUBMB Life. 63:419–429.

Crowe, J.H., Hoekstra, F.A., and Crowe, L.M. (1992). Anhydrobiosis. Annu. Rev. Physiol. 54(1): 579–599. doi:10.1146/annurev.ph.54.030192.003051.

Daviess G, Henrissat B. (1995). Structures and mechanisms of glycosyl hydrolases. Structure. 3:853–859.

Ding, X., Gopalakrishnan, B., Johnson, L. B., White, F. F., Wang, X.,

Morgan, T. D., Kramer, K. J. and Muthukrishnan, S. (1998). Insect resistance of transgenic tobacco expressing an insect chitinase gene. *Transgenic Res.* 7, 77-84.

Downer R.G.H. (1985). Lipid metabolism, pp. 77–113. In: Kerkut G.A. & Gilbert L.I. (eds) Comprehensive Insect Physiology, Biochemistry and Pharmacology. Pergamon Press, Oxford.

Downer, R G H (1978) Role of lipids in insects. In: Biochemistry of Insects, Rockstein M, ed. Academic Press, New York, pp 57-92

Elnitsky, M.A., Hayward, S.A.L., Rinehart, J.P., Denlinger, D.A., and Lee, R.E., Jr. (2008). Cryoprotective dehydration and the resistance to inoculative freezing in the Antarctic midge, Belgica antarctica. J. Exp. Biol. 211(4): 524–530. doi:10.1242/jeb.011874. PMID: 18245628.

Filho, B. P., Lemos, F. J., Secundino, N. F., Pascoa, V., Pereira, S. T. And Pimenta, P. F. (2002). Presence of chitinase and beta-*N*acetylglucosaminidase in the *Aedes aegypti*: a chitinolytic system involving peritrophic matrix formation and degradation. *Insect Biochem. Mol. Biol.* 32, 1723-1729.

Forcella M, Berra E, Giacchini R, Rossaro B, Parenti P. (2007). Increased alaninecconcentration is associated with exposure to fenitrothion but not carbamatesin *Chironomus riparius* larvae. Ecotoxicol. Environ. Saf. 66:326–334.

Forcella M, Cardona F, Goti A, Parmeggiani C, Cipolla L, Gregori M, Schirone R, Fusi P, Parenti P. (2010). A membrane-bound trehalase from *Chironomus riparius* larvae: Purification and sensitivity to inhibition. *Glycobiology.* 20:1186–1195.

Fukamizo, T., and Kramer, K. J. (1987). Effect of 20-hydroxyecdysone on chitinase and β-N-acetylglucosaminidase during the larval–pupal transformation of *Manduca sexta* (L.). *Insect Biochem.*, *17*, 547–550.

Gade G (1990) The adipokinetic hormone/red pigment-concentrating hormonepeptide family: structures, interrelationships and functions. *J Insect Physio!. 36,* 1-12.

Gade G and Kellner R (1989) The metabolic neuropeptides of the corpus cardiacum from the potato beetle and the American cockroach are identical. *Peptides* 10, 1287-1289.

Gade G, Auerswald L. (2002) Beetles' choice—proline for energy output: control by AKHs. *Comp Biochem Physiol* B. 132:117–29.

Gade, G (1997a) Hyperprolinaemia caused by novel members of the AKHIRPCH family of peptides isolated from corpora cardiaca of onitine beetles. *Biochem. 1.* 321,201-206.

Gade, G (1997b) Distinct sequences of AKHIRPCH family members in beetle *(Scarabaeus-species)* corpus cardiacum contain three aromatic amino acid residues. *Biochem. Biophys. Res. Commun.* 230, 16-21.

Ge LQ, Zhao KF, Huang LJ, Wu JC. (2011). The effects of triazophos on the trehalose content, trehalase activity and their gene expression in the brownplanthopper Nilaparvata lugens (Stål) (Hemiptera: Delphacidae). Pestic Biochem Physiol. 100:172–181.

Gibson RP, Gloster TM, Roberts S, Warren RA, Storch de Gracia I, Garcia A, Chiara JL, and Davies GJ. (2007). Molecular basis for trehalase inhibition revealed by the structure of trehalase in complex with potent inhibitors. Angew Chem. 46:4115–4119.

Goldsworthy G J (1983) The endocrine control of flight metabolism in locusts. In *Advances in Insect Physiology* 17 (ed. MJ. Berridge, lE. Treheme and V.B. Wigglesworth), pp. 149-204. New York: Academic Press

Gronke S, Muller G, Hirsch J, Fellert S, Andreou A, et al. (2007) Dual lipolytic control of body fat storage and mobilization in Drosophila. *PLoS Biol.*; 5:e137

Hansford R G and Johnson R N (1975) The nature and control of the tricarboxylate cycle in beetle flight muscle. *Biochem. J.* 148, 389-401

Hansford, R G and Johnson R N (1975) The nature and control of the tricarboxylate cycle in beetle flight muscle. *Biochem. J.* 148, 389-401

Hardie, D.G. (2007). AMP-activated/SNF1 protein kinases: conserved guardians of cellular energy. *Nat. Rev. Mol. Cell Biol.* 8(10): 774–785. doi:10.1038/nrm2249. PMID:17712357.

Hargrove, I .W. (1976) Amino acid metabolism during flight in tsetse flies. *J. Insect Physiol.* 22, 309-313.

Heinrich, B. (1993) The Hot Blooded Insects. Berlin Heidelberg, Germany: Springerm Verlag.

Heinrich, B., and Bartholojew, G. A. (1979) Roles of endothermy and size in interand intraspecific competition for elephant dung in an African dung beetle, *Scarabaeus laevistratus. Physiol. Zool.* 56,484-496.

Hoek, J .B., Pearson, D. J., and Olembo, N. K. (1976) Nicotineamide-adenine dinucleotide-linked "malic" enzyme in flight muscle of the tsetse fly *(Glossina)* and other insects. *Biochem.* 1. 160, 253-262.

Hogenkamp, D. G., Arakane, Y., Zimoch, L., Merzendorfer, H., Kramer, K. J., et al. (2005). Chitin synthase genes in *Manduca sexta*: Characterization of a gut-specific transcript and differential tissue expression of alternately spliced mRNAs during development. *Insect Biochem. Mol. Biol.*, *35*,529–540.

Homma T, Watanabe K, Tsurumaru S, Kataoka H, Imai K, Kamba M, et al (2006). G protein-coupled receptor for diapause hormone, an inducer of Bombyx embryonic diapause. Biochem Biophys Res Commun. 344(1):386–93

Hue, L., and Rider, M.H. (2007). The AMP-activated protein kinase: more than an energy sensor. *Essays Biochem.* 43(1): 121–137.

Irwin, J.T., and Lee, R.E., Jr. (2002). Energy and water conservation in frozen vs. supercooled larvae of the goldenrod gall fly, Eurosta solidaginis (Fitch) (Diptera: Tephritidae). J. Exp. Zool. 292(4): 345–350. doi:10.1002/jez.10082. PMID:11857468.

Izumida, H., Imamura, N. and Sano, H. (1996). A novel chitinase inhibitor from a marine bacterium, *Pseudomonas* sp. *J. Antibiot.* 49, 76-80.

Janssens, M. P. E. (1995) Hormonal control of flight metabolism in Odonata? M.Sc. thesis, University of Cape Town, South Africa.

Jaworski, E., Wang, L. and Margo, G. (1963). Synthesis of chitin in cellfree extracts of *Prodenia eridania. Nature* 198, 790.

Joanisse, D.R., and Storey, K.B. (1994) Mitochondrial enzymes during overwintering in two species of cold-hardy gall insects. *Insect Biochem. Mol. Biol.* 24(2): 145–150. doi:10.1016/0965- 1748(94)90080-9.

Jutsum, A R and Goldsworthy G J (1976) Fuels for flight in *Locusta.* 1. *Insect Physiol.* 22, 243-249.

Kageyama, T and Ohnishi E (1973) Carbohydrate metabolism in the eggs of the silkworm, Bombyx mori. II Anaerobiosis and Polyol formation. Development, growth and Differentiation, Vol 15 (1) 47-55

Kamei Y, Hasegawa Y, Niimi T, Yamashita O, Yaginuma T. (2011). Trehalase-2 contributes to trehalase activity enhanced by diapause hormonein developing ovaries of the silkworm, *Bombyx mori. J Insect Physiol.* 57:608–613.

Kamei Y, Hasegawa Y, Niimi T, Yamashita O, Yaginuma T. (2011). Trehalase-2 protein contributes to trehalase activity enhanced by diapause hormone in developing ovaries of the silkworm, *Bombyx mori. J Insect Physiol.* 57:608–613

Kawooya, J. K., and Law, J. H. (1988) Role of lipophorin in lipid transport to the insect egg. *J. Biol. Chem.* 263, 8748-8763

Kayukawa, T., Chen, B., Hoshizaki, S., and Ishikawa, Y. (2007). Upregulation of a desaturase is associated with the enhancement of cold hardiness in the onion maggot, Delia antiqua. *Insect Biochem. Mol. Biol.* 37(11): 1160–1167. doi:10.1016/j.ibmb. 2007.07.007. PMID:17916502.

Kenchington, W. (1976). Adaption of insect peritrophic membranes to form cocoon fabrics. In *The Insect Integument* (ed. H. R. Hepburn), pp. 497-513. Amsterdam: Elsevier Science

Knuesel I, Murao S, Shin T, Amachi T, Kayser H. (1998). Comparative studies of suidatrestin, a specific inhibitor of trehalases. *Comp Biochem Physiol BBiochem Mol Biol.* 120:639–646.

Koga, D., Funakoshi, T., Mizuki, K., Ide, A., Kramer, K. J., et al. (1992). Immunoblot analysis of chitinolytic enzymes in integument and molting fluid of the silkworm, *Bombyx mori*, and the tobacco hornworm, *Manduca sexta. Insect Biochem. Mol. Biol.*, *22*, 305–311.

Kondo, K., Matsumoto, M., Kojo, A. and Mauda, R. (2002). Purification and characterization of chitinase from pupae *Pieris rapae crucivora* (Boiduval). *J. Chem. Eng. Jap.* 35, 241-246.

Kostyukovsky, M., Chen, B., Atsmi, S. and Shaaya, E. (2000). Biological activity of two juvenoids and two ecdysteroids against three stored product insects. *Insect Biochem. Mol. Biol.* 30, 891-897.

Kramer, K. J. and Koga, D. (1986). Insect chitin: physical state, synthesis, degradation and metabolic regulation. *Insect Biochem.* 16, 851-877.

Kramer, K. J., & Koga, D. (1986). Insect chitin: Physical state,csynthesis, degradation and metabolic regulation. *Insect Biochem.*, *16*, 851–877.

Kramer, K. J., Corpuz, L., Choi, H. K., & Muthukrishnan, S. (1993). Sequence of a cDNA and expression of the gene encoding epidermal and gut chitinases of *Manduca sexta. Insect Biochem. Mol. Biol.*, *23*, 691–701.

Krishna R V and Leisinger T (1979) Biosynthesis of proline in *Pseudomonas aeruginosa.* Partial purification and characterisation ofy-glutamyl kinase. *Biochem. J* 181, 215-222.

Levin, D.B., Danks, H.V., and Barber, S.A. (2003). Variations in mitochondrial DNA and gene transcription in freezing-tolerant larvae of *Eurosta solidaginis* (Diptera: Tephritidae) and Gynaephora groenlandica (Lepidoptera: Lymantriidae). *Insect Mol. Biol.* 12(3): 281–289. doi:10.1046/j.1365-2583.2003.00413.x. PMID:12752662.

Liu, X.J, Zhang, H.H, Li, D.Q, Cui, M, Ma E.B, and Zhang J.Z. (2012). Sequence characterization and mRNA expression profiling of a soluble trehalase gene in Locusta migratoria (Orthoptera: Acrididae). *Acta Entomol Sin.* 55: 1264–1271

Locke, M. (1991). Insect epidermal cells. In K. Binnington, & A. Retnakaran (Eds.). *Physiology of the Insect Epidermis* (pp. 1–22). Melbourne, Australia: CRISCO Publications

Lopata, A., and Gade, G. (1994) Physiological action of a neuropeptide from the corpora cardiaca of the fruit beetle, *Pachnoda sinuata,* and its possible role in flight metabolism. *J Insect Physiol.* 40, 53-62.

Lorenz, J., Lenz, M. and Hoffmann, K. H. (1995). Effects of pharmacological agents on ecdysteroid synthesis *in vitro* in ovaries and abdominal integument from female adult crickets, *Gryllus bimaculatus* de Geer (Ensifera, Gryllidae). *Z. Naturforsch.* 50, 286-293.

Mc Cabe C. T., and Bursell, E. (1977) Interrelationships between amino acid and lipid metabolism in tsetse fly, *Glossina morsitans. Insect Biochem.* 5, 781-789.

Mitsumasu K, Azuma M, Niimi T, Yamashita O, Yaginuma T. (2008). Changes in the expression of soluble and integral-membrane trehalases in the midgut during metamorphosis in Bombyx mori. Zool Sci. 25:693–698.

Morin, P., Jr., McMullen, D.C., and Storey, K.B. (2005). HIF-1a involvement in low temperature and anoxia survival by a freeze tolerant insect. Mol. Cell. Biochem. 280(1–2): 99–106. doi:10. 1007/s11010-005-8236-x. PMID:16311910.

Moya, P., Castillo, M., Primo-Yufera, E., Couillaud, F., Martinez-Manez, R., Garcera, M. D., Miranda, M. A., Primo, J. And Martinez-Pardo, R. (1997). Brevioxime: a new juvenile hormone biosynthesis inhibitor isolated from *Penicillium brevicompactum. J. Org. Chem.* 62, 8544-8545.

Muthukrishnan, J., Seifert, K., Hoffmann, K. H. and Lorenz, M. W. (1999). Inhibition of juvenile hormone biosynthesis in *Gryllus bimaculatus* by *Glycosmis pentaphylla* leaf compounds. *Phytochemistry* 50, 249-254.

Muthukrishnan, S., Merzendorfer, H., Arakane, Y., Kramer, K.J., (2012). Chitin metabolism in insects. In: Gilbert, L.I. (Ed.), Insect Molecular Biology and Biochemistry. Elsevier B.V., London, pp. 193–235)

Nicolson, S W and Louw G N 1980 Pre-flight thermogenesis, conductance and thermoregulation in the Protea beetle, *Trichostetha fascicularis* (Scarabaeidae Cetoniinae). S. *Afr. J. Sci.* 76, 124-126.

Omura S, Arai N, Yamaguchi Y, Masuma R, Iwai Y, Namikoshi M, Turberg A, Kolbl H, Shiomi K. (2000). Argifin, a new chitinase inhibitor, produced by Gliocladium sp. FTD-0668. I. Taxonomy, fermentation, and biological activities. *J Antibiot* (Tokyo) 2000;53:603–608

Peters, W. (1992). Peritrophic membranes. In *Zoophysiology*, vol. 30 (ed. S. D. Bradshaw, W. Burggren, H. C. Heller, S. Ishii, H. Langer, G. Neuweiler and D. J. Randall), pp. 1-238. Berlin: Springer.

Post, L. C., and Vincent, W. R. (1973). A new insecticide inhibits chitin synthesis. *Naturwissenschaften, 60*, 431–432.

Rao, F. V., Houston, D. R., Boot, R. G., Aerts, J. M., Sakuda, S. and Van Aalten, D. M. (2003). Crystal structures of allosamidin derivatives in complex with human macrophage chitinase. *J. Biol. Chem.* 14, 20110- 20116

Retnakaran, A., Gelbic, I., Sundaram, M., Tomkins, W., Ladd, T., Primavera, M., Feng, Q., Arif, B., Palli, R. and Krell, P. (2001). Mode of action of the ecdysone agonist tebufenozide (RH-5992), and an exclusion mechanism to explain resistance to it. *Pest Manag. Sci.* 57, 951-957.

Reynolds, S. E. and Samuels, R. I. (1996). Physiology and biochemistry of insect moulting fluid. *Adv. Insect Physiol.* 26, 157-232.

Royer, V., Fraichard, S., & Bouhin, H. (2002). A novel putative insect chitinase with multiple catalytic domains: Hormonal regulation during metamorphosis. *Biochem. J., 366*, 921–928.

Rudall, K. M. and Kenchington, W. (1973). The chitin system. *Biol. Rev.* 48, 597-636.

Ruiz-Herrera, J., & San-Blas, G. (2003). Chitin synthesis as target for antifungal drugs. *Curr. Drug Targets Infect. Disord., 3*, 77–91.

Sacktor B and Childress C C (1967) Metabolism of proline in insect flight muscle and its significance in stimulating the oxidation of pyruvate. *Archs. Biochem. Biophys.* 120, 583-588.

Sacktor, B (1975) Utilisation of fuels by muscle. In *Insect Biochemistry and Function* (ed. DJ. Candy and B.A. Kllby), pp. 1-81. London: Chapman and Hall.

Sformo, T., Walters, K., Jeannet, K., Wowk, B., Fahy, G.M., Barnes, B.M., and Duman, J.G. (2010). Deep supercooling, vitrification and limited survival to –100 °C in the Alaskan beetle Cucujus clavipes puniceus (Coleoptera: Cucujidae) larvae. J. Exp. Biol. 213(3): 502–509. doi:10.1242/jeb.035758. PMID:20086136.

Silva, M.C.P., Ribeiro, A.F, Terra. W.R, and Ferreira C. (2009). Sequencing of *Spodoptera frugiperda* midgut trehalases and demonstration of secretion of soluble trehalase by midgut columnar cells. *Insect Mol Biol.* 18:769–784.

Singtripop T, Wanichacheewa S, Tsuzuki S, Sakurai S. (1999). Larval growth and diapause in a tropical moth, *Omphisa fuscidentalis* Hampson. *Zool Sci.* 16:725–733.

Storey, J.M., and Storey, K.B. (1985). Freezing and cellular metabolism in the gall fly larva, Eurosta solidaginis.*J.Comp.Physiol.BBiochem.Syst. Environ.Physiol.* 155(3): 333–337. doi:10.1007/BF00687475.

Storey, K.B. (1982). Phosphofructokinase from the overwintering gall fly larva, Eurosta solidaginis: control of cryoprotectant polyol synthesis. Insect Biochem. 12(5): 501–505.

Storey, K.B., and Storey, J.M. (1991). Biochemistry of cryoprotectants. In Insects at low temperature. Edited by D. Denlinger and R.E. Lee, Jr. Chapman and Hall, New York. pp. 64–93.

Storey, K.B., and Storey, J.M. (2010). Oxygen: stress and adaptation in cold hardy insects. In Low temperature biology of insects. Edited by D.L. Denlinger and R.E. Lee, Jr. Cambridge University Press, Cambridge. pp. 141–165.

Suarez RK, Darveau CA, Welch KC, O'Brien DM, Roubik DW, Hochachka PW. (2005) Energy metabolism in orchid bee flight muscles: carbohydrate fuels all. *J Exp Biol.*;208(Pt 18):3573-9

Takahashi, M., Kiuchi, M., & Kamimura, M. (2002). A new chitinase-related gene, BmChiR1, is induced in the *Bombyx mori* anterior silk gland at molt and metamorphosis by ecdysteroid. *Insect Biochem. Mol. Biol., 32*, 147–151.

Tellam, R. L., Vuocolo, T., Johnson, S. E., Jarmey, J., & Pearson, R. D. (2000). Insect chitin synthase cDNA sequence, gene organization and expression. *Eur. J. Biochem., 267*, 6025–6043.

Terwisscha van Scheltinga, A. C., Armand, S., Kalk, K. H., Isogai, A., Henrissat, B. and Dijkstra, B. W. (1995). Stereochemistry of chitin hydrolysis by a plant chitinase/lysozyme and X-ray structure of a complex with allosamidin: evidence for substrate assisted catalysis. *Biochemistry* 34, 15619-15623

Thompson SN. (2003) Trehalose: the insect 'blood' sugar. *Adv Insect Physiol.;* 31:205–85.

Thorat LJ, Gaikwad SM, Nath BB. (2012). Trehalose as an indicator of desiccation stress in Drosophila melanogaster larvae: A potential marker of anhydrobiosis. Biochem Biophys Res Commun. 419: 638–642.

van Aalten, D. M., Komander, D., Synstad, B., Gaseidnes, S., Peter, M. G. and Eijsink, V. G. (2001). Structural insights into the catalytic mechanism of a family 18 exo- chitinase. *Proc. Natl. Acad. Sci. USA* 98, 8979-8984.

van Eck, W. H. (1979). Mode of action of two benzoylphenyl ureas as inhibitors of chitin synthesis in insects. *Insect Biochem.* 9, 295-300.

Vandenborre G, Smagghe G, Ghesquière B, Menschaert G, Nagender Rao R, Gevaert K, et al. (2011) Diversity in Protein Glycosylation among Insect Species. PLoS ONE 6(2): e16682. doi:10.1371/journal.pone.0016682

Weeda E (1981) Hormonal regulation of proline synthesis and glucose release in the fat body of the Colorado potato beetle, *Leptinotarsa decemlineata. J. Insect Physiol.* 27,411-417.

Weeda E, De Kort CAD, and Beenakkers AM T (1980a) Oxidation of proline and pyruvate in flight muscle mitochondria of the Colorado beetle, *Leptinotarsa decemlineata* Say. *Insect Biochem.* 10, 305-311.

Weeda E, DE Kort, CAD and Beenakkers AM I (1979) Fuels for energy metabolism in the Colorado potato beetle, *Leptinotarsa decemlineata* Say. *J. Insect Physioi.* 25,951-955.

Weeda E, Koopmanshap, A.B, De Kort, CAD and Beenakkers AM I (1980b) Proline synth'esis in fat body of*Leptinotarsa decemlineata. Insect Biochem.* 10,631-636.

Wegener G, Tschiedel V, Schlöder P, Ando O. (2003). The toxic and lethal effects of the trehalase inhibitor trehazolin in locusts are caused by hypoglycaemia. J Exp Biol. 206:1233–1240.

Wharton DA. (2009). Life without water. In: Wharton DA, editor. Life at the limits: Organisms in extreme environments (online). Cambridge: Cambridge University Press, p. 92–128.

Wheeler, C.H (1989) Mobilisation and transport of fuels to the flight muscles. In *Insect Flight* (eds. G.J. Goldsworthy and C.H. Wheeler), pp. 273-303. Boca Raton, Florida: CRC Press Inc.

Worm R A A and Beenakkers AM T (1980) Regulation of substrate utilisation in the flight muscle of the locust, *Locusta migratoria,* during flight. *Insect Biochem.* 10,53-59.

Yamashita, O. (1996) Diapause hormone of silkworm, Bombyx mori: structure, gene expression and function. J.Insect Physiol. 42, 669-679

Yarema, C., McLean, H. and Caveney, **S.** (2000). L-Glutamate retrieved with the moulting fluid is processed by a glutamine synthetase in the pupal midgut of ***Calpodes ethlius.*** *J. Insect Physiol.* 46,1497 -1507

Zebe, E and Gade, G (1993) Flight metabolism in the African fluit beetle, *Pachnoda sinuata. J Comp. Physiol.* B 163, 107-112.

Zhang Q, Lu DH, Pu J, Wu M, Han ZJ. (2012). Cloning and RNA interference effects of trehalase genes in Laodelphax striatellus (Homoptera: Delphacidae). Acta Entomol Sin. 55:911–920.

Zhang, D., & Miller, M. J. (1999). Polyoxins and nikkomycins: Progress in synthetic and biological studies. *Curr. Pharm. Des., 5,* 73–99.

Zhao KF, Shi Z, Wu J. (2011). Insecticide-induced enhancement of flight capacity of the brown planthopper Nilaparvata lugens Stål (Hemiptera: Delphacidae). Crop Prot. 30:476–482.

Zheng, Y. P., Retnakaran, A., Krell, P. J., Arif, B. M., Primavera, M., & Feng, Q. L. (2003). Temporal, spatial and induced expression of chitinase in the spruce budworm, *Choristoneura fumiferana. J. Insect Physiol., 49,* 241–247.

Ziegler R (1991) Changes in lipid and carbohydrate metabolism during starvation in adult Manduca sexta. *J Comp Physiol* B.;161(2):125-31.

Ziegler, R and Schulz, M (1986) Regulation oflipid metabolism during flight in *Manduca sexta.* 1. *Insect Physiol.* 32, 903-908.

Ziegler, R., Ashida, M., Fallon, A.M., Wimer, L.T., Silver Wyatt, S., and Wyatt, G.R. (1979). Regulation of glycogen phosphorylase in fat body of Cecropia silkmoth pupae. *J. Comp. Physiol.* 131: 321–332.

Zimoch, L., & Merzendorfer, H. (2002). Immunolocalization of chitin synthase in the tobacco hornworm. *Cell Tissue Res., 308,* 287–297.

Muscular Co-ordination System 11

Synopsis

1. INTRODUCTION

We correlate animal movement with life and essentially all animal movement is accomplished by the most remarkable tissue, the muscles. Muscle is the excitable, contractile tissue of animals that is responsible for movement, posture and behaviour. The musculature of insects is composed of an external, multilayered array of body wall muscles (somatic musculature), an internal layer of visceral muscles that surround the digestive tract and gonads, and specialised myoepithelial tubes forming the vascular system. Although there is great variability in structure and performance among different insect muscles, many basic features of biochemical composition, ultrastructural organisation, and contractile performance are common among insect muscles and indeed are similar between muscles of insects and those of vertebrates. Insect muscles are mostly translucent, colourless or grey, though the flight muscles often show a yellowish or pinkish-brown tinge. All insect muscles are striated and they lack the smooth and typical cardiac muscles seen in vertebrates. In most skeletal muscles of insect, especially those of the appendages, one end of the muscle is attached to a movable part. Cuticular invaginations or apodemes, in the form of cords, bands or plate like structure, may provide the true sites of attachment. As muscles became better designed for faster contraction, to promote faster locomotion, especially that required to conquer the air, they needed evermore support systems for their greater energy demands.

Two types of fight muscles mechanisms are seen in insects- the **synchronous** and the **asynchronous** mechanisms. Synchronous flight muscle is anatomically most analogous to vertebrate skeletal (striated) muscle, whereas asynchronous (fibrillar) flight muscle is much like vertebrate cardiac muscle. The asynchronous mode of muscle action is a unique insect adaptation (in Diptera, Hymenoptera, Heteroptera and Coleoptera) for attaining high-frequency wing beats with concomitant high power output that are also related to molecular differences of protein isoforms- different myosins result in different shortening velocities, others affect passive stiffness.

One of the major variables for specific flight muscle performance includes mechanical hook-ups. The most obvious of these are probably the direct versus indirect attachment to the wings that differentiate between whether a muscle acts on the wing directly and whether the wing is indirectly depressed or elevated due to changes in shape of the thoracic box to harness resonant mechanical properties of the flight apparatus. In all insects, the muscles used for controlling the trajectory of the wings are direct muscles whereas in the vast majority of insects the power of the up-stroke and down-stroke of the wings results from two antagonistic sets of muscles that, unlike in all vertebrates, pull not on the wings themselves but on the thoracic box. The odonates are an exception. In them the power-producing wing muscles pull directly on the wings. Additionally, the two sets of wings can operate independently of one another for flight control.

Muscles themselves became adapted for use in different organ systems to facilitate the movement of nutrients through the gut, to facilitate nutrient distribution through the body to other muscles and other organs, and to help in gas exchange. The muscular system has the following functions:

1. Support of the body
2. Helps maintain posture
3. Movement of limbs, including ovipositor
4. Movement of wings
5. Movement of viscera
6. Locomotion
7. Closure of spiracles
8. Operation of various pumps such as cibarial pump and pumping of poison glands
9. Generation of heat by shivering.

2. TYPES OF MUSCLES

Insect muscles can be classified based on their location into skeletal, visceral muscles and cardiac muscles (Box 11.1). Again the skeletal muscles are further grouped into (i) Cephalic muscle (ii) Thoracic muscle (iii) Muscles of flight (iv) Abdominal muscles. The muscle of the viscera are grouped into (i) Alary muscle (ii) Dorsal blood vessel (iii) Accessory pulsatile organs and various diaphragms (iv) Alimentary canal including crop (v) Reproductive organs and duct (vi) Venom glands (vii) Repugnatorial glands (viii) Organs of defence and (ix) Malpighian tubules.

Box 11.1 Types of muscles in insects based on location

1. Skeletal muscles
   A. Cephalic muscle
   B. Thoracic muscle
   C. Muscles of flight
   D. Abdominal muscles
2. Visceral muscles
   A. Alary muscle
   B. Dorsal blood vessel
   C. Accessory pulsatile organs and various diaphragms
   D. Alimentary canal including crop
   E. Reproductive organs and duct
   F. Venom glands
   G. Repugnatorial glands
   H. Organs of defence
   I. Malpighian tubules
3. Cardiac muscles

2.1 Skeletal muscles

2.1.1 *Cephalic muscles*

The principal muscle of the head may be divided into:

1. **Cervical muscle:** These muscle control the movement of head and are classified as levators, depressor, retractors and rotators according to their function. They take their origin from the prothorax and cervix and are inserted into the tentorium and epicranium.
2. Muscles of the mouthparts. They are classified as:
   A. **The Labral compresssors**: Running between the dorsal and ventral surfaces of the labrum

   B. **The posterior labral muscles:** Run from the tormal sclerites of the labrum to the wall of the head.

   C. **The anterior labral muscles (Retractors):** Run from the anterior margin of the labral base to the wall of the head

   D. **The dorsal abductors:** Originate on the upper lateral part of the epicranium and insert each on an apodeme connected with the inner, basal region of the mandible.

   E. **The ventral abductors:** Present only in the Apterygotes and some lower Pterygotes.

   F. **Dorsal basal muscle:** Arising on the dorsal part of the head and forming the anterior and posterior rotators of the cardo and the crsnial flexure of the lacinia.

   G. **Ventral basal muscles:** Inserted on the cardo and stipes and originate on the tentorum in most pterygotes and on the tentorial apodemes in the apterygotes.

   H. **Stipital muscles:** Originate on the stipes and include the levator and depressor of the palp, flexor of the galea and the stipital flexor of the lacinia.

   I. **Extrinsic labial muscles:** Arise on the tentorium or cranial wall and insert on the prementum. They correspond to the ventral basal muscles of the maxilla.

   J. **Median labial muscles:** Run from the back of the prementum to the postmentum and have no homologues in the maxilla.

   K. **Labial salivary muscles:** Usually two pairs, arising on the prementum and convergin on the labiral wall of the salivarium near the opening of the salivary duct.

   L. **Muscles of the endites and palps:** From the prementum there run the levatora and depressor muscles of the palps and a flexor of each glossa and paraglossa.

M. **Intrinsic palp muscles:** Inserted on the suspensorium of the hypopharynx.

3. Muscles of Antenna. They are classified as follows:
   A. **Extrinsic antennal muscles:** A levator and usually two depressors are inserted on the base of the scape.
   B. **Intrinsic antennal muscles:** Pair of muscles arising in the scape and inserted on the base of the pedicel.

2.1.2 *Thoracic muscle*

The principal thoracic muscle may be divided as follows:

1. **Longitudinal:** Divisible into tergal and sterna groups, the former being important indirect flight muscle.
2. **Dorsoventral:** Two main groups here are: tergosternal muscles which are principle levators of wing (acting antagonistically to the longitudinal tergals) and tergocoxal muscles that act as the tergal promoters and remotors of the leg (Fig 11.1).
3. **Pleural:** Three main groups included here are - Tergopleural muscles that are variable in development and include the axillary muscle, pleurosternal muscles which are short fibres linking the pleural and sterna apophyses, and pleurocoxal muscles that act as abductors of the coxae.

2.1.3 *Abdominal muscles*

1. **Longitudinal:** Divided into i) tergal and ii) sterna longitudinal muscles. In each case they run between the intersegmental folds or antecostae of successive segment. Acting together the groups acts as retractors by telescoping the abdomen. Acting alone, the sternal muscles curve the abdomen downwards and the tergals straighten it or bend it upwards.
2. **Lateral:** Typically run dorsally and are both inter and intrasegmental in position. They are usually tergosternals, but when distinct pleurites are present there may also be tergopleural and sternopleural muscles. By contraction they tend to compress the segment and are therefore important in respiratory movements.
3. **Transverse:** Better known as the muscles of the dorsal and ventral diaphragms.

In addition there are special muscles for the movement of genitalia, cerci and spiracles.

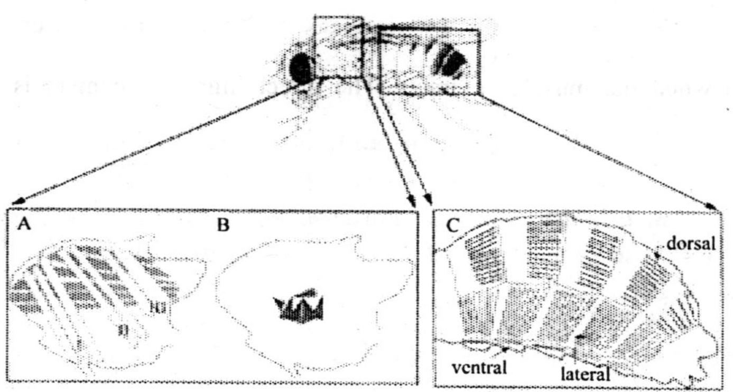

Fig 11.1 Sagittal view of flight muscles and abdominal muscles
A Indirect Flight Muscle* B- Direct Flight Muscle at wing hinge region,
C- Abdominal Muscles of segments 2-6
(*Grey: Dorsal-Longitudinal muscle consists of 6 myotubes; White: dorso-ventral muscle ; I,II,III -consists of three, two and one myotubes respectively)

2.1.4 *Muscles of flight*

Direct and Indirect Flight Muscles

Insect wings are driven up and down by flight muscles located in the animal's thorax. The muscles that power the oscillating wing movements of insect flight are customarily divided into direct flight muscles and indirect flight muscles (Fig. 11.1). Direct flight muscles insert on the base of a wing or on cuticular patches in the wing articulation that are in turn attached to the wings. The direct flight muscles lie ventral to the wings and their attachment to a wing is lateral to the wing s fulcrum. Consequently direct flight muscles are wing depressors, muscles whose contraction produces ventral movement of the wings. Indirect flight muscles induce wing movements by changing the position and shape of the tergum, the dorsal plate of the thorax. The suspension of the wings to the thorax is such that downward movement of the tergum, caused by contraction of the dorso-ventral thoracic muscles, causes wing elevation while upward bowing along the fore-and-aft axis, brought about by contraction of dorsal longitudinal muscles, causes wing depression (Box 11.2).

There are two types of insect flight mechanisms, based on how the flight muscles transmit force to the wings. They are called **direct** and **indirect flight mechanisms**. In the **direct flight mechanism**, at least one power muscle connects to the wing near the joint at one end and at the other end is attached to the thoracic segment. For example, a wing depressor muscle might attach to a special part of the wing near the wing base, and directly pull the wing down when the muscle contracts. The direct flight mechanism is more primitive than the indirect mechanism, and seen in dragonflies and mayflies. However, insects with the direct flight mechanism are not inherently worse flyers than those with the more advanced indirect mechanism. In fact, dragonflies are some of the most impressive fliers, due to their direct flight mechanism allowing them to independently control each of their four wings.

In the **indirect flight mechanism**, the flight muscles invest their energy into deforming the insect's thorax, which in turn causes the wings to move up and down. The muscles are not attached to the wings but to the notum dorsally and to the sternum ventrally (Fig 11.2). This mechanism in particular involves a complicated hinge system at the wing base. The majority of flying insects have indirect flight muscles, including butterflies and moths, beetles, grasshoppers and crickets, flies and bees. The big benefit of the indirect flight mechanism is that the thorax is able to store substantial elastic energy from each half-stroke, and return it to help propel the subsequent half-stroke, in a process that involves mechanical resonance. Such a system makes flapping flight more energy-efficient.

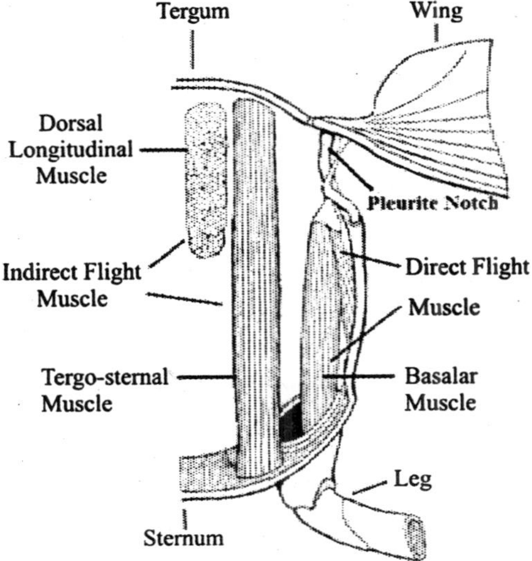

Fig. 11.2 Direct flight muscle (Modified from Snodgrass 1935)

In all insects, the upstroke is affected by contractions of the indirect flight muscles. This results in the notum being pulled down towards the sternum. A fulcrum point at the pleural process forces the wing up. In butterflies, moths, beetles, roaches, dragon flies, the down stroke is powered by the direct flight muscles in conjunction with the relaxation of the indirect muscles. In insects such as bees, wasps and flies, the down stroke is also achieved by indirect muscles called the **dorsal longitudinal muscles**.

Box 11.2 Flight muscles of Insects

1. Direct flight muscles attach to a wing directly.
2. Indirect flight muscles alter the shape of the thorax to cause wing movement.
3. All insects cause the upstroke with indirect muscles that pull the tergum downward.
4. Dragonflies and cockroaches contract direct muscles to pull the wing downward.
5. Bees, wasps and flies arch the tergum to cause the downward stroke indirectly.
6. Beetles and grasshoppers use a combination of direct and indirect muscles to move wings.

Muscles of flight can generally be distinguished from other muscles of the thorax, including control muscles of the wing, by their decidedly pinkish-brown colour. Flight is a metabolically expensive way to move about. The high metabolic rates of the flight muscles are supported by a high concentration of mitochondria within the muscle fibres. Mitochondria make up 20-40% of the volume of flight muscle fibres, and only 10% or less of equivalent muscles from non-flying forms. It is the high concentration of mitochondria, with their constituent cytochromes, that give flight muscle a distinct colour.

Synchronous and Asynchronous Muscles (or Fibrillar and Nonfibrillar Muscles)

Flight muscle fibers of some insects differ histologically from the usual striated muscle fibers of arthropods and vertebrates is being composed mainly of large, easily dissociable fibrils (Tiegs, 1955). Muscles which have large fibrils, measuring up to 2 μm in diameter and which are readily seen under light microscopy are identified as fibrillar muscles. The muscle fibers are sometimes unusually large in fibrillar muscles. The fibers of ordinary striated muscles are typically 0.05 to 0.1 mm in diameter; in contrast those of the fibrillar muscles of diptera are up to 2 mm in diameter (Tiegs, 1955).

The flight muscles of insects could be divided into two classes on the basis of the neural control of contraction- synchronous muscles and asynchronous muscles. In **asynchronous** muscles, the muscle fibres no longer require a nerve impulse for each and every contraction, as they do in typical **synchronous** muscles. Instead, a single nerve impulse elicits multiple consecutive muscle contractions. Insects with asynchronous flight muscles, such as beetles, bees and flies, have wing beat frequencies that are no longer limited by the firing rate of nerve cells, and many of these representatives have incredibly high flapping frequencies as a result (Box 11.3).

Synchronous muscles of insects are like striated muscles elsewhere in that each muscle contraction is initiated by impulses or bursts of impulses in one or more motor neurons to the muscle. The motor neuron impulses evoke depolarisation of the innervated fibres of the muscle termed muscle **action potentials**. An action potential in a fibre triggers release of calcium from an internal store, the sarcoplasmic reticulum (SR), within the fibre. The released calcium diffuses into the myofibrils, binds to control sites on the contractile filaments, and in binding turns on contractile activity of the filaments. Contraction is terminated as the **sarcoplasmic reticulum** re-sequesters the released calcium, reducing its cytoplasmic concentration to a level below that needed for contractile activity. The muscles are called **synchronous** because with them there is a 1:1 relationship between muscle electrical activity and contraction. **Asynchronous muscles** too are activated by muscle action potentials and triggered calcium release from the SR, but once activated asynchronous muscles can contract in an oscillatory manner if they are connected to an appropriate resonant load such as is offered, in life, by the wings and thorax of the insect. The oscillatory frequency of an active asynchronous muscle is typically substantially higher than the frequency of muscle action potentials that maintain activation of the muscle. It is because there is no congruence between action potentials and individual muscle contractions during oscillatory activity that the muscles are termed **asynchronous**.

Box 11.3 Highest wing beat frequency

The current world record holder for wingbeat frequency, weighing in at a ridiculous 1,046 beats/second, is the mighty midge:*Forcipomyia*!

Indirect flight mechanism can be either asynchronous or synchronous (the direct flight mechanism is always synchronous). Insect must alternately contract its flight muscles to drive the thorax and is considered the **driving force**. This system also exhibit mechanical vibration, with energy being stored and released during each stroke reversal to set up a natural oscillation. The insect's flight muscles contract at the right time in order to enhance the natural oscillatory motion of the thorax. This is what is meant by **mechanical resonance** in the indirect flight mechanism. In **synchronous muscle**, a nerve impulse is required for each and every contraction. This limits the rate at which the muscles are able to contract, because it takes time for the physiological processes associated with a nervous stimulus to run their course, and refresh for the next contraction. **Asynchronous muscles** do not require a nerve impulse for every contraction. Instead, a nerve impulse now and then is all that is needed to keep these muscles contracting with exquisite regularity (Box 11.4).

Box 11.4 Flight muscle contraction

1. Synchronous muscle control uses a single volley of nerve impulse to stimulate a wing stroke.
2. Asynchronous muscle stretches the antagonistic muscle and causes it to contract in response.
3. Asynchronous muscles only need occasional nervous stimulation.
4. Potential energy can be stored in resilient tissue.
5. Wing beats may vary from a slow 4/second in butterflies to over 1000/second in midges.

Power and Control Muscles

Power muscles drive the oscillations of the wings, while the control muscles are involved in the continuous adjustment of wing stroke amplitude and angle of attack needed to produce stable, directed flight and for turning during flight. In locusts there is a single control muscle for each wing, the pleuroaxillary muscle, which originates on the lateral body wall and attaches to a cuticular patch in the hinge region of each of the wings. The pleuroaxillary muscles are active during flight, and contraction of these muscles decreases wing pronation (= downward twisting of the leading edge of the wing) during the downstroke, and decreases wing supination (upward twisting of the leading edge) during the upstroke. In addition to and probably more important than the flight control offered by the pleuroaxillary muscles is that provided by differential activation and changes in the timing of activation among the many power muscles of the thorax. Of particular importance here are the basalar and subalar muscles; the former causing wing depression and **pronation**; the latter wing depression and **supination**.

The basalar (s) and subalar muscles of a wing are thus synergists for wing depression but antagonists for altering the angle of attack of the wing. Control muscles become more important in advanced fliers in which there is a reduction in the total number of power muscles able to be individually activated, and in which the power muscles are largely indirect muscles that act by distorting the entire thorax rather than on individual wings. The honeybee, has two pairs of power muscles and five pairs of muscles that probably act as flight control muscles, and in the dipteran *Drosophila,* with 4 pairs of power muscles, there are 18 pairs of small muscles in the thorax that contribute to the control of the one pair of wings.

Bifunctional Muscles

Some of the power muscles of insect flight, both direct and indirect flight muscles, attach ventrally to proximal leg segments. Contraction of such muscles could cause limb movements, especially if the wings are stationary, or wing movements, especially if the leg is held in a fixed position. Thus these muscles are anatomically bi-functional. There is some uncertainty about the extent to which these muscles are functionally as well as anatomically bi-functional. Based on electrical recordings from muscles and motor neurons, Wilson (1962) and Ramirez and Pearson (1988) concluded that the bifunctional muscles of locusts are active, but in different patterns, during both walking and flight. In contrast, Duch and Pfluger (1995) also working with locusts, found that two of the three bifunctional muscles that they examined were active during flight but not during walking, while one participated in both walking and flight but was turned on by different motor neurons during the two activities. It is actually a bit surprising that bifunctional muscles of locusts are sometimes used during walking. These muscles are large, produce twitches whose amplitude is a substantial fraction of the maximum force available from the muscle (ratio of twitch to tetanic tension = 60%), and have quite brief twitch contractions (approximately 40 ms, onset to 90% relaxation at 30°C 19). The muscle kinetics are appropriate for flight, during which the operating frequency is about 20 Hz, but the muscles seem ill-designed for walking movements which require low power and fine control, and which occur at an operating frequency an order of magnitude or so lower than that of flight.

2.2 Visceral muscles

These muscles differ in structure from skeletal muscles in several respects. Adjacent fibres are held together by desmosomes, which are absent from skeletal muscles. The sarcoplasmic reticulum is poorly developed;

mitochondria are small and often few in number. All insect muscles are striated, including the visceral muscles. Visceral muscles can be innervated from the stomodeal nervous system or from the ganglia of the ventral nerve cord, but are sometimes without innervations as in the heart of *Anopheles spp* larvae.

2.3 Cardiac muscles

The insect heart usually consists of a simple tube that contains a layer of contractile myocardial cells. They are usually mononucleate cells with striated longitudinal and cuticular myofibrils. The heart rate is influenced by the nerves that innervate it. These muscles are not homologous to the cardiac muscles of the vertebrates.

3. MUSCLE ULTRASTRUCTURE

3.1 Muscle Fibre Bundles

The skeletal muscles of insects are bundles of elongate, multinucleate cells called **muscle fibres**. The fibres attach at each end to the exoskeleton. The muscles typically span joints of the exoskeleton and, when active, cause bending of the joint or stabilisation of the joint against external forces. In addition to skeletal muscles, insects contain visceral muscles that cause movement of the gut, Malpighian tubules, and parts of the reproductive system; there are also cardiac muscles that cause contraction of tissue sheets and vessels associated with the circulatory system. The visceral and cardiac muscle cells are typically small, spindle shaped, and with a single nucleus.

An individual insect contains many morphologically identifiable skeletal muscles. The large number of muscles is a consequence of the segmental organisation of insects and metamerism, the serial replication of parts associated with segmentation. For instance, each of the wing-bearing segments of a cockroach contains about 50 separate muscles, while an abdominal segment has a lesser number.

In most insect muscles the fibres lie parallel to one another, and when the muscle contracts, it shortens along the long axis of the fibre bundle. Such muscles are spoken of as being **parallel-fibreed muscles** (Fig.11.3, left). In some muscles, in particular peripheral leg muscles, the fibres attach obliquely at one of their ends onto an internal, cuticular extension called an **apodeme** (Fig. 11.3, right). When these muscles are activated, the muscle as a whole shortens along the axis of the apodeme, oblique to the fibre axis. The oblique insertion of fibres onto the apodeme resemble the oblique junction between lateral filaments and the main shaft of a feather, hence muscles with such an oblique fibre arrangement are called **pinnate**. The force that a muscle can generate increases with increasing cross-sectional area. The pinnate arrangement of muscles increases the effective cross-sectional area and hence the force that the muscle can produce.

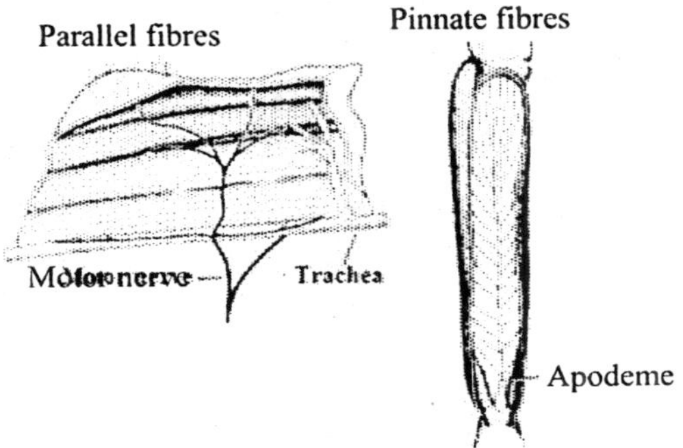

Fig. 11.3. Muscle with parallel and pinnate fibres

3.2 Filaments and Fibrils

Each muscle is made up of a number of fibres, which are long, usually multinucleate cells running the length of the muscle. The characteristic feature of the muscle fibres is the presence of myofibrils (Fig 11.4). These are embedded in the cytoplasm known as sarcoplasm and extend continuously from one end of the fibre to the other. The fibrils are long serial arrays of contractile units known as sarcomeres. Each sarcomere is composed of interdigitated molecular filaments, consisting mainly of two proteins- actin and myosin. Actin molecules constitute the thin filaments and the myosin molecules constitute the thick filaments. These proteins through their cyclic, ATP dependent interactions generate the contractile forces and movements. Each sarcomere is bounded by electron-dense Z-discs which connect neighbouring sarcomeres (Fig 11.4). From either side of each Z-disc (also called Z-line), thin filaments of actin extend toward, but do not reach the centre of the sarcomere. Each sarcomere comprises of an anisotropic region (the A-band) and two half isotropic regions (the I-band) containing the proximal parts of the thin filaments. The dense A-band further has a lighter region, the H-band. The thin I-band actin filaments (5nm in diameter) extend from the Z-disc to the edge of the H-band. While the thicker myosin filament (15 nm in diameter) run through the A-band. Actin and myosin filaments are linked temporary by cross-bridges. Each myosin filament is usually surrounded by six actin filaments.

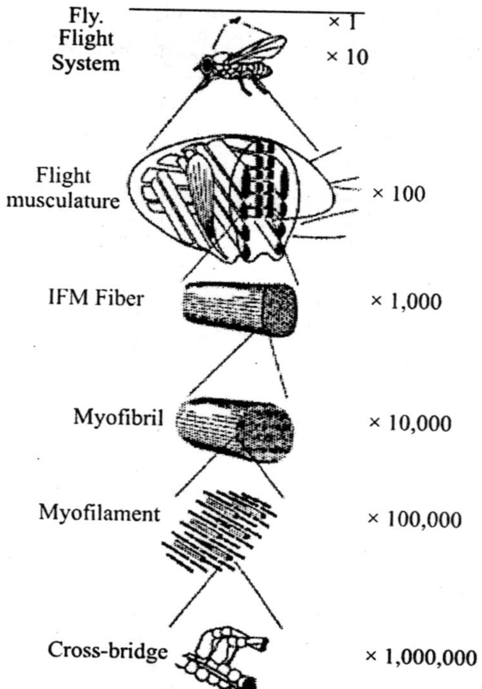

Fig. 11.4. Organisational levels of the flight system of *Drosophila melanogaster* (in powers of ten)

(IMF- Indirect Flight Muscle)
(modified from Maughan DW, Vigoreaux JO. 1992)

Muscle shortening in insects as in other animal's results from sliding movement between interdigitating thick and thin filaments contained within the muscle fibres. The force of contraction is a shearing force developed between these filaments. The thick filaments are made up largely of the protein myosin, the thin filaments of the protein actin. A single thick filament is composed of many individual myosin molecules and, similarly, a thin filament contains many actin molecules. Projections of the myosin molecules from the thick filaments toward the thin filaments, called **cross-bridges**, are the sites of interaction between the two and are the force generators for contraction.

The thick and thin filaments are grouped into longitudinal bundles called **fibrils**. The filaments within a fibril are grouped precisely, both longitudinally and transversely (Figs. 11.3 and 11.4). The thin filaments attach to and project from both sides of transverse structures called Z discs. The Z discs occur regularly along the length of the fibril. The interval from one Z disc to the next is called a **sarcomere**. The thick filaments lay side by side in the middle of the sarcomere. The sarcomere lengths in fibrils of fast muscles such as flight muscles are 2-4 μm; those in leg muscles, body wall muscles, and visceral muscles tend to be longer, up to 7-10 μm. The regular longitudinal arrangement of Z discs, thin filaments, and thick filaments creates a striped pattern along the length of a fibril (Figs. 11.5 and 11.6). The most obvious components of the striped pattern are (1) the Z discs; (2) the A bands, corresponding to that part of a sarcomere containing thick filaments; and (3) the I bands, corresponding to that part of a sarcomere without thick filaments. When a muscle shortens, the thin filaments slide toward the center of the sarcomere and the I bands become shorter. Because of their transverse banding pattern, muscles in insects (and skeletal muscle in vertebrates, which have a similar organization) are described as being striated muscles. The visceral muscles of insects are similar in function to vertebrate smooth muscles and in many ways similar in physiology as well. But although vertebrate smooth muscles lack striations, the visceral muscles of insects, like the skeletal muscles, are striated.

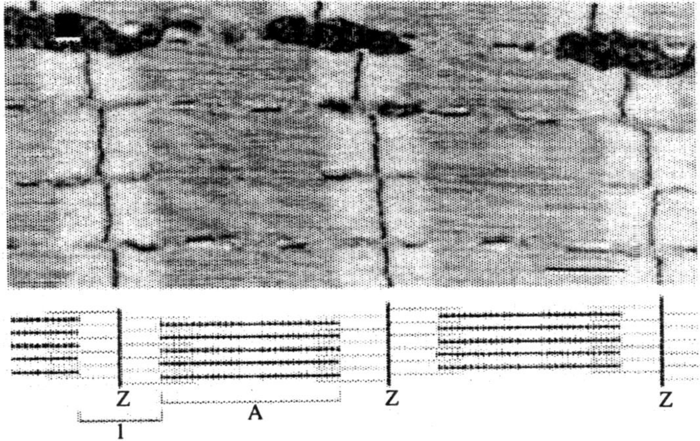

Fig 11.5 Transverse section of skeletal muscle
(M- mitochondria; Z – Z line; A- A band; I- I band)
(Adapted from http://what-when-how.com/insects/muscle-system-insects/)

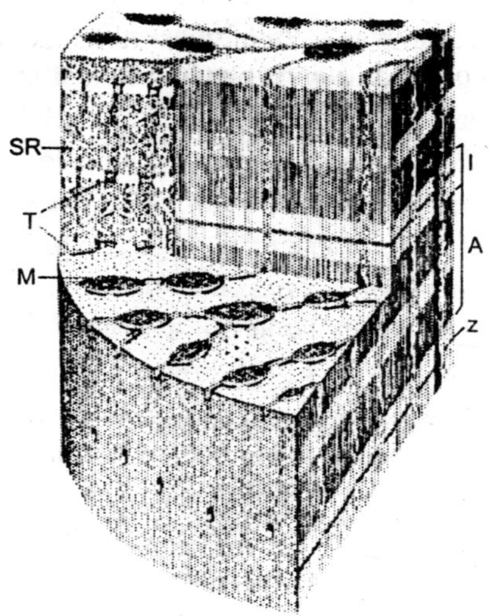

Fig 11.6 Structural organization of a fibre from an insect fast muscle.
(A, A band; I, I band; M, mitochondrion; SR, sarco-plasmic reticulum; T, transverse tubule; Z, Z disk).
(Adapted from http://what-when-how.com/insects/muscle-system-insects/)

The thick filaments of insect muscles, and of vertebrate striated muscles, occur in a regular, hexagonal array. In vertebrate muscles a thick filament is surrounded by six thin filaments, each of which lies at the midpoint between three adjacent thick filaments (Fig 11.7), and the overall ratio of thin-to-thick filaments is 2:1. In fast muscles of insects, for example flight muscles, there are also six thin filaments surrounding each thick filament, but these occur at the midpoint between two thick filaments and the thin-to-thick ratio is 3:1. In slower insect muscles, such as body wall muscles, the usual pattern is for each thick filament to be surrounded by a circle of up to 12 thin filaments.

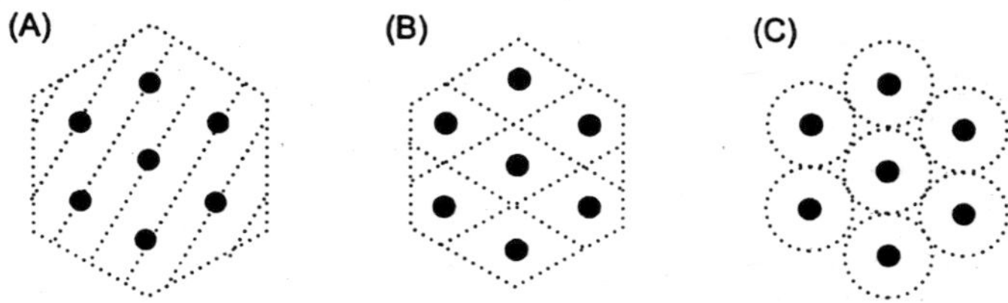

Fig. 11.7. Organisation of thick and thin filaments as seen in cross sections of fibres
(A) a vertebrate skeletal muscle (B) a fast insect muscle (C) a slow insect muscle

The fibrils of insect muscles occur in two basic patterns, cylindrical and radial-lamellar. In muscles with cylindrical fibrils the bundles of filaments forming the fibrils occur as elongate cylinders that are often polygonal in cross section. In radial-lamellar fibres the fibrils are ribbon-shaped structures arranged radially about the centre of the fibre.

3.3 Other Components

The cellular components of muscle fibres fall into four functional groups.

1. Structures directly involved in the generation of force and mechanical power. These structures are the thick and thin filaments that collectively form the fibrils.
2. Components involved in the control of contraction. The most obvious structures involved in the control of contraction are the transverse tubular system (T tubules) and the sarcoplasmic reticulum (SR). The T tubules are membrane-bound tubular structures oriented perpendicular to the fibre axis. The membrane of a T tubule is continuous with the surface membrane of the fibre, and the T tubule can be regarded as an inwardly directed extension of the surface membrane. In most insect muscles there are two tubules per sarcomere, lying in the overlap areas between thick and thin filaments, but in the fibres of some muscles there is a single, centrally located T tubule per sarcomere. Within the fibre the T tubules make specialised junctions with the SR, which is an internally closed, membrane-bound compartment within the fibre. Other elements involved in the control of contraction are the surface membrane of the muscle fibres and membrane specializations at the sites at which nerve processes contact muscle fibres.
3. Structural elements of the metabolic power supply. These are the mitochondria, which provide ATP, and glycogen granules. ATP is the immediate energy source for contraction; glycogen is a stored fuel for cellular metabolism. The mitochondria of insect muscles vary greatly in size, shape and distribution. They are most extensively developed in flight muscles, because of high metabolic rate of these contracting structures. They may be scattered throughout the sarcoplasm or arranged between fibrils opposite to z-discs. It would be appropriate to include among the elements involved in metabolic power the tracheoles, the terminal portions of the gas-exchange system that ramify throughout muscle fibres, even though topologically tracheoles are external to and not really part of the muscle fibres.
4. Structures involved in long-term maintenance of muscle, specifically the many nuclei of the fibres.

The relative abundance of different cellular components in muscle is tightly correlated with the functional capacity of the muscle fibres. SR and T tubules are particularly abundant in muscles that can produce brief contractions, that is, in muscles in which the contractile apparatus can be rapidly activated and inactivated. Muscles capable of sustained activity at high power output are particularly well supplied with mitochondria and tracheolar endings (Box 11.5). Mitochondria make up 30-40% of the muscle volume in wing muscle of active fliers and in sound-producing muscles that are active continuously and at high frequency. Such muscles are often pink, because of the cytochromes in the abundant mitochondria.

Box 11.5 Evolutionary problems with insect flight muscles

Because of their size (thickness) and mode of oxygen delivery, unlike oxygen delivery in vertebrates via the blood and capillaries, insects would have had a problem in getting oxygen to all of the muscle fibers and problems with the depolarization system. These constraints have been overcome by:

A. The tubular T-system and

B. The tracheal system

This tubular system is created by the tracheoles, which deliver the oxygen deep into the muscle fiber and the T-system that permits the depolarization reaction, taking place on the surface of the cell, to also penetrate deeper into the muscle fibers.

3.4 Contractile Proteins of Myofibrils

The myofibril is a structural complex of approximately twenty proteins that compose distinct but interacting filament systems and cross-linking structures organised in repeating units known as **sarcomeres** . The majority of these proteins (e.g., actin, myosin, tropomyosin) are highly conserved in striated muscles of invertebrates and vertebrates. The IFM also contain additional proteins, such as flightin, and variants of conserved proteins (arthrin, troponin H) that are unique to this tissue (Table 11.1).

3.4.1 Myosin- Structure and function

Myosin II of insect flight muscle, like those of other invertebrate and vertebrate muscles, is composed of two heavy chains, two essential light chains and two regulatory light chains (Fig. 11.8). Myosin heavy chain (MHC) (-200 kDa monomer) contains an N-terminal head domain and a C-terminal rod region. Initial dimerisation and subsequent multimerisation of myosin occurs through the rod region (-1150 amino acid residues). The multimerised rods serve as the main constituent of the thick filament. The head (-850 amino acid residues) contains the nucleotide-binding site, actin-binding site and the lever arm domain that binds the light chains. The light chains interact with the long alpha-helix of the heavy chain and this complex is called the **lever arm domain**. The head protrudes from the thick filament surface, forming a cross-bridge upon transient strong binding to actin, but spends the remainder of the contraction cycle in a detached or weakly bound state.

Table 11.1. Myofibrillar Proteins in IFM Development

Protein	Size (kDa)	Component of	Function
Myosin essential light chain	18	Thick filaments	Not known
Myosin regulatory light chain assembly	24-30	Thick filaments myofibril	Required for normal
Paramyosin	90-105	Thick filaments	Not known
Mini-paramyosin	55	Thick filaments/	Stretchin-mlck M line
Myostrandin	231, 225	Thick filaments	Not known
P165	165		
Myosin heavy chain	200-220	Thick filaments	Essential for thick filament assembly
Zetalin	210	Z band	Not known
α-actinin	97-107	Z band	Required for normal Morphogenesis of Z bands and terminal insertion sites
Troponin C	18	Thin filaments	Not known
Troponin H	70, 78	Thin filaments	Not known
Troponin T	47	Thin filaments integrity	Required for myofibril
Troponin I tractile force	25-29-35	Thin filaments	Required for sarcomere formation; regulate con
Arthrin	49-56	Thin filaments	Not known
Tropomodulin	45	Thin filaments	Required for thin filament length determination
Actin	43	Thin filaments	Essential for thin filament and Z band assembly
Tropomyosin	35-37	Thin filaments	Required for thin filament stability
ADP/ATP translocase	33	?	Not known
Glutathione-S- 2 transferase	32-35	Thin filaments	Not known
Flightin	20	Thick filaments	Required for thick filament length determination and stability
Projectin	1000	Connecting filaments	Not known
Kettin	540, 700, 1000	Connecting filaments	Mutations lead to defective terminal Z discs

Prior to assembling into the thick filament, myosin molecules dimerise via their alpha-helical rod regions to produces the hexameric complex. Partial digestion with a-chymotrypsin or papain cleaves the molecule just C-terminal to the regulatory light chain (RLC) (black oval) binding domain, producing the S1 head and the rod domain. The RLC can be digested away during α-chymotrypsin incubation, while the essential light chain (ELC) (white oval) is retained. Treatment of vertebrate myosin nnerva with trypsin cleaves the molecule on the C-terminal side of the hinge region resulting in heavy meromyosin (HMM) and light meromyosin (LMM), while prolonged incubation liberates the hinge region.

Significant conformational changes occur in myosin as it progresses through its biochemical cycle. These include reconfiguration of the nucleotide-binding pocket, opening and closing of a long cleft that runs between the actin-binding site and the active site, and movement of the lever arm. Thermodynamic coupling of the actin and nucleotide-binding domains ensures that the chemical energy released during ATP hydrolysis results in mechanical movement of the lever arm. Power generated from the lever arm stroke promotes muscle contraction by thin filament sliding. This force-generating step occurs concurrently with attachment to actin and is generally thought to take place just prior to or simultaneous with phosphate release. Summation of the forces generated by individual myosin powerstrokes provides the force for sarcomeric shortening. ADP is released after the powerstroke and produces the rigor state; subsequent binding of ATP to the rigor myosin conformation liberates the myosin head from actin and thus governs exit from the attached state.

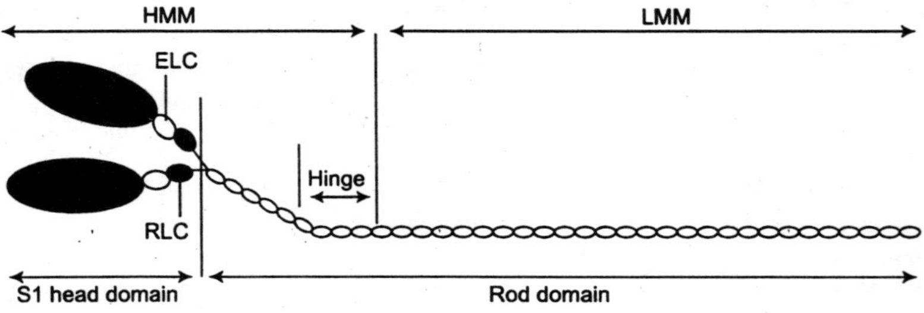

Fig. 11.8. Schematic sketch of mysoin molecule

(HMM- Heavy Mero Myosin; LMM- Light Mero Myosin; RLC- Regulatory Light Chain; ELC- Essential Light Chain)

Recent evidence in *Drosophila* suggests stage- and tissue-specific myosin isoforms are primary determinants of muscle contractile properties (Swank *et al.*, 2002). Myosin light chains bind to the alpha-helical lever arm of the heavy chain. The essential light chain (ELC), also known as MLC1 or the alkali light chain, is a protein of – 20 kDa. The role for the ELC in insect muscle contraction is not clear, and as yet there is no consensus regarding its function in vertebrates. The regulatory light chain (RLC), also designated MLC2, is a protein of 30-34 kDa and is important for regulation of muscle contraction. In *Drosophila,* RLC phosphorylation at Ser 66 and *67* is necessary for maximal ATPase activity, recruitment of cross-bridges, enhancement of stretch-activation of the indirect flight muscle and optimal flight function. The RLC has also been implicated in stabilizing the myosin neck and is important for maintaining the appropriate structural separation of the myosin heads of a given dimer. In *Drosophila* there is a unique RLC N-terminal extension of 46 amino acids, similar to the ELC extension in vertebrates that binds actin. This extension is thought to be important in maintaining a link between the thick and thin filament parallel to the myosin cross-bridge.

The heptad repeat of charged and hydrophobic amino acids within the C-terminal rod domain of myosin heavy chain (MHC) plays a role in assembly of monomers into nnerva through formation of an α-helical coiled-coil (Fig. 11.8). Subsequently, these myosin nnerva multimerise into hollow thick filaments through higher order interactions between neighbouring rod nnerva, α-helical coiled-coil paramyosin core proteins and flightin. Studies of the indirect flight muscle (IFM) *of Drosophila* show that the rod domain is critical for assembly and stability of the thick filament. Interestingly, the catalytic head domain also serves to fine tune the assembly of thick filaments and sarcomeres.

While myosin biochemical properties have been studied in only a few insects to date, the tools of molecular genetics have revealed that multiple isoforms of insect myosin exist in a single species, with specialised isoforms accumulating in flight muscles. In at least some insect species, isoforms of myosin heavy chain and the essential light chain arise from the process of alternative splicing of transcripts from a single gene. Mutations in *Drosophila* myosin, in conjunction with molecular modelling, implicate particular amino acid residues in thick filament assembly, sarcomere stability and ATPase activity. Molecular genetics approaches and transgenic technology in *Drosophila* are proving powerful in demonstrating how structural elements of myosin affect functional properties at the biochemical, fiber and whole organism levels. These integrative studies show that properties of the indirect flight muscle are critically dependent on the specific myosin isoform expressed.

3.4.2 Paramysoin and Miniparamyosin

Paramyosin and miniparamyosin are two proteins that are only found in association with thick filaments in invertebrate striated muscles. *Drosophila* paramyosin has a central α-helical coiled-coil rod flanked by two non α-helical terminal regions. The terminal regions of *Drosophila* paramyosin are very short and the regions are much less conserved than the rod portion when the sequences of different species are compared.

Paramyosin has a very similar structure to the myosin heavy chain rod tail. It contains 878 amino acids and is a rod-like molecule in which two 102 kDa monomers interact to form a coiled coil. In *Drosophila melanogaster* three paramyosin isoforms exist, which share a similar molecular mass. In general, the protein accumulates in the adult head, abdomen and thorax and in vivo, the appearance of the most acidic phosphorylated isoform of paramyosin has been specifically related to the functioning of the flight-related thoracic musculature. Moreover, as in the case of myosin light chains, and of the flightin protein, it appears that the phosphorylation of the paramyosin isoforms may modulate the assembly and/or function of thick filaments.

Paramyosin is produced during two different stages of development. It is detected at the late embryonic stages and remains present through the larval instars. It can then be detected again in mid-pupal stages through adulthood. The protein appears at around ten hours post-fertilisation and it accumulates progressively during middle and late embryogenesis, until reaching maximal levels in adults. The protein is present in oocytes and very early embryos (3 hours of development), indicating a possible maternal inheritance. By immunocytochemistry in whole vembryos and cryosections of adult flies, paramyosin protein can be seen to accumulate in all muscle groups: pharyngeal, somatic, visceral and specialized thoracic musculature. In electron microscopy images, paramyosin is distributed along the entire A-band of the sarcomere in *Drosophila* IFM and Tergal depressor of the trochanter (TDT) muscles.

The miniparamyosin dimer is a minor component of myofibrillar thick filaments. This protein is found mainly in the thoracic tubular jump muscles (TDT) while it is much less predominant in the fibrillar flight muscles. The miniparamyosin monomer has a predicted molecular weight of 54.887 daltons and it shares a common 363 amino acid C-terminal cc-helical domain with paramyosin. In addition, miniparamyosin has a unique 114 amino acid N-terminal domain that lacks homology to other known proteins. Six in vivo phosphorylated miniparamyosin isoforms have been identified in the head and thoracic musculature, over a very wide pH range (pH 6 to 8). As with paramyosin, it is possible that the phosphorylation/

dephosphorylation of miniparamyosin is involved in the acquisition of flight ability. In contrast to paramyosin, miniparamyosin is almost exclusively found in the adult musculature. A transient accumulation of miniparamyosin has been detected in third instars larvae but this decreases rapidly during pupation. Furthermore, miniparamyosin is differentially distributed in the sarcomeres of *Drosophila* muscle. In the IFM, the protein is found in the M-line and at the ends of the A-band whereas in the TDT, it is located throughout the A-band.

3.4.3 *Novel Mysoin Associated Proteins*

Flightin

Drosophila flightin is a 20 kDa protein that is expressed exclusively in insect flight muscles (IFM). Northern blot and microarray analyses showed that transcripts are present from late pupal stages through adulthood, a pattern consistent with flightin being specific to adult muscle. In situ hybridization and southern blot analyses indicated that flightin is encoded by a single gene located in polytene region *76* D/E. In the sarcomere, flightin is distributed rather homogeneously throughout the A band except at the M line and the edge of the A/I junction. This distribution suggests that flightin is associated with the thick filament. *Drosophila* flightin exists as two unphosphorylated and nine phosphorylated isovariants in adult muscle.

Flightin bears no sequence homology to any known protein. It also lacks motifs or modular domains that may allow its assignation to a protein family. There are several interesting features in the sequence of flightin that may be relevant to its functional properties. The N-terminal one third of the protein is highly acidic (pI ~3.8) while the C-terminal two thirds are basic (pI ~10.6). Separating these two regions is a stretch of 5 prolines. The tertiary structure of flightin is not known, but secondary structure prediction programs suggest the protein has two α-helical regions (residues 88-114 and 149-180), with the rest of the protein having an extended or random conformation. It suggests that flightin is essential for maintaining the structural integrity of the myofibril by linking or stabilizing the thick filaments in the lattice, sarcomere integrity in active muscle, and normal contractile function.

Zeelins

Zeelins are proteins found in *Lethocerus* leg and flight muscle. Zeelin 1 (35 kDa) is found in both leg and flight muscle, while zeelin 2 (23 kDa) is found only in flight muscle. Immunofluorescence microscopy of IFM myofibrils show that zeelin 1 is found predominantly at the ends of the A band (towards the H zone and Z disc) while zeelin 2 has a more broad distribution throughout the A band, on either side of the H zone.

Two dimensional gel electrophoresis revealed that zeelin 2 separates into two major and one minor isoform while the more basic zeelin 1 separates into two isoforms. Lower molecular mass forms of zeelin 1 were also found. The proportion of zeelin isoforms differ in leg muscle and IFM. The isoforms of zeelin 2 were phosphorylated in vitro with the catalytic subunit of cAMP dependent protein kinase, resulting in a uniform shift of the three isoforms. This suggests that the differences among the in vivo isoforms, which are of the same molecular weight, are not due to phosphorylation.

The role of zeelins is not known. Ferguson *et al.*, (1994) proposed that zeelins may be involved in maintaining the ordered structure of the thick filament and the regularity of the myofibrillar lattice. Given the similarities between zeelin 2 and flightin, these proteins are likely to play analogous roles.

Stretchin-MLCK

Stretchin-MLCK is a conceptual *Drosophila* protein that has been studied through the virtual translation of its coding region. The transcription unit is hypothesised to express seven different transcripts. None of the conceptual protein products have been shown to be expressed in IFM, although preliminary studies have identified a protein that may correspond to one of the small kinase isoforms and a second, kettin-like isoform that does not correspond to any of the predicted products.

3.4.4 *Actin and Arthrin*

Actin

Actin is a highly conserved protein found in all eukaryotic cells where it exists in two forms, the soluble monomeric G-actin and the filamentous polymeric F-actin. It contributes as a major cytoskeletal component to the structure of cells, and as actomyosin to generate force with members of the different families of the so-called nonconventional myosins. In all these activities the roles of F-actin are determined by its interactions with a vast array of cellular proteins known as actin-binding proteins (ABP). During myofibrillogenesis sarcomeric G-actin polymerises to F-actin to form the core of the thin filament and the Z-disc.

Monomeric G-actin consists of a single polypeptide chain of 375 residues with a mass of 42,000 that binds a single adenosine nucleotide (ATP or ADP) and a single divalent cation, usually magnesium in vivo. Actin has an ATPase activity. G-actin, ATP monomers are added to the 'barbed' end of the F-actin during polymerisation and, shortly afterwards, the actin ATP is converted to actin ADP.Pi with the subsequent release of Pi.

Polymerised, filamentous actin, F-actin, is helical with a diameter of about 10nm. Decoration of F-actin with myosin subfragment 1 (SI) reveals a **herringbone** pattern, which indicates that F-actin filaments are polar from one end to the other. All the actin monomers lie in similar orientations within a filament.

Arthrin

A novel heavy actin, arthrin, was first described in actin preparations from the IFM of the waterbug *Lethocerus*. This protein has very similar properties to actin: it polymerises to form F-arthrin, filaments which can be decorated with myosin SI subfragments. F-arthrin will activate the myosin Mg. ATPase activity and this activity is regulated by tropomyosin-troponin complexes from insect or mammalian muscle sources in the same way as F-actin. Arthrin has now been detected in a large number of insects including all the higher Diptera, except the primitive *Tipula spp.*, all Hymenoptera and a number of the Hemiptera. In all cases examined it has been found to be restricted to the IFM.

Studies of arthrin in *Drosophila* showed that it is not encoded by a novel gene but is a post-translationally modified form of the IFM-specific ACT88F actin isoform. Recent mass spectrometry studies have confirmed that arthrin is formed through conjugation of the carboxyl group of the C-terminal glycine of ubiquitin to one of the ε-amino groups of lysine 118 of actin.

3.4.5 *Troponin, Tropomyosin and GST-2*

Troponin complex

The regulation of the sliding of thin over thick filaments during muscle contraction depends to a great extent on the proper activity of the troponin complex. This is a tetrameric protein ensemble composed of three troponins: I, C and T, and tropomyosin. The troponin acronyms stand for their respective roles within the complex. In thin filament regulated muscle contraction, troponin I (Tnl) inhibits the interaction between actin and the myosin head at rest. Troponin C (TnC) is the calcium binding component when this ion is released from the storage compartments at the muscle activation point. Troponin T (TnT) anchors the complex to tropomyosin (Tm) providing a regular spacing pattern along the actin based thin filament (Fig 11.9).

A flight muscle specific form of tropomyosin, is the so called **troponin heavy** (TnH) because it co-purifies with the troponin complex. In *Drosophila,* TnH has been suggested to interact with both the thick and thin filaments thus allowing length changes to be directly transmitted to the Ca^{++} regulatory machinery. In *Lethocerus,* homologies between TnH and portions of Tnl have been proposed to be responsible for the effects of stretch on the troponin complex. However, the normal oscillatory work observed in IFM isolated from *Drosophila,* which are haploid for TnH fail to provide support for its role in stretch activation. Furthermore, TnH is not found in several insect species that have stretch activated IFM thus indicating that TnH is not a universal requirement for stretch activation.

Troponin I

Tnl is a globular protein about 269 amino acids long depending on isoforms or species. In *Drosophila,* it is encoded by a single gene, *wings up A (wupA),* located in chromosomal band 16F7, that yields up to ten protein isoforms by differential splicing. Tnl isoforms are muscle type specific. A given muscle, however, can express several Tnl isoforms.

The most characteristic role of Tnl is its inhibitory effect of muscle contraction by preventing the actin-myosin head interaction, that is, the acto-myosin MgATPase enzymatic activity. However, the TnT/TnC interaction plus the Tm binding to actin are also contributing factors to this enzymatic inhibition. Recent data using peptidic fragments of these proteins and NMR methods indicate that the binding of Tnl to actin causes a conformational change in the latter that result in allosteric effects at additional sites of the actin monomer. A further modulatory process on Tnl, at least the vertebrate cardiac isoform, is phosphorylation first at Ser24 and later at Ser23 by cAMP-dependent protein kinase. This modulation affects the Ca^{+}-binding activity by TnC and, perhaps, relates to the phenomenon of stretch activation.

Troponin C

TnC is a Ca^{2+}-binding protein about 155 amino acids long that represents a specialised form of calmodulin, to which it is evolutionary related. The crystal structure of vertebrate fast skeletal version shows two globular domains bound by an α-helix. Two EF-hand motifs for Ca^{2+} binding are present in each globular domain. Depending on muscle type and species, however, these four sites may be active, inactive, or mutated to a Mg^{2+}-binding activity. Also, their affinity constant for these cations differ according to isoforms. Contrary to vertebrates, in most insects and molluscs the two N-terminal EF-hand sites are either inactive, due to changes in cation coordinating key amino acids (site 1), or show low affinity for Ca^{2+} (site 2). The two C-terminal sites bind Mg^{2+} (site 3) and exhibit low affinity for Ca^{2+} (site4), in contrast to vertebrates where both sites have high affinity for Ca^{2+}. It is through the C-terminus that the interaction between TnC and Tnl takes place at rest, when the free Ca^{2+} concentration is very low. Thus, the two EF-hand sites at the C-terminus play a structural role in vertebrates (because of their high cation affinity) while they are regulatory in invertebrates (because of their low affinity). Also at rest, the N-terminus of vertebrate TnC interact withTnT. When mollusc muscle is activated, these interactions change drastically due, in part, to the binding of Ca^{2+} to the low affinity EF-hand sites at the N terminus of TnC. At this stage, the interaction between TnC and Tnl shifts towards the N-terminus of TnC and the C-terminus of Tnl liberating the inhibition on the actin-myosin head interaction.

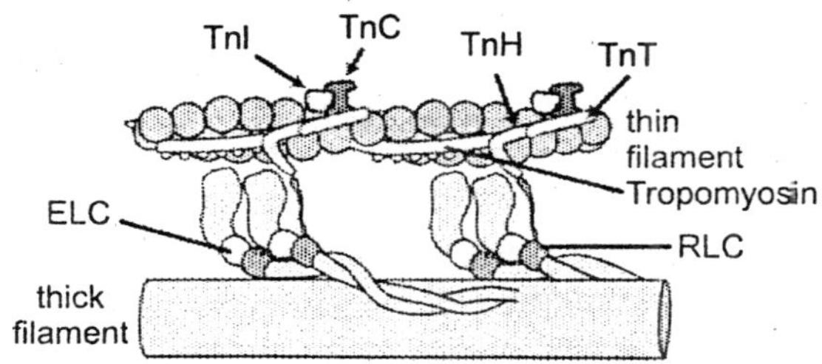

Fig. 11.9. Proteins in the thin filament

In addition to the structural and functional diversity of the four EF-hand motifs, insect TnC isoforms differ in the α-helix separating the two globular domains. Single amino acid insertions in this helix are frequent in certain isoforms of *Drosophila, Anopheles, Apis* and *Lethocerus*. Considering their restricted muscle type expression, these structural peculiarities are likely to sustain specialized functional properties, such as stretch activation.

Troponin T

TnT is a long molecule (18 nm) that interacts with Tm for most of its length and, as a result, serves to space the Tn/Tm complex at 38.5 nm intervals along the thin filament. Many different isoforms are expressed throughout development either from different genes (vertebrate skeletal) or by differential splicing (insects and vertebrate cardiac). The main functional difference among TnT isoforms is their effect on the actomyosin Mg-ATPase sensitivity to Ca^{2+} concentrations. TnT is a 396 amino acid long protein.

In contrast to vertebrates, *Drosophila* TnT has a negatively charged, Glu-rich, C-terminal extension that has been proposed to bind Ca^{2+} and influence the Ca^{2+} activation of the muscle. This suggests that the mechanism by which insect TnT influences Ca^{2+} activation differs from that of vertebrate TnT where the Glu-rich extension is very short.

Tropomyosin

Tropomyosins exhibit several (6-7 in vertebrates) actin-binding domains, each composed of about 40 amino acids. Two Tm monomers bind in opposite orientation to form the functional unit. Its two α helices arrange into a coiled-coil structure that is able to bind along the two grooves of the actin filament through clusters of Ala residues that produce local destabilization of the coiled-coil rigid structure, allowing the required flexibility to adapt to the filament grooves. Vertebrate α-tropomyosin cooperates with the actin-binding protein tropomodulin 1 in the process of thin filament capping and stabilization. At the other end of the thin filament, where it binds to the Z disk lattice, tropomyosins are competed out by kettin.

While vertebrate tropomyosins are encoded by four independent genes (α, β, γ and δ) yielding more than 20 protein isoforms, the *Drosophila* counterparts are encoded by two genes: *Tm1* and *Tm2,* located at chromosomal band positions 88E12 and 88E13 respectively with shared regulatory elements. Tropomyosin 1 is a 518 amino acids long protein that has been referred to as Tropomyosin II in previous literature. In addition to a regular muscle (mTml) and a cytoskeletal (cTml) protein isoforms, the gene *Tml* encodes two IFM specific isoforms, TnH33 and TnH34, characterised by a Pro-Ala rich extension of about 200 amino acids at the C-terminus. It has been proposed also, that these Pro-Ala extensions, following phosphorylation, could form connections between thin and thick filaments and be involved in the phenomenon of stretch activation.

Tropomyosin 2 is a 284 amino acids long protein that has been referred to as Tropomyosin I and Ifm(3)3 in earlier publications. Tml and Tm2 play distinct roles in muscle biology. While a single copy of the *Tm2* gene yields flightless adults with some defects in myofibrillogenesis, a single copy *of Tml* does not. In the first case, muscle power output is 32% of wild type and in the second is 73%. Also, flies nearly deprived of Tm2 yield 1% of power output only, though the core myofibrillar structure is rather normal. These data do not seem to support the key role suggested for Tml (TnH isoforms) unless some compensatory effects between both genes takes place, perhaps through the common regulatory elements.

GST-2

Glutathione S-transferase (GST) is a family of enzymes that play a general role in oxidative and detoxification processes. In insects, GSTs can be classified in three types (I-III). While types I and III (a.k.a. δ and ε classes respectively) are involved in insecticide resistance, type II (a.k.a. *σ* class) remain poorly studied. Currently known as DmGstSl-1 (E.C.:2.5.1.18), *Drosophila* GST-2 is a type II GST enzyme that is abundant in the indirect flight muscles. Since GST-2 has been shown to conjugate 4-hydroxynonenal, an aldehyde product from lipid peroxidation, it is thought to protect from oxidative stress during flight. Its crystal structure at 1.75Å resolution is known. It forms diers showing the canonical GST fold with glutathione bound to only one of the two binding sites. In addition, the alpha6 helix domain provides a novel electrophilic substrate-binding site (H-site) which is similar to that of other GSTs and may explain the selectivity for peroxidation products, thanks to the directionality of its polar and hydrophobic residues.

3.4.6 Z –Band Proteins

a-Actinin

Alpha actinin was discovered in 1964 by Ebashi and localised in vertebrate Z-bands by Masaki *et al*., in 1967. It has been found in Z-bands or related structures that anchor thin filaments in every muscle studied. *Drosophila* and vertebrate muscle α-actinin sequences have an average 68% identity. The α-actinin monomer has two N-terminal calponin homology domains that bind actin, a central rod segment with four spectrin-like repeats and a C-terminal calmodulin-like domain with two EF-hand motifs. Since a recombinant peptide containing C-terminal EF hand domain sequences of *Drosophila* α-actinin does not bind Ca^{2+}, the native protein is unlikely to be Ca^{2+} regulated. Two subunits combine to form an anti-parallel homodimer in which the four spectrin repeats of the anti-parallel partners are aligned. Since the configuration places actin binding domains at both ends of the molecule, α-actinin is presumed to participate in cross-linking actin filaments within the Z-band. This assumption has been supported in part by compelling electron microscope images of actin filaments linked in vitro by transversely oriented α-actinin molecules that form ladder-like structures with thin filaments. Cross linked actin filament pairs have been observed with opposite as well as common polarity.

Kettin

Kettin is a high molecular weight protein that was identified by monoclonal antibodies raised against isolated IFM Z-disc preparations. Kettin sequences have been found at the Z-band in all larval and adult tissues examined. Kettin is an actin binding protein with 35 immunoglobulin domains separated by linker sequences. The kettin/actin stoichiometry suggests that each Ig domain and one adjacent linker interacts with an actin monomer, and it is speculated that kettin may follow the genetic helix of actin. Immunoelectron microscopy studies place the N-terminal domain of the molecule near the center of the Z-band where it is thought to overlap somewhat with kettin from an adjacent sarcomere. The model predicts that actin is associated with 15 kettin Ig modules that span 45nm within the Z-band and with another 20 modules that extend 60 nm into the I-band where they might add structural support to the thin filament. Since a bacterially expressed kettin Ig domain with flanking spacer sequences binds to α-actinin, it may be that kettin binds to both actin and α-actinin within the Z-band and helps to reinforce cross links between thin filaments. Since kettin and α-actinin both bind noncompetitively to F actin, they must have different binding sites on the thin filament. In in-vitro assays tropomyosin has been found to compete with both kettin and α-actinin for actin binding. However, immunoelectron microscopy evidence indicates that tropomyosin is restricted from the Z-band.

Projectin

Insect projectin (mini-titin) is a ~700-1200kD protein that forms all or a part of the connecting filaments that link thick filaments to the Z-band in insect fibrillar flight muscle. Isolated projectin molecules from locust flight muscle have a contour length of about 250Å and are long enough to extend from the Z-band to the terminal regions of the thick filaments. Although the orientation of projectin in the IFM has not been established, its similarity to vertebrate titin in structure and function suggest that the N-terminus is likely to be toward the Z-band. This view is supported by preliminary immunoelectron microscopy studies showing that polyclonal antibodies, raised against N-terminal immunoglobulin domains of *Drosophila* projectin, bind within the Z-band.

Since connecting filaments in honeybee IFM remain distinct and have the same lattice position in the I-band as thick filaments in the A-band, they are assumed to follow a straight course into the Z-band where they form the hub of the C4 filaments within the Z-band lattice. Projectin is likely to be the protein at this lattice position, but the identity of the C4 components is not known. Projectin is expressed in all insect muscles, but in those outside the IFM it is limited to the A band where it cannot function as a connecting filament. Nonfibrillar muscles have longer I-bands than those in the IFM, and connecting filaments would likely be comprised of proteins considerably larger than projectin. Titin/connectin-like proteins of ~3 Mda have been detected in leg muscles of beetle, bumblebee and waterbug, and evidence suggests that they tether thick filaments to the Z-band. Thus, in nonfibrillar insect muscles projectin and titin/connectin-like proteins coexist, and titin/connectin may play the role in these fibers that projectin plays in the IFM.

Zetalin

Four monoclonal antibodies raised against *Drosophila* isolated Z-disc preparations recognize a protein that migrates just behind myosin on 5% polyacrylamide SDS gels of *Drosophila* IFM myofibrils and is found within the Z-band matrix. The protein was originally called Z(210) but was changed to zetalin (from the Greek 'zeta' or Z, and 'lin, short for line).

Zetalin appears during development at pupal stage P8, when myofibrils begin to be assembled. It is organised within the Z band in the IFM and in a subset of jump muscle cells, but it is not detected in any other somatic or visceral fibers. Of the 32 jump muscle fibers, the 28 that express zetalin are notably larger than the 4 anterior cells that lack the protein, and they use a different myosin isoform. The large jump muscle fibers also differ, however, from the IFMs not only in sarcomere structure (e.g., 6:1 vs. 3:1 thin/thick filament ratios), but in contractile protein isoforms and mechanical properties. Zetalin is not widely expressed in adult muscles.

Z(158) and Z(175)

Z(158) and Z(175) are immunologically similar proteins that are localised at the Z-line of honeybee IFM. Polyclonal antibodies directed against either antigen cross-react with the other, and both proteins are recognised by a monoclonal antibody raised against honeybee isolated Z-discs. Unlike α-actinin, kettin and projectin, which are found in all insect fiber types, Z(158) and Z(175) are expressed only in the IFM. In this respect, these honeybee proteins are similar to zetalin which has a restricted distribution in the fly. However, neither polyclonal nor monoclonal antibodies against Z(158/175) cross react with zetalin or any other *Drosophila* proteins.

Capping Protein

Drosophila capping protein (CP) is an αβ heterodimer that binds to the barbed end of actin and is found at the Z-band in adult thoracic muscle. It is related to the vertebrate Z-band protein, CapZ (36/32) which not only caps thin filaments, but serves as a Z band docking site for various nnervatio molecules. Since procedures that interfere with the CapZ/actin interactions in cultures of chick myotubes delay thin filament organisation into sarcomeres, Cap Z is considered an important player in myofibrillogenesis.

MSP-300

MSP-300 is a ~300kDa dystrophin-like *Drosophila* protein, located in the Z-line of larval fibers and expressed in differentiating embryonic muscle. In cultures of fusing myoblasts the protein is found at sites where actin filaments are connected to the sarcolemma, and it is concentrated at the leading edge of migrating myotubes. In embryos, during the early stages of muscle development, the protein is detected at regions of muscle-muscle and muscle-ectoderm attachment sites. Since the protein partially co-localizes with integrin, a transmembrane receptor for the extracellular matrix, and since integrin β-chain null mutations disturb the membrane targeting of MSP-300, the two proteins are likely to interact either directly or through other cytoplasmic integrin binding proteins. Although in differentiated larval muscle MSP-300 is predominately located at the Z-line,100 it retains its membrane association, and may help to tether the Z-band of peripheral myofibrils to the sarcolemma.

Non-muscle Myosin-II

Like MSP-300, non-muscle myosin-II heavy chain, a product of the *zipper (zip)* gene, co-localizes with integrin during *Drosophila* embryonic myogenesis. At stage 16 when muscle-specific myosin-II is diffusely distributed throughout the sarcoplasm, non-muscle myosin-II is concentrated with integrin at attachment sites of muscle termini where actin is linked to the membrane. At embryonic Stage 17 contractile proteins are organized into sarcomeres, and the process is dependent on non-muscle myosin-II. In muscles of *zip* null mutants I-bands are not observed, and normal myofibril formation is arrested. The *zip* gene product is assumed to have a role in organising arrays of thin filaments. This is further suggested by the finding that in 3rd instar larvae non-muscle myosin-II is localized at the Z-line. It is speculated that the protein functions as an actin cross-linker and supports α-actinin in stabilising the Z-band.

4. MUSCLE ATTACHMENTS

Skeletal muscles attach to the cuticle of the exoskeleton at one end through specialised epidermal cells. At the other end they are attached to the skeleton so that contraction of the muscle moves one part of the skeleton relative to the other. The muscle fibres are joined to the epidermal cells of the cuticle by specialized junctions called desmosomes. The terminal sarcomeres of the fibrils lack a final Z disc; instead, the thin filaments are attached to the muscle portion of the terminal desmosome through a band of what have been called junctional filaments. Within the epidermal cells, microtubules connect the desmosomes with cone like depressions of the outer epidermal plasma membrane. From each cone an electron dense muscle attachment fibre or tonofibrilla runs through the procuticle in a pore canal and finally inserts on the epicuticle. These tonofibrillae is slowly dissolved when the old cuticle is digested by moulting fluid. New tonofibrillae become attached to the epicuticle only when it is growing.

The muscle attachment fibres are not digested by moulting fluid. During moulting they retain their attachment to the old cuticle across the exuvial space between the new and old cuticles. As a result the insect is able to continue its activities after apolysis during the development of the new cuticle. Finally, at the time of ecdysis, the connections to the old cuticle are broken. Visceral muscles are frequently joined to one another by desmosomes, and cardiac muscle fibres are joined by structures resembling the intercalary disks of vertebrate cardiac muscle.

5. INNERVATION AND ACTIVATION

Nerve cells in the central nervous system send out long processes (motor axons) to the muscle fibres where they make specialised contacts termed **synapses**. A motor axon makes many synaptic contacts along the length of each muscle fibre that it innervates (multiterminal nnervations), and a single muscle fibre may receive inputs from more than one motor axon (polyneuronal nnervations). Basically each unit is innervated by a fast axon and a slow axon and also by an inhibitory axon. In the jumping muscles of locusts, about 40% of the fibres receive branches from both axons, but in the flight muscles of Odonata, Orthoptera, Diptera and Hymenoptera, only fast axons are present. Sometimes different part of the muscles has different functions and in this case the two parts have separate nerve supplies. For example, the posterior part of the basalar muscle of *Oryctes* is concerned only with wing depression and has only a fast innervations, but the anterior part also controls wing twisting and its nnervations is complex, consisting of up to four axons, one of which is inhibiting.

Insects, like other arthropods, manage their muscles using relatively few motor neurons. Some major muscles, for example tymbal muscles of cicadas, are innervated by a single motor neuron. Many muscles receive two to four motor neurons. The largest number of motor neurons yet described for an insect muscle is 16, to the flexor muscle in the leg of a locust.

There is an electrical potential across the surface membrane of a living, resting muscle fibre; the interior of the fibre is typically 30-70 mV electrically negative with respect to the extracellular solution. Impulses initiated in the central nervous system travel along the motor axons and cause the release of specific chemical signals (transmitters) from the motor axon terminals at the synapses. The transmitter released from the terminals of most motor axons leads to a reduction (depolarization) in the transmembrane potential of the muscle fibre in the vicinity of the nerve terminal. Muscle fibre depolarization initiates contraction of the fibre. Motor axons that depolarise muscle fibres and cause muscle contraction are called excitatory axons. Some axons, termed **inhibitory axons**, release transmitters that stabilize the transmembrane potential of the muscle fibre or even make it greater, thus antagonising excitatory inputs. In addition to excitatory and inhibitory neural inputs, many muscles receive inputs from modulatory motor neurons, activity that releases chemicals that modify muscle performance, for example, increasing muscle force and work output or speeding relaxation.

The processes linking membrane depolarisation and contractile activation have been little studied in insect muscles, but the ultrastructure, biochemistry, and contractile performance of insect muscle are so similar to those of the far better studied frog, cat, and rodent muscles that one can predict with confidence that the basic principles worked out for vertebrate muscles apply to insects as well. The expected scheme is as follows: Membrane depolarisation spreads inwardly into the fibre along the T tubules. Depolarisation of the T tubules, which are coupled to the SR through specialised junctions, leads to release of calcium from the SR. Released calcium reversibly binds to regulatory sites in the fibrils, turning on the contractile machinery. Relaxation occurs as the SR takes up the released calcium and reduces the calcium concentration in the cytoplasm below that needed for contractile activity.

The size of the twitch produced by the arrival of the stimulus varies depending on whether the stimulation occurs via the fast or slow axons. "Fast" and "Slow" refer to the type of post synaptic potential and hence the muscle twitch, which is produced. Stimulation via the fast axon produces a post-synaptic potential of constant size and a brief powerful contraction of the muscle. Contractions tend to fuse if the rate of stimulation exceeds 10/second and with 20-25 stimuli/second, the muscle undergoes tetanus contraction.

A single impulse from the slow axon, on the other hand, produces only a small post-synaptic potential and a very small twitch, but with successive stimuli, the potential and strength of the contraction increase due to summation. The response is thus said to be graded and is important in the production of slow and precise movements.

6. MUSCLE MECHANICS

6.1 Muscle Force and Muscle Length

The muscle contraction initiated by a single stimulus, or by a single impulse in an innervating motor neuron, is termed a **twitch**; that evoked by repetitive input at a frequency high enough to maintain full activation of the muscle is termed as **tetanus**. A response in which a stimulated muscle develops force while held at constant length is called an isometric contraction. The isometric force generated by a muscle stimulated to contract in tetanus is maximal at about the normal muscle length in the insect body and declines at longer and shorter lengths. The decline in force with increasing muscle length beyond the optimum is thought to be caused by a reduction in the overlap between the thick and the thin filaments and therefore in the number of myosin cross-bridges that can interact with the actin filaments. The decrease in force at short muscle lengths is probably a consequence of the thick filaments running into and being impeded by the Z discs, of collision of thin filaments in the middle of the sarcomere, and, at still shorter lengths, of overlap of thin filaments with portions of thick filaments of inappropriate polarity on the far side of the center of the sarcomere.

Some muscles in insects can shorten to a small fraction of their resting length, a response termed super-contraction. The capacity for super-contraction appears to involve modifications in the structure of the Z disc such that there is not a collision between the Z discs and the thick filaments at short muscle lengths. In the super-contracting muscles that have been examined, the Z disc becomes perforate at short muscle lengths and the thick filaments slide through the spaces in the discs.

The posterior, intersegmental, abdominal muscles of female locusts are of particular interest for the wide range of lengths over which they can operate. During oviposition, appendages on the end of the abdomen dig and pull the posterior abdomen down into a relatively deep hole. The intersegmental muscles become stretched to about nine times their resting length. During this stretch, called **super-extension**, the Z discs become broken up into discontinuous, nonaligned elements to which the thin filaments are attached. Muscle contractility is not lost, and contraction of intersegmental muscles returns the abdomen to its normal length following oviposition. The latter part of the recovery may be super-contraction, for in the resting state the posterior intersegmental muscles are normally super-contracted, with thick filaments protruding through gaps in the Z discs.

6.2 Force, Shortening Velocity, and Power

There is an inverse relationship between the force on a muscle and the velocity with which it can shorten, a relationship conveniently expressed in a force-velocity plot (Fig. 11.10). To facilitate comparison of muscles of differing size, force in a force-velocity plot is usually expressed as stress (force per unit cross-sectional area) and shortening velocity as strain rate (shortening velocity per unit muscle length). Two points on a force-velocity plot are frequently used to characterise a muscle's contractile properties: the maximum isometric stress of the muscle (Fmax, the intercept of the curve with the 0 velocity axis) and the maximum shortening velocity (Vmax, the intercept of the curve with the 0 force axis). Values for the maximum force in insect muscles are generally 5-35 N cm-2 (N, Newton; 1 N is approximately the downward force exerted by a mass of 100 g in the gravitational field at the earth's surface). The few available measurements of the maximum shortening velocity for insect muscles, all from fast muscles, are on the order of 3-15 lengths.

.The product of force and shortening velocity has dimensions of force X distance per time (work per time) and is the rate of doing work, that is, the mechanical power output. The product of stress (force per area) and strain rate (shortening distance per second per unit muscle length) is the mechanical power per unit volume of muscle, which is readily convertible to power output per unit muscle mass. Thus each point on a force-velocity plot (or a plot of stress against strain rate) represents a power output. The power predicted from a force-velocity curve is the instantaneous power output. The peak instantaneous power is substantially greater than the sustainable power from a muscle, for during maintained activity a muscle goes through repeated contraction-relaxation cycles and therefore shortens and produces power for only part of the total time. For fast muscle, including fast insect muscles, the peak power output is 100-500Wkg-1(lW= ljs-1 = lNms-1).

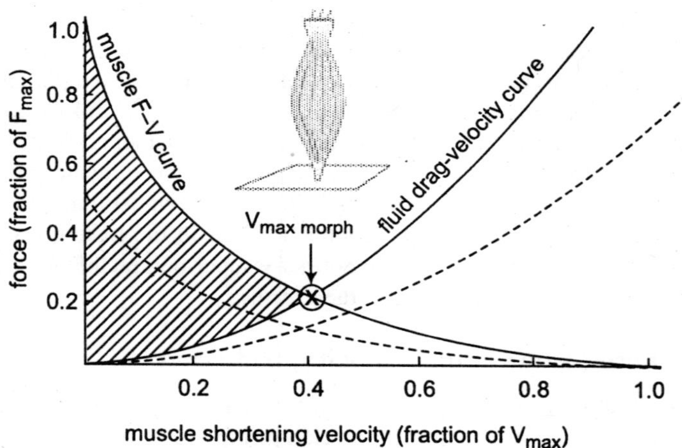

Fig. 11.10. Relationship between force and muscle shortening velocity

Muscles in insects may be divided into **synchronous** muscles and **asynchronous** muscles on the basis of the relationship between the patterns of neural activation and of contraction. The sustainable power available during repetitive, cyclic contraction has been determined for several synchronous insect muscles using the work loop approach, in which the muscle is subjected to length changes simulating those during normal activity and stimulated phasically during the length cycles. A plot of muscle force against muscle length for a full cycle produces a loop, the area of which is the net work output of the muscle for that cycle. The product of work per cycle and cycle frequency is the power output. The mechanical power available from synchronous flight muscles of several locusts and katydids and of a moth measured in this way ranges from 50 to 120 Wkg-1 at normal operating temperature.

Most insect muscles are like vertebrate skeletal muscles in that each contraction is initiated by depolarisation of muscle fibres, and there is a 1:1 relationship between muscle electrical activity and muscle contraction. Such muscles may be termed **synchronous muscles**, reflecting the congruence between electrical and mechanical activities. The major flight muscles in several insect groups are different in that there is no synchrony between electrical and mechanical events. These muscles are known as asynchronous muscles. Neural input and fibre depolarisation are needed to activate an asynchronous muscle, but when it is activated an asynchronous muscle can contract in an oscillatory manner if it is attached to a mechanically resonant load. The resonant loads for the flight muscles are the wings, which may be regarded as small, somewhat dampened tuning forks. The frequency of the oscillatory contraction is the mechanically resonant frequency of the load, which is greater than the neural input frequency required to keep the muscle fully activated. The contraction frequency of asynchronous wing muscles during flight is typically 3-10 times higher than the neural input frequency in each of the motor neurons activating the muscle. The main singing muscles in some but not all cicadas are asynchronous muscles; the resonant load here is the cuticular tymbal to which the muscle is attached and whose inward movement produces the sound pulses.

The features of asynchronous muscle that allow oscillatory contraction are stretch activation and shortening deactivation. When allowed to shorten rapidly an asynchronous muscle becomes deactivated, and while deactivated it can be stretched out to its original length, developing less force than it did while shortening. Stretching the muscle, in turn, reactivates it. Because of shortening deactivation, less work is required to restretch an asynchronous muscle than is produced by the muscle during shortening, and there is net work output when the muscle undergoes a shortening-lengthening cycle. It is this net work that is available to drive the wings and power flight.

Asynchronous muscles occur in several of the most successful insect groups. They power flight in beetles, flies, bees, and wasps and many of the true bugs. The distribution of asynchronous muscles among insect taxa suggests that this mode of muscle control has evolved independently as many as 7-10 times. It is likely that asynchronous muscle has been favoured by evolution because it is more powerful and more efficient than is synchronous muscle for operation at the high frequencies characteristic of insect flight. It is more powerful, in part, because asynchronous control does not demand rapidity in the rate at which muscle is turned on and off by neural input. High-frequency contraction is achieved without hypertrophy of the SR, leaving more room in muscle fibres for fibrils, which are the power-producing component. It is more efficient because a relatively low-frequency neural input is needed to maintain full activation, which reduces the amount of calcium that is released and rebound during activity and the associated metabolic costs of calcium cycling.

6.3 Determinants of muscle power

When a muscle shortens, the repeated power stroke of several hundred active cross-bridges cause the overlapping thick and thin filaments to slide past each other (Alexander, 2003). However the ability of these contractile proteins to interact is complicated by both the spatial and temporal dynamics of force development. Because of these dynamics, morphological and neurological characteristics can greatly affect muscle performance. Extending from the idea of filament geometry and position dependent cross-bridge cycling, muscle performance is a function of its length, velocity, and activation characteristics.

6.3.1 *Force-length relationship*

Muscle force depends on the ability of cross-bridges to interact with actin binding sites, a trait directly attributable to myofilament overlap. Isometric contraction experiments on isolated fiber segments revealed the now classic length-tension relationship; demonstrating muscle force is directly related to sarcomere length (Gordon *et al.*, 1966). At the optimum sarcomere length for force production, myosin and actin filaments overlap such that the maximum number of cross-bridges is within binding range. At longer lengths force production declines due to reduced filament overlap and available binding sites. At shorter muscle lengths, thin filaments from the two halves of the sarcomere begin to overlap, resulting in obstructed binding sites and incorrect cross-bridge binding (Gordon *et al.*, 1966). Not surprisingly, the actual relationship between length and tension depends, in part, on the specific lengths of the thin filaments in individual sarcomeres, a factor that markedly varies across species.

6.3.2 *Force-velocity relationship*

Muscle has an approximate inverse relationship between force and shortening velocity: when muscle is stimulated and allowed to shorten under isotonic conditions, muscle velocity declines with increased loads, until a maximum load where muscle fibers can no longer shorten. In contrast, under increasing isotonic stretch conditions, muscle velocity continues to rise until approaching a yield point. This relationship between the force (F) and velocity (V) is well described by Hill's equation:

$$(F + a) * (V + b) = (F_o + a) * b$$

Where F_0 is the isometric force produced, and a and b are constants that vary by individual muscle (Alexander, 2003). Thus, there is a tradeoff between muscles that are particularly fast and muscles that generate large forces. The mechanical power output of a muscle is the product of its force and velocity. Thus, every point along the force-velocity curve has a corresponding power output value. Power output has a unique maximum at some intermediate velocity between 0 and the maximum shortening rate (Alexander, 2003).

6.3.3 Phase of activation

During the time course of muscle activation, force produced by the muscle rises rapidly following stimulation and then declines gradually during relaxation. As such, muscle activation characteristics, described by the timing, frequency, and rate of activation following stimulation, are additionally responsible for modifying power output and muscle performance. Varying the time at which activation occurs, commonly termed the phase of activation, can lead to significant differences in power output (Tu and Daniel, 2004). In the hawkmoth *Manduca sexta*, advancing the phase of activation from 0.36 (during the lengthening phase) to 0.7 (during the shortening phase) actually reduces power output from maximal (~90 W kg-1), to significantly negative, (~ –50 W kg-1) (Tu and Daniel, 2004). This variation in power output is the result of the previously described length-tension relationship, where force output varies with myofilament overlap

In addition, the timing of a contraction, defined by the time to peak force and duration of relaxation, is essential in determining locomotor speeds. Internal fiber dynamics, including variation in cytoplasm Ca^{2+}levels, ATPase rates, and myofilament overlap, dictate a muscles contraction times. As with other biological rate processes, contractile rates also significantly depend on temperature. Both vertebrates and invertebrates benefit from the accelerated contractile rates associated with an increase in temperature. Underlying this temperature dependence are the enzymatically catalysed mechanisms associated with muscle contraction (Bennett, 1985). Decreased contraction times allow organisms to attain higher locomotor speeds and reaction rates, essential for high frequency cyclic motions such as synchronous insect flight and ballistic movements including escape maneuvers or prey capture.

6.4 Functional diversity of muscles

Muscles perform a broad range of specialised functions, including providing power for rapid motions, braking, storing and releasing energy, and aiding in energy transmission.

6.4.1 *Muscles that function as motors*

Muscles that operate as motors predominantly drive locomotion, including ballistic movements such as escape jetting by scallops and oscillatory motions such as wing movement in flight (Marsh *et al.*, 1992; Biewener, 1998). These muscles generate force as they shorten and produce positive mechanical power throughout a contraction cycle. Power producing muscles are generally expected to have long parallel muscle fibers, a trait which allows them to create larger amplitude displacements as they shorten to move limb or body segments. For instance, in a comparison of two differently specialized muscles, Biewener (1998) measured extensively greater shortening in the muscle responsible for generating high mechanical power output, ~35% of the rest length versus 2% in the nonpower producing muscle. Because the timing of muscle activation significantly affects power output, there is also a time dependent nature to muscle function. Muscles that generate positive mechanical power in rapid oscillatory movements require an early phase of activation in order to adequately develop force prior to the shortening phase of the cycle (Altringham and Johnston, 1990). This is exemplified by the power-phase curve of the dominant flight muscle of the hawkmoth. Maximal power output, ~90 W kg-1, occurs when the muscle is activated ~6 ms before shortening, this is followed by a significant decrease, with power output actually transitioning to highly negative values shortly after shortening (Tu and Daniel, 2004).

6.4.2 *Muscles that function as brakes*

Muscles may also produce negative power output, absorbing energy during the contraction cycle, and therefore function as a brake. In this case, muscles produce large forces during the lengthening phase of the cycle to actively slow limb or body segment movement. In cyclic motions, such as in insect flight, muscles producing negative power can dampen oscillations that become unsteady. In addition, muscles that absorb work facilitate changes in motion requiring decelerations. Therefore, it appears that negative power output is essential in maintaining stability and control in locomotion. By modulating extrinsic factors such as the phase of activation and shortening velocity, the muscular system can selectively absorb energy when the mechanics demand it. In the example of a running cockroach, select leg extensor muscles actually absorb energy (–19 W kg-1) under in vivo stimulus and strain conditions (Ahn and Full, 2002). Thus, in the cockroach leg there appears to be a division of labour, with some leg extensor muscles acting to extend the leg while others act as brakes to slow the leg during the swing phase and therefore stabilize motion.

6.4.3 *Muscles that function as springs*

Elastic energy savings is a crucial aspect of locomotion, allowing organisms to minimize their metabolic costs and increase efficiency. Although the sites of elastic energy storage are generally believed to be in non-muscular structures such as tendons and cuticle, muscle fibers can also act as springs to store and release energy (Alexander and Bennet-Clark, 1977; Alexander, 1984; Dickinson et al., 2005; Patek *et al*., 2011). This recoil energy could be stored in compliant cross-bridges and in the axial portions of the myofilament lattice. A recent study measuring changes in the muscle lattice structure using X-ray diffraction techniques, observed elastic deformation of thick filaments in asynchronous insect flight muscle (Dickinson *et al*., 2005). This elastic stretch and recoil of the myosin filaments suggests that a portion of energy storage takes place within the myofilaments themselves. Elastic energy stored in stretched muscle filaments and cross-bridges at the end of one phase of the contraction cycle would be released during the second phase, passively contributing to reduce the total energy requirements of locomotion.

6.4.4 *Muscles that function as struts*

Lastly, muscles can also serve as struts, stiffening to play a force-transmitting role. These muscles are commonly short and pinnate, allowing them the ability to generate large forces with little change in length. The well-described force-velocity curve highlights this tradeoff between work rate and force output (Hill, 1938). By operating isometrically (v = 0), strut-like muscles transfer energy from one part of the body to another and can also function in series with tendons to store and recover energy. In turkeys, for example, the isometrically contracting lateral gastrocnemius muscle functions as a strut, transferring energy to the spring-like tendon, which stretches and recoils to provide 60% of the work needed to move the body up and down (Roberts, 1997). In fish, muscle function can actually vary based on the timing of muscle activation with respect to the bending wave travelling across the body during undulatory swimming. It appears that early in the cycle, the anterior muscle shortens to produce positive power. Later in the cycle, and further along the body position, the muscle becomes stiff and passively transmits power to the tail fin, where the majority of the hydrodynamic forces are generated.

7. MUSCLES INVOLVED IN WALKING AND JUMPING

7.1 Movements of the legs and their constituent joints

Insects walk by moving their six legs in reproducible patterns. It involves two phases, the stance phase and the swing phase that alternate with each other. In the **stance phase**, the tarsus is on the ground and the animal is pushed forward, and in the **swing phase**, the tarsus is moved forward through the air. At slow speeds, the legs follow a metachronal pattern, moving from the hind legs to the middle and then to the front legs on either side. However, to move more rapidly, some legs must be moved at the same time; therefore, the insect shifts into a modification of the metachronal pattern called the **tripod gait**. Here the front and rear legs on one side of the animal move as a unit with the middle leg on the opposite side. This tripod alternates between swing and stance with the tripod made up of the remaining legs. The tripod gait is very stable, because at most speeds, the animal's center of mass remains within the base of support. However, at very high speeds, many insects make dynamic postural adjustments to stay upright. The American cockroach *Periplaneta americana* has been seen to rise up on its hind legs and run with a bipedal gait during high speed walking. This gait is not statically stable but represents a balance of forces that produces dynamic stability. In cockroaches, the tripod gait can be further subdivided into a slower "amble", a faster "trot" which has much less variability in leg coordination, and an even faster escape run which is used only about 1% of the time.

All the three pairs of insect legs are made up of segments that are similar from leg to leg but differ in dimensions. The leg segments are the coxa, trochanter, femur, and tibia, and a series of tarsal segments ending in a retractable claw (Fig. 11.11). In the cockroach, the most important joints for walking are the coxa-trochanter (CTr) joint and the femur-tibia (FTi) joint. The CTr joint actually moves the femur relative to the coxa because the trochanter-femur (TrF) joint makes only small movements. Although flexion of the TrF joint can effectively rotate the tarsus, during many movements it acts mechanically as a fused joint. In other insects, relative proportions of these leg segments are changed to match the needs of specialized forms of locomotion. Thus, the hind leg of a locust has a very short coxa and a long muscular femur, making a powerful jumping leg.

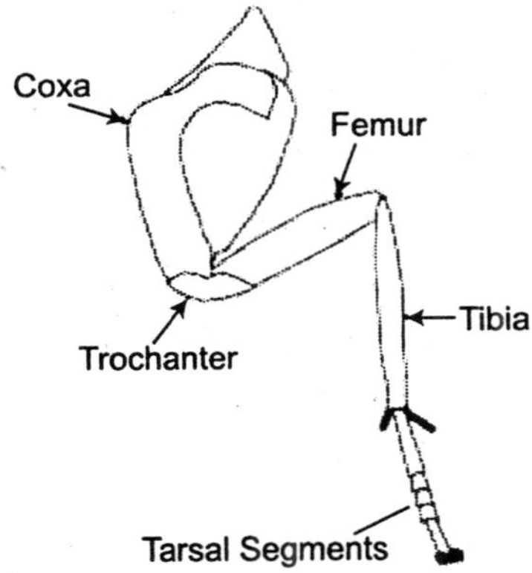

Fig. 11.11. Insect leg

Although the legs within a tripod move their feet as a unit, the joint movements and resulting forces are unique for each pair of legs. In cockroaches, the hind legs are specialised to propel the animal forward in walking. To accomplish this, the CTr joint and the FTi joint move in near synchrony. This action allows these rotary joints to direct the movements of the tarsi in a line nearly parallel to the long axis of the animal's body. The middle legs make similar movements, but with smaller joint (CTr) excursions. The orientation of this leg causes it to first brake and then accelerates forward movements of the animal.

The movements of the front legs are very different. Unlike the other two pairs of legs, the front legs make much greater use of the thorax-coxa (ThC) joint, which attaches the leg to the thorax. This joint has three degrees of freedom, similar to the ball-and-socket joint in a human shoulder. Movement of the ThC joint swings the front legs far forward much like an extending human arm. The CTr and FTi joints then move out of phase with each other as the foot is drawn back toward the body and then is pushed backward. The resulting ground reaction forces slow the forward movement of the body and keep the animal from losing control. The neural control of the front leg is much different from the control of the other two pairs of legs. Nevertheless, the sum of the ground reaction forces of all three pairs of legs is similar to that seen in the bipedal leg movements of a human. Cockroaches can run extremely rapidly (up to 25 steps per second), ranking them as the fastest terrestrial animal (in steps per second) (Box 11.6). Running, which follows wind stimulation of the cerci (abdominal appendages) or tactile stimulation of antennae or body cuticle, is used to escape from predators. Although fast running shows a number of similarities to walking, it occurs so rapidly that it may actually be a separate behaviour. Passive elastic structures in the legs can play an important role in generating such rapid locomotion, prompting cockroach running to be modelled as a mass on a spring.

Box 11.6 Fascinating facts about insect locomotion

1. Cockroaches can run 2.9 mph (= 70mph if scaled to size of lion
2. Fly wings can beat 200x/s
3. Bumble bee wings can beat 160x/s
4. Average flight speed of fly is 4.5 mph
5. Fastest flier is the desert locust Schistocerca gregaria @ 33km/h
6. Fastest runner is the tiger beetle Cicindela hudsoni @ 5.57 mph

The largest and most powerful muscles are closer to the body while the smaller ones are in the more distal segments of the leg. This helps in reducing the inertia in the distal part of the limb and allows controlled movement. Also, the movements of the proximal joints act on the tarsus through the relatively large lever arm of the intervening leg segments to the tarsus. Thus, for example, relatively small movements at the ThC joint can greatly alter the orientation of the tarsal end point.

The most distal segment, the tarsus, is actually made up of a series of segments ending in a retractable claw that have again remarkable mobility. However, as with human fingers, the muscle for these segments is found in more proximal segments and imparts its movements via a long apodeme that serves the role of a tendon. Although the claw can be engaged by this muscle, there is no antagonistic muscle. Rather, the claw is disengaged and lifted from the substrate by a remarkably efficient elastic protein (resilin) found in the joints of the tarsal segments. The resilin is stretched when the claw is engaged, and causes the tarsus to be lifted automatically when the muscle is relaxed.

7.2 Neuromuscular System

Insect muscles are innervated by very few motor neurons. In the stance phase muscles there are only two motor neurons. These can be readily distinguished as either fast or slow motor neurons depending on the types of muscle contraction they produce. **Slow motor neurons** need a series of action potentials to generate significant movement, whereas **fast motor neurons** generate a reasonable twitch with a single action potential. Typically, extracellular recordings indicate that fast motor neurons generate larger action potentials. In insects most muscle fibers produce graded potentials when motor neurons fire action potentials. Simply altering the frequency of action potentials in a single motor neuron can control tension.

The leg movements that occur during walking on a horizontal surface are associated with typical patterns of motor activity. In both the middle and hind legs of cockroaches, CTr extension is generated by a burst of activity in the slow depressor motor neuron (Ds). At faster speeds, one or more muscle potentials from the fast depressor motor neuron (Df) occur at specific times in the leg cycle. These actions generate the stance phase of the leg cycle and alternate with activity from several flexor motor neurons that produce the swing phase. The simultaneous extension at the FTi joint is generated by the slow extensor of the tibia (SETi) motor neuron and the fast extensor of the tibia (FETi), and is again opposed by activity in several flexor motor neurons that generate the swing phase of that joint. As expected, the motor patterns of the front legs are more complicated, matching the joint movements described earlier.

The cockroach controls the speed of walking by altering the frequency of motor action potentials. Motor frequency is positively correlated with stepping frequency and joint velocity. Thus, the animal can move faster by increasing the frequency of Ds and SETi in each leg. It can turn by increasing frequency in one or more legs of the tripod while decreasing the frequency in the leg or legs located on the opposite side. This change creates stronger forces on the outer leg of the turn and weaker lateral forces on the inner leg, thereby turning the animal.

7.3 Control of Basic leg movements: Central Pattern Generation

Despite the morphological variation in insect legs, common neural control systems are found even in insects as different as cockroaches and stick insects. The muscles that extend leg joints are typically innervated by very few motor neurons. Most of the muscles that extend the CTr joints of cockroaches are innervated by two excitatory motor neurons: one fast and one slow motor neuron. In slow movements, the slow neuron fires in burst that lead to a series of facilitating muscle potentials that generate joint extension (Tryba and Ritzmann, 2000). At faster speeds, the fast motor neurons are added (Watson and Ritzmann, 1998). These neurons generate few spikes, but each causes a rapid extension of the joint greatly increasing the joint velocity. Flexion of the joint is generated by antagonistic muscles that are typically controlled by a larger number of motor neurons.

The local control systems found in the thoracic ganglia must be able to generate basic swing-stance patterns but still be open to adjustments in both timing and force. First, the individual joints must be made to alternate between extension and flexion. In many neural systems, this basic timing function is often performed by central pattern generation (CPG) circuits. Then the joints of each leg must be coordinated to create an effective movement. It would be possible to have one overarching CPG that imposes a rhythm on all joints of each leg or even all of the legs. However, observations in stick insects strongly suggest that no single CPG exists for all legs or even for all joints of a single leg (Büschges *et al.*, 1995). This turns out to be an important design feature, since intra-joint coordination can differ in front, middle and hind legs. Inter-leg coordination must also change from time to time (e.g. turning or walking backward). On top of this timing pattern, the forces generated by stance phase muscles must also adjust as the insect walks up and down inclines or walls, carries items, or even climbs over substantial barriers.

Although both stance and swing phase motor neurons increase burst duration with increasing walking speed, the change in stance phase motor activity is much greater. This observation led to a model for insect walking referred to as the **flexor burst generator**. In this model, a set of interneurons within the central nervous system called a central pattern generator controls the flexor motor neurons directly and the stance phase motor neurons indirectly through inhibition from swing phase interneurons.

Investigators have recorded traces from oscillatory neurons in cockroaches and stick insects that could be part of such a central pattern generation circuit. For example, in the cockroach, neurons have been found that undergo membrane potential oscillations in time with levator motor activity. Moreover, stimulation of these cells activates the flexor motor neurons. Thus, these cells could be part of the flexor burst pattern generator. However, they by no means represent the entire circuit.

Observations in stick insects strongly suggest that there are separate burst generators that activate motor neurons for each joint in the leg. Patterned activity can be recorded in motor neurons for each joint of a single leg even after removing all sensory inputs. However, the timing of oscillations is not correlated between joints.

7.4 Leg sensors

In intact insects, muscle contractions and joint movements are tuned and coordinated by sensory inputs. For example, sensory receptors that monitor joint angles (called chordotonal organs) affect the timing of phase changes of motor neurons serving the leg joints and segments that they monitor as well as adjacent joints (Hess and Büschges, 1997; Akay *et al.*, 2004). Campaniform sensilla are embedded in specific areas of the cuticle and respond to changes in strain that allow the insect's nervous system to adjust to changing load (Zill *et al.*, 2004; Zill and Seyfarth, 1996). Signals of loads and muscle forces detected by campaniform sensilla can influence groups of muscles at different joints (e.g. leg extensors) to insure that they are tuned to work together as synergists

Even where central pattern generation circuits contribute to the control of locomotion, sensory systems still play important roles in generating behaviourally meaningful motor activity. This concept was particularly well demonstrated in the locust flight system. Locust preparations that have been isolated from sensory structures in the periphery (deafferented) are capable of generating bursts of activity that could move the wings in an appropriate manner. However, the frequency of the **wingbeat cycle** is much lower than normal and the pattern quickly dies out, whereas in the intact animal, flight can go on for very long periods of time. If appropriately timed activity from wing stretch receptors is added in, a more normal flight pattern occurs that is maintained for long periods of time. Thus, the central circuits are capable of generating part of the cyclical pattern but not a complete, behaviourally relevant one.

Similarly, normal walking behaviours in all insects that have been studied require input from sensory structures on the legs. However, in rapid running movements associated with escape, feedback from sensory inputs probably is too slow to influence walking within a single step, and the animal may be running on autopilot. The sensors that play an important role in walking provide precise bits of feedback information to the walking system through structures that are most often found in the same locations in all three pairs of legs. Internal sensors such as the chordotonal organs span the joints and monitor joint angle. Hair plates are positioned near joints where adjacent segments will touch during maximal flexion. Campaniform sensilla are located in strategic locations near joints and muscle insertions and detect strain in the cuticle. These structures are made up of sensory nerve endings positioned in sockets with a flexible dome. As strain increases in the surrounding cuticle, the dome is deformed and activates the sensory neuron. There are groups of campaniform sensilla near each of the leg joints, including the segments of the tarsus.

Each peripheral sensor provides information that is used at discrete phases of the step cycle as the insect walks. Some sensory inputs in walking are essential in making transitions between the stance and swing phases while others adapt muscle activities to reach levels needed for support and propulsion. The effects of leg sense organs are demonstrated particularly well in a computer simulation that was developed by Ekeberg, Bumel, and Buschges for stick insect walking. The simulation is controlled by calculations similar to those done by the nervous system. Signals from chordotonal organs that monitor joint angles are used to determine the times of transition from swing to stance, when the leg is first placed on the ground. Inputs from campaniform sensilla that detect load aid in generating extensor muscle activities in a self-regenerating loop known as positive feedback. Extensor activities continue until chordotonal inputs trigger the start of swing.

Beyond helping to set the normal locomotor pattern, these sensors are, of course, readily available to make rapid adjustments to the motor pattern as the animal moves away from horizontal surfaces and climbs inclines or vertical surfaces, or even walks on ceilings and negotiates obstacles in its path. Experiments have shown that the tibial campaniform sensilla can aid in this process by detecting the proportion of body weight that is supported by a leg. In these studies, load was varied by using magnets attached to the body and an electrical coil under the floor on which the insect stood. Some campaniform sensilla discharged when the load was increased while others fired when the body load was decreased. Those signals could be important in detecting when a leg slips. Arrays of neurons with diverse properties receive sensory inputs and could mediate appropriate effects. Within each thoracic ganglion, the sensory neurons from these leg sensors project to populations of spiking local interneurons (interneurons that are isolated to a single ganglion).

The projections of many receptors follow a topographic pattern that reflects the location of the sensors in the leg. The spiking interneurons project both to motor neurons and to another set of local interneurons called **nonspiking interneurons**. These interneurons never generate action potentials. Rather, they act through graded potentials to control and coordinate motor activity to various leg muscles within a thoracic segment. Coordination among joints of each leg (intraleg coordination) occurs through interactions in these local circuits. Finally, spiking interganglionic interneurons project between thoracic ganglia, influencing local circuits for adjacent legs. Through these circuits, the sensory activity directs appropriate adjustments in individual muscles, coordinates segments within a single leg, and influences the coordination between legs.

The notion of feedback circuits influencing timing of leg movements suggests a solution to coupling of joint pattern generators alluded to earlier. Leg sensors influence timing of extension both in the motor neurons that control the joint that is monitored and in motor neurons serving adjacent joints. These weaker connections provide coupling for the joint pattern generators. Moreover, that coupling can be altered to provide for considerable flexibility. In this way, the joint pattern generators that produce walking could also be used in other rhythmic behaviours, such as righting (when an animal falls over and uses its legs to regain an upright posture), turning, or grooming, by changing the coupling between the oscillators. Indeed, in the cockroach, reflex effects in adjacent leg joints can be reversed by altering signals descending from the brain.

In stick insects, the influences of some sense organs are changed when the direction of walking is reversed. Stick insects readily walk forward or backward, as is needed when negotiating branches of trees. However, backward walking involves changing the activities of leg muscles. For example, protractor muscles are active in swing in forward walking but fire during stance when the animal walks backward. The effects of load-detecting sense organs on motor activities also change when the animal reverses direction. Trochanteral sensilla that excite the retractor in forward walking instead activate the rectractor when the animal walks backward. The nervous system, therefore, changes the effects of the sense organs so that they counter the perturbation in the direction in which the animal is walking.

7.5 Interleg co-ordination

To generate an effective tripod gait, insects must coordinate movements of joints not only within a single leg but also between legs. This interleg coordination on the surface appears to be a daunting task. The tarsi of the three legs that make up an effective tripod must move in synchrony. However, as described earlier, the legs of each segment make different movements. The patterns of the motor neurons that control those movements also differ in each leg. Thus, the animal must coordinate the tarsi through joint movements and neural activities that are specific to each pair of legs.

Observations on stick insects suggest that these potentially complex connections can be functionally formulated as a set of fairly straightforward rules indicating that events in one leg that are readily detected by leg sensors influence the actions of other legs in discrete ways. For example, one rule is that when one leg is in swing, the adjacent anterior leg cannot enter into swing phase. Alternatively, the start of the stance phase promotes the start of swing in adjacent anterior legs and in the contralateral leg of the same segment. More subtle influences also exist to account for transient changes. For example, an increased load on one leg prolongs the power stroke on the contra-lateral leg of the same segment. This would result in co-activation of legs when the animal is encountering increased resistance, such as during climbing.

7.6 Negotiating obstacles

Most insects effortlessly negotiate barriers. There is considerable variability in the strategies used. However, some common themes are evident. In climbing over a block, for example, insects must move their centre of mass (CoM) upward to surmount the obstacle. For cockroaches, the strategies used to accomplish this task vary with the size of the obstacle and the speed that the insect is moving at when the block is encountered. The front legs are normally lifted up fairly high during normal walking. If the block that is encountered is lower than the normal front leg trajectory, the insect hardly needs to change its movement at all. The front leg will be placed on top of the obstacle and pushed down to move the animal's CoM upward as a natural consequence of the encounter with the block.

Larger blocks, especially at slower speeds, require a redirection of leg movements that now push the body upward upon joint extension. This rearing movement allows the front legs to be readily placed on top of the block so that climbing can commence. These redirected leg movements often require intervention from descending activity originating in the brain. Lesion of a region of the brain called the central complex (CC) compromises the cockroach's ability to negotiate barriers. Although little is known regarding how the CC generates appropriate descending commands, neurons in this region receive sensory inputs from a variety of structures including mechanoreceptors on the antennae and visual systems.

For very large obstacles, the insect must actually climb up the front face of the object. Here the insect must attach itself as if the face of the block were a wall. This problem is solved by a combination of claws and pads that adhere to various substrates. Cockroach claws are normally held up during horizontal walking, like the claws of a cat. However, when a cockroach is walking up a vertical surface, the claw is pulled down by a muscle in the tibia, which attaches to the claw through a long apodeme. There is no return muscle. Rather, as we have described, the claw pops back up passively with the aid of strategically placed resilin ligaments. For smoother surfaces, including those as polished as glass, some cockroaches have cuticular pads between their claws that can adhere to the substrate for stability but be readily lifted when needed in walking.

Many insects have developed efficient jumping behaviours to move long distances with a single jump. The jump requires a powerful and rapidly accelerating movement of the jumping leg. Orthopterans such as grasshoppers and locusts have very large hind legs that can generate powerful jumping movements. However, even the large femur extensor muscles of these jumping legs cannot generate the quick extension necessary for an efficient jump without some mechanical modification within the leg structure for storing energy. In locusts, the tendon of the tibial flexor muscle moves over a stop that allows the extensor to contract without moving the leg when the muscles are co-activated. As a result, a considerable amount of energy is stored in the mechanical distortion of the femur, tibia, and extensor tendon (like the bow of an archer). Inhibition of the tibial flexor muscle releases locking mechanisms and produces a very rapid and powerful extension that propels the animal upward. The timing of these events is critical, and neural circuits have been identified in association with kicking movements that provide the appropriate control.

Fleas generate remarkable jumps relative to their tiny size. Again, a proportionally very large hind leg is used to generate the jumping movement, with a hook on the hind leg preventing movement until a large isometric force has been achieved. The isometric force is stored in strategically located resilin ligaments and released during the jump. One of the most remarkable jumping strategies in insects does not even involve legs. Click beetles can jump from a standstill to four times their body length by rapidly accelerating the joint between the prothoracic and mesothoracic segments of the body. Here again a mechanical stop prevents movement until a large isometric force has been achieved, this time in thoracic muscles, and then is released suddenly to shoot the animal upward.

8. TYMBAL MUSCLES: SOUND PRODUCTION

The song of male cicadas is a fine example of a rhythmic motor output involving specialised cuticular, muscular and sensory structures. Pulses of sound are produced by the regular buckling of a pair of cuticular membranes, the tymbals. Each tymbal is buckled inwards by the twitch contraction of a single, large muscle, and springs back to the resting position after the muscle relaxes. The form of the sound pulses is adjusted by accessory muscles, especially by the tonic contraction of the tensor muscle which inserts on the anterior rim of the tymbal.

The sound-producing structures of cicadas are located mainly in the anterior of the abdomen and consist of a pair of tymbals, tymbal muscles, various accessory muscles and a tracheal air-sac.

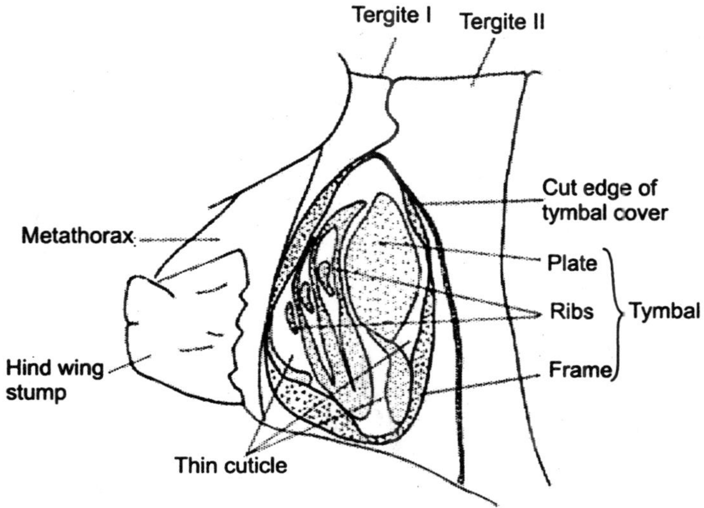

Fig. 11.12. Tymbal structure

The tymbal is located on the side of the 1st abdominal segment. It comprises a thin, flexible membrane (frame), biconvex in shape and with its posterior surface turned sharply inward (Figs.11.12). The curved part of the tymbal bears a number of sclerotised ribs. There are usually seven long ribs, diminishing in size from posterior to anterior, but sometimes there is an incomplete eighth rib anteriorly. Interspersed between the long ribs are seven short ribs, arranged in a row and also diminishing in size from posterior to anterior. The long ribs are slightly raised above the membranous area but the short ribs are flush. Dorsally the long ribs fuse with the sclerotised tymbal rim but ventrally they stop short of the rim. The long ribs are assigned numbers according to the order in which they buckle so that the largest, most posterior rib is rib 1 and the most anterior is rib 7, or sometimes rib 8. The inward-sloping, posterior face of the tymbal bears an irregularly shaped area of sclerotised cuticle which is termed the tymbal plate. The region of membrane between the tymbal plate and the vertical bar is especially flexible. The tymbal is bounded ventrally by the tymbal bar, which projects forward from the 2nd abdominal segment.

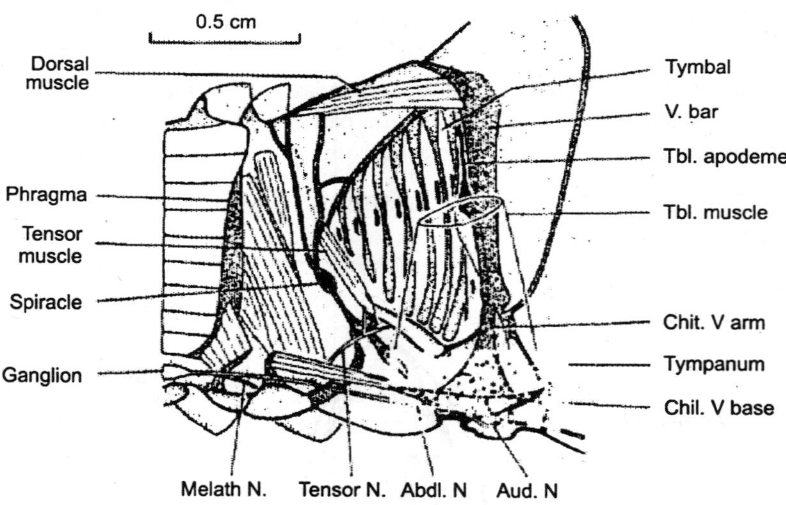

Fig. 11.13. Arrangement of sound producing structures

(Abdl.N- Abdominal nerve; Aud. N- Auditory nerve; Chit.V.arm- Lateral arm of chitinous V; Chit.V base- base of chitinous V; Metath. N- Metathoracic nerve; Tbl- tymbal)
(Taken from Simpson and Young, 1978)

Dorsally on the tymbal plate there is a small indentation of less sclerotised cuticle and at the top of this is the insertion of the tymbal apodeme. The tymbal apodeme is a slender shaft at this point of insertion but it expands internally to form a broad plate upon which the tymbal muscle inserts. Tymbal muscle fibres travel some way up within the shaft of the apodeme. The broad part of the tymbal apodeme is approximately kidney-shaped and this allows the distal part of the tymbal muscle to curve round the vertical bar. The tymbal apodeme touches the vertical bar in the resting position and at this point of contact there is a pronounced knob on the vertical bar and a corresponding dip in the broad part of the tymbal apodeme (Fig 11.13). The knob on the vertical bar is slightly hollowed on its anterior surface and there is a corresponding small bulge on the shaft of the tymbal apodeme, about level with the limit of muscle insertion. This combination of raised and hollowed surfaces produces a precise fit between the tymbal apodeme and the knob on the vertical bar in the resting position. This arrangement strongly suggests the function of a return stop to prevent the elasticity of the tymbal from causing overshoot during the relaxation of the tymbal. Each tymbal muscle has its origin ventrally, close to the midline on a specialized sclerite, the chitinous V (Myers, 1928) (Fig 11.13). The left and right tymbal muscles are opposed to one another in the midline so that the tymbal muscles slope outwards to the tymbals. A large surface for muscle attachment is provided by the anterior and posterior extension of the base of the chitinous V. On each side the lateral arm of the chitinous V extends from the base to the vertical bar. The two arms form the compression members balancing tension developed by the tymbal muscles (Pringle, 1954).

The tensor muscle, situated at the anterior rim of the tymbal, is an important part of the sound-producing mechanism (Pringle, 1954; Hagiwara, 1955). It has its origin on a pronounced knob below the spiracle at the border of the metathorax and 1st abdominal segment (Fig 11.13) and inserts above the spiracle upon a distinct oval area of cuticle immediately adjacent to the anterior rim of the tymbal. This point of insertion is in line with the row of short ribs on the tymbal (Young, 1975). In *C. saundersii* the tensor muscle is comparatively slender and pale in colour. Two chordotonal organs (mechanoreceptors) are located near the tensor muscles (Young, 1975). The larger of the two, the tymbal chordotonal organ, inserts on the anterior rim of the tymbal and this point of insertion is discernible on the external surface. Another muscle considered to have a possible role in sound production (Pringle, 1954) is the dorsal muscle (Fig. 11.13). This flat muscle extends longitudinally between the anterior and posterior margins of its abdominal segment, passing across the dorsal rim of the tymbal. All the muscles and sense organs involved in sound production are innervated from the thoracic-abdominal nervous system. In the species of cicadas studied by Pflugfelder (1937) and Pringle (1954), all the thoracic and abdominal ganglia are fused into a single mass, with at most a slight waist between prothoracic and mesothoracic regions. In *C. saundersii* the thoracic nervous system is not so condensed, the prothoracic ganglion being separated from the mesothoracic by short connectives and the mesothoracic ganglion being distinguishable from the metathoracic by a distinct waist and central hole. The metathoracic and abdominal ganglia are entirely fused together and this combined ganglion is located in the mesothorax (Fig. 11.13).

The nerves concerned with sound production and hearing emanate from the posterior part of this metathoracic-abdominal ganglion. The most medial of this posterior group of nerves is the abdominal nerve, which supplies all organs posterior to the 2nd abdominal segment. Next to it is the auditory auditory nerve is the most dorsal of these nerves in its origin on the ganglion. In *C. saundersii* the auditory nerve carries both the sensory fibres from the auditory organ and the motor innervation to the tymbal muscle (Young, 1972), which consists of a single motor axon about 20 μm in diameter. In other cicadas the tymbal motor neurone axon is carried in a separate tymbal nerve which travels dorsally (Pringle, 1954). The auditory nerve travels with the abdominal nerve through the channel beneath the base of the chitinous V and then it turns peripherally along the edge of the tympanum to innervate the tymbal muscle from the peripheral side. The tensor nerve leaves the ganglion slightly anteriorly and ventrally with respect to the auditory nerve. It travels closely with the auditory and abdominal nerves for a short distance and then turns laterally near the anterior margin of the 1st abdominal segment (Fig. 11.13). The majority of its target organs are in the region of the tensor muscle (Fig. 11.14), but one branch continues on round the ventral rim of the tymbal, travels up the vertical bar and around the tymbal apodeme to innervate the dorsal muscle. There are three axons which innervate the tensor muscle. About the same number of axons goes to the spiracle muscle and to another small muscle below the spiracle. The motor axons to the accessory tensor muscle could not be found by any of the methods employed. A small sensory branch carries axons from a group of innervated hairs situated just anterior to the insertion of the tensor muscle. The tensor nerve also receives sensory branches from the tymbal and tensor chordotonal organs (Young, 1975).

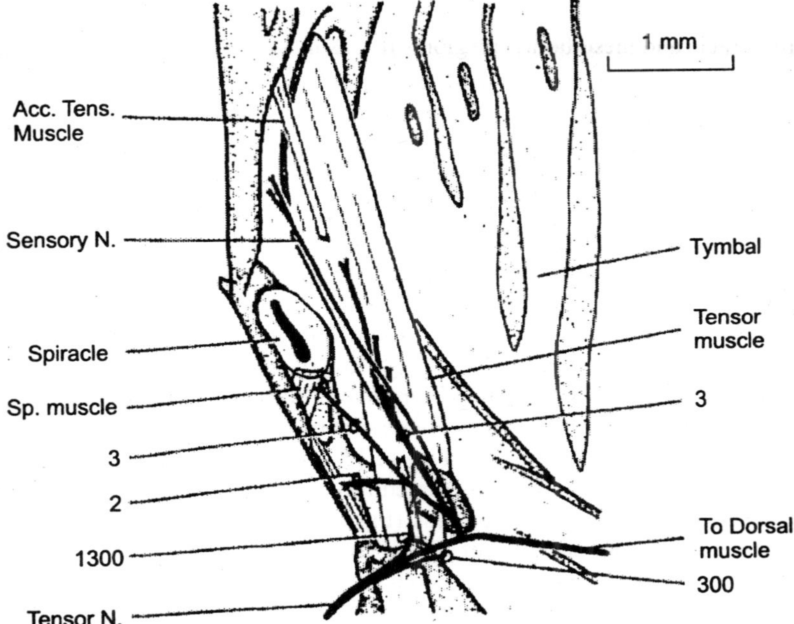

Fig. 11.14. Innervation of the tensor muscle region of *Cystosoma saundersii* (1300 and 300 indicate number of sensillae in chordotonal organs; 3and 2 indicate the number of axons in the motor branch of the nerve; Sp-Spiracular muscle; Acc. Tens.- Accessory tensor; N- nerve (Taken from Simpson and Young, 1978)

8.1 The mechanism of tymbal action

In an intact animal the inward buckling of a tymbal is caused by a twitch contraction of its tymbal muscle. The delay between the spike and the buckling of the tymbal, as judged by the onset of a sound pulse, is 8-10 ms. The tymbal muscle develops maximum tension 12 ms after a spike invades it, as measured by a straingauge attached to its apodeme. This means that the tymbal begins to buckle and produce sound before the peak in the force that buckles it is attained. Following the inward buckling, the resting shape of a tymbal is usually regained by inherent elasticity. Sometimes a tymbal remains collapsed, particularly when high frequency electrical shocks are applied to its muscle. Even in intact, freely singing males a tymbal occasionally remains collapsed for several seconds or even for some minutes. Three sound pulses are produced following each spike in the myogram. The peak in sound intensity is reached rapidly in each and is followed by an exponential decay. Usually the second pulse has a greater amplitude than the first and follows it by 5-7 ms. The third pulse is only sometimes observed; it follows the second by 15-20 ms and is always much smaller. In both intact and experimentally manipulated specimens, sound produced by the tymbal has a dominant frequency of about 850 Hz. Removal of the abdomen greatly reduces the intensity but does not significantly alter the form or carrier frequency of sound pulses.

Only three tymbal shapes were ever seen: the resting shape; a shape where only one rib had buckled inwards; and a shape where other ribs in addition to rib 1 had buckled inwards (Fig. 11.14). Initially, following contraction of the tymbal muscle, the tymbal plate swings forward and inwards, presenting a much broader profile to the observer. This movement of the tymbal plate brings about the buckling of rib 1, which is adjacent to it, thereby effecting the transition from shape 1 to shape 2. The transition to the third shape is effected by a smaller arc of movement of the tymbal plate as it continues to swing inward (Fig. 11.15). In the third shape ribs 2 and 3 always buckled and other ribs sometimes buckled. The line of buckling coincides with the row of short ribs (Fig.11.14). The first sound pulse is coincident with the buckling of rib 1. The remaining ribs buckle almost synchronously and their buckling coincides with the second sound pulse. All the ribs are restored to their original shapes almost simultaneously and this movement produces the third sound pulse.

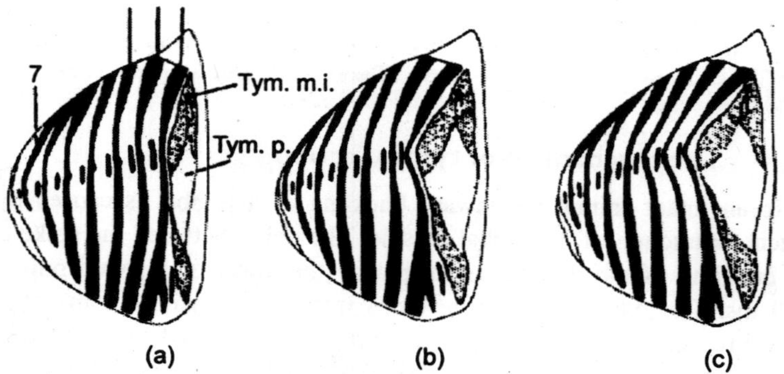

Fig. 11.15. The tymbal of *Cystosoma saundersii* in different stages of buckling Resting, (b) with rib 1 bucked (c) ribs 1-3 bucked)
(Numbers refers to ribs; Tym.m.i- Tymbal muscle insertion; Tym.p-Tymbal plate)
(Taken from Simpson and Young, 1978)

8.2 The action of the tensor muscle

The form of the sound pulses produced by a single tymbal is modified by the tensor muscle, which has very different physiological properties from the tymbal muscle. A single shock applied to the tensor nerve produces little or no contraction of the tensor muscle. Tension develops gradually when a train of shocks is applied to its nerve and the muscle relaxes slowly when stimulation ceases. The rate at which it develops tension depends on the frequency of stimulation and is maximal at 40 Hz. In an intact animal, the tensor muscle, immediately following insertion of silver electrodes into the tensor muscle, was still developing tension 2s after onset of a train of shocks applied to it at a frequency of 20 Hz. Contraction of the tensor muscle increases the convexity of the tymbal. The first two sound pulses occur closer together in time and often merge. Rib 1 buckle sooner before the other ribs. The latency between excitation of the tymbal muscle and the production of sound is lengthened, indicating that the structural alteration to the tymbal increases its stiffness. Also the first two sound pulses increase in amplitude (i.e. loudness) when the tensor muscle contracts, which is readily understood in view of the tymbal's increased stiffness.

REFERENCES

Ahn, A. N. and Full, R. J. (2002). A motor and a brake: two leg extensor muscles acting at the same joint manage energy differently in a running insect. *J. Exp. Biol.* 205, 379-89.

Akay, T., Haehn, S., Schmitz, J. and Büschges, A. (2004). Signals from load sensors underlie interjoint coordination during stepping movements of the stick insect leg. J Neurophysiol 92, 42-51

Alexander, R. M. (1984). Elastic Energy Stores in Running Vertebrates. *Integ. Comp. Biol.* 24, 85-94.

Alexander, R. M. (2003). Principles of Animal Locomotion. Princeton University Press: Princeton.

Alexander, R. M. and Bennet-Clark, H. C. (1977). Storage of elastic strain energy in muscle and other tissues. *Nature* 265, 114-117.

Altringham, J. D. and Johnston, I. A. (1990). Scaling effects on muscle function: power output of isolated fish muscle fibers performing oscillatory work. *J. Exp. Biol.* 151, 453- 467.

Bennett, A. F. (1985). Temperature and muscle. *J. Exp. Biol.* 115, 333-344.

Biewener, A. A. (1998). Muscle-tendon stresses and elastic energy storage during locomotion in the horse. *Comp. Biochem. Physiol.* B 120, 73–87

Büschges, A., Schmitz, J. and Bässler, U. (1995). Rhythmic patterns in the thoracic nerve cord of the stick insect induced by pilocarpine. *J. Exp. Biol.* 198, 435-456.

Dickinson, M., Farman, G., Frye, M., Bekyarova, T., Gore, D., Maughan, D., and Duch C, Pfluger HJ. (1995) Motor patterns for horizontal and upside-down walking and vertical climbing in the locust. *J Exp Biol* 198:1963-1976.

Ferguson C, Lakey A, Hutchings A et al. (1994) Cytoskeletal proteins of insect muscle: Location of zeelins in *Lethocerus* flight and leg muscle. *J Cell Science* 107:1115-1129.

Gordon, A. M., Huxley, A. F., and Julian, F. J. (1966). The variation in isometric tension with sarcomere length in vertebrate muscle fibres. *J. Physiol.* 184, 170-92.

Hagiwara , S. (1955). Neuro-muscular mechanism of sound production in the cicada. *Phytiol. Comp. Oecol,* 4, 142-153.

Hess, D. and Büschges, A. (1997). Sensorimotor pathways involved in interjoint reflex action of an insect leg. J. Neurobiol. 33, 891 - 913.

Hill, A. (1938). The heat of shortening and the dynamic constants of muscle. *Proc. R. Soc. Lond. B* 126, 136–195.

Irving, T. (2005). Molecular dynamics of cyclically contracting insect flight muscle inlocomotion in the horse. *Comp. Biochem. and Physiol. Part B* 120, 73-87.

Marsh, R. L., Olson, J. M., and Guzik, S. K. (1992). Mechanical performance of scallop adductor muscle during swimming. *Nature*. 357, 411-413.

Myers, J. G. (1928). The morphology of the Cicadidae. *Proc. Zool. Soc. Land.* pp. 365-472.

Patek, S. N., Dudek, D. M., and Rosario, M. V. (2011). From bouncy legs to poisoned arrows: elastic movements in invertebrates. *J. Exp. Biol.* 214, 1973-80.

Pfluofelder, O. (I937). Vergleichendanatomische, erperimentelleund embryologische Untersuchungen Uber das Nervensystem und die Sinnesorgane der Rhynchoten. *Zoologica, Stuttgart* 34, 1—102

Pringlb, J. W. S. (1954). A physiological analysis of cicada song. *J. exp. Biol.* **33,** 525-560.

Ramirez JM, Pearson KG.1988. Generation of walking patterns for walking and flight in motoneurons supplying bifunctional muscles in the locust. J Neurobiol 19:257-282.

Roberts, T. J. (1997). Muscular Force in Running Turkeys: The Economy of Minimizing Work. *Science* 275, 1113-1115.

Simpson, P and Young, D (1978) The tymbal mechanism and song patterns of the bladder cicada *Cystosoma saundersii J.Exp.Biol.* 76, 27-45

Swank DM, Knowles AF, Suggs JA et al. (2002) The myosin converter domain modulates muscle performance. *Nat Cell Biol*; 4:312-317.

Tryba, A. K. and Ritzmann, R. E. (2000). Multi-joint coordination during walking and foothold searching in the Blaberus cockroach. II. Extensor motor neuron pattern. *J. Neurophysiol.* 83, 3337-3350.

Tu, M. S. and Daniel, T. L. (2004). Submaximal power output from the dorsolongitudinal flight muscles of the hawkmoth *Manduca sexta. J. Exp. Biol.* 207, 4651-62.

Watson, J. T. and Ritzmann, R. E. (1998). Leg kinematics and muscle activity during treadmill running in the cockroach, Blaberus discoidalis: II. Fast running. *J. Comp. Physiol.* A 182, 23-33.

Wilson DM. 1962 Bifunctional muscles in the thorax of grasshoppers. *J Exp Biol;* 39:669-677.

Young, D. (1975). Chordotonal organs associated with the sound producing apparatus of cicadas (Insecta, Homoptera). *Z. Morph. Tiere* 81, m-135.

Zill, S. N. and Seyfarth, E.-A. (1996). Exoskeletal sensors for walking. *Sci. Am.* 275, 86-90.

Zill, S. N., Schmitz, J. and Büschges, A. (2004). Load sensing and control of posture and locomotion. *Arth. Struct. Dev.* 33, 273-286.

Insect Sensory Physiology and Behaviour 12

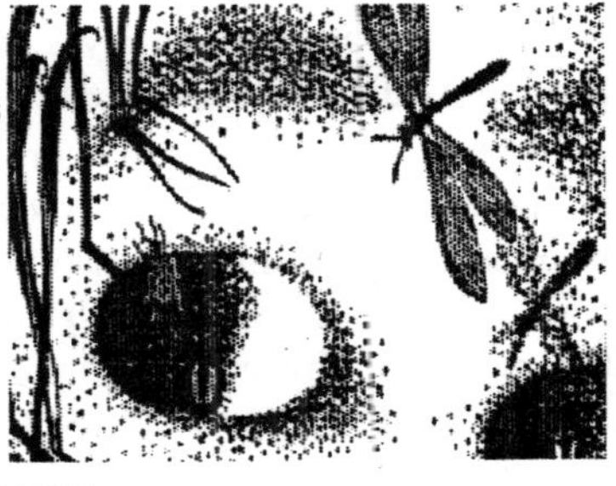

Synopsis

1. Introduction
2. Classification of insect sense organs
3. Mechanoreceptors
    - 3.1 The tactile organs or trichoid sensilla
    - 3.2 Proprioceptors
    - 3.3 Chordotonal organs
        - 3.3.1 Subgenual organs
        - 3.3.2 Johnston's organ
        - 3.3.3 Auditory or tympanal organ
        - 3.3.4 Non-tympanal sound perception
        - 3.3.5 Sound-producing mechanisms
        - 3.3.6 Behaviours associated with insect sound
            - A. Intraspecific communication: rivalry and territorial behaviour
            - B. Detection and avoidance of predators
            - C, Host finding by parasitoid insects
4. Chemoreceptors
    - 4.1 Olfactory (odorant) receptor
    - 4.2 Gustatory receptors
    - 4.3 Ionotropic receptors (IRs)
    - 4.4 Common chemical sense
    - 4.5 Olfaction in insects
5. Temperature and humidity receptors
    - 5.1 Mechanism of hygroreception
    - 5.2 Thermoregulatory behaviour in insects
6. Photoreceptors
    - 6.1 Compound eyes
    - 6.2 Ocelli - simple eyes
    - 6.3 Extra-ocular photoreception
    - 6.4 The mechanism of image formation
    - 6.5 Mechanism of vision
7. Industrial application of knowledge of insect sensory system
    - 7.1 Sensory system application in crickets
    - 7.2 Robot control
    - 7.3 Controlling insect behaviours

1. INTRODUCTION

Entomological research has revealed that the insect sensory systems are crucial for their success. Compared to human brains, the insect central nerve systems are extremely primitive and simple, both structurally and functionally, and are of minimal learning ability. Faced with these constraints, insects have evolved a set of extremely effective sensory systems that are structurally simple, functionally versatile and powerful, and highly distributed, as well as noise and fault tolerant.

There are a variety of factors accounting for the success of insects and the following four are among the most frequently cited: (1) the high level of organisation of insect sensory and neuromotor systems, (2) the minimal learning capacity and short generation time, (3) their small size, and(4) the nature of the co-evolutionary interactions that involve insects. The first factor is the essential and insects are only matched by vertebrates in this aspect. A high level of organisation enhances the ability to perceive small environmental differences and react to them; it effectively maximizes the apparent heterogeneity. Insects have minimal learning ability and vertebrates are superior without the slightest dispute. However, the minimal learning capability forces insects to depend on instincts that are genetically programmed. In order to adapt to constantly changing environments, they have to reprogram , i.e., to change their genetic code rapidly and to adapt to local environment perturbations. Their short lifetimes are actually favourable for evolving the rapid genetic changes. Therefore, the deficiency in learning ability forces insects to diverge genetically and evolve more species. The small size of insects is an advantage in that it makes them less likely to become extinct than larger animals. Insects' co-evolution, e.g. with plants, is the most interesting arms race in evolution and has undoubtedly contributed to their success by evolving feeding strategies that make them adaptable to the immediate environment.

Sensory complexity allows for both simple and complex behaviours of insects. For example, to control flight, the aerial environment must be sensed and appropriate responses made. Because much insect activity is nocturnal, orientation and navigation cannot rely solely on conventional visual cues, and in many night-active species, odours and sounds play a major role in communication. The range of sensory information used by insects differs from that of humans. We rely heavily on visual information, and although many insects have well-developed vision, most insects make greater use of olfaction and hearing than humans do. Insects perceive their immediate environment by using sense organs that protrude from the cuticle, or sometimes lie within or beneath it. Specialised cells detect stimuli that may be categorized as mechanical, thermal, chemical and visual. Other cells (the neurons) transmit messages to the central nervous system, where they are integrated. The nervous system instigates and controls appropriate behaviours, such as posture, movement and feeding, as well as behaviours associated with mating and oviposition.

All sense organs (receptors) act as transducers, converting light energy, chemical energy, or mechanical energy from the environment into electrical energy of nerve impulses in sensory neurons. Signals generated by insect sensory receptors travel to the brain or ventral nerve cord where they stimulate appropriate behavioural responses such as finding resources (e.g. food, mate, etc.), avoiding danger, or reacting to changes in the environment. All sensory receptors are derived from embryonic ectoderm and are integral parts of the insect's exoskeleton. Box 12.1 provides some different types of sensillae seen in insects.

Box 12.1. Select types of sensillae seen in insects

1. ***Trichoid sensilla*** have long hairs and 1-3 un-branched dendrites covered by a thick cuticle.
2. ***Basiconic sensilla*** are shorter with 1-3 branched dendrites and a thicker cuticle.
3. ***Placoid sensilla*** are plate-like, with 2-50 branched dendrites.
4. ***Coeloconic sensilla*** have a peg-like structure and are often located within a cuticular pit, innervated by 3-5 unbranched dendrites.
5. ***Campaniform sensilla*** refer to the bell shape appearance of the sensory structure which has only a single neuron.

2. CLASSIFICATION OF INSECT SENSE ORGANS

The sense organs in an insect body are distributed on different parts and respond to a given stimulus such as light, sound, touch, chemicals etc. The sense organs may be classified as:

1. Mechanoreceptors
2. Chemoreceptors
3. Temperature and humidity receptor
4. Photoreceptors or visual organ

3. MECHANORECEPTORS

Mechanoreceptors are the sense organs of insect, which respond to the sense of touch due to contact with external solid objects, current of air and water or even because of internal body pressure. Insect mechanoreceptors can be found almost anywhere on the surface of an insect's body. They may act as tactile receptors, detecting movement of objects in the environment, or they may provide proprioceptive cues (sensory input about the position or orientation of the body and its appendages). These receptors are innervated by one or more sensory neurons that fire in response to stretching, bending, compression, vibration, or other mechanical disturbance. Some mechanoreceptors produce a **phasic response** when stimulated, that is, they fire once when activated and again when deactivated. Other receptors generate a **tonic response**, firing repeatedly as long as a stimulus persists (Box 12.2). Neural processing centres in the brain or segmental ganglia interpret the combinations of tonic and phasic signals sent from nearby receptors.

Box 12.2 Basic forms of receptors according to adaptation

1. Phasic receptors
   - adapt rapidly to steady stimuli
   - react to short but strong stimuli
2. Tonic receptors
   - adapt slowly to steady stimuli
   - react to long lasting but less strong stimuli

The principal mechanoreceptors are:

1. The tactile organs or trichoid sensilla.
2. Proprioceptors - Hair plate, Stretch receptors and the campaniform organs.
3. The chordotonal organs.

3.1 The tactile organs or trichoid sensilla

The trichoid sensilla (Trichoform sensilla) are the simple articulated sensory hairs and distributed on the entire body surface commonly called **sensilla**. Trichoform sensilla are probably the simplest mechanoreceptors. These are **tactile hairs** (setae) that are innervated by a sensory neuron. Dendrites of the neuron attach near the base of the hair and generate a nerve impulse whenever they detect movement. **Hair beds** (clusters of tactile setae) are often found behind the head, on the legs, or near joints where they respond to movements of the body.

3.1.1 *Structure of tactile organ*

They are made up of cuticle and seen in the socket within the body wall. The trichoid sensillum is a sensory nerve cell. It is a modified epithelial cell with a single immotile cilium which is modified as a sensor and called **dendrite**. The sensory cell is accompanied by several accessory, supporting or sheath cells. The **tormogen cell** secretes the socket, the **trichogen cell** secretes the seta, and the **thecogen cell** (or sheath cell) secretes the **scolopale sheath**. A **glial cell** also wraps around the axon, forming an insulating **tunic**(Fig 12.1). The sensory cell is bipolar - with a dendrite at one pole and an **axon** at the other. The dendrite carries signals towards the cell body, the axon away from the cell body and towards the central nervous system. The **cell body** is the rounded region of the sensory cell containing the nucleus. The signals are electrical in nature (strictly electrochemical).The whole structure rests, along with the rest of the epithelium, on the **basement membrane** - a supporting skeleton of extracellular matrix materials.

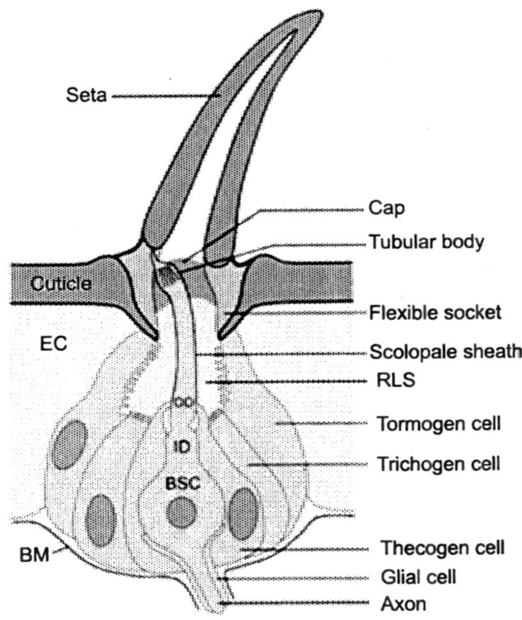

Fig. 12.1. Trichoid sensillum

(BM; basement membrane; BSC: the cell body of the bipolar sensory cell; EC, epithelial cell; ID, inner segment of sensory cell dendrite, inner dendrite; OD: outer segment of sensory cell dendrite, outer dendrite; RLS, receptor lymph space)

3.1.2 Location and function of tactile organ

Trichoid sensillae are present on the antennae, tarsi, tibia and cerci. These organs of the insects sub-serve the sense of touch. The functions of the tactile organs are:

1. There are some trichoid hair plates at the joints of various appendages and function as proprioceptors during sliding of the segments over each other.
2. There are tactile hair- beds on the facial region of the head of locusts and Lepidoptera, on the neck of dragonflies, on the wing margins of Lepidoptera which responds to the air movements during flight.
3. The tactile hairs of the antennae and lower segments of legs perceive earth-born vibrations in terrestrial insects and water surface vibrations in aquatic insects.

3.2 Proprioceptors

Insects require continuous knowledge of the relative position of their body parts, such as limbs or head, and need to detect how the orientation of the body relates to gravity. This information is conveyed by proprioceptors (self-perception receptors). Three types are commonly seen and descried namely the **hair plate, stretch receptors** and **campaniform sensilla.**

3.2.1 The hair plate

These are trichoid sensilla that form a bed of grouped small hairs, a **hair plate**, at joints or at the neck, in contact with the cuticle of an adjacent body part (Fig 12.2). The degree of flexion of the joint gives a variable stimulus to the sensilla, thereby allowing monitoring of the relative positions of different parts of the body.

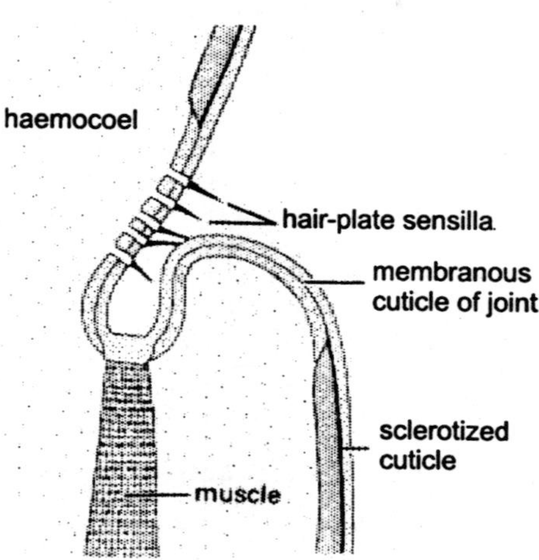

Fig. 12.2. Hair plate

3.2.2 *The stretch receptors*

The stretch receptors comprise internal proprioceptors associated with muscles such as those of the abdominal and gut walls. Alteration of the length of the muscle fibre is detected by multiple-inserted neuron endings, producing variation in the rate of firing of the nerve cell. Stretch receptors monitor body functions such as abdominal or gut distension, or ventilation rate. Stretch receptors are multi-polar neurons. They are commonly embedded in intersegmental membranes and in the muscular walls of the digestive system. Some insects may stop feeding when the gut expands enough to stimulate stretch receptors in the crop. In others, oviposition may begin when mature eggs stimulate stretch receptors in the reproductive system.

3.2.3 *The Campaniform sensilla*

Campaniform sensilla are flattened oval discs that usually serve as **flex receptors** in the exoskeleton. The campaniform sensilla cannot be seen externally but recognised from the dome-shaped cuticular areas. They are elevated above or depressed below the general body surface. Cell structure and arrangement is similar to that of trichoid sensilla. They respond whenever mechanical stress causes the exoskeleton to bend. Campaniform sensilla are found throughout the body, especially on the legs, near the base of the wings, and along sutures where two sclerites of the exoskeleton meet (Fig 12.3).

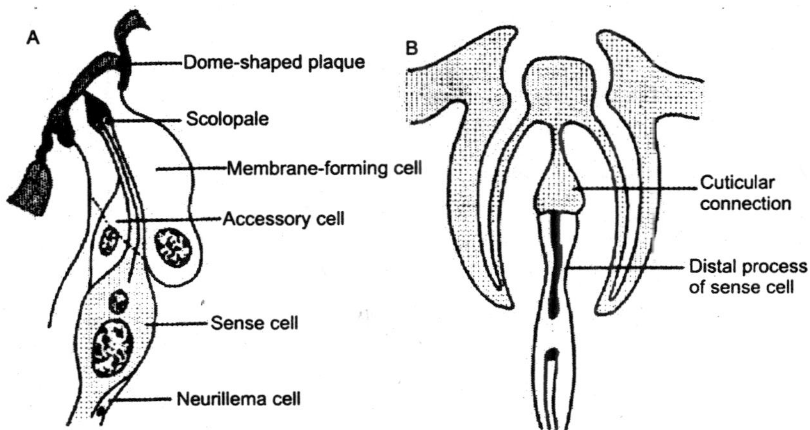

Fig. 12.3 A: Campaniform sensilla, B: Section showing stiffening rod of cuticle running along the cuticular plate

The cuticular stretches and the forces exerted on the cuticle by muscles and gravitational forces stimulate the campaniform sensilla. The campaniform sensilla functions as the proprioceptors. They respond to the mechanical stimuli in terrestrial insects, water pressure in aquatic insects and air pressure in flying insects. Pressure receptors provide sensory information about an aquatic insect's depth in the water. These receptors are usually associated with a cushion of air against the body or within the tracheal system. Increasing water pressure deflects hair-like processes within the receptor and stimulates tonic and phasic impulses.

3.3 Chordotonal organs

Chordotonal organs include several types of mechanoreceptors in which one or more bipolar neurons bridge a gap between two internal surfaces of the exoskeleton. Chordotonal organ consists of single unit or group of similar unit is called **scolopidia**. They are sub-cuticular and are attached to the cuticle at one or both end often no sign of their presence. Each scolopidia consists of three cells:

1. Neuron
2. Scolopale cell or enveloping cell
3. Cap cell

Each neuron is usually accompanied by the other two cells which form a sheath (the scolopale cell) and a point of attachment (the cap cell). Together, these three cells create a unit (called **scolopidia**) that may occur singly or in groups (Fig 12.4). The chordotonal organs occur in legs at femoral, distal tibial and tibio-tarsal regions, in abdomen and wing base. They are stimulated by passive movement of segments, tension of muscles, internal pressure changes due to blood and tracheae.

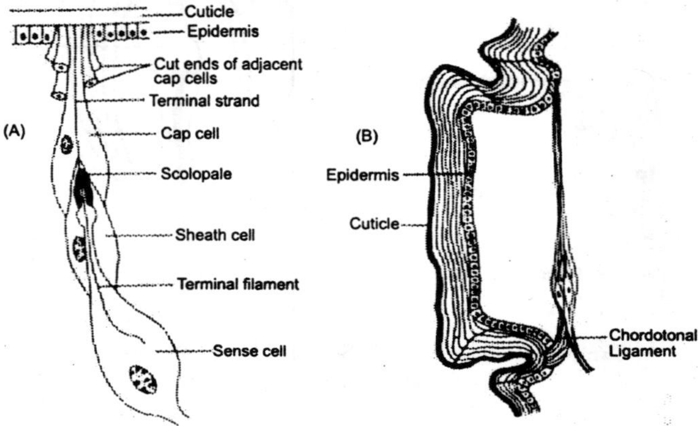

Fig. 12.4. A: Single chordotonal sensillum; B: Chordotonal organ

Box 12.3 Sense of vibrations in insects

1. Properties:
   - vibrations can be perceived on the boundaries of mediums e.g. solid - air, air water
   - relatively low frequency periodic stimuli
   - can be perceived only from a short distance
   - generally has alarming effect or plays role in communication
2. Insects are classified according to vibration sense:
1. Sensitive species: nerve impulses are synchronous with the frequency of stimuli e.g.: Lepidoptera, Hymenoptera
   - typical sense organ: subgenual organ is a complex chordotonal organ in the tibia of each leg with 10-40 sensilla to perceive substrate vibrations
2. Less sensitive or indifferent species:
   - nerve impulses are asynchronous with the frequency of stimuli e.g.: Heteroptera
   - typical sense organs: trichoid sensilla and campaniform sensilla

Common types of chordotonal organs include:

1. Subgenual organs
2. Johnston's organ
3. Auditory or tympanal organ

3.3.1 *Subgenual organs*

These are located in the legs of many insects. These receptors contain relatively few scolopidia yet they appear to be very sensitive to substrate vibrations. Insects may lack specialized sound receptors, yet they can still **hear** vibrations transmitted through the substrate (Box 12.3).

3.3.2 *Johnston's organ*

Johnston's organ is a specialised chordotonal organ in the 2nd antennal segment (pedicel), and occurs in all adult insect except Collembola and Diplura. It consists of single mass or several groups of scolopidia and is highly developed in Culicidae, where the pedicel is enlarged. Axon of sense cells run back and enter the antennal nerve. It perceives movement of antennal flagellum and flight speed indicator.In mosquitoes and midges, they respond to certain frequencies of airborne sound by detecting resonant vibrations in antennal hairs. Shorter hairs near the tip of the antennae resonate to higher frequencies than longer hairs near the base.

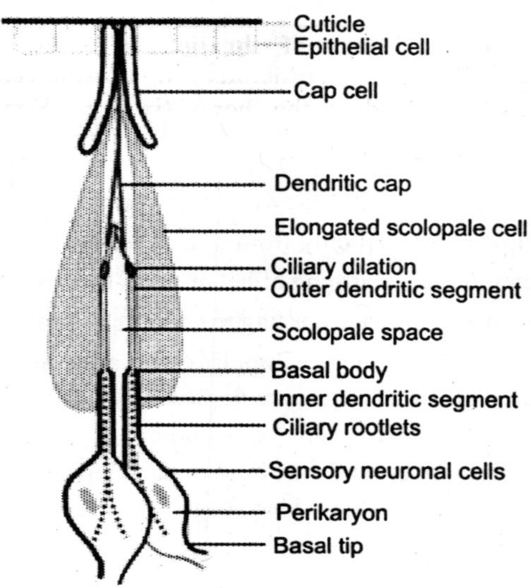

Fig. 12.5. Diagrammatic sketch of the scolopidia of Johnston's organ

In *Drosophila*, the Johnston's organ is composed of about two hundred functional units called **scolopidia** (Fig.12.5). Each scolopidium is formed of five different types of cells: scolopale cell, cap cell, ligament cell, dendritic cell and neuronal cell. The sensory neuron has axon at the proximal region and apically it has a dendrite with outer segment having characteristic features of non-motile cilia. The cilia contain a dilated structure known as **ciliary dilation** and are formed of paracrystalline like material. The basal part of the cilia contains ciliary rootlet, which extends from the base of the dendritic outer segment through inner segment into the perikarya. The scolopale cell forms an endolymph around the cilia and produces actin based rod, which provide strength to the scolopidia. Apical to the scolopale cell is the cap cell, which is surrounded by attachment cell.

3.3.3 Auditory or tympanal organ

Tympanal organs are present in the adult of many insect species. There is tremendous morphological diversity of insect ears. The multitude of different ear designs and locations reflects the unique physical and behavioural challenges faced by each insect. Yet despite their many differences, most ears follow a similar morphological plan. Tympanal organ consists of three identifiable structures namely, (1) a thin layer of cuticular structure, called **tympanic membrane**, (2) a tracheal air sac and (3) a group ofchordotonal organ. Tympanic membrane and air sac form drum; sound waves that strike the drum causes it to vibrate and therefore the sensilla can be stimulated. These **ears** may be located on the thorax (in some Hemiptera), on the abdomen (in grasshoppers, cicadas, and some moths), on the front tibia (in crickets and katydids), on the wing bases of certain moths and radial vein of some lacewings. In tiger beetles they are seen under the elytra and they hear when in flight. In scarab beetles they are seen in the cervical membranes and in flies in the prosternum.

The **tympanal membrane** is a thinned region of exoskeleton, typically supported by a chitinous and stretched across an enlarged air-filled cavity. Sound impinges upon the membrane, setting it and its associated nerve cells into motion. The thickness of the membrane can vary from 1 to over 100 μm. The ultra-sound-sensitive ears of many nocturnal Lepidoptera (butterflies and moths), for example, are so thin that they are transparent. Such fragile membranes are typically protected within body cavities or by external flaps of cuticle. In contrast, the thicker, opaque tympanal membranes of some diurnal butterflies or grasshoppers are conspicuously positioned on the outer surface of the body.

The internal face of the eardrum backs onto an enlarged **air-filled chamber**, which forms part of the tracheal respiratory system. In some ears, the air chambers are connected directly to other sound input sources (spiracles or contralateral ears) or resonating chambers via the tracheal system. In a few rare cases [e.g., green lacewings (Chrysopidae) and some water bugs (Corixidae)] the tympanic chambers are largely fluid filled.

Associated with the inner surface of the tympanal membrane is one to several **chordotonal organs.** Each chordotonal organ comprises one or more individual sensory units called **scolopidia**, and each scolopidium consists of three cells arranged in a linear array: a sensory cell, a scolopale cell, and an attachment cell. The total number of scolopidia in an ear ranges from one in some moths (Notodontidae) to almost 2000 in the bladder grasshopper (*Bullacris membracioides*). A chordotonal organ may attach directly to the inner surface of the tympanic membrane or to tracheal air sacs indirectly associated with the tympanum. The axons of the sensory neurons collectively make up the auditory nerve that forms the neural link between the mechanosensory stimulation of the eardrum and its **perception** by the central nervous system. Vibrations of the eardrum and/or air chamber cause bioelectric currents to flow in the sensory cell, initiating action potentials in the auditory nerves and signalling neurons in the auditory pathways of the nervous system.

There are two distinct types of insect tympanal ears: pressure receivers and pressure difference receivers (or pressure gradient receiver). In a

pressure receiver, the tympanal membrane forms one side of an otherwise enclosed air chamber. The membrane vibrates in response to sound waves arriving only to the external surface of the tympanal membrane. In insects whose ears are sensitive to high-frequency sounds (like many bat-detecting moth ears), the sound shadow of the body is sufficient to cause a difference in sound amplitudes arriving to each ear, which provides useful directional cues to the animal. Many insects using lower frequency sounds are faced with the difficulty of being able to localsze sounds that arrive almost simultaneously to both ears. They overcome this problem using pressure-difference receivers in which the sound impinges on both the outer and the inner surfaces of the membrane, and the membrane vibrates in response to the difference in pressure between the two sides.

A third and very unusual way to localise sounds has been invented by certain parasitic tachinid and sarcophagid flies. These flies localise their hosts (field crickets, katydids, and cicadas) by hearing and then homing in on the mating calls. The ears possessed by these flies are conventional chordotonal organs, containing 70 or more scolopidia. However, the fly's eardrums are mechanically connected, which confers an acute sense of directionality such that they can detect and localise their calling hosts/prey from a distance of tens of meters, even from high in the air. Such ears appear to be unique to these parasitoid flies.

The frequencies detectable by insect tympanal ears range from a few kilohertz (e.g., the water boatman Corixa) to over 100 kHz (e.g., some tettigoniid species). This broad bandwidth of frequency sensitivity is used by some insects for mate calling and in others it enables them to detect their predators (e.g., echolocating bats). Some insects have the ability to discriminate between different sound frequencies, which may be achieved at the level of the tympanal membrane or the chordotonal organ.

Tympanal sound reception is particularly well developed in orthopterans, especially the crickets and katydids. In most of these ensiferan Orthoptera, the tympanal organs are located in the tibia of each foreleg. Behind the paired tympanal membranes lies an acoustic trachea that runs from a prothoracic spiracle down each leg to the tympanal organ.

The acoustic tracheae of katydids form a system completely isolated from the ventilator tracheae, opening via a separate pair of acoustic spiracles. In many katydids, the tibial base has two separated longitudinal slits, each of which leads into a tympanic chamber. The acoustic trachea, which lies centrally in the leg, is divided in half by a membrane at this point, such that one half closely connects withthe anterior and the other half with the posterior tympanal membrane. The primary route of sound to the tympanal organ is usually from the acoustic spiracle and along the acoustic trachea to the tibia. The change in the cross-sectional area from the enlargement of the trachea behind each spiracle to the tympanal organ in the tibia functions as a horn in amplifying the sound. Although the slits of the tympanic chambers allow the entry of sound, their exact function is debatable. They may allow directional hearing, because very small differences in the time of arrival of sound waves at the tympanum can be detected by pressure differences across the membrane.

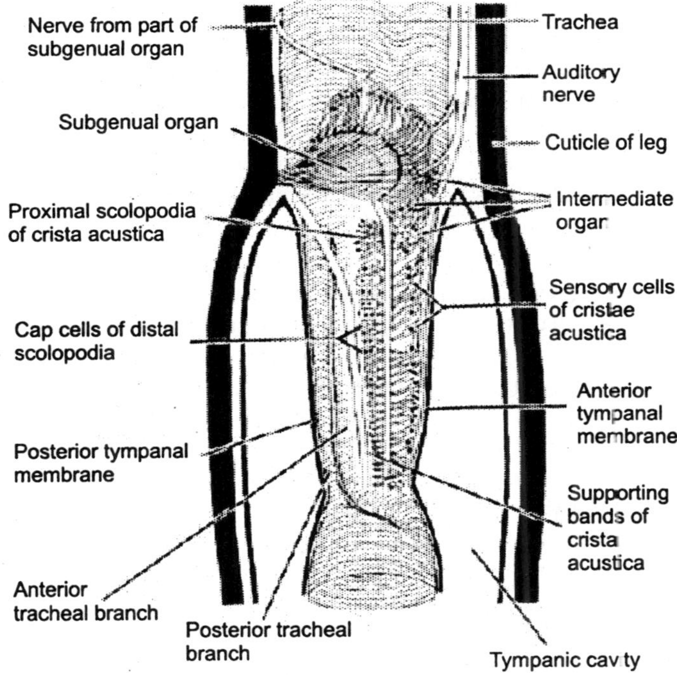

Fig. 12.6 Tympanal organs of katydid
(After Schwabe 1906 in Michelsen & Larsen 1985)

Entry of sound to the tympanal organs by air- and substrate-borne signals causes the tympanal membranes to vibrate. Vibrations are sensed by three chordotonal organs: the **subgenual organ**, the **intermediate organ** and the **crista acustica** (Fig. 12.6). The subgenual organs,which have a form and function similar to those in non-orthopteroid insects, are present on all legs, but the crista acustica and intermediate organs are foundonly on the forelegs inconjunction with the tympana. This implies that the tibial hearing organ is a serial homologue of the proprioceptor units of the mid and hind legs. The crista acustica consists of a row of up to 60 scolopidial cells attached to the acoustic trachea and is the main sensory organ for airborne sound in the 5–50 kHz range. The intermediate organ, which consists of 10–20 scolopidial cells, is posterior to the subgenual organ and is virtually continuous with the crista acustica. The role of the intermediate organ is uncertain but it may respond to airborne sound of frequencies from 2 to 14 kHz. Each of the three chordotonal organs is innervated separately, but the neuronal connections between the three imply that signalsfrom the different receptors are integrated.

Hearing insects can locate a point source of sound, but exactly how they do so varies between taxa. Localisation of sound directionality clearly depends upon detection of differences in the sound received by one tympanum relative to another, or in some orthopterans by a tympanum within a single leg. Sound reception varies with the orientation of the body relative to the sound source, allowing some precision in locating the source.

Night activity is common, as shown by the abundance and diversity of insects attracted to artificial light, especially at the ultraviolet end of the spectrum, and on moonless nights. Night flight allows avoidance of visual-hunting predators, but exposes the insect to specialist nocturnal predators—the insectivorous bats (Microchiroptera). These bats employ a **biological sonar system**, using ultrasonic frequencies that range (according to species) from 8to 215 kHz, for navigating and for detecting and locating prey, predominantly flying insects. Although bat predation on insects occurs in the darkness of night and high above a human observer, it is evident that a range of insect taxa can detect bat ultrasounds and take appropriate evasive action. The behavioural response to ultrasound, known as the **acoustic startle response**, involves very rapid and co-ordinated muscle contractions. This leads to reactions such as freezing, unpredictable deviation in flight, or rapid cessation of flight and plummeting towards the ground. Instigation of these reactions, which assist in escape from predation, obviously requires that the insect hears the ultrasound produced by the bat.

Physiological experiments show that the response can be within a few milliseconds of the emission of such a sound, which would precede the detection of the prey by a bat. Certain insects belonging to at least six orders detect and respond tc ultrasound: praying mantids (Mantodea), locusts, katydids and crickets (Orthoptera), lacewings (Neuroptera), moths (Lepidoptera), beetles (Coleoptera), and some parasitoid flies (Diptera). Tympanal organs occur in different sites amongst these insects, showing that ultrasound reception has several independent origins amongst these orders. The orthopteran ability to hear bat ultrasounds is an **exaptation**—in this case, amorphological–physiological predisposition that has been modified to permit sensitivity to ultrasound. The crickets, katydids and acridid grasshoppers that communicate intraspecifically and also hear ultrasound have sensitivity to high- and low-frequency sound— and perhaps limit their discrimination to only two discrete frequencies. The ultrasound elicits aversion; the low-frequency sound (under suitable conditions) elicits attraction.

3.3.4 *Non-tympanal sound perception*

Sound is a pressure fluctuation transmitted in a wave form via movement of the air or the substrate, including water. Human ears perceive sounds from 20 to 20,000 Hz (1 hertz (Hz) is a frequency of one cycle per second). Amongst insects some receive vibrations ranging from as low as 1–2 Hz to ultrasound frequencies perhaps as high as 100 kHz.

Two types of vibration or sound reception that do not involve tympana are the detection of substrate-borne signals and the ability to perceive the relatively large translational movements of the surrounding medium (air or water) that occur very close to a sound. The latter, referred to as near-field sound, is detected by either sensory hairs or specialised sensory organs.

A simple form of sound reception occurs in species that have very sensitive, elongate, trichoid sensilla that respond to vibrations produced by a near-field sound. For example, caterpillars of the noctuid moth *Barathra brassicae* have thoracic hairs about 0.5 mm long that respond optimally to vibrations of 150 Hz. Although in air this system is effective only for local sounds, caterpillars can respond to the vibrations caused by the audible approach of parasitic wasps.

The cerci of many insects, especially crickets, are clothed in long, fine trichoid sensilla (filiform setae or hairs) that sense air currents, which can convey information about the approach of predatory or parasitic insects or a potential mate. The direction of approach of another animal is indicated by which hairs are deflected; the sensory neuron of each hair is tuned to respond to movement in a particular direction. The dynamics (the time-varying pattern) of air movement gives information on the nature of the stimulus (and thus on what type of animal is approaching), and is indicated by the properties of the mechanosensory hairs. The length of each hair determines the response of its sensory neuron to the stimulus: neurons that innervate short hairs are most sensitive to high-intensity, high-frequency stimuli, whereas long hairs are more sensitive to low-intensity, low-frequency stimuli. The responses of many sensory neurons innervating different hairs on the cerci are integrated in the central nervous system to allow the insect to behave appropriately to detected air movement.

For low-frequency sounds in water (a medium more viscous than air), longer distance transmission is possible. Although few aquatic insects communicate through underwater sounds, "drumming" sounds are produced by some aquatic larvae to assert territory, and noises are produced by underwater diving hemipterans such as corixids ·and nepids. Many insects can detect vibrations transmitted through a substrate at a solid–air or solid– water boundary or along a water–air surface. The perception of substrate vibrations is particularly important for ground-dwelling insects, especially nocturnal species, and social insects living in dark nests. Some insects living on plant surfaces, such as sawflies(Hymenoptera: Pergidae), communicate with each other by tapping the stem. Various plant feeding bugs (Hemiptera), such as leafhoppers, planthoppers and pentatomids, produce vibratory signals that are transmitted through the host plant. Water-striders (Hemiptera:Gerridae) create pulsed waves across the water surface film to communicate in courtship and aggression. Moreover, they can detect the vibrations produced by the struggles of prey that fall onto the water surface. Whirligig beetles (Coleoptera: Gyrinidae) can navigate using a form of echolocation: waves that move on the water surface ahead of them are reflected from obstacles and are sensed by their antennae in time to take evasive action.

Some species of buprestid beetles (*Melanophila* spp.) live on trees killed in forest fires. These insects depend on paired thoracic pit organs (Fig 12.7) for infrared reception (Box 12.4).

Box 12.4 Infrared reception

1. Infrared irradiation can be detected by ***paired thoracic pit organs*** in the species of buprestid beetles (Melanophila spp.) that breed only in trees that have been recently killed by fire.
2. Beetles respond to the infrared radiation emitted from forest fires using dome-shaped sensilla located in paired pits near the mesothoracic coxa that are exposed during flight.
3. Each pit organ contains 50 to 100 sensilla that change in volume with exposure to infrared radiation and deform the dendrite of the underlying mechanoreceptor.
4. These sensilla are activated specifically by the infrared wavelengths (360-410 nm) from burning forests, and other sensilla on the antennae respond to olfactory cues from phenolic compounds present in the smoke from these recent fires.

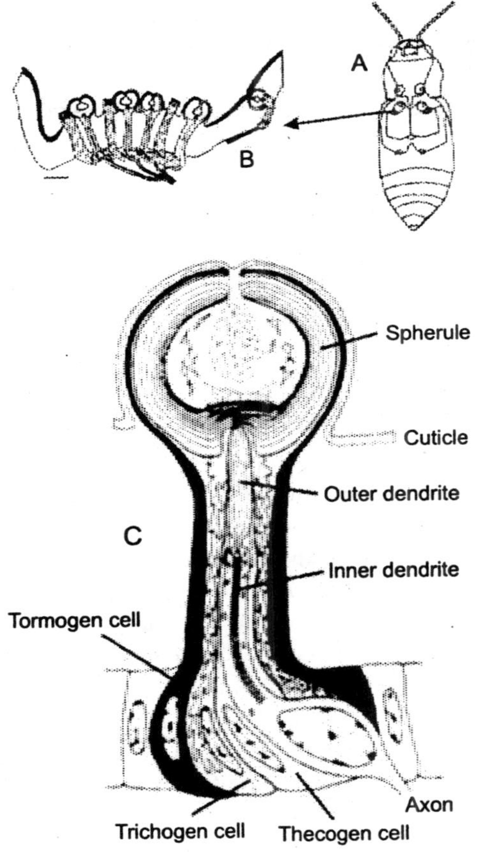

Fig. 12.7. Infrared receptor (C) from *Melanophila* located in a cuticular pit (A,B)

3.3.5 *Sound-producing mechanisms in insects*

Many groups have specialised mechanisms of sound production and hearing which are used in intraspecific acoustic communication. Ewin (1989) recognised five categories of sound producing mechanisms.

Vibration (including Tremulation)

In some insects sound emissions results from vibrations of relatively unspecialised body parts of the insect, generally oscillations of the abdomen, either dorso-ventrally or laterally. The oscillatory movement of the wings of an insect sets up regions of compression and rarefaction and a vibrational sound is produced. Tremulation produces bending waves in the substrate. As the insect vibrates its abdomen vigorously in the vertical plant, the lightweight stem or leaf upon which it is standing is shaken up and down at low frequencies ranging typically between 10 and 150 Hz.

Tremulation sound is transmitted through the legs to the substrate on which the insect is walking or standing. The flight sound, made by the wings, in swarming mosquitoes is considered to be used for species-specific recognition (e.g. Gibson, et al, 2010; Pennetier, et al 2010). Wing vibration is also used in the courtship dances of *Drosophila* species (Von Schilcher, 1976).

Percussion

Percussive (drumming) signals results from some part of the insect's body, usually the abdomen or wings which usually strike the substrate such that the compression waves are propagated through the material. The tone of the signal is largely dependent on the resonant properties of the substrate.

Striking one part of the body against another as a communication system for pair formation, is known for example in the Australian moth (Lepidoptera); males produce ultrasonic acoustical long distance signals to attract sexually receptive females and to establish territorial residency in competition with other males (Alcock & Bailey, 1995).

Striking the substrate with the tip of the abdomen is also considered as a **percussion signal**. An example of an insect that uses percussion as a sound producing mechanism is a species of the suborder Arctoperlaria (Plecoptera), which produce drumming signals on the substrate as a mating system. The male emits a species-specific call searching for the female, Once the female answers, a duet is established with vibrational signals. Then she becomes stationary and the male starts localising the stationary female until location and mating are accomplished (Stewart & Sandberg, 2005).

Stridulation

Stridulation consists of sounds produced by frictional mechanisms, involving the movements of two specialised body parts against each other in a systematic patterned manner.

For stridulation to work effectively a cricket must have two specialised body parts that together make up the tegminal, a **scraper** and a **file** (Gullan and Cranston, 2010). These are located on the wings of the insect; one wing will be the file while the other will be the scraper (Fig 12.8.). To generate their call the scraper strikes file and rasps against the teeth that are present

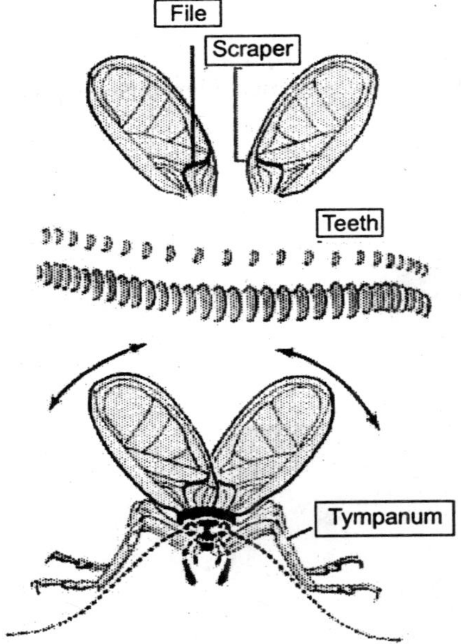

Fig. 12.8. The stridulating organs of cricket

in both the file and the scraper. Each of these strikes of the scraper on the file generates a single pulse **(Silbe)** (Huber, 1962). The length of the call and the frequency of the song that is created by these strikes is dependent on the morphological and mechanical structures of the tegminal and the speed that the strikes are occurring (Huber, 1962). Thus song formation is controlled by six main things; (1) the position of the both the tegman relative to each other when the strike occurs, (2) the number of teeth used during each strike (length of the pulse), (3) the number of pulses per song, (4) the sequence of pulses (long and short), (5) the intensity of the pulses (how loud the cricket makes them) and (6) the frequency at which the sound that is produced (Huber, 1962).

In Figure 12.8 the wing on the right makes the file section of the tegminal while the left wing makes the scraper section. When the wings move back and forth (bottom of the figure) the file strikes the scraper which generates the call that we hear from the cricket. The amplification of calls in crickets can be done in two ways, changing the environment or using the subalar space for resonance (Stephen & Hartley, 1995). Using the subalar space to amplify calls is quite different to the altering the surroundings and allows for better control over the frequency of the calls released, it is the relative position of the wings to the abdomen that allows for this frequency control (Stephen and Hartley, 1995). By changing the amount of space that is between the wings and abdomen the amount of echo that occurs and the

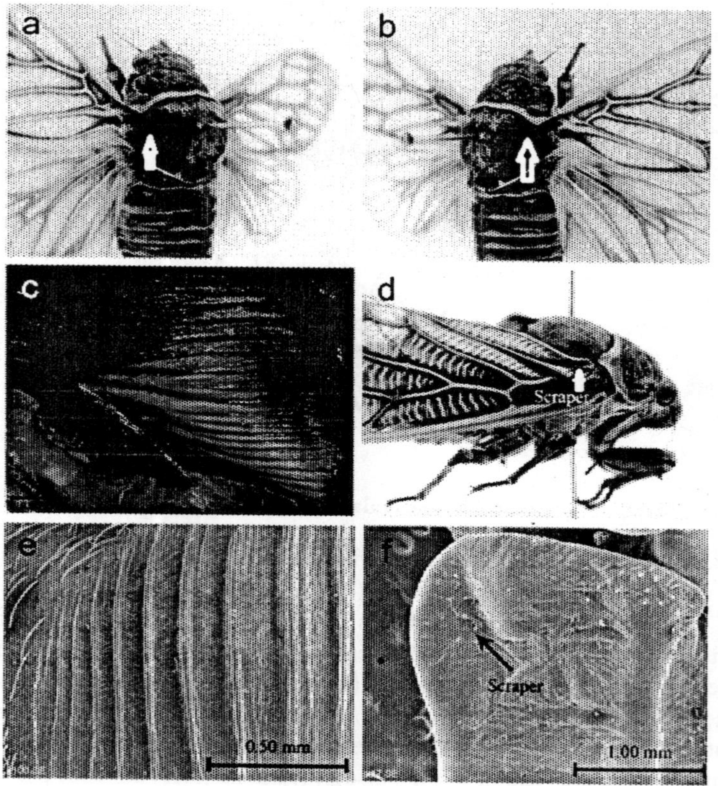

Fig. 12.9. The paired stridulatory organs of female *Subpsaltria yangi*.

(a) A file (arrow) is situated at the left anterior angle of the mesonotum. (b) A file (arrow) is situated at the right anterior angle of the mesonotum. (c) A micrograph of a stridulatory file. (d) The arrow indicates the scraper of a stridulatory organ. (e) A scanning electron micrograph showing the ridges of a file. (f) A scanning electron micrograph showing the scraper of a stridulatory organ (Adapted from Changqing Luo and Cong WePLoS One. 2015; 10(2): e0118667)

length of the sound waves produced is also altered (Stephen and Hartley, 1995). The amount of echo controls how loud the call is (how far it will travel) and the length of the sound wave is the frequency at which the call is produced.

Reception of the sound is the same in both the male and the female crickets; they have sense organs that are specialised for the pickup of sound that if generated by another cricket and they are able to distinguish that sound and the sound that they produce themselves (Kowalski and Lakes-Harlan, 2009). They have 27 of these sense organs on the crista acustica, 18 located on the mid-leg and 14 on the hind-leg, this allows the insect to identify the direction that they sound came from, this characteristic is especially important for the female crickets as it is how they locate a mate (Kowalski and Lakes-Harlan, 2009).

The stridulating organs of the cicada are vibrating membranes made use of for the purpose of producing sound. In cicadas, the main method of sound production is the tymbal mechanism, and the tymbal system is only developed in males, except that in two species from the family Tettigarctidae, *Tettigarcta crinita* and *T. tomentosa*, both sexes possess rudimentary tymbals with reduced apparent tymbal muscles. As in most other cicada species, the females of *Subpsaltria yangi* lack tymbal organs, but they have evolved well-developed stridulatory structures. In cicada, *Subpsaltria yangi,* females have a stridulatory organ on each side of the body. The left and right stridulatory organs are similar in morphology. The stridulatory file is a conspicuous oval area on the anterior angle of the mesonotum (Fig. 12.9 a,b). The file is yellowish in colour, and bears a series of ridges (Fig. 12.9 c). The ridges are highly sclerotised and almost parallel to each other (Fig. 12.9 c,e). The base of the inner margin of the forewing, serving as the scraper, is thickened, sclerotised and slightly curved outwards.

Click mechanisms

These sounds depend on the deformation of a modified area of cuticle, generally by contraction and relaxation of specialised musculature within the insect body. This movement results in a succession of clicks which may be repeated quickly in distinctive patterns. **Tymbal** is the name of the specialised area of cuticle, as exemplified in the loud singing cicadas (Hemiptera). For example, the males of *Tibicina* (Hemiptera) cicada species produce a sustained and monotonous calling song by tymbal activity. Tymbalis an organ that they click in and out extremely quickly. A tymbal is on the cicada's abdomen, and the cicada uses its large cavernous body to amplify the noise, much like a large concert hall amplifies the sounds of an individual musician (Fig 12.10) This acoustic signal constitutes the first step in pair formation, attracting females at long distances, and is involved in male-male interactions (Sueur and Aubin, 2003).

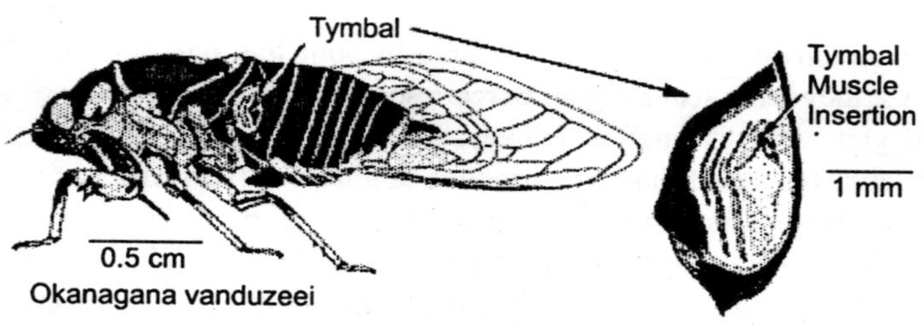

Fig. 12.10. Tymbal of cicada, *Okanagana vanduzeei*

Air expulsion

This is an unusual sound-producing mechanism within insects. This sound is described as an **exhalatory sound**, frequently expelled via the tracheal spiracles, however little is known about its function (Ewing, 1989). The Madagascar hissing cockroach *Gromphadorhina portentosa,* is able to produce audible hisses from a pair of modified spiracles. Adult males hiss in three social contexts: during courtship, during copulation, and during aggressive encounter. Adults and nymphs of both sexes also hiss when disturbed (Nelson and Fraser, 1980).

3.3.6 Behaviours associated with insect sound

Acoustic signals (sounds) can be associated with several insect behaviours, mainly mate attraction, courtship and territorial behaviour. For instance, the mature unmated female cricket *Gryllus campestris* and long-horn grasshopper *Thamnotrizon apterus*, are attracted by the singing (chirping) of the males. Mosquitoes also produce noises that are used for mate attraction. The following are major behaviours associated with sound production in insects:

Intraspecific communication: rivalry and territorial behaviour

Some grasshoppers have developed one of the most highly intricate song behaviour amongst insects, called the **rival duet** behaviour. If a male grasshopper, singing his courtship song to a female, is interrupted by the appearance of a second male who also tries to court the female; the first male leaves the female, faces the intruder, and sings the **rivalry song**. This behaviour produces an abrupt and similar response from the interloper, and the two males sing against one another. The contest may be brief, when one or other leaves the field and the remaining male returns to his courtship, or so long that the female abandons the scene herself (Haskell, 1961).

Busnel (1953) described the territorial behaviour of the cricket *Oecanthus pellucens*. The male lives in bushes, about 50 cm above the ground and stays for the great majority of its life restrained to a small area approximately 50 sq cm. The male moves around his territory, patrolling, and singing its normal song. If it encounters another interfering male, it sings a type of song called the **warning song.** This interesting behaviour appears analogous to the territory marking of certain song-birds, which is established by the male by singing at the territorial boundaries prior to the commencement of the mating season (Haskell, 1961).

Detection and avoidance of predators

Numerous predators utilise a combination of sensors to hunt insects: vision, olfaction, and hearing are the main ones; with a bias toward vision in diurnal animals and toward hearing and olfaction in the nocturnal predators. Nocturnal acoustic hunters include felids, canids, owls, rodents, which are all passive listeners. Bats are also passive listeners, but also active echolocators. According to Conner (2014), all traits including visual, chemical tactile, electric, and acoustic cues, that minimize the probability of a prey being detected when potentially detectable to an observer, can be called Crypsis. For example, certain moths interrupt their sexual acoustic calls in the presence of acoustic predators, to avoid detection by the predator (Greenfield, 2014; Spangler, 1984). Similar adaptive silence behaviours have been reported in crickets and katydids (Bailey and Haythornthwaite, 1998; Conner, 2014; Faure and Hoy, 2000; Spangler, 1984).

The cricket *Teleogryllus oceanicus* depicts one of the most remarkable examples of adaptive silence (Hedwig & Robert, 2014; Zuk, et al 2006). The male field cricket stridulates by scissoring the wings; females are attracted to the males' songs. Interestingly, an Australian and Pacific Island species of *T. oceanicus*, have been forced to alter that strategy because of an acoustically orienting parasitoid fly *Ormia ochracea.* Like the female cricket, this fly is attracted to the male cricket's calling **song.** After locating the male cricket, the fly deposits her offspring on or near the male. The larvae burrow into the male, killing him as they develop. Within approximately 20 generations of intense selection has resulted in a morphological change in the wings of the male crickets on Kauai. Mute males, called **flat-wing males**, have wings similar in appearance to those of females. Mute males have lost their file and scraper and are thereby silenced. Flat-wing males produce no calling song and do not attract the parasitoid. However, it is not clear how the silent flat-wing males now attract mates, although Cade (1980) suggested that they function as satellite males, waiting close to calling males, and intercepting females as they move toward them.

Whispering moths exemplify another antidetection strategy (Nakano *et al.*, 2008). *Ostrinia furnicalis*, male Asian corn borer moths, use specialised very low intensity courts cm) produced by sex-specific scales on the forewings and mesothorax. In order to protect the pair from conspecific competitors and predators, this male moth produces the sounds in the immediate vicinity of the female's ear, which provide a private communication channel between the male singer and the female listener (Conner, 2014).

Host finding by parasitoid insects

Animal acoustic signalling uncommonly constitutes a confidential communication channel. Sound tends to propagate in all directions, broadcasting towards hardly predictable destinations and, sometimes, is unintentionally received by predators and parasitoids. The Dipteran parasitic family Tachinidae and the flesh flies family Sarcophagidae take advantage of acoustic signalling in Orthoptera (crickets) and Hemiptera (cicadas), enabling the parasitic exploitation. Guided by acoustic communication signals, these flies identify and locate their singing target, depositing their larvae on or close the host. Larvae then develop as endoparasites, eventually killing the host (Hedwig and Robert, 2014).

4. CHEMORECEPTORS

Insects have the ability to sense various chemical substances in their environment. When these chemicals are present in gaseous form at relatively low concentrations, they may be detected as odours by **olfactory receptors**. When they are in solid or liquid form (usually at higher concentrations) they are perceived as tastes by **gustatory receptors**. In general, the sense of taste involves direct contact with a substrate (contact chemoreception) whereas olfaction usually implies detection of compounds in gaseous or airborne form (remote chemoreception).

Chemosensory systems play an important role in shaping the life of an insect, such that fundamental behaviours such as mating, food choice and seeking, predator and parasitoid avoidance, and egg-laying, are strongly regulated by external chemical stimuli. The recent focus on the molecular mechanisms of chemosensory detection in insect chemical ecology research has identified canonical chemosensory receptors in insects that consist of **odorant receptors (ORs), gustatory receptors (GRs), and ionotropic receptors (IRs).**

The fundamental chemoreceptive organ of the insect is the sensillum commonly called **sensory hair, sensory peg, sense organ, or sensory receptor**. Sensilla are distributed on a range of insect structures including the antenna, mouthparts, and limbs especially tarsi. The early classification of insect sensilla was based on external morphology as revealed by light microscopy. This recognised five categories of sensilla to which a further five were subsequently added giving 10 principal categories ordered in terms of the extent of their cuticular extensions as follows:

- S. trichodea, long, slender and hair-like; s. chaetica, long, heavy and thick-walled ;
- S. basiconica, shorter and peg-like and s. auricillia, similar but 'rabbit-eared';
- S. styloconica, sharply-tipped pegs on a stylus; s. coeloconica, pegs in shallow pits;
- S. ampullacea, pegs in deep pits; s. campaniformia, dome- or button-shaped;
- S. placodea, flattened plates; s. scolopalia and scolopophora, subcuticular;
- S. squamiformia, scale-like (Fig. 12.11).

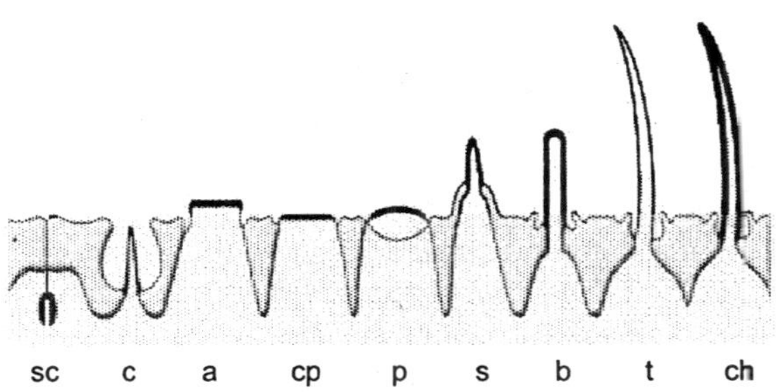

Fig. 12.11. Principal types of sensilla (sc, scolopophora; c, coeloconic; a, ampullaceous; cp, campaniform; p, placoid; s, styloconic; b, basiconic; t, trichoid; ch, chaetica)

The advent of the electron microscope facilitated more detailed information on sensillar ultrastructure and a steady stream of reports on a wide range of insects indicated a basic, common pattern of construction comprising: (1) cuticular components most usually expressed as hair-like extensions; (2) sensory neurons; and (3) associated cells.

Cuticular components

Typically, sensilla with no pores, aporous (AP) and with one neuron are sensitive to mechanical stimuli; however, some no-pore sensilla may be hygro- or thermosensitive. Sensilla with one pore (uniporous, UP) or with multiple pores (multiporous, MP) are chemoreceptive with contact and olfactory functions respectively. The single pore, when brought into physical contact with a chemical stimulus, funnels molecules to relatively unbranched dendrites. In addition and more widespread are dual-function gustatory/mechanoreceptive organs, the first detailed account of which was provided by Lewis in 1954. Essentially, this cuticular hair is thick walled and grooved, tapering to a blunt tip with a single terminal or sub-terminal pore which may have a closure apparatus. A viscous sticky substance exuding from the tip has been reported as α- glucosidase. Distally, the cuticular sheath or scolopale divides the lumen of the peg into channels isolating mechanoreceptive from chemoreceptive dendrites. At the pore, the scolopale appears as an invagination of the cuticle and is shed during ecdysis.

In contrast, the walls of olfactory sensilla are perforated by n x 10^3pores (10 to 100 μm in external diameter) that filter molecules from the airstream. Pore densities vary from 3 to 8/μm^2, in the pheromone sensitive sensilla of the silk moths*Bombyx mori* and *Antheraea pernyi*, to 35/ μm^2 in the olfactory peg of the face fly*Musca autumnalis.* Such pores lead to a chamber, the **pore kettle**, where puffs of stimuli may be accommodated and concentrated. Pore tubules, 100 to 200 A° diameter extend toward the lumen of the sensillum that contains much-branched dendrites. Three pore systems may be distinguished:

1. The pore funnel of the surface leads to a narrow pore canal that widensinto a pore kettle from which are subtended several pore tubules;
2. Pores give rise to spoke canals leading to the lumen, usually in longitudinally-grooved sensilla. If there is more than one peg lumen, pore tubules connect with the outermost one, and radial canals penetrate to the innermost one;
3. Dense osmiophilic pore strands extend from the pores to the lumen,perhaps as an adaptation to dry climates.

The sensory neuron

Usually bipolar, this comprises three distinct regions, **dendrite, cell body** or **perikaryon,** and the **axon** that conveys impulses to the central nervous system. The dendrite is the source of much attention on account of its primary position in chemoreception. It comprises an outer segment or distal dendritic process (DDP); the constricted ciliary region; and a thicker inner segment or proximal dendriticprocess (PDP) (Fig. 12.12).

The typical DDP contains filaments, vesicles, and only longitudinal microtubules that increase in number from dendrite termination to about mid-length. There they decrease in number to multiples of about eight in the narrower ciliary region, so called because its microtubules are frequently arranged in the ciliary pattern as nine peripheral doublets with or without a central pair, i.e. 9 x 2 +2 or 9 x 2 + 0 but 36, 45 or 54 x 2 and 0 also are recorded. A thin, porous, electron dense membrane or ciliary collar, that in fly larvae may be a fibrillar mass, lines the circumference of peripheral tubules.

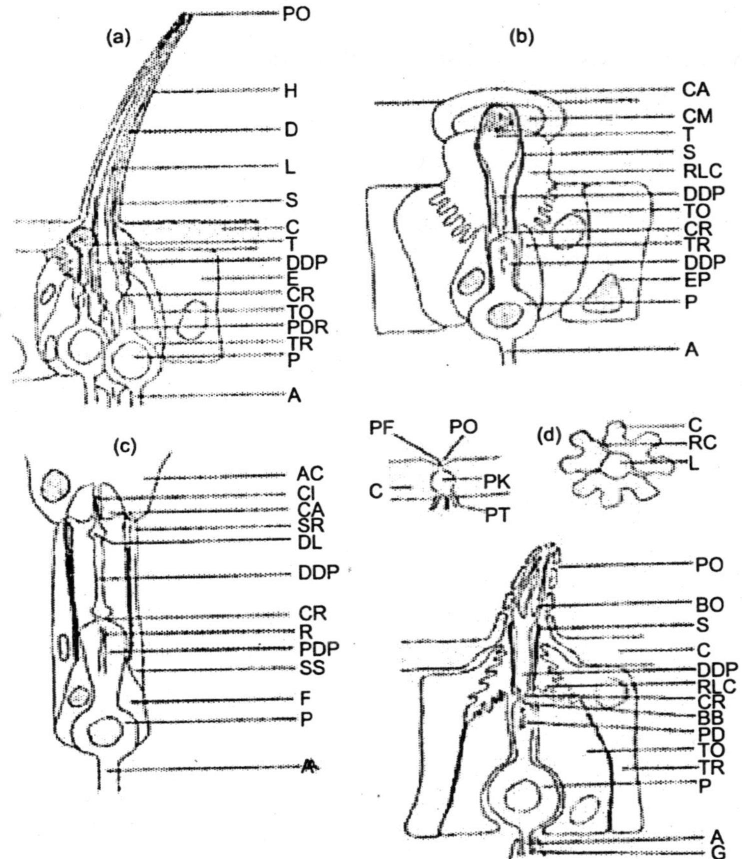

Fig 12.12 Schematic diagram of ultra structure of : (a) UP dual function gustatory/mechanoreceptive trichoid sensillum, (b) AP campaniform mechanosensitive sensillum, (c) UP coeloconic sensillum, (d) MP trichoid sensillium with inset of pore tubule system (left) and end-on view

A, axon; AC, attachment cell; BB, basal bodies; BD, branched dendrites; C, cuticle; CA, cap; CI, cilium; CM, cap membrane; CR, ciliary region; D, dendrite; DL, dilations; DDP, distal dendritic process; E, epidermal cell; F, fibrous sheath cell; G, glial cell; H, hair; L, lumen; P, perikaryon; PF, pore funnel; PK, pore kettle; PO, pore; PT, pore tubule; PDP, proximal dendritic process; R, ciliary rootlets; RC, radial canal; RLC, receptor lymph cavity; S, scolopale; SR, scolopale rods; SS, scolopale sheath cell; T, tubular body; TO, tormogen cell; TR, trichogen cell

(Taken from Michael F Ryan 2002 Insect Chemoreception- Fundamental and Applied, Kluwer Academic Publishers, N.York)

The outer member of each doublet lacks the arms that are associated with motile cilia. Ciliary microtubules lead to a centriole-like structure the basal body (of which there may be two, one above the other), comprising nine sets of peripheral triplet tubules. This body gives rise to banded rootlets, the ciliary rootlets. The similarity of this ultrastructure to that found in cilia of lower Phyla has promoted the suggestion, unverified and perhaps unverifiable, that insect sensilla are derivable from cilia either intrinsic to arthropods, or originally associated with ectoparasitic Protozoa.

The role of microtubules is unambiguous in mechanoreceptors in which they form a tubular body, possibly derived from folding of the DDP, of 50-100 microtubules but sometimes much fewer, arranged in parallel and embedded in an electron dense matrix (Fig.12.12). Physical contact with the sensillum compresses the tubular body, deforms the transducer membrane, and initiates nerve impulses. The cytoplasm of the PDP may have a distal concentration of mitochondria and many small vesicles.

The perikaryon located just below the surface of the cuticle has a large spherical nucleus with finely dispersed chromatin that has been associated with high metabolic activity. Abundant cytoplasmic organelles are: mitochondria; longitudinal microtubules; rough endoplasmic reticulum (RER); free ribosomes; vesicles; Golgi and onion bodies. Mitochondria and microfibrils are also abundant in the axons. The number of sensory neurons is usually in the range of one to 10 but placoid sensilla may be innervated by 35 bundles of two to 35 neurons, each bundle ensheathed by a scolopale.

Associated cells

Surrounding the neuron are two or more sheath cells that have been named **trichogen** and **tormogen** if two in number. A Schwann cell surrounds the nerve fibre, glial cells surround the perikarya and axons, and a neurilemma cell surrounds the PDP region. Alternatively, cells additional to trichogen and tormogen cells (numbered I and II respectively) are designated accessory cells III, IV and V as appropriate. Usually however, the innermost cell is termed the **thecogen cell** as it enwraps the bundle of sensory neurons from their perikarya to the base of the scolopale. Individual neurons are also enwrapped by cytoplasmic extensions of this cell as is the proximal part of each individual dendrite, presumably for electrical insulation.

The contents of the thecogen cell of the uniporous, gustatory chemosensilla (UP GCS) of the ventral sensory field on the maxillary palps of the American cockroach *Periplaneta americana*have been studied in detail. This cell is conspicuous for the abundance of granules that are scarce in the other enveloping cells. The structure of these electron dense granules indicated that they were lysosomes which was confirmed by acid phosphatise cytochemistry. Coated pits occurred on the thecogen cell plasma membrane where it borders the receptor lymph cavity. Use of the fluorescent dye Lucifer yellow indicated very extensive fluid-phase endocytosis from the cavity into the cell, followed by digestion in the endosome-lysosome pathway. Possible functions attributed to lysosomes and endocytosis included: purging the receptor lymph of physical impurities ingested following probing by the maxillary palps; participation in catabolic processes associated with embryonic development and moulting; and direct involvement in the chemoreceptive process by removal and metabolism of stimulus molecules from the receptor lymph to preclude continuous stimulation. Such endocytosis would be highly efficient if receptor-mediated.

The trichogen cell forms the hair wall, the hair, and the scolopale; the tormogen cell forms the socket and enwraps the trichogen cell from near the perikaryon to the cuticle. After forming the hair, the trichogen cell withdraws from the hair lumen down to the hair base or below, filling the spaces with the liquid or sensillum liquor that it secretes. Processes of this cell surround the cilium and extend microvilli into the receptor lymph cavity. This vacuole lies within the tormogen cell and is formed by that cell's withdrawal after it has formed the socket. The receptor lymph cavity extends up to the hair base and is continuous with the sensillum liquor in the hair peg. Labyrinthine connecting canals and lamellae are characteristic of both enveloping cells. Septate desmosomes connect them to each other and to epidermal cells. In the moth *A. polyphemus*a very tight contact, sufficient to preclude entry of ionic lanthanum, is formed between the apical membrane of the tormogen cell and the socket cuticle of olfactory s. trichodea. The contact is characterised by numerous hemidesmosome-like structures that could electrically isolate the receptor lymph cavity and might also attach the sensillum to the cuticle.

4.1 Olfactory (odorant) receptor

Olfactory receptors are usually thin-walled pegs, cones, or plates with numerous pores through which airborne molecules diffuse. Dendrites of sensory neurons branch profusely within these pores and may respond to very low concentrations of detectable compounds (*e.g.* sex pheromones). Some receptors respond to a wide range of substances while others are highly specific. Olfactory receptors are most abundant on the antennae, but may also be associated with the mouthparts or external genitalia.

The basiconic peg is an extension of the tough exoskeletal cuticle of the antenna (Fig 12.13). The cuticular wall of the peg is punctured by rows of pores. These pores vary in diameter, depending upon insect species and sense organ, but are usually between 6 and 65 nanometres in diameter. Various accessory or auxiliary cells are associated with the nerve cell and the peg - the **trichogen cell** synthesises the peg, whilst the **tormogen cell** secretes the surounding socket. The tormogen cell encircles the trichogen cell. A **thecogen cell** ensheaths the nerve cell (covering part of its dendrite and the cell body) and a glial cell forms a neurolamella, or neural sheath, around the axon. This provides electrical insulation for the axon.

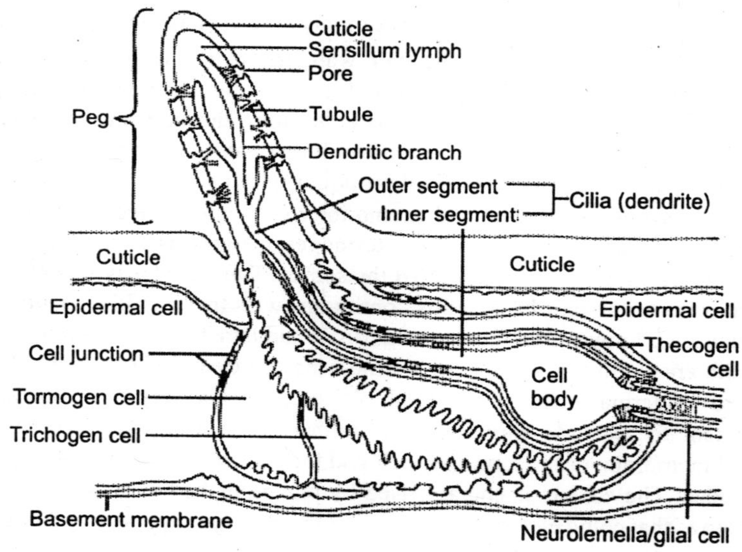

Fig 12.12 A Basiconic olfactory sensillum
(fromhttp://cronodon.com/BioTech/Insects_antenna.html)

Figure 12.14 shows a cross section through the basiconic peg. The dendrites are bathed by the sensillum lymph. Each pore opens into a tiny expansion, called the **pore kettle**, and from this pore kettle extends a number of minute pore tubules. These ar1e tubules, presumed to be hollow, numbering 3-20 per pore (3-4 are visible per pore in these sections) and only 10 to 20 nanometres in diameters. These pore tubules sometimes appear to extend into the sensillum lymph for a short distance.

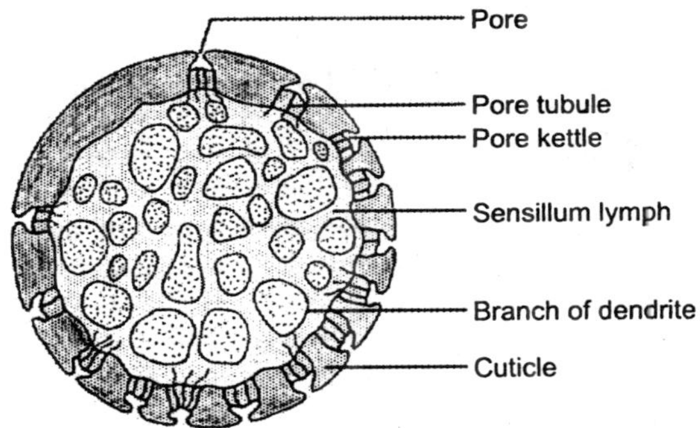

Fig 12.14 Cross section through the basiconic peg
(from http://cronodon.com/BioTech/Insects_antenna.html)

Odours consist of gas molecules, usually very dilute gases that get carried in air. These molecules are extremely tiny - even smaller than the pores and pore tubules and they find their way into the pores and are thought then to pass down the pore tubules into the sensillum lymph and then to attach (temporarily) to one of the dendritic branches. The dendrite then generates an electrical signal that travels down the dendrite to the cell body, and, if the signal is strong enough, it will then travel down the axon to the brain of the insect. The hair (peg) is covered in a very thin film of oily material. When an odour molecule strikes the hair it gets trapped in this oily layer. It is now only free to move across the surface of the hair, it cannot escape, and so sooner or later it will most probably fall into one of the pores, which also contain oil. Once in the pore it has few places to go and so will often end up inside the pore tubule. In this way, odorant molecules get funnelled into the pore tubules. Once inside the pore tubules, the molecules dissolve into the sensillum lymph. Some researchers suggests that the tubules may actually end on or very close to the surface of a dendritic branch. The odorants bind to the surface of the dendrites and are then detected. Carrier proteins (odorant-binding proteins or OBPs) bind tightly to specific odour molecules, trapping them and then delivering them to the dendritic branches. Different sensilla, which may otherwise look identical, may have different types of OBPs, each specific to one type of odour, so that different sensilla specialise in detecting different odours. Some insects have olfactory pegs sunken into pits, so-called **coeloconic sensilla**.

4.2 Gustatory receptors

Gustatory receptors are commonly described as thick-walled hairs, pegs, or pits where the dendrites of several (usually up to five) sensory neurons are exposed to the environment through a single opening (pore) in the cuticle. Each neuron appears to respond to a different range of compounds (*e.g.* sugar, salt, water, protein, acid, etc.). Taste receptors are most abundant on the mouthparts, but may also be found on the antennae, tarsi, and genitalia (especially near the tip of the female's ovipositor).

Lepidopteran insects harbour two main types of sensilla on their maxilla, the lateral and the medial sensilla known as **sensilla styloconica**. Additionally, some species exhibit a pair of epipharyngeal sensilla located on the ventral side of the labrum (Schoonhoven and Van Loon, 2002). These sensilla styloconica have four gustatory receptor neurones (GRNs) and one mechanosensory neurone, which are involved in sensing taste and other related responses. A typical GRN has GRs on its dendritic end with a specific affinity towards a diverse range of ligands such as sugars, salt (NaCl), inositol and plant alkaloids. These GRs also help to discriminate between the host plant nutrients and deterrents. GR proteins belong to the G-protein coupled receptor (GPCR) family, having their signature seven transmembrane domain topology (Xu et al., 2012). Binding of various ligands to GRs on the extracellular domain activates a downstream signalling cascade in either a G-protein dependent or independent manner (Yao and Carlson, 2010; Zhang *et al.*, 2011). Thus, effective functioning by this signalling cascade depends on the differential binding of ligands to the extracellular domain of GRs.

4.3 Ionotropic receptors (IRs)

Insects can detect a wide range of environmental chemicals: bitter, sweet, and salty tastants, odours, pheromones, humidity, carbon dioxide, and carbonated water. Most of these chemosensory stimuli are recognised by members of two evolutionarily related insect-specific chemosensory receptor families, the **odorant receptors** (ORs) and **gustatory receptors** (GRs). These proteins contain seven predicted transmembrane domains but are evolutionarily unrelated to G protein-coupled receptor (GPCRs) and adopt a distinct membrane topology. Recent analysis has indicated that insect ORs function as odour-gated ion channels, setting them mechanistically apart from metabotropic vertebrate ORs (Benton, *et al.*, 2009).

Recently, a third family of *Drosophila* chemosensory receptors-the ionotropic receptors (IRs) has been identified besides ORs and GRs in coeloconic sensilla, and was named for their homology to ionotropic glutamate receptors (iGluRs) (Albuin *et al.*, 2011). Importantly, IRs contains divergent ligand-binding domains that lack glutamate-interacting residues which suggest that they recognise distinct ligands. IRs are expressed in a combinatorial fashion in sensory neurons that respond to many distinct odours but do not express either insect odorant receptors (ORs) or gustatory receptors (GRs) (Benton *et al.*, 2009). So IRs may be equally important with ORs in chemoreception.Comparison of chemosensory receptors in insects and mammals showed that IRs are only found in insects but not mammals.

Functional analyses of IRs indicated that IRs affect environment exploration and sexual behaviours of insects. IRs are specific for selecting amines and acids, which complements the broader tuning of ORs for esters and alcohols. Olfactory sensory neurons in the fruitfly, *Drosophila melanogaster*, express an ionotropic receptor IR64a that is highly selective for acidity. Besides, IRs also affects insect sexual behaviours. IR84a, an olfactory receptor for food-derived odours, the aromatic odours phenylacetic acid and phenylacetaldehyde, was found to promote male courtship in Drosophila.

4. 4 Common chemical sense

High concentrations of irritant compounds (*e.g.* ammonia, chlorine, acids, essential oils, etc.) simulate avoidance reactions and cleaning behaviour. Insects can detect these compounds even when all known chemoreceptors have been covered or destroyed. The irritants evidently trigger a generalised response from other types of sensory neurons.

4.5 Olfaction in insects

Insects perceive odorant mainly with sensory organs called **sensilla** on their antennae. Olfactory sensilla are generally characterised by bearing tiny pores on their cuticular surface and occur in a bewildering variety of shapes, of which the most common one is probably the hair. There are several olfactory receptor cells in the sensillum. Their dendrites extend into a sensillar lumen and their axons project to the first order olfactory information processing center of the brain, the antennal lobe. An odorant can be adsorbed on the cuticular surface of the sensillum, diffuse towards a pore and through it into the inside of the sensillum. Since the sensillum is filled with sensillum lymph, it is thought that volatile and insoluble odorants such as pheromones bind to specific odorant binding proteins (OBP), are thereby solubilised and transferred to the olfactory receptors on the dendrites of the olfactory receptor cells. Recently it has been reported for a special class of OBPs, the pheromone-binding proteins (PBP) that they are necessary for the activation of pheromone-sensitive odorant receptors. When the solubilised odorant binds to odorant receptors in the dendritic membrane of the olfactory receptor cell, the cell is depolarised and generates action potentials, which transmit the olfactory signal to the antennal lobe. In this chapter, we introduce the current knowledge concerning the molecular mechanisms of olfaction.

Insects olfactory receptor neurons (ORNs) can roughly be classified into two types: those are responsive to general odors such as floral odors or food-related odors and those that are responsive to pheromone, i.e. species-specific odorants. General odorant receptor proteins (ORs) show low specificity and respond to various odorants hence they are called generalists. They have partially overlapping odorant response spectra. On the contrary pheromone receptor proteins show high specificity and respond to only one ligand (a specific pheromone) hence they are called specialists. Genera ORs are well investigated in fruit fly, cockroach and honeybee and categorized into a number of types based on their response spectra.

The domestic silkmoth (*Bombyx mori*) has become a model organism of pheromone reception for which much information has been accumulated. Pheromone (specific) receptor neurons are housed in a type of hair sensillum, the long sensillum trichodeum, of which a large number covers the side branches of the antennae of the male silkmoth. Silkmoth pheromone is composed of bombykol, the major component [(E, Z)-10, 12-hexadecadien-1-ol], and bombykal, a minor (or accessory) component [(E, Z)-10, 12-hexadecadien-1-al]. Bombykol alone is sufficient to elicit mating behaviour in the male silkmoth. On the other hand bombykal raises the threshold for the induction of male mating behaviour by bombykol and can thus effectively inhibit mating behaviour. Bombykol and bombykal receptor neurons are housed in pairs in the same type of trichoid sensillum. These receptor cells respond with high specificity to their pheromone ligands.

4.5.1 *Transduction mechanism of odour reception*

Insects perceive odorants through olfactory sensilla mostly located on the antennae. There is also a smaller number of olfactory sensilla on the maxillary palp and labial palp. An olfactory sensillum has a characteristic structure generally with numerous pores on its cuticular surface. The lumen of the sensillum is filled with sensillum lymph surrounding the dendrites of olfactory receptor neurons (ORNs). ORN soma is surrounded by support cells (thecogen, thecogen, and trichogen cell) that secrete odorant binding proteins (OBP) or pheromone binding protein (PBP) (Steinbrecht et al., 1995; Shanbhag et al., 1999; Shanbhag et al., 2000). The ORN is a bipolar cell that extends its single dendrite into the sensillum and its single axon projects to the brain. The ORN is excited and can transmit an electrical signal to the brain when an odorant binds to the receptor protein in its dendritic membrane. Olfactory sensilla are categorized into several types according to their morphology and size (Shanbhag *et al*., 1999) and the odour response of each cell type can be recorded by single sensilum recording under favourable conditions in some types of sensilla (de Bruyne *et al*., 2001). In *Drosophila* the olfactory response spectra of most types of sensilla have been recorded and each type has a distinct olfactory spectrum (de Bruyne *et al.*, 2001; Hallem et al., 2004; Yao *et al.*, 2005; van der Goes van Naters and Calrson, 2007). Each sensillum houses two to four ORNs with different odorant receptor and the olfactory response spectrum of each sensillum represents these ORN spectra. It has been shown that these response spectra of ORNs are derived from the ligand affinity of their olfactory receptor proteins.

Recently, peculiar molecular mechanisms have been discovered in insect odorant receptors. G-proteins are present in insect antennae (Laue et al., 1997), and IP3, a second messenger activated via a trimeric G-protein, transiently increases in response to odorant (Boekhoff *et al.*, 1993). Also, IP3-activated ion channels are known in insects. From these lines of evidence it has been hypothesised that insect olfactory transduction occurs though a trimeric G-protein-coupled receptor (Krieger and Breer, 1999). This resembles the scheme also found in vertebrates. However, the function of Or83b, a Drosophila odorant receptor family protein, indicates that olfactory signal transduction mechanisms in insects are different.

5. TEMPERATURE AND HUMIDITY RECEPTORS

Thermoreceptors are temperature sensors, whilst hygroreceptors are moisture-sensors. Very often the two are combined in insects, into dual-function thermohygroreceptors. The mode of operation of these remarkable sensors is poorly understood.

The thermohygroreceptors occur in different forms but are usually pointed/trichoid hairs (Altner *et al.*, 1983), or blunt pegs/cones which may be enclosed inside a pit (coeloconic sensilla) (Altner and Prillinger, 1980). These sensilla are non-porous, except for a terminal moulting pore (which may lead to confusion with gustatory sensilla which have a pore in their tip to allow liquids to contact the sensory dendrites inside); some thermo-hygroreceptive sensilla also contain olfactory sensory cells, in which case the walls may be multi-porous, with the pores allowing diffusable gases to reach the sensory cells within.

A typical thermo-hygroreceptive sensillum contains 2-5 sensory cells (though three cells is the most common), with four striking features: (1) The dendritic outer segments of only two of the cells extend into the lumen of the peg. (2) The dendrite which does not enter the peg is usually flattened and forms characteristic lamellae (layered sheets). (3) The outer dendrites which are found within the lumen of the peg, fill the lumen completely (there is no receptor lymph cavity) and tightly. (4) These latter dendrites usually contain many prominent microtubules.

The two dendrites occupying the lumen correspond to hygroreceptor units identified electrophysiologically, one for dry-air and one for moist-air sensitive cell. The fact that these dendrites are well-packed with microtubules and often surrounded by electron-dense material is suggestive of a mechanoreceptive role, as in the tubular body of other mechanoreceptors. It has been suggested that a swelling process, in the presence of moisture, could lead to the exertion of mechanical forces on the dendritic membranes of the hygroreceptors, in such a way that the inner lymph-cavity shrinks and the wall constricts around the sensory dendrites. In this state the inner wall squeezes and places pressure on the dendrites, which get squeezed against the expanded inner wall. This squeezing of the dendrites may activate the dendrites to send signals to the sensory cell body and, if the stimulus is strong enough, the cell body will relay the signals to the central nervous system (as action potentials travelling in sensory axons). Thus, these moisture sensitive dendrites are thought to function as pressure sensors and so are fundamentally mechanoreceptive. Therefore, this sensillum is a device called a **hygromechanical transducer**.

Box 12.5 Specialised Thermo-hygro receptors

1. Typical sense organ: sensillum coeloconicum with 1 thermosensitve and 2 hygrosensitive receptor cells.
2. Because they are poreless, hygroreceptors clearly cannot operate in the same manner as chemoreceptors, that is, by water molecules binding to receptors in the dendritic membrane. In fact, it seems that they function as mechanoreceptors, behaving like a hair hygrometer.
3. The sensillar cuticle or a substance associated with it is hygroscopic and undergoes deformations as humidity changes, leading to physical stimulation of the dendrites.
4. There are two types of hygroreceptors: moist receptors respond to increasing humidity, while dry receptors are stimulated by a decrease in humidity.

Of the two dendrites which enter the lumen of the peg, and branch, one is thought to be **moisture-sensitive**,the other **dry-sensitive**. It is the moisture-sensitive dendrite which responds to pressure as the inner wall expands and constricts around it. The dry receptor, responds when it expands as the wall dries, shrinks and moves outwards (Tichy and Kallina, 2010). The moisture-sensitive receptor also responds to increasing air pressure, the dry receptor to decreasing air pressure, supporting the hygromechanical transducer model (Box 12.5). However, the exact mechanism of operation of these sensors is not firmly established.

Externally, hygroreceptive sensilla can be described as small, peg-like cuticular structures with a cap, like a mushroom, or with a rounded tip (Altner *et al.*, 1983). The first type includes the sensillum capitulum of the cockroach (Fig 12.15A), the second type the stick insect's peg-in-pit sensillum (Fig 12.15B). Their position on the antennae appears to be a trade-off between exposure and protection. While the occurrence on a body appendage enhances the contact with ambient air, abrasion by encountered objects demands protection by elevated surrounding structures. The cockroach's sensillum capitulum projects from the antenna surface but is overtowered by neighbouring hairs. A tightly fitting protection wall encloses the mushroom-shaped stem, so tight that it can be taken for a thin outer wall of the sensillum with a ring-shaped opening just under the cap. In the stick insect, on the other hand, the peg-in-pit sensillum is located in a cavity and surrounded by an inflated collar.

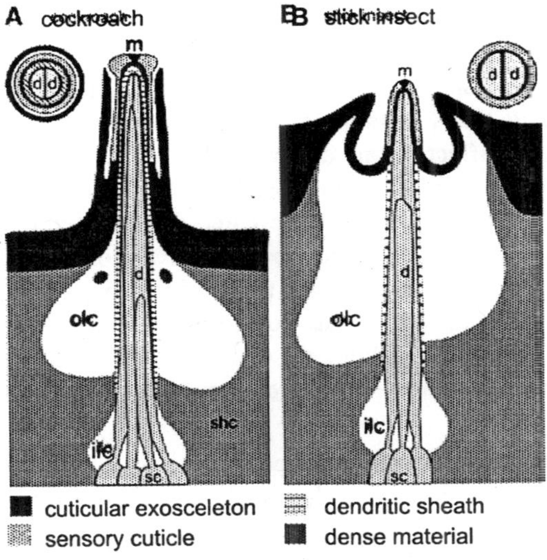

Fig. 12.15 Hygroreceptive sensilla on antenna of (A) cockroach and (b)stick insect
[moulting pore (m); outer dendritic segments (d); inner receptor lymph cavity (ilc); outer receptor lymph cavity (olc) ; sheath cells (shc); sensory cell (sc)] (after Tichy and Loftus 1996).

The sensillum wall typically lacks pores except for a single terminal moulting pore (m) plugged by electron-dense material. The wall consists of cuticle and is discarded at moult together with the exuvium. Thickness varies but is generally <1 μm—in the cockroach and in the stick insect 0.6 μm. The dendrites of two of the three or four sensory cells are quite long, closely adjoining, and extend far out into the peg. The third dendrite branches and sends numerous slender extensions into the peg (cockroach, Fig 12.14A) or is folded and terminates below the cuticle (stick insect, Fig 12.14B). Two sensory cells have been identified as hygroreceptive (a moist and a dry cell) and one as thermoreceptive (a cold cell). Their total of three dictates that at least one of the two long dendrites is a hygroreceptor. However, the two long dendrites probably have similar function and belong to the two hygroreceptors. Consequently, the third is considered to be thermoreceptive.

A dendritic sheath encloses the dendrites in a small lumen filled with finely granular lymph. In the cockroach, the sheath extends up to the sensillum tip, where it narrows to form an apex just inside the tip. In the stick insect, the sheath terminates deeper, near the base of the cuticular peg. Here the sheath merges with the inside of the peg wall so that the outer region of the dendrites seems to be enclosed directly within the wall rather than within the sheath. The dendritic membranes need to be intimately associated with the wall. Too much lymph in between could cushion the volume changes in the wall and impede their effect on the dendrites. But if the wall is so hygroscopic that it withdraws sufficient water vapour from the air to produce a graded effect on the dendrites, it could as well withdraw water from the inside. Studying the resolving power of the stick insect's moist and dry cells for steady-state humidity revealed that both cell types withstood several 6-min periods of adaptation in an air stream of 0% relative humidity (Tichy 1978). Because the air stream travelled at 2.5 ms^{-1} and the sensillum diameter was <25 µm, the dry air flowed at $>10^6$ sensillum diameters per second. Relative to the sensillum, the mass of air contacting it per second is very great. Lack of effective insulation would have immediately dried the sensilla out.

The cuticular wall swells or expands due to its tendency to take up water. This, in turn, increases the discharge rate of the moist cell and decreases that of the dry cell ; conversely, the cuticular wall would shrink or contract because it tends to lose water. This increases the discharge rate of the moist cell and lowers that of the dry cell. The effects of swelling and shrinking cannot be realised from moist and dry cell responses during humidity changes. Swelling of the cuticular wall would be expected to increase the internal pressure, shrinking to decrease it. Just as the forces caused by swelling and shrinking stimulate the moist and dry cells, it is proposed that the internal pressure changes due to external changes in air pressure have similar effects on the discharge rates of both types of cells. However, the mechanical model considers the thin cuticular wall (<1 µm) to be hygroscopic, thereby balancing out its moisture content with the surrounding air. In both insects, the moist and dry cells respond with transient discharge rates when an adapting air stream aimed at the sensillum was replaced within 15 ms by another air stream having a different humidity (Tichy 1978; Tichy and Loftus 1990).

5.1 Mechanism of hygroreception

The mechanisms underlying the impressive ability of many insect species to react behaviourally to humidity gradients are still poorly understood. Experiments on humidity transduction present many technical difficulties beyond those involved in the electrophysiological study of the function of the dual system (i.e., moist cells and dry cells) of hygroreceptors. The various notions concerning the transduction process have been summed up in three models (Tichy and Loftus 1996). The first model is based on a mechanical mode of action similar to a hair hygrometer in which activity is initiated by swelling and shrinking of hygroscopic sensillum structures. The second model holds that hygroreceptors operate as psychrometers in which the degree of cooling due to evaporation of water from the sensillum surface is used to measure the humidity (or the dryness) of the air. In the third model, humidity affects the electrolyte concentration just outside the dendritic processes of the hygroreceptive cells. The drier the air, the greater the evaporation rate of water out of the sensillum, and the greater the quantitative changes in the electrolyte concentration.

5.2 Thermoregulatory behaviour in insects

Insects are **poikilothermic**, that is they lack the means to maintain homeothermy – a constant body temperature that is independent of fluctuations in their surroundings. Although the temperature of an inactive insect tends to track ambient, many insects can alter their temperature, both upwards and downwards, even if only for a short time. The temperature of an insect can be varied from ambient either behaviourally using external heat (ectothermy) or by physiological mechanisms (endothermy). Endothermy relies on internally generated heat, predominantly from metabolism associated with flight. Some 94% of flight energy is generated as heat (only 6% is directed to mechanical force on the wings), thus flight demands much energy and also produces a great deal of heat.

In insects, as in other animals, body temperature strongly affects the rate of energy expenditure, the rate at which food can be located and harvested, growth, the facility with which mates can be acquired and predators avoided, and sometimes also the susceptibility to disease organisms. Thermoregulation refers to the ability to regulate that body temperature which best serves survival and reproduction, and it encompasses numerous conflicting constraints and selective pressures. In insects, major considerations involve body mass and access to either external or internal heat. Thermoregulation operates through behaviour, physiology, and morphology. For the most part, insects are too small to be able to appreciably elevate, or regulate their body temperature by internal heat production, although some are large enough and that, coupled with their high flight metabolism, could easily cause them to overheat. In numerous insects, elaborate mechanisms of thermoregulation have evolved both for heating and for cooling the body that possibly rival those of the typically endothermic vertebrates.

5.2.1 Endothermy

Endothermy is physiological temperature regulation. The ability to maintain an optimum constant core temperature means endothermic insects are less vulnerable to fluctuating temperatures. Honeybees (*Apis mellifera*) are thought to be endothermic via controlled blood circulation. Schmaranzer (1999) found evidence that heat flow from the thorax into the head and into the abdomen are two independently controlled processes. This is achieved by regurgitating nectar and holding it in the mouthparts. Evaporation of water from the nectar cools the head. When the temperature gradient between the head and thorax becomes significant, heat is transferred from the thorax to the head. This allows the honeybee to maintain temperatures in each segment within favourable ranges (Stabentheiner, 2001).

Flight is one of the most metabolically active and energy dense processes of the class insect. During flight heat is released as an inevitable by-product of cellular metabolism whenever muscle contracts. Close to 94% of the energy expended by muscles during contraction is degraded to heat, while approximately 6% appears as mechanical force on the wings. Insect flight is one of the most energetically demanding activities known, and thus most insects produce more heat per unit muscle mass when they fly than almost any organism on earth. Most insects exist under conditions somewhere in between the cold-blooded crawler and the hot- blooded flyer.

Small insects have much lower body temperatures in flight than large insects, not because they produce less heat—in fact, they may have higher rates of heat production than larger insects, but because they have much greater conductance because of their large relative surface area. In bees, for example, only the large species heat up in flight and generate an appreciable elevation of body temperature even though metabolic cost of flight per unit weight declines approximately 230% for a 10-fold increase in mass. A mosquito in flight may heat up 5°C, and a honey bee heats up its thorax about 15°C. Having a much larger thorax, and hence a smaller relative surface area, means that the internally generated heat during flight is not lost by convection at the same rate that it is produced until a much higher temperature gradient has been generated.

Warm-up by shivering

Bees are able to raise their thoracic temperature by shivering, which can work their flight muscles harder than flight itself does. During preflight warm-up, there are synchronous contractions of groups of muscles that normally contract alternately in flight. That is, the main wing-depressor muscles, the dorsal longitudinal muscles, are excited simultaneously—in other words, in synchrony—with the dorsoventral wing elevator muscles. Some insects, such as dragonflies and moths, do have visible external wing vibrations that were originally called **wing whirring**; these were once thought to pump air into the animal before the true function was elucidated.

Honey bees and bumble bees have a physiological sophistication either not existing or not yet observed in other insects, and they exploit shivering behaviour to an unprecedented extent and in a variety of ways. Like flies and beetles, bees are myogenic flyers in which the wing-beat cycle runs in part on automatic; as the downstroke muscles contract they stretch the upstroke muscles. This stretching *by* itself causes the upstroke muscles to contract. The downstroke of the wing therefore automatically causes the upstroke muscles to contract and vice versa, in a repeating cycle that is sparked by neural commands that are at a much lower frequency than the wing beats and are no longer specific to a single wing beat. (This system permits some of the smallest insects, such as midges, to achieve the unprecedented coordination required for wing-stroke cycles of over 1000 beats per second.) But the stretching of opposing muscle groups that maintains the myogenic contraction cycle can occur only if the wings are actually beating—namely,during flight. When the wings are not in use, as when they are folded back dorsally and the clutch-like wing hinge is engaged, then the muscle groups are in near tetanus. That is the reason for the bees' shivering.

5.2.2 Ectothermy

Ectothermy is the behavioural regulation of temperature. This can take the form of basking, crowding, tent making, and limiting activity to particular times of day. These are practices used by both adults and juveniles. Effective thermoregulatory behaviours by larvae can assist in more rapid growth. Less time spent at the vulnerable larval stage can result in reduced mortality rate, as well as higher population growth rate. Larvae of the codling moth (*Cydia pomonella*) complete life stages from the first-instar to the fourth-instar inside apples, and are termed cryptic baskers in that they move through temperature gradients within the fruit during development.

First-instar larvae feed directly beneath the surface, and bore towards the centre to feed until reaching the fifth-instar life stage. At this point, larvae leave to find cocooning sites under the bark of the tree (Kuhrt *et al.*, 2005).

Warm-up by basking

Like warm-up by shivering, warm-up by basking occurs in all major orders of insects that have fast flyers large enough to heat up from their flight metabolism. In its simplest form, behavioural warm-up is merely heat-seeking. A basking insect usually takes specific postures that simultaneously maximize solar input and minimize convective heat loss. Heat input is maximised by exposing the maximum surface area to the sun, while convective heat loss is minimised by using body parts (such as the spread wings) as baffles to retard air movement around the body. Orienting the body parallel to the air stream (as a wingless insect might do) would reduce the effect of convective cooling, but orienting the body perpendicular to the sun's rays to facilitate heating should take precedence, because no heat loss can be minimised until heat is first gained. Grasshoppers, beetles, and flies use these basking methods. For some dragonflies and butterflies the wings are especially important during warm-up in their role of reducing convective cooling.

Behaviorally distinct types of basking have been described in butterflies, although some (tropical) butterflies do not bask at all. In one type, called **lateral basking**, the butterfly closes its wings dorsally and then tilts to present either the right or the left wing and body surface to the sun. The lower portions of the wings wrap around the body and touch it, and warming the lower wing portions in sunshine then causes heat to be conducted directly through them and into the body.

Many species of small-bodied butterflies, primarily pierids and some lycaenids bask by opening their wings partially in a V so that the body is directly available to the sun's rays at the bottom of the V. The wings then serve as convection baffles to reduce cooling in the breezy environment. **Dorsal baskers** hug a solid substrate, such as the ground, and pull their wings down around them. They thereby expose the dorsal body surface to the sun while simultaneously capturing heat from the sun- heated substrate.

5.2.3 *Heat Loss Mechanisms*

Insects near the size of a honey bee (approximately 200 mg) or larger may experience body temperatures during forced flight exercise that are potentially lethal to them. Such insects resort to one or more of the following mechanisms for heat loss:

Harnessing Convection

The rate of convective heat loss from a body is determined by the conductance of the body, i.e., its intrinsic rate of heat loss (which is a function of body size, shape, and insulation). Conductance, in turn, is a function of the wind speed (a hot body cools more quickly in wind than in still air). However, no convective heat loss is possible, regardless of conductance, if body and ambient temperature are equal, and at any one conductance and wind speed the amount of heat loss is directly proportional to the temperature difference between the body and the ambient surroundings.

Small insects that are endothermic in flight are sufficiently air-cooled such that they almost never reach the potentially dangerous high-temperature ceiling of near 45°C that is common to most animal tissues at normal atmospheric pressures. These small insects thus have no need of a specialised cooling system: they lose sufficient heat passively. Theoretically, larger insects could cool themselves by increasing flight speed and thus increasing convective heat loss, but flying faster would generally increase metabolic heat production, which would cancel out the increased heat loss, unless the internally generated heat is redistributed.

Heat Radiators

Radiators are utilised by many large insects from the very diverse orders Lepidoptera, Odonata, Dipera, and Hymenoptera. The animals have a fluid-transfer cooling mechanism that dissipates heat through an abdominal radiator, while small members of the same groups that are not strong or continuous flyers lack the heat-transfer response. When we humans exercise in the heat, blood is pumped to the skin or extremities to facilitate heat loss, but this is done at the expense of pumping blood and oxygen to the muscles instead. Therefore, work capacity is compromised. Insects, on the other hand, do not need to compromise aerobic work capacity at higher air temperatures because of thermoregulation. In insects, the total separation of respiratory and heat-transfer functions makes it possible for them to continue working, even when the fluid flow is interrupted.

The radiator tube that conducts the haemolymph to the abdominal heat radiator in insects also serves as a pump, which operates by peristaltic contractions along its entire length. Although sphinx moths in whom surgery has rendered this heart inoperative (by tying it shut) can still fly until reaching near-lethal thoracic temperatures, removal of their insulating layer of thoracic scales makes continuous flight again possible.

Evaporative Cooling

One of the extraordinary examples of an evaporative cooling mechanism specifically for thermoregulation is that found in the workers of honey bees, *Apis mellifera*, and yellow jackets, *Vespula* spp. These insects use the head as a radiator, but they do so with a difference. Nectar-gathering honey bees normally fly with flight-motor temperatures near 15°C above air temperature. They are capable of the astounding feat of flying even at ambient temperatures near 45°C while maintaining the thorax at the same or only slightly lower temperature. They do so by regurgitating nectar from the honeycrop, and while the nectar is held on the mouthparts and the head, water evaporates from it. Because of the physical contact between the head and the thorax, thermoregulation of one effectively results in thermoregulation of the other. Thus, the head is cooled by evaporation of water until there is a large temperature difference between the head and the metabolically heated thorax, at which point heat from the thorax follows the temperature gradient and is transmitted to the head. Head temperature is actively regulated, with thoracic temperature passively following, because artificial heating of the thorax alone does not result in the heat-dissipation response so long as head temperature remains low. However, artificial heating of the head (as with a narrow beam of light from a heat lamp) almost immediately results in nectar regurgitation and evaporative cooling, even while thoracic temperature is still (momentarily) low.

Some insects cool evaporatively from the back. In the hot Australian deserts, the larvae of the sawfly *Perga dorsalis*, in response to solar heat stress, first raise their abdomen to the sun to shade the body and to increase convective heat loss. In an emergency, when this response is insufficient, they also emit rectal fluid and spread it over their ventral surface to cool themselves by evaporation.

Diceroprocta apache, a species of cicada of the Sonoran desert of the southwestern United States, employs a third evaporative cooling mechanism which is analogous to sweating. These cicadas are plant-sap feeders, and despite living in a dry environment, they have access to a large fluid supply by inserting their sucking mouthparts into the xylem of deep-rooted shrubs, such as mesquite. They thus indirectly tap water from deep underground stores. Cicadas sing when ambient temperatures in the shade reach 40°C, and the repetitive contractions of their tymbal muscles result in internal heat production that adds to the already considerable external heat load. Body temperature during this exercise in the heat is reduced to tolerable levels by evaporative cooling from fluid shed through large pores distributed over their dorsal body surfaces. The release of this fluid, and the consequent evaporative cooling, occurs only in response to very high body temperature. Most insects, especially those of desert environments, are instead highly resistant to water loss when alive, and upon death, there results an immediate increase in water loss as the spiracles are no longer actively maintained shut. Killing of the cicada, in contrast, immediately stops the sweating response, therefore showing that it is under metabolic control. The cooling response is mediated, ironically enough, by aspirin-like substances produced in their bodies in response to heat stress.

6. PHOTORECEPTORS

Photoreceptors are primary sense cells that are sensitive to light and convert light energy into electrical signals. These impulses are then summed, processed and integrated by interneurons in the central nervous system and eventually the insect has an image of the world around it. Three types of photoreceptive structures are found in insects:

1. Compound eyes
2. Dorsal ocelli
3. Lateral ocelli (stemmata)

6.1 Compound Eyes

A pair of compound eyes is the principle visual organs of most insects; they are found in nearly all adults and in many immatures of ametabolous and hemimetabolous orders. As the name suggests, compound eyes are composed of many similar, closely-packed facets (called ommatidia) which are the structural and functional units of vision. The number of ommatidia varies considerably from species to species: some worker ants have fewer than six while some dragonflies may have more than 25,000. Compound eyes are of two distinct and optically different kinds: **apposition eyes**, in which each receptor cluster has its own lens, and **superposition eyes**, in which the image at any point on the retina is the product of many lenses (Box 12.6).

Box 12.6 Two main types of Compound eyes

1. Appositional – photopic – eyes
   - typical of diurnal insects; ommatidia are optically isolated by accessory pigment cells ; photoreceptor cells and rhabdom are long reaching the crystalline cone; ingoing strait axial light beams have importance

- two possible paths of a beam:
  - (a) focused on the top of the rhabdom (focal dioptrics)
  - (b) in a wider beam starting from the focal point within the crystalline cone (afocal dioptrics)

2. Superpositional – scotopic – eyes
   - typical of nocturnal species
   - especially able to photo-adaptation to dim light conditions
   - pigment cells isolate only in the upper region
   - adjacent (as good as 30) ommatidia can take part in the vision owing to the light diffraction
   - there is a fine crystalline tract from the rhabdom to the crystalline cells
   - on the bottom of the ommatidium there is a special light reflective layer = *(iris)tapetum layer*

Externally, each ommatidium is marked by a convex thickening of transparent cuticle, the corneal lens (Fig 12.16). Beneath the lens, there is often a crystalline cone secreted by a pair of semper cells. According to existence and properties of cone, the ommatidia can be a) eucone- cone exists intracellular, b) acone- cone absent and substituted with crystalline cells and c) pseduocone- cone absent and substituted with intracellular liquid or get. Together, the lens and the crystalline cone form a dioptric apparatus that refracts incoming light down into a receptor region containing visual pigment.

The light-sensitive part of an ommatidium is called the **rhabdom.** It is a rod-like structure, secreted by an array of 6-8 specialised neurons (retinula cells), and centred on the optical axis just below the crystalline cone. The rhabdom contains an array of closely packed microtubules where light-sensitive pigments (e.g. rhodopsin, etc.) are stored. Pteridin pigments (yellow) and Ommochrom pigments (red, brown) can move within the pigment cells. These pigments absorb certain wavelengths of incident light and generate nerve impulses through a photochemical process similar to that of vertebrates.

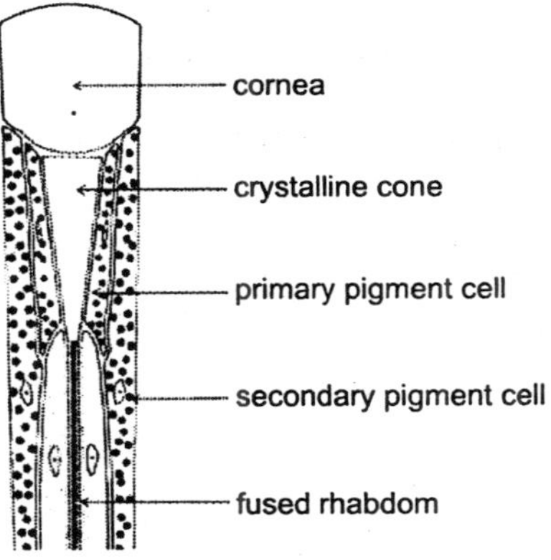

Fig. 12.16. Structure of an ommatidium
(from https://desperatelyseekingstrepsiptera.wordpress.com/)

Optically, each ommatidium works as follows. The inverted image is focused onto the distal tip of the rhabdom. Having a slightly higher refractive index than its surroundings, the rhabdom behaves as a light guide, so that the light that enters its distal tip travels down the structure, trapped by total internal reflection. Any spatial information in the image that enters the rhabdom tip is lost, scrambled by the multiple reflections within the light guide, so that the rhabdom itself acts as a photocell that averages all the light that enters it. Its field of view is defined, in geometric terms, by the angle that the tip subtends at the nodal point of the corneal lens and in a typical apposition eye this acceptance angle is approximately the same as the angle between the ommatidial axes. Thus the field of view of one rhabdom abuts (or "apposes," hence the name) the field of its neighbour, producing an overall erect image made up of a mosaic of adjacent fields of view.

Most diurnal insects have pigment cells surrounding each ommatidium. These cells limit a facet's field of view by absorbing light that enters through adjacent corneas. Each facet points toward a slightly different part of the visual field – in composite, they render a mosaic-like impression of the environment. Nocturnal and crepuscular insects have pigment cells that do not completely isolate each facet. Their ommatidia are stimulated by light from larger fields of view. This produces a brighter but theoretically less distinct mosaic image.

Since insects cannot form a true (i.e. focused) image of the environment, their visual acuity is relatively poor compared to that of vertebrates. On the other hand, their ability to sense movement, by tracking objects from ommatidium to ommatidium, is superior to most other animals.

Temporal resolution of flicker is as high as as 200 images/second in some bees and flies (in humans, still images blur into constant motion at about 30 images/second). They can detect polarisation patterns in sunlight, and discriminate wavelengths in a range from ultraviolet to yellow (but not red).

6.2 Ocelli - Simple eyes

Two types of **simple eyes** can be found in insects: **dorsal ocelli** and **lateral ocelli** (= stemmata) (Fig 12.17). Although both types of ocelli are similar in structure, they are believed to have separate phylogenetic and embryological origins.

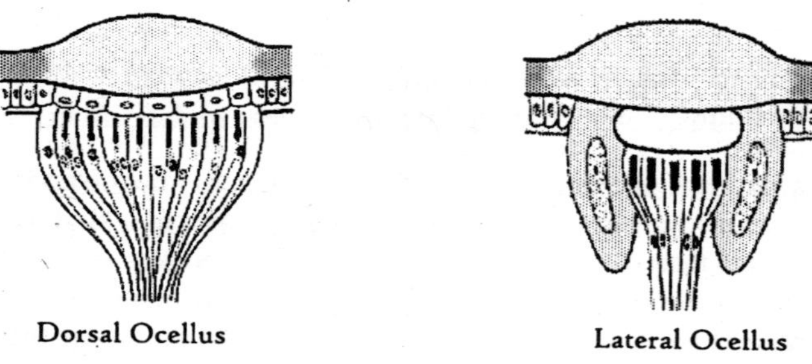

Fig. 12.17. Simple eyes of insects

Dorsal ocelli are commonly found in adults and in the immature stages (nymphs) of many hemimetabolous species. They are not independent visual organs and never occur in species that lack compound eyes. Whenever present, dorsal ocelli appear as two or three small, convex swellings on the dorsal or facial regions of the head. They differ from compound eyes in having only a single corneal lens covering an array of several dozen rhabdom-like sensory rods. These simple eyes do not form an image or perceive objects in the environment, but they are sensitive to a wide range of wavelengths, react to the polarization of light, and respond quickly to changes in light intensity. No exact function has been clearly established, but many physiologists believe they act as an **iris mechanism**- adjusting the sensitivity of the compound eyes to different levels of light intensity.

Lateral ocelli (=stemmata) are the sole visual organs of holometabolous larvae and certain adults (e.g. Collembola, Thysanura, Siphonaptera, and Strepsiptera). Stemmata always occur laterally on the head, and vary in number from one to six on each side. Structurally, they are similar to dorsal ocelli but often have a crystalline cone under the cornea and fewer sensory rods. Larvae use these simple eyes to sense light intensity, detect outlines of nearby objects, and even track the movements of predators or prey. Covering several ocelli on each side of the head seems to impair form vision, so the brain must be able to construct a coarse mosaic of nearby objects from the visual fields of adjacent ocelli.

6.3 *Extra-ocular Photoreception*

Some (perhaps most) insects respond to changes in light intensity even when all known photoreceptive structures are rendered inoperative. This dermal light sense has been attributed to the response of individual neurons in the brain and/or ventral nerve cord. There is also convincing evidence that some insects can perceive infrared radiation (heat), although specific receptors for this ability have not been described.

In insects able to detect light through their body surface, there are sensory receptors below the body cuticle but no optical system with focusing structures. Evidence for this general response to light comes from the persistence of photic responses after covering all visual organs, for example in cockroaches and certain beetle and lepidopteran larvae. Some blind cave insects, with no recognizable visual organs, respond to light, as do decapitated cockroaches. In most cases, the sensitive cells and their connection with the central nervous system have yet to be discovered. However, within the brain itself, aphids have light-sensitive cells that detect changes in day length—an environmental cue that controls the mode of reproduction (i.e. either sexual or parthenogenetic). The setting of the biological clock relies upon the ability to detect photoperiod.

6.4 The mechanism of image formation

1. **The apposition mechanism:** The mechanism of image formation in the bright day light by most of the diurnal insects. The pigment cells envelop the ommatidia completely and thus the rays of light can enter from the central point of the dioptric apparatus. Different ommatidia finally produced a single complete, erect image of an object.
2. **The super position mechanism:** The mechanism of image formation in poor light generally during night, mostly the nocturnal insects. At night time the pigment cells become contracts so that the light from the wide visual field can enter to the rhabdom of retinula cells obliquely. So light can enter centrally and obliquely. Each rhabdom therefore receives the light from several cones of neighboring ommatidia. Hence there is an overlapping of points of light. The image thus formed from a group of rays refracted by neighboring cones, called the **superposition image**. It is erect in

position and represents merely a part of an image. The compound image is formed after an amalgamation of all such images and it is the final form of a complete superposition image.

6.5 Mechanism of vision

6.5.1 *Visual acuity*

Visual acuity is the actual spatial resolution that the eye can see, and can be measured, for example, by using a grating of alternating black and white vertical lines and seeing how close together the striped must be when viewed at a set distance before the stripes merge and can no longer be distinguished as a series of lines. Visual acuity is determined in part by the maximum possible resolution of the eye, as determined by the spatial density of sensors - retinal cells or ommatidia which determines the number of points that make up the final image. It is also determined by the optical limits and imperfections of the lens system. Even if the optics are as perfect as possible, light will still diffract (spread-out) by some degree as it passes through the lenses, such that a single point of light coming from an object becomes an extended or blurred spot of light in the image. Insect eyes are limited by this diffraction.

Ommatidial facets are very small, about 10 micrometres (or one hundredth of a millimetre) in diameter. This allows many points to compose the final image. Diurnal insects, such as flies (Diptera), wasps and bees (Hymenoptera), many beetles (Coleoptera), dragonflies and damselflies (Odonata) and day flying butterflies (Lepidoptera) have this type of eyes. This type of eye is adapted for bright light and is called **apposition compound eye** because the final image is made up of discrete points, each point formed by a single ommatidium, placed side-by-side (apposed to one another) to form an image which is a mosaic of points.

Many insects are **crepuscular** (meaning that they are most active in twilight). Moths and beetles in particular, but also some flies, some dragonflies and some butterflies fly at light levels comparable to Moonlight. These insects may have apposition eyes with wider facets and they may collect light over a longer time period (up to about 0.5 seconds) before integrating the signal to produce the final image. Moths and beetles, in particular, may have a different type of compound eye, called the **superposition eye**. In this type of eye the iris cells only ensheath the top part of the ommatidium, around the facet and cone. A translucent light-conducting rod connects the bottom of the crystalline cone to the rhabdom which is now far beneath the cone. Now each rhabdom not only receives light from its own facet and cone lens system, but it also receives light from neighbouring ommatidia, since there is no screening pigment to prevent light leaking between adjacent rhabdoms.

In this way, light from as many as 30 ommatidia may overlap and focus onto the same point. Clearly this intensifies the image, improving sensitivity in dim light. However, the trade-off is that visual acuity is reduced - 30 or so ommatidia are now working as one large ommatidium, so the final image will be made from 30 times fewer points and spatial resolution will be reduced. However, some insects have optical apposition eyes that superimpose their signals neurologically, so-called **neural superposition.** In neural superposition, it is the electrical signals from neighbouring ommatidia that are added together by the nervous system, even though the light illuminates separate ommatidia by apposition.

6.5.2 *Visual cascade*

The biochemical pathway starts with the activation of **rhodopsin** and the resulting membrane depolarisation. The unstable cis-retinal is converted to its trans-form (Fig 12.18) when it absorbs a photon of light. Its transformation to the stable form of metarhodopsin activates G-proteins that ultimately cause the membrane depolarisation. The metarhodopsin loses its ability to activate the G-proteins after it is phosphorylated and binds with arrestin proteins to assume its unstable form. Light converts this metarhodopsin to an inactive rhodopsin that releases its arrestin and again becomes capable of responding to light (Fig 12.19).

6.5.3 *Dark Adaptation*

The eyes of most insects are capable of adapting to light and dark. In diurnal insects with apposition eyes, the pigment in the iris cells moves upward in the dark, exposing the rhabdom to light from neighbouring ommatidia - effectively turning the eye from an optical apposition eye into an optical superposition eye. Neural changes can further

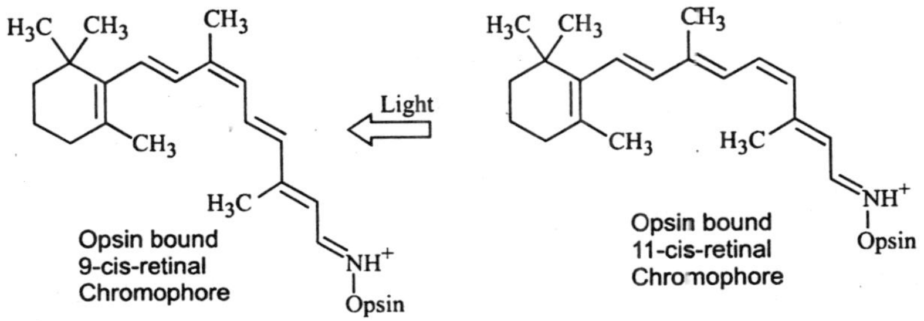

H_3C CH_3 CH_3 CH_3 H_3C NH^+ Opsin

Opsin bound
9-cis-retinal
Chromophore

Light

H_3C CH_3 CH_3 CH_3 H_3C NH^+ Opsin

Opsin bound
11-cis-retinal
Chromophore

Fig. 12.18. Two forms of retinal that combines with opsin to create rhodopsin

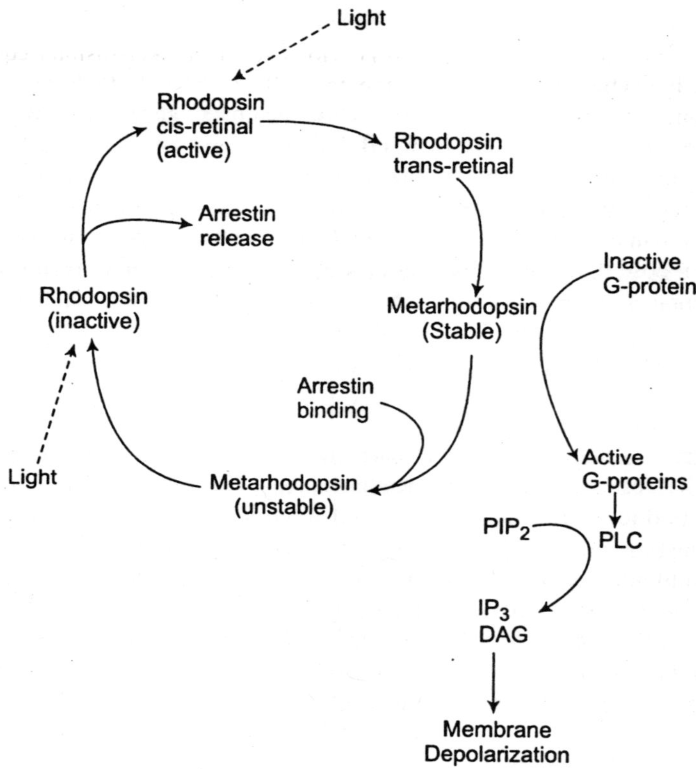

Fig. 12.19. Biochemical pathway of vision

increase the sensitivity of dark-adapted insect vision. Nocturnal insects show a similar pattern, but with greater ranges in sensitivity, with the eye becoming about 1000 times more sensitive to light in the dark. Thus, though insects may have the geometry of apposition or superposition type eyes, most can change in functionality to some degree. Clearly, however, the range of light intensities which best suites each type of eye is restricted and best suited to the life habits of the species.

6.5.4 *Flicker-fusion Frequency*

The **flicker fusion threshold** (or **flicker fusion rate**) is a concept in the psychophysics of vision. It is defined as the frequency at which an intermittent light stimulus appears to be completely steady to the average human observer. Flicker fusion threshold is related to persistence of vision. For any visual stimulus that blinks faster than the **flicker-fusion frequency** of the visual system, the flickers fuse into a single continuous image and the flickering cannot be perceived. The flicker-fusion frequency of human vision is 15-20 times a second. However, the flicker-fusion frequency for a honeybee is about 300, so the bee will see the light flickering. Thus, although the spatial visual acuity of the honeybee visual system is only 1/100 to 1/60 that of the human eye, its **temporal resolution** is much greater. This helps account for the very fast reflexes of many insects. The dragonfly can intercept a flying prey insect on the wing because its vision responds much faster than humans. Fast-flying diurnal insects have very high flicker-fusion frequencies. Slow fliers, like the stick insect, *Carausius*, have flicker-fusion frequencies of about 40 per second.

6.5.5 *Colour Vision*

Nobel Laureate Karl von Frisch (1914) conducted classical experiments with honeybees, and demonstrated that honeybees do see colour. He trained bees to associate the colour blue with food. He placed out a checkerboard series of paper squares, one blue and the others varying shades of grey. This was to demonstrate that bees did not simply recognise the blue square by seeing it as a particular shade of grey. He also covered the papers in a glass plate to rule out any odours associated with the blue paper in particular, in case the bees could smell that the paper was different. He placed an identical clean and empty dish on each square, but only the dish on the blue square contained food. This ruled out any visual cues that the bees might use to find the food. The bees quickly learned that the blue square contained food. The position of this square in the checkerboard was changed every 20 minutes, to prevent the bees remembering its position, but they still flew straight to the blue square, no matter where it was. Even when no food was provided, the bees would fly to the blue square initially, expecting to find food there. This demonstrated that bees have true colour vision, and also that they were capable of learning. Furthermore, the bees could not be trained to respond to a grey, black or white square - so the bees really see the colour blue and do not see it as a shade of grey. The reason why so many insect-pollinated flowers are large and brightly coloured is to advertise their presence to insects. Flowers provide both pollen and often nectar for food for the insect, in return for pollen dispersal and delivery to recipient flowers. Kuhn (1927) confirmed that worker bees can discriminate colours across the visible spectrum from orange to violet wavelengths, but they are red-blind. He also discovered that bees respond to ultraviolet light.

There are three receptor types in the worker bee retina belonging to three principle classes of insect receptor cells: **ultraviolet, violet and green sensitive**. The quantitative analysis of colour vision in bees shows it to be similar to human, being trichromatic, but it is shifted to about 100 nm toward shorter wavelength. *Formica polyctena, Periplaneta americana* and

Drosophila melanogaster have bichromatic colour vision while *Manduca sexta* has tetrachromatic colour vision.

Colour vision is presumably widespread among insects, as multiple rhodopsin types have been identified in most species investigated. Extensive knowledge exists of the colour vision system of the honeybee, where three photoreceptor classes (UV, B and G) form the standard set of photoreceptors underlying colour vision. The photoreceptor cells mediating colour vision in the blowfly Lucilia, the central photoreceptors R7 and R8, exist in two pairs with different visual pigments. The two pairs are randomly distributed in the retina, i.e. in a heterogeneous pattern.

For papilionid butterflies, which have a specifically well-developed colour vision system,the responsible photoreceptors have not yet been identified. At least the photoreptors in the distal part of the retina must participate; as it is here that the UV, B and G receptors are located (Kitamoto et al. 2000). However, the proximal photoreceptors must also be involved, because the yellow and red photoreceptor screeningpigments modify the sensitivity spectra of the cells in the proximal part of the rhabdom (Arikawa *et al.*, 1999), and red receptors dominate the proximal tier of the retina in Papilio (Arikawa and Uchiyama 1996). It is highly likely that the short basal receptors play a role in the discrimination of red light. In ommatidia with red-filtering screening pigment in the distal photoreceptors, red sensitivity of the basal receptors is relatively enhanced. This red sensitivity will be further enhanced when a tapetum exists which reflects well in the long-wavelength range. Outstanding examples are the eyes of Pieridae, where the two types of red filters are combined with two tuned types of tapeta. Ample behavioural evidence indicates that colour discrimination in the red is well-developed in Pieris and other butterflies.

6.5.6 Perception of polarised light by insects

The eyes of insects are sensitive to a natural phenomenon that man is blind to: the polarised light of the daytime sky. It is this capacity that underlies the remarkable navigational ability of many insect species. The light radiated by the sun is unpolarised, that is, its waves vibrate in all directions at right angles to the line of sight. In travelling through the earth's atmosphere, however, it is scattered by molecules and other particles that are small with respect to the wavelength of light, so that at each point in the sky its waves tend to vibrate in a specific direction.

The pattern of polarisation in the sky varies with the position of the sun or, more exactly, with the orientation of the plane of a triangle formed by the sun, the observer and the point observed. At any point on the celestial sphere the direction of polarisation is always perpendicular to the plane of such a triangle. By disregarding a few exceptions and applying this general rule to all points in the sky one can determine the entire pattern of polarisation for any given position of the sun.

The ability of honeybees to navigate by the polarised light of the sky was first described some 25 years ago by Karl von Frisch. His finding came as a surprise; even though the polarization of skylight had been known since the 19th century, no one had really considered the possibility that the phenomenon could serve any navigational purpose.

The ability of the visual cells to analyse the axial orientation of the polarised light is the result of a molecular oddity. In all animals, invertebrates and vertebrates alike, the visual pigment rhodopsin is present in the photoreceptor membrane of the visual cells in the form of dipolar molecules, that is, molecules with a distinct axis. As a result the pigment absorbs a maximum of the incoming polarised-light energy when the direction of polarization is parallel to the dipole axis of the molecule. In insects the photoreceptor membranes are bent into arrays of narrow tubes, the microvilli. The rhodopsin molecules in the microvillar membrane are preferentially aligned parallel to the axis of the microvilli. Such an orientation would of course result in the maximum absorption of polarized light when, and only when, the axis of polarization coincided with the microvillar axis. (This, incidentally, is part of the reason the human eye is blind to polarized skylight. Among vertebrates, man included the rhodopsin molecules are free to rotate in the photoreceptor membrane, so that their axial orientation is random. There is an equal chance that any pigment molecule will maximally absorb light with any direction of polarisation, and so there is no particular sensitivity to the polarisation of skylight.)

7. INDUSTRIAL APPLICATION OF KNOWLEDGE OF INSECT SENSORY SYSTEM

At present two main lines of research are underway to exploit the knowledge of the mechanisms of insect sensory systems, neural information processing in the brain and behaviour of insects for industrial application. The first is to learn from the simplicity and efficiency of information processing mechanisms of insect sensory and control systems that often function with amazing speed, sensitivity, and precision and to apply them to improve conventional technology sensors and control systems. The second is to utilise the insects directly though targeted manipulations of sensory and central nervous systems that leads to behavioural control, providing a simple readout of such a system. These two approaches are also being merged, using both genetic manipulations and interfacing of insects to robots. Such hybrid systems are being used to gain deeper insight into the functions of insect nervous systems with the goal of reconstructing a fully featured insect brain. This information can in turn be applied, especially in autonomous robots. Some examples of applications that have emerged from studies of insect nervous systems and behaviour are listed below.

7.1 Sensory system application

7.1.1 Odour sensors

Insects have olfactory systems that set landmarks in terms of sensitivity and selectivity. Intensive research in the past decades has recently led to the elucidation of the primary processes of olfaction in insects. The key advantage of odorant sensors in insects is direct coupling between chemical detection and electrical transduction in a single protein complex. As genetic tools are readily available for making these sensors, the development of fast, highly sensitive and highly selective odorant sensors is currently under way.

Marques and de Almeida (2000) developed insect behaviour-based algorithms in order to follow the chemical gradient of an odour. The insect world demonstrates that the laying and detection of chemical trails can be useful as an aid for navigation and to help organise large groups of workers. The excellent performance of insect antennae for chemical sensing make them ideal candidates to be used as sensitive elements for the development of insect antenna-based biomimetic smell biosensors, which have many potential applications in a wide range of fields such as plant protection, agricultural production, environmental monitoring, and fire detection. Biologically functional components that originated from the insects, which include isolated insect antenna, olfactory sensory neurons, and olfactory receptors, can be utilised as sensing components in smell sensors. Upon exposure to the specific volatile compounds, these biological functional components can generate corresponding responses. Knowledge on the sensitivity and selectivity pattern of insect antennae for chemical sensing allows the better design of biomimetic smell sensors by combination of amazing capability of insect antennae with various reliable transducers. Fast advancements in micro fabrication technologies and nanotechnologies also promote the research of insect-antenna based smell sensors by providing novel transducers and coupling technologies to enhance the performance of chemical sensing applications.

The powerful sense of smell that insects possess is being put to use in applications from detecting rotten tomatoes to controlling one of the deadliest of diseases in Africa. Studying the nerve response when insects are exposed to different smells has allowed scientists to detect trace volatiles that are sure signs of fish or fruit deterioration. Insect-based assays have been used to detect compounds as diverse as cocaine and Semtex (a plastic explosive containing RDX and PETN). Using insects' smell proteins, a method to control malaria could be developed.

This important natural resource has immediate applications in areas ranging from insect pest control products to commercial pharmaceutical production and design, disease control products and sensor products for manufacturing, quality control, military, and homeland security applications. Insects are keenly aware of their environment because of their extremely sensitive chemosensory systems, and many important aspects of their behaviour rely on tastes and odours. For example, feeding, mating, and selecting a place to deposit eggs all rely on chemosensory stimuli. In fact, the variety of insects is surpassed only by the sheer number of tastes and odours they can detect. The rich molecular diversity present in the chemosensory systems of millions of insect species represents an untapped resource with lucrative applications in areas of pharmaceutical production and chemical detection. Since odours and tastes profoundly influence insect behaviour, the molecules and processes involved in the chemosensory pathways are also ideal targets for the development of novel, more effective insect control products.

7.1.2 Odour source localisation robot

The French naturalist Henri Fabre, observed that moths are able to localise an odour source, a specific female, from a distance up to several kilometers. This obviously requires not only very sensitive and specific odorant sensors but also a suitable strategy to cope with environmental conditions in the natural surroundings. Copying the behavioural strategy used by moths and other insects in localising odour sources is a promising approach currently being employed to develop robust odour-source localisation robots.

Honey bees forage on a wide variety of flowers and remember the smell of any new variety that they find to be a good source of nectar. This behavioural flexibility is being exploited to develop insect-based technology for detecting illicit or dangerous materials.

7.1.3 *A flow sensor using an artificial sensory hair*

Many arthropods are highly sensitive to minute air currents. Particularly well-studied examples are crickets, which bear a number of filiform hairs on their cerci as well as spiders (*Cupiennius salei* and *Agelena* sp.), and whip scorpions, that have trichobothria. All these hair sensilla which are specialised for the detection of small air currents are very fine and relatively long hair sensilla with specially adapted socket structures. Taking advantage of the ultrastructural information available for such sensilla, efforts are being made to implement corresponding sensors.

7.1.4 *Visual navigation in insects*

The remarkable navigational abilities of social insects are proof that small brains can produce exquisitely efficient, robust navigation in complex environments. Because social insects produce specialist foragers that are amenable to field and laboratory studies, they have been productive model systems for studies of navigation. Ideas derived from these studies of insect navigation have shown how simple mechanisms can produce robust and seemingly complex behaviour. This is important for a general understanding of spatial cognition as these 'insect-like' navigational behaviours are probably phylogenetically widespread. Current insect research is helping to show how simple panoramic views, without the need for cognitive processes such as object identification or labelling, can provide explanations for many findings, including behaviours that in the vertebrate literature have traditionally drawn conclusions about sophisticated high-level spatial modules or learning rules.

The interaction of sunlight with the atmosphere results in a polarization pattern in the sky that varies with the position of the sun. Polarisation is strongest opposite to the sun. If the sun is not directly visible, the polarisation pattern can be used to infer the position of the sun. The dorsal part of the compound eye of many insects contains specialised photoreceptors that are polarisation-sensitive. Polarisation sensitivity (dichroism) results from the arrangement of membrane evaginations called **microvilli** that are oriented in one direction with respect to the long axis of the photoreceptor that coincides with the axis along which light travels as projected by the lens of the ommatidium. Polarisation-sensitive receptors are short and each has all its microvilli oriented in one direction. All other polarisation-insensitive receptors in the rest of the eye are longer and twisted such that microvilli are oriented in all possible directions, abolishing dichroism. The detection of the orientation of polarised light by receptors with differently oriented microvilli (and scanning the sky if there are only receptors with two orientations) and the overall polarization pattern in the sky thus enables these insects to infer sun position without direct visibility of the sun. In bees, it has been shown that they also have a clock to compensate for the course of the sun over the time of the day enabling them to infer the direction to the hive from the position of the sun at all times. As a result, homing is always possible using these external and internal references. Recently, it has also been shown that polarised light orientation is possible in moonlight in a dung beetle (Warrant and Dacke, 2010). Studies of insect navigation could be a useful complement to understanding the philosophy of comparative cognition.

7.1.5 Sound orientation in crickets

The cricket has tympanal organs on the tibiae of the front legs. The left and right tympanal organs are connected through an acoustic trachea. Sound from both sides can interact to result in amplification or attenuation though this structure. The female cricket uses the tympanal organs in particular for localising a calling male. Sound localisation under natural environmental conditions is not a trivial task. Sound can be reflected, other sounds can interfere, etc. Thus, sound localisation in crickets is a very useful system to investigate how to solve this task under natural conditions, conditions under which it is difficult for man-made systems to operate reliably.

7.2 Robot control

7.2.1 A visually guided robot inspired by fly

A prominent feature of insect eyes is their enormously high flicker fusion frequency. In some insects, time resolution goes up to over 200Hz, but spatial resolution of compound eyes is limited. For insects, rapid detection motion is of utmost importance for course control. In seminal experiments using a curculionid beetle walking on a *Spangenglobus*, scientists have described behavioural reactions to visual motion and proposed an elegant model for motion detection (based on autocorrelation of visual inputs) that has been one of the biggest successes of theoretical neuroscience. Later studies recorded neurons from the fly's lobula plate that show responses that are exactly predicted by the autocorrelation detector model. The success of this model is also shown by a number of robotic implementations that use these detectors for motion detection and course control.

A lot of research has been conducted on robots inspired by the nervous system of arthropods, including insects. For example, both insects and humans can be regarded as vehicles capable of autonomously moving around in a complex environment. One challenge in treating insects from a robotic viewpoint is to clearly demonstrate how movement functions under the guidance of vision. Insects have an extensive behavioural repertoire, and using their excellent sensors and limited resources, they can respond to an unpredictable environment. Flying insects, such as flies and dragonflies in particular, are far more agile than vertebrates or modern mobile robots. The sensory–motor control system of an insect is an engineering masterpiece.

Franceschini *et al.*, (2007) reported the creation of a fully autonomous mobile robot that detected optic flow (OF) with planar compound eyes and an elementary motion detector (EMD) array, inspired by the fly. This robot, named **Fly**, detects the outside world with ring-shaped ommatidia. A given pair of adjacent ommatidia drives a single EMD, and 114 EMDs as a whole analyze of in the azimuthal plane. The compound eyes of Fly have a built-in resolution gradient that increases on the basis of a certain rule. The resolution gradient incorporated in the anatomical structure of the eyes ensures the detection of the contrast of objects entering the robot's field of vision during translational motion (a distance that adjacent ommatidia predict for a point a certain distance away). "Fly" could process analog signals in parallel, just like the brain. The robot could weave its way around inanimate objects.

There is a visually guided flight robot inspired by the fly, which uses the same motion detection principles mentioned above to guide flight agents. This robot can fly and land on a rough terrain and follow the terrain without any information on the ground speed or surface contours. This robot is FANIA, a semi-constrained miniature helicopter with a variable pitch rotor. It has eyes equipped with 19 EMDs, and it uses signals from the eyes to avoid the ground. It was able to fly over contrasting obstacles in its trial in a circular arena.

Franceschin *et al.*, (2007) also formatted OF perceived by flying organisms and construct an autopilot called **OF regulator**. A miniature helicopter (OCTAVE), equipped with this autopilot, can take off, fly, and land smoothly in trials over various terrains in a circular arena. It also showed a very good approximation to an insect response when buffeted by the wind.

The fly-inspired autopilot differed greatly from the conventional autopilot designed by humans. A conventional autopilot is designed to allow an airplane to maintain a constant speed and constant altitude, and it requires many expensive, large sensors. In contrast, OCTAVE does not collide with the ground by only regulating its altitude.

The retina of a fly in flight is microscanning. It moves back and forth by a few micrometers 5 times a second. Franceschin *et al.*, hypothesised that microscanning is involved in motion detection, and on this basis, they built 2 types of robots with visual sensing (SCANIA and OSCAR) using only circular-motion OF activity. SCANIA has only low-resolution eyes; however, using its vision, it could avoid walls surrounding a rectangular area in which SCANIA was. OSCAR and robots equipped with OSCAR had 2 propellers suspended from the ceiling of the test chamber and could accordingly move their heads freely. They were equipped with microscanning and could fix a line of site on bars or other nearby objects. This robot could track at an angle up to 30° above the target. Even if buffeted by the pendulum motion of a hanging strap, floor vibrations, or wind gusts, the robot could hold its gaze. This feature would be very useful in real helicopters for detecting transmission lines without relying on radar or lasers.

The bio-robot approach outlined here paves the way for designing new devices and machinery, particularly inexpensive sensorimotor control systems for micro-vehicles. It also provides useful feedback in the fields of neuroethology and neurophysiology.

7.2.2 *A flight-stabilising device based on ocelli*

Ocelli are visual sensory organs of insects separate from the compound eyes. The resolution of the ocelli is too low for detailed images and the visual fields covered are large but contrast detection is possible. Depending on body tilt, the distribution of light intensity received by the different ocelli changes and this information is being used as a gyroscope. A flight stabiliser equipped with 3 light sensors that mimics the mechanism of the ocelli in controlling the flight attitude is being sold as an industrial product. It is used in radio-controlled airplanes. The stabiliser functions to correct the attitude if the radio-controlled airplane falls in an abnormal manner, such as in a spin.

7.2.3 *Robots using insect design principle*

Insects may have tiny brains, but they are capable of a wide variety of behaviours. Most striking and obvious are their capabilities in walking and flying. Even these tasks are difficult to reproduce by machines built according to traditional design principles. Therefore, numerous investigators are working on attempts to emulate the insects' highly elaborate motor systems. On the one hand, insects are used to steer robots through their behaviour to be able to investigate movement patterns in detail under defined conditions and even in the presence of perturbations which can be chosen by the experimenter. This approach is very promising for characterising insect locomotion or flight in great detail. On the other hand, a chosen physiological parameter, the activity of well-defined neurons for example is used to steer a robot. This makes it possible to go beyond phenomenological analysis and allows direct testing of candidate neural codes for generating behaviour.

7.2.4 *An insect-based flying robot*

Bees and wasps have excellent flying ability. By examining their high ability of flight control and the shape of the insect body and wings enabling the flight, we may be able to develop small-scale, lightweight, and stable flying machines. Small and lightweight aircrafts are demanded as information-acquiring tools.

Insects show extremely agile flight control performance. This surprising level of flight performance is attained by a wide range of factors such as strong sensory feedback, complex muscle systems for wing beats, a diversity of movements, and unsteady aerodynamic mechanisms. For example, insects have a surprisingly diverse sensory system. It includes simple eyes (ocelli), compound eyes, antennae, wind receptor sensillae, and distortion sensors buried in the wing surface. Flies have well-developed inertia-sensory organs derived from the hind wings. One wing is controlled by approximately 20 muscles; therefore, its control is incredibly complex. In flight, the wing not only beats upward and downward but also moves along a precise trajectory by bending, twisting, and turning. The wing beat motion inevitably changes the size, direction, and site of action of aerodynamic force, which means that there is a fine interaction between supporting the body, propelling, and controlling the flight. Although this situation complicates flight control, it also enables the insect to enhance its flight control ability and momentum by changing the direction of lift and thrust. The hydrodynamic forces working on the wing are the inertial force and viscous force. For millimeter-sized insects, the viscous force is as large as the inertial force. This means that the aerodynamic force applied to the wing has strong unsteadiness. Insects can fly by flapping their wings because of this unsteady air flow.

The above descriptions indicate that flight control in insects is exceedingly complex; however, it has many possibilities. Insects produce force and momentum for flight control using their wings, and they use the same wings for lifting and thrusting. Compared with artificial flying machines, insects can thus generate a large amount of power. This means that small-scale models based on the insect wing beat pattern should generate the same level of controllability and mobility. In recent years, research has been actively conducted to develop small-scale flying machines, particularly Micro Air Vehicles (MAVs) with a length of less than 15 cm and weight of less than 150 g and Nano Air Vehicles (NAVs) with a length of less than 7.5 cm and weight of 10 g, for information collection. It is highly expected that airfoil imitating insect wing flap could be applicable to small flying vehicles.

7.2.5 *Insect walking and a walking-controlled robot*

Characteristic of walking insects is the trip gait in which three legs (two one side) swing and the remaining legs have ground contact to support the body. The most important properties of the neural networks generating walking patterns have been elucidated and as a result, insect-like hexapod robots have been successfully implemented. An advantage of hexapod robots (and their insect models) is the high stability and robustness to disturbances of locomotion.

Walking insects can cope with a rough terrain. They can also use their legs for sensing and manipulating objects. The control system for these activities is composed of what appear to be low-precision circuit elements - neurons. However, this control system is robust enough to tolerate noises and demonstrate the ability of self-organisation. The control system has a distributed processing structure, relying not only on sensory feedback (feedback loops from environmental interaction) but also on internal feedback.

Machines that move on land usually use wheels or caterpillar tracks; however, their use is generally restricted to a flat ground. In contrast, insects can move on natural land that has not undergone artificial preparation. Insects can walk around rugged and sandy places, climb tree branches, and track over obstacles of various sizes and shapes on the ground. Insect legs are also used in complex operations such as swimming, food gathering, grooming, and moving objects. In insects and crustaceans, their leg actions are controlled by relatively small brains. The processing speed and accuracy of neurons constituting their brains are vastly inferior to those of modern computers. However, the leg control system must be reliable and robust and must continue to function even if some legs are missing. It must also function with low energy consumption. Therefore, an artificial walking system based on the insect control mechanism may be interesting not only to biologists but also to engineers.

At first glance, complex behavioural control would seem to require a complex brain. However, it has been shown that the information obtained from the use of physical properties of the body and its interactions with environment can drastically simplify the controlling mechanisms. Research on insect walking supports this idea. An artificial neural network (ANN) model called **Walknet** has been constructed using results from research on stick insect locomotion. This ANN model is conceptually close to biological systems. In addition, it can perform learning using online and offline information and is robust to input noises.

7.3 Controlling insect behaviours

7.3.1 *Remote control of insect behaviours*

Research is underway presently at the University of California, Berkeley, aimed to control insect behaviours though a tiny remote-controlled stimulation apparatus carried by the insects. The controlled movement of rhinoceros beetle wings and some other parts using radio signals sent to six electrodes on its brain and muscles was established. Beetles were equipped with a module incorporating a circuit to send signals to the electrodes, wireless circuit, microcontroller and battery. The researchers used rhinoceros beetles in this experiment because they can carry a weight of up to 3g. They can fly carrying the module weighing about 1.3g on their backs.

7.3.2 *Architectural insecticide method*

Physical technologies of insect control to protect against insect invasion of production facilities (factories) and logistic facilities (warehouses) are being developed. This technology is not based on using insecticides or similar chemical treatments in principle but based on architectural means such as designing structures and features when designing new buildings or restoring existing ones to protect against insect invasion. By considering insect behavioural traits and by intervening possible entrance pathways and breeding grounds in each room with appropriate countermeasures, it is possible to prevent or reduce insect invasions. It is well known that insects have a positive phototaxis, and the choice of lighting regime and light source are a major factor to reduce insect invasions. Several light sources have been developed that exploit such a trait of insects.

REFERENCES

Abuin L et al. (2011) Functional Architecture of Olfactory Ionotropic Glutamate Receptors. Neuron: 69, 44-60

Altner H, Schaller-Selzer L, Stetter H, Wohlrab I. (1983). Poreless sensilla with inflexible socket. *Cell Tissue Res.*234:279–307

Altner, H. and Prillinger, L. (1980). Ultrastructure of invertebrate chemo-, thermo-, and hygroreceptors and its functional significance. *International Review of Cytology* 67: 69-139.

Arikawa K, Mizuno S, Scholten DG, Kinoshita M, Seki T, Kitamoto J, Stavenga DG (1999) An ultraviolet absorbing pigment causes a narrow-band violet receptor and a singlepeaked green receptor in the eye of the butterfly Papilio. *Vision Res* 39:1–8

Arikawa K, Uchiyama H (1996) Red receptors dominate the proximal tier of the retina in the butterfly *Papilio xuthus*. *J Comp Physiol* A 178:55–61

Bailey, W.F. & Haythornthwaite, S. (1998). Risks of calling by the field cricket *Teleogryllus oceanicus*: potential predation by Australian long-eared bats. *Journal of Zoology*, 244, 505-513.

Benton, R., Vannice, K. S., Gomez-Diaz, C. and Vosshall, L. B. (2009). Variant ionotropic glutamate receptors as chemosensory receptors in Drosophila. *Cell* 136(1): 149-62.

Boekhoff I., Seifert E., Goggerle S., Lindemann M., Kruger B.W. & Breer H. (1993) Pheromone-induced second-messenger signaling in insect antrnnae. *Insect Biochem. Mol. Biol.* 23, 757-762.

Busnel, M.C. (1953). Contribution à l'étude des émissions acoustiques des Orthoptères. Recherches sur les spectres de fréquence et sur les intensités. *Annales des épiphyties,* 3, 333-421

Cade, W.H. (1980). Alternative male reproductive behaviors. *Florida Entomologist*, 63, 30-45.

Conner, W.E. (2014). Adaptive Sounds and Silences: Acoustic Anti-Predator Strategies in Insects. In: B. Hedwig (Ed.), Insect Hearing and Acoustic Communication (pp. 65-79). London, UK: Springer

de Bruyne M., Foster K. & Carlson J.R. (2001) Odor coding in the Drosophila antenna. Neuron 30, 537-552.

Ewing, A.W. (1989). Arthropod Bioacoustics: Neurobiology and Behaviour. Ithaca, New York: Comstock Publishing Associates.

Faure, P.A. and Hoy, R.R. (2000). The sounds of silence: cessation of singing and song pausing are ultrasound-induced acoustic startle behaviour in the katydid *Neoconocephalus ensiger* (Orthoptera: Tettigoniidae). *Journal of Comparative Physiology* A, 186, 129-142

Franceschini N, F.Ruffier, and J Serres (2007) A bio-inspired flying robot sheds light on insect piloting abilities. *Current Biology* 17: 329-335

Frisch K von (1914) Der Farbensinn und Formensinn der Biene. Jb Abt Allg *Zool Physiol* 37, 1-238 G

Greenfield, M.D. (2014). Acoustic Communication in the Nocturnal Lepidoptera. In: B. Hedwig (Ed.), Insect Hearing and Acoustic Communication (pp. 81-100). London, UK: Springer.

Gullan, P.J and Cranston, P.S. (2010). The Insects an Outline of Entomology Fourth Edition. Wiley-Blackwell, West Sussex

Hallem E.A., Ho M.G. & Carlson J.R. (2004) The molecular basis of odor coding in the Drosophila antenna. *Cell* 117, 965-979.

Haskell, P.T. (1961). Insect sounds (p. 189). Chicago: Quadrangle Books.

Hedwig, B. & Robert, D. (2014). Auditory Parasitoid Flies Exploiting Acoustic Communication of Insects. In: B. Hedwig (Ed.), Insect ·Hearing and Acoustic Communication (pp. 44-63).· London, UK: Springer

Huber F. 1962. Central Nervous Control of Sound Production in Crickets and Some Speculations on Its Evolution. *Evolution* 16, 429-442.

Kowalski K and Lakes-Harlan R. (2009). Sound, behaviour, and auditory receptors of the armoured ground cricket,***Acanthoplus longipes***. *Journal of Insect Science* 10, 1-15.

Stephen, R.O and Hartely J.C. (1995). Sound Production in Crickets. *The Journal of Experimental Biology* 198, 2138-2152.

Kitamoto J, Ozaki K, Arikawa K (2000) Ultraviolet receptors and violet receptors express identical mRNA encodingan ultraviolet-absorbingopsin: identification and histological localization of two mRNAs encodingshort-waveleng thabsorbingopsins in the retina of the butterfly *Papilio xuthus*. *J Exp Biol* 203:2887–2894

Krieger, J., and Breer, H., (1999). Olfactory Reception in Invertebrates. *Science* 286, 720–723.

KühnA.(1927)ÜberdenFarbensinnderBienen.*ZverglPhysiol*;5(4):762–800.

Kuhrt U, Samietz J and Dorn S. (2005). Thermoregulation behavior in codling moth larvae. *Physiological Entomology* 30, 54-61.

Laue M.L., Maida R. & Redkozubov A. (1997) G-protein activation, identification and immunolocalization in pheromone sensitive sensilla trichodea of moth. *Cell Tissue Res.* 288, 149-158.

Marques, L. and de Almeida, A.T. (2000), "Electronic nose based odour source localization", Proc. of 6th International Workshop on Advanced Motion Control, Nagoya, Japan, March 30-April 1, pp. 36-40.

Nakano, R.; Skals, N.; Takanashi, T.; Surlykke, A.; Koike, T.; Yoshida, K.; Maruyama, H. et al. (2008). Moths produce extremely quite ultrasonic courtship songs by rubbing specialize scales. *Proceedings of the National Academy of Sciences,* 105, 11812-11817

Schmaranzer S. 1999. Thermoregulation of water collecting honey bees (***Apis mellifera***). ***Journal of Insect Physiology*** 46, 1187-1194.

...nhoven, L.M. and Van Loon, J.J.A. (2002) An inventory of taste in caterpillars: each species its own key. *Acta Zool Acad Sci Hungaricae* 48: 215–263.

Shanbhag S.R., Muller B. & Steinbrecht R.A. (1999) Atlas of olfactory organs of Drosophila melanogaster 1. Types, external organization, innervation and distribution of olfactory sensilla. *Int. J. Insect Morphol. Embryol.* 28, 377-397.

Shanbhag S.R., Muller B. and Steinbrecht R.A. (2000) Atlas of olfactory organs of Drosophila melanogaster 2. Internal organization and cellular architecture of olfactory sensilla. *Arthropod Struct.* Dev. 29, 211-229.

Spangler, H. (1984). Silence as a defense against predatory bats in two species of calling insects. *The Southwestern Naturalist,* 29, 481-488.

Stabentheiner A. (2001). Thermoregulation of dancing bees: thoracic temperature of pollen and nectar foragers in relation to profitability of foraging and colony need. ***Journal of Insect Physiology*** **47**, 385-392.

Steinbrecht R.A., Laue M. and Ziegelberger G. (1995) Immunolocalization of pheromone-binding protein and general odorant-binding protein in olfactory sensilla of the silk moths Antheraea and Bombyx. *Cell Tissue Res.* 282, 203-217.

Tichy H and Loftus R. (1990) Response of moist-air receptor on antenna of the stick insect, *Carausius morosus*, to step changes in temperature. J Comp Physiol A. 166:507–516.

Tichy H and Loftus R. (1996) Hygroreceptors in insects and a spider: humidity transduction models. *Naturwissenschaften*. 83:255–263.

Tichy H. 1978 Hygroreceptor identification and response characteristics in the stick insect ***Carausius morosus***. *J Comp Physiol* A;160:43–5

Tichy, H. and Kallina, W. 2010. Insect Hygroreceptor Responses to Continuous Changes in Humidity and Air Pressure. *Neurophysiol* 103: 3274–3286.

van der Goes van Naters W.M. and Carlson J.R. (2007) Receptors and neurons for fly odors in Drosophila. *Curr. Biol.* 17, 606-612.

Warrant E, and Dacke M. 2010 Visual orientation and navigation in nocturnal arthropods. *Brain Behav Evol.* 2010;75(3):156-73.

Xu, W., Zhang, H.J. and Anderson, A. (2012) A sugar gustatory receptor identified from the foregut of cotton bollworm Helicoverpa armigera. *J Chem Ecol* 38: 1513–1520.

Yao C.A., Ignell R. and Carlson J.R. (2005) Chemosensory coding by neurons in the coeloconic sensilla of the Drosophila antenna. *J. Neurosci.* 25, 8359-8367.

Yao, C.A. and Carlson, J.R. (2010) Role of G-proteins in odorsensing and CO2-sensing neurons in Drosophila. *J Neurosci* 30: 4562–4572.

Zhang, H.J., Anderson, A.R., Trowell, S.C., Luo, A.R., Xiang, Z.H. and Xia, Q.Y. (2011) Topological and functional characterization of an insect gustatory receptor. *PLoS ONE* 6: e24111.

Zuk, M.; Rotenberry, J.T. & Tinghitella, R.M. (2006). Silent night: adaptive disappearance of a sexual signal in a parasitized population of field crickets. *Biology Letters*, 2, 521-525.